# THE ELEMENTS

|  |  |  |  |  | 17 VIIA | 18 O |
|---|---|---|---|---|---|---|
| 13 IIIA | 14 IVA | 15 VA | 16 VIA | | 1 **H** 1.00794 | 2 **He** 4.00260 |

| | | | 13 IIIA | 14 IVA | 15 VA | 16 VIA | | |
|---|---|---|---|---|---|---|---|---|
| | | | 5 **B** 10.811 | 6 **C** 12.011 | 7 **N** 14.0067 | 8 **O** 15.9994 | 9 **F** 18.9984 | 10 **Ne** 20.1797 |
| 10 | 11 IB | 12 IIB | 13 **Al** 26.9815 | 14 **Si** 28.0855 | 15 **P** 30.9738 | 16 **S** 32.066 | 17 **Cl** 35.4527 | 18 **Ar** 39.948 |
| 28 **Ni** 58.69 | 29 **Cu** 63.546 | 30 **Zn** 65.39 | 31 **Ga** 69.723 | 32 **Ge** 72.61 | 33 **As** 74.9216 | 34 **Se** 78.96 | 35 **Br** 79.904 | 36 **Kr** 83.80 |
| 46 **Pd** 106.42 | 47 **Ag** 107.868 | 48 **Cd** 112.411 | 49 **In** 114.82 | 50 **Sn** 118.710 | 51 **Sb** 121.75 | 52 **Te** 127.60 | 53 **I** 126.904 | 54 **Xe** 131.29 |
| 78 **Pt** 195.08 | 79 **Au** 196.967 | 80 **Hg** 200.59 | 81 **Tl** 204.383 | 82 **Pb** 207.2 | 83 **Bi** 208.980 | 84 **Po** (209) | 85 **At** (210) | 86 **Rn** (222) |

Metals ← → Nonmetals

| 63 **Eu** 151.965 | 64 **Gd** 157.25 | 65 **Tb** 158.925 | 66 **Dy** 162.50 | 67 **Ho** 164.930 | 68 **Er** 167.26 | 69 **Tm** 168.934 | 70 **Yb** 173.04 | 71 **Lu** 174.967 |
|---|---|---|---|---|---|---|---|---|

| 95 **Am** (243) | 96 **Cm** (247) | 97 **Bk** (247) | 98 **Cf** (251) | 99 **Es** (252) | 100 **Fm** (257) | 101 **Md** (258) | 102 **No** (259) | 103 **Lr** (262) |
|---|---|---|---|---|---|---|---|---|

# INTRODUCTION TO
# GENERAL, ORGANIC & BIOCHEMISTRY

# INTRODUCTION TO
# GENERAL, ORGANIC & BIOCHEMISTRY

## FOURTH EDITION

**Frederick A. Bettelheim**
**Jerry March**
*Adelphi University*

Saunders College Publishing

**Harcourt Brace College Publishers**
Fort Worth · Philadelphia · San Diego · New York
Orlando · Austin · San Antonio
Toronto · Montreal · London · Sydney · Tokyo

*To our wives:*
*Vera S. Bettelheim and Beverly March.*

Text Typeface: Caledonia
Compositor: Progressive Information Technologies
Vice President; Publisher: John Vondeling
Developmental Editor: Beth Rosato
Managing Editor: Carol Field
Project Editor: Beth Ahrens
Copy Editor: Janis Moore
Manager of Art and Design: Carol Bleistine
Art Director: Anne Muldrow
Art and Design Coordinator: Sue Kinney
Cover Designer: Lou Fuiano
Text Artwork: J & R Technical Services, Inc. and Rolin Graphics, Inc.
Layout Artist: Julie Anderson
Director of EDP: Tim Frelick
Production Manager: Charlene Squibb
Vice President, Marketing: Marjorie Waldron


Cover Credit: Uncompahgre National Forest, Colorado. © Willard Clay/FPG International
Frontispiece: California coast. Comstock, Inc., Mike and Carol Warner © 1991.

Printed in the United States of America


INTRODUCTION TO GENERAL, ORGANIC & BIOCHEMISTRY, Fourth Edition

0-03-001068-3

Library of Congress Catalog Card Number: 94-065433

567890123   032   10 9876543

# PREFACE

*USDA*

".  .  .  . matter is matter, neither noble nor vile, infinitely transformable, and its proximate origin is of no importance whatsoever. Nitrogen is nitrogen, it passes miraculously from the air into plants, from these into animals and from animals to us; when its function in the body is exhausted, we eliminate it, but it still remains nitrogen, aseptic, innocent" so says Primo Levi in his book *The Periodic Table.* In transforming a life-long love affair toward his chosen profession, chemistry, into literature, he succeeded in expressing his enthusiasm for all to see.

In writing the Preface for this, the fourth edition of our textbook, we hope that somewhat similarly we manage to convey our delight in observing the chemical processes in the core of life sciences. The increasing use of our textbook enables this new edition, and we wish to thank our colleagues who adopted previous editions for their courses. Testimony from colleagues and students indicates that we have conveyed our enthusiasm to students who find this book a great help in studying difficult concepts.

Thus, this fourth edition intends to be even more readable and understandable than earlier editions. While maintaining the overall organization of the textbook, we strive to produce more integration of the three domains of the text: general, organic, and biochemistry. Chemistry, especially biochemistry, is a fast-developing discipline, and we include new relevant material in the text. This is done not just by upgrading information but also by enlarging the scope of the book, both in the text and in the boxes containing medical applications of chemical principles. At the same time, we are aware of the necessity to keep the book to a manageable size and proportion. Twenty percent of the problems are new; there is an increase in the number of more challenging, thought-provoking problems (marked by asterisks).

## AUDIENCE

As were the previous editions, this book is intended for nonchemistry majors, mainly those entering health science and related fields (such as nursing, medical technology, physical therapy, and nutrition). It also can be used by students in environmental studies. In its entirety it can be used for a one-year (two-semester or three-quarter) course in chemistry, or parts of the book can be used in a one-term chemistry course.

We assume that the students using the book have little or no chemistry background. Therefore, we introduce the basic concepts slowly at the beginning, increasing the tempo and the level of sophistication as we go on. We progress from the basic tenets of general chemistry to organic chemistry and finally to biochemistry. We consider this progression an ascent in terms of both practical importance and sophistication. However, the three parts are integrated by keeping a unified view of chemistry. We do not consider the general chemistry sections to be the exclusive domain of inorganic compounds, so we frequently use organic and biological substances to illustrate general principles.

While teaching the chemistry of the human body is our ultimate goal, we try to show that each subsection of chemistry is important in its own right, besides being required for future understanding.

## BOXES  (Medical and Other Applications of Chemical Principles)

The boxes contain applications of the principles discussed in the text. Comments from users of the earlier editions indicate that these are especially well received, providing a much requested relevance to the text. The large number of boxes deal mainly with health-related applications, including many also related to the environment. A list of these medically relevant applications follows the Contents. Five boxes from the third edition have been dropped. Two others have been incorporated into the text. Nine new boxes have been added dealing with diverse topics such as oncogenes, the indoor radon problem, taxol as an anticancer agent, nitric oxide and long term memory, sex hormones and old age, excess vitamins, hydrates and air pollution, rubber, sunglasses, and deodorants. Many boxes have been enlarged and updated. For example, boxes on Parkinson's disease, lipid storage disease, anesthetics, birth control, goiter, and AIDS now include recent material.

The presence of boxes allows a considerable degree of flexibility. If an instructor wants to assign only the main text, the boxes will not interrupt continuity and the essential material will be covered. However, most instructors will probably wish to assign at least some of the boxes, since they enhance the core material. In our experience, students are eager to read the relevant boxes even without assignments, and they do so with discrimination. From such a large number of boxes, the instructor can select those that best fit the particular needs of the course and of the students. Problems are provided for nearly all of the boxes.

## ORGANIZATION

We have maintained the organization of the previous edition. Nine chapters deal with general chemistry, six with organic chemistry and eleven with biochemistry. The presentation of nuclear chemistry is in Chapter 9 at the end of the general chemistry sections. Some instructors think that this topic is better presented immediately after Chapter 2 on Atoms. There are no difficulties in moving up nuclear chemistry, if the instructor so desires. We think, however, that the complexity of this topic is better appreciated after the exploration of equilibrium and kinetics.

In organic chemistry, we concentrate on the structure and properties and only the most important reactions of each class of compounds. As for the mechanisms of the reactions, we provide only one example: carbocation intermediates in addition reactions. We do this deliberately because we think that in the relatively brief portion of the course devoted to organic chemistry, students do not have time to learn a large number of reactions or anything substantial about mechanisms. As stated before, we consider the progression from general to organic and to biochemistry an ascent. Therefore, we selected mainly organic compounds and reactions which have physiological activity and biological importance. In order to help students learn the reactions, we include summaries of reactions at the ends of the chapters.

Within the biochemistry chapters, we maintain the traditional order. We find this a pedagogical imperative. Even though most of the important new developments in biochemistry occur in molecular biology (Chapter 23 on Nucleic Acids and Protein Synthesis) and neurochemistry and immunology (Chapter 24 on Chemical Communication: Neurotransmitters, Hormones, and Immunoglobulins), these chapters come late in the book. We think that the appreciation of these topics requires a previous acquaintance with carbohydrate, lipid, and protein chemistry and metabolism. We hope that each instructor, to his or her taste, will judiciously appropriate time to discuss the exciting developments presented in the late chapters.

## NEW MATERIAL

In addition to several new boxes, we include new material in the text. Some examples are

- We provide a new section on the basic types of reactions.
- We discuss environmental problems, such as the Clean Air Act and the oxygenation of gasoline as well as the threat to the ozone layer by CFCs.
- We include the fullerenes in the discussion of carbon.
- We enlarge the discussion on cis-trans isomerism.
- We add new material on the role of chaperones in protein folding and denaturation.
- We extend the discussions on transcription factors and gene therapy.
- We describe the important polymerase-chain-reaction (PCR) technique in molecular biology.

## METABOLISM; COLOR CODE

The biological functions of chemical compounds are explained in each of the biochemistry chapters and in many of the organic chapters. The emphasis is on chemistry rather than physiology. We received much positive feedback regarding the way in which we organized the topic of metabolism (Chapters 20, 21, and 22). We maintained this organization.

First, we introduce the common metabolic pathway through which all food will be utilized (citric acid cycle; oxidative phosphorylation), and only after that do we discuss the specific pathways leading to the common pathway. We find this a useful pedagogical device, and it enables us to sum up the caloric values of each type of food because their utilization through the common pathway has already been learned. Finally, we separate the catabolic pathways from the anabolic pathways by treating them in different chapters, emphasizing the different ways the body breaks down and builds up different molecules.

The topic of metabolism is a difficult one for most students. We try to explain it as clearly as possible. As in the previous edition, we enhance the clarity of presentation by the use of a color code for the most important biological compounds discussed in Chapters 20, 21, and 22. Each type of compound is shown in a specific color, which remains the same throughout the three chapters. These colors are as follows:

ATP and other nucleoside triphosphates

ADP and other nucleoside diphosphates

The oxidized coenzymes $NAD^+$ and FAD

The reduced coenzymes NADH and $FADH_2$

Acetyl coenzyme A

The circled numbers in the figures showing the steps involved in the various metabolism pathways are always in yellow.

In addition to this main use of a color code, other figures in various parts of the text are color coded, so that the same color is used for the same entity throughout the text. For example, in Chapter 19, enzymes are always shown in blue and substrates in orange in all of the figures that show enzyme substrate interactions.

## INTERVIEWS

Each of the three sections—general, organic, and biochemistry—opens with an interview with an individual who has made significant contributions in that particular field. Roald Hoffman, Nobel Laureate in 1981, not only enhanced our understanding of chemical bonding through his theoretical work but also regularly teaches general chemistry at Cornell University. Carl Djerassi of Stanford University is a master of steroid chemistry, both in synthesis and analyses of organic compounds, and he is often cited as the inventor of the birth control pill. Jacqueline K. Barton of the California Institute of Technology has made important discoveries relating to the structure and conformation of the DNA double helix.

These interviews are intended to give the student a human face of science and an insight of how science affects our lives.

## FEATURES

One of the main features of this book, as in earlier editions, is the number of applications of chemical concepts presented in the boxes. Another important feature is the Glossary-Index. The definition of each term is given along with the index entry and the page number. Another feature is the list of key terms at the end of each chapter, with notation of the section number in which the term is introduced. Many students find these lists to be helpful study guides.

Other features are the summaries at the end of each chapter (including summaries of organic reactions in Chapters 11 to 15) and the substantial number of margin notes. We subsectioned the end-of-chapter problems, to guide students to the section of the chapter where they can find the relevant material.

## STYLE

Feedback from colleagues and students alike indicates that the style of the book, which addresses the students directly in simple and clear phrasing, is one of its major assets. We continue to make special efforts to provide clear and concise writing. Our hope is that this eases the understanding and the absorption of the difficult concepts.

## PROBLEMS

About one fifth of the problems are new in this fourth edition. The number of starred problems, which contain the more challenging, thought-provoking questions, has increased. The end-of-chapter problems are grouped and sub-headed in the order of topic coverage. The last group headed as Additional Problems is not arranged in any specific order. The answers to all the in-text problems and to the odd-numbered end-of-chapter problems are given at the end of the book. Answers to the even-numbered problems are included in the Instructor's Manual and the Study Guide.

## ANCILLARIES

This textbook is accompanied by a number of ancillary publications to help support your teaching and your students' learning:

**1.**  Flash Cards by Hugh Akers (Lamar University). 200 bi-directional cards provide handy and convenient reference to the important reactions, terms, structures, and classifications of general, organic, and biological chemistry.
**2.**  Study Guide by W. Scovell (Bowling Green State University). Includes review of chapter objectives, important terms and comparisons, focused review of concepts, self tests, and answers to the even-numbered problems in the text.
**3.**  Instructor's Manual and Test Bank by F. A. Bettelheim and J. March. Contains suggested course outlines, 90% newly revised exam questions organized chapter by chapter, answers to the exam questions, and answers to the even-numbered problems.

**4.** Computerized Test Bank available in IBM and Macintosh versions. The format permits modification of questions in the test bank and the addition of new ones.

**5.** Approximately 160 overhead transparencies in full color. Figures and tables are taken from the text.

**6.** Laboratory Experiments for *Introduction to General, Organic & Biochemistry,* 2e by F. A. Bettelheim and J. Landesberg. Forty-six experiments illustrate important concepts or principles in general, organic, and biochemistry. Simple equipment and inexpensive, common, and environmentally safe chemicals are used. The large number of experiments allows sufficient flexibility for the instructor to select the usual 24 experiments the students can perform in a two-semester course.

**7.** Instructor's Manual to accompany Laboratory Experiments for *Introduction to General, Organic & Biochemistry,* 2e by F. A. Bettelheim and J. Landesberg. This will help instructors in grading the answers to the questions as well as in assessing the range of the experimental results obtained by the students.

**8.** Saunders Chemistry of Life Videodisc. Includes nearly all the still images from *Introduction to General, Organic & Biochemistry*, 4e, with large type for better classroom viewing, and contains live-action footage of chemical demonstrations.

**9.** Chemistry of Life LectureActive™ Software. This outstanding, convenient package is noted for its ease of use and economy of time because all video clip and still frame data from the videodisc are entered and listed on the software. This unique program for both IBM Windows and Macintosh formats enables the instructor to swiftly access every image and film clip on the videodisc and to easily prepare a customized presentation of disc images.

## ACKNOWLEDGMENTS

The publication of a book such as this requires the efforts of many more people than merely the authors. A number of reviewers have read all or significant portions of the manuscript in various stages. We thank the following for their constructive criticism and helpful suggestions in revising the text for this fourth edition:

Hugh Akers, Lamar University

Karen Eichstadt, Ohio University

Leland Harris, University of Arizona

James T. Johnson, Sinclair Community College

Margaret G. Kimble, Indiana-Purdue Universities at Fort Wayne

Kenneth E. Miller, Milwaukee Area Technical College

Thomas Pynadath, Kent State University

Guy Rosenthal, The University of Vermont

Robert VanLanen, Saint Xavier University

Reviewers of the first three editions:

David Adams, North Shore Community College

Michael Bobrick, North Virginia Community College

Lorraine Brewer, University of Arkansas

Ana A. Ciereszko, Miami-Dade Community College South Campus

William D. Covey, Los Angeles Pierce College

Charles Cowell, Rancho Santiago College

Adrienne Curtis, Orange Coast College

Andrew Dachauer, University of San Francisco

Jane Dant, Florida Community College

Bruce Dunlap, University of South Carolina

Stanley Grenda, University of Nevada

John Griswold, Cedar Crest College

Marvin Hackert, University of Texas at Austin

Leland Harris, University of Arizona

Robert Harris, University of Nebraska

Larry Jackson, Montana State University

Eugene Kline, Tennessee Technological University

Herman Knoche, University of Nebraska

Joseph Landesberg, Adelphi University

Jessie Lee, Philadelphia Community College

William Masterton, University of Connecticut

Melvin Merken, Worcester State College

Norman Meyers, Bowling Green State University

Robert Midden, Bowling Green State University

Paul O'Brien, West Valley College

Pat O'Day, Pasadena City College

Alan Price, University of Michigan

Jacqueline Scholar, Bellevue College

Ronald M. Scott, Eastern Michigan University

William Scovell, Bowling Green State University

Brahama Sharma, Los Angeles Pierce College

David Tuleen, Vanderbilt University

William H. Voige, James Madison University

Shan Wong, University of Lowell

Albert Zabady, Montclair State College

William Zuber, Memphis State University

We wish to also thank the many respondents to a survey of questions on our text and the market that Saunders commissioned for our consideration as we made plans for the fourth edition: Hugh Akers, Lamar University; Kim Caluo, University of Akron; Thomas Cogdell, University of Texas at Arlington; Carolyn T. Connelly, Waynesburg College; Thomas D. Crute, Augusta College; James Deavor, College of Charleston; Lawrence K. Duffy, University of Alaska, Fairbanks; Amal Girgis, Pierce College; Marie Hankins, University of Southern Indiana; Lillian M. Hansen, Tarrant County Junior College Northeast; Karen Harding, Pierce College; Chu-ngi Ho, East Tennessee State University; Vanessa Hull, Tarrant County Junior College South; Eric Johnson, Ball State University; Warren Johnson, University of Wisconsin-Green Bay; M. G. Kimble, Indiana Purdue University at Ft. Wayne; Richard Kletsch, Centralia College; M. L. Morris, North Dakota State University; Don Naschert, Illinois Central College; Mary O'Sullivan, Indiana State University; Helen Place, Washington State University; Bruce P. Ronald, Idaho State University; L. Rosenhein, Indiana State University; J. Scholar, Bellevue College; Daniel R. Tauber, Villa Julie College; Chris Vogel, Sacramento City College; Craig Weaver, Brevard Community College; Jack Wolf, El Banuaru College.

Many of our reviewers pointed out inadvertent errors or certain weaknesses in the previous editions. We attempted to correct these, and thus we hope that the fourth edition will prove even more useful than the earlier editions.

We also wish to thank several of our colleagues at Adelphi University for their useful advice. These include Stephen Goldberg, Joseph Landesberg, Sung Moon, Donald Opalecky, Reuben Rudman, Charles Shopsis, Kevin Terrance, and Stanley Windwer. We are grateful for the support of John Vondeling, Vice President and Publisher, Saunders College Publishing. We thank Beth Rosato, Developmental Editor, and Beth Ahrens, Project Editor, for their congenial steady assistance. We would like to express our appreciation to Charlene Squibb for supervising the production of this edition and to Anne Muldrow for supervising the art. Rolin Graphics, Inc. transformed our crude drawings into pieces of art. Last but not least we want to thank Beverly March and Charles D. Winters for their many excellent photographs.

*Frederick A. Bettelheim and Jerry March*
*Adelphi University*
*June 1994*

# CONTENTS OVERVIEW

*Beverly March*

**Part I   General Chemistry      1**
**Interview:   Roald Hoffmann      2**

1   Matter, Energy, and Measurement      7
2   Atoms      33
3   Chemical Bonds      65
4   Chemical Reactions      95
5   Gases, Liquids, and Solids      129
6   Solutions and Colloids      163
7   Reaction Rates and Equilibrium      195
8   Acids and Bases      219
9   Nuclear Chemistry      251

**Part II   Organic Chemistry      279**
**Interview:   Carl Djerassi      280**

10   Organic Chemistry. Alkanes      285
11   Alkenes, Alkynes, and Aromatic Compounds      323
12   Alcohols, Phenols, Ethers, and Halides      361
13   Aldehydes and Ketones      391
14   Carboxylic Acids and Esters      415
15   Amines and Amides      443

**Part III   Biochemistry   473**
**Interview:   Jacqueline K. Barton   474**

16   Carbohydrates   479
17   Lipids   517
18   Proteins   547
19   Enzymes   573
20   Bioenergetics. How the Body Converts Food to Energy   591
21   Specific Catabolic Pathways: Carbohydrate, Lipid, and Protein Metabolism   613
22   Biosynthetic Pathways   637
23   Nucleic Acids and Protein Synthesis   649
24   Chemical Communication: Neurotransmitters, Hormones, and Immunoglobulins   685
25   Body Fluids   707
26   Nutrition and Digestion   725

**Appendix A**   Exponential Notation   A.1
**Appendix B**   Significant Figures   A.5

Answers to In-Text and Odd-Numbered End-of-Chapter Problems   A.9

Glossary-Index   I.1

# CONTENTS

*USDA*

**Interview:** Roald Hoffmann    2

1.1  Introduction    7
1.2  The Scientific Method    9
1.3  Measurements    11
1.4  Unit Conversions. The Factor-Label Method    18
1.5  The States of Matter    20
1.6  Density and Specific Gravity    21
1.7  Energy    23
1.8  Heat    25
     Summary    28
     Key Terms    28
     Problems    29

**Box 1A**    The Scientific Method in Medicine    10
**Box 1B**    Drug Dosage and Body Mass    15
**Box 1C**    Laser Surgery    25
**Box 1D**    Hypothermia and Hyperthermia    26
**Box 1E**    Cold Compresses    27

PART I    1
General Chemistry

**CHAPTER 1**    7
**Matter, Energy, and
Measurement**

**CHAPTER 2**
**Atoms**
**33**

2.1   Introduction    33
2.2   Classifications of Matter    34
2.3   Dalton's Atomic Theory    39
2.4   Inside the Atom    41
2.5   The Periodic Table    44
2.6   The Electronic Structure of Atoms    49
2.7   Electronic Configuration and the Periodic Table    54
2.8   How Small Are Atoms?    57
        Summary    60
        Key Terms    60
        Problems    61

**Box 2A**   Elements Necessary for Human Life    36
**Box 2B**   Strontium-90    46
**Box 2C**   The Use of Metals as Historical Landmarks    48
**Box 2D**   Abundance of Elements on Earth and in People    58

*Beverly March*

**CHAPTER 3**
**Chemical Bonds**
**65**

3.1    Introduction    65
3.2    Ions    65
3.3    Ionic Bonds    69
3.4    Covalent Bonds    72
3.5    Coordinate Covalent Bonds    77
3.6    The Shapes of Molecules    77
3.7    Electronegativity and Dipoles    82
3.8    How to Predict the Kinds of Bonds that Form    85
3.9    What Bonds to What?    86
3.10   Polyatomic Ions    87
3.11   Naming of Simple Inorganic Compounds    88
        Summary    90
        Key Terms    90
        Problems    91

**Box 3A**   Biologically Important Ions    67
**Box 3B**   Ionic Compounds in Medicine    70
**Box 3C**   Hemoglobin    78

*C. D. Winters*

**CHAPTER 4**
**Chemical Reactions**
**95**

4.1   Introduction    95
4.2   Formula Weight    97
4.3   The Mole    98
4.4   Chemical Equations    101
4.5   Weight Relationships in Chemical Reactions    104
4.6   Percentage Yield    107
4.7   Basic Types of Reactions    109
4.8   Reactions Between Ions in Aqueous Solution    110
4.9   Oxidation–Reduction    115

4.10    Heat of Reaction    119
        Summary    122
        Key Terms    122
        Problems    123

        **Box 4A**    Industrial Production and Percentage Yield    108
        **Box 4B**    Voltaic Cells    116
        **Box 4C**    Oxidizing Antiseptics    117
        **Box 4D**    The Caloric Value of Foods    121

5.1     The Organization of Matter    129
5.2     Gases    131
5.3     Pressure    132
5.4     Gas Laws    135
5.5     Avogadro's Law and the Ideal Gas Law    140
5.6     Dalton's Law and Graham's Law    142
5.7     Intermolecular Forces    145
5.8     Liquids    149
5.9     Evaporation and Condensation. Boiling Point    149
5.10    Surface Tension    153
5.11    Solids    154
5.12    Freezing and Boiling. Phase Changes    156
        Summary    158
        Key Terms    159
        Problems    160

        **Box 5A**    The Uses of Atmospheric Gases    131
        **Box 5B**    Mercury as a Poison    133
        **Box 5C**    Blood Pressure Measurement    134
        **Box 5D**    Breathing and Boyle's Law    136
        **Box 5E**    Hyperbaric Medicine    139
        **Box 5F**    Breathing and Dalton's Law    144
        **Box 5G**    The Densities of Ice and Water    147
        **Box 5H**    Ethyl Chloride    151
        **Box 5I**    Diamonds    157

6.1     Introduction    163
6.2     Types of Solutions    164
6.3     Characteristics of Solutions    165
6.4     Solubility    166
6.5     Concentration Units    168
6.6     Water as a Solvent    175
6.7     Colloids    181
6.8     Colligative Properties. Freezing-Point Depression    184
6.9     Osmotic Pressure    185

**CHAPTER 5**    **129**
**Gases, Liquids, and**
**Solids**

*Beverly March*

**CHAPTER 6**    **163**
**Solutions and**
**Colloids**

© *Ray Pfortner/Peter Arnold, Inc.*

6.10   Dialysis   188
       Summary   190
       Key Terms   190
       Problems   191

**Box 6A**   Acid Rain   165
**Box 6B**   The Bends   169
**Box 6C**   Hydrates and Air Pollution: The Decay of Buildings
             and Monuments   178
**Box 6D**   Hypoglycemia and Potassium Balance   180
**Box 6E**   Emulsions and Emulsifying Agents   183
**Box 6F**   Magnesium Salts in Hypertonic Solution   189
**Box 6G**   Hemodialysis   189

**CHAPTER 7**          **195**
**Reaction Rates and**
**Equilibrium**

7.1   Introduction   195
7.2   Molecular Collisions   196
7.3   Activation Energy and Energy Diagrams   198
7.4   Factors Affecting Rates of Reaction   200
7.5   Reversible Reactions and Equilibrium   204
7.6   Equilibrium Constants   206
7.7   Equilibrium and Reaction Rates   209
7.8   Le Chatelier's Principle   209
      Summary   214
      Key Terms   214
      Problems   215

**Box 7A**   Fire Hazards in Hospitals   200
**Box 7B**   Why High Fever is Dangerous   201
**Box 7C**   The Effects of Lowering Body Temperature   202
**Box 7D**   Sunglasses and Le Chatelier's Principle   211
**Box 7E**   The Haber Process   213

**CHAPTER 8**          **219**
**Acids and Bases**

8.1    Introduction   219
8.2    Acid and Base Strength   220
8.3    Brønsted-Lowry Acids and Bases   223
8.4    Acid Dissociation Constants   225
8.5    Some Properties of Acids and Bases   226
8.6    Reactions of Acids   227
8.7    Self-Ionization of Water   230
8.8    pH   231
8.9    The pH of Aqueous Salt Solutions   236
8.10   Buffers   238
8.11   Titration, Equivalents, and Normality   242
       Summary   246
       Key Terms   246
       Problems   247

**Box 8A**    Some Important Acids and Bases    221
**Box 8B**    Acid and Alkali Burns of the Cornea    227
**Box 8C**    Drugstore Antacids    228
**Box 8D**    Rising Cakes and Fire Extinguishers    229
**Box 8E**    Acidosis and Alkalosis    237
**Box 8F**    The Henderson-Hasselbalch Equation    240

9.1    Introduction    251
9.2    What Happens When a Nucleus Emits Radioactivity?
       Natural Transmutation    255
9.3    Half-Life    257
9.4    Characteristics of Radiation    259
9.5    Radiation Dosimetry and the Effects of Radiation
       on Human Health    261
9.6    Medical Uses of Radioactive Materials    265
9.7    Nuclear Fusion. Artificial Transmutation    269
9.8    Nuclear Fission. Atomic Energy    270
       Summary    274
       Key Terms    274
       Problems    274

**Box 9A**    Rutherford's Experiment    254
**Box 9B**    Radioactive Dating    258
**Box 9C**    The Indoor Radon Problem    263
**Box 9D**    Radium in Wristwatch Dials    265
**Box 9E**    CAT and Nmr Scans    266
**Box 9F**    Isotope Generators for Medical Use    269
**Box 9G**    Energy from Mass    270
**Box 9H**    Radioactive Fallout from Nuclear Accidents    272

Interview:   Carl Djerassi    280

10.1    Introduction    285
10.2    Sources of Organic Compounds    287
10.3    Structures of Organic Compounds    289
10.4    Hybrid Orbitals    291
10.5    Hydrocarbons    293
10.6    Alkanes    293
10.7    IUPAC Nomenclature of Alkanes and Alkyl Halides    298
10.8    Cycloalkanes    304
10.9    Stereoisomerism in Cyclic Compounds    305
10.10   Physical Properties    307
10.11   Chemical Properties    308

**CHAPTER 9**    **251**
**Nuclear Chemistry**

*Atomic Industrial Forum*

**PART II**    279
Organic Chemistry

**CHAPTER 10**    **285**
**Organic Chemistry**
**Alkanes**

10.12  Functional Groups    312
       Summary    315
       Key Terms    315
       Problems    316

**Box 10A**  The Actual Shape of the Cyclohexane Ring    306
**Box 10B**  The Biological Effects of Alkanes    308
**Box 10C**  Petroleum    309
**Box 10D**  Carbon Dioxide and Carbon Monoxide    310
**Box 10E**  The Greenhouse Effect    311
**Box 10F**  Taxol, A New Anticancer Drug and the
             Environment    314

**CHAPTER 11**          **323**    11.1   Introduction    323
**Alkenes, Alkynes,**              11.2   Nomenclature of Alkenes    323
**and Aromatic**                   11.3   $sp^2$ Hybridization and $\pi$ Orbitals    327
**Compounds**                      11.4   Geometry and Stereoisomerism of Alkenes    328
                                   11.5   Physical Properties of Alkenes    332
                                   11.6   Chemical Properties of Alkenes: Addition Reactions    333
                                   11.7   Addition Polymers    338
                                   11.8   Alkynes    341
                                   11.9   Aromatic Hydrocarbons    344
                                   11.10  Nomenclature of Benzene Derivatives    346
                                   11.11  Reactions of Aromatic Compounds    349
                                   11.12  Fused Aromatic Rings    350
                                   11.13  Heterocyclic Compounds    350
                                          Summary    352
                                          Key Terms    353
                                          Problems    353

                                   **Box 11A**  Terpenes    324
                                   **Box 11B**  Pheromones    329
                                   **Box 11C**  Cis-Trans Isomerism in Vision    330
                                   **Box 11D**  The Mechanism of HX Addition    336
                                   **Box 11E**  Rubber    339
                                   **Box 11F**  Carcinogenic Fused Aromatics and Smoking    351

**CHAPTER 12**          **361**    12.1   Introduction    361
**Alcohols, Phenols,**             12.2   Nomenclature of Alcohols    362
**Ethers, and Halides**            12.3   Chemical Properties of Alcohols    364
                                   12.4   Some Important Alcohols    368
                                   12.5   Phenols    372
                                   12.6   Ethers    375
                                   12.7   Physical Properties of Alcohols, Phenols, and Ethers    378
                                   12.8   Thiols, Thioethers, and Disulfides    380
                                   12.9   Alkyl and Aryl Halides    381

Summary    384
Key Terms    384
Problems    385

**Box 12A**    Alcoholic Beverages    369
**Box 12B**    Alcoholism and Disulfuram    370
**Box 12C**    The Poisonous Effects of Methanol    371
**Box 12D**    Nitroglycerine, an Explosive and a Drug    373
**Box 12E**    Phenols as Antiseptics    374
**Box 12F**    Anesthetics    376
**Box 12G**    Deodorants    377
**Box 12H**    Marijuana    379
**Box 12I**    Chlorinated Hydrocarbon Insecticides    381
**Box 12J**    Iodine and Goiter    383

*USDA*

13.1    Introduction    391
13.2    Nomenclature    392
13.3    Physical Properties    394
13.4    Preparation of Aldehydes and Ketones by Oxidation    395
13.5    Oxidation of Aldehydes    395
13.6    Other Chemical Properties of Aldehydes and Ketones    397
13.7    Some Important Aldehydes and Ketones    405
        Summary    407
        Key Terms    408
        Problems    408

**Box 13A**    Chloral Hydrate    398
**Box 13B**    Biological Reduction of Aldehydes and Ketones    404
**Box 13C**    Formaldehyde in Plastics and Plywood    406
**Box 13D**    Some Important Naturally Occurring Aldehydes
        and Ketones    407

**CHAPTER 13**    391
**Aldehydes and
Ketones**

14.1    Introduction    415
14.2    Nomenclature of Carboxylic Acids    416
14.3    Physical Properties of Carboxylic Acids    417
14.4    Some Important Carboxylic Acids    418
14.5    Preparation of Carboxylic Acids    419
14.6    The Acidity of Carboxylic Acids    419
14.7    Properties of Carboxylic Acid Salts    420
14.8    Nomenclature of Carboxylic Esters    422
14.9    Physical Properties of Carboxylic Esters    424
14.10    Preparation of Carboxylic Esters    425
14.11    Hydrolysis of Carboxylic Esters    431
14.12    Esters and Anhydrides of Phosphoric Acid    434
        Summary    436
        Key Terms    437
        Problems    438

**CHAPTER 14**    415
**Carboxylic Acids
and Esters**

**Box 14A** Some Uses of Carboxylate Salts     421
**Box 14B** Carboxylic Esters as Flavoring Agents     425
**Box 14C** Aspirin     429
**Box 14D** Polyesters     430
**Box 14E** The Use of Absorbable Staples in Hysterectomy     432

**CHAPTER 15**     **443**     15.1     Introduction     443
**Amines and Amides**                 15.2     Nomenclature of Amines     443
                                                  15.3     Physical Properties of Amines     446
                                                  15.4     The Basicity of Amines. Quaternary Salts     448
                                                  15.5     Amides: Nomenclature and Physical Properties     451
                                                  15.6     Acid–Base Properties of Organic Compounds. The Neutrality
                                                               of Amides     453
                                                  15.7     Preparation of Amides     454
                                                  15.8     Hydrolysis of Amides     456
                                                  15.9     Alkaloids     459
                                                              Summary     465
                                                              Key Terms     465
                                                              Problems     466

**Box 15A** Amphetamines (Pep Pills)     446
**Box 15B** The Solubility of Drugs in Body Fluids     449
**Box 15C** Nitrites in Meat and the Carcinogenicity
                    of *N*-Nitrosamines     450
**Box 15D** Nonprescription Pain-Relievers     452
**Box 15E** Nylon     455
**Box 15F** Urea     456
**Box 15G** Barbiturates     457
**Box 15H** Tranquilizers     459
**Box 15I** Cocaine     462

**PART III**     **473**     **Interview:**  Jacqueline K. Barton     474
**Biochemistry**

**CHAPTER 16**     **479**     16.1     Introduction     479
**Carbohydrates**                 16.2     Enantiomers and Chirality     480
                                                  16.3     Properties of Enantiomers. Optical Activity     483
                                                  16.4     Compounds with More Than One Chiral Carbon     487
                                                  16.5     Glucose     489
                                                  16.6     Monosaccharides     491
                                                  16.7     Physical Properties of Monosaccharides     494
                                                  16.8     Chemical Properties of Monosaccharides     495
                                                  16.9     Derivatives of Monosaccharides     496
                                                  16.10    Disaccharides     498
                                                  16.11    Polysaccharides     500

16.12   Acidic Polysaccharides   506
16.13   Chiral Compounds in Biological Organisms   508
Summary   510
Key Terms   510
Problems   511

**Box 16A**   Amount of Rotation   485
**Box 16B**   The Chair Form of Pyranose Rings   490
**Box 16C**   Blood Groups   500
**Box 16D**   Galactosemia   503
**Box 16E**   Polysaccharides as Fillers in Drug Tablets   505
**Box 16F**   Retina Detachment   507
**Box 16G**   Thalidomide   509

18.6   Some Properties of Peptides and Proteins    555
18.7   The Primary Structure of Proteins    557
18.8   The Secondary Structure of Proteins    560
18.9   The Tertiary and Quaternary Structure of Proteins    563
18.10  Denaturation    565
18.11  Glycoproteins    567
       Summary    568
       Key Terms    568
       Problems    569

Box 18A   Aspartic Acid Racemization and Aging    550
Box 18B   Artificial Sweeteners    553
Box 18C   Glutathione    555
Box 18D   The Use of Human Insulin    559
Box 18E   Sickle Cell Anemia    560
Box 18F   Collagen Abnormalities    562
Box 18G   The Structure of Collagen in Bones and Teeth    565
Box 18H   Some Applications of Protein Denaturation    566
Box 18I   Permanent Waves    567

*Brookhaven Data Bank*

**CHAPTER 19**    **573**    19.1   Introduction    573
**Enzymes**                 19.2   Naming and Classifying Enzymes    574
                            19.3   Common Terms in Enzyme Chemistry    577
                            19.4   Factors Affecting Enzyme Activity    578
                            19.5   Mechanism of Enzyme Action    579
                            19.6   Enzyme Regulation and Allosterism    583
                            19.7   Enzymes in Medical Diagnosis and Treatment    585
                                   Summary    586
                                   Key Terms    587
                                   Problems    587

Box 19A   Muscle Relaxants and Enzyme Specificity    575
Box 19B   Meat Tenderizers    577
Box 19C   Active Sites    580
Box 19D   Sulfa Drugs as Competitive Inhibitors    581
Box 19E   Noncompetitive Inhibition and Heavy Metal Poisoning    583
Box 19F   Penicillin. War of Enzyme Against Enzyme    586

**CHAPTER 20**        **591**    20.1   Introduction    591
**Bioenergetics. How**          20.2   Mitochondria    592
**the Body Converts**           20.3   The Principal Compounds of the Common Catabolic Pathway    594
**Food to Energy**              20.4   The Citric Acid Cycle    598
                                20.5   Electron and $H^+$ Transport (Oxidative Phosphorylation)    603
                                20.6   Phosphorylation and the Chemiosmotic Pump    605
                                20.7   The Energy Yield    606
                                20.8   Conversion of Chemical Energy to Other Forms of Energy    607

Summary    609
Key Terms    609
Problems    610

**Box 20A**    Nicotine and Nicotinic Acid    597
**Box 20B**    2,4-Dinitrophenol as an Uncoupling Agent    604
**Box 20C**    Overdose of Barbiturates    606

21.1    Specific Pathways and Their Convergence to the Common Pathway    613
21.2    Glycolysis    615
21.3    The Energy Yield from Glucose    619
21.4    Glycerol Catabolism    621
21.5    β-Oxidation of Fatty Acids    622
21.6    The Energy Yield from Stearic Acid    624
21.7    Ketone Bodies    625
21.8    Catabolism of the Nitrogen of Amino Acids    626
21.9    Catabolism of the Carbon Skeleton of Amino Acids    629
21.10    Catabolism of Heme    631
Summary    633
Key Terms    633
Problems    634

**Box 21A**    Fructose Intolerance    618
**Box 21B**    Lactate Accumulation    619
**Box 21C**    Glycogen Storage Diseases    621
**Box 21D**    Diphtheria    622
**Box 21E**    Ketoacidosis in Diabetes    626
**Box 21F**    MSG and Chinese Restaurant Syndrome    631
**Box 21G**    Hereditary Defects in Amino Acid Catabolism. PKU    632
**Box 21H**    Jaundice    633

**CHAPTER 21**    **613**
**Specific Catabolic Pathways: Carbohydrate, Lipid, and Protein Metabolism**

© *CNRI/Phototake NYC*

22.1    Introduction    637
22.2    Biosynthesis of Carbohydrates    638
22.3    Biosynthesis of Fatty Acids    641
22.4    Biosynthesis of Membrane Lipids    643
22.5    Biosynthesis of Amino Acids    645
Summary    646
Key Terms    647
Problems    647

**Box 22A**    Photosynthesis    640
**Box 22B**    Phospholipid Synthesis and Hyaline Membrane Disease    644
**Box 22C**    Amino Acid Transport and Blue Diaper Syndrome    645

**CHAPTER 22**    **637**
**Biosynthetic Pathways**

**CHAPTER 23**    **649**    23.1    Introduction    649
**Nucleic Acids and**                23.2    Components of Nucleic Acids    650
**Protein Synthesis**                23.3    Structure of DNA and RNA    654
23.4    DNA Replication    660
23.5    RNA    662
23.6    Transmission of Information    663
23.7    Transcription    664
23.8    The Role of RNA in Translation    665
23.9    The Genetic Code    666
23.10    Translation and Protein Synthesis    667
23.11    Genes, Exons, and Introns    669
23.12    Gene Regulation    672
23.13    Mutations, Mutagens, and Genetic Diseases    674
23.14    Recombinant DNA    676
Summary    680
Key Terms    681
Problems    681

**Box 23A**    Anticancer Drugs    653
**Box 23B**    Okazaki Fragments in DNA Replication    661
**Box 23C**    The Action of Antibiotics    664
**Box 23D**    Viruses    670
**Box 23E**    Mutations, Biochemical Evolution, and Molecular
Disease    674
**Box 23F**    Mutagens and Carcinogens    675
**Box 23G**    Oncogenes    676
**Box 23H**    DNA Fingerprinting    678

*Drs. Salunke, Caspar, and Garcea*

**CHAPTER 24**    **685**    24.1    Introduction    685
**Chemical**                24.2    Neurotransmitters    687
**Communication:**                24.3    Acetylcholine    687
**Neurotransmitters,**                24.4    Monoamines and Amino Acids: Adrenergic Neurotransmitters    690
**Hormones, and**                24.5    Peptidergic Neurotransmitters    694
**Immunoglobulins**                24.6    Hormones    696
24.7    Immunoglobulins    699
24.8    How Antibodies Work    702
Summary    703
Key Terms    704
Problems    704

**Box 24A**    Alzheimer's Disease and Acetylcholine Transferase    688
**Box 24B**    Nerve Gases and Antidotes    689
**Box 24C**    Fugu (Puffer Fish) Poison and Communication
Between Neurons    690
**Box 24D**    Parkinson's Disease—Depletion of Dopamine    692
**Box 24E**    Chemically Induced Headaches    693

**Box 24F**   Nitric Oxide and Long-Term Memory    694
**Box 24G**   Diabetes    698
**Box 24H**   Immunization    699
**Box 24I**   AIDS    700
**Box 24J**   Using Antibodies to Locate Specific Cell
Components    702

25.1   Introduction    707
25.2   Functions and Composition of Blood    709
25.3   Blood as a Carrier of Oxygen    713
25.4   Transport of Carbon Dioxide in the Blood    714
25.5   Blood Buffers    714
25.6   Blood Cleansing—A Kidney Function    715
25.7   Buffer Production—Another Kidney Function    719
25.8   Water and Salt Balance in Blood and Kidneys    719
25.9   Blood Pressure    720
Summary    721
Key Terms    722
Problems    722

**Box 25A**   Using the Blood-Brain Barrier to Eliminate
Undesirable Side Effects of Drugs    709
**Box 25B**   Blood Clotting    712
**Box 25C**   Composition of Body Fluids    716
**Box 25D**   Sex Hormones and Old Age    718
**Box 25E**   Hypertension and Its Control    720

**CHAPTER 25**    **707**
**Body Fluids**

*Dr. Peter T. Fox*

26.1   Introduction    725
26.2   Nutrition    725
26.3   Calories    727
26.4   Carbohydrates, Fats, and Proteins in the Diet    728
26.5   Vitamins, Minerals, and Water    730
26.6   Digestion    731
26.7   Digestion of Carbohydrates    731
26.8   Digestion of Lipids    737
26.9   Digestion of Proteins    737
Summary    738
Key Terms    738
Problems    738

**Box 26A**   Iatrogenic Malnutrition    728
**Box 26B**   Parenteral Nutrition    729
**Box 26C**   Too Much Vitamin Intake?    730
**Box 26D**   Hypoglycemia and Hyperglycemia    737

**CHAPTER 26**    **725**
**Nutrition and
Digestion**

**APPENDIX A**        **A.1**
**Exponential**
**Notation**


**APPENDIX B**        **A.5**
**Significant Figures**


**ANSWERS**           **A.9**


**GLOSSARY-INDEX**    **I.1**

# HEALTH-RELATED TOPICS

*Drs. Dieppe, Bacon, Bamji, and Watt; Gower Medical Publishing Co.*

| | |
|---|---|
| Absorbable staples in surgery | Box 14E |
| Acidic polysaccharides in the body | Sec. 16.12 |
| Acidosis and alkalosis | Box 8E |
| Aging and racemization | Box 18A |
| AIDS | Box 24I |
| Alcoholic beverages | Box 12A |
| Alcoholism and disulfuram | Box 12B |
| Alcohols in medical use | Sec. 12.4 |
| Alkaloids | Sec. 15.9 |
| Alzheimer's disease | Box 24A |
| Amphetamines | Box 15A |
| Anabolic steroids | Box 17H |
| Anesthetics | Box 12F |
| Antacids | Box 8C |
| Anti-inflammatory drugs | Box 17J |
| Antibiotic's action | Box 23C |
| Anticancer drugs | Box 23A |
| Antiseptics | Box 4C |
| Aspirin | Box 14C |
| Barbiturates | Boxes 15G; 20C |
| Bends | Box 6B |
| Bile salts | Sec. 17.11 |
| Biological effects of alkanes | Box 10B |

| | |
|---|---|
| Biologically important ions | Box 3A |
| Blood buffers | Secs. 8.10, 25.5 |
| Blood circulation | Sec. 1.2 |
| Blood cleansing | Sec. 25.6 |
| Blood clotting | Sec. 16.12, Box 25B |
| Blood groups | Box 16C |
| Blood pressure | Sec. 25.9 |
| Blood pressure measurement | Box 5C |
| Blood-brain barrier | Box 25A |
| Blue diaper syndrome | Box 22C |
| Body fluids | Box 25C |
| Breathing | Boxes 5D; 5F |
| Bubonic plague | Sec. 1.1 |
| Caloric values of foods | Box 4D |
| Calories | Sec. 26.3 |
| Carbon monoxide as a poison | Box 10D |
| Carboxylate salts | Box 14A |
| Carboxylic acids | Sec. 14.4 |
| CAT and Nmr scans | Box 9E |
| Chaperones | Secs. 18.9; 18.10 |
| Chinese restaurant syndrome | Box 21F |
| Chiral compounds in the body | Sec. 16.13 |
| Chloral hydrate | Box 13A |
| Cholesterol | Sec. 17.9 |
| Cholesterol and heart attacks | Box 17G |
| Cis-trans isomerism in vision | Box 11C |
| Cocaine | Box 15I |
| Cold compresses | Box 1E |
| Collagen abnormalities | Box 18F |
| Collagen in bones | Box 18G |
| Complex lipids | Secs. 17.4; 17.6–17.8 |
| Corneal burns | Box 8B |
| Crenation | Sec. 6.9 |
| Diabetes | Box 24G |
| Dialysis | Sec. 6.10 |
| Digestion | Secs. 26.6–26.9 |
| Diphtheria | Box 21D |
| Drug dosage and body mass | Box 1B |
| Elements in people | Boxes 2A; 2D |
| Enzymes in medicine | Sec. 19.7 |
| Essential amino acids | Secs. 18.2; 26.4 |
| Essential fatty acids | Secs. 17.3; 26.4 |
| Ethyl chloride spray | Box 5H |
| Fever | Box 7B |
| Fire hazards in hospitals | Box 7A |
| Formalin | Sec. 13.7 |
| Fructose intolerance | Box 21A |
| Galactosemia | Box 16D |
| Gene regulation | Sec. 23.12 |

| | |
|---|---|
| Genes | Sec. 23.11 |
| Genetic diseases | Sec. 23.13 |
| Glucose | Sec. 16.5 |
| Glutathione | Box 18C |
| Glycogen storage diseases | Box 21C |
| Glycoproteins | Sec. 18.11 |
| Headaches, chemically induced | Box 24E |
| Heavy metal poisoning | Box 19E |
| Hemodialysis | Box 6G |
| Hemoglobin | Box 3C |
| Hemolysis | Sec. 6.9 |
| Hemorrhoids and sitz bath | Box 6F |
| High energy phosphates | Sec. 14.12 |
| Hormones | Sec. 24.6 |
| Hyaline membrane disease | Box 22B |
| Hyperbaric medicine | Box 5E |
| Hypertension and its control | Box 25E |
| Hypoglycemia and hyperglycemia | Boxes 6D; 26D |
| Hypothermia and hyperthermia | Box 1D |
| Iatrogenic malnutrition | Box 26A |
| Immunization | Box 24G |
| Immunoglobulins | Sec. 24.7 |
| Insulin | Box 18D |
| Iodine and goiter | Box 12J |
| Ionic compounds in medicine | Box 3B |
| Isotope generators in medicine | Box 9F |
| Jaundice | Box 21H |
| Ketoacidosis in diabetes | Box 21E |
| Ketone bodies | Sec. 21.7 |
| Lactate accumulation | Box 21B |
| Laser surgery | Box 1C |
| Leukotrienes | Sec. 17.12 |
| Lipid storage diseases | Box 17F |
| Lipoproteins: HDL and LDL | Sec. 17.9 |
| Lowering body temperature | Box 7C |
| Marijuana | Box 12H |
| Membranes | Sec. 17.5 |
| Menstrual cycle | Box 17I |
| Mercury as a poison | Box 5B |
| Methanol as a poison | Box 12C |
| Molecular diseases | Box 23E |
| Monoamine oxidase inhibitors | Sec. 24.4 |
| Multiple sclerosis | Box 17E |
| Muscle relaxants | Box 19A |
| Mutagens and carcinogens | Box 23F |
| Myelin sheath | Box 17E |
| Nerve gases | Box 24B |
| Neurotransmitters | Secs. 24.2–24.5 |
| Nicotine and nicotinic acid | Box 20A |

Nitric oxide and memory — Box 24F
Nitroglycerine — Box 12D
Nitrosamines as carcinogens — Box 15C
Nutrition — Chapter 26
$O_2$ and $CO_2$ in blood — Box 5F
Oncogenes — Box 23G
Oral contraception — Box 17I
Oxytocin — Sec. 18.7
Pain relievers — Box 15D
Parenteral nutrition — Box 26B
Parkinson's disease — Box 24D
Penicillin — Box 19F
pH of body fluids — Sec. 8.8
Phenols as antiseptics — Box 12E
Phenylketonuria (PKU) — Box 21G
Plastics in surgery — Sec. 11.7
Polyesters in surgery — Box 14D
Protein denaturation — Box 18H
Prostaglandins — Sec. 17.12
Radiation dosimetry — Sec. 9.5
Radiation sickness — Sec. 9.5
Radioactive fallout — Box 9H
Radioactive isotopes in medicine — Sec. 9.6
Recombinant DNA — Sec. 23.14
Retina detachment — Box 16F
Scientific method in medicine — Box 1A
Sex hormones and old age — Sec 17.10; Box 25D
Sickle cell anemia — Box 18E
Smoking and carcinogens — Box 11F
Solubility of drugs in body — Box 15B
Steroids — Sec. 17.10
Strontium-90 — Box 2B
Sulfa drugs — Box 19D
Teratogen — Box 16G
Thalidomide — Box 16G
Tranquilizers — Box 15H
Urea — Box 15F
Vasopressin — Sec. 18.7
Viruses — Box 23D
Vitamins — Sec. 26.5; Box 26C
Water and life — Sec. 6.6

Genes                                      Sec. 23.11
Genetic diseases                           Sec. 23.13
Glucose                                    Sec. 16.5
Glutathione                                Box 18C
Glycogen storage diseases                  Box 21C
Glycoproteins                              Sec. 18.11
Headaches, chemically induced              Box 24E
Heavy metal poisoning                      Box 19E
Hemodialysis                               Box 6G
Hemoglobin                                 Box 3C
Hemolysis                                  Sec. 6.9
Hemorrhoids and sitz bath                  Box 6F
High energy phosphates                     Sec. 14.12
Hormones                                   Sec. 24.6
Hyaline membrane disease                   Box 22B
Hyperbaric medicine                        Box 5E
Hypertension and its control               Box 25E
Hypoglycemia and hyperglycemia             Boxes 6D; 26D
Hypothermia and hyperthermia               Box 1D
Iatrogenic malnutrition                    Box 26A
Immunization                               Box 24G
Immunoglobulins                            Sec. 24.7
Insulin                                    Box 18D
Iodine and goiter                          Box 12J
Ionic compounds in medicine                Box 3B
Isotope generators in medicine             Box 9F
Jaundice                                   Box 21H
Ketoacidosis in diabetes                   Box 21E
Ketone bodies                              Sec. 21.7
Lactate accumulation                       Box 21B
Laser surgery                              Box 1C
Leukotrienes                               Sec. 17.12
Lipid storage diseases                     Box 17F
Lipoproteins: HDL and LDL                  Sec. 17.9
Lowering body temperature                  Box 7C
Marijuana                                  Box 12H
Membranes                                  Sec. 17.5
Menstrual cycle                            Box 17I
Mercury as a poison                        Box 5B
Methanol as a poison                       Box 12C
Molecular diseases                         Box 23E
Monoamine oxidase inhibitors               Sec. 24.4
Multiple sclerosis                         Box 17E
Muscle relaxants                           Box 19A
Mutagens and carcinogens                   Box 23F
Myelin sheath                              Box 17E
Nerve gases                                Box 24B
Neurotransmitters                          Secs. 24.2–24.5
Nicotine and nicotinic acid                Box 20A

| | |
|---|---|
| Nitric oxide and memory | Box 24F |
| Nitroglycerine | Box 12D |
| Nitrosamines as carcinogens | Box 15C |
| Nutrition | Chapter 26 |
| $O_2$ and $CO_2$ in blood | Box 5F |
| Oncogenes | Box 23G |
| Oral contraception | Box 17I |
| Oxytocin | Sec. 18.7 |
| Pain relievers | Box 15D |
| Parenteral nutrition | Box 26B |
| Parkinson's disease | Box 24D |
| Penicillin | Box 19F |
| pH of body fluids | Sec. 8.8 |
| Phenols as antiseptics | Box 12E |
| Phenylketonuria (PKU) | Box 21G |
| Plastics in surgery | Sec. 11.7 |
| Polyesters in surgery | Box 14D |
| Protein denaturation | Box 18H |
| Prostaglandins | Sec. 17.12 |
| Radiation dosimetry | Sec. 9.5 |
| Radiation sickness | Sec. 9.5 |
| Radioactive fallout | Box 9H |
| Radioactive isotopes in medicine | Sec. 9.6 |
| Recombinant DNA | Sec. 23.14 |
| Retina detachment | Box 16F |
| Scientific method in medicine | Box 1A |
| Sex hormones and old age | Sec 17.10; Box 25D |
| Sickle cell anemia | Box 18E |
| Smoking and carcinogens | Box 11F |
| Solubility of drugs in body | Box 15B |
| Steroids | Sec. 17.10 |
| Strontium-90 | Box 2B |
| Sulfa drugs | Box 19D |
| Teratogen | Box 16G |
| Thalidomide | Box 16G |
| Tranquilizers | Box 15H |
| Urea | Box 15F |
| Vasopressin | Sec. 18.7 |
| Viruses | Box 23D |
| Vitamins | Sec. 26.5; Box 26C |
| Water and life | Sec. 6.6 |

# PART I

*Photograph by Beverly March.*

# General Chemistry

# INTERVIEW

# Roald Hoffmann

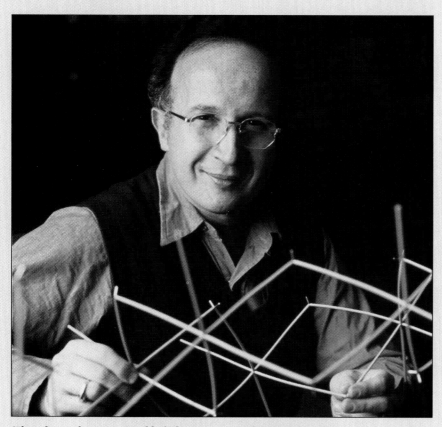

*(Photo from video series "World of Chemistry," Annenberg/CPB Project.)*

Roald Hoffmann is a remarkable individual. When he was only 44 years old, he shared the 1981 Nobel Prize in Chemistry with Kenichi Fukui of Japan for work in applied theoretical chemistry. In addition, he has received awards from the American Chemical Society in both organic chemistry and inorganic chemistry, the only person to have achieved this honor. And, in 1990, he was awarded the Priestley Medal, the highest award given by the American Chemical Society.

The numerous honors celebrating his achievements in chemistry tell only part of the story of his life. He was born to a Polish Jewish family in Zloczow, Poland, in 1937, and was named Roald after the famous Norwegian explorer Roald Amundsen. Shortly after World War II

began in 1939, the Nazis first forced him and his parents into a ghetto and then into a labor camp. However, his father smuggled Hoffmann and his mother out of the camp, and they were hidden for more than a year in the attic of a schoolhouse in the Ukraine. His father was later killed by the Nazis after trying to organize an attempt to break out of the labor camp. After the war, Hoffmann, his mother, and his stepfather made their way west to Czechoslovakia, Austria, and then Germany. They finally emigrated to the United States, arriving in New York on Washington's birthday in 1949. That Hoffmann and his mother survived these years is our good fortune. Of the 12,000 Jews living in Zloczow in 1941 when the Nazis took over, only 80 people, three of them children, survived the Holocaust. One of those three children was Roald Hoffmann.

On arriving in New York, Hoffmann learned his sixth language, English. He went to public schools in New York City and then finally to Stuyvesant High School, one of the city's select science schools. From there he went to Columbia University and then to Harvard University, where he earned his Ph.D. in 1962. Shortly thereafter, he began the work with Professor R. B. Woodward that eventually led to the Nobel Prize. Since 1965 he has been a professor at Cornell University, where he regularly teaches first-year general chemistry.

In addition to his work in chemistry, Professor Hoffmann also writes popular articles on science for the American Scientist and other magazines, and he has published two volumes of his poetry. Fi-

*nally, he appears in a series of 26 half-hour television programs for a chemistry course called "World of Chemistry," which began airing on public television and cable channels in 1990.*

## From medicine to cement to theoretical chemistry

We visited Professor Hoffmann in his office at Cornell University, an office full of mineral samples, molecular models, and Japanese art. When asked what brought him into chemistry he said that "I came rather late to chemistry; I was not interested in it from childhood." However, he clearly feels that one can come late to chemistry, and that it can be a very positive thing. "I am always worried about fields in which people exhibit precocity, like music and mathematics. Precocity is some sort of evidence that you have to have talent. I don't like that. I like the idea that human beings can do anything they want to. They need to be trained sometimes. They need a teacher to awaken the intelligences within them. But to be a chemist requires no special talent, I'm glad to say. Anyone can do it, with hard work."

He took a standard chemistry course in high school. He recalls that it was a fine course, but apparently he found biology more enjoyable because, in his high school year book, "under the picture of me with a crew cut, it says 'medical research' under my name." Indeed, he says that "medical research was a compromise between my interest in science and the typical Jewish middle class family pressures to become a medical doctor. The same kind of pressures apply to Asian-Americans today."

When he went to Columbia University, Hoffmann enrolled as a pre-med student, but says that there were several factors that shifted him away from a career in medicine. One of these was his work at the National Bureau of Standards in Washington, D.C., for two summers and then at Brookhaven

National Laboratory for a third summer. He says that these experiences gave him a feeling for the excitement of chemical research. Nonetheless, during his first summer at the Bureau of Standards, he "did some not very exciting work on the thermochemistry of cement." During his second summer there he visited the National Institutes of Health to find out what medical research was about. "To my amazement," he says, "most of the people had Ph.D.'s and not M.D.'s. I just didn't know. Young people do not often know what is required for a given profession. Once I found that out, and found that I did well in chemistry, it made me feel that I didn't really have to do medicine, that I could do some research in chemistry or biology. Later, what influenced me to decide on theoretical chemistry was an excellent instructor. Had I had some really good instruction in organic chemistry, I'm pretty sure I would have become an organic chemist."

"At the very same time I was being exposed to the humanities, in part because of Columbia's core curriculum —which I think is a great idea—that had so-called contemporary civilization and humanities courses. I took advantage of the liberal arts education to the hilt, and that has remained with me all my life. The humanities teachers have remained permanently fixed in my mind and have changed my ways of thinking. These were the people who really had the intellectual impact on me and helped to shape my life."

"To trace the path, I was a late-comer to chemistry and was inspired by research. I think *research* is the way in. It just gives you a different perception."

## A love for complexity

Having discussed what brought him into chemistry, we were interested in his view of the qualities that a student should possess to pursue a career in the field. He said that "one thing one

needs to be a chemist is a love for complexity and richness. To some extent that is true of biology and natural history, too. I think one of the things that is beautiful about chemistry is that there are 10,000,000 compounds, each with different properties. What's beautiful when you make a molecule is that you can make derivatives in which you can vary substituents, the pieces of a molecule, and we know that those substituents give a molecule function, give it complexity and richness. That's why a protein or nucleic acid with all its variety is essential for life. That's why to me, intellectually, isomerism and stereochemistry in organic chemistry are at the heart of chemistry. I think we should teach that much earlier. It requires no mathematics, only a little model building; you can do this without theory. I think it is no accident that organic chemistry drew to it the intellects of its time."

## Experiment and theory

Professor Hoffmann has spent his career immersed in the theories of chemistry. However, he believes that fundamentally "chemistry is an experimental science, in spite of some of my colleagues saying otherwise. However, the educational process certainly favors theory. It's in the nature of things for teacher and student both to want to understand and then give primacy to the soluble and the understood at the expense of other things. We also have this reductionist philosophy of science, the idea that the social sciences derive from biology, that biology follows from chemistry, chemistry from physics, and so on. This notion gives an inordinate amount of importance to theoretical thinking, the more mathematical the better. Of course this is not true in reality, but it's an ideology; it is a religion of science."

There is of course a role for theory. "You can't report just the facts and nothing but the facts; by themselves they are dull. They have to be woven into a framework so that there is understanding. That's accomplished usually by a theory. It may not be mathematical, but a qualitative network of relationships." Indeed, Hoffmann believes that the incorporation of theory into chemistry "is what made American science better than that in many other countries. The emphasis in chemistry on theory and theoretical understanding is very important, but not nearly as important as the syntheses and reactions of molecules."

"Although I think chemists need to like to do experiments, that doesn't mean there is no role for people like me. It turns out that I am really an experimental chemist hiding as a theoretician. I think that is the key to my success. That is, I think I can empathize with what bothers the experimentalists. In another day I could have become an experimentalist."

## Major issues in chemistry and science today

Professor Hoffmann has worked, and is presently working, at the forefront of several major areas of chemistry. He is currently quite intrigued by surface science. "For instance, there is the Fischer-Tropsch process, a pretty incredible thing in detail. Carbon monoxide and hydrogen gas come onto a metal catalyst, a surface of some sort, and off come long-chain hydrocarbons and alcohols. The richness of all these things happening is intriguing, and we are on the verge of understanding. We now have structural information on surfaces that's reliable, and we are just beginning to get kinetic information. Surface science is at a crossing of chemistry, physics, and engineering. The field is in some danger of being spun off on its own, but I would like to keep it in chemistry."

"Bioinorganic chemistry is another such field. In my group we are doing some work trying to understand the mechanism of oxygen production in photosynthesis, the last steps. What is known is very little. There is an enzyme in photosystem II that involves 3 to 4 manganese atoms. And they somehow take oxide or hydroxide to peroxide and eventually to molecular oxygen. That's all we know. Experimentally, not theoretically, I think bioinorganic chemistry is a very interesting field."

Finally, Hoffmann remarked that "there are going to be finer and finer ways of controlling the synthesis of molecules, the most essential activity of chemists. If I were to point to a single thing that chemists do, it would be that they make molecules. Chemistry is the science of molecules and their transformations. The transformations are the essential part. I think there are exciting possibilities for chemical intervention into biological systems with an ever finer degree of control. We need not be afraid of nature. We can mimic it, and even surpass its synthetic capabilities. And find a way to cooperate with it."

## Scientific literacy and democracy

Roald Hoffmann is very concerned not only about science in general and chemistry in particular but also about our society. One of his concerns is scientific literacy because "some degree of scientific literacy is absolutely necessary today for the population at large as part of a democratic system of government. People have to make intelligent decisions about all kinds of technological issues." He recently offered his comments on this important issue in the *New York Times*. He wrote that "What concerns me about scientific, or humanistic, illiteracy is the barrier it poses to rational democratic governance. Democracy occasionally gives in to *technocracy*—a reliance on experts on matters such as genetic engineering, nuclear waste disposal, or the cost of medical care. That is fine, but the people must be able to vote intelligently on these issues. The less we know as a nation, the more we must rely on experts, and the more likely we are to be misled by demagogues. We must know more."

## The responsibility of scientists

Our discussion of the importance of scientific literacy led to a conversation about the broader responsibilities of scientists. "Scientists have a great obligation to speak to the public," Hoffmann says. "We have an obligation as educators to train the next generation of people. We should pay as much attention to those people who are *not* going to be chemists, and sometimes need to make compromises about what is to be taught and what is the nature of our courses. I think scientists have an obligation to speak to the public broadly, and here I think they have been negligent. I think society is paying scientists money to do research, and can demand an accounting in plain language. That's why I put in a lot of time on that television show [*The World of Chemistry*]."

## A teacher of chemistry—and proud of it

In the Nobel Yearbook, Professor Hoffmann wrote that the technical description of his work "does not communicate what I think is my major contribution. I am a teacher, and I am proud of it. At Cornell University I have taught primarily undergraduates—I have also taught chemistry courses to nonscientists and graduate courses in bonding theory and quantum mechanics. To the chemistry community at large, and to my fellow scientists, I have tried to teach "applied theoretical chemistry": a special blend of computations stimulated by experiment and coupled to the construction of general models—frameworks for understanding." His success in this is unquestioned.

Interview conducted by John C. Kotz, SUNY Distinguished Teaching Professor, State University of New York, College at Oneonta.

*Charles Winters*

# FLUORITE
## *by* **Roald Hoffmann**

I was asked about my hobbies.
"Collecting minerals," I said
and stopped to think.
"Minerals in their matrix
are what I like best."

   Fluorite wears a variable habit.
Colorless when pure, it is vodka
in stone. More commonly
it brandishes shades of rose to blue,
an occasional yellow. A specimen I have
tumbles in inch-long cubes,
superimposed, interpenetrating,
etched on all their faces.
The cubes have a palpable darkness,
a grainy darkness, texture
blacker than black.

Solid yet fragile when held
up to light, the darkness
deposited in this ordered
atomic form a million years ago
allows some rays through.
But only on the thin edges,
in sinister violet.

   Struck with a chisel
the cubes cleave and
octahedra emerge.
I have seen it done,
but my hands tremble.
I know why it cleaves so,
but why destroy what took
centuries to grow, then
rested aeons in a cool
fissure in the rock?

   Eerie crystal.
Were a Martian photograph
enlarged to reveal such polyhedral
regularity, it would be deemed
a sign of intelligence at work.
But the only work here, and it is free,
   is that of entropy.

Roald Hoffmann, the John A. Newman professor of physical science at Cornell, won the Nobel Prize in chemistry in 1981. He has published two books of poems: *The Metamict State* (1987), and *Gaps and Verges* (1990).

5

*The Earth from space. (Courtesy of NASA.)*

# Matter, Energy, and Measurement

## 1.1 INTRODUCTION

There was a time—only a few hundred years ago—when physicians were powerless to treat many diseases. Cancer, tuberculosis, smallpox, typhus, plague, and many other sicknesses struck people seemingly at random, and doctors, who had no idea how any of these diseases were caused, could do little or nothing about them. Between 1348 and 1350 the disease known as the Black Death (bubonic plague) wiped out about one third of the population of Europe (Fig. 1.1). Doctors treated it with magic as well as by such measures as bleeding (Fig. 1.2), laxatives, hot plasters, and pills made from powdered stag horn, saffron, or gold. None of these were of any use, and the doctors, because they came into direct contact with a highly contagious disease, died at a much higher rate than the general public.

"Bring out your Dead"

**Figure 1.1**
This old engraving shows a town crier asking people to bring out the bodies of plague victims. (Courtesy of the National Library of Medicine.)

Doctors of those times did have remedies for a few conditions, such as the quinine bark that was used to treat malaria, but all such remedies were discovered by trial and error. For most illnesses, human beings, including the doctors themselves, were entirely at the mercy of body processes and microorganisms they knew nothing about.

**Figure 1.2**
A woman being bled by a leech on her left forearm; a bottle containing more leeches is on the table. From a 1639 woodcut. (Courtesy of the National Library of Medicine.)

Medicine has made great progress since those times. Although people still get sick, they do on the average live much longer, and many once-feared diseases have been essentially eliminated or made easily curable. Smallpox has been completely eradicated, and polio, typhus, bubonic plague, diphtheria, and other diseases that once killed millions no longer pose a serious problem, at least not in the developed countries.

How has this medical progress come about? The answer is that diseases could not be cured until they were understood, and this understanding has come about through a knowledge of how the body functions. The study of bodily functions is an important part of the science of biology, but biologists know that because most body processes are chemical reactions it is also necessary to study chemistry. It is progress in our understanding of the principles of biology, chemistry, and physics that has led to these advances in medicine. When further advances are made—for example, the curing of cancer—they too will come from an increased understanding of biology and chemistry.

Because so much of modern medicine depends on chemistry, it is essential that students who intend to enter the health professions have some understanding of basic chemistry. This book was written to help you achieve that goal. Even if you choose a different profession, you will find that the chemistry you learn in this course will enrich your life.

### Chemistry as the Study of Matter

The universe consists of matter, energy, and empty space. **Matter** is the stuff that everything is made of. It can be defined as **anything that has mass and takes up space. Chemistry is the science that deals with matter:** the structure and properties of matter and the transformations from one form of matter to another. We will discuss energy in Section 1.7.

## 1.2 THE SCIENTIFIC METHOD

Scientists learn what they know by using a tool called the **scientific method.** The heart of the scientific method is the *testing* of theories. It was not always so. Before about 1600, philosophers believed statements because they sounded right. For example, the great philosopher Aristotle (384–322 BC) believed that if you took the gold out of a mine it would grow back. He believed this because it fitted in with a more general picture that he had about the workings of nature. In ancient times, this was the way most thinkers behaved. If a statement sounded right, they believed it without testing it.

Before about 1600, science as such did not exist except as a branch of philosophy.

About 1600 the scientific method came into use. Let us look at an example to see how the scientific method operates. The Greek physician Galen (AD 130–200, Fig. 1.3) recognized that the blood on the left side of the heart somehow gets to the right side. This is a fact. A **fact** is a statement that is obvious to anyone who cares to look; that is, a statement based on direct experience. Having observed this fact, Galen then proposed a hypothesis to explain it. A **hypothesis** is a statement that is not obvious but is offered to explain the facts. Since Galen could not actually see how the blood got from the left side to

For example, it is a fact that humans must breathe in order to live.

**Figure 1.3**
Galen. (Courtesy of the National Library of Medicine.)

the right side of the heart, he came up with the hypothesis that there must be tiny holes present in the muscular wall that separates the two halves.

Up to this point, modern scientists and ancient philosophers would behave the same way. Each offers a hypothesis to explain the facts. From this point on, however, their methods differ. To Galen, his explanation sounded right and that was enough to make him believe it, even though he couldn't see any holes. His hypothesis was in fact believed by virtually all physicians for more than 1000 years. When we use the scientific method, we do not believe a hypothesis just because it sounds right. **We test it,** using the most rigorous testing we can think of.

Around 1600 William Harvey (1578–1657) tested Galen's hypothesis by dissecting human and animal hearts and blood vessels. He discovered that there are one-way valves separating the upper chambers of the heart from the lower chambers. He also discovered that the heart is a pump that, by contracting and expanding, pushes the blood out. Harvey's teacher, Fabricius (1537–1619), had

---

## Box 1A

## The Scientific Method in Medicine

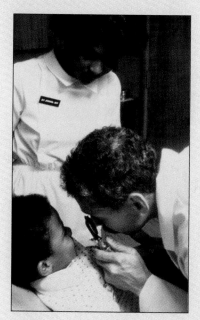

**Figure 1A**
A physician examining a child. (© Superstock, Inc.)

The use of the scientific method is not confined to scientific laboratories. It can be used profitably anywhere a person wants to understand something that isn't immediately obvious. Good detectives use the scientific method in investigating a crime, and doctors use it in diagnosing illnesses. The following is an example of the use of the scientific method in a hospital.

A two-year-old child was brought to the hospital. The child had been vomiting for several weeks and suffered from constipation. No apparent abnormalities were visible. Blood tests showed low hemoglobin levels: 7 g/100 mL against the normal range of 14 to 18 g/100 mL. The red blood cell count was also low: $3.3 \times 10^{12}$/L against $5 \times 10^{12}$/L in normal blood. These tests indicated anemia of unspecified origin. X-rays showed some opaque material in the intestinal lining and along the growth areas of the bones. With this information the doctors were ready to advance a hypothesis: that the opaque material accumulated in the intestines and in the growth areas of the bones was a heavy metal and that the child was suffering from heavy metal poisoning.

It was now necessary to test the hypothesis. If there was a heavy metal, the child must somehow have ingested it. In this case, the first test consisted of asking the parents if there was any way this could have happened. Upon questioning the parents, the doctors learned that the family lived in a rundown house that had peeling paint and that the child had a habit of chewing on window sills and furniture. Since house paint can contain significant amounts of lead, a heavy metal, this lent support to the hypothesis, but it still needed further testing. The next test was for lead in the blood, and it was found that almost two and a half times the normal lead concentration was present.

The hypothesis could now be regarded as confirmed. As a final test, treatment for lead poisoning was administered. If successful it would not only prove the hypothesis but also cure the patient. This was done by administering calcium EDTA intravenously. EDTA is a chemical that extracts heavy metals, binds them tightly, and carries them through the kidney membranes into the urine. One day after the injection a 25-fold increase in lead concentration was found in the child's urine. A few days later the poisoning symptoms disappeared, and after a week the patient was discharged, cured (Fig. 1A).

previously observed that there are one-way valves in the veins, so that blood in the veins can travel only toward the heart and not the other way.

Harvey put all these facts together to come up with a better hypothesis: blood circulates throughout the body, pumped by the heart. This was a better hypothesis than Galen's because it fitted the facts more closely. Even so, it was still a hypothesis and, according to the scientific method, had to be tested further. One important test took place in 1661, four years after Harvey died. According to Harvey's hypothesis, there had to be a way for the blood to get from the arteries to the veins, and Harvey predicted that there must be tiny blood vessels connecting them. In 1661 the Italian anatomist Malpighi (1628–1694), using the newly discovered microscope, found these tiny vessels, now called **capillaries.**

Malpighi's discovery supported the blood circulation hypothesis by fulfilling Harvey's prediction. When a hypothesis passes the tests, we have more confidence in it and call it a **theory.** A theory, in this sense, is the same as a hypothesis except that we have a stronger belief in it because it is supported by more evidence. No matter how much confidence we have in a theory, however, if new facts are discovered that conflict with it or if new tests are devised that it does not pass, the theory must be altered or rejected. In the history of science many firmly established theories have eventually been thrown out because they could not pass new tests.

The scientific method is thus very simple. We don't accept a hypothesis or a theory just because it sounds right. We devise tests, and only if the hypothesis or theory passes the tests do we accept it. The enormous progress made since 1600 in chemistry, biology, and the other sciences is a testimony to the value of the scientific method.

The word "theory" has another meaning as well: a model that explains many interrelated facts and can be used to make predictions about natural phenomena. Examples are Newton's theory of gravitation and the kinetic-molecular theory of gases, which we will meet in Section 5.2. This type of theory is also subject to testing and will be discarded or modified if it is contradicted by new facts.

## 1.3 MEASUREMENTS

In our daily lives we are constantly making measurements. We measure ingredients for recipes, driving distances, gallons of gasoline, weights of fruits and vegetables, the timing of TV programs. Doctors and nurses measure pulse rates, blood pressures, temperatures, and drug dosages. Chemistry, like other sciences, is based on measurements.

A measurement consists of two parts: a number and a unit. A number without a unit is usually meaningless. If you were told that a person's weight is 57, the information would be of very little use. Is it 57 pounds, which would indicate that the person is very likely a child or a midget, or 57 kilograms, which is the weight of an average woman or a small man? Or is it perhaps some other unit? Because there are so many units, a number by itself is not enough; the unit must also be stated.

In the United States most measurements are made with the English system of units: pounds, miles, gallons, and so on. In most of the rest of the world, however, this is not the case. In Germany, France, Russia, Japan, and most other countries, almost nobody could tell you what a pound or an inch is. These countries use the **metric system,** a system that was begun in France about 1800 and has since spread throughout the world. Even in the United States,

**Figure 1.4**
A highway sign in Missouri. (Photograph by Beverly March.)

metric measurements are slowly being introduced (Fig. 1.4). For example, many soft drinks and most alcoholic beverages now come in metric sizes (Fig. 1.5). *Scientists* in the United States have been using metric units all along.

Around 1960, international scientific organizations adopted another system, called the **International System of Units** (abbreviated **SI**). The SI is based on the metric system and uses some of the metric units. The main difference is that the SI is more restrictive: it discourages the use of certain metric units and favors others. Although the SI has advantages over the older metric system, it also has significant disadvantages, and for this reason U.S. chemists have been very slow to adopt it. At this time, approximately 35 years after its introduction, not many U.S. chemists use the entire SI, although some of its preferred units are gaining ground.

In this book we shall use the metric system. Occasionally we shall mention the preferred SI unit where it appears that that unit might have a chance of gaining favor.

*"SI" is actually an abbreviation of the French title Système Internationale.*

## Length

The key to the metric system (and the SI) is that there is one basic unit for each kind of measurement and that other units are related to the basic unit only by powers of ten. For example, let us look at measurements of length. In the English system we have the inch, the foot, the yard, and the mile (not to mention such older units as the league, furlong, ell, and rod). If you want to convert one to another, you must memorize or look up these conversion factors:

$$12 \text{ inches} = 1 \text{ foot}$$
$$3 \text{ feet} = 1 \text{ yard}$$
$$1760 \text{ yards} = 1 \text{ mile}$$

*The official definition of the meter is the distance that light travels in*
$$\frac{1}{299\,792\,458} \text{ of a second.}$$

All this is unnecessary in the metric system (and the SI). In both systems the basic unit of length is the **meter** (m). To convert to larger or smaller units we do

**TABLE 1.1   The Most Common Metric Prefixes**

| Prefix | Symbol | Value |
|--------|--------|-------|
| giga | G | $10^9$ = 1 000 000 000 (one billion) |
| mega | M | $10^6$ = 1 000 000 (one million) |
| kilo[a] | k | $10^3$ = 1000 (one thousand) |
| deci | d | $10^{-1}$ = 0.1 (one tenth) |
| centi | c | $10^{-2}$ = 0.01 (one hundredth) |
| milli | m | $10^{-3}$ = 0.001 (one thousandth) |
| micro | $\mu$ | $10^{-6}$ = 0.000001 (one millionth) |
| nano | n | $10^{-9}$ = 0.000000001 (1 billionth) |

[a] The prefixes shown in color are those we encounter most frequently in this book.

not use arbitrary numbers like 12, 3, and 1760, but only 10, 100, 1/10, 1/100, or other powers of ten. This means that **to convert from one metric or SI unit to another, we only have to move the decimal point.** Furthermore, the other units are named by putting prefixes in front of "meter," and **these prefixes are the same throughout the metric system and the SI.**

The most important of these prefixes are shown in Table 1.1. If we put some of these prefixes in front of "meter," we have

$$1 \text{ kilometer (km)} = 1000 \text{ meters (m)}$$

$$1 \text{ centimeter (cm)} = 0.01 \text{ meter}$$

$$1 \text{ nanometer (nm)} = 10^{-9} \text{ meter}$$

For people who have grown up using English units, it is helpful to have some idea of the size of metric units. A meter is equal to 39.37 in, which makes it a little longer than a yard (36 in). An inch is about 2.5 cm. Some conversion factors between the English and metric systems are shown in Table 1.2.

Now that we have discussed length, we shall look at the metric units for some other measurements important in chemistry.

## Volume

**Volume is space.** The volume of a liquid, solid, or gas is the space occupied by that substance. The basic unit of volume in the metric system is the **liter** (L).

**TABLE 1.2   Some Conversion Factors Between the English and Metric Systems**

| Length | Mass | Volume |
|--------|------|--------|
| 1 in = 2.54 cm | 1 oz (avdp) = 28.35 g | 1 qt = 0.946 L |
| 1 m = 39.37 in | 1 lb (avdp) = 453.6 g | 1 gal = 3.785 L |
| 1 mile = 1.609 km | 1 kg = 2.205 lb | 1 L = 33.81 fl oz |
| | 1 g = 15.43 grains | 1 fl oz = 29.57 mL |
| | | 1 L = 1.057 qt |

**Figure 1.5**
Soft drinks in metric sizes. (Photograph by Beverly March.)

This unit is a little larger than a quart (Table 1.2). The only other common metric unit is the milliliter (mL), which is of course equal to $10^{-3}$ L:

$$1000 \text{ mL} = 1 \text{ L}$$

One milliliter is exactly equal to one cubic centimeter (cc or $cm^3$):

$$1 \text{ mL} = 1 \text{ cc}$$

which means that there are 1000 cc in 1 L.

Volume measurements are very important in chemistry and in medicine, and there are many devices for this purpose (Fig. 1.6). Graduated cylinders are used for approximate measurements; burets, pipets, and volumetric flasks are used for more exact ones. In a hypodermic syringe, the means for injecting a drug (the needle) is combined with the means for measuring the volume injected (the barrel).

## Mass

**Mass is the quantity of matter in an object.** The basic unit of mass in the metric system is the **gram** (g). As always in the metric system, larger and smaller units are indicated by prefixes. The ones in common use are

$$1 \text{ kilogram (kg)} = 1000 \text{ g}$$
$$1 \text{ milligram (mg)} = 0.001 \text{ g}$$

The gram is a small unit; there are 453.6 g in one pound (Table 1.2).

The official kilogram is defined by the mass of a particular platinum-iridium bar kept at the International Bureau of Standards in Paris.

**Figure 1.6**
Some instruments for measuring volume.

Graduated cylinder     Pipet     Buret     Volumetric flask     Hypodermic syringe

## Drug Dosage and Body Mass

In many cases, drug dosages are prescribed on the basis of body mass. For example, the recommended dosage of a drug may be 3 mg of drug for each kilogram of body weight. In this case, a 50-kg (110-lb) woman would receive 150 mg and an 82-kg (180-lb) man would get 246 mg. This is especially important for children, since a dose suitable for an adult will generally be too much for a child, who has much less body mass. For this reason, manufacturers package and sell smaller doses of certain drugs, such as aspirin, for children.

In chemistry we use balances to measure mass. As with devices for measuring volume, some balances are used for approximate measurements and others for more exact measurements. Some laboratory balances are shown in Figure 1.7. Most balances operate on the same very simple principle, shown in Figure 1.8. A beam is balanced on a pivot. The object whose mass is to be determined is placed on one side, and some known masses are placed on the other. When enough known masses have been placed to make the beam horizontal, the unknown mass equals the known mass.

There is a basic difference between mass and weight. Mass is independent of location. The mass of a stone, for example, is the same whether we measure it at sea level, on top of a mountain, or in the depths of a mine. In contrast, weight is not independent of location. **Weight is the force a mass experiences under the pull of gravity.** This was dramatically demonstrated when the

Both mass and weight are independent of temperature.

**Figure 1.7**
Two laboratory balances. (Courtesy of Brinkman Instruments Co.)

**Figure 1.8**
Principle of the balance. When the beam is horizontal, the masses are equal.

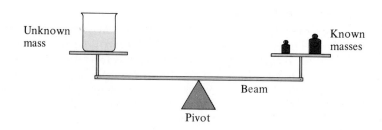

astronauts walked on the surface of the moon. The moon, being a smaller body than the earth, exerts a weaker gravitational pull. Consequently the astronauts, although they wore space suits and equipment that would be heavy on earth, felt lighter on the moon and could execute great leaps and bounces during their walks.

Although mass and weight are different concepts, they are related to each other by the force of gravity. We frequently use the words interchangeably because we weigh objects by comparing their masses to standard reference masses (weights) on a balance, and the gravitational pull is the same on the unknown object and on the standard masses (Fig. 1.8). Because the force of gravity is essentially constant, mass is always directly proportional to weight.

## Time

Time is the one quantity for which the units are the same in all systems: English, metric, and SI. The basic unit is the **second** (s), and

$$60 \text{ s} = 1 \text{ min}$$
$$60 \text{ min} = 1 \text{ h}$$

## Temperature

Most people in the United States are familiar with the Fahrenheit scale of temperature. In the metric system the **centigrade,** or **Celsius,** scale is used. In this scale, the boiling point of water is set at 100°C and the freezing point at 0°C. We can convert from one scale to the other by using the formulas

$$°F = \tfrac{9}{5}°C + 32$$
$$°C = \tfrac{5}{9}(°F - 32)$$

**EXAMPLE**

Normal body temperature is 98.6°F. Convert this to °C.

**Answer**

$°C = \tfrac{5}{9}(98.6 - 32) = \tfrac{5}{9}(66.6) = 37°C$

---

**PROBLEM 1.1**  Convert: (a) 64.0°C to Fahrenheit  (b) 47°F to Celsius.

---

The relationship between the Fahrenheit and Celsius scales is shown in Figure 1.9.

There is also another temperature scale, called the **Kelvin (K)** scale or the **absolute** scale. The *size* of the Kelvin degree is the same as that of the Celsius degree. The only difference is the zero point. The temperature −273°C is taken as the zero point on the Kelvin scale. This makes conversions between Kelvin and Celsius very easy. To go from Celsius to Kelvin, just *add* 273; to go from Kelvin to Celsius, *subtract* 273:

$$K = °C + 273$$

The relationship between the Kelvin and Celsius scales is also shown in Figure 1.9. Note that we don't use the degree symbol in the Kelvin scale: 100°C equals 373 K, not 373°K.

Why was −273°C chosen as the zero point on the Kelvin scale? The reason is that **−273°C, or 0 K, is the lowest possible temperature.** Because of this, 0 K is called **absolute zero.** Temperature depends on how fast molecules move. The more slowly they move, the colder it gets. At absolute zero molecules stop moving altogether, and it is not possible for them to move more slowly than that. Therefore the temperature cannot get any lower. For some purposes it is convenient to have a scale that begins at the lowest possible temperature; the Kelvin scale fulfills this need.

Scientists in laboratories have been able to attain temperatures as low as 0.001 K, but a temperature of 0 K is impossible.

**Figure 1.9**
Three temperature scales.

## 1.4   UNIT CONVERSIONS. THE FACTOR-LABEL METHOD

We frequently need to convert a measurement from one unit to another. The best and most foolproof way to do this is the **factor-label method.** In this method we follow the rule that *when multiplying numbers we also multiply units, and when dividing numbers we also divide units.*

A conversion factor is a ratio of two different units.

For conversions between one unit and another it is always possible to set up two fractions, called **conversion factors.** Suppose we wish to convert 381 g to pounds. We can see in Table 1.2 that there are 453.6 g in 1 lb. The conversion factors between grams and pounds therefore are

$$\frac{1\ lb}{453.6\ g} \quad and \quad \frac{453.6\ g}{1\ lb}$$

To convert 381 g to pounds, all we need do is multiply by the proper conversion factor. But which one? Let us try both and see what happens. First let us multiply by 1 lb/453.6 g:

$$381\ \cancel{g} \times \frac{1\ lb}{453.6\ \cancel{g}} = 0.840\ lb$$

Following the procedure of multiplying and dividing units when we multiply and divide numbers, we find that dividing grams by grams cancels out the grams and we are left with pounds, which is the answer we want. This is the correct method because it converts grams to pounds. Suppose we had done it the other way, multiplying by 453.6 g/1 lb:

$$381\ g \times \frac{453.6\ g}{1\ lb} = 173\ 000\ \frac{g^2}{lb}$$

When we multiply grams by grams, we get $g^2$ (grams squared). Dividing by pounds gives $g^2/lb$. Since this is not the unit we want, this must be the incorrect method.

The advantage of the factor-label method is that it lets us know when we have made a wrong calculation. *If the units of the answer are not the ones we are looking for, the calculation must be wrong.* Incidentally, this principle works not only in unit conversions but in all problems where we make calculations using measured numbers. Here are some additional examples.

---

**EXAMPLE**

The distance between Paris and Marseilles (the two largest cities in France) is 777 km. How many miles is this?

**Answer**

We want to convert kilometers to miles. Table 1.2 shows that 1 mile = 1.609 km. From this we get two conversion factors:

$$\frac{1\ mile}{1.609\ km} \quad and \quad \frac{1.609\ km}{1\ mile}$$

Which one should we use? We use the one that gives the answer in miles:

$$777 \text{ km} \times \text{conversion factor} = \text{? miles}$$

This means that the kilometers must cancel, and so the conversion factor $\dfrac{1 \text{ mile}}{1.609 \text{ km}}$ is appropriate.

$$777 \text{ k\cancel{m}} \times \frac{1 \text{ mile}}{1.609 \text{ k\cancel{m}}} = 483 \text{ miles}$$

**PROBLEM 1.2**   Convert 132 lb to kilograms.

**EXAMPLE**

The label on a container of olive oil says 1.752 gal. How many milliliters is this?

**Answer**

Table 1.2 shows no factor converting gallons to milliliters, but it does show that 1 gal = 3.785 L. Since we know that 1000 mL = 1 L, we can solve this problem by multiplying by two conversion factors, making certain that all units cancel except milliliters:

$$1.752 \text{ g\cancel{al}} \times \frac{3.785 \text{ \cancel{L}}}{1 \text{ g\cancel{al}}} \times \frac{1000 \text{ mL}}{1 \text{ \cancel{L}}} = 6631 \text{ mL}$$

**PROBLEM 1.3**   Convert 2.61 miles to meters.

**EXAMPLE**

The maximum speed limit on many roads in the United States is 55 mph. How many meters per second is this?

**Answer**

Here we have essentially a double conversion problem. We must convert miles to meters and hours to seconds. We use as many conversion factors as necessary, always making sure that we use them in such a way that the proper units cancel:

$$55 \frac{\text{mil\cancel{es}}}{\text{\cancel{h}}} \times \frac{1.609 \text{ k\cancel{m}}}{1 \text{ mil\cancel{e}}} \times \frac{1000 \text{ m}}{1 \text{ k\cancel{m}}} \times \frac{1 \text{ \cancel{h}}}{60 \text{ mi\cancel{n}}} \times \frac{1 \text{ mi\cancel{n}}}{60 \text{ s}} = 25 \frac{\text{m}}{\text{s}}$$

As shown in these examples, when canceling units we do *not* cancel the numbers. The numbers are multiplied and divided in the ordinary way.

**PROBLEM 1.4**   Convert the speed of sound, 332 m/s, to miles per hour.

## 1.5  THE STATES OF MATTER

**Matter can exist in three states: gas, liquid, and solid.** *Gases* have no definite shape or volume. They expand to fill whatever container they are put into. On the other hand, they are highly compressible and can be forced into small containers. *Liquids* also have no definite shape. However, they do have a definite volume that remains the same when they are poured from one container to another. Liquids are only slightly compressible. *Solids* have definite shapes and definite volumes. They are essentially incompressible.

Whether a substance is a gas, a liquid, or a solid depends on its temperature. On a cold winter day a puddle of liquid water turns to *ice;* it becomes a solid. If we boil water, the liquid becomes a gas—we call it *steam.* Most substances can exist in the three states: they are gases at high temperature, liquefy at a lower temperature, and solidify when their temperature becomes low enough. Figure 1.10 shows a single substance in the three different states.

We do not regard a substance as having changed its identity when it is converted from one state to another. Water is still water whether in the form of ice, steam, or liquid water. We discuss the three states of matter, and the changes between one state and another, at greater length in Chapter 5.

**Figure 1.10**
The three states of matter for the compound $NO_2$, which is a gas at room temperature. (a) The brown gas. (b) The gas is frozen to a solid. (c) When the solid is warmed, it melts to a liquid. (Photographs by Charles D. Winters.)

(a)                              (b)                              (c)

## 1.6 DENSITY AND SPECIFIC GRAVITY

One of the many pollution problems that the world faces is the spillage of petroleum in the oceans from oil tankers or from offshore drilling (Fig. 1.11). When oil spills into the ocean, it floats on top of the water. Why doesn't it sink? The reason is that water has a higher *density* than oil, and when two liquids are mixed (assuming that one does not dissolve in the other), the one of lower density floats on top (Fig. 1.12).

The **density** of any substance is defined as its **mass per unit volume.** Not only do all liquids have a density, but so do all solids and gases, too. Density is calculated by dividing the mass of a substance by its volume:

$$d = \frac{m}{V} \qquad d = \text{density}, \ m = \text{mass}, \ V = \text{volume}$$

---

**EXAMPLE**

If 73.2 mL of a liquid has a mass of 61.5 g, what is its density in g/mL?

**Answer**

$$d = \frac{m}{V} = \frac{61.5 \text{ g}}{73.2 \text{ mL}} = 0.840 \ \frac{\text{g}}{\text{mL}}$$

---

**Figure 1.11**
The spillage of more than 10 million gallons of petroleum in Prince William Sound, Alaska, in March 1989 caused a great deal of environmental damage. (Wide World Photos.)

**Figure 1.12**
Two separatory funnels containing water and another liquid. In each case the liquid with the lower density is on top.

Water

Carbon tetrachloride

Ether

Water

### EXAMPLE

The density of iron is 7.86 g/mL. What is the volume, in mL, of an irregularly shaped piece of iron that has a mass of 524 g?

### Answer

Here we are given the mass. In this type of problem, it is useful to derive a conversion factor that we can obtain from the density. We know that the density is 7.86 g/mL. This means that 1 mL has a mass of 7.86 g. From this we can get two conversion factors:

$$\frac{1\ mL}{7.86\ g} \quad \text{and} \quad \frac{7.86\ g}{1\ mL}$$

As usual, we multiply the mass by whichever conversion factor results in the cancellation of all but the correct unit:

$$524\ \cancel{g} \times \frac{1\ mL}{7.86\ \cancel{g}} = 66.7\ mL$$

**PROBLEM 1.5** The density of titanium is 4.54 g/mL. What is the mass, in g, of 17.3 mL of titanium?

**PROBLEM 1.6** An unknown substance has a mass of 56.8 g and occupies a volume of 23.4 mL. What is its density, in g/mL?

See Section 2.2 for a discussion of physical properties.

In old movies, villains frequently used a chloroform-wetted handkerchief to subdue their victims.

This is one of the many discoveries made by Benjamin Thompson (1753–1814), who was born in Massachusetts but later became Count Rumford in Munich.

The density of any liquid or solid is a physical property that is constant, which means that it always has the same value at a given temperature. We use physical properties to help identify a substance. For example, the density of chloroform (a liquid formerly used as an anesthetic) is 1.483 g/mL at 20°C. If we want to find out if an unknown liquid is chloroform, one thing we might do is measure its density at 20°C. If the density is, say, 1.355 g/mL, we know the liquid isn't chloroform. If the density does come out to be 1.483 g/mL, we cannot be sure the liquid is chloroform, because other liquids might also have this density, but we can then measure other physical properties (the boiling point, for example). If all the physical properties we measure match those of chloroform, we can be reasonably sure the liquid *is* chloroform.

We have said that the density of a liquid or solid is a constant at a given temperature. Density does change when the temperature changes. Almost always, *density decreases with increasing temperature.* This is true because *mass* does not change when a substance is heated, but volume almost always increases because atoms and molecules tend to get farther apart as the temperature increases. Since $d = m/V$, if $m$ stays the same and $V$ gets larger, $d$ must get smaller.

The most common liquid, water, provides a partial exception to this rule. As we go from 4°C to 100°C the density of water does decrease, but from 0°C to 4°C the density *increases.* That is, water has its maximum density at 4°C.

Since density is equal to mass divided by volume, it always has units, most commonly g/mL or g/cc (g/L for gases). **Specific gravity** is numerically the same as density, but it has no units (it is dimensionless). The reason is that specific gravity is defined as a comparison of the density of a substance with that of water, which is taken as a standard. For example, the density of copper at 20°C is 8.92 g/mL. The density of water at the same temperature is 1.00 g/mL. Therefore, copper is 8.92 times as dense as water, and its specific gravity at 20°C is 8.92. Because water is taken as the standard and because the density of water is 1.00 g/mL at 20°C, the specific gravity of any substance is always numerically equal to its density, provided that the density is measured in g/mL or g/cc.

---

**EXAMPLE**

The density of pure ethyl alcohol at 20°C is 0.789 g/mL. What is its specific gravity?

**Answer**

$$\text{Specific gravity} = \frac{0.789 \text{ g/mL}}{1.00 \text{ g/mL}} = 0.789$$

---

**PROBLEM 1.7** The specific gravity of a urine sample is 1.016. What is its density, in g/mL?

---

### Intensive and Extensive Properties

There is an important difference between such properties as density and specific gravity on the one hand and mass and volume on the other. The latter depend on how much we have; the former do not. For example, if we have 25 g of butter and add more butter, we have a larger mass and a larger volume but the density remains the same. Whenever we add more, we increase both mass and volume, so that density, which is mass divided by volume, is not affected. **A property that does not depend on the quantity of a substance** is called an **intensive property.** Density is thus an intensive property. **A property that does depend on the quantity of a substance** is called an **extensive property.** Mass and volume are extensive properties.

## 1.7 ENERGY

Matter constantly undergoes changes. In order to understand these changes, we must know something about energy. **Energy** is defined as **the capacity to do work.** Energy can be described as being either kinetic energy or potential energy.

 **Kinetic energy** is the **energy of motion.** Any object that is moving possesses kinetic energy. We can calculate how much energy by the formula $KE = \frac{1}{2}mv^2$, where $m$ is the mass of the object and $v$ is its velocity. This means that kinetic energy increases (1) when an object moves faster and (2) when it is a heavier object that is moving. When a truck and a bicycle are moving at the same velocity, the truck has more kinetic energy.

You impart kinetic energy to a baseball or a football when you throw it.

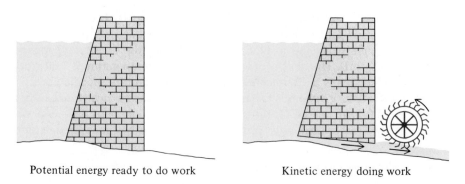

Potential energy ready to do work          Kinetic energy doing work

**Figure 1.13**
The water held back by the dam possesses potential energy, which is converted to kinetic energy when the water is released.

Hydroelectric plants use the kinetic energy of falling water.

**Potential energy** is **stored energy.** The potential energy possessed by an object arises from its capacity to move or to cause motion. For example, a book balanced at the edge of a table contains potential energy—it is capable of doing work. If given a slight push, it will fall. The potential energy the book has while it is on the table is converted to the kinetic energy it has while it is falling. Another way in which potential energy is converted to kinetic energy is shown in Figure 1.13.

In chemistry the most important form of potential energy is **chemical energy.** This is the energy stored within chemical substances. The energy is given off when they react. For example, a piece of paper possesses chemical energy. When the paper is ignited, the chemical energy (potential) is turned into energy in the form of heat and light.

An important principle in nature is that things have a tendency to seek their lowest possible potential energy. We all know that water always flows downhill and not uphill.

There are several forms of energy, of which the most important are mechanical energy (the kinetic energy possessed by moving objects), heat, light, electrical energy, chemical energy, and nuclear energy. The various forms of energy can be converted from one to another. In fact, we do this all the time. A power plant operates either on the chemical energy derived from burning fuel or on nuclear energy. This energy is converted to heat, which is converted to the electricity that is sent over transmission wires into houses and factories. Here we convert the electricity to light, heat (in an electrical heater, for example), or mechanical energy (in the motors of refrigerators, vacuum cleaners, and so on).

Although one form of energy can be converted to another, the *total amount* of energy in any system does not change. **Energy can neither be created nor destroyed.** This statement is called **the law of conservation of energy.***

---

* This statement is not completely true. As discussed in Box 9G, it is possible to convert matter to energy, and vice versa. Therefore a more correct statement would be **matter-energy can neither be created nor destroyed.** However, the law of conservation of energy is valid for most purposes and is highly useful.

**B o x   1C**

## Laser Surgery

An example of conversion of energy from one form to another is demonstrated in laser surgery. When surgery is required on a particular organ of the body, a highly coherent light beam of a single wavelength, a laser beam, can be delivered to that organ through fiber optics. These are tiny fibers that guide the light with only a small loss of energy. Ten thousand such thin fibers can be fitted into a tube only 1 mm in diameter. The instruments containing such bundles of fibers are called **fiberscopes.**

Depending on the wavelength of the laser beam, certain tissues absorb the energy of the laser beam and convert it to heat. In this way the laser beam can seal wounds or join blood vessels without the necessity of cutting through healthy tissues. Laser beams can also cauterize incisions so that a minimum amount of blood is lost during the operation. A routine use of laser beams is to cut films that occasionally form after cataract surgery on an implanted lens in an eye, restoring vision without additional invasive surgery. Laser energy has also been used to shatter kidney stones so that they can be expelled from the body through the urine. Fiberscopes have been used successfully to diagnose and treat many bleeding ulcers in the stomach, intestines, and colon.

Fiberscopes are also used for diagnostic purposes. With these instruments a physician can detect the location of bleeding in ulcers, a polyp in the colon, or a tumor in the esophagus. Fiberscopes can also take samples for analysis without surgery.

## 1.8   HEAT

One form of energy that is particularly important in chemistry is heat. This is the form of energy that most frequently accompanies chemical reactions. Heat is not the same as temperature. Heat is a form of energy, but temperature is not.

We can understand the difference between heat and temperature if we realize that heat is an extensive property and temperature is an intensive property (Section 1.6). If we have two beakers, one containing 100 mL of water and the other 1 L of water at the same temperature, the heat content of the water in the larger beaker is ten times that of the water in the smaller beaker, even though the temperature is the same in both. If you were to dip your hand accidentally into a liter of boiling water, you would be much more severely burned than if only one drop fell on your hand. Even though the water is at the same temperature in both cases, the liter of boiling water has much more heat.

As we saw in Section 1.3, temperature is measured in degrees. Heat can be measured in various units, the most common of which is the **calorie,** which is defined as the **amount of heat necessary to raise the temperature of 1 g of water by 1°C.** This is a small unit, and chemists more often use the **kilocalorie** (kcal):

$$1 \text{ kcal} = 1000 \text{ cal}$$

Nutritionists use the word "Calorie" (with a capital C) to mean the same thing as "kilocalorie"; that is, 1 Cal = 1000 cal = 1 kcal. The calorie is not part of the

## Hypothermia and Hyperthermia

The human body cannot tolerate temperatures that are too low. A person outside in very cold weather (say, −20°F, or −29°C) who is not protected by heavy clothing will eventually freeze to death because the body loses heat. Normal body temperature is 37°C, and when the outside temperature is lower than that, heat flows out of the body. When the air temperature is moderate (10 to 25°C), this poses no problem and is, in fact, necessary because the body produces more heat than it needs and must lose some. At extremely low temperatures, however, too much heat is lost and body temperature drops, a condition called **hypothermia.** A drop in body temperature of 1 or 2°C causes shivering, which is an attempt by the body to increase its temperature by the heat generated through muscular action. A greater drop than that results in unconsciousness and eventually death.

The opposite condition is **hyperthermia.** This can be caused either by high outside temperatures or by the body itself when an individual develops a high fever. A sustained body temperature as high as 41.7°C (107°F) is usually fatal.

Calories and joules are units not only of heat, but of other forms of energy as well.

SI. The official SI unit for heat is the **joule (J),** which is about one fourth as big as the calorie:

$$1 \text{ cal} = 4.184 \text{ J}$$

### Specific Heat

As we noted, it takes 1 cal to raise the temperature of 1 g of water by 1°C. **The amount of heat (in calories) necessary to raise the temperature of 1 g of any substance by 1°C** is called the **specific heat** of that substance. Each substance has its own specific heat, which is a physical property of that substance, like density or melting point. Specific heats for a few common substances are shown in Table 1.3. For example, the specific heat of copper is 0.092 cal/g·deg. Therefore, if we had 1 g of copper at 20°C, it would require only 0.092 cal to increase the temperature to 21°C. Note from Table 1.3 that ice and steam do not have the same specific heat as water.

Water has the highest specific heat of any common substance.

**TABLE 1.3  Specific Heat Values for Some Common Substances**

| Substance | Specific Heat (cal/g·deg) | Substance | Specific Heat (cal/g·deg) |
|-----------|---------------------------|-----------|---------------------------|
| Water | 1.00 | Wood (typical) | 0.42 |
| Ice | 0.48 | Glass (typical) | 0.12 |
| Steam | 0.48 | Rock (typical) | 0.20 |
| Iron | 0.11 | Ethyl alcohol | 0.59 |
| Aluminum | 0.22 | Methyl alcohol | 0.61 |
| Copper | 0.092 | Ether | 0.56 |
| Lead | 0.038 | Carbon tetrachloride | 0.21 |

The high specific heat of water is useful in cold compresses and makes them last a long time. For example, consider two patients with cold compresses, one compress made by soaking a towel with water and the other made with ethyl alcohol, both at 0°C. Each gram of water in the water compress requires 25 cal to make the temperature of the compress rise to 25°C (after which it must be changed). Since the specific heat of ethyl alcohol is 0.59 cal/g deg (Table 1.3), each gram of the alcohol requires only 15 cal to reach 25°C. If the two patients give off heat at the same rate, the alcohol compress is less effective because it will reach 25°C a good deal sooner than the water compress and will need to be changed sooner.

It is easy to make calculations involving specific heats. The equation is

$$\text{Heat absorbed} = \text{specific heat} \times \text{mass} \times \text{change in temperature}$$
$$\text{Heat} = SH \times m \times (T_2 - T_1)$$

**EXAMPLE**

How many calories are required to heat 352 g of water from 23°C to 95°C?

**Answer**

$$\text{Heat} = SH \times m \times (T_2 - T_1)$$
$$= \frac{1.0 \text{ cal}}{\cancel{g} \cdot \cancel{\deg}} \times 352 \cancel{g} \times (95 - 23) \cancel{\deg}$$
$$= 2.5 \times 10^4 \text{ cal} = 25 \text{ kcal}$$

**PROBLEM 1.8**  How many calories are required to heat 731 g of water from 8°C to 74°C?

**EXAMPLE**

How many calories are required to heat 755 g of iron from 23°C to 175°C?

**Answer**
According to Table 1.3, 0.11 cal/g·deg is the specific heat of iron.

$$\text{Heat} = SH \times m \times (T_2 - T_1)$$
$$= \frac{0.11 \text{ cal}}{\cancel{g} \cdot \cancel{\deg}} \times 755 \cancel{g} \times (175 - 23) \cancel{\deg}$$
$$= 1.3 \times 10^4 \text{ cal} = 13 \text{ kcal}$$

**PROBLEM 1.9**  How many calories are required to heat 26 g of aluminum from 20°C to 115°C?

---

**EXAMPLE**

If we add 450 cal of heat to 37 g of ethyl alcohol at 20°C, what is the final temperature?

**Answer**

The specific heat of ethyl alcohol is 0.59 cal/g·deg (Table 1.3).

$$\text{Heat} = SH \times m \times (T_2 - T_1)$$

$$450 \text{ cal} = \frac{0.59 \text{ cal}}{\cancel{g} \cdot \text{deg}} \times 37 \cancel{g} \times (T_2 - T_1)$$

$$(T_2 - T_1) = \frac{450 \cancel{cal}}{0.59 \cancel{cal}/\text{deg} \times 37} = 21 \text{ deg}$$

Since the starting temperature is 20°C, the final temperature is 41°C.

---

**PROBLEM 1.10**  It required 88.2 cal to heat 13.4 g of an unknown substance from 23°C to 176°C. What is the specific heat of the unknown substance?

## SUMMARY

In order to understand modern medicine, it is necessary to know some chemistry. **Chemistry** is the science that deals with the structure of matter and the changes it can undergo. There are three states of matter: **solid, liquid, and gas.**

The **scientific method** is a tool used in science and medicine. The heart of the scientific method is the testing of **hypotheses** and **theories.**

In chemistry we use the **metric system** for measurements. The main units are the meter for length, the liter for volume, and the gram for mass. Other units are indicated by prefixes that represent powers of ten. **Temperature** is measured in degrees Celsius or in kelvins. Conversions from one unit to another are best done by the **factor-label method,** in which units are multiplied and divided.

**Density** is mass per unit volume. **Specific gravity** is relative density and thus has no units. Density usually decreases with increasing temperature.

**Kinetic energy** is energy of motion; **potential energy** is stored energy. Energy can neither be created nor destroyed, but it can be converted from one form to another.

**Heat** is measured in calories. A calorie is the amount of heat necessary to raise the temperature of 1 g of water by 1°C. Every substance has a **specific heat,** which is a physical constant. The specific heat is the number of calories required to raise the temperature of 1 g of the substance by 1°C.

Heat is not the same as temperature. Heat, mass, and volume are **extensive properties.** Temperature and density are **intensive properties.** An intensive property is one that does not change when more matter is added.

## KEY TERMS

Chemistry (Sec. 1.1)
Conversion factor (Sec. 1.4)
Density (Sec. 1.6)
Energy (Sec. 1.7)
Extensive property (Sec. 1.6)

Fact (Sec. 1.2)
Gas (Sec. 1.5)
Heat (Sec. 1.8)
Hyperthermia (Box 1D)
Hypothermia (Box 1D)

Hypothesis (Sec. 1.2)
Intensive property (Sec. 1.6)
Kinetic energy (Sec. 1.7)
Laser (Box 1C)
Liquid (Sec. 1.5)
Mass (Sec. 1.3)
Matter (Sec. 1.1)
Metric system (Sec. 1.3)
Potential energy (Sec. 1.7)

Scientific method (Sec. 1.2)
SI (Sec. 1.3)
Solid (Sec. 1.5)
Specific gravity (Sec. 1.6)
Specific heat (Sec. 1.8)
Temperature (Secs. 1.3, 1.8)
Theory (Sec. 1.2)
Weight (Sec. 1.3)

## PROBLEMS

Difficult problems are designated by an asterisk.

### Chemistry and the Scientific Method

**1.11** What subject does the science of chemistry deal with?

**1.12** Why is the study of chemistry necessary for students planning a career in the health sciences?

**1.13** What is the difference between a hypothesis and a fact?

**1.14** What is the difference between a theory and a hypothesis?

**1.15** What does it take to get scientists to accept a theory? If they do accept it, does that make it correct?

### Measurements

**1.16** In the metric system, what is the basic unit of (a) mass  (b) length  (c) volume?

**1.17** How many meters are there (a) in 1 km  (b) in 1 mm?

**1.18** How many milliliters are there (a) in 1 L  (b) in 1 cc?

**1.19** For each of these, tell which answer is closest:
  (a)  a man weighs 75 mg or 75 g or 75 kg
  (b)  a paper clip weighs 50 mg or 50 g or 50 kg
  (c)  an audiocassette weighs 40 mg or 40 g or 40 kg
  (d)  a baseball bat has a length of 100 mm or 100 cm or 100 m
  (e)  your hand has a width of 100 mm or 100 cm or 100 m
  (f)  a glass of milk holds 23 cc or 230 mL or 23 L
  (g)  a tablespoon contains 15 mL or 150 mL or 1.5 L

**1.20** What is the difference between mass and weight?

**1.21** Convert to Celsius and to Kelvin:
  (a)  320°F          (c)  0°F
  (b)  212°F          (d)  −250°F

**1.22** Convert to Fahrenheit and to Kelvin:
  (a)  25°C           (c)  250°C
  (b)  40°C           (d)  −273°C

### Unit Conversions

**1.23** Make the following conversions (conversion factors are given in Table 1.2):
  (a)  2.73 gal to L      (f)  43.7 L to gal
  (b)  34 in to cm        (g)  1.62 lb to g
  (c)  34.9 mL to fl oz   (h)  1.1 miles to km
  (d)  42.6 kg to lb      (i)  37.2 km to miles
  (e)  62 g to oz (avdp)  (j)  33.61 qt to L

**1.24** Make the following metric conversions:
  (a)  711 g to kg        (g)  96.4 mL to L
  (b)  45.7 kg to g       (h)  3.29 L to cc
  (c)  0.073 kg to mg     (i)  361 mg to g
  (d)  83.4 m to mm       (j)  275 mm to cm
  (e)  475 cm to m        (k)  21.64 cc to mL
  (f)  63.7 mL to cc      (l)  0.044 L to mL

**1.25** What is your weight in kilograms? Your height in meters?

**1.26** The speed limit in some European cities is 80 km/h. How many miles per hour is this?

**1.27** The speed of light is 186 000 miles per second. What is it in cm/s?

### Density and Specific Gravity

**1.28** The density of a solid is 7.5 g/mL. What is its density in kg/L?

**1.29** The density of manganese is 7.21 g/mL; that of calcium chloride is 2.15 g/mL; and that of sodium acetate is 1.528 g/mL. If you place these three solids in a liquid that has a density of 2.15 g/mL, which will sink to the bottom, which will stay on the top, and which will stay in the middle of the liquid?

**1.30** The density of uranium is 18.95 g/mL. What is the volume, in mL, of 63 g of uranium?

**1.31** A 335.0-cc sample of urine has a mass of 342.6 g. What is the density, in g/mL, to three decimal places?

**1.32** The density of methanol at 20°C is 0.791 g/mL. What is the mass, in g, of a 280-mL sample?

**1.33** Boiling point is a physical property. Name three others.

**1.34** Which of the following properties are extensive and which are intensive?
   (a) volume        (f) density
   (b) heat           (g) temperature
   (c) mass          (h) melting point
   (d) specific heat   (i) length
   (e) kinetic energy

**1.35** The density of chloroform, a liquid insoluble in water, is 1.48 g/cc. If chloroform and water are placed in a separatory funnel, which will be the upper layer?

**\*1.36** A sample of 10.00 g of oxygen has a volume of 6702 mL. The same weight of carbon dioxide occupies 5058 mL. (a) What are the densities in g/L? (b) Carbon dioxide is used as a fire extinguisher to cut off the fire's supply of oxygen. Do the densities of these two gases explain the fire-extinguishing ability of carbon dioxide?

**\*1.37** Two one-quart containers of butter are for sale. Butter A has a density of 1.2 g/mL and costs $5.00; butter B has a density of 0.96 g/mL and costs $4.00. Which is a better buy?

## Energy

**1.38** What is the difference between kinetic and potential energy?

**1.39** How many calories are required to heat 250.0 g of water from 60°F to 120°F?

**1.40** How many calories are required to heat the following (specific heats are given in Table 1.3)?
   (a) 52.7 g of aluminum from 100°C to 285°C

   (b) 93.6 g of methyl alcohol from −35°C to 55°C
   (c) 3.4 kg of lead from −33°C to 730°C
   (d) 71.4 g of ice from −77°C to −5°C

**\*1.41** If 168 g of an unknown liquid requires 2750 cal of heat to raise its temperature from 26°C to 74°C, what is the specific heat of the liquid?

## Boxes

**1.42** (Box 1A) Suppose that in examining the patient described in Box 1A the doctors had found normal lead levels in the blood. What would they conclude about their hypothesis of lead poisoning?

**1.43** (Box 1A) How did the physicians mentioned in Box 1A test their hypothesis?

**1.44** (Box 1B) If the recommended dose of a drug is 3.7 mg for each kilogram of body weight, how many mg should be prescribed for a 114-lb woman?

**1.45** (Box 1B) If the recommended dose of calcium is 1150 mg for a 123-lb woman, what is this per kilogram of body weight?

**1.46** (Box 1B) If the recommended dose of a drug is 825 mg for a 190-lb man, what would be a suitable dose for a 145-lb man?

**1.47** (Box 1B) The average lethal dose of heroin is 150 mg per kilogram of body weight. Estimate how many grams of heroin would be lethal for a 200-lb man.

**1.48** (Box 1C) What is the advantage of being able to shatter kidney stones with a laser beam?

**1.49** (Box 1D) How does the body react to hypothermia?

**1.50** (Box 1D) Low temperatures often cause people to shiver. What is the function of this involuntary body action?

**1.51** (Box 1E) Which would make a more efficient cold compress, ethyl alcohol or methyl alcohol? (Refer to Table 1.3.)

## Appendix A: Exponential Notation

**1.52** Write in exponential notation:
   (a) 75.2         (c) 0.082
   (b) 35 600     (d) 0.00000057

**1.53** Write out in full:
   (a) $4.03 \times 10^5$    (c) $7.13 \times 10^{-5}$
   (b) $3.2 \times 10^3$     (d) $5.55 \times 10^{-10}$

**1.54** Multiply:
  (a) $(4.73 \times 10^5)(1.37 \times 10^2)$
  (b) $(6.49 \times 10^7)(7.22 \times 10^{-3})$
  (c) $(2.7 \times 10^{-4})(5.9 \times 10^8)$
  (d) $(3.4 \times 10^{-5})(8.2 \times 10^{-11})$

**1.55** Divide:
  (a) $\dfrac{6.02 \times 10^{23}}{3.01 \times 10^5}$    (d) $\dfrac{5.8 \times 10^{-6}}{6.6 \times 10^{-8}}$

  (b) $\dfrac{3.14}{2.30 \times 10^{-5}}$    (e) $\dfrac{7.05 \times 10^{-3}}{4.51 \times 10^5}$

  (c) $\dfrac{7.08 \times 10^{-8}}{300}$

**1.56** Add:
  (a) $(6.7 \times 10^3) + (3.8 \times 10^3)$
  (b) $(6.17 \times 10^5) + (7.4 \times 10^4)$
  (c) $(4.44 \times 10^{-5}) + (3.289 \times 10^{-4})$

**1.57** Subtract:
  (a) $(7.00 \times 10^5) - (4.20 \times 10^5)$
  (b) $(5.82 \times 10^3) - (7.9 \times 10^2)$
  (c) $(2.134 \times 10^{-7}) - (5.62 \times 10^{-8})$

**\*1.58** Solve:
$$\frac{(3.14 \times 10^3) \times (7.80 \times 10^5)}{(5.50 \times 10^2)}$$

**\*1.59** Solve:
$$\frac{(9.52 \times 10^4) \times (2.77 \times 10^{-5})}{(1.39 \times 10^7) \times (5.83 \times 10^{-3})}$$

## Appendix B: Significant Figures

**1.60** How many significant figures are there in
  (a) 657            (e) 3.108
  (b) 0.0052         (f) 0.0001
  (c) 0.0860         (g) 0.00000700
  (d) 6.4007

**1.61** How many significant figures are there in
  (a) $5.71 \times 10^{13}$      (d) $4.000 \times 10^{-11}$
  (b) $4.4 \times 10^5$          (e) $5.5550 \times 10^{-3}$
  (c) $3 \times 10^{-6}$

**1.62** Round off to two significant figures:
  (a) 73.82          (d) 0.005118
  (b) 2.811          (e) 3.77
  (c) 0.899

**1.63** Multiply these numbers, using the correct number of significant figures in your answer:
  (a) $3630.15 \times 6.8$
  (b) $512 \times 0.0081$
  (c) $5.79 \times 1.85825 \times 1.4381$

**1.64** Divide these numbers, using the correct number of significant figures in your answer:
  (a) $\dfrac{3.185}{2.08}$    (c) $\dfrac{0.0035}{7.348}$

  (b) $\dfrac{6.5}{3.0012}$

**1.65** Add these groups of measured numbers:

| (a) | 37.4083 | (c) | 51.51 |
|---|---|---|---|
| | 5.404 | | 100.27 |
| | 10916.3 | | 16.878 |
| | 3.94 | | 3.6817 |
| | 0.0006 | | |

| (b) | 84 |
|---|---|
| | 8.215 |
| | 0.01 |
| | 151.7 |

**1.66** A student had to solve the following equation, where all the quantities were measured numbers:
$$x = \frac{19.227 \times 4.300}{0.000450 \times 370.00}$$

Her calculator gave the answer $x = 496.5531532$. What answer should she report to her chemistry teacher?

## Additional Problems

**1.67** Which of the following is the largest mass and which the smallest?
  (a) 41 g              (c) $8.2 \times 10^6$ $\mu$g
  (b) $3 \times 10^3$ mg      (d) $4.1 \times 10^{-8}$ kg

**1.68** Classify these as kinetic or potential energy:
  (a) Water held by a dam
  (b) A speeding train
  (c) A book on its edge before falling
  (d) A falling book
  (e) Electric current in a light bulb

**1.69** A brain weighing one pound occupied a volume of 620 mL. What is the specific gravity of the brain?

**1.70** What is the difference between heat and temperature?

**1.71** The *meter* is a measure of length. Tell what each of the following units measures:
    (a) kg           (e) cal
    (b) mL          (f) cm/s
    (c) °C           (g) g/cc
    (d) cm$^3$        (h) joule

**1.72** If 80 g of an unknown substance released 3000 cal of heat when it was cooled from 100°C to 50°C, what was the specific heat of the substance?

**1.73** Make the following conversions (conversion factors are given in Table 1.2):
    (a) 1.57 km to inches
    (b) 328 mL to quarts
    (c) 232 oz (avdp) to kilograms
    (d) 836 grains to pounds

**\*1.74** The specific heats of some elements at 25°C are: aluminum = 0.215 cal/g·deg; carbon (graphite) = 0.170 cal/g·deg; iron = 0.107 cal/g·deg; mercury = 0.0331 cal/g·deg. (a) Which element would require the smallest amount of heat to raise the temperature of 100 g of the element by 10°C? (b) If the same amount of heat needed to raise the temperature of 1 g of aluminum by 25°C were applied to 1 g of mercury, by how many degrees would its temperature be raised? (c) If a certain amount of heat is used to raise the temperature of 1.6 g of iron by 10°C, the temperature of 1 g of which element would also be raised by 10°C, using the same amount of heat?

**1.75** Which scientist found evidence that Harvey's theory about the circulation of the blood was correct?

**\*1.76** If the density of air is $1.25 \times 10^{-3}$ g/cc, what is the mass in kilograms of the air in a room that is 5.3 m long, 4.2 m wide, and 2.0 m high?

**\*1.77** The kinetic energy possessed by an object with a mass of 1 gram moving with a velocity of 1 cm/sec is called **1 erg.** What is the kinetic energy, in ergs, of an athlete with a mass of 127 lb running at a velocity of 14.7 miles/hr?

**1.78** One quart of milk costs 80 cents and one liter costs 86 cents. Which is the better buy?

**1.79** The density of gasoline is 0.71 g/mL. What is this in lb/gal?

**1.80** Which of these answers is most correct? The density of a substance decreases with increasing temperature (a) always (b) almost always (c) sometimes (d) never.

**\*1.81** (a) If 1.00 mL of butter, density 0.860 g/ml, is thoroughly mixed with 1.00 mL of sand, density 2.28 g/ml, what is the density of the mixture? (b) What would be the density of the mixture if 1.00 g of the same butter were mixed with 1.00 g of the same sand?

**1.82** When the astronauts walked on the moon they could make giant leaps in spite of their heavy gear. (a) Why were their weights on the moon so low? (b) Were their masses different on the moon than on the earth?

**1.83** In Japan, high-speed bullet trains move with an average speed of 220 km/h. If Chicago and New York were connected by such a train, how long would it take to travel nonstop between these cities (830 miles)?

**1.84** Which speed is the fastest?
    (a) 80 miles/h       (c) 5.0 km/s
    (b) 160 km/h       (d) 50 miles/min

**\*1.85** Is it possible to convert 83 mL to grams?

# CHAPTER 2

*Antoine Laurent Lavoisier and his wife, by Jacques David. (Courtesy of the Metropolitan Museum of Art.)*

# Atoms

## 2.1 INTRODUCTION

What is matter made of? This question was discussed for thousands of years, long before human beings had any reasonable way of getting an answer. In ancient Greece two possible answers were given, each having its own group of supporters. One group, led by a scholar named Democritus (about 470–380 BC), believed that all matter was made of very small particles—much too small to see. Democritus called these particles *atoms.* Some of his followers developed the idea that there were different kinds of atoms, with different properties, and that the properties of the atoms caused ordinary matter to have the properties we all know.

But these ideas were not accepted by all ancient thinkers. Another group, led by Zeno of Elea (born about 450 BC), did not believe in atoms at all. They insisted that matter was infinitely divisible; that is, if you took any object, such as a coin or a piece of wood, or a liquid such as water or olive oil, you could cut it or otherwise divide it into two parts, divide each of these into two more parts, and continue the process forever. According to Zeno and his followers, you would never reach a particle of matter that could no longer be divided.

They said that water is a liquid because its atoms are smooth and round and can easily glide over each other, whereas atoms of iron are rough and hard.

Today we know that Democritus was right and Zeno was wrong, but there is a great difference in the way we look at this question. Today our ideas are based on *evidence*. Democritus had no evidence to prove that matter cannot be divided an infinite number of times, just as Zeno had no evidence for his claim that matter can be divided infinitely. Both claims were based not on evidence but on visionary beliefs: one in unity, the other in diversity.

In Section 2.3 we will discuss the evidence for the existence of atoms, but first we need to look at the diverse forms of matter.

## 2.2 CLASSIFICATIONS OF MATTER

Matter can be divided into three classes: elements, compounds, and mixtures.

### Elements

An **element** is **a pure substance that cannot be broken down into simpler substances by chemical reactions.** At this time, 109 elements are known. Of these, 88 occur in nature; the others have been made by chemists. The elements are the building blocks of all matter. A list of the known elements is given on the inside back cover of this book, along with their symbols. These symbols, consisting of one, two, or three letters, are the same in all countries. Most of the symbols correspond directly to the name in English (for example, C for carbon, H for hydrogen, Li for lithium), but a few come from the Latin or German names. Some of the more important symbols are listed in Table 2.1. Photographs of some elements are shown in Figure 2.1.

For elements with three-letter symbols, see the periodic table on the inside front cover of this book.

The symbol for tungsten, W, comes from the German *wolfram*. The symbol for iron, Fe, comes from the Latin *ferrum*.

Students will do well to learn these symbols.

**TABLE 2.1  Names and Symbols of Some Important Elements**

| Symbol | Element | Symbol | Element |
|--------|---------|--------|---------|
| Al | Aluminum | Hg | Mercury |
| As | Arsenic | Mo | Molybdenum |
| B | Boron | Ne | Neon |
| Br | Bromine | N | Nitrogen |
| Ca | Calcium | O | Oxygen |
| C | Carbon | P | Phosphorus |
| Cl | Chlorine | Pt | Platinum |
| Cr | Chromium | Pu | Plutonium |
| Co | Cobalt | K | Potassium |
| Cu | Copper | Se | Selenium |
| F | Fluorine | Si | Silicon |
| Au | Gold | Ag | Silver |
| He | Helium | Na | Sodium |
| H | Hydrogen | Sr | Strontium |
| I | Iodine | S | Sulfur |
| Fe | Iron | Sn | Tin |
| Pb | Lead | U | Uranium |
| Mg | Magnesium | V | Vanadium |
| Mn | Manganese | Zn | Zinc |

**Figure 2.1**
Some elements. (Photographs by Charles D. Winters.)

Phosphorus (P)

Arsenic (As) (left), antimony (Sb) (right), bismuth (Bi) (top)

Magnesium (Mg) (left), calcium (Ca) (right)

Nitrogen ($N_2$)

Clockwise from the top: lead (Pb), tin (Sn), carbon (C), silicon (Si)

Zinc (Zn) (left), mercury (Hg) (right)

**Box 2A**

## Elements Necessary for Human Life

Only a few elements are necessary for human life. The four most important of these—carbon, hydrogen, oxygen, and nitrogen—make up the subjects of organic chemistry and biochemistry (Chapters 10 to 26), so we need not consider them here. Seven other elements are also quite important, and besides these there are at least nine additional ones (trace elements) that our bodies use in very small quantities. The table shows these 20 elements and their functions in the human organism. Many of these are more fully discussed later in the book. For the average daily requirements, sources in foods, and symptoms of deficiencies, see Table 26.2.

| Element | Function |
|---------|----------|
| **The Big Four** | |
| Carbon | |
| Hydrogen | Discussed in Chapters 10 to 26 |
| Oxygen | |
| Nitrogen | |
| **The Next Seven** | |
| Calcium | Strengthens bones and teeth; aids in blood clotting |
| Phosphorus | Present in phosphates, which regulate neutrality of body fluids and are involved in energy transfer |
| Potassium | Helps regulate electrical balance of body fluids |
| Sulfur | An essential component of proteins; important in protein structure |
| Chlorine | Helps regulate electrical balance of body fluids |
| Sodium | Helps regulate electrical balance of body fluids |
| Magnesium | Helps nerve and muscle action; present in bones |
| **The Trace Elements** | |
| Chromium | Increases effectiveness of insulin |
| Cobalt | A part of vitamin $B_{12}$ |
| Copper | Strengthens bones; assists enzyme action |
| Fluorine | Reduces incidence of dental cavities |
| Iodine | Necessary for thyroid function |
| Iron | An essential part of such proteins as hemoglobin, myoglobin, cytochromes, and FeS protein. |
| Manganese | Present in bone-forming enzymes; aids fat and carbohydrate metabolism |
| Molybdenum | Necessary for activity of certain enzymes |
| Zinc | Necessary for normal growth |

### Compounds

A **compound** is **a pure substance made up of two or more elements in a fixed proportion by weight.** For example, water is a compound made up of hydrogen and oxygen, and table salt is a compound made up of sodium and chlorine. There are millions of known compounds, many of which we will meet in this book.

## Mixtures

Elements and compounds are pure substances. The third category of matter consists of **mixtures** of pure substances. In our daily life most of the matter we meet (including our own bodies) consists of mixtures rather than pure substances. For example, blood, butter, gasoline, soap, the metal in a wedding ring, the air we breathe, and the earth we walk on are all mixtures of pure substances.

The difference between mixtures and compounds can be illustrated by a simple experiment. The elements zinc and sulfur are shown in Figure 2.2(a) and (b). If we mix them, Figure 2.2(c), we have a *mixture* of zinc and sulfur. We can put the mixture in a blender, so as to make it a very thorough mixture, with the particles evenly distributed, but it still consists of tiny pieces of sulfur and tiny pieces of zinc. On the other hand, if in this case we heat the mixture of zinc and sulfur, we get the *compound* zinc sulfide, shown in Figure 2.2(d). This compound is an entirely different substance, with properties that are completely different from those of either zinc or sulfur. An important difference between a compound and a mixture is that the proportions by weight of the elements in a compound are fixed (see Sec. 2.3), whereas, for example, in a mixture of zinc and sulfur, the two elements can be mixed in any proportion.

**Figure 2.2**
(a) Zinc; (b) sulfur; (c) a mixture of zinc and sulfur. Both yellow sulfur and gray zinc particles are present; (d) the compound zinc sulfide. Neither gray nor yellow particles are present. The compound is white. (Photographs by Beverly March.)

(a)

(b)

(c)

(d)

Note that elements present in a compound cannot be separated by such methods.

One of the most important tasks of chemists is the separation of the components of a mixture. Many methods are known, all of which rely on some difference in the properties of the components. For example, one such method is **distillation,** used to separate the components of a mixture of liquids. Here we take advantage of the difference in boiling points. The mixture is heated until the liquid with the lower boiling point evaporates. The vapor is then condensed to form a liquid. The liquid with the higher boiling point remains in the flask. A distillation apparatus is shown in Figure 2.3. You will probably encounter other methods of separation in the laboratory.

## Chemical and Physical Changes

It has long been known that matter can change, or be made to change, from one form to another. In a **chemical change,** substances are used up (disappear) and others are formed to take their places. An example is the burning of propane ("bottled gas"). When this chemical change takes place, propane and oxygen from the air are converted to carbon dioxide and water. Another chemical change is shown in Figure 2.4. Chemical changes are more often called *chemical reactions.* Many thousands of them are known, and the study of these reactions is the chief business of most chemists.

Digestion of food also involves chemical changes.

Matter also undergoes other kinds of changes, called **physical changes.** These changes differ from chemical reactions in that substances do not change

**Figure 2.3**
A laboratory distillation apparatus. The liquid mixture in the flask on the left is heated until it boils. The component with the lower boiling point evaporates first, and the resulting vapor is condensed in the water-cooled condenser and collected in the flask on the right.

**Figure 2.4**
A chemical change. The beaker on the left contains silver nitrate dissolved in water. A piece of copper wire is added. In the beaker on the right, the copper is reacting with the silver compound to produce silver, which is coating the surface of the copper, and a copper compound that dissolves in water. (Photograph by Charles Steele.)

their identity. Most physical changes are changes of state—for example, the melting of solids and the boiling of liquids. Water remains water whether in the liquid state or in the form of ice or steam. Conversion from one state to another is a physical, not a chemical, change. Another important type of physical change involves making or separating mixtures. Dissolving sugar in water is a physical change.

The mixing of vinegar, salad oil, and eggs to make mayonnaise is a physical change.

When we talk of the *chemical properties* of a substance, we mean the chemical reactions it undergoes. *Physical properties* are all properties that do not involve chemical reactions; for example, density, color, melting point, and physical state (liquid, solid, gas) are all physical properties.

## 2.3 DALTON'S ATOMIC THEORY

The atomic theory of Democritus (Sec. 2.1) was vague. Around 1805 an English chemist, John Dalton (1766–1844), put forth a scientific atomic theory. The big difference between Dalton's theory and that of Democritus was that Dalton based his on evidence. First let us state the theory and then see what kind of evidence supported it.

**1.** All matter is made up of very tiny indivisible particles, which Dalton called **atoms,** using the same name that Democritus had used.

An atom is the smallest particle of an element that shows the chemical behavior of that element.

**2.** All atoms of the same element are identical to each other.
**3.** Atoms of any one element are different from those of any other element.
**4.** Atoms combine to form molecules. A molecule is a tightly bound group of atoms that acts as a unit.

### The Evidence for Dalton's Atomic Theory

**A.** The great French chemist Antoine Laurent Lavoisier (1743–1794) discovered the *law of conservation of mass.* This says that **matter can neither be created nor destroyed.** Lavoisier proved this by many experiments in which he showed that, no matter what chemical changes took place, the total weight of the matter at the end was exactly the same as at the beginning. Dalton's theory explained this fact very nicely. If all matter is made up of indestructible atoms, then any chemical change just changes the attachments between atoms but does not destroy the atoms themselves. For example, carbon monoxide reacts with lead oxide to give carbon dioxide and lead. This is how Dalton might have explained what happens:

| A carbon monoxide molecule | A lead oxide molecule | A carbon dioxide molecule | A lead atom |

In this figure, the carbon atom is shown as a red ball, the lead atom as a black ball, and the oxygen atoms as blue balls. All of the original atoms are still there in the end; they have only changed partners. This explains why the total weight at the end is the same as at the beginning.

**B.** Before Dalton presented his theory, another French chemist, Joseph Proust (1754–1826), had demonstrated the *law of constant composition,* which states that **any compound is always made up of elements in the same proportion by weight.** For example, the compound water contains 89 percent oxygen and 11 percent hydrogen by weight. Proust said that all pure water contains these two elements in these exact proportions, whether the water comes from the Atlantic Ocean or the Seine River or is collected as rain, squeezed out of a watermelon, or distilled from urine. Water is never formed in any other proportions. This fact was also evidence for Dalton's theory. If a water molecule is made up of two atoms of hydrogen and one of oxygen, and if an oxygen atom weighs 16 times as much as a hydrogen atom (as today we know it does), water must always contain 89 percent oxygen and 11 percent hydrogen and can never be found in any other proportions.

### Monatomic, Diatomic, and Polyatomic Elements

Some elements—for example, helium and neon—consist of single atoms not connected to each other. In contrast, oxygen, in its most common form, contains two atoms in each molecule, connected to each other by chemical bonds. We write this as $O_2$, the little number (subscript) telling us how many atoms are in the molecule. Other elements that occur as diatomic molecules are hydrogen ($H_2$), nitrogen ($N_2$), fluorine ($F_2$), chlorine ($Cl_2$), bromine ($Br_2$), and iodine ($I_2$). It is important to understand that under normal conditions the free atoms O, H, N, F, Cl, Br, and I do not exist. These elements occur only as diatomic molecules.

Some elements have even more atoms in each molecule. In one form of phosphorus, each molecule has four atoms ($P_4$); sulfur forms $S_8$ molecules. There are even elements whose molecules are much larger than this. For example, a molecule of carbon (a diamond) has millions of carbon atoms all bonded together in a gigantic cluster. Just the same, all of these are elements because they all contain only one kind of atom.

## 2.4  INSIDE THE ATOM

Today we know that matter is more complex than Dalton believed. A wealth of evidence obtained over the last 100 years or so has convinced us that atoms are not indivisible, but are made up of even smaller particles. There are **three elementary particles that make up all atoms:** *protons, electrons,* and *neutrons.* The main things we know about these three particles are their charge and their mass (Table 2.2).

There are many other subatomic particles, but we do not deal with them in this book.

Protons and electrons carry electric charges. There are two kinds of charges, which we call *positive* and *negative.* The only way we can tell that they exist is that positive charges repel other positive charges, negative charges repel other negative charges, and positive charges attract negative charges. In other words, **like charges repel and unlike charges attract.**

A **proton** has a positive charge. By convention we say that the magnitude of the charge is 1. Thus, the proton has a charge of $+1$ (two protons have a total charge of $+2$, and so forth). The mass of a proton is $1.673 \times 10^{-24}$ g, but this number is so small that it is more convenient to use another unit, called the atomic mass unit (amu):

$$1 \text{ amu} = 1.67 \times 10^{-24} \text{ g}$$

The proton has a mass of approximately 1 amu.

The **electron** has a charge of $-1$, equal in magnitude to the charge on the proton but opposite in sign. However, the mass of the electron is much less, about 1/1835 that of the proton. It takes about 1835 electrons to equal the mass of one proton.

To be more exact, the mass of the proton is 1.0073 amu.

---

**TABLE 2.2   Properties and Location of the Elementary Particles**

|          | Mass (g)              | Mass (amu) | Charge | Found in          |
|----------|-----------------------|------------|--------|-------------------|
| Proton   | $1.673 \times 10^{-24}$ | 1          | $+1$   | Nucleus           |
| Electron | $9.110 \times 10^{-28}$ | 1/1835     | $-1$   | Outside of nucleus |
| Neutron  | $1.675 \times 10^{-24}$ | 1          | 0      | Nucleus           |

The exact mass of the neutron is 1.0087 amu.

**Neutrons** have no charge at all. Therefore they do not attract or repel each other or any other particle. The mass of the neutron is slightly greater than that of the proton: $1.675 \times 10^{-24}$ g. This mass is so close to that of the proton, however, that for most purposes we say that these two particles have the same mass: 1 amu.

These are the three particles that make up atoms, but where are they found? Protons and neutrons are found in a tight cluster in the center of the atom (Fig. 2.5). This part of the atom is called the nucleus, and both protons and neutrons are called **nucleons.** Electrons are found as a diffuse cloud outside the nucleus. We will discuss the nucleus in greater detail in Chapter 9.

## Mass Number

Each atom has a fixed number of protons, electrons, and neutrons. **The total number of protons and neutrons (nucleons) in any atom** is called the **mass number.** For example, an atom with 5 protons, 5 electrons, and 6 neutrons has a mass number of 11 (electrons are not counted in determining mass number).

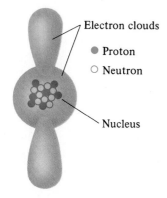

**Figure 2.5**
A typical atom. Protons and neutrons make up the nucleus. The electrons are found as clouds outside the nucleus. This figure is not drawn to scale.

---

### EXAMPLE

What is the mass number of (a) an atom with 58 protons, 58 electrons, and 78 neutrons and  (b) an atom with 17 protons, 17 electrons, and 20 neutrons?

### Answer
(a)  The mass number is $58 + 78 = 136$.
(b)  The mass number is $17 + 20 = 37$.

In each case we add the number of protons and neutrons and ignore the electrons.

---

**PROBLEM 2.1**   What is the mass number of (a) an atom with 47 protons, 47 electrons, and 62 neutrons and  (b) an atom with 91 protons, 91 electrons, and 139 neutrons?

---

## Atomic Number

The number of protons contained in any atom is very important. *It determines what element the atom is.* For example, any atom with 7 protons is a nitrogen atom; if an atom has 24 protons it is a chromium atom, and if there are 80 protons it is an atom of mercury. This is true no matter how many electrons or neutrons are present. The identity of an element is established only by the number of protons in the atom. We call this number the **atomic number** of the element. At the present time 109 elements have been discovered, and these elements have atomic numbers from 1 to 109. The smallest atomic number belongs to the element hydrogen, which has only 1 proton, and the largest (so far) to the element unnilennium (Une), with 109 protons.

Atomic numbers for all the elements are given in the atomic weight table on the inside back cover.

**EXAMPLE**

Name the elements mentioned in the previous example.

**Answer**

(a) Since there are 58 protons, we see from the atomic weight table on the inside back cover that the atom is a cerium atom (Ce), atomic number 58.

(b) This atom has 17 protons, making it a chlorine (Cl) atom.

**PROBLEM 2.2** Name the elements mentioned in Problem 2.1.

## Ions

Atoms are electrically neutral; that is, they have no charge. This means that the number of electrons in any atom must be equal to the number of protons. The mercury atom, for example, with 80 protons, also has 80 electrons. So why don't we say that the number of electrons establishes the identity of the atom? The reason is that atoms can lose or gain electrons but do not thereby change to other elements. If a mercury atom, with 80 protons and 80 electrons, loses two electrons, it becomes a mercury *ion*, with 80 protons and 78 electrons. We no longer call it an atom, but it is still mercury because it still has 80 protons. We call it an **ion,** which is defined as **a particle with unequal numbers of protons and electrons.** Atoms do not normally lose or gain protons or neutrons, only electrons. The mercury ion we just mentioned has 78 electrons and 80 protons. The total charge of the 78 electrons ($-78$) is canceled by the charge on 78 of the protons ($+78$), but there are no electrons present to cancel the other two positive charges. Therefore this particular mercury ion has a charge of $+2$, and we write it as $Hg^{2+}$.

If in an ion there are more protons than electrons, the ion has a positive charge. If there are more electrons than protons, the ion has a negative charge. We will discuss ions again in Section 3.2.

A sulfide ion, $S^{2-}$, has 16 protons and 18 electrons.

## Isotopes

Although we can say that a mercury atom or ion always has 80 protons and a neutral mercury atom has 80 electrons, we cannot say that a mercury atom (or ion) must have any particular number of neutrons. Some of the mercury atoms found in nature have 122 neutrons; the mass number of these atoms is 202, and they are designated as either $^{202}Hg$ or mercury-202. Other mercury atoms have only 120 neutrons (and therefore a mass number of 200). Still others have 118, 119, or 121 neutrons, and others as many as 124. **Atoms with the same number of protons but different numbers of neutrons** are called **isotopes.** All isotopes of mercury contain 80 protons (or they wouldn't be mercury). If they are neutral atoms, they all contain 80 electrons. Each isotope contains a different number of neutrons, however, and therefore has a different mass number.

The fact that isotopes exist means that the second statement of Dalton's theory (Sec. 2.3) is not correct.

The properties of isotopes of the same element are almost identical, and for most purposes we regard them as identical. (They differ in radioactivity properties, which we will discuss in Chapter 9.)

## Atomic Weight

Most elements are found in nature as mixtures of isotopes, in a more or less constant ratio. For example, any sample of the element chlorine contains 75.5 percent chlorine-35 (18 neutrons) and 24.5 percent chlorine-37 (20 neutrons). Silicon is composed of a fixed ratio of three isotopes, with 14, 15, and 16 neutrons. For some elements these ratios may vary slightly, but for most purposes the slight variations can be ignored. The **atomic weight** of an element is a **weighted average of the masses of the isotopes** (the mass of an isotope is approximately the same as its mass number). Thus the atomic weight of chlorine is 35.5 amu, which is a weighted average of the masses of the chlorine atoms, 75.5 percent of which have a mass number of 35 and 24.5 percent of which have a mass number of 37:

$$\left(\frac{75.5}{100} \times 35.0 \text{ amu}\right) + \left(\frac{24.5}{100} \times 37.0 \text{ amu}\right) = 35.5 \text{ amu}$$

Some elements—for example, gold, fluorine, and aluminum—occur naturally as only one isotope. The atomic weights of these elements are of course close to whole numbers (gold, 196.97; fluorine, 18.998; aluminum, 26.98).

The atomic weight is a very important property of an element, and we shall make much use of it in this course. A table of atomic weights is found on the inside back cover of this book.

## 2.5   THE PERIODIC TABLE

In the 1860s a number of scientists, most notably the Russian Dmitri Mendeleev (1834–1907, Fig. 2.6), produced periodic tables. In its modern form (see the inside front cover) the periodic table consists of all the elements arranged in order of atomic number. When this is done we find that the elements fall into rows, in such a way that elements in the same vertical column (*group*) have similar properties. This is best understood by looking at specific examples in the table. The elements fluorine (atomic number 9), chlorine (17), bromine (35), iodine (53), and astatine (85) are all in the same column in the table. These elements are all called *halogens*. They are all colored substances, and the colors deepen as we go down the table (see Fig. 2.7). All form compounds with sodium that have the formula NaE (for example, NaCl, NaBr) and not $NaE_2$, $Na_2E$, $Na_3E$, or anything else. Elements not in this column do not share these properties.

At this point we must say a word about the numbering of the columns of the periodic table. Mendeleev gave them Roman numerals, and added the letter A for some columns and B for others. These numbers have been used for many years. Recently, however, because U.S. and European chemists used the A and B in different ways, international and U.S. chemistry organizations have

---

Strictly speaking, the atomic weight of an element is a ratio of the weighted average mass of the atom compared with the mass of the carbon isotope whose mass number is 12 and whose mass is taken to be 12.0000 amu. Thus the amu is defined as $\frac{1}{12}$ the mass of a $^{12}$C atom.

This illustrates how to calculate a weighted average.

Some biochemists use **daltons** as a unit of atomic weight.

1 amu = 1 dalton

Note that hydrogen (H) appears in columns IA and VIIA but is neither a halogen nor an alkali metal.

"E" is a general symbol meaning element.

The elements of group 0 were not yet discovered when Mendeleev first made his table, so this column does not have a Roman numeral.

**Figure 2.6**
Dmitri Mendeleev. (Courtesy of E. F. Smith Memorial Collection, Special Collection Department, Van Pelt Library, University of Pennsylvania.)

agreed to recommend a different set of numbers, 1 to 18, going from left to right. Thus, by Mendeleev's numbering, the halogens are in group VIIA, but in the new numbering they are in group 17. Although in this book we use the traditional numbering, both kinds are shown on the periodic table in the inside front cover.

**Figure 2.7**
(a) Chlorine, a greenish-yellow gas. (From W. L. Masterton, et al., *Chemical Principles,* 6th ed. Philadelphia: Saunders College Publishing, 1985.) (b) Bromine, a deep red liquid. (Photograph by Charles D. Winters.) (c) Iodine, a shiny solid, which, on gentle heating, sublimes to a purple gas. (Photograph by Beverly March.)

(a)                              (b)                   (c)

The similarity of properties shown by elements in the same column of the periodic table can have biological consequences. One important, and troubling, example comes about because the properties of the element strontium are fairly similar to those of calcium, which is just above it on the table. Calcium is an important element for humans, since our bones and teeth consist largely of calcium compounds. We need some of this mineral in our diet every day, and we get it mostly from milk, cheese, and other dairy products, where it is present in the form of calcium ions.

Unfortunately, one of the products released by nuclear explosions is an isotope of the element strontium, strontium-90. This isotope is radioactive, with a half-life of 28.1 years. (Half-life is discussed in Section 9.3.) Strontium-90, present in the fallout from above-ground nuclear test explosions, was carried all over the earth and slowly settled to the ground, where it was eaten by cows and other animals. By this means it got into milk and eventually into our bodies. If it were not similar to calcium, our bodies would eliminate all of it within a few days, but because it is similar to calcium, some of the strontium-90 ions were deposited in bones and teeth (especially of children), subjecting all of us to a small amount of radioactivity for long periods of time. Fortunately, above-ground nuclear testing was pretty much stopped some years ago by a treaty between the United States and the former Soviet Union, although a few other countries still conduct occasional tests.

Not only do the elements in any one column have similar properties, but the properties vary in some fairly regular way as we go up or down a column. As shown in Table 2.3, the melting and boiling points regularly increase as we go down a column. Another example is found in group IA. Once again, the elements here (called *alkali metals*) all share similar properties. All are metals with low melting points and are soft enough to be cut with a knife. All react with water (Fig. 2.8), and all form compounds with fluorine, whose formula is EF. Also, as we go down the column, the melting and boiling points of the alkali metals change in a regular way, the violence of the reaction with water increases, and so does the softness of the metal. The elements in the last column (group 0) provide still another example. All are gases at room temperature and form either no compounds at all or very few.

**TABLE 2.3   Melting and Boiling Points of Group VIIA Elements**

| Element | Melting Point (°C) | Boiling Point (°C) |
|---|---|---|
| Fluorine | −220 | −188 |
| Chlorine | −101 | −35 |
| Bromine | −7 | 59 |
| Iodine | 114 | 184 |
| Astatine | 302 | 337 |

The same type of regularity is found in all the columns of the periodic table. The table is so useful that it hangs in nearly all chemistry classrooms and in most chemical laboratories throughout the world. What makes it so useful is that it correlates a vast amount of data about the elements and their compounds and allows us to make many predictions about physical as well as chemical properties. For example, if you were told that the boiling point of germane ($GeH_4$) is $-88°C$ and that of methane ($CH_4$) is $-164°C$, could you predict the boiling point of silane ($SiH_4$)? The position of silicon in the table, between germanium and carbon, might lead you to a prediction of about $-120°C$. The actual boiling point of silane is $-112°C$, not far from this prediction.

If we look at the periodic table on the inside front cover of this book, we see that columns IA and IIA extend above the rest, and so do columns IIIA to 0. The elements in these eight columns are called **representative elements.** The elements in the B columns and those in group VIII are called **transition elements.** Notice that elements 58 to 71 and 90 to 103 are not included in the main body of the table but are shown separately at the bottom. These elements (called **inner transition elements**) actually belong in the main body, between columns IIIB and IVB (the place is between La and Hf). As is customary, we put them outside the main body only in order to make a neater and more convenient presentation. So, if you will, you may mentally take a pair of scissors, cut through the heavy line between columns IIIB and IVB, move them apart, and insert the inner transition elements. You will now have a table with 32 columns.

**Figure 2.8**
A piece of metallic sodium reacting with water. (Photograph by Beverly March.)

## Metals, Nonmetals, and Metalloids

Another way to classify the elements is as metals or nonmetals. **Metals** are shiny substances whose atoms tend to give up electrons. They are usually good conductors of electricity and heat and are **malleable,** which means that they can be hammered into flat sheets, and **ductile,** which means that they can be drawn into thin wires. Most elements are metals, including all the transition and inner transition elements. In addition, the elements in columns IA and IIA are all metals, as are those in columns IIIA to VIA *below the heavy step-like line.*

**Nonmetals** do not conduct electricity, and their atoms do not give up electrons but rather tend to accept them. The nonmetals are the ones *above the heavy step-like line* in the periodic table. The elements right next to the heavy line share some of the properties of metals and nonmetals and are called **metalloids.** The following elements are classified as metalloids: B, Si, Ge, As, Sb, Te, Po, and At. In general, representative elements behave more like metals in going down a column of the periodic table, and in going from right to left.

The six elements in group 0 (called **noble gases**) are a special group, with properties different from those of the other nonmetals. They neither give nor take electrons. Apart from them, the nonmetals are H, C, N, O, F, P, S, Cl, Se, Br, and I. Although there are only 11 of these elements, they make up nearly all the compounds we shall study when we get to organic chemistry and biochemistry later in this book. Figure 2.9 shows the distribution of the elements in the periodic table.

| H | | | | | | | | | | | | | | | | | H | He |
|---|---|---|---|---|---|---|---|---|---|---|---|---|---|---|---|---|---|---|
| Li | Be | | | | | | | | | | | | B | C | N | O | F | Ne |
| Na | Mg | | | | | | | | | | | | Al | Si | P | S | Cl | Ar |
| K | Ca | Sc | Ti | V | Cr | Mn | Fe | Co | Ni | Cu | Zn | Ga | Ge | As | Se | Br | Kr |
| Rb | Sr | Y | Zr | Nb | Mo | Tc | Ru | Rh | Pd | Ag | Cd | In | Sn | Sb | Te | I | Xe |
| Cs | Ba | La | Hf | Ta | W | Re | Os | Ir | Pt | Au | Hg | Tl | Pb | Bi | Po | At | Rn |
| Fr | Ra | Ac | Unq | Unp | Unh | Uns | Uno | Une | | | | | | | | | |

| Ce | Pr | Nd | Pm | Sm | Eu | Gd | Tb | Dy | Ho | Er | Tm | Yb | Lu |
|---|---|---|---|---|---|---|---|---|---|---|---|---|---|
| Th | Pa | U | Np | Pu | Am | Cm | Bk | Cf | Es | Fm | Md | No | Lr |

☐ Metals

☐ Nonmetals

☐ Metalloids

☐ Noble gases

**Figure 2.9**
Classification of the elements.

▶ **Box 2C**

## The Use of Metals as Historical Landmarks

**Figure 2C**
King Solomon's copper mines, Timna, Israel. (Courtesy of Holy Views, Ltd., Jerusalem.)

The malleability of metals played an important role in the development of human society. In the early days, tools were made from stone, which has no malleability. This period is called the Stone Age. Later it was discovered that copper found on the surface of the earth could be hammered into sheets, which made it suitable for vessels, utensils, and religious or artistic objects. This practice started about 9000 or 10 000 years ago, and this period is sometimes called the Copper Age. However, pure copper on the surface of the earth is scarce. Around 5000 BC human beings found a way to obtain copper from its ore by putting the green copper-containing stone, malachite, into a fire. Malachite yielded pure copper at the relatively low temperature of 200°C. Copper is a soft metal made of layers of copper crystals. It can easily be drawn into wires as the layers slip past each other. When hammered, the large crystals break into smaller ones with rough edges. The layers cannot slip, so hammered copper sheets were harder. As a result, the ancient profession of coppersmith was born, and beautiful plates, pots, and ornaments were produced.

Around 4000 BC, it was discovered that by mixing molten copper with tin a greater hardness could be achieved. Copper, even when hammered, does not take an edge, but a copper-tin mixture (an alloy called bronze) does take an edge, and knives and swords could be manufactured. Thus the Bronze Age was born somewhere in the Middle East and quickly spread to China and all over the world.

An even harder metal was soon to come. The first raw iron was found in meteorites. (The ancient Sumerian name of iron is "metal from heaven.") Later, around 2500 BC, it was smelted from its ores and the Iron Age began. Higher technology was needed because iron melts at a higher temperature (about 1500°C) than copper (about 1100°C). Thus, it took a longer time to perfect this process and to learn how to manufacture steel (about 90 to 95 percent iron and 5 to 10 percent carbon). Steel objects appeared first in India around 100 BC.

Modern historians look back at ancient cultures and use the discovery of a new metal as a landmark for that age.

## 2.6   THE ELECTRONIC STRUCTURE OF ATOMS

In discussing the electronic structure of atoms, we will begin with hydrogen because it is the simplest, with only one electron, but before we do that it is necessary to mention the work of Niels Bohr (1885–1962). In 1913 Bohr, a Danish physicist, found that the electron in a hydrogen atom could have one of a number of energies. An electron is always moving around the nucleus and so possesses energy. What Bohr proposed is that certain values were possible for this energy, but not others. This was a very surprising proposal. If you were told that you could drive your car at 23.4 mph or 28.9 mph or 34.2 mph, but never at any speed in between these values, you wouldn't believe it. Yet this is just what Bohr said about electrons in atoms. There is a lowest possible energy level; we call it the **ground state** (electrons in atoms cannot stop moving, so the ground-state energy is not zero). If an electron is to have more energy than in the ground state, however, there are only certain values allowed; values in between are not permitted. We can liken the situation to walking up a flight of stairs (Fig. 2.10). You can put your foot on any step, but you cannot stand anyplace between two steps. Bohr was unable to explain why these levels exist, but the accumulated evidence forced him to the conclusion that they do. We say that the energy of the electron is **quantized.**

Electrons in the lowest energy state of the hydrogen atom—the ground state—occupy a spherical cloud surrounding the nucleus (Fig. 2.11). Clouds of this type are called **orbitals.** They are designated by numbers and letters. This one is called the **1s orbital.** There are also orbitals of higher energy. Some of them (2s and 2p) are shown in Figure 2.11. There are three 2p orbitals (designated $2p_x$, $2p_y$, and $2p_z$), which have equal energy. These three orbitals look like dumbbells (but are still clouds) and are at right angles to each other.

Niels Bohr was awarded the Nobel prize for this work in 1922.

**Figure 2.10**
The energy stairway. Unlike our stairways, the spaces between levels are not equal, but get smaller as they go up.

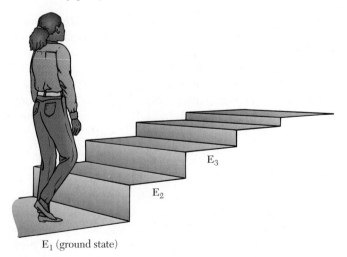

$E_3$

$E_2$

$E_1$ (ground state)

**Figure 2.11**
The 1s, 2s, and 2p orbitals.

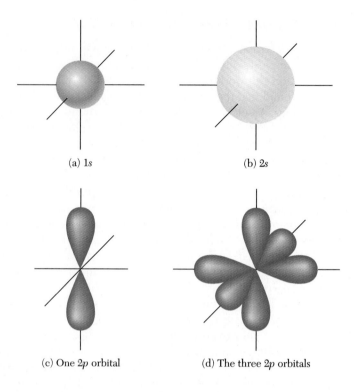

(a) 1s

(b) 2s

(c) One 2p orbital

(d) The three 2p orbitals

The *d* and *f* orbitals are less important to us, and we shall not discuss their shapes.

There are no 1*p*, 1*d*, or 1*f* orbitals.

A subshell can also be called a **sublevel.**

There are also orbitals of still higher energy. After 2p come 3s, 3p, 4s, and others even higher. All *s* orbitals are spherical; all *p* orbitals look like dumbbells and come in sets of three. Besides *s* and *p* orbitals there are two other kinds, called *d* and *f*. All *d* orbitals come in sets of five and *f* orbitals in sets of seven.

The number used in the designation of orbitals is called the **principal energy level.** The first principal energy level (the 1 level) contains only one orbital, the 1s orbital. The second principal energy level (the 2 level) contains four orbitals (one 2s and three 2p). The number of orbitals in each level increases as the principal energy level increases.

All the orbitals in any principal energy level constitute a **shell.** Thus, the 1 shell contains only the 1s orbital, and the 2 shell contains one 2s and three 2p orbitals. A **subshell** consists of all the orbitals with the same number and letter. For example, there are two subshells in the 2 shell: One of them consists of the three 2p orbitals, and the other contains only the single 2s orbital.

Note that the orbital designations show both the energy aspect and the spatial aspect of each orbital. The number tells the energy level and the size of the orbital; the letter tells its shape.

### The Hydrogen Atom

We now have a good picture of the electronic structure of the hydrogen atom. In its lowest possible energy state (1s), the atom looks like the picture shown in Figure 2.11(a), with the nucleus in the middle. This is the ground state, but it is not the only possible state. The electron can also exist in any one of the higher energy levels, called *excited states.* This is how it works: A hydrogen atom is in its ground state, looking like Figure 2.11(a). It then receives additional energy from some outside source (usually in the form of heat or light). This additional energy, when added to the energy it already had, equals the amount of energy the electron needs to reach some higher orbital—the $2p_x$, for example. The $1s$ orbital is now vacant, and the electron appears in the $2p_x$ orbital. The hydrogen atom now looks like Figure 2.11(c) and no longer like Figure 2.11(a). We can imagine a whole collection of hydrogen atoms, some having the electron in the $1s$ orbital, some in the $2p_x$, some in the $3d$, and some in other orbitals. In any individual hydrogen atom, only one orbital is occupied at any time. At room temperature this orbital is overwhelmingly the $1s$.

### Rules for Other Atoms

Now that we have seen the electronic structure of hydrogen, what about other atoms? The situation is similar except that these atoms have more than one electron. The orbitals available to all atoms are the same as in hydrogen: $1s$, $2s$, $2p$, and so on. In the ground state of each atom, only the orbitals of lowest energy are occupied; all higher ones are empty. The electrons fill in according to the following rules:

**1.   Each orbital can hold two electrons, but no more.** Furthermore, if there are two electrons in an orbital, they must have opposite *spin*. This is a statement of the **Pauli exclusion principle.** When there is only one electron in an orbital, we call the electron **unpaired** and say that the orbital is singly occupied.

*Electrons can spin either clockwise or counterclockwise.*

**2.**   When we have a set of orbitals of equal energy (such as the $2p_x$, $2p_y$, and $2p_z$ orbitals), each orbital becomes half filled before any is completely filled. This is called **Hund's rule.**

**3.**   The orbitals can be likened to a set of boxes. The electrons in any atom go into whichever boxes have the lowest energies, until there are no more electrons left. If we want to know which boxes are occupied, we have to know the order of energies. This is shown in Figure 2.12. Note that the lowest $p$ orbitals are $2p$, the lowest $d$ orbitals are $3d$, and the lowest $f$ orbitals are $4f$. Within any shell the energy levels increase in the order $s, p, d, f$. That is, $s$ is the lowest and $f$ the highest.

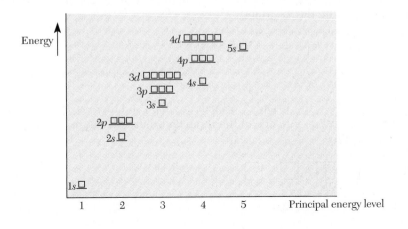

**Figure 2.12**
Energy levels of atomic orbitals up to the $4d$ level.

### The Electronic Buildup

Using the rules just given, we now determine the ground-state electronic configuration for the first 36 elements.

**1. Hydrogen** As we have already seen, the single hydrogen electron is in the 1s orbital in the ground state.

Arrow indicates direction of spin

H configuration: $1s^1$

This digit tells us that there is one electron in this orbital

**2. Helium** There are two electrons in this atom. Since any orbital can hold two electrons, both are able to go into the lowest-energy orbital, the 1s.

The 2 tells us that there are two electrons in this orbital

He configuration: $1s^2$

All electrons at the 1 level (the first principal energy level) constitute the first *shell*. This shell can hold only two electrons, so helium *has a complete shell*.

**3. Lithium** The first two of lithium's three electrons fill the 1s orbital. The other must go into the orbital of next-lowest energy, the 2s.

| ↿⇂ | ↑ |
|---|---|
| 1s | 2s |

Li configuration: $1s^2 2s^1$

In lithium the first shell (the 1 shell) is complete and there is one electron in the 2 shell. The 1 shell is the *inner shell,* and the 2 shell is the incomplete *outer shell.*

**4. Beryllium** The first two electrons fill the 1s orbital (and the 1 shell). The remaining two fill the 2s orbital.

| ↿⇂ | ↿⇂ |
|---|---|
| 1s | 2s |

Be configuration: $1s^2 2s^2$

**5. Boron** Now the 1s and 2s orbitals are filled and there is one electron left over, which must go into one of the three 2p orbitals. It doesn't matter which; we shall say $2p_x$.

| ↿⇂ | ↿⇂ | ↑ | | |
|---|---|---|---|---|
| 1s | 2s | $2p_x$ | $2p_y$ | $2p_z$ |

B configuration: $1s^2 2s^2 2p^1$

**6. Carbon** We now come to our first application of Hund's rule. The sixth electron of carbon obviously must go into one of the $2p$ orbitals, but which one? Hund's rule says that it cannot fill the $2p_x$ orbital while either of the other two $2p$ orbitals is vacant, and so it goes into the $2p_y$ or $2p_z$ (it doesn't matter which; we shall say $2p_y$)..A carbon atom thus has *two* unpaired electrons.

$1s$ $\quad$ $2s$ $\quad$ $2p_x$ $\quad$ $2p_y$ $\quad$ $2p_z$

C configuration: $1s^2 2s^2 2p^2$

**7. Nitrogen** Hund's rule puts the seventh electron into the $2p_z$ orbital. A nitrogen atom has three unpaired electrons.

$1s$ $\quad$ $2s$ $\quad$ $2p_x$ $\quad$ $2p_y$ $\quad$ $2p_z$

N configuration: $1s^2 2s^2 2p^3$

There is experimental proof for Hund's rule. Magnetic measurements show that a nitrogen atom has three unpaired electrons.

**8–10. Oxygen, fluorine, neon** Now that all three $2p$ orbitals are half filled, the next three elements fill these orbitals, one at a time.

8  O  $1s^2 2s^2 2p^4$

9  F  $1s^2 2s^2 2p^5$

10  Ne  $1s^2 2s^2 2p^6$

Oxygen has two unpaired electrons, and fluorine has one. The tenth electron of neon fills not only the $2p_z$ orbital but also the entire 2 shell. Neon is the second element with all its electrons in completely filled shells (helium was the first).

**11. Sodium** Since the 1 and 2 shells are now complete, the eleventh electron of sodium goes into the $3s$ orbital and begins the 3 shell. The 2 shell is now an inner shell, and the 3 shell becomes the outer shell.

$1s$ $\quad$ $2s$ $\quad$ $2p_x$ $\quad$ $2p_y$ $\quad$ $2p_z$ $\quad$ $3s$ $\quad$ $3p_x$ $\quad$ $3p_y$ $\quad$ $3p_z$

Na configuration: $1s^2 2s^2 2p^6 3s^1$ $\quad$ or $\quad$ $[\text{Ne}]3s^1$

Since the configuration of the two inner shells is the same as that of the neon atom, we use the abbreviation [Ne] to show the configuration of the two inner shells.

**12–18.** The next seven elements follow the same pattern as elements 4 through 10, but at the 3 rather than the 2 level.

12  Mg  $[\text{Ne}]3s^2$

13  Al  $[\text{Ne}]3s^2 3p^1$

14  Si  $[\text{Ne}]3s^2 3p^2$

15  P  $[\text{Ne}]3s^2 3p^3$

16  S  $[\text{Ne}]3s^2 3p^4$

17  Cl  $[\text{Ne}]3s^2 3p^5$

18  Ar  $[\text{Ne}]3s^2 3p^6$

Argon is the third element to have all its electrons in complete shells. This is true even though the 3*d* orbitals are vacant. We say this because the behavior of argon is very similar to that of helium and neon (none of the three form any compounds). This is what makes us regard argon as an element with no incomplete shells.

**19–20.   Potassium, calcium**   The next two electrons go into the 4*s* orbital.

19   K   [Ar]4*s*$^1$

20   Ca   [Ar]4*s*$^2$

**21.   Scandium**   If scandium followed the previous pattern, the twenty-first electron would go into the 4*p* orbital. It does not do so, however, because the 3*d* orbital has a lower energy (Fig. 2.12).

21   Sc   [Ar]4*s*$^2$3*d*$^1$

**22–30.**   The next nine elements fill the five 3*d* orbitals; the last of these is zinc.

30   Zn   [Ar]4*s*$^2$3*d*$^{10}$

Zinc does not have a complete outer shell. The 3 shell *is* complete, but the 3 shell is an inner shell here. **The occupied shell with the highest number is always the outer shell.** In the case of zinc, the occupied shell with the highest number is the 4 shell, and that one is not complete since the 4*p* orbitals are vacant.

**31–36.**   The next six elements fill the 4*p* orbitals.

36   Kr   [Ar]4*s*$^2$3*d*$^{10}$4*p*$^6$

Krypton is the fourth element with a complete outer shell (even though the 4*d* and 4*f* orbitals are vacant).

We have now described the electronic configuration of the first 36 elements (they are summarized in Table 2.4). The remaining elements fill up the next orbitals (5*s*, 4*d*, and so on) in a similar manner, in the order shown in Figure 2.12.

## 2.7   ELECTRONIC CONFIGURATION AND THE PERIODIC TABLE

When Mendeleev published his first periodic table in 1869, he had no explanation for why it worked, and indeed nobody else had any good explanation either. It was not until after the discovery of electronic structure that chemists finally understood why the table works. The answer is very simple. *The periodic table works because elements in the same column have the same configuration of electrons in the outer shell.* For example, let us look at the elements in column IA. We already know the configuration for lithium, sodium, and potassium. We now add rubidium and cesium:

Li   [He]2*s*$^1$

Na   [Ne]3*s*$^1$

K   [Ar]4*s*$^1$

Rb   [Kr]5*s*$^1$

Cs   [Xe]6*s*$^1$

| TABLE 2.4 | Electronic Configurations of the First 36 Elements | | |
|---|---|---|---|
| 1 | H | $1s^1$ | |
| 2 | He | $1s^2$ | |
| 3 | Li | $1s^22s^1$ | |
| 4 | Be | $1s^22s^2$ | |
| 5 | B | $1s^22s^22p^1$ | |
| 6 | C | $1s^22s^22p^2$ | |
| 7 | N | $1s^22s^22p^3$ | |
| 8 | O | $1s^22s^22p^4$ | |
| 9 | F | $1s^22s^22p^5$ | |
| 10 | Ne | $1s^22s^22p^6$ | |
| 11 | Na | $1s^22s^22p^63s^1$ | or $[\text{Ne}]3s^1$ |
| 12 | Mg | $1s^22s^22p^63s^2$ | or $[\text{Ne}]3s^2$ |
| 13 | Al | $1s^22s^22p^63s^23p^1$ | or $[\text{Ne}]3s^23p^1$ |
| 14 | Si | $1s^22s^22p^63s^23p^2$ | or $[\text{Ne}]3s^23p^2$ |
| 15 | P | $1s^22s^22p^63s^23p^3$ | or $[\text{Ne}]3s^23p^3$ |
| 16 | S | $1s^22s^22p^63s^23p^4$ | or $[\text{Ne}]3s^23p^4$ |
| 17 | Cl | $1s^22s^22p^63s^23p^5$ | or $[\text{Ne}]3s^23p^5$ |
| 18 | Ar | $1s^22s^22p^63s^23p^6$ | |
| 19 | K | $1s^22s^22p^63s^23p^64s^1$ | or $[\text{Ar}]4s^1$ |
| 20 | Ca | $1s^22s^22p^63s^23p^64s^2$ | or $[\text{Ar}]4s^2$ |
| 21 | Sc | $1s^22s^22p^63s^23p^64s^23d^1$ | or $[\text{Ar}]4s^23d^1$ |
| 22 | Ti | $1s^22s^22p^63s^23p^64s^23d^2$ | or $[\text{Ar}]4s^23d^2$ |
| 23 | V | $1s^22s^22p^63s^23p^64s^23d^3$ | or $[\text{Ar}]4s^23d^3$ |
| 24 | Cr | $1s^22s^22p^63s^23p^64s^13d^{5\,\text{a}}$ | or $[\text{Ar}]4s^13d^5$ |
| 25 | Mn | $1s^22s^22p^63s^23p^64s^23d^5$ | or $[\text{Ar}]4s^23d^5$ |
| 26 | Fe | $1s^22s^22p^63s^23p^64s^23d^6$ | or $[\text{Ar}]4s^23d^6$ |
| 27 | Co | $1s^22s^22p^63s^23p^64s^23d^7$ | or $[\text{Ar}]4s^23d^7$ |
| 28 | Ni | $1s^22s^22p^63s^23p^64s^23d^8$ | or $[\text{Ar}]4s^23d^8$ |
| 29 | Cu | $1s^22s^22p^63s^23p^64s^13d^{10\,\text{a}}$ | or $[\text{Ar}]4s^13d^{10}$ |
| 30 | Zn | $1s^22s^22p^63s^23p^64s^23d^{10}$ | or $[\text{Ar}]4s^23d^{10}$ |
| 31 | Ga | $1s^22s^22p^63s^23p^64s^23d^{10}4p^1$ | or $[\text{Ar}]4s^23d^{10}4p^1$ |
| 32 | Ge | $1s^22s^22p^63s^23p^64s^23d^{10}4p^2$ | or $[\text{Ar}]4s^23d^{10}4p^2$ |
| 33 | As | $1s^22s^22p^63s^23p^64s^23d^{10}4p^3$ | or $[\text{Ar}]4s^23d^{10}4p^3$ |
| 34 | Se | $1s^22s^22p^63s^23p^64s^23d^{10}4p^4$ | or $[\text{Ar}]4s^23d^{10}4p^4$ |
| 35 | Br | $1s^22s^22p^63s^23p^64s^23d^{10}4p^5$ | or $[\text{Ar}]4s^23d^{10}4p^5$ |
| 36 | Kr | $1s^22s^22p^63s^23p^64s^23d^{10}4p^6$ | |

[a] The electronic configurations of chromium and copper are slightly irregular, because a filled (10 electrons) or half-filled (5 electrons) $d$ subshell has a special stability.

All these atoms have complete inner shells and an outer shell consisting of one $s$ electron. The properties of elements largely depend on the electronic configuration of their outer shell. This being so, it is not surprising that these elements, all of which have the same outer-shell configuration, should have such similar properties.

We could not expect *identical* properties because the number of protons in the nucleus is different and so is the number of inner shells. These numbers also have an effect on the properties and allow us to explain why properties change regularly as we move down a column of the periodic table. For example, all of the elements in group IA are metals, which means they tend to give up electrons. In this case they give up only the single electron in the outer shell, but

cesium gives it up most easily, rubidium next, and so on up to lithium. We explain this by saying that the 2s electron is closer to the positive nucleus in lithium than is the 3s electron in sodium, and so the 2s electron is more attracted to its nucleus. Furthermore, sodium has one more filled inner shell than lithium, and this filled inner shell (made up of negatively charged electrons) "shields" the 3s electron from the nucleus more effectively than the fewer inner electrons of lithium. Since atoms get bigger as we go down the table, these effects explain why loss of the outer electron becomes easier as we go down a column.

### Ionization Energy

The energy necessary to remove an electron from an atom in the gaseous state is called the *ionization energy*. We can actually measure its numerical value for each atom. Table 2.5 shows the ionization energy for atoms of some of the representative elements. We can see from Table 2.5 that ionization energy decreases as we go down any column. We have already seen why this is so: Electrons are easier to remove the farther they get from the nucleus. Table 2.5 also shows that, with a few exceptions, ionization energy increases regularly as we go from left to right across any row. This is consistent with the change in metallic character mentioned in Section 2.5. As we go from left to right in the table, elements become less metallic. Metals tend to give up electrons more readily than nonmetals, and the information in Table 2.5 largely bears this out.

In all cases these regularities exist because elements in the same column have the same outer-shell electronic configuration.

The regularity that we see in Table 2.5 is not unusual. We could construct similar tables for many other properties, and we would find similar regularities, going both down and across.

Electronic configuration also explains the presence of the transition elements (p. 47). Let us look at the electronic structure of iron, a typical transition element:

Two incomplete shells

$$26 \quad \text{Fe} \quad [\text{Ar}]4s^2 3d^6$$

In this atom, the outer shell—the 4 shell—contains two electrons and is obviously incomplete because the 4p orbitals, which can hold six more electrons, are empty. The 3 shell, however, an inner shell, is also incomplete. There are six

| TABLE 2.5 | Ionization Energy of Some Representative Elements | | | | | | |
|-----------|-----------|-----------|-----------|-----------|-----------|-----------|-----------|
| IA | IIA | IIIA | IVA | VA | VIA | VIIA | 0 |
| Li | Be | B | C | N | O | F | Ne |
| 124[a] | 215 | 191 | 260 | 335 | 314 | 402 | 497 |
| Na | Mg | Al | Si | P | S | Cl | Ar |
| 119 | 176 | 138 | 188 | 242 | 239 | 299 | 363 |
| K | Ca | Ga | Ge | As | Se | Br | Kr |
| 100 | 141 | 138 | 182 | 226 | 225 | 272 | 323 |

[a] All values are kilocalories/mole, a unit of energy.

electrons in the five $3d$ orbitals, and these orbitals, of course, can hold a total of ten electrons. So iron has not only an incomplete outer shell but also an incomplete inner shell. We define a **transition element** as **an element whose atoms have one or more incomplete inner shells.** (As you might have guessed, inner transition elements have two incomplete inner shells.)

The presence of unfilled inner shells gives transition elements special properties. For example, most of them form more than one kind of ion.

## 2.8  HOW SMALL ARE ATOMS?

In the previous section we examined the electronic structure of atoms. It is useful to get some idea of how small atoms and their components really are.

### The Mass of an Atom

A typical heavy atom (though not the heaviest) is lead-208, a lead atom with 82 protons, 82 electrons, and 126 neutrons. It has a mass of $3.5 \times 10^{-22}$ g. You would need $1.3 \times 10^{24}$ lead-208 atoms to make one pound of lead. This is a very large number. There are about five billion people on earth right now. If you divided one pound of these atoms among all the people on earth, each person would get about $2.6 \times 10^{14}$ atoms. This is still a large number. If these atoms had a value of 1 cent per 1000 atoms, every person on earth could get more than two billion dollars by selling his or her share.

### The Size of an Atom

An atom of lead has a diameter of about $3.5 \times 10^{-8}$ cm. If you could line them up with the atoms just touching, it would take 73 million lead atoms to make a line 1 inch long. Despite their tiny size, we can actually see atoms, in certain cases, by the use of special microscopic techniques. An example is shown in Figure 2.13.

**Figure 2.13**
Atoms of silicon on a specially reconstructed surface of a silicon crystal as seen by a scanning tunneling microscope. (Courtesy of Dr. Sang-Il Park, Applied Physics, Stanford University, Stanford, CA.)

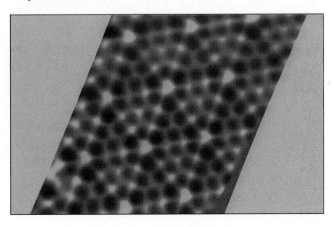

## Abundance of Elements on Earth and in People

Although 88 elements are found in the earth's crust (we know very little about the interior of the earth because we have not been able to penetrate very far), they are not present in anything like equal amounts. Table 2D.1 shows that the chief elements are oxygen, silicon, aluminum, and iron, and only six other elements make up even as much as one half of 1 percent by weight of the earth's crust.

**TABLE 2D.1    Relative Abundance of Elements in the Earth's Crust, Including Atmosphere and Oceans**

| Element | Percent by Weight | Element | Percent by Weight |
|---|---|---|---|
| O | 49.3 | Cl | 0.2 |
| Si | 25.8 | P | 0.12 |
| Al | 7.6 | Mn | 0.09 |
| Fe | 4.7 | C | 0.08 |
| Ca | 3.4 | S | 0.06 |
| Na | 2.7 | Ba | 0.05 |
| K | 2.4 | Cr | 0.03 |
| Mg | 1.9 | N | 0.03 |
| H | 0.9 | All others | <0.1 |
| Ti | 0.6 | | |

Table 2D.2 shows the elements present in the human body. Oxygen is still the most abundant element by weight, but otherwise the picture is greatly different. Apart from oxygen, the most abundant elements in the body are hydrogen, carbon, and nitrogen. If we go by number of atoms instead of by weight, hydrogen is even more abundant than oxygen. Note that such common elements (on earth) as silicon, aluminum, and iron are not even listed in Table 2D.2 (although iron is an important trace element in the human body, as mentioned in Box 2A).

**TABLE 2D.2    Relative Abundance of Elements in the Human Body**

| | Percent | |
|---|---|---|
| Element | By Number of Atoms | By Weight |
| H | 63.0 | 10.0 |
| O | 25.4 | 64.8 |
| C | 9.4 | 18.0 |
| N | 1.4 | 3.1 |
| Ca | 0.31 | 1.8 |
| P | 0.22 | 1.4 |
| K | 0.06 | 0.4 |
| S | 0.05 | 0.3 |
| Cl | 0.03 | 0.2 |
| Na | 0.03 | 0.1 |
| Mg | 0.01 | 0.04 |
| Others | 0.01 | |

**Figure 2D**
Only oxygen is a principal element in both the sand, gravel, and water, and in the water buffaloes. (Photograph by Beverly March.)

### The Mass of an Electron

An electron has a mass of $9.1 \times 10^{-28}$ g. The sun is very big; it weighs about 333 000 times as much as the earth. The mass of the sun is to the mass of a 3-lb cantaloupe as the mass of the cantaloupe is to the mass of an electron (Fig. 2.14).

### The Mass of the Nucleus

As shown in Table 2.2, the mass of an electron is very much smaller than that of a proton or neutron. Since protons and neutrons are in the nucleus and electrons are outside, this means that almost all the mass of an atom is in the nucleus. The total mass of the 82 electrons in a lead-208 atom is $7.5 \times 10^{-26}$ g, and the mass of the whole atom is $3.5 \times 10^{-22}$ g.

This means that 99.98 percent of the mass of this atom is in the nucleus. A similar percentage holds for all other atoms.

### The Size of the Nucleus

Although the mass of the nucleus is very great, the size of the nucleus is very small. The nucleus of a lead-208 atom has a diameter of about $1.6 \times 10^{-12}$ cm. When we compare this with the diameter of the whole atom, which is $3.5 \times 10^{-8}$ cm, we see that the nucleus occupies only a tiny fraction of the atom. If the nucleus of a lead-208 atom were the size of a baseball, then the whole atom would be much larger than a baseball stadium. In fact, it would be a sphere about one mile in diameter.

**Figure 2.14**
The mass of the sun is to that of a cantaloupe as the mass of a cantaloupe is to that of an electron.

**Figure 2.15**
If a paper clip this size were made entirely of atomic nuclei packed together, it would weigh about ten million tons.

## The Density of the Nucleus

With such a large mass in such a small space, it is obvious that the nucleus must be very dense. For example, the density of the lead-208 nucleus is $1.8 \times 10^{14}$ g/cc. This is an extremely high density. Nothing in our daily life has a density anywhere near as high. If a paper clip had this density, it would weigh about ten million ($10^7$) tons (Fig. 2.15).

## SUMMARY

Matter can be classified as **elements** (substances all of whose atoms are the same), **compounds** (substances made up of at least two elements), or **mixtures.** In **chemical changes (chemical reactions)** compounds or elements are used up and others are formed. In **physical changes** substances do not change their identity. Dalton's atomic theory says that all atoms of a given element are identical. The theory is based on the law of conservation of mass (matter cannot be created or destroyed) and on the law of constant composition (any compound is always made up of elements in the same proportion by weight). An **atom** is the smallest unit of an element that shows the chemical behavior of the element. A **molecule** is a cluster of two or more atoms connected by chemical bonds.

Atoms consist of **protons** and **neutrons** inside the nucleus and **electrons** outside it. Electrons are very light and have a charge of $-1$. Protons and neutrons are much heavier; their masses are about equal. Protons have a charge of $+1$; neutrons have no charge. The **atomic number** of an element is the number of protons; this determines the identity of the element. The sum of the number of protons and neutrons is the **mass number.** Atoms have an equal number of electrons and protons; but **ions,** which are atoms that have lost or gained electrons, have an unequal number of electrons and protons and therefore are charged. **Isotopes** are atoms with the same number of protons but different numbers of neutrons. The **atomic weight** of an element is a weighted average of the masses of the isotopes as they occur in nature.

The **periodic table** arranges elements into columns of elements with similar properties, which gradually change as we move down a column. Metallic elements, which tend to lose electrons, are below the heavy line in the periodic table. Nonmetals are above the line.

Electrons in atoms can exist in a number of **energy levels.** The energy levels are clouds called **orbitals.** All $s$ orbitals are spherical, and all $p$ orbitals are shaped like dumbbells. Each orbital can hold a maximum of two electrons, of opposite spin. Electrons in the **ground state** (lowest energy state) of atoms fill up the available orbitals of lowest energy. Atoms are very tiny, with a very small mass, almost all of which is in the nucleus. The nucleus is extremely tiny, with a high density.

The periodic table works because elements in the same column have the same outer-shell configuration. **Ionization energy,** which is the energy necessary to remove an electron from an atom, decreases as we go down the periodic table and increases as we go from left to right. Transition elements have one or more incomplete inner shells.

## KEY TERMS

Alkali metal (Sec. 2.5)
Atom (Sec. 2.3)
Atomic number (Sec. 2.4)
Atomic weight (Sec. 2.4)
Chemical change (Sec. 2.2)
Chemical property (Sec. 2.2)
Chemical reaction (Sec. 2.2)

Compound (Sec. 2.2)
Distillation (Sec. 2.2)
Electric charge (Sec. 2.4)
Electron (Sec. 2.4)
Element (Sec. 2.2)
Excited state (Sec. 2.6)
Ground state (Sec. 2.6)

Hund's rule (Sec. 2.6)
Inner transition element (Secs. 2.5, 2.7)
Ion (Sec. 2.4)
Ionization energy (Sec. 2.7)
Isotope (Sec. 2.4)
Mass number (Sec. 2.4)
Metal (Sec. 2.5)
Metalloid (Sec. 2.5)
Mixture (Sec. 2.2)
Molecule (Sec. 2.3)
Neutron (Sec. 2.4)
Noble gas (Sec. 2.5)
Nonmetal (Sec. 2.5)
Nucleon (Sec. 2.4)

Nucleus (Sec. 2.4)
Orbital (Sec. 2.6)
Outer shell (Sec. 2.6)
Pauli exclusion principle (Sec. 2.6)
Periodic table (Sec. 2.5)
Physical change (Sec. 2.2)
Physical property (Sec. 2.2)
Principal energy level (Sec. 2.6)
Proton (Sec. 2.4)
Quantized (Sec. 2.6)
Representative element (Sec. 2.5)
Shell (Sec. 2.6)
Subshell (Sec. 2.6)
Transition element (Secs. 2.5, 2.7)

## PROBLEMS

**2.3** Why are the theories of Democritus and Zeno considered nonscientific?

### Classifications of Matter

**2.4** Try to classify each of the following as an element, a compound, or a mixture:
- (a) silver
- (b) wine
- (c) milk
- (d) table salt
- (e) oxygen
- (f) diamond
- (g) a pebble
- (h) sea water
- (i) carbon dioxide
- (j) air
- (k) gasoline

**2.5** Classify each as a chemical or physical change:
- (a) burning gasoline
- (b) making ice cubes
- (c) boiling oil
- (d) melting lead
- (e) rusting iron
- (f) making ammonia from nitrogen and hydrogen
- (g) digesting food

**2.6** Name these elements (try not to look at a table):
- (a) C
- (b) Cu
- (c) Ag
- (d) O
- (e) Na
- (f) Fe
- (g) K
- (h) Sn
- (i) P

**2.7** Explain how a mixture of two liquids can be separated by distillation.

### Dalton's Atomic Theory

**2.8** How does Dalton's atomic theory explain (a) the law of conservation of mass (b) the law of constant composition?

**2.9** When 2.16 g of mercury oxide is decomposed to yield 2.00 g of mercury and 0.16 g of oxygen, which law is supported by the experiment?

**2.10** Which, if any, of the four statements of Dalton's atomic theory are still regarded as completely valid today?

### Inside the Atom

**2.11** What is the charge of (a) a proton (b) an electron (c) a neutron?

**2.12** What is the mass in amu of (a) a proton (b) an electron (c) a neutron?

**2.13** Where in the atom are these particles located:
- (a) protons
- (b) electrons
- (c) neutrons

**2.14** What is the mass number of an atom with
- (a) 40 protons, 40 electrons, 51 neutrons
- (b) 12 protons, 10 electrons, 12 neutrons
- (c) 35 protons, 36 electrons, 44 neutrons
- (d) 86 protons, 86 electrons, 136 neutrons
- (e) 92 protons, 90 electrons, 146 neutrons

**2.15** What is the name of each element in Problem 2.14?

**2.16** Given the mass number and the number of neutrons, what is the name of each of the following elements?
(a) mass number 107; 60 neutrons
(b) mass number 36; 18 neutrons
(c) mass number 48; 26 neutrons
(d) mass number 45; 24 neutrons
(e) mass number 246; 156 neutrons

**2.17** If each of the atoms in Problem 2.16 acquired two more neutrons, what would their names be?

**2.18** How many neutrons are there in
(a) a boron atom of mass number 11
(b) an arsenic atom of mass number 73
(c) a mercury atom of mass number 199
(d) a molybdenum atom of mass number 97

**2.19** How many protons and how many neutrons does each of these isotopes of radon contain: Rn-210, Rn-218, and RN-222?

**2.20** How many nucleons are there in
(a) $^{35}Cl$      (d) lithium-7
(b) $^{22}Ne$      (e) tellurium-128
(c) $^{104}Pd$      (f) uranium-238

**2.21** State the number of protons and neutrons in each isotope in Problem 2.20.

**2.22** Define nucleon.

**2.23** Barium-134 is one of the isotopes of barium. Name the other isotopes of barium that contain two, four, and five more neutrons than barium-134.

**2.24** Define: (a) ion (b) isotope.

**2.25** Write the symbol for each atom or ion, given the number of protons (p) and electrons (e) [example: 3p, 2e = $Li^+$]:
(a) 11p, 10e      (e) 30p, 30e
(b) 36p, 36e      (f) 26p, 23e
(c) 9p, 10e      (g) 20p, 18e
(d) 52p, 54e      (h) 1p, 0e

**2.26** For each atom or ion, tell how many electrons, protons, and neutrons are present:
(a) Ca with a mass number of 40
(b) $Br^-$ with a mass number of 79
(c) $Cr^{3+}$ with a mass number of 52
(d) Bi with a mass number of 210
(e) $Rh^{2+}$ with a mass number of 102

**2.27** If copper occurred in only two isotopic forms, $^{63}Cu$ and $^{65}Cu$, and if the naturally occurring mixture contained 69.00% $^{63}Cu$ and 31.00% $^{65}Cu$, what would be the atomic weight of copper?

**2.28** If the three isotopes of hydrogen occurred in nature as follows: 98.00% $^1H$, 1.800% $^2H$, and 0.2000% $^3H$, what would be the atomic weight of hydrogen?

**2.29** If there are only two naturally occurring isotopes of silver, $^{107}Ag$ and $^{109}Ag$, and the atomic weight of silver is 107.87, what are the abundances of these two isotopes?

**The Periodic Table**

**2.30** Which of the following elements would you expect to have fairly similar properties (look at the periodic table): As, I, Ne, F, Mg, K, Ca, Ba, Li, He, N, P.

**2.31** Which are transition elements?
(a) Pd      (d) Ce
(b) K      (e) Br
(c) Co      (f) Cr

**2.32** Which element in each pair is more metallic?
(a) arsenic or germanium
(b) selenium or sulfur
(c) silicon or phosphorus
(d) aluminum or boron

**2.33** Classify as metal, nonmetal, or metalloid:
(a) potassium      (g) silicon
(b) nitrogen      (h) boron
(c) iodine      (i) sulfur
(d) argon      (j) lead
(e) vanadium      (k) antimony
(f) arsenic

**Electronic Structure of Atoms**

**2.34** What is the shape of (a) an $s$ orbital (b) a $p$ orbital?

**2.35** What is the maximum number of electrons that can go into (a) an orbital (b) an outer shell (c) the 3 shell?

**2.36** State (a) the Pauli exclusion principle (b) Hund's rule.

**2.37** Write ground-state electronic configurations for
(a) Li      (e) P
(b) He      (f) Ne
(c) Be      (g) Mn
(d) C      (h) Cl

**2.38** What is common and what is different in the electronic configuration of: (a) Rb and Li (b) O and Se (c) Ge and Si?

**2.39** Positive ions are obtained from neutral atoms by removing the number of electrons indicated by the charge on the ion. Give the electronic configuration of (a) Mg and $Mg^{2+}$  (b) Na and $Na^+$ (c) Al and $Al^{3+}$.

**2.40** Negative ions are obtained from neutral atoms by adding the number of electrons indicated by the charge on the ion. Give the electronic configuration of (a) F and $F^-$  (b) S and $S^{2-}$  (c) P and $P^{3-}$.

**2.41** The electronic configurations for the elements with atomic numbers higher than 36 follow the same rules as given in the text for the first 36. Write the ground-state electronic configuration for (a) Sr  (b) Cs  (c) I.

**\*2.42** The element radon (atomic number 86) has the electronic configuration $[Xe]6s^2 5d^{10} 4f^{14} 6p^6$. The element francium (atomic number 87) has the same configuration plus one more electron. Which orbital does this extra electron occupy?

**2.43** What is the total number of electrons that can fit into the $6s$, $6p$, $6d$, and $6f$ orbitals?

## Electronic Configuration and the Periodic Table

**2.44** Why does the periodic table work?

**2.45** Why do the elements in column IA of the periodic table (the alkali metals) have similar but not identical properties?

**2.46** List in order of increasing ionization energy: Al, Li, N, Ar, B, O, K, Cl, C.

## Boxes

**2.47** (Box 2A) Which four elements are the most important for human life?

**2.48** (Box 2A) Why does the body need sulfur, calcium, and iron?

**2.49** (Box 2B) Why is strontium-90 more dangerous to human beings than most other radioactive isotopes that are present in fallout?

**2.50** (Box 2C) Historically the Copper Age preceded the Iron Age. Why was that so?

**2.51** (Box 2C) Bronze is an alloy of which two metallic elements?

**2.52** (Box 2D) Which are the two most abundant elements, by weight, in (a) the earth's crust  (b) the human body?

**2.53** (Box 2D) Which elements constitute more than 5 percent by weight in the human body but less than that in the earth's crust?

## Additional Problems

**2.54** Define the excited state of an atom. Can an atom be in more than one unique excited state?

**\*2.55** After the $s$, $p$, $d$, and $f$ orbitals, the next higher series is the $g$ series. (a) How many electrons does each $g$ subshell hold?  (b) What is the lowest principal energy level in which $g$ orbitals can exist?

**2.56** Define molecule.

**2.57** How many protons, electrons, and neutrons are present in
(a) $^{35}Cl$ (d) $^{127}I^-$
(b) $^{98}Mo$ (e) $^{158}Gd$
(c) $^{44}Ca^{2+}$ (f) $^{212}Bi^{3+}$

**\*2.58** Write the electronic structure of an excited state of the Be atom.

**2.59** Write the electronic configurations of the ground states of (a) S  (b) $S^{2-}$.

**\*2.60** The mass of a proton is $1.67 \times 10^{-24}$ g. The mass of a grain of salt is $1.0 \times 10^{-2}$ g. How many protons would it take to have the same mass as a grain of salt?

**2.61** How many neutrons are there in
(a) radium-223 (c) plutonium-242
(b) thorium-231 (d) berkelium-247

**2.62** What is the symbol for (try not to look at a table)
(a) nitrogen (f) zinc
(b) sodium (g) silver
(c) phosphorus (h) bromine
(d) potassium (i) carbon
(e) tin (j) calcium

**\*2.63** The natural abundance of titanium isotopes is as follows: titanium-46 = 7.95%, titanium-47 = 7.75%, titanium-48 = 73.45%, titanium-49 = 5.51%, and titanium-50 = 5.34%. Calculate the atomic weight of titanium.

**2.64** Give the designation of all subshells in
(a) the 1 shell (c) the 3 shell
(b) the 2 shell (d) the 4 shell

**\*2.65** The mass of a single atom of vanadium-51 is $8.5 \times 10^{-23}$ g. How many vanadium-51 atoms make up 1.0 lb of vanadium-51?

**2.66** Write the symbol for each atom or ion, given the number of protons (p) and electrons (e) [example: 3p, 2e = Li$^+$]:

(a)  10p, 10e          (d)  56p, 54e
(b)  19p, 18e          (e)  1p, 2e
(c)  35p, 36e          (f)  21p, 18e

**2.67** How many electrons are there in the outer shell of

(a)  Ca          (f)  Se
(b)  O           (g)  Pb
(c)  Si          (h)  He
(d)  Br          (i)  P
(e)  K           (j)  Kr

**2.68** What is the outer-shell electronic configuration of the elements in (a) group IIIA  (b) group VIIA (c) group VA?

**\*2.69** Rubidium has two natural isotopes: rubidium-85 and rubidium-87. What is the natural abundance of each of these isotopes if the atomic weight of rubidium is 85.47?

**2.70** Which element of group IVA gives up an electron least easily?

**2.71** This state of C

$$1s^2 2s^2 2p^1 3s^1$$

represents  (a) a  C$^+$ ion   (b) a  C$^-$ ion   (c) the ground state of a C atom   (d) an excited state of a C atom.

# CHAPTER 3

*The Space Shuttle* Discovery *fires thrusters. Two important rocket fuels are hydrogen, $H_2$, and hydrazine, $N_2H_4$, both small covalent molecules. (Courtesy of NASA.)*

# Chemical Bonds

## 3.1  INTRODUCTION

In Chapter 2 we mentioned that molecules are tightly bound clusters of atoms. In this chapter we shall see what it is that holds the atoms together. Atoms are held together by powerful attractions called **chemical bonds.** There are two main types: *ionic bonds* and *covalent bonds.* In order to talk about ionic bonds, we must first discuss ions.

## 3.2  IONS

In Section 2.4 we defined *ions* and mentioned that there are two kinds.

An ion is a particle with an unequal number of protons and electrons.

65

### Cations

Many atoms have a tendency to lose one or more electrons from the electron cloud that surrounds the nucleus. An atom that loses one or more electrons becomes an ion. It now has a positive charge. **Positive ions** are called **cations.** Some examples are

The smallest cation is the hydrogen ion, $H^+$, which is a bare proton.

$$Li \longrightarrow e^- + Li^+$$ (the lithium ion has three protons and two electrons; the charge is $+1$)

$$Mg \longrightarrow 2e^- + Mg^{2+}$$ (the magnesium ion has 12 protons and 10 electrons; the charge is $+2$)

$$Al \longrightarrow 3e^- + Al^{3+}$$ (the aluminum ion has 13 protons and 10 electrons; the charge is $+3$)

### Anions

If an atom gains one or more electrons, it becomes an ion that contains more electrons than protons. It now has a negative charge because it has extra electrons. **Negative ions** are called **anions.** Some examples are

$$Br + e^- \longrightarrow Br^-$$ (the bromide ion has 35 protons and 36 electrons; the charge is $-1$)

$$S + 2e^- \longrightarrow S^{2-}$$ (the sulfide ion has 16 protons and 18 electrons; the charge is $-2$)

$$N + 3e^- \longrightarrow N^{3-}$$ (the nitride ion has seven protons and ten electrons; the charge is $-3$)

In Section 3.10 we will meet some polyatomic ions.

Table 3.1 lists the names of some of the more important ions that are formed from a single atom.

### Atoms and Their Ions

As you might have guessed from our discussion of metals and nonmetals in Section 2.5, metals generally form cations, and nonmetals, if they form ions at all, form anions. Two important elements that do not form ions are carbon and boron. Also, nitrogen is reluctant to form $N^{3-}$ ions, and those ions are not very stable.

The element chlorine exists not as free chlorine atoms but as chlorine molecules, $Cl_2$.

A **chemical species** is any type of particle: atom, ion, or molecule.

**It is important to understand that there is an enormous difference between the properties of atoms and the properties of their corresponding ions.** For a common example, we can look at the elements sodium and chlorine. Sodium is a soft metal made of sodium atoms and reacts violently with water. Chlorine atoms are very unstable and even more reactive than sodium atoms. Both sodium and chlorine are poisonous. Yet common table salt, NaCl, is made up entirely of sodium *ions* and chloride *ions*. These two ions are quite stable and unreactive. Sodium ions do not react violently with water; in fact, they do not react with water at all. Neither do chloride ions. In all cases ions and atoms are completely different chemical species with completely different chemical and physical properties.

Sometimes people are not careful to distinguish between atoms and ions. One example is the drug many people call "lithium," which is used to treat

**TABLE 3.1   Some Common Cations and Anions[a]**

| Group | Ion Name | Ion Symbol |
|---|---|---|
| | Hydrogen | $H^+$ |
| | Hydride | $H^-$ |
| IA | Lithium | $Li^+$ |
| | Sodium | $Na^+$ |
| | Potassium | $K^+$ |
| IIA | Magnesium | $Mg^{2+}$ |
| | Calcium | $Ca^{2+}$ |
| | Strontium | $Sr^{2+}$ |
| | Barium | $Ba^{2+}$ |
| IIIA | Aluminum | $Al^{3+}$ |
| IVA | Lead | $Pb^{2+}$ |
| VIA | Oxide | $O^{2-}$ |
| | Sulfide | $S^{2-}$ |
| VIIA | Fluoride | $F^-$ |
| | Chloride | $Cl^-$ |
| | Bromide | $Br^-$ |
| | Iodide | $I^-$ |
| Transition elements | Copper(I) | $Cu^+$ |
| | Copper(II) | $Cu^{2+}$ |
| | Iron(II) | $Fe^{2+}$ |
| | Iron(III) | $Fe^{3+}$ |
| | Mercury(II) | $Hg^{2+}$ |
| | Silver(I) | $Ag^+$ |

[a] See also Table 3.3.

**Box 3A**

**Biologically Important Ions**

The human body uses several ions to fulfill important functions. For example, electrical neutrality (Sec. 3.3) must be maintained both inside and outside body cells. Inside the cells the main positive ion used for this purpose is potassium ($K^+$) and the main negative ion is hydrogen phosphate ($HPO_4^{2-}$). In body fluids outside the cells, different ions are used to accomplish this task, chiefly sodium ($Na^+$) and chloride ($Cl^-$).

The calcium ion ($Ca^{2+}$) is a major component of bones and teeth. It also has other functions, including assistance in blood clotting, muscle action, and heartbeat control. About 90 percent of the calcium in the body is present in bones as calcium phosphates and carbonates. The iron(II) ion ($Fe^{2+}$) is a part of hemoglobin, which transports oxygen from the lungs to the cells (see Box 3C).

Another important ion is magnesium ($Mg^{2+}$), which is involved in the action of nerves and muscles and is necessary for the activity of certain enzymes. It is also present in bones and teeth, along with $Ca^{2+}$.

All of these ions, as well as certain others present in lesser amounts, must be obtained from the diet. Lack of some of them causes diseases; for example, a deficiency of $Fe^{2+}$ results in anemia. However, too much of them can also be a problem. Too much table salt, NaCl, for example, can cause water retention (edema) and high blood pressure in some people. See also Table 26.2.

manic-depressive symptoms (a form of mental illness). The element lithium, like sodium, is a soft metal that reacts with water. It would certainly be poisonous, probably fatal, if ingested. The drug is not lithium (Li) but lithium *ions* ($Li^+$), usually given in the form of lithium carbonate ($Li_2CO_3$) (Fig. 3.1). Another example comes from the fluoridation of water. Fluorine ($F_2$), an extremely reactive element and a deadly poison, is *not* used for this. Water is fluoridated with fluoride ions ($F^-$), which are unreactive and not poisonous in the low concentrations used.

Small amounts of tin(II) fluoride are added to water to prevent tooth decay.

### The Octet Rule

In Table 3.1 we see that $K^+$ is listed, but not $K^{2+}$ or $K^{3+}$ or $K^-$. In fact, $K^{2+}$, $K^{3+}$, and $K^-$ are extremely unstable and do not normally exist. Is there any way to predict which ions are stable? Yes, there is, although our predictions will not be perfect. We use the principle that **atoms and ions are most stable when they have a complete outer shell of electrons.** Since a complete outer shell of an atom or ion has eight electrons (Sec. 2.6)—except for the very first shell, which has only two—we call this the **octet rule.**

The outer shell of an atom is called the **valence shell.**

The octet rule readily explains why $K^+$ is stable but the other potassium ions are not. Potassium is in column IA of the periodic table, so a potassium atom has one electron in its outer or valence shell ($4s^1$). If it loses this electron

$$K \longrightarrow e^- + K^+$$

then there are no electrons left in the 4 shell, and the 3 shell becomes the outer shell in the $K^+$ ion. If the $K^+$ were to lose another electron (to form $K^{2+}$), the electron would have to come from the 3 shell, leaving only seven electrons in it. Because $K^+$ has a complete octet but $K^{2+}$ does not, $K^{2+}$ is much less stable than $K^+$ and is not found in normal matter. Once an atom or ion has formed a complete octet, it is usually difficult to get it to lose or gain any electrons.

That explains why $K^{2+}$ and $K^{3+}$ are not stable ions, but what about $K^-$? This ion does not exist because $K^-$ has only two electrons in its outer shell, and this is a long way from an octet.

**Figure 3.1**
Lithium (a) and lithium carbonate (b) are entirely different substances. (Photograph by Beverly March.)

(a)                                    (b)

We can use the octet rule with a reasonable degree of success to predict the stable ions of the representative elements. All the elements in column IA, like potassium, have one electron in the valence shell, and so all form only $E^+$ ions. The elements in column IIA have two outer-shell electrons and lose these to form $E^{2+}$ ions. For example,

As in Section 2.5, "E" is a general symbol for any element.

$$Ba \longrightarrow 2e^- + Ba^{2+}$$

The elements in column VIIA have seven electrons in the outer shell. They can achieve a complete octet by gaining an electron, but only one, so they form $E^-$ ions. For example,

$$Cl + e^- \longrightarrow Cl^-$$

In column VIA each element has six electrons in the outer shell and so gains two electrons to form an $E^{2-}$ ion. For example,

$$Se + 2e^- \longrightarrow Se^{2-}$$

One test of the octet rule is column 0. The elements in this column (the noble gases) already have a complete outer shell, so the octet rule predicts they will form no ions at all. This prediction is amply borne out: No stable ions of any group 0 elements have ever been found.

The octet rule can help us to predict the charges of many ions. However, it is not perfect, for two reasons:

**1.** Concentrated charges are unstable. Boron has three valence-shell electrons. If it lost these three, it would, as $B^{3+}$, have a complete outer shell. It seems, however, that this is too concentrated a charge for such a tiny ion, and this ion is not found in normal matter. This also explains why the upper group IVA elements (C, Si) do not form any $E^{4+}$ or $E^{4-}$ ions and why an atom like chlorine, which forms $Cl^-$ ions by gaining one electron, does not form $Cl^{7+}$ ions by losing seven electrons.
**2.** The octet rule cannot be applied to transition elements because they are too far removed from the noble-gas structure. Table 3.1 shows that iron forms $Fe^{2+}$ and $Fe^{3+}$ ions. This behavior is typical of transition elements, which frequently form more than one type of ion.

In spite of these weaknesses, the octet rule is useful in those cases where it does apply.

## 3.3  IONIC BONDS

Since sodium chloride (NaCl), also known as table salt, is the most common ionic compound, let us look at its structure rather closely. It is made up of sodium ions, $Na^+$, and chloride ions, $Cl^-$. An **ionic bond** is the **attraction between positive and negative ions,** but this attraction is not on a one-to-one basis. In its solid (crystalline) form, sodium chloride consists of a three-dimensional array of $Na^+$ and $Cl^-$ ions arranged as shown in Figure 3.2. Note that, for any particular $Na^+$ ion, there is no individual $Cl^-$ ion that is closer than any other. Rather, each $Na^+$ has six $Cl^-$ ions as "nearest neighbors," as shown in Figure 3.3(a). Since the six $Cl^-$ ions are equidistant from the $Na^+$, each is

**Figure 3.2**
The structure of the sodium chloride crystal.

$Cl^-$          $Na^+$

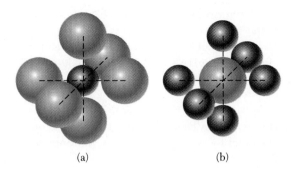

(a)                    (b)

**Figure 3.3**
(a) The environment around one Na$^+$ ion. (b) The environment around one Cl$^-$ ion.

Ionic compounds are compounds that contain ionic bonds.

attracted to the same degree. Likewise, each Cl$^-$ ion is surrounded by six Na$^+$ ions, each an equal distance away, as shown in Figure 3.3(b). This situation is typical of ionic compounds in the solid state. There are no discrete molecules in these compounds. The entire crystal consists of ions arranged as shown in Figure 3.2.

There are many other ionic compounds, some of which are shown in Box 3B. All are solids at room temperature, and all consist of ions arranged in some kind of regular three-dimensional array. The arrays are different in each case since they depend on the size of the ions and on the relative number of negative

---

**B o x  3B**

## Ionic Compounds in Medicine

**Figure 3B**
A plaster cast being removed from a patient. (FourByFive/John Wang.)

There are many ionic compounds that have medical uses. Some of them are shown here.

| Formula | Name | Medical Use |
|---------|------|-------------|
| $AgNO_3$ | Silver nitrate | Astringent, styptic in veterinary medicine |
| $BaSO_4$ | Barium sulfate | Radiopaque medium in x-ray work |
| $CaSO_4$ | Calcium sulfate | Plaster casts |
| $FeSO_4$ | Iron(II) sulfate | For iron deficiency |
| $KMnO_4$ | Potassium permanganate | Anti-infective (external) |
| $KNO_3$ | Potassium nitrate (saltpeter) | Diuretic |
| $Li_2CO_3$ | Lithium carbonate | Antidepressive |
| $MgSO_4$ | Magnesium sulfate (epsom salts) | Cathartic |
| $NaHCO_3$ | Sodium bicarbonate (baking soda) | Antacid |
| $NaI$ | Sodium iodide | Source of iodide for thyroid |
| $NH_4Cl$ | Ammonium chloride | To acidify the digestive system |
| $(NH_4)_2CO_3$ | Ammonium carbonate | Expectorant |
| $SnF_2$ | Tin(II) fluoride | To strengthen teeth (external) |
| $ZnO$ | Zinc oxide | Astringent (external) |

and positive ions, but in every case there is some regular array. No ionic compounds contain discrete molecules.

Ions are charged particles, but the matter we see all around us and deal with every day is electrically neutral (uncharged). Nature does not allow a large concentration of charge, either positive or negative, to build up in any one place. We can assume that all the substances we deal with are uncharged. If there are ions present in any sample of matter, the total number of positive charges must equal the total number of negative charges. Therefore, we cannot have a sample containing only $Na^+$ ions. Any sample that contains $Na^+$ ions must also contain negative ions, which may be $Cl^-$, $Br^-$, $S^{2-}$, or other anions, and the sum of the positive charges must equal the sum of the negative charges. This is called electrical neutrality.

## Predicting Formulas of Ionic Compounds

The principle of electrical neutrality allows us to predict the formulas of ionic compounds.

The naming of ionic compounds is discussed in Section 3.11.

### EXAMPLE

What is the formula for the compound formed from lithium ion and bromide ion?

### Answer

The charge on $Li^+$ is +1; the charge on $Br^-$ is −1. The formula for lithium bromide is LiBr (one to one).

Ionic charges are given in Tables 3.1 and 3.3.

### EXAMPLE

What is the formula for the compound formed from barium ion and iodide ion?

### Answer

The charge on $Ba^{2+}$ is +2, and $I^-$ has a charge of −1. Two $I^-$ ions are required to neutralize one $Ba^{2+}$ ion. Therefore, the formula for barium iodide is $BaI_2$.

### EXAMPLE

What is the formula for the compound formed from aluminum ion and sulfide ion?

### Answer

The charge on $Al^{3+}$ is +3, and $S^{2-}$ has a charge of −2. It takes three $S^{2-}$ ions to neutralize the positive charge of two $Al^{3+}$ ions, so the formula is $Al_2S_3$.

$Al_2S_3$, like all other ionic compounds, obeys the principle of electrical neutrality:

$$2Al^{3+} = +6$$
$$\underline{3S^{2-} = -6}$$
$$\text{Net charge} = \quad 0$$

It was Gilbert N. Lewis (1875–1946; Fig. 3.4) who first proposed that two atoms can form a bond by sharing a pair of electrons.

**Figure 3.4**
G. N. Lewis. (AIP Niels Bohr Library; photograph by Francis Simon.)

---

**PROBLEM 3.1**   Write formulas for compounds formed from (a) potassium ion and chloride ion   (b) calcium ion and fluoride ion   (c) iron(III) ion and oxide ion.

---

Note that we can generally arrive at the formula for ionic compounds by "crossing" the charges; for example, for aluminum sulfide,

$$Al^{3+} \qquad S^{2-}$$
$$Al_2S_3$$

Remember that the subscripts in the formulas for ionic compounds represent the *ratio* of the ions. Thus, a crystal of $BaI_2$ has twice as many $I^-$ as $Ba^{2+}$ ions. (This ratio is, of course, necessary to keep the substance neutral.)

## 3.4   COVALENT BONDS

The second type of chemical bond—the covalent bond—is very different from the ionic bond. Because of this the properties of compounds containing covalent bonds are very different from those of ionic compounds.

**A covalent bond is one in which two atoms share a pair of electrons.** Let us examine a typical case, that of the element fluorine. Each fluorine atom has seven outer-shell electrons and needs one more to complete the octet. Of course, the fluorine atom could gain an electron and thereby become an $F^-$ ion, but it can do this only if there is a nearby atom (or molecule) from which to take the electron. It cannot get an electron from nowhere. If we have a sample that contains only fluorine atoms, then the only nearby atom would be another fluorine atom. Would a fluorine atom take an electron from another fluorine atom? A little thought shows that the answer is no. The first atom would then have a complete octet and be a stable $F^-$ ion, but the second atom would have only six outer-shell electrons and be an extremely unstable $F^+$ ion. It turns out that fluorine atoms do not have to take electrons from each other. They can (and do) **share** electrons to form a covalent bond. When they do this, *both* atoms have a complete octet. We can write it this way (the dots stand for outer-shell electrons):

$$:\!\ddot{F}\!\cdot \quad + \quad \cdot\!\ddot{F}\!: \quad \longrightarrow \quad :\!\ddot{F}\!\cdot\!\cdot\!\ddot{F}\!:$$
$$\text{F atom} \qquad \text{F atom} \qquad F_2 \text{ molecule}$$

As we can see, each fluorine atom still has possession of six of its original seven electrons. The other two (one contributed by each atom) are now *shared* between them. Each fluorine atom acts as if those two electrons are part of its outer shell, that is, as if each had a complete octet.

Another way to look at it is to remember that the electrons of fluorine (and all other) atoms are in orbitals, which are clouds of electrons (Sec. 2.6). A fluorine atom has the electron configuration $1s^2 2s^2 2p_x^2 2p_y^2 2p_z^1$, which means that its seventh outer-shell electron is in a $2p$ orbital, which is shaped like a dumbbell. When two fluorine atoms come together to form a covalent bond,

2$p_z$ orbital
of F atom

2$p_z$ orbital
of F atom

Molecular orbital

**Figure 3.5**
When two F atoms form a molecule of $F_2$, their two 2$p_z$ orbitals overlap to form a molecular orbital (called a sigma orbital). The other orbitals are not affected. (The 1$s$ and 2$s$ orbitals are not shown in this figure.)

these orbitals overlap to form a new cloud that surrounds both nuclei (Fig. 3.5). This new cloud is the covalent bond. We call it a **molecular orbital** because it binds two atoms together to form a molecule. The other orbitals (containing the 1$s$, 2$s$, 2$p_x$, and 2$p_y$ electrons) are not much affected by the bonding and remain virtually unchanged. Molecular orbitals are given Greek letters, and this type is called a **sigma ($\sigma$) orbital.** In Section 11.3 we will meet another type of molecular orbital, called pi ($\pi$).

Clouds in the sky can overlap, and so can electron clouds.

A fluorine atom can achieve an outer-shell octet by sharing *one* of its electrons. The same is true for the other elements in group VIIA. Because of this, each of these atoms forms only *one* covalent bond. The same principle holds for the other nonmetallic elements. **Each nonmetallic atom has a strong tendency to form a particular number of covalent bonds.** This number is equal to the number of electrons that must be shared to achieve a noble-gas electronic configuration. For example, carbon, which has four electrons in its outer shell, forms four covalent bonds; oxygen, with six electrons, forms two covalent bonds; and nitrogen, with five electrons, forms three covalent bonds.

Any orbital, atomic or molecular, can hold a maximum of two electrons.

When an atom forms more than one bond, do all the bonds have to be to the same atom? The answer is no. The bonds can be formed in many different ways. As an example, let us look at carbon, which forms four bonds. Here are some examples of stable, well-known compounds that contain carbon:

Elements in groups VA, VIA, and VIIA (except O, N, and F) may form different numbers of bonds in different compounds. For example, phosphorus forms three bonds in some compounds ($PCl_3$) and five in others ($PCl_5$).

Carbon
tetrachloride

Methanol

Vinyl chloride

Acetylene

Formaldehyde

Ethane

Bromochloroiodomethane

Carbon dioxide

In these formulas, a covalent bond (a pair of electrons in a molecular orbital) is shown as a straight line connecting two atoms. We understand that, when we draw them this way, we are showing that the two atoms involved are sharing the electrons.

## Double and Triple Bonds

An inspection of these formulas shows that a carbon atom can use its four bonds to connect with four atoms of the same element or with four different atoms, and can also form *double* and *triple* bonds. For example, in vinyl chloride the carbon on the left is bonded to two H atoms and, by a double bond, to another C atom. In the carbon-carbon double bond, *two* pairs of electrons are shared. The total number of bonds for each C atom must be four, but they can be made with any combination of single, double, or triple bonds adding up to four. In virtually all carbon compounds, every carbon atom has four bonds, and stable compounds containing carbon atoms with three or five bonds, or any other number, are practically unknown (though a few curiosities do exist).

The formulas given above are called structural formulas. A **structural formula** shows all the atoms in a molecule and all the bonds connecting them. A **molecular formula** gives less information, since it shows only the number of atoms and not the bonds. For example, the molecular formulas for methanol and vinyl chloride are $CH_4O$ and $C_2H_3Cl$, respectively.

*The structural formulas for these compounds are shown above.*

## Lewis Structures

We have now seen two kinds of formulas for molecules, structural and molecular. There is still another kind, one that gives even more information than a structural formula. This kind, called a **Lewis structure,** shows not only all atoms and covalent bonds but also all other outer-shell electrons, including those that are unshared.

Lewis structures can be drawn for atoms as well as molecules. They show all outer-shell electrons as dots. We have already shown the fluorine atom in this way on page 72. Some other examples are

$$Li\cdot \quad \cdot \dot{B}\cdot \quad \cdot \dot{C}\cdot \quad \cdot \ddot{N}\cdot \quad \cdot \ddot{O}\cdot \quad :\ddot{Ne}:$$

As an example of a Lewis structure of a molecule, consider ammonia, whose molecular formula is $NH_3$. Because we know that a hydrogen atom forms only one covalent bond, it is clearly not possible for a hydrogen atom *in this molecule* to be connected to another hydrogen atom. Each hydrogen atom must be bonded to the nitrogen. Since nitrogen forms three bonds and since there are three hydrogen atoms, this works out all right.

$$\underset{\displaystyle |}{H-N-H} \qquad \text{Structural formula for ammonia}$$
$$H$$

But a nitrogen atom has five outer-shell electrons (Sec. 2.6). It must use three of them to bond to the hydrogens (one each), and so two remain. We show the two unshared electrons as a pair of dots:

*Unshared pairs are also called **nonbonding pairs** or **lone pairs.***

$$H-\ddot{N}-H \qquad \text{Lewis structure for ammonia}$$
$$H$$

Similarly, the formula we gave for $F_2$ is also a Lewis structure:

$$:\ddot{F}-\ddot{F}:$$

Lewis structures are usually easy to draw if we follow a few steps:

**1.** Find out how many atoms of each kind are in the molecule and which are connected to which. Sometimes we can figure out the connections (as we did for $NH_3$), but this is not always possible.

**2.** Calculate the total number of outer-shell electrons brought into the molecule by all the atoms in it. For any given atom, this number is always the same—it is the number of outer-shell electrons in the free atom. For example, every time we write a Lewis structure for an oxygen-containing compound, we can be certain that each oxygen atom brings six outer-shell electrons.

**3.** Insert these electrons in pairs (either bonds or unshared pairs) so that every atom has an octet if possible (two for hydrogen). Where necessary, use double or triple bonds.

The procedure is best illustrated by examples.

---

**EXAMPLE**

Draw the Lewis structure for hydrogen peroxide, $H_2O_2$.

**Answer**
(Step 1) From the fact that a hydrogen atom forms only one covalent bond, we know the connections must be H—O—O—H. (Step 2) An oxygen atom has six outer-shell electrons, so the two O atoms bring in a total of 12 electrons. Each H atom brings in 1 electron, so the total is 14 outer-shell electrons, or 7 pairs. (Step 3) We can begin by putting in the three bonds we know are there:

$$\text{H—O—O—H}$$

This leaves four electron pairs to be accounted for. If we want to give each O atom an octet, we need to put the four pairs in as follows:

$$\text{H—}\overset{\cdot\cdot}{\underset{\cdot\cdot}{\text{O}}}\text{—}\overset{\cdot\cdot}{\underset{\cdot\cdot}{\text{O}}}\text{—H} \qquad \textit{Lewis structure for hydrogen peroxide}$$

$$
\begin{array}{ll}
\text{O} & 2 \times 6 = 12 \\
\text{H} & 2 \times 1 = \underline{\phantom{0}2} \\
\text{Total} & \phantom{2 \times 1 =}14
\end{array}
$$

---

**PROBLEM 3.2** Draw the Lewis structure for phosphorus tribromide, $PBr_3$.

---

**EXAMPLE**

Draw the Lewis structure for formaldehyde, $H_2CO$. The atoms are connected like this:

$$\text{H} \quad \text{C} \quad \text{H}$$
$$\text{O}$$

**Answer**
(Step 1) Given. (Step 2) The C brings in four electrons, the O brings in six, and each H brings in one. The total is thus 12 electrons, or 6 pairs. (Step 3) As usual, we begin by putting in pairs to bond all the atoms:

$$
\begin{array}{c}
\text{H—C—H} \\
| \\
\text{O}
\end{array}
$$

$$
\begin{array}{ll}
\text{C} & \phantom{2 \times 1 =}4 \\
\text{O} & \phantom{2 \times 1 =}6 \\
\text{H} & 2 \times 1 = \underline{\phantom{0}2} \\
\text{Total} & \phantom{2 \times 1 =}12
\end{array}
$$

Three pairs remain. If we put one of them between C and O, making this a double bond, the carbon has a complete octet:

$$H—C—H$$
$$\overset{\parallel}{O}$$

Two pairs of electrons remain, and these go to fill the octet of the O:

$$H—C—H$$ Lewis structure for formaldehyde
$$\overset{\parallel}{\underset{\displaystyle ..}{O}}:$$

**PROBLEM 3.3**  Draw the Lewis structure for acetic acid, $C_2H_4O_2$. The connections are

$$
\begin{array}{ccccc}
& & H & & \\
& & & & O \quad H \\
H & C & C & & \\
& & & & O \\
& & H & &
\end{array}
$$

**EXAMPLE**

Draw the Lewis structure for the sulfate ion, $SO_4^{2-}$. The connections are

$$
\begin{array}{ccc}
& O & \\
O & S & O \\
& O & \\
\end{array}
$$

**Answer**

This time we have an ion rather than a molecule, so our counting procedure has an extra step. A sulfur atom has 6 outer-shell electrons and each oxygen has 6, so the total brought in by these five atoms is 30 electrons. However, the $SO_4^{2-}$ ion has a charge of $-2$, meaning that it must have two more electrons than this, or a total of 32 (16 pairs). Note that if the ion had a positive charge, we would subtract a number equal to the charge. (Step 3) As usual, we begin by putting in four electron pairs to bond all the atoms:

$$
\begin{array}{ccc}
& O & \\
& | & \\
O & —S— & O \\
& | & \\
& O & \\
\end{array}
$$

This leaves 12 pairs. If we put three pairs around each O, we give all atoms an octet and use up all the electrons:

$$
\left[
\begin{array}{c}
:\ddot{O}: \\
| \\
:\ddot{O}—S—\ddot{O}: \\
| \\
:\ddot{O}: \\
\end{array}
\right]^{2-}
$$

The $-2$ charge means that this ion has two more electrons than protons. It is clear that the atoms involved have not brought in any more electrons than protons, since they were neutral. The two extra electrons therefore came from some outside source, not from the atoms that went into it. The extra electrons are there, though, so we must count them.

$$
\begin{array}{ll}
S & 6 \\
O \quad 4 \times 6 = & \underline{24} \\
& 30 \\
-2 \text{ charge} & \underline{2} \\
\text{Total} & 32 \\
\end{array}
$$

When we write the Lewis structure of an ion, it is customary to use brackets and to show the total charge outside the brackets.

**PROBLEM 3.4** Draw the Lewis structure for the sulfite ion, $SO_3{}^{2-}$. The connections are

O   S   O

O

## 3.5 COORDINATE COVALENT BONDS

We have seen that a covalent bond is formed when a single electron from one atom and a single electron from another atom are both shared by the two atoms. There is another way a covalent bond can form, though it is less common. In this case one atom supplies two electrons and the other none. This type of bond is called a **coordinate covalent bond.** As an example, consider the compound ammonia, which has an unshared pair of electrons.

Ammonia    Boron    Adduct
           trifluoride

Note that atoms in coordinate covalent bonds do not form their normal number of bonds. The nitrogen atom in the adduct has four bonds rather than its usual three.

It can share this pair with a boron atom in the compound boron trifluoride to give a compound in which both the boron and the nitrogen have a complete octet. This type of bond can form only between two species that have the right characteristics: One of them must have an unshared pair of electrons, and the other must have space for a pair. In $BF_3$ the boron has only six electrons in its outer shell and thus lacks an octet. Therefore, it makes a fine partner for the ammonia because it needs two more to make eight.

Boron trifluoride is one of the compounds that violate the octet rule.

*A coordinate covalent bond is just like any other covalent bond once it has been formed.* The only difference is that both electrons are supplied by one atom rather than each atom supplying one electron.

Certain metal ions, most notably those of the transition metals, form large numbers of compounds that contain coordinate covalent bonds. Many of these ions can take two, four, or even six pairs of electrons. Such compounds are called **coordination compounds,** and thousands of them are known. Some examples follow (coordinate covalent bonds are shown in color).

The coordination number is the number of bonds connected to a central atom or ion.

$[H_3N—Ag—NH_3]^+$

CO
|
OC—Fe—CO
|      |
OC    CO

Cl      NH$_3$
  \    /
    Pt
  /    \
NH$_3$    Cl

## 3.6 THE SHAPES OF MOLECULES

The atoms in a molecule are not arranged randomly. **Molecules have definite three-dimensional shapes.** These shapes are important because they help to determine the properties of the molecules. For molecules with only two atoms, only one shape is possible: The two atoms must be in a straight line. Thus

The function of the red cells of the blood is to carry oxygen, $O_2$, to the cells of the body. The actual carrier of the $O_2$ is a protein called **hemoglobin.** The $O_2$ attaches itself to the hemoglobin by a coordinate covalent bond. Hemoglobin is a very large molecule (molecular weight about 68 000) whose structure is discussed in Section 18.9, but of all the atoms in this molecule, the only ones that concern us here are four iron atoms. Each of these four atoms is imbedded in a portion of the hemoglobin called **heme,** which has the following structural formula:

At this point we need not concern ourselves with the meaning of all these symbols. All that is relevant here is that the iron atom in the center is held in place by covalent bonds to two nitrogen atoms and by coordinate covalent bonds to two other nitrogen atoms (shown in color). Every hemoglobin molecule has four of these heme units, each containing one iron atom. When hemoglobin picks up oxygen in the lungs, each $O_2$ molecule bonds to one of the Fe atoms by a coordinate covalent bond, with both electrons supplied by the oxygen, because the Fe atom, though bonded to four N atoms, still has room for additional electrons.

$$:\ddot{O}\!\!=\!\!\ddot{O}\!-\!Fe$$

The hemoglobin now carries the $O_2$ to the cells, where it is released as needed.

The coordinate covalent bonding ability of the Fe in hemoglobin is not restricted to $O_2$. There are many other species that are also able to use an electron pair to form such a bond with the Fe in hemoglobin. Among these is the poison carbon monoxide, CO.

$$:O\!\!\equiv\!\!C\!-\!Fe$$

Carbon monoxide is poisonous because the bond it forms with the Fe in hemoglobin is stronger than the $O_2$—Fe bond. When a person breathes in CO the hemoglobin combines with this molecule rather than with $O_2$. The cells, deprived of $O_2$, can no longer function, and the person dies.

diatomic molecules, such as $F_2$, HCl, and CO, can only be linear (two points determine a line). With three or more atoms, however, different possibilities arise. Three atoms could be *linear* (all three in a straight line) or *angular* (the three atoms forming an angle). An example of a linear three-atom molecule is carbon dioxide, $CO_2$; an angular one is water, where the H—O—H angle is about 105° (Fig. 3.6). These shapes can be determined in laboratories through the use of instruments called x-ray and electron diffractometers.

Is there any way we can predict that $CO_2$ is linear and $H_2O$ angular? The answer is yes. There is a simple method that allows us to predict accurately the shapes of almost all molecules. The method, called the *valence-shell electron pair repulsion theory* (VSEPR), is based on Lewis structures.

Very simply, **VSEPR** says that **the electron pairs in the outer shell of an atom try to get as far away from each other as possible.** This is quite understandable because they are all negatively charged, and like charges repel. What we do is draw the Lewis structure and look for an atom connected to two or more other atoms (this is the *central atom*). We then simply count the number of electron pairs in the outer shell of the central atom while keeping in mind these two rules:

**1.** It doesn't matter whether an electron pair is unshared or bonded to another atom. It counts just the same.
**2.** For the purposes of VSEPR, a double or triple bond counts as *one* pair because it occupies one region of space.

**The number of pairs tells us the shape.** We will consider cases in which there are two, three, or four pairs. Each of these numbers corresponds to a different shape, and in each case this shape is adopted by the molecule because that is how the electrons can get farthest away from each other (Fig. 3.7). With *two pairs,* the shape is linear:

With *three pairs,* the shape is trigonal (planar triangular):

With *four pairs,* the shape is tetrahedral:

In compounds with more than one central atom, these shapes apply to each central atom. Let us look at some examples.

**Figure 3.6**
Carbon dioxide is a linear molecule; water is an angular one.

These are the simplest cases. Atoms with five or more pairs are not considered in this book.

A regular tetrahedron is a geometric figure made up of four equilateral triangles. The central atom is *inside* the tetrahedron, and the four electron pairs point to the four corners.

(a)

(b)          (c)

**Figure 3.7**
When balloons are tied together, they naturally assume the proper shapes, because that is how they can keep farthest apart. (a) Two balloons are linear. (b) Three balloons are trigonal. (c) Four balloons are tetrahedral. (Photographs by Charles D. Winters.)

### Two Pairs

There are not many examples in this category, because most molecules obey the octet rule, which means that they have four outer-shell pairs. However, remember that we count double or triple bonds as one pair, so that a molecule can obey the octet rule and still have two pairs for the purposes of VSEPR. Two important molecules in this category are carbon dioxide and acetylene, $C_2H_2$, whose Lewis structures are

$$:\ddot{O}{=}C{=}\ddot{O}: \qquad H{-}C{\equiv}C{-}H$$

In $CO_2$ the central atom, carbon, has two VSEPR pairs. In $C_2H_2$ there are two central atoms, each with two VSEPR pairs. The VSEPR theory predicts that both should be linear, and both are.

### Three Pairs

As mentioned earlier, boron trifluoride, $BF_3$, does not obey the octet rule. The Lewis structure shows that there are only six electrons in the outer shell of the boron atom:

$$\underset{\substack{\ddot{}\\ :\ddot{F}:}}{\overset{\substack{:\ddot{F}\qquad\ddot{F}:\\ \diagdown\,\nearrow \\ B\;\;\;\;120°}}{}}$$

Since there are three electron pairs around the B atom, VSEPR says that this molecule should be triangular, with all four atoms in a plane and F—B—F angles of 120°. This is indeed the case. Another example is ethylene, $C_2H_4$, whose Lewis structure is

$$\underset{H\qquad\qquad H}{\overset{H\qquad 120°\;\; H}{C{=}C}}$$

The octet rule is obeyed here, but we count the double bond as one, so each central atom has three pairs, and this compound too is planar, with 120° angles.

## Four Pairs

This is the most important case, since all molecules in which atoms have complete octets and no double or triple bonds are in this category. Some of these atoms have four bonds, some three, and some two. The most common four-bonded atom is carbon, and VSEPR says that **all carbon atoms with four single bonds have tetrahedral shapes.** Some examples are $CH_4$, $CCl_4$, and $CHCl_3$ (Fig. 3.8). The angle between any two groups in this geometry is 109.5°.

Nitrogen, which forms three bonds, also has a complete octet (four pairs), and thus most nitrogen-containing compounds belong in this category. Ammonia (p. 74) is a typical example, as well as the simplest. VSEPR says that the four pairs point to the corners of a tetrahedron, even if one of these pairs is not connected to another atom:

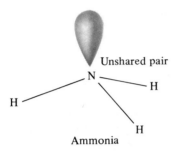

If we look only at the atoms and not at the unshared pair, we see that ammonia (as well as any other nitrogen-containing compound with no double or triple bonds) has a *pyramidal* shape.

$$\begin{array}{c} N \\ H \;|\; H \\ H \end{array}$$

**Figure 3.8**
Some simple carbon compounds with tetrahedral shapes.

109.5°

Methane
$CH_4$

Carbon tetrachloride
$CCl_4$

Chloroform
$CHCl_3$

Finally, in this category, we may look at oxygen. In most of its compounds, oxygen has a complete octet (four pairs) but only two bonds. When these are single bonds, the four pairs again mean a tetrahedral structure. The simplest example is water:

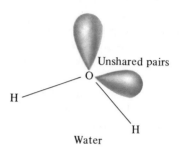

Water

Again, if we look only at the three atoms (and ignore the unshared pairs), water is predicted by VSEPR to have an angle of 109.5° (the tetrahedral angle). The actual angle, 105°, is not very far off. Water is thus an angular molecule.

Though simple, VSEPR theory is very powerful. Without it we would have a hard time explaining what would otherwise be some very puzzling facts. Why should two such similar molecules as HOH and OCO have such different geometries, one angular and the other linear? The same question can be asked about $NH_3$ and $BF_3$. Why should one be pyramidal and the other planar triangular? VSEPR theory gives us simple and satisfying answers.

## 3.7   ELECTRONEGATIVITY AND DIPOLES

In Section 3.4 we discussed the electronic configuration of $F_2$. We saw that the covalent bond is actually a cloud of electrons (containing two electrons), as shown in Figure 3.5. The electrons in this cloud are equally distributed around the two fluorine nuclei because there is no reason for them to prefer one side over the other. But that is not the case when a covalent bond connects two different atoms. For example, the Lewis structure for hydrogen fluoride, HF, is

In this case the two electrons in the bond do not remain equidistant from the two nuclei; they are closer to the F atom than to the H atom. That is, the electron cloud is distorted, as shown in Figure 3.9, because the F atom attracts the electron pair of the bond more than the H atom does.

The name given to this attraction is **electronegativity.** Fluorine has a higher electronegativity than hydrogen and so attracts the electrons of the bond more. Note that we are not talking about attracting anything outside the molecule. Electronegativity refers only to the attraction that an atom has for a pair of electrons *in a covalent bond to itself.*

We now know that fluorine has a higher electronegativity than hydrogen, but how much higher? Linus Pauling (1901–  ) set up a scale in which each element is assigned a number that measures its electronegativity. Part of this scale is shown in Table 3.2. Fluorine has the highest electronegativity, 4.0, and cesium the lowest, 0.7. With this scale we can tell in which direction the electrons of any covalent bond are shifted.

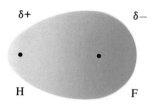

**Figure 3.9**
The covalent bond in HF is distorted toward the fluorine atom.

**TABLE 3.2 Electronegativities of Some Elements**

| | | | | | | |
|---|---|---|---|---|---|---|
| H<br>2.1 | | | | | | |
| Li<br>1.0 | Be<br>1.5 | B<br>2.0 | C<br>2.5 | N<br>3.0 | O<br>3.5 | F<br>4.0 |
| Na<br>0.9 | Mg<br>1.2 | Al<br>1.5 | Si<br>1.8 | P<br>2.1 | S<br>2.5 | Cl<br>3.0 |
| Rb<br>0.8 | | | Ge<br>1.8 | As<br>2.0 | Se<br>2.4 | Br<br>2.8 |
| Cs<br>0.7 | | | | | Te<br>2.1 | I<br>2.5 |

**EXAMPLE**

Toward which atom are the electrons shifted in (a) sulfur dichloride, $SCl_2$ (b) iodine bromide, IBr?

**Answer**

(a) The electronegativity of Cl is 3.0, and that of S is 2.5. The electrons are shifted toward the chlorine atoms.   (b) The electrons are shifted toward the Br, but since the difference is small, the distortion is not very much.

**PROBLEM 3.5**   Toward which atom are the electrons shifted in (a) oxygen difluoride, $OF_2$   (b) phosphorus tribromide, $PBr_3$?

Table 3.2 shows that electronegativity, like ionization energy (Sec. 2.7), changes with position in the periodic table. We can see two trends:

**1.**   Electronegativity increases from left to right in the table (except for the noble gases).
**2.**   Electronegativity decreases down any column.

Thus electronegativity correlates with metallic character: Nonmetals have high electronegativity; metals have low electronegativity.

Let us now return to HF. The electron cloud is distorted so that it covers the fluorine atom more than the hydrogen atom (Fig. 3.9). This means that there is a partial negative charge on the fluorine atom and a partial positive charge on the hydrogen, which we can show as

$$\delta^+ H - F^{\delta-}$$

The Greek letter $\delta$ (delta) means "partial."

There are two things we should remember:

**1.** These are *partial* charges, that is, less than $+1$ or $-1$. If the electron pair were transferred *completely*, the charges would be $+1$ and $-1$. Since the cloud is not completely transferred, the charges are less than that. How much less varies from case to case and depends on the difference between the electronegativities of the two atoms (the greater the difference, the greater the partial charges).

**2.** Whatever the magnitude of $\delta+$, it must be the same as $\delta-$ because the sum of $\delta+$ and $\delta-$ must be zero (the total number of electrons in the whole molecule is equal to the total number of protons, and so the net charge must be zero).

What are the consequences of these partial charges? Because each end of the H—F molecule has a charge, the molecule behaves like a little bar magnet. If you put two bar magnets together, the north poles repel each other and attract the south poles (Fig. 3.10). The molecules behave exactly the same way. The H end of one repels the H end of the other and attracts the F end (Fig. 3.10).

Because the H—F molecule has two poles, like a magnet, we call it a **dipole** or a **polar molecule.** We have already seen how we can tell which is the negative and which the positive end. **The *extent* of charge separation** is given by a number called the **dipole moment,** which can be measured in laboratories by a special technique. For molecules with two atoms, the dipole moment is related to the electronegativities. For example, we would expect HF to have a higher dipole moment than HCl (and it does), whereas $F_2$ and $H_2$ should have dipole moments of zero (and they do).

Dipole moment depends not only on electronegativities but also on the three-dimensional geometry of a molecule. For example, let us consider $H_2O$, which has an angular geometry:

*The greater the charge separation, the greater the dipole moment.*

**Figure 3.10**
Dipoles, like magnets, attract and repel each other.

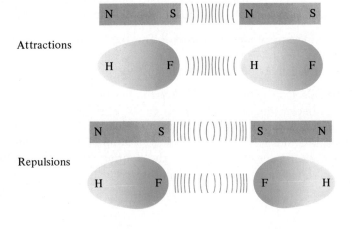

The electronegativity of oxygen is 3.5 and that of hydrogen is 2.1. Thus the oxygen must have $\delta-$ and the hydrogen $\delta+$. Molecules behave like *dipoles*, however, not like tripoles. Each polar molecule has only one negative end and one positive end. We know where the negative end is: on the oxygen. But where is the positive end? The answer is, halfway between the hydrogens:

$$
\underset{H \quad \overset{}{\underset{\delta+}{|}} \; H}{\overset{\overset{|}{O}\,\delta-}{}}
$$

*Laboratory methods can give only an overall dipole moment, not a moment for each bond.*

Sometimes the geometry is such that the bond moments completely cancel. For example, consider carbon dioxide, which, as we saw, is a linear molecule:

$$
\overset{\delta-}{O}=\overset{\delta+}{C}=\overset{\delta-}{O}
$$

Each C=O bond is polar. Since the electronegativities are 2.5 for carbon and 3.5 for oxygen, the electrons are shifted away from the central carbon and toward the two oxygens. Yet the entire molecule is nonpolar (has no dipole moment). Because the two polar bonds point in opposite directions (180° apart) and because the dipole moment of each bond is the same, the polarities cancel.

Another such case is the tetrahedral molecule carbon tetrachloride, where four C—Cl bond polarities cancel:

$$
\begin{array}{c}
Cl^{\delta-} \\
\overset{}{\underset{}{C^{\delta+}}} \\
{}^{\delta-}Cl \diagup \quad \diagdown Cl^{\delta-} \\
Cl^{\delta-}
\end{array}
$$

This compound is also nonpolar.

The dipole moment (or lack of one) has a major effect on the behavior of molecules, especially their boiling points and solubilities. We shall see the consequences of this in Chapter 5 and in several subsequent chapters.

## 3.8   HOW TO PREDICT THE KINDS OF BONDS THAT FORM

In Section 3.7 we learned that, in a bond between two different atoms, the electrons are not usually halfway between the two but are shifted toward the atom of higher electronegativity. When the difference in electronegativities is great, we can say that the electron pair is (almost) completely displaced to one side, and an ionic bond is formed. For example, consider a bond between cesium and chlorine. The electronegativity of Cl is 3.0; that of Cs is 0.7.

$$
Cs\cdot \; + \; \cdot\ddot{\underset{..}{Cl}}: \; \longrightarrow \; Cs^+ \quad :\ddot{\underset{..}{Cl}}:^-
$$

This is a very large difference; it means that Cs has very little attraction for the electron pair and essentially allows the Cl to have it completely. The Cs, having contributed one electron to the bond, no longer has it and has become a positive

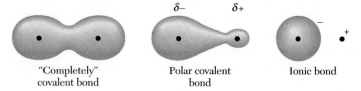

$\delta-$ $\delta+$

"Completely" covalent bond    Polar covalent bond    Ionic bond

**Figure 3.11**
The electron distribution in three types of bonds.

cesium ion. The chlorine atom has also contributed one electron but now has both, so that it has become a negative chloride ion. In this case we don't write $\delta+$ or $\delta-$, but $+$ or $-$, because each ion has acquired a full charge. Cesium chloride, CsCl, is therefore an ionic compound. (We must remember that in ionic compounds the bonding is not one to one. Instead, the ions are arranged in a lattice like the one shown in Figure 3.2.)

We can thus have basically three kinds of bonds between two atoms: a "completely" nonpolar covalent bond (when the electronegativities are identical or fairly similar), a polar covalent bond, or an ionic bond (when the electronegativities are very different) (see Fig. 3.11). We shall say that a difference of 1.8 or less gives a polar covalent bond, and a difference of 1.9 or more gives an ionic bond. Although this is a purely arbitrary rule, we shall not go very wrong in using it.

**EXAMPLE**

Classify as nonpolar, polar covalent, or ionic: (a) Cl—Cl   (b) C—N
(c) Rb—F.

**Answer**
To answer this question we look at the difference in electronegativities (Table 3.2). (a) This is a nonpolar bond because the electronegativity difference is zero.   (b) The electronegativity difference is $3.0 - 2.5 = 0.5$, which is less than 1.9, so this is a polar covalent bond.   (c) The electronegativity difference is $4.0 - 0.8 = 3.2$, which is greater than 1.8, so this is an ionic bond.

**PROBLEM 3.6**   Classify as nonpolar, polar covalent, or ionic: (a) N—N
(b) S—Cl   (c) Na—F.

## 3.9   WHAT BONDS TO WHAT?

We have now learned to predict the polarity of bonds and the shape of molecules (Sec. 3.6). In this section, let us consider one more thing: What bonds to what? In Section 2.5 we saw that most elements are either metals or nonmetals (or somewhere between). With respect to bonding, the main tendencies of metals and nonmetals can be summarized:

| Metals | Form **positive** ions | Usually reluctant to form covalent bonds |
|---|---|---|
| Nonmetals | Form **negative** ions | Form covalent bonds |

We can then state the following:

**1.** Metals bond to nonmetals, usually by ionic bonds but by covalent bonds if the difference in electronegativity is moderate.
**2.** Metals do not form compounds with other metals.
**3.** Nonmetals bond to nonmetals, usually by covalent bonds.

These rules are broad and general and have many exceptions in specific cases. Still, they serve as a rough guide and are valid for most cases.

## 3.10  POLYATOMIC IONS

Many compounds have both ionic and covalent bonds. In such cases the compounds contain ions with more than one atom. These ions are called **polyatomic ions.** An example is the important compound sodium hydroxide, NaOH. This is an ionic compound containing the positive $Na^+$ ion and the negative $OH^-$ ion. The structure is not very different from that of NaCl (Fig. 3.2). The difference is that $Cl^-$ is a monatomic ion and $OH^-$ is a polyatomic ion. The O and H are connected by a covalent bond, and the Lewis structure is

$$[:\ddot{O}-H]^-$$

Another notable polyatomic ion, one important in biochemistry, is the phosphate ion:

$$\begin{bmatrix} & :\ddot{O}: & \\ & | & \\ :\ddot{O}-&P&-\ddot{O}: \\ & \| & \\ & O: & \end{bmatrix}^{3-}$$

A list of some important polyatomic ions is given in Table 3.3. As with $OH^-$, the atoms in each ion are connected by covalent bonds, and Lewis structures can be drawn for each (we did this for $SO_4^{2-}$ in Sec. 3.4). Most of the important polyatomic ions are negatively charged, but there are some positively charged ones also, most notably the ammonium ion, $NH_4^+$:

$$\begin{bmatrix} & H & \\ & | & \\ H-&N&-H \\ & | & \\ & H & \end{bmatrix}^+$$

**TABLE 3.3  Some Important Polyatomic Ions**

| | |
|---|---|
| $NH_4^+$ | Ammonium |
| $OH^-$ | Hydroxide |
| $NO_3^-$ | Nitrate |
| $NO_2^-$ | Nitrite |
| $CH_3COO^-$ | Acetate |
| $CN^-$ | Cyanide |
| $MnO_4^-$ | Permanganate |
| $CO_3^{2-}$ | Carbonate |
| $HCO_3^-$ | Bicarbonate |
| $SO_3^{2-}$ | Sulfite |
| $HSO_3^-$ | Bisulfite |
| $SO_4^{2-}$ | Sulfate |
| $PO_4^{3-}$ | Phosphate |
| $HPO_4^{2-}$ | Hydrogen phosphate |
| $H_2PO_4^-$ | Dihydrogen phosphate |

Note that the phosphorus atom has ten electrons in its outer shell. This is an exception to the octet rule.

## 3.11 NAMING OF SIMPLE INORGANIC COMPOUNDS

Inorganic compounds are those that do not contain the element carbon. The naming of organic compounds will be discussed beginning in Chapter 10.

There are a large number of inorganic compounds. It would be convenient to be able to name them just by looking at their formulas and, conversely, to be able to write their formulas by looking at their names. This is relatively easy to do for the simple ones, which include most of the inorganic compounds you are ever likely to come across.

### Binary Ionic Compounds

A **binary** compound contains just two elements.

The rule for binary ionic compounds is that the positive ion (cation) is written first and then the negative ion (anion), in both the formula and the name. The name of the compound is simply the name of the positive ion followed by that of the negative ion.

---

**EXAMPLE**

Name these compounds: $LiBr$, $Ag_2S$, $SrF_2$.

**Answer**
Lithium bromide, silver sulfide, strontium fluoride.
   Note that the *number* of ions of each type is *not* given in the name. $SrF_2$ is called strontium fluoride, and not strontium difluoride. Note also that negative ions of a single element usually end in **ide;** for example, sulfide, iodide, oxide.

---

**PROBLEM 3.7**   Name these compounds: $MgO$, $BaI_2$, $NaBr$.

---

**EXAMPLE**

Give formulas for barium hydride and sodium sulfide.

**Answer**
$BaH_2$, $Na_2S$.

---

**PROBLEM 3.8**   Give formulas for strontium sulfide and sodium fluoride.

---

It is mostly transition elements that form ions with more than one charge.

In Table 3.1 we showed that some metals can form positive ions with more than one charge (for example, $Cu^+$, $Cu^{2+}$). In such cases we use Roman numerals in the name to show the charge. We do this because we must have some way of distinguishing them.

---

**EXAMPLE**

Name these compounds: $CuCl$, $CuCl_2$.

**Answer**
Copper(I) chloride, copper(II) chloride.

An older way to do this was to use the suffix **ic** for the higher charge and **ous** for the lower one. By this method CuCl is cuprous chloride and $CuCl_2$ is cupric chloride.

The older method is still in wide-spread use. Note that the Latin forms *cupr* and *ferr* are used in the older method, but not the newer one.

---

**EXAMPLE**

Name $FeBr_2$ and $FeBr_3$ by the two methods.

**Answer**
$FeBr_2$: ferrous bromide; iron(II) bromide.
$FeBr_3$: ferric bromide; iron(III) bromide.

---

**PROBLEM 3.9**  Name CuBr, $HgBr_2$, and $Fe_2O_3$ (see Table 3.1).

---

## Ionic Compounds that Contain Polyatomic Ions

The rule for these cases is exactly the same (that is, the positive ion comes first), except that now one or both of the ions contain two or more elements.

---

**EXAMPLE**

Name these compounds: $NaNO_3$, $AlPO_4$, $(NH_4)_2SO_3$, $NaH_2PO_4$.

**Answer**
Sodium nitrate, aluminum phosphate, ammonium sulfite, sodium dihydrogen phosphate.

---

**PROBLEM 3.10**  Name these compounds: $K_2HPO_4$, $Al_2(SO_4)_3$, $FeSO_3$.

---

## Binary Molecular Compounds

There are two rules we use for these compounds. (1) The less-electronegative element (see Table 3.2) comes first, in both the formula and the name; and (2) prefixes *di*, *tri*, and so on, are used to show the number of atoms of each element.

Compounds that do not contain ions are called *molecular compounds*.

Prefixes up to 8 are:

| 1 | mono | 5 | penta |
|---|------|---|-------|
| 2 | di | 6 | hexa |
| 3 | tri | 7 | hepta |
| 4 | tetra | 8 | octa |

---

**EXAMPLE**

Name these compounds: NO, $SF_2$, $P_2O_3$.

**Answer**
Nitrogen oxide, sulfur difluoride, diphosphorus trioxide.

---

**PROBLEM 3.11**  Name these compounds: $NO_2$, $PBr_3$, $SCl_2$.

We will not deal with the names of molecular compounds that contain more than two elements in this chapter. Some of them will be given in the chapter on acids and bases (Chapter 8) and others in the organic and biochemistry chapters.

## SUMMARY

Atoms form positive ions (**cations**) by losing electrons or negative ions (**anions**) by gaining electrons. An **ionic bond** is the attraction between positive and negative ions. Atoms usually lose or gain enough electrons to convert them to ions with complete octets. Atoms and ions have completely different properties. **Ionic compounds** are not composed of molecules. In the solid state they consist of regular arrays of ions.

**Covalent bonds** are formed when atoms share electrons. Atomic orbitals overlap to give **molecular orbitals.** Here too the atoms usually try to form octets and share enough electrons to do so. Each atom has a strong tendency to form a certain number of covalent bonds. Carbon always forms four bonds; nitrogen forms three; oxygen forms two. The four bonds of carbon can be made from single, double, or triple bonds. In **coordinate covalent bonds,** one atom supplies both electrons.

A **molecular formula** shows all the atoms in the molecule. A **structural formula** also shows all the bonds connecting them. A **Lewis structure** also shows all unshared outer-shell electrons.

Molecules have definite three-dimensional shapes, and these shapes can be determined by **VSEPR theory,** which is based on counting all electron pairs in the outer shell of each atom. In this theory, double and triple bonds are treated as single pairs. Atoms with two pairs are **linear,** and those with three pairs are **planar triangular.** With four pairs the shape is **tetrahedral.**

**Electronegativity** is a measure of how much an atom attracts the electrons in a covalent bond. When electronegativities of the two atoms in a bond differ, the atom with the higher electronegativity preferentially attracts the electron pair. This gives rise to **polar covalent bonds.** Molecules with polar covalent bonds may be **dipoles.** The **dipole moment** measures the extent of charge separation. When a very large difference in electronegativities (1.9 or greater) exists, we say that the bond is ionic.

Simple ionic inorganic compounds are named by giving the name of the positive ion followed by the name of the negative ion. Binary molecular compounds are named by giving the less-electronegative element first, followed by the more-electronegative one. In this case, prefixes are used to show the number of atoms of each element.

## KEY TERMS

Anion (Sec. 3.2)
Binary compound (Sec. 3.11)
Cation (Sec. 3.2)
Chemical bond (Sec. 3.1)
Chemical species (Sec. 3.2)
Coordinate covalent bond (Sec. 3.5)
Covalent bond (Sec. 3.4)
Dipole (Sec. 3.7)
Dipole moment (Sec. 3.7)
Double bond (Sec. 3.4)
Electronegativity (Sec. 3.7)
Inorganic compound (Sec. 3.11)
Ionic bond (Sec. 3.3)
Ionic compound (Sec. 3.3)
Lewis structure (Sec. 3.4)
Linear shape (Sec. 3.6)

Molecular compound (Sec. 3.11)
Molecular formula (Sec. 3.4)
Molecular orbital (Sec. 3.4)
Octet rule (Sec. 3.2)
Polar covalent bond (Sec. 3.8)
Polar molecule (Sec. 3.7)
Polyatomic ion (Sec. 3.10)
Pyramidal shape (Sec. 3.6)
Sigma ($\sigma$) orbital (Sec. 3.4)
Structural formula (Sec. 3.4)
Tetrahedral shape (Sec. 3.6)
Trigonal shape (Sec. 3.6)
Triple bond (Sec. 3.4)
Valence shell (Sec. 3.2)
VSEPR (Sec. 3.6)

## PROBLEMS

**3.12** Name the two main types of chemical bonds.

### Ions and Ionic Bonding

**3.13** Calcium is a soft metal that reacts with water. When a physician prescribes calcium for osteoporosis, what species will the patient actually get?

**3.14** How many electrons does each of these atoms have to gain or lose to reach a complete outer shell?
(a) Li      (f) S
(b) Cl      (g) Si
(c) Ne      (h) O
(d) Al      (i) Kr
(e) Sr      (j) P

**3.15** Write the formula for the most stable ion of
(a) Li      (d) Na
(b) Cl      (e) Al
(c) Ca      (f) Ba

**3.16** Predict which of these ions are stable:
(a) $Na^+$      (g) $Ba^{3+}$
(b) $Li^{2+}$      (h) $Se^{2+}$
(c) $Br^{2-}$      (i) $Na^-$
(d) $I^-$      (j) $C^{4-}$
(e) $Ca^+$      (k) $Cs^+$
(f) $S^{2-}$      (l) $Ar^+$

**3.17** Why is $Li^-$ not a stable ion?

**3.18** The ion $O^{6+}$ has a complete outer shell. Why is this ion never found in normal matter?

**3.19** Why are carbon and silicon reluctant to form ionic bonds?

**3.20** Do any ions violate the octet rule? Explain.

**3.21** Is there such a thing as a KF molecule? Explain.

**3.22** Describe the structure of sodium chloride in the solid state.

**3.23** Since sodium sulfate, $Na_2SO_4$, is an ionic compound, there are no discrete $Na_2SO_4$ molecules. What does the "2" in $Na_2SO_4$ mean?

**3.24** Complete the chart by writing formulas:

|  | $Br^-$ | $ClO_4^-$ | $O^{2-}$ | $NO_3^-$ | $SO_4^{2-}$ | $PO_4^{3-}$ | $OH^-$ |
|---|---|---|---|---|---|---|---|
| $Li^+$ |  |  |  |  |  |  |  |
| $Ca^{2+}$ |  |  |  |  |  |  |  |
| $Co^{3+}$ |  |  |  |  |  |  |  |
| $K^+$ |  |  |  |  |  |  |  |
| $Cu^{2+}$ |  |  |  |  |  |  |  |

**3.25** Write formulas for ionic compounds formed from these elements:
(a) sodium and bromine
(b) potassium and oxygen
(c) aluminum and chlorine
(d) barium and chlorine
(e) calcium and oxygen

**3.26** The compound $Na_2SO_4$ is made up of $Na^+$ and $SO_4^{2-}$ ions. Write the formulas for the ions in
(a) NaBr      (d) $KH_2PO_4$
(b) $FeSO_3$      (e) $NaHCO_3$
(c) $Mg_3(PO_4)_2$      (f) $Ba(NO_3)_2$

**3.27** What are the charges on the ions in each of the following?
(a) $CaCl_2$      (c) $Al_2S_3$
(b) MgO      (d) $Rb_2O$

### Covalent Bonding

**3.28** Explain how a covalent bond forms.

**3.29** How many covalent bonds are normally formed by
(a) N      (d) Br
(b) F      (e) O
(c) C

**3.30** What is a double bond? A triple bond?

**3.31** Draw the structural formula for
(a) $CH_4$      (d) $BF_3$
(b) $C_2H_2$      (e) $CH_2O$
(c) $C_2H_4$      (f) $C_3Cl_8$

**3.32** Some of the following formulas are incorrect (that is, they do not represent any real compound) because they contain atoms that do not have their normal number of covalent bonds. Which compounds are they, and which atoms have the incorrect number of bonds?

(a)

$$Cl-\underset{\underset{\displaystyle H}{|}}{\overset{\overset{\displaystyle H}{|}}{C}}=\underset{}{\overset{\overset{\displaystyle H}{|}}{C}}-H$$

(b)

(c)

(d)

(e) H—C≡C—C—O—C—H with Br, H, H below

(f) H—C≡C—C=C—H with H, H, H below

**3.33** Draw structural formulas for molecules in which a carbon atom is connected by a double bond to (a) another carbon atom (b) an oxygen atom (c) a nitrogen atom.

**3.34** Why can't a carbon atom be connected to an oxygen atom by a triple bond?

**3.35** Explain why argon does not form (a) ionic bonds (b) covalent bonds.

## Lewis Structures

**3.36** What is the difference between a molecular formula, a structural formula, and a Lewis structure?

**3.37** Draw the Lewis dot structure for
(a) C               (e) O
(b) S               (f) P
(c) N               (g) Br
(d) Na

**3.38** Give the total number of outer-shell electrons in
(a) $NH_3$          (e) $CCl_4$
(b) $C_3H_6$        (f) $NO_3^-$
(c) $C_2H_4O_2$     (g) $PO_4^{3-}$
(d) $C_2H_6O$

**3.39** Draw the Lewis structure for the following (in each case, there is only one way the atoms can be connected):
(a) $Br_2$          (d) $CN^-$
(b) $H_2S$          (e) $NH_4^+$
(c) $N_2H_2$        (f) $BF_4^-$

**3.40** Draw the Lewis structure for
(a) HCN    H    C    N
(b) $HNO_2$    H    O    N    O
                              O
(c) $H_2SO_4$    H    O    S    O    H
                                O

(d) $C_2H_3O_2^-$    H    C    C    O
                                        H
                                  H    O
(e) $CO_3^{2-}$    O    C    O
                          O
                          O
(f) $ClO_4^-$    O    Cl    O
                        O

**3.41** What is the difference between a bromine atom, a bromine molecule, and a bromide ion? Write the formula for each.

## Coordinate Covalent Bonds

**\*3.42** Draw a Lewis structure for carbon monoxide, CO. Is a coordinate covalent bond present?

**3.43** The reaction between ammonia, $NH_3$, and the hydrogen ion, $H^+$, produces the ammonium ion, $NH_4^+$. Show how this reaction involves the formation of a coordinate covalent bond.

## The Shapes of Molecules

**3.44** What is the shape of a molecule whose central atom contains (a) two (b) three (c) four electron pairs?

**3.45** Predict the shape of
(a) $CH_4$          (g) $C_2Cl_4$
(b) $PH_3$          (h) $NO_2^-$
(c) $BF_3$          (i) $H_3O^+$
(d) $C_2F_2$        (j) $NO_3^-$
(e) $SO_2$          (k) $SeO_2$
(f) $SO_3$          (l) $PCl_3$

## Electronegativity and Dipoles

**3.46** Which atom are the electrons shifted toward in a bond between
(a) K and Cl        (e) Na and O
(b) S and N         (f) P and S
(c) C and O         (g) H and O
(d) Cl and Br

**3.47** Why is a molecule of HCl attracted to another molecule of HCl?

**3.48** Which of these bonds is the most polar? The least polar?
(a) C—N  (b) C—C  (c) C—O

**3.49** Carbon disulfide, $CS_2$, is a linear molecule (S=C=S). What is its dipole moment? Show the partial charges.

**3.50** The $H_2S$ molecule has an angular geometry similar to that of water. Is the $H_2S$ molecule a dipole? Where are the partial charges located?

**\*3.51** Like ethylene, the chlorinated ethylenes are planar (flat). Which, if any, of these molecules have a dipole moment?

### Kinds of Bonds

**3.52** In each case, tell whether the bond is ionic, polar covalent, or nonpolar covalent:
(a) $Br_2$           (e) $SiH_4$
(b) BrCl           (f) CO
(c) HCl            (g) $N_2$
(d) $SrF_2$          (h) CsCl

**3.53** Predict whether a bond will form and, if so, whether it will be ionic or covalent, between
(a) Cl and I        (e) Na and Cr
(b) Li and K        (f) Ar and S
(c) O and K         (g) Na and Ca
(d) Al and N        (h) Li and Ne

### Polyatomic Ions

**3.54** Draw Lewis structures for the following, each of which contains at least one ionic bond (show the ions as separate structures):
(a) $Na_2SO_3$        (d) $NH_4OH$
(b) $KNO_3$          (e) $K_2HPO_4$
(c) $Cs_2CO_3$

**3.55** The perchlorate ion is $ClO_4^-$ (a) Draw its Lewis structure. (b) Write the molecular formula for calcium perchlorate.

### Naming Inorganic Compounds

**3.56** Write formulas for the following ionic compounds (see Tables 3.1 and 3.3):
(a) potassium bromide
(b) calcium oxide
(c) magnesium acetate
(d) lithium sulfate
(e) ammonium bisulfite
(f) copper(II) phosphate
(g) silver(I) carbonate
(h) lead(II) hydroxide
(i) barium permanganate
(j) strontium chloride
(k) iron(III) sulfide
(l) mercury(II) oxide
(m) strontium dihydrogen phosphate

**3.57** Name these ionic compounds:
(a) NaF            (g) $Sr_3(PO_4)_2$
(b) MgS            (h) $Fe(OH)_2$
(c) $Al_2O_3$         (i) $NaH_2PO_4$
(d) $BaCl_2$         (j) $Pb(CH_3COO)_2$
(e) $Ca(HSO_3)_2$     (k) $BaH_2$
(f) KI             (l) $(NH_4)_2HPO_4$

**3.58** Write molecular formulas for these compounds:
(a) carbon monoxide
(b) bromine chloride
(c) phosphorus pentachloride
(d) silicon tetrabromide
(e) dinitrogen trioxide
(f) sulfur hexafluoride

**3.59** Name these molecular compounds:
(a) $SO_2$           (e) $SeO_3$
(b) $H_2S$           (f) NO
(c) IF             (g) $N_2O_3$
(d) $IF_7$

### Boxes

**3.60** (Box 3A) What is the main use of $Ca^{2+}$ in the body?

**3.61** (Box 3A) In what biological molecule do we find $Fe^{2+}$?

**3.62** (Box 3A) What are the main ions inside the body cells?

**3.63** (Box 3A) What are the main ions in the body fluids outside the cells?

**3.64** (Box 3B) What is the medical use of barium sulfate?

**3.65** (Box 3B) Why is sodium iodide often present in the table salt we buy in the supermarket?

**3.66** (Box 3B) What is the medical use of $KMnO_4$?

**3.67** (Box 3C) How is the central iron(II) ion bound to the nitrogens in heme?

## Additional Problems

**3.68** Which of these magnesium compounds are ionic compounds? (a) $MgF_2$ (b) $Mg_2C$ (c) $MgS$

**\*3.69** What is the shape of the $NH_4^+$ ion?

**3.70** Which type of element (metal or nonmetal) is more likely to form (a) anions (b) cations?

**3.71** Why is each of the following ions never found in normal matter?
(a) $O^{3-}$ (b) $C^{4-}$ (c) $S^+$

**\*3.72** Which of these have an atom that does not obey the octet rule (not all of these are stable molecules)?
(a) $BF_3$             (e) $CH_3$
(b) $CF_4$             (f) $N_2$
(c) $BeF_2$            (g) $NO$
(d) $C_2H_4$

**3.73** What type of bond is formed between $BCl_3$ and $NH_3$?

**3.74** Name these ionic compounds:
(a) $Na_2SO_3$         (e) $(NH_4)_2CO_3$
(b) $KMnO_4$           (f) $LiNO_2$
(c) $Sr(CN)_2$         (g) $NaHCO_3$
(d) $FeO$

**\*3.75** The correct Lewis structure for ethylene is

$$\text{H—C=C—H}$$
$$\quad\ |\quad |$$
$$\quad\ \text{H}\quad\text{H}$$

Two other structures that might be drawn are incorrect:

(a) $\text{H—}\overset{..}{\text{C}}\text{—C—H}$     (b) $\text{H—}\overset{..}{\text{C}}\text{—}\overset{..}{\text{C}}\text{—H}$
$\qquad\ |\quad |$                $\qquad\ \ |\quad |$
$\qquad\ \text{H}\quad\text{H}$          $\qquad\ \ \text{H}\quad\text{H}$

Explain why each of these is incorrect.

**3.76** Write formulas for the compounds formed between
(a) potassium ion and sulfide ion
(b) aluminum ion and nitrate ion
(c) lead(II) ion and sulfite ion
(d) calcium ion and bicarbonate ion
(e) mercury(II) ion and dihydrogen phosphate ion
(f) barium ion and hydride ion

**3.77** Show how $BF_3$ forms a coordinate covalent bond with phosphine, $PH_3$.

**\*3.78** Predict which of these molecules will be polar (that is, will have a dipole moment):
(a) $CBr_4$            (d) $CH_4$
(b) $CHCl_3$           (e) $CCl_2I_2$
(c) $CH_2Br_2$         (f) $CH_3Cl$

**3.79** A covalent bond is directed in space from one particular atom to another. Can you say the same thing for an ionic bond? Explain.

**\*3.80** Draw the three-dimensional shape of the ethane molecule, $C_2H_6$.

**\*3.81** In the ferricyanide ion $Fe(CN)_6^{3-}$ there are six coordinate covalent bonds. Draw the Lewis structure of the ferricyanide ion.

**\*3.82** The two compounds $AlCl_3$ and $NCl_3$ would seem to be very similar in structure. Yet one has a flat planar geometry, while the other has a pyramidal geometry. Explain why these compounds have such different structures.

**3.83** In analogy to Figure 3.5, show how the two $1s$ orbitals overlap to form a molecular orbital in the $H_2$ molecule.

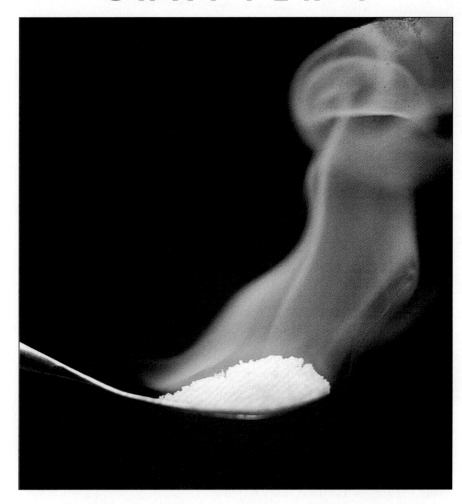

*In this chemical reaction, ammonium chloride, NH$_4$Cl, is converted to ammonia, NH$_3$, and hydrogen chloride, HCl. (Photograph by Charles D. Winters.)*

# Chemical Reactions

## 4.1   INTRODUCTION

In Chapter 1 we learned that chemistry is mainly concerned with two things: the structure of matter and the transformations of one form of matter to another. In Chapters 2 and 3 we discussed the first of these topics ("what is matter made of"), and now we are ready to turn our attention to the other. In a chemical change, also called a **chemical reaction,** one or more original substances (called the **starting materials** or **reactants**) are converted to one or more new substances (called the **products**). Chemical reactions are all around us. They fuel and keep alive the cells of living tissues; they occur when we light a match, cook dinner, start a car, listen to a portable radio, or watch television

(a)

(b)

(c)

**Figure 4.1**
Examples of chemical reactions: (a) fireworks,
(b) baking (photographs by Beverly March),
(c) rocket blast-off (courtesy of NASA).

(Fig. 4.1). Most of the world's manufacturing processes involve chemical reactions (Fig. 4.2); they include petroleum refining, metal-smelting operations, and food processing, as well as the manufacture of drugs, plastics, artificial fibers, explosives, and many other materials.

In this chapter we discuss three aspects of chemical reactions: (a) weight relationships, (b) types of reactions, and (c) heat gains and losses. In later chapters we will discuss many reactions in greater detail, especially those taking place in living organisms.

**Figure 4.2**
A chemical factory that makes isopropyl alcohol, as well as other compounds. (Photograph by Beverly March.)

## 4.2 FORMULA WEIGHT

We begin our study of weight relationships with formula weight. In Section 2.4 we learned that every atom has an atomic weight. The **formula weight** of any substance is simply **the sum of the atomic weights of all the atoms in the molecular formula.**

From our previous discussions it is evident that molecular formulas mean different things for molecular compounds (those that contain only covalent bonds) and ionic compounds. Molecular compounds are made of molecules, and the molecular formula tells how many atoms of each element are present in that molecule. For example, a $CCl_4$ molecule contains one C and four Cl atoms. Since there are no molecules in an ionic compound, however, its "molecular" formula represents the ratio of ions. For example, $Na_2SO_4$ means that two $Na^+$ ions are present for every one $SO_4^{2-}$ ion.

The term "formula weight" can be used for both ionic and molecular compounds and tells nothing about whether the compound is ionic or molecular. There is an older term, **molecular weight,** that is strictly correct only when used for molecular compounds. In this book we use it only for them. Strictly speaking, a more accurate term than "weight" would be "mass" (Sec. 1.3), but the expressions "molecular weight" and "formula weight" are more commonly used.

A table of atomic weights is given on the inside back cover. Atomic weights can also be found in the periodic table on the inside front cover.

Many people use "molecular weight" for both molecular and ionic compounds.

---

**EXAMPLE**

What is the formula weight of (a) $CO_2$ (b) $(NH_2)_2CO$ (urea)?

**Answer**

(a)

|   | AW (amu) | Total weight |
|---|---|---|
| C | $12.0 \times 1 =$ | 12.0 |
| O | $16.0 \times 2 =$ | 32.0 |
| Formula weight of $CO_2$ | | 44.0 amu |

(b)

|   | AW (amu) | Total weight |
|---|---|---|
| N | $14.0 \times 2 =$ | 28.0 |
| H | $1.0 \times 4 =$ | 4.0 |
| C | $12.0 \times 1 =$ | 12.0 |
| O | $16.0 \times 1 =$ | 16.0 |
| Formula weight of $(NH_2)_2CO$ | | 60.0 amu |

Note that in (b) we include all the atoms. The formula is $(NH_2)_2CO$. The $(NH_2)_2$ tells us that there are two nitrogen atoms and four hydrogen atoms.

In this book we take formula weights to one decimal place.

---

**PROBLEM 4.1** What is the formula weight of (a) aspirin, $C_9H_8O_4$ (b) barium phosphate, $Ba_3(PO_4)_2$?

## 4.3    THE MOLE

Atoms and molecules are so tiny (Sec. 2.8) that chemists are seldom able to deal with them one at a time (very special and expensive apparatus is necessary for that). When a chemist weighs out even a very small quantity of a molecular compound, huge numbers of molecules (perhaps $10^{19}$) are present. To overcome this problem, chemists long ago defined a unit called the mole. Quite simply, one **mole** of any substance is its **formula weight expressed in grams.** For instance, the formula weight of $CO_2$, as we learned in our last example, is 44.0 amu; therefore 44.0 g of $CO_2$ is 1 mole of $CO_2$. Likewise, the formula weight of $(NH_2)_2CO$ is 60.0 amu, and so 60.0 g of $(NH_2)_2CO$ is 1 mole of this compound. For atoms, 1 mole is the atomic weight expressed in grams: 12 g of carbon is 1 mole of carbon; 32 g of sulfur is 1 mole of sulfur.

One-mole quantities of several compounds are shown in Figure 4.3.

More than 60 years ago, scientists found out that 1 mole of any compound always contains $6.02 \times 10^{23}$ molecules. In other words, 44.0 g of $CO_2$ always contains $6.02 \times 10^{23}$ molecules of $CO_2$, and 60.0 g of $(NH_2)_2CO$ always contains $6.02 \times 10^{23}$ molecules of $(NH_2)_2CO$. Thus we can say, as a secondary definition, that a **mole is $6.02 \times 10^{23}$ of anything:** A mole of hydrogen atoms is $6.02 \times 10^{23}$ hydrogen atoms, a mole of sugar molecules is $6.02 \times 10^{23}$ sugar molecules, and a mole of apples is $6.02 \times 10^{23}$ apples. As you can see, there is nothing mysterious about the mole. Just as we call 12 of anything a dozen, 20 a score, and 144 a gross, we call $6.02 \times 10^{23}$ of anything a mole. The number $6.02 \times 10^{23}$ is called **Avogadro's number,** after the Italian physicist Amedeo Avogadro (1776–1856).

One mole of pennies placed side by side would stretch for more than one million lightyears, a distance far outside our solar system and even our own galaxy. Six moles of freshman chemistry books would weigh as much as the earth.

Let us see why the fact that 1 mole always contains $6.02 \times 10^{23}$ molecules is very useful information. An atomic weight table tells us that the mass of a

**Figure 4.3**
One-mole quantities of five compounds (clockwise from upper left): $CuSO_4 \cdot 5H_2O$ (249.68 g); $K_2Cr_2O_7$ (294.19 g); $(NH_2)_2CO$ (urea, 60.0 g); $KMnO_4$ (158.0 g); $C_2H_5NO_2$ (glycine, 75.1 g). (Photograph by Beverly March.)

silicon atom is 28 amu and the mass of a nitrogen atom is 14 amu; that is, a silicon atom has twice the mass of a nitrogen atom (to the nearest integer). Suppose we had ten of each. Ten Si atoms weigh 280 amu, and ten N atoms weigh 140 amu. The ratio is still 2:1. In fact, the weight ratio is 2:1 no matter how many atoms we have, as long as we have the same number of N and Si atoms. Now suppose we have a mole of each. Avogadro's number of N atoms weighs 14 g. Therefore, the same number of Si atoms *must* weigh 28 g.

The same argument can be made for any two atoms or molecules. In 1 mole each of any two substances, the weight ratios are the same as they are for the individual atoms or molecules. This fact allows chemists to measure out a given *number* of atoms or molecules by *weighing out* a sample. It is impossible for a chemist to weigh out one molecule or even 100 molecules, but it is very easy to weigh out one mole of molecules, or half a mole, or 2.7 moles, or any convenient fraction or multiple desired.

Even 100 molecules represent such a small mass that no balance is sensitive enough to detect it.

Let's look at some examples. In these examples we use the factor-label method discussed in Section 1.4.

---

**EXAMPLE**

We have 27.5 g of stannous fluoride, $SnF_2$. How many moles is this?

**Answer**

The formula weight of $SnF_2 = 118.7 + 2(19.0) = 156.7$ amu. This means that each mole of $SnF_2$ has a mass of 156.7 g, allowing us to use the conversion factor 1 mole $SnF_2 = 156.7$ g $SnF_2$:

$$27.5 \text{ g } \cancel{SnF_2} \times \frac{1 \text{ mole } SnF_2}{156.7 \text{ g } \cancel{SnF_2}} = 0.175 \text{ mole } SnF_2$$

Note that such calculations can be performed for ionic compounds, such as $SnF_2$, as well as molecular compounds, such as $CO_2$ and urea.

---

**PROBLEM 4.2**  We have 756.3 g of $ZnCl_2$. How many moles is this?

---

**EXAMPLE**

We wish to weigh out 3.41 moles of ethyl alcohol, $C_2H_5OH$. How many grams is this?

**Answer**

The formula weight of $C_2H_5OH$ is $2(12.0) + 6(1.0) + 16.0 = 46.0$ amu, and so the conversion factor is 1 mole $C_2H_5OH = 46.0$ g $C_2H_5OH$. In order for the proper units to cancel, we multiply as follows:

$$3.41 \text{ moles } \cancel{C_2H_5OH} \times \frac{46.0 \text{ g } C_2H_5OH}{1 \text{ mole } \cancel{C_2H_5OH}} = 157 \text{ g } C_2H_5OH$$

---

**PROBLEM 4.3**  We wish to weigh out 2.84 moles of sodium sulfide, $Na_2S$. How many grams is this?

TNT (trinitrotoluene) is a powerful explosive.

**EXAMPLE**

How many moles of nitrogen atoms and oxygen atoms are there in 21.4 moles of TNT, $C_7H_5(NO_2)_3$?

**Answer**

The formula $C_7H_5(NO_2)_3$ tells us that each molecule of TNT contains three nitrogen atoms. Therefore each mole of TNT must contain 3 moles of nitrogen atoms. The number of moles of N atoms in 21.4 moles of TNT is therefore

$$21.4 \text{ moles TNT} \times \frac{3 \text{ moles N}}{1 \text{ mole TNT}} = 64.2 \text{ moles N}$$

Similarly, each mole of TNT contains 6 moles of oxygen atoms, so that the total number of moles of oxygen atoms is

$$21.4 \text{ moles TNT} \times \frac{6 \text{ moles O}}{1 \text{ mole TNT}} = 128.4 \text{ moles O}$$

**PROBLEM 4.4**   How many moles of C atoms, H atoms, and O atoms are there in 2.5 moles of glucose, $C_6H_{12}O_6$?

**EXAMPLE**

How many moles of sodium ions, $Na^+$, are there in 5.63 g of $Na_2SO_4$?

**Answer**

First we find out how many moles of $Na_2SO_4$ we have. As in the $SnF_2$ example, we obtain the conversion factor by calculating the formula weight: $2(23.0) + 32.1 + 4(16.0) = 142.1$ amu. Then

$$5.63 \text{ g Na}_2\text{SO}_4 \times \frac{1 \text{ mole Na}_2\text{SO}_4}{142.1 \text{ g Na}_2\text{SO}_4} = 0.0396 \text{ mole Na}_2\text{SO}_4$$

The formula $Na_2SO_4$ tells us that there are two $Na^+$ ions in each mole of $Na_2SO_4$. The number of moles of $Na^+$ ions in 5.63 g of $Na_2SO_4$ is therefore

$$0.0396 \text{ mole Na}_2\text{SO}_4 \times \frac{2 \text{ moles Na}^+}{1 \text{ mole Na}_2\text{SO}_4} = 0.0792 \text{ mole Na}^+$$

**PROBLEM 4.5**   How many moles of copper(I) ions, $Cu^+$, are there in 0.062 g of copper(I) nitrate, $CuNO_3$?

**EXAMPLE**

A typical aspirin tablet, $C_9H_8O_4$, contains 0.36 g aspirin. How many aspirin molecules are present?

**Answer**

The formula weight of aspirin is $9(12.0) + 8(1.0) + 4(16.0) = 180.0$ amu. First we find out how many moles of aspirin there are in 0.36 g:

$$0.36 \text{ g aspirin} \times \frac{1 \text{ mole aspirin}}{180.0 \text{ g aspirin}} = 0.0020 \text{ mole aspirin}$$

$$= 2.0 \times 10^{-3} \text{ mole aspirin}$$

Each mole contains $6.02 \times 10^{23}$ molecules, and so the number of molecules is

$$(2.0 \times 10^{-3} \text{ mole}) \left( 6.02 \times 10^{23} \frac{\text{molecules}}{\text{mole}} \right) = 1.2 \times 10^{21} \text{ molecules}$$

A single aspirin tablet contains $1.2 \times 10^{21}$ molecules of $C_9H_8O_4$. Chemists seldom need to find out the number of molecules in anything. They work with grams and moles. If they ever do need this information, however, it is very easy to do, as the last example shows.

---

**PROBLEM 4.6** How many molecules of water, $H_2O$, are there in a glass of water (235 g)?

---

## 4.4 CHEMICAL EQUATIONS

A typical chemical reaction is the burning of propane (bottled gas). In this reaction, propane reacts with the oxygen in the air; both of these substances are converted to the products: carbon dioxide and water. We could write this

Propane plus oxygen gives carbon dioxide and water

If we didn't know the molecular formulas, we would in fact be forced to write it this way. We do know the formulas, however, and so a better way to write it is

$$C_3H_8 \quad + \quad O_2 \quad \longrightarrow \quad CO_2 \quad + \quad H_2O$$

Propane    Oxygen Yields Carbon    Water<br>dioxide

The arrow indicates that the starting materials (on the left) are converted to the products (on the right).

This kind of expression is called a **chemical equation.**

The equation we have written is incomplete, however. Though it does tell us the formulas for all the starting materials and all the products (which every chemical equation must do), it does not give the amounts correctly. That is, it is qualitatively right but quantitatively wrong. *It is not balanced.* In chemical reactions atoms are never destroyed or created; they are only shifted from one substance to another. This means that all the atoms present at the start (on the left side of the arrow) must still be there at the end (on the right side), as seen in Figure 4.4. In the equation we have written, there are three carbon atoms on the left and only one on the right. This means the equation cannot be correct. In order to make it correct, we must balance it.

**Figure 4.4**
In the combustion of propane, one $C_3H_8$ molecule and five $O_2$ molecules are converted to three $CO_2$ and four $H_2O$ molecules. No atoms are created or destroyed. They merely change partners, and so the total mass is unchanged.

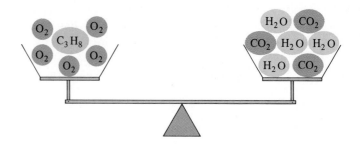

## Balancing Equations

Most equations are easily balanced. It only takes a few minutes. Basically, all we do is place numbers in front of the formulas until the equation is balanced. These numbers are called **coefficients** and represent numbers of molecules. As an example, let us balance our propane equation:

$$C_3H_8 + O_2 \longrightarrow CO_2 + H_2O$$

How do we choose our coefficients? Let us begin with carbon. There are three carbon atoms on the left and one on the right. If we put a 3 in front of the $CO_2$ (indicating that three $CO_2$ molecules are formed), there will be three carbons on each side and the carbons will be balanced:

$$C_3H_8 + O_2 \longrightarrow 3CO_2 + H_2O$$

Next we look at the hydrogens. There are eight on the left and two on the right. We therefore put a 4 in front of the $H_2O$ to make a total of eight hydrogens on the right ($4 \times 2$). We now have

$$C_3H_8 + O_2 \longrightarrow 3CO_2 + 4H_2O$$

The only thing still unbalanced is oxygen. There are two oxygen atoms on the left and ten on the right (make sure you can find ten). If we put the coefficient 5 in front of the oxygen on the left, we balance the oxygens and the whole equation:

$$C_3H_8 + 5O_2 \longrightarrow 3CO_2 + 4H_2O \qquad \text{Balanced equation}$$

At this point the equation ought to be balanced, but we should always check, just to make sure. **In a balanced equation there must be the same number of atoms of each element on both sides.** A check of our work shows three C, ten O, and eight H atoms on each side. The equation is indeed balanced.

Note that the numbers we put in to balance an equation are always coefficients. We are not permitted to change formulas. For example, when we started to balance $C_3H_8 + O_2 \rightarrow CO_2 + H_2O$, we saw that we needed three C atoms on the right. Why not change $CO_2$ to $C_3O_2$? We cannot because the product of the reaction is carbon dioxide and the formula for carbon dioxide is $CO_2$, not $C_3O_2$. In balancing equations we are permitted to say that three $CO_2$ molecules are formed (or whatever number is necessary for a balanced equation), but we are *not* permitted to write incorrect formulas. Here are two other examples.

**EXAMPLE**

Balance:

$$Ca(OH)_2 + HCl \longrightarrow CaCl_2 + H_2O$$

**Answer**

The calcium is already balanced: There is one on each side. There is one Cl on the left and two on the right. In order to balance them, we add the coefficient 2 to HCl. The equation now is

$$Ca(OH)_2 + 2HCl \longrightarrow CaCl_2 + H_2O$$

In the next step we note that there are four hydrogens on the left but only two on the right. Placing the coefficient 2 in front of $H_2O$ remedies the situation:

$$Ca(OH)_2 + 2HCl \longrightarrow CaCl_2 + 2H_2O \qquad \text{Balanced equation}$$

Note that the last step also balanced the oxygen atoms.

**PROBLEM 4.7**   Balance: $As_2O_5 + H_2O \rightarrow H_3AsO_4$

---

**EXAMPLE**

Balance:

$$C_3H_6 + O_2 \longrightarrow H_2O + CO_2$$

**Answer**

This equation, for the combustion of cyclopropane, is very similar to the first one we did. As before, we put a 3 in front of the $CO_2$ and this time a 3 in front of the $H_2O$ (since there are six hydrogens on the left):

$$C_3H_6 + O_2 \longrightarrow 3H_2O + 3CO_2$$

But now when we balance the oxygens, we find 2 on the left and 9 on the right. The only way to balance the equation is to put a $4\frac{1}{2}$ in front of the $O_2$:

$$C_3H_6 + 4\tfrac{1}{2}O_2 \longrightarrow 3H_2O + 3CO_2$$

Although there are times when chemists have good reasons to write equations with fractional coefficients, it cannot be completely correct because there is no such thing as half an oxygen molecule. We get around the difficulty by multiplying everything by 2:

$$2C_3H_6 + 9O_2 \longrightarrow 6H_2O + 6CO_2 \qquad \text{Balanced equation}$$

A check will show that the equation is balanced.

Propane, $C_3H_8$, and cyclopropane, $C_3H_6$, are different compounds.

**PROBLEM 4.8**   Balance: $C_6H_{14} + O_2 \rightarrow H_2O + CO_2$

One more thing. If

$$C_3H_8 + 5O_2 \longrightarrow 3CO_2 + 4H_2O$$

is a correctly balanced equation (and indeed it is), would it be correct if we doubled all the coefficients?

$$2C_3H_8 + 10O_2 \longrightarrow 6CO_2 + 8H_2O$$

The answer is that it is mathematically and scientifically correct, but chemists do not normally write equations with multiple coefficients. A correctly balanced equation is almost always written with the coefficients expressed "in lowest terms."

## 4.5   WEIGHT RELATIONSHIPS IN CHEMICAL REACTIONS

As we saw in Section 4.4, a balanced chemical equation tells us not only which substances react and which are formed, but also in what proportions. This is very useful information when we want to carry out a chemical reaction. For example, we can easily calculate how much starting materials to weigh out if we want to produce a particular weight of a product. **The study of weight relationships in chemical reactions** is called **stoichiometry.**

Let us look once again at the balanced equation for the burning of propane:

$$C_3H_8 + 5O_2 \longrightarrow 3CO_2 + 4H_2O$$

This equation not only tells us that propane and oxygen are converted to carbon dioxide and water; it also tells us that **1 mole** of propane combines with **5 moles** of oxygen to produce **3 moles** of carbon dioxide and **4 moles** of water; that is, we know the quantities involved. The same is true for any balanced equation.

This fact allows us to answer questions such as

**1.**   How many moles of any particular product are produced if we start with a given mass of any starting material?
**2.**   How many grams of one starting material are necessary to react completely with a given number of grams of another starting material?
**3.**   If we want to produce a certain number of grams (or moles) of a certain product, how many grams (or moles) of starting materials are needed?
**4.**   If a certain amount of one product is produced, how much of another product is also produced?

It sounds as if we have four different types of problems here, and yet they are all done by the same very simple procedure, which involves three steps.

"Stoichiometry" comes from the Greek *stoicheion*, element, and *metron*, measure.

In Section 4.4 we saw that the coefficients in an equation represent numbers of molecules. Since moles are proportional to molecules (Sec. 4.3), the coefficients in an equation also represent numbers of moles.

Ammonia, $NH_3$, which is used to make fertilizer, is produced by this reaction (the Haber process).

**EXAMPLE**

Consider the reaction

$$N_2 + 3H_2 \longrightarrow 2NH_3$$

How many grams of $N_2$ are necessary to produce 7.5 g of $NH_3$?

**Answer**

**STEP 1**   The coefficients in an equation refer to moles, not grams. There-fore we must first find out how many moles of $NH_3$ there are in 7.5 g. As before, we multiply the 7.5 g by the proper conversion factor (1 mole $NH_3$ = 17.0 g $NH_3$). At this stage we don't do the actual multiplication, but just leave it set up:

$$7.5 \text{ g } NH_3 \times \frac{1 \text{ mole } NH_3}{17.0 \text{ g } NH_3} = \text{number of moles of } NH_3$$

**STEP 2**   Next we turn to the balanced equation, which tells us that *two* moles of $NH_3$ are produced from *one* mole of $N_2$. We multiply the number of moles of $NH_3$ by this fraction:

$$7.5 \text{ g } NH_3 \times \frac{1 \text{ mole } NH_3}{17.0 \text{ g } NH_3} \times \frac{1 \text{ mole } N_2}{2 \text{ moles } NH_3} = \text{number of moles of } N_2$$

This gives us the number of *moles* of $N_2$ that are necessary.

The fraction obtained from the bal-anced equation is shown in color.

**STEP 3**   We now find the number of grams of $N_2$, as usual, by multiplying by the proper conversion factor (28.0 g $N_2$ = 1 mole $N_2$):

$$7.5 \text{ g } NH_3 \times \frac{1 \text{ mole } NH_3}{17.0 \text{ g } NH_3} \times \frac{1 \text{ mole } N_2}{2 \text{ moles } NH_3} \times \frac{28.0 \text{ g } N_2}{1 \text{ mole } N_2} = 6.2 \text{ g } N_2$$

At this point we do the arithmetic to get our answer, 6.2 g $N_2$.

In all such problems, we are given a weight (or number of moles) of one compound and asked to find the weight (or number of moles) of another com-pound. The two compounds can be on the left or the right, on the same side of the equation, or on opposite sides. All such problems can be done by the three simple steps we have just used:

**1.**   If you are given a mass, convert to moles by multiplying by the proper conversion factor. If you are given the number of moles, just write it down.
**2.**   Multiply by a fraction consisting of the coefficient of the compound whose mass is unknown divided by the coefficient of the compound whose mass is known. Since this information is obtained from the balanced equation, we can-not do such problems without a balanced equation.
**3.**   If you are asked for the mass of a compound, multiply by the appropriate conversion factor. If you are asked for moles, this step is unnecessary and the answer is already present at the end of step 2.

Note that we look at the coefficients of the balanced equation only in step 2. These coefficients play no part in steps 1 and 3.

The simplicity of the method is shown in the following examples.

**EXAMPLE**

How many moles of $Cl_2$ are required to produce 11.6 moles of $PCl_3$ by the reaction

$$2P + 3Cl_2 \longrightarrow 2PCl_3$$

**Answer**

**STEP 1**   Since we are given moles, we merely write the information down: 11.6 moles $PCl_3$.

**STEP 2**   Multiply by the fraction obtained from the equation:

$$11.6 \text{ moles } PCl_3 \times \frac{3 \text{ moles } Cl_2}{2 \text{ moles } PCl_3} = 17.4 \text{ moles } Cl_2$$

**STEP 3**   Unnecessary, since we were asked for moles.

---

**PROBLEM 4.9**   In the reaction

$$CH_4 + 2O_2 \longrightarrow CO_2 + 2H_2O$$

how many moles of $CH_4$ are required to produce 16.6 moles of $H_2O$?

---

**EXAMPLE**

When urea, $(NH_2)_2CO$, is acted on by the enzyme urease in the presence of water, ammonia and carbon dioxide are produced:

$$(NH_2)_2CO + H_2O \longrightarrow 2NH_3 + CO_2$$

If excess water is present (more than necessary for the reaction), how many grams each of $CO_2$ and $NH_3$ are produced from 0.83 mole of urea?

**Answer**
For $CO_2$:

**STEP 1**   Once again, we merely write down the number of moles: 0.83 mole urea.

**STEP 2**   Multiply by the fraction obtained from the equation:

$$0.83 \text{ mole urea} \times \frac{1 \text{ mole } CO_2}{1 \text{ mole urea}}$$

**STEP 3**   Multiply by the conversion factor (44.0 g $CO_2$ = 1 mole $CO_2$):

$$0.83 \text{ mole urea} \times \frac{1 \text{ mole } CO_2}{1 \text{ mole urea}} \times \frac{44.0 \text{ g } CO_2}{1 \text{ mole } CO_2} = 37 \text{ g } CO_2$$

For $NH_3$:
Step 1 is the same, but in step 2 the equation gives a different fraction:

$$0.83 \text{ mole urea} \times \frac{2 \text{ moles } NH_3}{1 \text{ mole urea}} \times \frac{17.0 \text{ g } NH_3}{1 \text{ mole } NH_3} = 28 \text{ g } NH_3$$

**PROBLEM 4.10** In the reaction

$$C_2H_2 + 2Br_2 \longrightarrow C_2H_2Br_4$$

how many grams of $C_2H_2Br_4$ are produced from 7.24 moles of $Br_2$ (assuming excess $C_2H_2$)?

Excess means more than is necessary for complete reaction.

---

**EXAMPLE**

In the reaction

$$2CH_3OH + PCl_5 \longrightarrow 2CH_3Cl + POCl_3 + H_2O$$

how many grams of $PCl_5$ are necessary to react with 137 g of methanol, $CH_3OH$?

**Answer**

The formula weight of methanol is 32.0 amu and that of $PCl_5$ is 208.2 amu. Steps 1, 2, and 3 give us

$$137 \text{ g } CH_3OH \times \frac{1 \text{ mole } CH_3OH}{32.0 \text{ g } CH_3OH} \times \frac{1 \text{ mole } PCl_5}{2 \text{ moles } CH_3OH} \times \frac{208.2 \text{ g } PCl_5}{1 \text{ mole } PCl_5}$$

$$= 446 \text{ g } PCl_5$$

---

**PROBLEM 4.11** In the reaction

$$C + 2H_2 \longrightarrow CH_4$$

how many grams of $H_2$ are necessary to react with 51.4 g of C?

---

## 4.6 PERCENTAGE YIELD

When carrying out chemical reactions in the laboratory, we often get less of a product than expected from the weight-weight calculation. For example, if we start with 1 mole (32.0 g) of $CH_3OH$ and add excess $PCl_5$ in the reaction

$$2CH_3OH + PCl_5 \longrightarrow 2CH_3Cl + POCl_3 + H_2O$$

we should get 1 mole (50.5 g) of $CH_3Cl$. However, we often don't; we get less. Does this mean that the law of conservation of mass is being violated? No, it does not. We get less than 50.5 g of $CH_3Cl$ either because some of the $CH_3OH$ does not react or because some of it reacts in some other way, or perhaps because our laboratory technique is not perfect and we lose a little in transferring from one container to another. Whatever the reason, the weight of product we get in the laboratory is called the **actual yield.** The expected weight of the product—the amount calculated by the weight-weight method—is called the **theoretical yield.** The **percentage yield** is actual yield divided by theoretical yield multiplied by 100:

Reactions that do not give the main product are called *side reactions*.

$$\% \text{ yield} = \frac{\text{actual yield}}{\text{theoretical yield}} \times 100$$

### Box 4A

### Industrial Production and Percentage Yield

The percentage yield of a chemical product is of great economic importance. Because starting compounds cost money, and any compounds produced by side reactions must be removed from the product, which also costs money, manufacturers of chemical compounds have a strong incentive to try to get the highest possible percentage yields. Many patents and new manufacturing processes deal with this problem. A typical case is the manufacture of urea. Urea was the first organic compound synthesized (Sec. 10.1). It is used mainly as a nitrogen source in fertilizers, as a supplemental feed for cattle, as a raw material for plastics, and, in its purest form, as starting compound in the pharmaceutical industry to synthesize barbiturates (Box 15G).

Urea is manufactured from ammonia (Box 7E) and carbon dioxide in two steps:

$$2NH_3 + CO_2 \longrightarrow NH_2{-}\underset{\underset{O}{\|}}{C}{-}O^-\,NH_4{}^+$$

Ammonium carbamate

which decomposes above 150°C to form

$$NH_2{-}\underset{\underset{O}{\|}}{C}{-}O^-\,NH_4{}^+ \longrightarrow NH_2{-}\underset{\underset{O}{\|}}{C}{-}NH_2 + H_2O$$

Urea

The two reactions run simultaneously. The percentage yield of a single passage of the reaction is about 62 percent under the best temperature–pressure conditions. This means that the main product, urea, should be separated and the ammonia, carbon dioxide, and ammonium carbamate recycled. This is what is accomplished in many different patented industrial processes, some of them reaching a 72 to 74 percent yield of urea.

---

**EXAMPLE**

In an experiment using the reaction just described, the actual yield of $CH_3Cl$ was 46.8 g (theoretical yield is 50.5 g). What is the percentage yield?

**Answer**

$$\frac{46.8 \text{ g}}{50.5 \text{ g}} \times 100 = 92.7\%$$

Most chemists consider this a pretty good percentage yield for this reaction.

---

**PROBLEM 4.12** In one method for making aspirin, the theoretical yield is 153.7 g. If the actual yield obtained by a student was 124 g, what is the percentage yield?

## 4.7   BASIC TYPES OF REACTIONS

There are many thousands of chemical reactions; yet chemists are able to understand nearly all of them and are even able to predict new ones. Chemists can do this because many reactions are similar, so it is possible to classify most reactions into a relatively small number of categories.

For example, it is easy to see that these two reactions have something in common:

$$2Na + 2H_2O \longrightarrow 2NaOH + H_2$$

and

$$Zn + Cu(NO_3)_2 \longrightarrow Zn(NO_3)_2 + Cu$$

In both cases, one element displaces another element from a compound. In the first case the element sodium reacts with water to replace hydrogen. The hydrogen that was formerly in the water becomes a free element (though some of the hydrogen does remain combined with the oxygen). In the second case, zinc displaces copper from copper nitrate. This general equation covers these two reactions as well as many others:

$$A + BC \longrightarrow B + AC$$

We call them **single displacement reactions.**

In this section we will discuss four general types of reactions. Many examples of each are known, though there are many other reactions that cannot be fitted neatly into these categories.

### Combination Reactions

In this type two (or sometimes more) compounds or elements come together to form larger compounds. The general equation is

$$A + B \longrightarrow C$$

Some examples are

$$2H_2 + O_2 \longrightarrow 2H_2O$$
$$2SO_2 + O_2 \longrightarrow 2SO_3$$
$$SO_3 + H_2O \longrightarrow H_2SO_4$$

These two are steps in the commercial manufacture of sulfuric acid.

### Decomposition Reactions

These reactions are the opposite of those in the previous category. Here a compound splits into two (or sometimes more) pieces. They can be represented as

$$C \longrightarrow A + B$$

These are less common than the former, and in most cases some outside agent, such as heat, light, or a catalyst is required to make them happen, though some

| TABLE 4.1   Major Types of Reactions | |
|---|---|
| Combination reactions | $A + B \longrightarrow C$ |
| Decomposition reactions | $C \longrightarrow A + B$ |
| Single displacement reactions | $A + BC \longrightarrow B + AC$ |
| Double displacement reactions | $AB + CD \longrightarrow AC + BD$ |

compounds are inherently unstable and will decompose without an outside agent. Some examples are:

*The conversion of limestone ($CaCO_3$) to quicklime (CaO) to make cement has been carried out for hundreds of years.*

$$2HgO \xrightarrow{\text{heat}} 2Hg + O_2$$

$$CaCO_3 \xrightarrow{\text{heat}} CaO + CO_2$$

Even water can be decomposed, though in this case electricity is needed:

$$2H_2O \xrightarrow{\text{electrolysis}} 2H_2 + O_2$$

Note that this is the reverse of the combination reaction referred to above.

### Single Displacement Reactions

These reactions were discussed at the beginning of this section.

### Double Displacement Reactions

In this very common category, also called *exchange* or *metathesis reactions,* two compounds "exchange partners." The general equation is

$$AB + CD \longrightarrow AC + BD$$

Some examples are

$$HCl + NaOH \longrightarrow NaCl + H_2O$$
$$Ba(NO_3)_2 + K_2SO_4 \longrightarrow BaSO_4 + 2KNO_3$$
$$3K_2S + 2FeCl_3 \longrightarrow Fe_2S_3 + 6KCl$$

We will see in the next section that it is possible to write simpler equations for many of these reactions, when they take place in water solution.

The four types of reactions are summarized in Table 4.1.

*An aqueous solution is a water solution.*

## 4.8   REACTIONS BETWEEN IONS IN AQUEOUS SOLUTION

Many ionic compounds are soluble in water. As we saw in Section 3.3, ionic compounds always consist of both positive and negative ions. When they dissolve in water, positive and negative ions are separated from each other by the water molecules (Sec. 6.6). We can say that a **dissociation** has taken place; for example,

*The (s) stands for solid and the (aq) for aqueous solution.*

$$NaCl(s) \xrightarrow{H_2O} Na^+(aq) + Cl^-(aq)$$

What happens when we mix two different ionic solutions? Does a reaction take place between the ions? The answer depends on what the ions are. If negative and positive ions can come together to form a compound, then a reaction takes place; otherwise it does not. For example, if we mix a solution of NaCl with one of silver nitrate, $AgNO_3$, four ions are present in the solution: $Ag^+$, $Na^+$, $Cl^-$, and $NO_3^-$. It so happens that two of these, $Ag^+$ and $Cl^-$, form a compound, AgCl (silver chloride), that is insoluble in water. A reaction therefore takes place,

$$Ag^+(aq) + Cl^-(aq) \longrightarrow AgCl(s)$$

and a white precipitate forms that slowly sinks to the bottom of the container. What about the $Na^+$ and $NO_3^-$ ions? They do nothing at all and merely remain dissolved in the water. Ions like these, which do not participate in a reaction, are called **spectator ions,** certainly an appropriate name.

In general, ions in solution react with each other only when one of these things can happen:

**1.**   Two of them form a solid that is insoluble in water (AgCl is one example; Fig. 4.5).

**2.**   Two of them form a gas. In this case the reaction takes place because the gas is not very soluble in water. An example (Fig. 4.6) is

$$HCO_3^-(aq) + H^+(aq) \longrightarrow CO_2(g) + H_2O(l)$$

Bicarbonate
ion

The (g) stands for gas and the (l) for liquid.

**3.**   An acid neutralizes a base. Acid-base reactions are so important that we devote a whole chapter to them (Chapter 8).

**4.**   One of the ions can oxidize another. This type of reaction is also important enough to be discussed separately (Sec. 4.9).

**Figure 4.5**
Adding $Cl^-$ ions to a solution of $Ag^+$ ions produces a white precipitate of silver chloride. (Photograph by Charles Steele.)

(a)

(b)

**Figure 4.6**
(a) The left beaker contains a solution of $HCO_3^-$ ions; the right beaker a solution of $H^+$ ions. (b) When the beakers are mixed the reaction between $HCO_3^-$ ions and $H^+$ ions produces the gas $CO_2$, which can be seen as bubbles. (Photographs by Beverly March.)

In many cases no reaction takes place when we mix solutions of ions because none of these situations holds. For example, if we mix a solution of copper nitrate, $Cu(NO_3)_2$, and potassium sulfate, $K_2SO_4$, we merely have a mixture containing $Cu^{2+}$, $K^+$, $NO_3^-$, and $SO_4^{2-}$ ions dissolved in water. None of these react with each other, and we will see nothing happening (Fig. 4.7).

The equations we write for ions in solution are called ionic equations. Like all other chemical equations, they must be balanced. We balance them the same way as any other equation, except that we must make sure that charges balance as well as atoms. For example, the precipitation of arsenic sulfide takes place according to this equation:

$$2As^{3+}(aq) + 3S^{2-}(aq) \longrightarrow As_2S_3(s)$$

**Figure 4.7**
(a) The beakers contain copper nitrate (blue) and potassium sulfate (colorless). (b) When they are mixed, no reaction takes place. The blue color gets lighter because the copper nitrate is less concentrated, but nothing else happens. (Photographs by Beverly March.)

(a)

(b)

Not only are there two As and three S atoms on each side, but the total charge on the left side is the same as the total charge on the right side: They are both zero. *Ionic equations show only the ions that react.* Spectator ions are not shown.

---

**EXAMPLE**

When a solution of barium nitrate, $Ba(NO_3)_2$, is added to a solution of potassium sulfate, $K_2SO_4$, a white precipitate of barium sulfate, $BaSO_4$, forms. Write the ionic equation for this reaction.

**Answer**

Because they are ionic compounds, both barium nitrate and potassium sulfate exist in water as the dissociated ions:

$$Ba^{2+} + 2NO_3^- + 2K^+ + SO_4^{2-}$$

We are told that a precipitate of barium sulfate forms:

$$Ba^{2+} + 2NO_3^- + 2K^+ + SO_4^{2-} \longrightarrow BaSO_4(s) + 2K^+ + 2NO_3^-$$

Since $K^+$ and $NO_3^-$ appear on both sides (nothing happens to them), we cancel them (they are spectator ions) and are left with

$$Ba^{2+}(aq) + SO_4^{2-}(aq) \longrightarrow BaSO_4(s) \qquad \text{Ionic equation}$$

In Section 4.7 we treated this as a double displacement reaction and wrote the equation:

$$Ba(NO_3)_2 + K_2SO_4 \longrightarrow BaSO_4 + 2KNO_3$$

When this reaction takes place in aqueous solution, however, it is the ionic equation that more accurately describes what is happening.

---

**PROBLEM 4.13**  When a solution of copper(II) chloride, $CuCl_2$, is added to a solution of potassium sulfide, $K_2S$, a black precipitate of copper(II) sulfide, $CuS$, forms. Write the ionic equation for the reaction.

---

Of the four ways for ions to react in water, one of the most common is the formation of an insoluble compound. We can predict when this will happen if we know the solubilities of the ionic compounds. There is no simple way to remember which ionic solids are soluble in water and which are not, but there are some useful generalizations.

The solubilities of compounds are found in handbooks.

1.  All compounds containing $Na^+$, $K^+$, or $NH_4^+$ ions are soluble in water.
2.  All nitrates ($NO_3^-$) are soluble in water.
3.  Most chlorides ($Cl^-$) and sulfates ($SO_4^{2-}$) are soluble. Some important exceptions are silver chloride, $AgCl$, barium sulfate, $BaSO_4$, and lead sulfate, $PbSO_4$, which are insoluble.
4.  Most carbonates ($CO_3^{2-}$), phosphates ($PO_4^{3-}$), sulfides ($S^{2-}$), and hydroxides ($OH^-$) are insoluble in water (Fig. 4.9). Important exceptions are those of $Na^+$, $K^+$, and $NH_4^+$, as well as barium hydroxide, $Ba(OH)_2$.

Some of these insoluble compounds are shown in Figure 4.8.

(a)                          (b)                          (c)

**Figure 4.8**
Some insoluble compounds: (a) barium sulfate, (b) ferric hydroxide, (c) copper carbonate, (d) lead phosphate, (e) cobalt sulfide. (Photographs by Beverly March.)

(d)                          (e)

**Figure 4.9**
A conch shell. Sea animals of the mollusc family often use insoluble $CaCO_3$ to construct their shells. (Photograph by Beverly March.)

## 4.9 OXIDATION–REDUCTION

One of the most important, as well as one of the most common, types of reaction is oxidation-reduction. We define **oxidation** as **the loss of electrons.** Reduction is defined as **the gain of electrons.** An oxidation-reduction reaction (often called a **redox reaction**) is the transfer of electrons from one species to another. An example is the oxidation of zinc by copper ions:

$$Zn + Cu^{2+} \longrightarrow Cu + Zn^{2+}$$

If you put a piece of zinc metal into a beaker containing copper ions in aqueous solution, you will see, after a short time, three things happening (Fig. 4.10):

1. Some of the zinc metal goes into solution.
2. Copper metal deposits on the zinc metal.
3. The blue color of $Cu^{2+}$ ions gradually disappears.

The zinc metal is giving electrons to the copper ions. Since it is losing electrons, we say that the zinc is being oxidized:

$$Zn \longrightarrow Zn^{2+} + 2e^- \qquad \text{Oxidation of Zn}$$

At the same time, the $Cu^{2+}$ ions are taking the electrons from the zinc. We say that they are being reduced:

$$Cu^{2+} + 2e^- \longrightarrow Cu \qquad \text{Reduction of } Cu^{2+}$$

It is evident that oxidation and reduction are not independent reactions. A species cannot take electrons from nowhere; a species cannot give electrons to nothing. **There can be no oxidation without an accompanying reduction, and vice versa.** In this reaction, the $Cu^{2+}$ is oxidizing the Zn. Thus we call it an **oxidizing agent.** Similarly, the Zn is reducing the $Cu^{2+}$, and so we call it a **reducing agent.**

Although the definitions we have given for oxidation (loss of electrons) and reduction (gain of electrons) are easy to apply in many redox reactions, they are not so easy to apply in others. For example, another redox reaction is the combustion (burning) of methane, $CH_4$, in which the $CH_4$ is oxidized to $CO_2$ while $O_2$ is reduced to $CO_2$ and $H_2O$:

$$CH_4 + 2O_2 \longrightarrow CO_2 + 2H_2O$$

There must also be negative ions present, such as $Cl^-$ or $NO_3^-$, for example, but we do not write them in the equation because they are spectator ions.

The oxidizing agent is the species that gets reduced. The reducing agent is the species that gets oxidized.

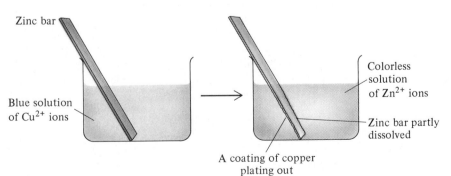

Blue solution of $Cu^{2+}$ ions

Zinc bar

A coating of copper plating out

Colorless solution of $Zn^{2+}$ ions

Zinc bar partly dissolved

**Figure 4.10**
When a piece of zinc is added to a solution containing $Cu^{2+}$ ions, the zinc is oxidized by the $Cu^{2+}$ ions.

## B o x  4B

# Voltaic Cells

In Figure 4.10 we see that when a piece of zinc metal is put in a solution containing $Cu^{2+}$ ions, the zinc atoms give electrons to the $Cu^{2+}$ ions. If we change the experiment by putting the zinc metal in one beaker and the $Cu^{2+}$ ions in another and connect the two beakers by a piece of wire and a salt bridge (Fig. 4B), *the reaction still takes place.* That is, zinc atoms still give electrons to the $Cu^{2+}$ ions, but now the electrons must flow through the wire to get from the Zn to the $Cu^{2+}$. **This flow of electrons is an electric current,** and the current keeps flowing until either the zinc or the $Cu^{2+}$ is used up. This apparatus is therefore a device that generates an electric current by making use of a redox reaction; in other words, it is a **voltaic cell** or, as we commonly say, a **battery.**

It is obvious that this particular battery would not be useful for flashlights, portable radios, and so on. The batteries that we do use for such purposes use different redox reactions, in many cases ones that do not require aqueous solutions, but the principle is the same in every case: A reducing agent transmits electrons to an oxidizing agent in an apparatus that prevents it from doing so directly. The electrons must go through an outside circuit and in doing so produce the electric current of the battery. (Electricity is the flow of electrons.) To see why a salt bridge is necessary, we must look at the $Cu^{2+}$ solution. Since we cannot have positive charges in any place without an equivalent number of negative charges, there must be negative ions in the beaker also, which may be sulfate, $SO_4^{2-}$, nitrate, $NO_3^-$, or some other ion. When electrons come over the wire, the $Cu^{2+}$ is converted to Cu:

$$Cu^{2+} + 2e^- \longrightarrow Cu$$

This diminishes the number of $Cu^{2+}$ ions, but the number of negative ions is unaffected. The salt bridge is necessary to carry some of these negative ions to the other beaker, where they are needed to balance the $Zn^{2+}$ ions being produced by the reaction

$$Zn \longrightarrow 2e^- + Zn^{2+}$$

**Figure 4B**
A voltaic cell. The electron flow, over the wire from Zn to $Cu^{2+}$, is an electric current that causes the light bulb to glow.

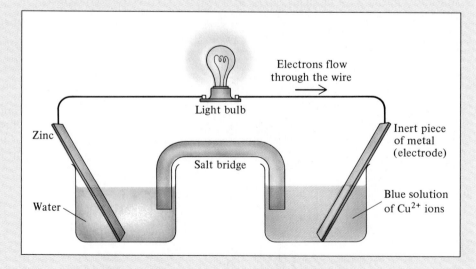

It is not easy to see the electron loss and gain in cases like this, and so chemists have another definition of oxidation and reduction, one that is easier to apply in many cases, especially where organic compounds are involved:

> Oxidation is the gain of oxygen or the loss of hydrogen.
> Reduction is the loss of oxygen or the gain of hydrogen.

In fact, this second definition is much older than the one involving electron transfer; it is the definition given by Lavoisier when he first discovered oxidation and reduction more than 200 years ago. Note that here, too, oxidation and reduction must occur together. Note also that we could not apply this definition to our zinc-copper example.

We have said that redox reactions are extremely common. Some important categories are the following:

**1. Combustion** All combustion (burning) reactions are redox reactions in which the compounds or mixtures that are burned are oxidized by the oxygen in the air. This includes the burning of gasoline, diesel oil, fuel oil, natural gas, coal, wood, and paper. All these materials contain carbon and most of them hydrogen also. If the combustion is complete, the carbon is oxidized to carbon dioxide and the hydrogen to water. In an incomplete combustion, these elements are oxidized to other compounds, many of which cause air pollution. It is unfortunate that much of the combustion that takes place in gasoline engines and in furnaces is incomplete and so contributes to air pollution.

An important product of incomplete combustion is carbon monoxide, CO, a significant air pollutant.

**Box 4C**

**Oxidizing Antiseptics**

An **antiseptic** is a compound that kills bacteria. Antiseptics are used to treat wounds—not to heal them any faster but to prevent them from becoming infected by bacteria. Some antiseptics operate by oxidizing (hence destroying) compounds essential to the normal functioning of the bacteria. One example is iodine, $I_2$, which was used as a household antiseptic for minor cuts and bruises for many years, not in the pure form but as a dilute solution in alcohol, called a *tincture*. Pure $I_2$ is a steel-gray solid that gives a purple vapor when heated. The tincture is a brown liquid. Other examples of oxidizing antiseptics are dilute solutions of hydrogen peroxide, $H_2O_2$, and potassium permanganate, $KMnO_4$, both of which oxidize a large number of organic compounds. Sodium hypochlorite (a bleaching agent) was also once used as an antiseptic. It was one of the main ingredients of *Dakin's solution,* which was used to treat the wounded in World War I.

Oxidizing antiseptics are often regarded as too harsh, however. They not only kill bacteria but also harm skin and other normal tissues. For this reason, they have largely been replaced by phenolic antiseptics (Box 12E). *Disinfectants* are also used to kill bacteria, but on inanimate objects rather than on living tissues. Many disinfectants are oxidizing agents. Two important examples are chlorine, $Cl_2$, a pale green gas, and ozone, $O_3$, a colorless gas. Both of these gases are added in small quantities to municipal water supplies to kill any harmful bacteria that may be present. Both gases must be handled carefully because they are very poisonous.

**2.   Respiration**   Humans and animals get their energy by respiration. As in combustion, the oxygen in the air we breathe oxidizes carbon-containing compounds in our cells to produce $CO_2$ and $H_2O$. In fact, this oxidation reaction is really the same as combustion, though it takes place more slowly and at a much lower temperature. We discuss these reactions more fully in Chapter 20. The important product of these reactions is not the $CO_2$ (which the body gets rid of anyhow) or the $H_2O$, but the energy.

**3.   Rusting**   We all know that when iron or steel objects are left out in the open air, they eventually rust (steel is mostly iron, but contains certain other elements also). In rusting (Fig. 4.11), iron is oxidized to a mixture of oxides. We can represent the main reaction by the equation

$$4Fe + 3O_2 \longrightarrow 2Fe_2O_3$$

**4.   Bleaching**   Most bleaching is oxidation, and common bleaches are oxidizing agents. The colored compounds are usually organic compounds; oxidation converts them to colorless compounds.

**5.   Batteries**   A voltaic cell (Box 4B) is a device in which electricity is generated from a chemical reaction. Such cells are often called **batteries.** We are all familiar with them in our cars and in such portable devices as radios and flashlights (Fig. 4.12). In all cases, the reaction taking place in the battery is a redox reaction.

**Figure 4.11**
The rusting of iron and steel can be a serious problem in an industrial society. (Diane Schiumo, Fundamental Photographs, New York.)

(a)

Anode

Cathode

Negative plates: lead grills filled with spongy lead.

Positive plates: lead grills filled with $PbO_2$

(b)

**Figure 4.12**
(a) Three types of dry cell battery. (Photograph by Beverly March.) (b) A lead storage battery. (From J. C. Kotz and K. F. Purcell, *Chemistry and Chemical Reactivity,* 2nd ed. Philadelphia: Saunders College Publishing, 1991.)

## 4.10   HEAT OF REACTION

In almost all chemical reactions, not only are starting materials converted to products, but there is also a gain or loss of heat. For example, when 1 mole of pure carbon is oxidized by oxygen to give $CO_2$, 94 000 cal of heat are given off:

$$C + O_2 \longrightarrow CO_2 + 94\ 000\ cal$$

The heat lost or gained in a reaction is called the **heat of reaction** (Fig. 4.13). A reaction that gives off heat is said to be **exothermic;** a reaction that requires heat is **endothermic** (Fig. 4.14). The *amount* of heat given off or taken in is proportional to the amount of material. For example, when two moles of pure carbon are oxidized, $2 \times 94\ 000 = 188\ 000$ cal of heat are given off.

The energy changes accompanying a chemical reaction are not limited to heat. In some reactions, as in voltaic cells (Box 4B), the energy output is in the form of electricity. In other reactions, such as photosynthesis (the reaction whereby plants convert water and carbon dioxide to sugar and oxygen), the energy absorbed is in the form of light. In most cases, however, the energy change is in the form of heat. An example of an endothermic reaction is the decomposition of mercury(II) oxide:

$$2HgO + 43\ 400\ cal \longrightarrow 2Hg + O_2$$

What this equation tells us is that if we want to break down 2 moles of mercury(II) oxide into the elements Hg and $O_2$, we must add 43 400 cal of energy.

The heat given off increases the temperature of the reaction mixture.

An **exergonic** reaction is one that gives off energy in any form, not necessarily heat. An **endergonic** reaction is the opposite: It uses up energy.

**Figure 4.13**
Two reactions that give off a large quantity of heat. (a) The elements bromine, in the beaker, and aluminum, on the watch glass. (b) When the aluminum is added to the bromine, the reaction is so vigorous that the aluminum melts and glows white hot. (c) The thermite reaction. After the reaction is started with a fuse of burning magnesium wire, iron(III) oxide reacts with aluminum powder. (d) The reaction generates so much heat that the product, iron, is produced in the molten state. (Photographs by Charles D. Winters.)

(a)

(b)

(c)

(d)

**Figure 4.14**
In an endergonic reaction (a) the products have *more* energy than the starting compounds. Energy has been absorbed. In an exergonic reaction (b) the products have *less* energy than the starting compounds. Energy has been given off.

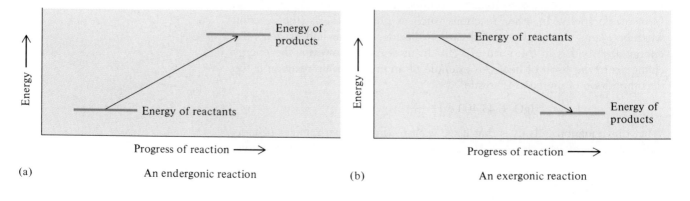

(a)      An endergonic reaction      (b)      An exergonic reaction

All of us have heard such expressions as "a teaspoon of sugar contains 16 Calories." What does this mean? We have pointed out (p. 118) that combustion is essentially the same reaction as the metabolism of food. In both cases the starting materials are converted to $CO_2$ and $H_2O$, although metabolism takes place at 37°C and by a series of reactions, whereas combustion is a high-temperature process that takes place directly. Nevertheless, **the heat of reaction depends only on the starting compounds and the final products, not on the pathway.** (This is called **Hess's Law.**) This means that if a chemist takes a given weight of any food (such as sugar) and burns it in order to measure the amount of heat given off, then the chemist knows that the same amount of heat is given off in the body when that weight of that food is metabolized.

By this simple method, the energy content of many foods has been determined (Table 4D). Of course, we all know that it works two ways. If we don't take in enough calories in our food to replace the heat we lose or use for energy, our bodies begin to "burn" some of our normal tissues and we lose weight. On the other hand, if we eat food with more than enough calories to match our losses, our bodies turn that extra energy to fatty tissue and we gain weight. That is why many people carefully count the number of Calories they eat. Remember (Sec. 1.8) that a nutritionist's Calorie is equal to 1000 chemist's calories.

**Box 4D**

# The Caloric Value of Foods

**TABLE 4D   Caloric Values of Some Foods**

| Food | Calories | Food | Calories |
|---|---|---|---|
| Milk, 8 oz | 165 | Apple, 5 oz | 70 |
| Swiss cheese, 1 oz | 105 | Banana, 5 oz | 85 |
| Egg, boiled, large | 80 | Grapefruit, $\frac{1}{2}$ medium | 50 |
| Hamburger patty, 3 oz | 245 | Orange juice, 8 oz | 100 |
| Beef, lean, 2 oz | 115 | Bread, rye, 1 slice | 55 |
| Lamb, shoulder, lean, 2 oz | 125 | Bread, white, 1 slice | 60 |
| Haddock, fried, 3 oz | 135 | Fudge, 1 oz | 115 |
| Shrimp, 3 oz | 110 | Honey, 1 tablespoon | 60 |
| Peanuts, roasted, 2 oz | 210 | Cookie, 3-in | 110 |
| Carrot, raw, 8 oz | 45 | Cornflakes, 1 oz | 110 |
| Corn, 5-in ear | 65 | Noodles, 8 oz | 200 |
| Potato, 5 oz | 90 | Butter, 1 tablespoon | 100 |
| Tomato, 5 oz | 30 | Margarine, 1 tablespoon | 100 |

Incidentally, the law of conservation of energy tells us that the reverse reaction, the oxidation of mercury, must give off exactly the same amount of heat:

$$2Hg + O_2 \longrightarrow 2HgO + 43\ 400\ cal$$

Especially important are heats of reaction for combustion reactions. As we saw in the previous section, combustion reactions are the most important heat-producing reactions, since most of our energy is derived from them. All combustions are exothermic. The heat given off in a combustion reaction is called the **heat of combustion.** Accurate heat of combustion values are available for thousands of compounds.

## SUMMARY

A **chemical equation** tells which starting materials are converted to which products. All equations must be balanced.

The **formula weight** of a compound is the sum of the atomic weights of all atoms in the molecular formula. A **mole** of a substance is defined as **Avogadro's number** $(6.02 \times 10^{23})$ of particles. The weight of this number of molecules of any compound is numerically equal to the formula weight expressed in grams.

A **balanced equation** tells how many moles of each starting material are converted to how many moles of each product. By the use of balanced equations we can calculate the required number of grams (or moles) of any starting material or product if we know the number of grams (or moles) of any other starting material or product (weight-weight calculations). **Percentage yield** equals actual yield divided by **theoretical yield** (calculated by a weight-weight calculation) multiplied by 100.

Most reactions can be divided into four general types: **combination reactions, decomposition reactions, single displacement reactions,** and **double displacement reactions.** When ions are mixed in water solution, they react with each other only if (a) a precipitate is formed, (b) a gas is formed, (c) an acid neutralizes a base, or (d) an oxidation-reduction takes place. Otherwise, no reaction takes place. Ions that do not react are called **spectator ions.** All compounds containing $Na^+$, $K^+$, $NH_4^+$, or $NO_3^-$ ions are soluble in water, as are most compounds containing $Cl^-$ or $SO_4^{2-}$ ions. Most compounds containing $CO_3^{2-}$, $PO_4^{3-}$, $S^{2-}$, or $OH^-$ are insoluble. **Ionic equations** show only the ions that react. In an ionic equation the charges as well as the atoms must be balanced.

In an **oxidation,** electrons are lost; in a **reduction,** electrons are gained. These two processes must take place together. The joint process is often called a **redox** reaction. Oxidation can also be defined as the gain of oxygen or loss of hydrogen; reduction is the reverse. Important examples of oxidation are combustion, respiration, bleaching, and rusting. Electric batteries use redox reactions to generate electricity.

Almost all reactions are accompanied by a loss or gain of heat. This is called the **heat of reaction.** Reactions that give off heat are **exothermic;** those that absorb heat are **endothermic.** The heat of a combustion reaction is called the **heat of combustion.**

## KEY TERMS

Antiseptic (Box 4C)
Avogadro's number (Sec. 4.3)
Chemical equation (Sec. 4.4)
Combination reaction (Sec. 4.7)
Decomposition reaction (Sec. 4.7)
Disinfectant (Box 4C)
Dissociation (Sec. 4.8)
Double displacement reaction (Sec. 4.7)
Endergonic (Sec. 4.10)
Endothermic (Sec. 4.10)
Exergonic (Sec. 4.10)
Exothermic (Sec. 4.10)
Formula weight (Sec. 4.2)
Heat of combustion (Sec. 4.10)
Heat of reaction (Sec. 4.10)
Hess's law (Box 4D)
Ionic equation (Sec. 4.8)

Mole (Sec. 4.3)
Molecular weight (Sec. 4.2)
Oxidation (Sec. 4.9)
Oxidizing agent (Sec. 4.9)
Percentage yield (Sec. 4.6; Box 4A)
Products (Sec. 4.1)
Reactants (Sec. 4.1)
Redox (Sec. 4.9)
Reducing agent (Sec. 4.9)
Reduction (Sec. 4.9)
Single displacement reaction (Sec. 4.7)
Spectator ion (Sec. 4.8)
Starting materials (Sec. 4.1)
Stoichiometry (Sec. 4.5)
Theoretical yield (Sec. 4.6)
Voltaic cell (Box 4B)
Yield (Sec. 4.6)

## PROBLEMS

### Formula Weight and Moles

**4.14** Calculate the formula weight of
(a) $Br_2$                (d) $P_4$
(b) $Ne$                 (e) $S_8$
(c) $I_2$

**4.15** Calculate the formula weight of
(a) $KCl$
(b) $Na_3PO_4$
(c) $Fe(OH)_2$
(d) sucrose, $C_{12}H_{22}O_{11}$
(e) $NaAl(SO_3)_2$
(f) glycine, $C_2H_7O_2N$
(g) $Al_2(SO_4)_3$
(h) $(NH_4)_2CO_3$
(i) DDT, $CCl_3CH(C_6H_4Cl)_2$

**4.16** How many
(a) moles of $CO_2$ are there in 83.2 g of $CO_2$?
(b) moles of glycerol, $C_3H_8O_3$, are there in 428 g of glycerol?
(c) moles of $NaH_2PO_4$ are there in 14 g of $NaH_2PO_4$?
(d) moles of quinine, $C_{20}H_{24}N_2O_2$, are there in 51.6 g of quinine?

**4.17** How many
(a) grams of nitrogen dioxide, $NO_2$, are there in 1.77 moles of $NO_2$?
(b) grams of 2-propanol, $C_3H_8O$ (rubbing alcohol), are there in 0.84 mole of 2-propanol?
(c) grams of $WCl_5$ are there in 3.69 moles of $WCl_5$?
(d) grams of galactose, $C_6H_{12}O_6$, are there in 0.348 mole of galactose?
*(e) grams of vitamin C, $C_6H_8O_6$, are there in $4.9 \times 10^{-2}$ mole of vitamin C?

**4.18** How many
(a) moles of O atoms are there in 18.1 moles of formaldehyde, $H_2CO$?
(b) moles of Br atoms are there in 0.41 mole of bromoform, $CHBr_3$?
*(c) moles of O atoms are there in $3.5 \times 10^3$ moles of $Al_2(SO_4)_3$?
(d) moles of Hg atoms are there in 87 g of HgO?
*(e) moles of N atoms are there in $2.1 \times 10^{-3}$ g of $K_3Fe(CN)_6$?

**4.19** How many
(a) moles of $Ag^+$ ions are there in 3.71 moles of $AgNO_3$?
(b) moles of $Na^+$ ions are there in 144 g of $Na_2CO_3$?
*(c) moles of $SO_4^{2-}$ ions are there in 84 g of $Al_2(SO_4)_3$?

**4.20** A single atom of cerium weighs just about twice as much as a single atom of gallium. What is the weight ratio of 25 atoms of cerium to 25 atoms of gallium?

**4.21** What is the mass in grams of (a) 20   (b) 300 (c) 100 000   (d) $1 \times 10^{18}$ molecules of acetic acid, $C_2H_4O_2$?

**4.22** What is the mass in grams of a single molecule of sucrose, $C_{12}H_{22}O_{11}$?

**4.23** How many
(a) molecules of TNT, $C_7H_5N_3O_6$, are there in 2.9 moles of TNT?
(b) molecules are in one drop (0.0500 g) of water?

**4.24** A typical cholesterol deposit in an artery might have a mass of 3.9 mg of cholesterol ($C_{27}H_{46}O$). How many molecules of cholesterol are in this mass?

**4.25** Protein molecules are very large. The molecular weight of hemoglobin is about 68 000. What is the mass in grams of a single molecule of hemoglobin?

### Balancing Equations

**4.26** Balance:
(a) $Cl_2 + H_2 \longrightarrow HCl$
(b) $H_2 + O_2 \longrightarrow H_2O$
(c) $K + Br_2 \longrightarrow KBr$
(d) $MgCO_3 \longrightarrow CO_2 + MgO$

**4.27** Balance:
(a) $HI + NaOH \longrightarrow NaI + H_2O$
(b) $Ba(NO_3)_2 + H_2S \longrightarrow BaS + HNO_3$
(c) $CH_4 + O_2 \longrightarrow CO_2 + H_2O$
(d) $C_4H_{10} + O_2 \longrightarrow CO_2 + H_2O$
(e) $Fe + CO_2 \longrightarrow Fe_2O_3 + CO$
(f) $Al + HBr \longrightarrow AlBr_3 + H_2$
(g) $P_4 + O_2 \longrightarrow P_4O_{10}$

## Weight Relationships in Chemical Reactions

**4.28** Define:
(a) stoichiometry
(b) aqueous solution

**4.29** For the equation

$$2N_2 + 3O_2 \longrightarrow 2N_2O_3$$

(a) how many moles of $N_2$ are required to react completely with 1 mole of $O_2$?
(b) how many moles of $N_2O_3$ are produced from the complete reaction of 1 mole of $O_2$?
(c) how many moles of $O_2$ are required to produce 8 moles of $N_2O_3$?

**4.30** For the equation

$$2Al + 6HCl \longrightarrow 2AlCl_3 + 3H_2$$

how many moles of $H_2$ are produced by the complete reaction of 132 g of Al?

**4.31** For the equation

$$CH_4 + 3Cl_2 \longrightarrow CHCl_3 + 3HCl$$
Chloroform

how many grams of $Cl_2$ are required to produce 6.74 moles of chloroform?

**4.32** Acetaldehyde, $C_2H_4O$, is produced commercially from acetylene by the reaction

$$C_2H_2 + H_2O \longrightarrow C_2H_4O$$

How many grams of $C_2H_4O$ can be produced from 81.7 g of $C_2H_2$, assuming enough water is present?

**4.33** Ethyl alcohol, $C_2H_6O$, can be added to gasoline to produce "gasohol," a fuel for automobile engines. The equation for the combustion of ethyl alcohol is

$$C_2H_6O + 3O_2 \longrightarrow 2CO_2 + 3H_2O$$

How many grams of $O_2$ are required for the combustion of 421 g of $C_2H_6O$?

**4.34** In the process called photosynthesis, plants convert $CO_2$ and $H_2O$ to glucose, $C_6H_{12}O_6$, by the equation

$$6CO_2 + 6H_2O \longrightarrow C_6H_{12}O_6 + 6O_2$$

How many grams of $CO_2$ are required to produce 5.1 g of glucose?

**4.35** Iron ore is converted to iron by a process that can be represented as

$$2Fe_2O_3 + 6C + 3O_2 \longrightarrow 4Fe + 6CO_2$$

If the process is run until 3940 g of Fe is produced, how many grams of $CO_2$ will also be produced?

**4.36** Given the reaction in Problem 4.35, how much C (in grams) is necessary to react completely with 0.58 g of $Fe_2O_3$?

**4.37** Aspirin can be made in the laboratory by the reaction

$$C_7H_6O_3 + C_2H_3ClO \longrightarrow C_9H_8O_4 + HCl$$
Salicylic    Acetyl              Aspirin
acid         chloride

How many grams of aspirin are produced if 85.0 g of salicylic acid is treated with excess acetyl chloride?

**4.38** The compound aniline reacts with chlorine as follows:

$$C_6H_7N + 3Cl_2 \longrightarrow C_6H_4Cl_3N + 3HCl$$
Aniline                    Trichloroaniline

How many grams of trichloroaniline are produced if 71.1 g of aniline is treated with excess $Cl_2$?

**\*4.39** If 29.7 g of $N_2$ is added to 3.31 g of $H_2$ in the reaction

$$N_2 + 3H_2 \longrightarrow 2NH_3$$

(a) which reactant is completely used up?
(b) how many grams of the other reactant are left over?
(c) how many grams of $NH_3$ are formed if the reaction goes to completion?

## Percentage Yield

**4.40** A chemist calculated the theoretical yield in a particular reaction to be 60.4 g. The mass of the product actually obtained was 51.7 g. What is the percentage yield?

**\*4.41** Diethyl ether can be made by the reaction

$$2C_2H_5OH \longrightarrow (C_2H_5)_2O + H_2O$$

If a chemist using 517 g of ethanol, $C_2H_5OH$, produced 391 g of ethyl ether, $(C_2H_5)O$. What is the percentage yield?

## Basic Types of Reactions

**4.42** Classify each of these as a combination, a decomposition, a single displacement, or a double displacement reaction:

(a) $Ca + 2H_2O \longrightarrow Ca(OH)_2 + H_2$

(b) $HNO_2 + LiOH \longrightarrow LiNO_2 + H_2O$

(c) $2P + 5S \longrightarrow P_2S_5$

(d) $2KClO_3 \xrightarrow[\text{heat}]{\text{catalyst}} 2KCl + 3O_2$

(e) $C_2H_2 + H_2O \xrightarrow{\text{catalyst}} C_2H_4O$

(f) $3BaCl_2 + 2Na_3PO_4 \longrightarrow$
$$Ba_3(PO_4)_2 + 6NaCl$$

(g) $2Cr + 3HgSO_4 \longrightarrow Cr_2(SO_4)_3 + 3Hg$

**4.43** The explosion of nitroglycerine takes place according to the following equation:

$$4C_3H_5N_3O_9 \longrightarrow 12CO_2 + 6N_2 + 10H_2O + O_2$$

Which class does this reaction belong to?

**4.44** Complete and balance each of these combination equations:

(a) $K + Br_2 \longrightarrow$

(b) $Al + O_2 \longrightarrow$

(c) $H_2 + Cl_2 \longrightarrow$

**4.45** Complete and balance each of these single displacement equations:

(a) $K + H_2O \longrightarrow$

(b) $Sr + AgBr \longrightarrow$

(c) $Al + Fe(NO_3)_3 \longrightarrow$

**4.46** Complete and balance each of these double displacement equations:

(a) $HF + Ba(OH)_2 \longrightarrow$

(b) $Na_2CO_3 + Pb(NO_3)_2 \longrightarrow$

(c) $NaOH + FeCl_2 \longrightarrow$

## Reactions Between Ions in Aqueous Solution

**4.47** In the equation

$$K^+ + SO_4^{2-} + Ba^{2+} + Cl^- \longrightarrow$$
$$K^+ + BaSO_4(s) + Cl^-$$

(a) which are the spectator ions?   (b) Write the balanced ionic equation.

**4.48** Balance:

(a) $Ag^+ + Br^- \longrightarrow AgBr$

(b) $Cd^{2+} + S^{2-} \longrightarrow CdS$

(c) $Sc^{3+} + SO_4^{2-} \longrightarrow Sc_2(SO_4)_3$

*(d) $Sn^{2+} + Fe^{2+} \longrightarrow Fe^{3+} + Sn$

(e) $K + H_2O \longrightarrow K^+ + OH^- + H_2$

**4.49** When a solution of ammonium chloride, $NH_4Cl$, is added to a solution of lead nitrate, $Pb(NO_3)_2$, a white precipitate, $PbCl_2$, is formed. Write a balanced ionic equation for this reaction. Both ammonium chloride and lead nitrate exist as dissociated ions in aqueous solution.

**4.50** When a solution of ammonium sulfide, $(NH_4)_2S$, is added to a solution of lead nitrate, $Pb(NO_3)_2$, a black precipitate of lead sulfide, $PbS$, is formed. Write a balanced ionic equation for this reaction. Both $(NH_4)_2S$ and $Pb(NO_3)_2$ exist as dissociated ions in aqueous solution.

**4.51** When a solution of hydrochloric acid, $HCl$, is added to a solution of sodium sulfite, $Na_2SO_3$, sulfur dioxide gas, $SO_2$, is released from the solution. Write an ionic equation for this reaction. An aqueous solution of $HCl$ contains $H^+$ and $Cl^-$ ions, and $Na_2SO_3$ exists as dissociated ions in aqueous solution.

**4.52** When a solution of sodium hydroxide, $NaOH$, is added to a solution of ammonium carbonate, $(NH_4)_2CO_3$, ammonia gas, $NH_3$, is released when the solution is heated. Write an ionic equation for this reaction. Both $NaOH$ and $(NH_4)_2CO_3$ exist as dissociated ions in aqueous solution.

**4.53** Using the solubility generalizations given in Section 4.8, predict whether each of the following ionic compounds is soluble in water.

(a) $KCl$         (d) $Na_2SO_4$

(b) $NaOH$         (e) $Na_2CO_3$

(c) $BaSO_4$         (f) $Fe(OH)_2$

**4.54** Using the solubility generalizations given in Section 4.8, predict whether each of the following ionic compounds is soluble in water.

(a) $CaCO_3$         (d) $AgCl$

(b) $Mg_3(PO_4)_2$         (e) $Fe(NO_3)_3$

(c) $(NH_4)_2CO_3$         (f) $PbS$

**4.55** Predict whether or not a solid precipitate will come out of solution when aqueous solutions of the following compounds are mixed. If a precipitate will form, write its formula.

(a) $CaCl_2 + K_3PO_4$

(b) $KCl + Na_2SO_4$

(c) $(NH_4)_2CO_3 + Ba(NO_3)_2$

(d) $FeCl_2 + KOH$

(e) $Ba(NO_3)_2 + NaOH$

(f) $Na_2S + SbCl_3$

(g) $Pb(NO_3)_2 + K_2SO_4$

## Oxidation-Reduction

**4.56** Give two definitions of oxidation and two definitions of reduction.

**4.57** Can a reduction take place without an oxidation? Explain.

**4.58** In the reaction

$$Pb + 2Ag^+ \longrightarrow 2Ag + Pb^{2+}$$

(a) which species gets oxidized and which gets reduced?
(b) which is the oxidizing agent and which the reducing agent?

**4.59** In the reaction

$$C_7H_{12} + 10O_2 \longrightarrow 7CO_2 + 6H_2O$$

(a) which species gets oxidized and which reduced?
(b) which is the oxidizing agent and which the reducing agent?

## Heat of Reaction

**4.60** Define: (a) exothermic (b) endothermic.

**4.61** Which of these reactions are endothermic and which are exothermic?
(a) $2NH_3 \longrightarrow N_2 + 3H_2 - 22\ 000$ cal
(b) $H_2 + F_2 \longrightarrow 2HF + 124\ 000$ cal
(c) $C + O_2 \longrightarrow CO_2 + 94\ 000$ cal
(d) $H_2 + CO_2 \longrightarrow H_2O + CO - 9800$ cal
(e) $C_3H_8 + 5O_2 \longrightarrow$
$$3CO_2 + 4H_2O + 531\ 000 \text{ cal}$$

**4.62** The reaction

$$2SO_2 + O_2 \longrightarrow 2SO_3 + 23\ 800 \text{ cal}$$

gives off 23 800 cal. What is the heat of reaction of

$$2SO_3 \longrightarrow 2SO_2 + O_2$$

**4.63** The equation for the combustion of acetone is

$$2C_3H_6O + 8O_2 \longrightarrow$$
Acetone $\qquad 6H_2O + 6CO_2 + 853\ 600$ cal

How much heat will be given off if 0.37 mole of acetone is completely burned?

**\*4.64** To convert one mole of iron(III) oxide to its elements requires 196 500 cal:

$$Fe_2O_3 + 196\ 500 \text{ cal} \longrightarrow 2Fe + 3O_2$$

How many grams of iron can be produced if 156 000 calories of heat are absorbed by a large-enough sample of iron(III) oxide?

## Boxes

**4.65** (Box 4A) Why is it important to get as high a percentage yield as possible in an industrial process?

**4.66** (Box 4B) The equation

$$Fe + Zn^{2+} \longrightarrow Zn + Fe^{2+}$$

represents a redox reaction that can be used to make a voltaic cell. Draw a voltaic cell that uses this reaction, showing the direction of electron flow.

**4.67** (Box 4C) Hydrogen peroxide is not only an antiseptic but also an oxidizing agent. A typical equation is

$$4H_2O_2 + CH_4 \longrightarrow CO_2 + 6H_2O$$

In this reaction, what is oxidized and what is reduced? To what is it being reduced?

**4.68** (Box 4D) The caloric value of an apple is given in Table 4D. If this quantity of heat were added to 1.00 kg of water at 25°C, what would be the final temperature of the water? (Sec. 1.8.)

**4.69** For a person on a reducing diet, which breakfast would furnish a lower number of calories: (a) 8 oz of milk, 1 oz of cornflakes, and 5 oz of banana slices, or (b) 8 oz of orange juice, two slices of white bread spread with $\frac{1}{2}$ tablespoon of butter and 1 tablespoon of honey?

## Additional Problems

**4.70** Define: (a) exergonic (b) endergonic.

**4.71** The active ingredient in one analgesic pill is 488 mg of aspirin, $C_9H_8O_4$. How many moles of aspirin does the pill contain? How many molecules?

**\*4.72** Chlorophyll, the compound responsible for the green color of leaves and grass, contains one atom of magnesium in each molecule. If the percentage by weight of magnesium in chlorophyll is 2.72%, what is the molecular weight of chlorophyll?

**4.73** Explain this statement: Our modern civilization depends on the heat obtained from chemical reactions.

**\*4.74** If 7 g of $N_2$ is added to 11 g of $H_2$ in the Haber reaction,

$$N_2 + 3H_2 \longrightarrow 2NH_3$$

which reactant is in excess?

**4.75** Which has more mass, a mole of tin (Sn) or a mole of platinum (Pt)?

**4.76** Ammonium sulfide, $(NH_4)_2S$, and arsenic nitrate, $As(NO_3)_3$, are both ionic compounds and dissociate in water to give the corresponding ions. When aqueous solutions of these compounds are mixed, arsenic sulfide, $As_2S_3$, precipitates. Write the ionic equation and identify the spectator ions.

**\*4.77** A test tube contains some calcium oxide, CaO. The total weight of the test tube and the CaO is 10.860 g. The CaO reacts with water vapor in the air to give calcium hydroxide, by this equation:

$$CaO + H_2O \longrightarrow Ca(OH)_2$$

After completely reacting with the water vapor, the test tube and its contents weigh 11.149 g. What is the weight of the empty test tube?

**4.78** Define spectator ion.

**\*4.79** In the reaction

$$3Pb(NO_3)_2 + 2AlCl_3 \longrightarrow 3PbCl_2 + 2Al(NO_3)_3$$

8.00 grams of lead nitrate interact with 2.67 grams of aluminum chloride. The reaction yields 5.55 g of lead chloride. (a) Which reagent is in excess? (b) What is the percentage yield?

**4.80** In the equation

$$Fe_2O_3 + 3CO \longrightarrow 2Fe + 3CO_2$$

(a) how many moles of $Fe_2O_3$ are required to produce 38.4 moles of Fe?

(b) how many grams of CO are required to produce 38.4 moles of Fe?

**\*4.81** The 1992 Clean Air Act mandates that an additive such as methyl *tert*-butyl ether, $CH_3-O-C(CH_3)_3$, be added to gasoline. (a) How many moles of $CO_2$ are produced in the complete combustion of 100 g of methyl *tert*-butyl ether? (b) To the extent that this ether is used in gasoline, it replaces, weight-for-weight, the alkanes normally present. A typical one is heptane, $C_7H_{16}$. How many moles of $CO_2$ are produced in the complete combustion of 100 g of heptane?

**\*4.82** Assume that the average red blood cell has a mass of $2 \times 10^{-8}$ g and that 20 percent of its content is hemoglobin (a protein whose molecular weight is 68 000). How many molecules of hemoglobin are there in one red blood cell?

*One of the first hot-air balloon ascents, flown by the Montgolfier brothers at Versailles, France, in 1783, in the presence of the King, Queen, and thousands of spectators. (Musée Carnavalet, Paris.)*

# Gases, Liquids, and Solids

## 5.1 THE ORGANIZATION OF MATTER

There are various forces that hold matter together and cause it to take different forms. In the atomic nucleus, very strong forces keep the protons and neutrons together (Chapter 9). In the atom itself, there are attractions between the positive nucleus and the negative electrons. Within molecules, atoms are attracted to each other by covalent bonds, the arrangement of which causes the molecules to assume a particular shape. Within ionic crystals, three-dimensional shapes are brought about by electrostatic attractions between ions.

In addition to all these forces, there are also attractive forces *between* molecules. These forces are weaker than any already mentioned, but they nevertheless help to determine whether a particular compound or element is a solid, a liquid, or a gas at room temperature. Because these are **forces between one molecule and another** (rather than within a molecule), they are called **intermolecular forces.**

All these attractive forces provide the potential energy that holds matter together and keeps it from falling apart, by counteracting another form of energy, kinetic energy, which tends to disorganize matter. In the absence of attractive forces, the kinetic energy that particles possess keeps them constantly moving, mostly in a random, disorganized way. This kinetic energy increases with increasing temperature. Therefore, **the higher the temperature, the greater the tendency of particles to fly randomly in every direction.**

The physical state of matter thus depends on a balance between the kinetic energy of particles and the attractive forces between them. At high temperatures, molecules possess a large amount of kinetic energy and move so fast that intermolecular forces are too weak to hold them together. This is the **gaseous state.** At lower temperatures, the molecules slow down. When the temperature is sufficiently reduced, the gas condenses to form the **liquid state.** The molecules are still moving past each other, but much more slowly. The attractive forces now become important. When the temperature goes even lower, the molecules no longer have enough velocity to move past each other. This is the **solid state** (Fig. 5.1). In this state each molecule has a certain number of nearest neighbors, and these neighbors do not change.

The attractive forces between molecules in the solid state are the same as in the gaseous and liquid states. The difference is that in the gaseous (and to a lesser degree in the liquid) state the kinetic energy of the molecules is great enough to overcome the attractive forces.

Most substances can exist in any of the three states. Typically a solid, when heated to a sufficiently high temperature, melts and becomes a liquid. This temperature, called the **melting point,** is different for each solid. Further heating causes the temperature to rise to the point at which the liquid boils and

**Figure 5.1**
Molecules in the three states of matter.

Gas

Molecules far apart and disordered
Negligible interactions between
molecules

Liquid

Intermediate situation

Solid

Molecules close together and
ordered
Strong interactions between
molecules

becomes a gas. However, not all substances can exist in all three states. For example, wood and paper cannot be melted. On heating, they decompose or burn (depending on whether or not air is present) but they do not melt. Another example is sugar, which does not melt when heated, but forms a dark substance called caramel.

## 5.2 GASES

Of the three states of matter, the gaseous state is the one most studied and best understood. The most important characteristic of a gas is that it completely fills whatever container it is in. In doing so, the gas exerts a pressure on the walls of the container. This behavior, and all other properties of gases, can be explained by the **kinetic molecular theory,** which assumes that

*The kinetic molecular theory explains the behavior of ideal gases.*

**1.** Gases consist of molecules that are constantly moving through space in straight lines, randomly, and with varying speeds.
**2.** The average kinetic energy of the gas molecules is proportional to the absolute (Kelvin) temperature. This means that as the temperature increases the gas molecules move faster.
**3.** The gas molecules collide with each other, as billiard balls do, bouncing off each other and changing direction. Each time they collide they also may exchange kinetic energies (one moves faster, the other slower than before), but the total kinetic energy is conserved (assuming that they do not react with each other).
**4.** The molecules of a gas are very tiny compared to the distances between the molecules, so tiny that the kinetic molecular theory assumes that the molecules have no volume at all. The **volume of a gas** is thus the **volume of the container** it is in.

---

**Box 5A**

## The Uses of Atmospheric Gases

Four gases, three of them major components of our atmosphere, serve important roles in many industrial and manufacturing processes. Nitrogen ($N_2$), the main constituent of air, is separated from the other gases at low temperatures. Nitrogen is used in oil fields to maintain pressure in the wells and in recovery of oil from old oil fields. Liquefied nitrogen, which maintains a temperature of $-196°C$, is used in refrigeration and to provide a neutral blanketing atmosphere in metal production.

Oxygen ($O_2$) is also separated from air at low temperatures. It is used primarily as an oxidizing agent in metal manufacturing, mainly in the steel industry. It is also important in medicine, where it is administered for heart attack treatment, resuscitations, and other situations.

Carbon dioxide ($CO_2$) is obtained mostly as a by-product of various industrial processes. The fermentation industry that produces ethanol is one source; ammonia-producing plants are another. The solid form of $CO_2$ (known by the name Dry Ice) is used in food freezing and other refrigeration processes. It is also used to manufacture carbonated beverages.

The fourth gas, hydrogen ($H_2$), is not an important component of the atmosphere. It is produced by the reaction of natural gas with steam. Its use is largely in chemical industries, in the manufacture of plastics and pesticides. In Chapters 11 and 17 we shall see its reactions with alkenes. Another use is in metal processing where, mixed with $N_2$, it forms a blanketing atmosphere.

**5.** The molecules do not stick together after collision because there are no attractions between them.

**6.** The molecules collide with the walls of the container. The collisions with the walls constitute the pressure of the gas. The greater the number of collisions per unit time, the greater the pressure.

These six assumptions of the kinetic molecular theory obviously give an idealized picture of what is going on. In real gases there are intermolecular attractions and the molecules really do occupy some volume. Because of this, a gas described by these six assumptions is called an **ideal gas.** There cannot actually be an ideal gas; all gases are real. However, under normal environmental conditions, the behavior of real gases does approximate that of an ideal gas, and we can safely use these concepts.

## 5.3   PRESSURE

The barometer was invented in 1643 by Evangelista Torricelli (1608–1647).

On the earth we live under a blanket of air that presses down on us, on the ground, and on everything around us. As we all know from weather reports, the pressure of the atmosphere is not constant, but varies from day to day. A device called a **barometer** is used to measure atmospheric pressure. In this instrument, as shown in Figure 5.2, there is a vacuum inside the tube, and the atmosphere presses down on the liquid in the dish. Since the vacuum has zero pressure, there is no pressure on the mercury inside the tube. On the other hand, there is a pressure (the pressure of the air) on the mercury in the dish.

**Figure 5.2**
The mercury barometer. A long glass tube completely filled with mercury is inverted into a pool of mercury in a dish. There is no air at the top of the mercury column in the tube because there is no way air could get in. Thus there is no gas pressure exerted on the mercury column. However, on the open dish of mercury the whole atmosphere exerts its pressure. The difference in the heights of the two mercury levels is the atmospheric pressure.

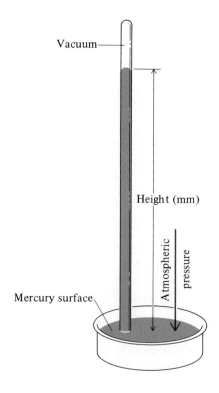

Vacuum

Height (mm)

Atmospheric pressure

Mercury surface

**Box 5B**

## Mercury as a Poison

Mercury is used in most manometers and barometers because it is a liquid over a wide temperature range and so can be used for pressure measurements between −30 and +300°C. However, mercury is poisonous. Mercury spills should be collected and stored under water. Frequent and prolonged contact with mercury should be avoided (use gloves). Prolonged inhalation of mercury vapor can cause teeth to fall out; later, nervous disorders develop.

Industrial mercury poisoning was prevalent among people who made felt hats, beginning in the middle of the seventeenth century. Felt hats are made of the fine hair of rabbits and hares. The workers treated the hairs with mercury salts, which twisted them and made them limp, aiding the felting process. This contaminated the air with mercury vapors, which the hat makers breathed in. As a consequence they developed twitches, a symptom of nervous disorders. The term "mad hatter" (as in *Alice in Wonderland*) originates from this industrial disease.

The atmospheric pressure is measured as the difference in height between the two liquid levels. This difference in height is used directly as a unit of pressure. **Pressure is measured in millimeters of mercury** (mm Hg). (The weather bureau usually gives it in inches of mercury.) In honor of Torricelli, we define another unit, the **torr,** as being equal to 1 mm Hg. At sea level the average pressure of the atmosphere is 760 mm Hg, and we use this number to define still another unit of pressure, the **atmosphere** (atm). There are several other units used for pressure. The official SI unit is the pascal, defined as 1 atm = 101 325 pascals. In this book we use only mm Hg and atm.

1 atm = 760 torr = 760 mm Hg

A barometer is fine for measuring the pressure of the atmosphere, but to measure the pressure of a gas in a container, we use a simpler instrument called a **manometer.** One form of manometer consists of a U-shaped tube containing mercury (Fig. 5.3) in which one arm (A) has been evacuated and sealed. This arm has zero pressure. The other arm (B) is connected to the container in which the gas sample is enclosed. The pressure of the gas depresses the level of the mercury in the B arm of the U tube, and **the difference between the two mercury levels gives the pressure directly in mm Hg.** If more gas is added to the bulb in Figure 5.3, the mercury level in B will be pushed down and that in A will rise, because the pressure in the bulb has now increased.

**Figure 5.3**
A manometer.

**Box 5C**

## Blood Pressure Measurement

Blood pressure is the result of the blood's pushing against the walls of the blood vessels. When the heart ventricles contract, pushing blood out into the arteries, the blood pressure is high **(systolic blood pressure);** when the ventricles relax, the blood pressure is low **(diastolic pressure).**

Blood pressure varies with age. In young adults the normal range is 100 to 120 mm Hg systolic and 60 to 80 mm Hg diastolic. In older people the corresponding normal pressures are 115 to 135 and 75 to 85. Blood pressure is usually expressed as a fraction showing systolic over diastolic pressure, for instance, 120/80. Normal blood pressure is lower for women than for men. Blood pressures higher than normal may lead to heart attack, stroke, or kidney failure. Early diagnosis is important because blood pressure can be reduced by proper diet (low in sodium ions) and drugs. The measurement for blood pressure is routinely performed by nurses and physicians.

The *sphygmomanometer,* or aneroid blood pressure measurement device, consists of a bulb, a cuff, a manometer, and a stethoscope (Fig. 5C). The cuff is wrapped around a bare arm and the bulb repeatedly squeezed. This increases the amount of air in the cuff and thus the pressure. This increased pressure is transmitted to the cuff, which transmits it to the upper arm. The pressure applied can be read on the manometer. At the beginning, the cuff is inflated by repeated squeezing of the bulb to a pressure of 200 mm Hg. Then a release valve is opened slowly and the pressure is allowed to drop 2 or 3 mm Hg per second. While the pressure is dropping, a stethoscope is placed over the brachial artery above the elbow, and one listens for a clear tapping sound. At the start no sound is heard because, in most cases, 200 mm Hg applied pressure is greater than the blood pressure and so no pulsating blood flows under the stethoscope. When the first faint tapping sound is heard, the pressure on the manometer is read. This is the systolic pressure. The applied pressure just matches the blood pressure when the ventricle contracts, allowing pulsating blood to flow into the lower arm. The clear tapping sound becomes louder as the applied cuff pressure drops and eventually fades away. When the last tapping sound is heard, the manometer is read again. This is the diastolic pressure. The applied cuff pressure now matches the blood pressure when the ventricle is relaxed, allowing continuous blood flow into the lower arm. Repeated blood pressure measurements can be taken, but the cuff should be deflated completely for two or three minutes before attempting the next reading.

**Figure 5C**
A sphygmomanometer.

$P = 1\ \text{atm}$
$V = 4\ \text{L}$

$P = 2\ \text{atm}$
$V = 2\ \text{L}$

$P = 4\ \text{atm}$
$V = 1\ \text{L}$

**Figure 5.4**
Boyle's law illustrated in an automobile cylinder. When the piston moves up, the volume occupied by the gas (a mixture of air and gasoline vapor) decreases. Since the temperature is constant, the pressure must increase.

## 5.4 GAS LAWS

In observing the behavior of gases, a number of relationships have been established:

1. **Boyle's law**  At constant temperature, the volume of any sample of gas decreases as its pressure increases, and vice versa. This means that for a fixed mass of gas at a constant temperature, the volume is inversely proportional to the pressure. This is what happens in the cylinder of a car engine; as the pistons move up, the volume of gas decreases and the pressure increases (Fig. 5.4).

2. **Charles's law**  At constant pressure, the volume of any sample of gas increases as its temperature increases, and vice versa. Putting it more precisely, the volume of a fixed mass of gas at a constant pressure is directly proportional to the absolute temperature. This is the basis of the hot-air balloon (Fig. 5.5). The air in the balloon expands as it is heated from below; this makes the air inside the balloon less dense than the surrounding air, providing the lift (Charles was one of the first balloonists).

3. **Gay-Lussac's law**  At constant volume, as the temperature increases, the pressure of any sample of gas increases, and vice versa. For a fixed mass of gas at

These gas laws hold not only for pure gases but also for any mixture of gases.

Although Jacques Charles (1746–1823) had earlier formulated a crude version of his law, it was first stated mathematically by Joseph Gay-Lussac (1778–1850) in 1802.

**Figure 5.5**
Charles's law illustrated in a hot-air balloon. Because the balloon can stretch, the pressure inside it is constant. When the air in the balloon is heated, its volume increases, expanding the balloon. Since the volume of the air has increased but not the mass, its density has decreased. The air inside the balloon being less dense than the air outside provides the lift, and the balloon can ascend.

## Breathing and Boyle's Law

Robert Boyle (1627–1691) discovered the pressure-volume relationship in 1662, but humans and animals were using the process described by the law long before then, in order to breathe. We breathe about 12 times a minute, each time taking in and exhaling about 500 mL of air. When we breathe, we lower the diaphragm or raise the rib cage (Fig. 5D). Either of these motions increases the volume of the chest cavity. In accord with Boyle's law, the pressure is therefore decreased and becomes lower than the outside pressure. Air thus flows from the higher-pressure area into the lungs. The difference is only about 3 mm Hg, but this is enough to make the air go in. In breathing out we reverse the process: We raise the diaphragm or lower the rib cage. The resulting decrease in volume increases the pressure inside the chest cavity, causing air to flow out.

In certain diseases the chest becomes paralyzed, and the patient cannot move the diaphragm or the rib cage. In such cases artificial respirators are used, pushing down on the chest and forcing the air inside the patient to go out the only way possible, through the lungs. The pressure in the respirator is then lowered below atmospheric pressure, causing the patient to breathe in, and the cycle is constantly repeated.

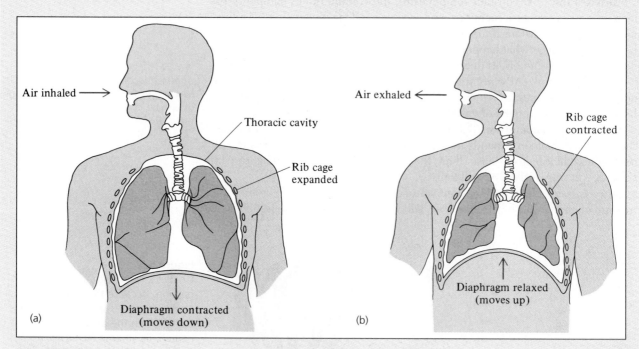

**Figure 5D**
Schematic drawings of the chest cavity. In (a) the lungs fill with air. In (b) air empties from the lungs.

constant volume, the pressure is directly proportional to the absolute temperature. This is what happens inside an autoclave. Autoclaves are used to sterilize hospital and laboratory equipment (Fig. 5.6). Steam generated at 1 atm pressure has a temperature of 100°C. When this steam is further heated in an enclosed vessel (autoclave) at constant volume, the increased temperature of the steam increases its pressure. A valve controls the maximum pressure for which the autoclave is designed; if the pressure exceeds this, the valve opens and the steam is released. At maximum pressure the temperature may be as high as 120 to 150°C. At these high temperatures all microorganisms in the equipment are destroyed.

Mathematical expressions of these three laws are shown in Table 5.1. The three gas laws can be combined and expressed by a mathematical equation called the **combined gas law:**

$$\frac{P_1 V_1}{T_1} = \frac{P_2 V_2}{T_2}$$

In this equation, $P_1$, $V_1$, and $T_1$ are the pressure, volume, and **absolute (Kelvin) temperature** of any sample of gas. When any change is made to the sample, $P_2$, $V_2$, and $T_2$ are the new pressure, volume, and temperature. The following examples illustrate calculations using the combined gas law equation.

**TABLE 5.1   Mathematical Expressions of the Three Gas Laws for a Fixed Mass of Gas**

| Name | Expression | Constant |
|---|---|---|
| Boyle's law | $P_1 V_1 = P_2 V_2$ | $T$ |
| Charles's law | $\dfrac{V_1}{T_1} = \dfrac{V_2}{T_2}$ | $P$ |
| Gay-Lussac's law | $\dfrac{P_1}{T_1} = \dfrac{P_2}{T_2}$ | $V$ |

**EXAMPLE**

A gas occupies 3.00 L at 1.50 atm pressure. What is its volume at 10.00 atm at the same temperature?

**Answer**

First we identify the known quantities:

$$P_1 = 1.50 \text{ atm}; \qquad V_1 = 3.00 \text{ L}; \qquad P_2 = 10.00 \text{ atm}; \qquad T_1 = T_2$$

$$\frac{P_1 V_1}{T_1} = \frac{P_2 V_2}{T_2}$$

Since we want the new volume, we solve our equation for $V_2$. Because $T_1$ and $T_2$ are the same in this example and consequently cancel each other, we do not even have to know what the temperature is. Thus,

$$V_2 = \frac{P_1 V_1 \cancel{T_2}}{\cancel{T_1} P_2} = \frac{(1.50 \text{ } \cancel{\text{atm}})(3.00 \text{ L})}{10.00 \text{ } \cancel{\text{atm}}} = 0.45 \text{ L}$$

The new volume is 0.45 L, or 450 mL.

In this type of problem, we first identify the quantities that remain constant, and then solve the equation for the property ($P$, $V$, or $T$) we are interested in and substitute the numbers.

**PROBLEM 5.1** A gas occupies 3.8 L at 0.7 atm pressure. If we expand the volume at constant temperature to 6.5 L, what is the new pressure?

**EXAMPLE**

In an autoclave, steam at 100°C is generated at 1.00 atm. After the autoclave is closed, the steam is heated at constant volume until the pressure gauge indicates 1.13 atm. What is the new temperature in the autoclave?

**Answer**

Since all temperatures in gas law calculations are Kelvin, we must first convert the Celsius temperature to Kelvin:

$$100°\text{C} = 100 + 273 = 373 \text{ K}$$

We then identify the known quantities:

$$P_1 = 1.00 \text{ atm}; \quad T_1 = 373 \text{ K}; \quad P_2 = 1.13 \text{ atm}; \quad V_1 = V_2$$

$$\frac{P_1 V_1}{T_1} = \frac{P_2 V_2}{T_2}$$

Since we want a new temperature, we solve this equation for $T_2$:

$$T_2 = \frac{P_2 \cancel{V_2} T_1}{P_1 \cancel{V_1}}$$

$$T_2 = \frac{(1.13 \text{ } \cancel{\text{atm}})(373 \text{ K})}{(1.00 \text{ } \cancel{\text{atm}})} = 421 \text{ K}$$

The new temperature is 421 K, or $421 - 273 = 148°\text{C}$. In this example the two volumes cancel out.

Ordinary air contains 20.9 percent oxygen. Under certain conditions, the cells of tissues are starved for oxygen (hypoxia), and there is a need for quick oxygen delivery. Increasing the percentage of oxygen in the air supplied to the patient is one way to remedy the situation, but sometimes even breathing pure (100%) oxygen may not be enough. For example, in carbon monoxide poisoning, the hemoglobin in the blood, which usually carries most of the $O_2$ from the lungs to the tissues, binds CO and cannot take up any $O_2$ in the lungs. Without any help, the tissues would soon be starved for oxygen and the patient would die. If oxygen is administered, especially under high pressure (2 or 3 atm), oxygen dissolves in the plasma to such a degree that the tissues receive enough to recover. Many large hospitals now have special *hyperbaric units* where pure oxygen under pressure is given to patients. The entire unit is under high pressure, and the patient and the accompanying medical staff must undergo gradual compression and, at the end of the treatment, decompression.

Besides treating carbon monoxide poisoning, hyperbaric units are also used along with other treatment. For example, in radiation necrosis, when tissues are damaged by x-rays during cancer treatment or accidents, hyperbaric medicine is beneficial in 80 to 100 percent of the cases. Chronic refractory osteomyelitis, a bone infection that does not heal, also responds well to hyperbaric medicine. Other conditions for which hyperbaric medicine is used are gas gangrene, smoke inhalation, cyanide poisoning, skin grafts, and thermal burns.

However, breathing pure oxygen for prolonged periods is toxic. For example, if $O_2$ is administered at 2 atm for more than six hours, the lung tissues begin to be destroyed, and the central nervous system can also be damaged. Therefore, recommended exposures to $O_2$ are two hours at 2 atm and 90 minutes at 3 atm.

The benefits of hyperbaric medicine must be carefully weighed against the contraindications. Reports have been published showing that this treatment can cause nuclear cataract formation, thus necessitating postrecovery eye surgery.

---

**PROBLEM 5.2** A constant volume of oxygen gas is heated from 120°C to 212°C. The final pressure is 20.3 atm. What was the initial pressure?

---

**EXAMPLE**

A gas is under 1.0 atm pressure in a flexible container with a volume of 0.50 L at a temperature of 393 K. When the gas is heated to 500 K and the volume expanded to 3.0 L, what will the new pressure be?

**Answer**
The known quantities are

$$P_1 = 1.0 \text{ atm}; \quad V_1 = 0.50 \text{ L}; \quad T_1 = 393 \text{ K};$$
$$V_2 = 3.0 \text{ L}; \quad T_2 = 500 \text{ K}$$

$$\frac{P_1 V_1}{T_1} = \frac{P_2 V_2}{T_2}$$

Solving for $P_2$, we get

$$P_2 = \frac{P_1 V_1 T_2}{T_1 V_2} = \frac{(1.0 \text{ atm})(0.50 \text{ L})(500 \text{ K})}{(3.0 \text{ L})(393 \text{ K})} = 0.21 \text{ atm}$$

The new pressure will be 0.21 atm, or

$$0.21 \ \cancel{\text{atm}} \times \frac{760 \ \text{mm Hg}}{1 \ \cancel{\text{atm}}} = 161 \ \text{mm Hg}$$

**PROBLEM 5.3** If 3.0 L of a gas has an initial pressure of 2.8 atm at 280 K, what will be its final pressure if it is compressed to 0.466 L and cooled to 200 K?

## 5.5 AVOGADRO'S LAW AND THE IDEAL GAS LAW

The combined gas law is valid only for a given constant sample of gas. That is, it does not work if any more gas is added or any taken away. However, it is often necessary to work with changing amounts of gas. Fortunately, there is an easy way to do this. The relationship between the amount of gas present and its volume is described by **Avogadro's law,** which states that **equal volumes of gases at the same temperature and pressure contain equal numbers of molecules.** For example, Figure 5.7 shows two tanks of equal volume, one containing hydrogen and the other carbon dioxide. Avogadro's law tells us that, if the temperature and pressure are the same, the two tanks contain the same number of molecules, regardless of what the gases are. The actual temperature and pressure don't matter. Avogadro's law is valid no matter what the gases are, as long as they don't become liquids. However, it is convenient to select one temperature and one pressure as standard, and chemists have chosen 1 atm to be the standard pressure and 0°C (273 K) to be the standard temperature. These conditions are called **STP (standard temperature and pressure).**

Just as at any other temperature and pressure, all gases at STP contain the same number of molecules in any given volume. But how many is that? In Chapter 4 we saw that the *mole* (which contains $6.02 \times 10^{23}$ molecules) is a convenient quantity of matter. What volume of gas at STP contains 1 mole of molecules? This quantity is very easy to measure, and it was done long ago. It turns out to be 22.4 L:

Avogadro's law holds only for gases, not liquids or solids.

STP = 1 atm and 273 K

1 mole of any gas at STP occupies 22.4 L.

**Figure 5.7**
Two tanks of gas of equal volume at the same temperature and pressure contain the same number of molecules.

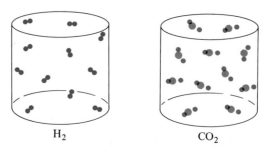

H₂                    CO₂

Avogadro's law allows us to write a gas law that is valid not only for any pressure, volume, and temperature but also for any quantity of gas. This law, called the **ideal gas law,** is

$$PV = nRT$$

where $P$ = pressure of the gas in atm
$V$ = volume of the gas in L
$n$ = number of moles of the gas
$T$ = temperature of the gas in K
$R$ = a constant for all gases, for all conditions of $n$, $P$, $V$, and $T$.

We can find out what the constant is by using the fact that 1 mole at STP occupies 22.4 L:

$$PV = nRT$$

$$R = \frac{PV}{nT} = \frac{(1 \text{ atm})(22.4 \text{ L})}{(1 \text{ mole})(273 \text{ K})} = 0.0821 \frac{\text{L} \cdot \text{atm}}{\text{mole} \cdot \text{K}}$$

Thus $R$, called the **universal gas constant,** is equal to 0.0821 L·atm/mole·K.

The **ideal gas law, $PV = nRT$, is a universal law.** It holds for all gases, including all mixtures of gases, at any temperature, pressure, volume, or amount, as long as they are ideal gases. As we mentioned before, real gases, while not exactly ideal, in most cases behave as ideal gases, and so we can use this equation with little trouble. Using $PV = nRT$, we can calculate any one quantity—$P$, $V$, $T$, or $n$—if we know the other three. Furthermore, since the number of moles of any substance is its mass in grams divided by its molecular weight (Sec. 4.3),

Real gases behave most like ideal gases at low pressures (1 atm or less) and high temperatures (300 K or higher).

$$n = \frac{m}{MW}$$

$MW$ stands for molecular weight.

we can replace $n$ in the ideal gas equation to give

$$PV = \frac{mRT}{MW}$$

This equation allows us to calculate the molecular weight of a gas if we know the other quantities. Some examples of the use of the ideal gas law follow.

---

**EXAMPLE**

One mole of $CH_4$ gas occupies 20.0 L at 1.00 atm pressure. What is the temperature of the gas in K?

**Answer**
Solve the ideal gas law for $T$:

$$T = \frac{PV}{nR}$$

$$T = \frac{(1.00 \text{ atm})(20.0 \text{ L})}{(1 \text{ mole})(0.0821 \text{ L} \cdot \text{atm/mole} \cdot \text{K})} = 244 \text{ K}$$

Note that the answer would be the same if the gas were $CO_2$, $N_2$, $NH_3$, or *any other gas*.

**PROBLEM 5.4**   If 3.4 moles of CO occupy 10.8 L at 300 K, what is the pressure?

---

**EXAMPLE**

If there is 5.0 g of $CO_2$ gas in a 10-L vessel at 350 K, what is the pressure?

**Answer**

Here we are given a mass, and so we use $PV = mRT/MW$. The molecular weight of $CO_2 = 44$:

$$P = \frac{mRT}{V(MW)} = \frac{(5.0 \text{ g})(0.0821 \text{ L} \cdot \text{atm/mole} \cdot \text{K})(350 \text{ K})}{(10 \text{ L})(44 \text{ g/mole})}$$

$$= 0.33 \text{ atm}$$

---

**PROBLEM 5.5**   A certain quantity of neon gas is under 1.05 atm pressure at 303 K in a 10.0-L vessel. How many moles of neon are present?

---

**EXAMPLE**

If 1.0 L of an unknown gas has 1.15 atm pressure at 40°C and the mass of the gas is 3.3 g, what is the molecular weight?

**Answer**

Convert 40°C to absolute temperature: $273 + 40 = 313$ K.

$$MW = \frac{mRT}{PV} = \frac{(3.3 \text{ g})(0.0821 \text{ L} \cdot \text{atm/mole} \cdot \text{K})(313 \text{ K})}{(1.15 \text{ atm})(1.0 \text{ L})}$$

$$= 74 \frac{\text{g gas}}{\text{mole gas}}$$

Therefore, the molecular weight is 74 amu.

---

**PROBLEM 5.6**   An unknown gas weighs 5.0 g and occupies 1.50 L at 1.00 atm and 300 K. What is its molecular weight?

---

## 5.6   DALTON'S LAW AND GRAHAM'S LAW

In a mixture of gases each molecule acts independently of all the others, assuming that the gases do not react with each other. Unlike liquids, where molecules are close together and strongly influence each other, the molecules in a gas are so far apart that they have little or no effect on each other. This is why the ideal gas law works for mixtures as well as for pure gases. In particular, the collisions with the walls of the container made by each molecule contribute independently to the total pressure. **Dalton's law of partial pressures** states that the total pressure of a mixture of gases is the sum of the individual partial pressures:

$$P_T = P_1 + P_2 + P_3 + \cdots$$

The **partial pressure of a single gas** in a mixture is defined as **the pressure the gas would exert if it were alone in the container.** The equation $PV = nRT$ holds separately for each gas in the mixture, as well as for the total mixture.

---

**EXAMPLE**

To a tank already containing $N_2$ at 2.0 atm and $O_2$ at 1.0 atm, we add an unknown quantity of $CO_2$ until the total pressure is 4.6 atm. What is the partial pressure of the $CO_2$?

**Answer**

Dalton's law tells us that the addition of $CO_2$ does not affect the $N_2$ and $O_2$ that are already there. Thus the partial pressures of $N_2$ and $O_2$ remain at 2.0 atm and 1.0 atm, respectively, even after the $CO_2$ is added. The sum of the partial pressures of $N_2$ and $O_2$ is

$$2.0 \text{ atm} + 1.0 \text{ atm} = 3.0 \text{ atm}$$

Since the rest of the total pressure of 4.6 atm must be caused by the $CO_2$, we can determine it by subtracting:

$$\underset{\substack{\text{Total} \\ \text{pressure}}}{4.6 \text{ atm}} - \underset{\substack{\text{Pressure of} \\ N_2 \text{ and } O_2}}{3.0 \text{ atm}} = \underset{\substack{\text{Pressure} \\ \text{of } CO_2}}{1.6 \text{ atm}}$$

---

**PROBLEM 5.7** A vessel under 2.015 atm pressure contains nitrogen and water vapor. The partial pressure of the nitrogen is 1.908 atm. What is the partial pressure of the water vapor?

---

**Figure 5.8**
When open containers of $NH_3$ (left) and HCl (right) in water solution are placed side by side, molecules of $NH_3$ and HCl gases escape and react with each other in the air to form $NH_4Cl$, which appears in this picture as white fumes. (Photograph by Charles D. Winters.)

When a bottle of ammonia, $NH_3$, is opened in one corner of a room, within minutes its odor spreads to other parts of the room (Fig. 5.8). The process whereby two gases mix is called **diffusion.** In this case, the ammonia molecules diffuse into the air. When a bottle of perfume is opened, its fragrance also fills the room, but this takes longer than with ammonia. Thomas Graham (1805–1869) observed that molecules with small masses diffuse faster than heavy molecules and formulated what is now known as **Graham's law:**

$$\frac{(\text{Diffusion rate of A})^2}{(\text{Diffusion rate of B})^2} = \frac{MW \text{ of B}}{MW \text{ of A}}$$

**B o x  5F**

## Breathing and Dalton's Law

A gas naturally flows from an area of higher pressure to one of lower pressure. This fact is responsible for respiration, though we must look not at the total pressure of air but only at the partial pressure of $O_2$ and $CO_2$. Our cells use $O_2$ to oxidize carbon compounds to $CO_2$ and thus obtain energy. Respiration is the process by which the blood carries $O_2$ from the lungs to the tissues and collects the $CO_2$ manufactured by the cells and brings it to the lungs, where it is breathed out (Fig. 5F).

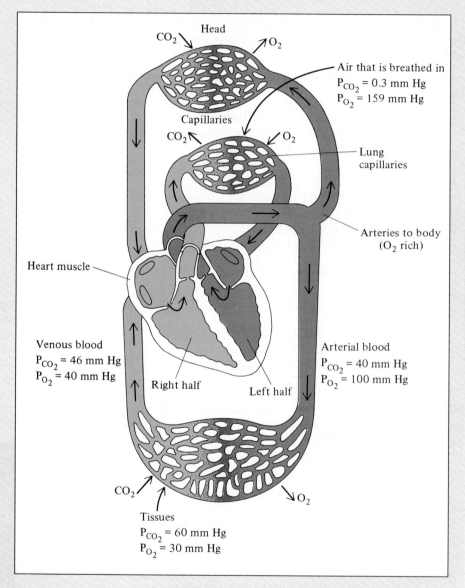

**Figure 5F**

Blood circulation showing partial pressures of $O_2$ and $CO_2$ in the different parts of the system.

The air we breathe contains about 21 percent $O_2$; the partial pressure of $O_2$ is about 159 mm Hg. The partial pressure of $O_2$ in the alveoli (tiny air sacs) in the lungs is about 100 mm Hg. Because the partial pressure is higher in the inhaled air, $O_2$ flows to the alveoli. At this point, venous blood (blood from the veins) flows past the alveoli. The partial pressure of $O_2$ in the venous blood is only 40 mm Hg, and so $O_2$ flows from the alveoli to the venous blood. This new supply of $O_2$ raises the partial pressure to about 100 mm Hg, and the blood that now has had its $O_2$ pressure increased becomes arterial blood and flows to the body tissues. There it finds that the partial pressure of $O_2$ is about 30 mm Hg or less (because the cells of the body tissues have used up most of their oxygen to provide energy). Oxygen now flows from the arterial blood to the body tissues, giving them a fresh supply. This naturally lowers the partial pressure of $O_2$ in the blood (to about 40 mm Hg), so that the blood now becomes venous blood again and returns to the lungs for a fresh supply of $O_2$. At each step of the cycle, $O_2$ flows from a region of higher partial pressure to one of lower partial pressure.

Meanwhile, the $CO_2$ goes the opposite way, for the same reason. The partial pressure of $CO_2$ in oxidized body tissue is about 60 mm Hg. Thus, $CO_2$ flows from the tissues to the arterial blood, which of course is changed by the process to venous blood and now has a $CO_2$ pressure of 46 mm Hg. When the venous blood reaches the lungs, it finds that the $CO_2$ pressure in the alveoli is 40 mm Hg, and so it flows to the alveoli. Since the partial pressure of $CO_2$ in air is only 0.3 mm Hg, $CO_2$ flows from the alveoli and is exhaled.

## 5.7 INTERMOLECULAR FORCES

If the temperature of a gas is sufficiently decreased, the gas condenses and becomes a liquid. If the temperature is lowered still more, the liquid freezes and becomes a solid. The great difference between the gaseous state and condensed states (liquid and solid) is the effect of **intermolecular forces.** In gases the individual molecules are far apart, so much so that we can ignore intermolecular attraction and treat most gases as ideal. This is valid when the temperature is high (room temperature and above) and the pressure low (1 atm or less), so that the molecules are far apart. When the temperature decreases and/or the pressure increases, distances between molecules become less, and intermolecular forces can no longer be ignored. As a matter of fact, it is these forces that cause condensation and solidification.

Therefore, before discussing the properties and structures of liquids and solids, we must look at the nature of intermolecular forces.

### Dipole–Dipole Attractions

As mentioned in Section 3.7, many molecules are dipoles, and the positive end of one molecule attracts the negative end of another. These forces are called

**dipole-dipole attractions** and can exist between two identical molecules or between two different molecules:

$$\delta+ \overline{(H-Cl)}\, \delta- \quad \delta+ \overline{(H-Cl)}\, \delta-$$

$$\delta+ \overline{(H-Br)}\, \delta- \quad \delta+ \overline{(H-Cl)}\, \delta-$$

## Hydrogen Bonds

**Figure 5.9**
Two water molecules joined by a hydrogen bond.

An especially strong attraction exists between molecules of water. The polarity of the O—H bond shifts the electrons toward the oxygen within the molecule so much that the hydrogen acquires a partial positive charge. Such a hydrogen is very strongly attracted toward the electron cloud of an oxygen in a neighboring water molecule. This attraction, indicated by the dotted line in Figure 5.9, is called a **hydrogen bond.**

Hydrogen bonds are not restricted to water. They form whenever two molecules, the same or different, come together, provided that

1.   **One molecule has a hydrogen atom attached by a covalent bond to an atom of oxygen, nitrogen, or fluorine.**
2.   **The other molecule has an oxygen, nitrogen, or fluorine atom.**

Therefore, we expect hydrogen bonds to form here:

but not here:

$$\begin{matrix} H \\ \diagdown \\ \quad O \\ \diagup \\ H \end{matrix} \qquad H-\overset{\displaystyle H}{\underset{\displaystyle H}{\overset{\displaystyle |}{\underset{\displaystyle |}{C}}}}-H$$

Although hydrogen bonds are among the strongest of the intermolecular forces, it must be kept in mind that they are still much weaker than the covalent bonds within the molecules. Covalent bonds are about ten times stronger than hydrogen bonds (Table 5.2). As we shall see in later chapters, hydrogen bonds play an important role in biological molecules.

| TABLE 5.2   Approximate Energy of Atomic and Molecular Interactions | |
|---|---|
| Interaction | Energy (kcal/mole) |
| Covalent bond | 50–100 |
| Hydrogen bond | 5–10 |
| Dipole-dipole attraction | 0.1–1 |
| London dispersion force | 0.001–0.2 |

The hydrogen-bonded superstructure of ice (Fig. 5G.1) has empty spaces in the middle. Therefore the molecules in ice are not so closely packed as those in liquid water. Because of this, ice has a lower density than liquid water. In the latter, more of the hydrogen bonds are broken, and the hexagonal superstructure is collapsed into a more densely packed organization (Fig. 5G.2). This is why ice floats on top of water instead of sinking to the bottom. Such behavior is highly unusual. Most solids are denser than the corresponding liquids. The lower density of ice keeps fish and microorganisms alive in many rivers and lakes that would freeze solid each winter if the ice sank to the bottom. The presence of ice on top insulates the remaining water and keeps it from freezing.

The fact that ice has a lower density than water means it has a greater molar volume (that is, any particular mass of ice takes up more space than the same mass of water). This explains the damage done to biological tissues by freezing. When parts of the body (usually fingers, toes, noses, and ears) are subjected to extreme cold, they develop a condition called **frostbite.** The water in the cells freezes despite the blood's attempt to keep the temperature at 37°C. The frozen water expands and in doing so ruptures the cell walls, causing damage. In some cases, frostbitten fingers or toes must be amputated.

Cold weather can also damage plants in a similar way. Many plants are killed when the air temperature drops below the freezing point of water for several hours. Trees can survive cold winters because they have a low water content inside their trunks and branches.

Many times, slow freezing is more damaging to plant and animal tissues than quick freezing. In slow freezing only a few crystals are formed, and these can grow to large sizes, rupturing the cells. In quick freezing, such as can be achieved in liquid nitrogen (at a temperature of −196°C), many tiny crystals are formed. Since these do not grow much, tissue damage may be much less.

# The Densities of Ice and Water

**Figure 5G.1**
The structure of ice. (Reprinted by permission from *Chemistry, Matter and the Universe* by R. E. Dickerson and I. Geis (1976), Benjamin/Cummings, Menlo Park, CA, Publishers.)

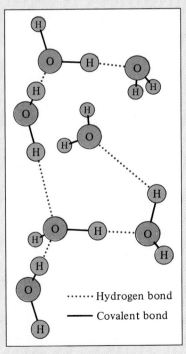

**Figure 5G.2**
The structure of liquid water.

147

**EXAMPLE**

(a) Can a hydrogen bond form between two molecules of methanol, $CH_3OH$?   (b) between two molecules of formaldehyde, $H-C-H$?

$$\overset{\|}{O}$$

(c) between one molecule of methanol and one of formaldehyde?

**Answer**

(a) Yes. One molecule has a hydrogen atom attached to an oxygen atom (actually, both have, but only one is necessary), and the other molecule has an oxygen atom:

(b) No. Neither molecule has a hydrogen attached to an oxygen, nitrogen, or fluorine.

(c) Yes. One molecule (methanol) has a hydrogen atom attached to an oxygen atom; the other (formaldehyde) has an oxygen atom.

**PROBLEM 5.8**   (a) Will a molecule of water form a hydrogen bond with a molecule of methanol?   (b) Will two molecules of methane ($CH_4$) form hydrogen bonds with each other?

## London Dispersion Forces

Not all gases contain polar molecules. Yet even nonpolar gases, such as helium and hydrogen, liquefy if the temperature is lowered sufficiently. Are there intermolecular forces even between molecules that are not dipoles? The answer is yes, though such forces, called **London dispersion forces,** are *very* weak (Table 5.2). Even though London forces are very weak, in large molecules they contribute significantly to the attractive forces between the molecules, because they act over large surface areas.

London forces were first explained by Fritz London (1900–1954) in 1930.

Even London forces have their origin in electrostatic interaction. We can explain the attraction between nonpolar atoms by assuming that, when the two atoms get close to each other, the motion of the electron clouds becomes synchronized. For a very small fraction of a second, the negative charge is located on one side of each spherical cloud. If the motion of the electrons in two helium clouds becomes synchronized during that tiny fraction of a second, the charges will be at the same relative position, as shown in Figure 5.10. Thus, the electrons of atom B attract the nucleus of atom A during this short interval. A fraction of a second later, when the electrons have moved to the 3 o'clock position, the electrons of A interact with the nucleus of B. Thus there are very short-lived and constantly changing interactions between the two atoms—the London dispersion forces.

**Figure 5.10**
The relative positions of the electron clouds of two helium atoms result in London dispersion forces.

## 5.8  LIQUIDS

We have seen that the behavior of gases under most circumstances can be described by the ideal gas law. As pressure increases, however, the molecules of a gas are squeezed into a smaller space. The distance the average molecule travels between collisions becomes smaller. Over short distances intermolecular forces become significantly stronger, and so molecules have a greater tendency to stick together.

If the contact between molecules increases to the extent that almost all the molecules touch (or almost touch) each other, the gas becomes a liquid. In contrast to gases, liquids do not fill all the available space. They have a definite volume, irrespective of the container. Because gases have a lot of empty space between molecules, it is easy to compress them into a smaller volume. In liquids there is very little empty space; consequently, liquids are hard to compress. A great increase in pressure is needed to cause a very small decrease in the volume of a liquid. Liquids are, for all practical purposes, incompressible. The density of liquids is much greater than that of gases because the same mass occupies a much smaller volume in liquids than in gases.

The position of the molecules in the liquid phase is random, and there is some irregular empty space into which molecules can slide. The molecules thus constantly change their position with respect to neighboring molecules. This property causes the *fluidity* of liquids; that is, they do not have a constant shape but only a constant volume.

## 5.9  EVAPORATION AND CONDENSATION. BOILING POINT

An important property of liquids is that they evaporate. A few hours after a heavy rain, most of the puddles have dried up. The water has evaporated, gone into the air. The same thing occurs if we leave a glass of water out in the open. Why does this happen?

There is a distribution of velocities among molecules. In our water example, some of the molecules have high energy and are moving rapidly. Others have low energy and are moving slowly. Whether fast or slow, a molecule in the center or at the bottom of the liquid cannot go very far before it hits another molecule and has its speed and direction changed by the collision. A molecule at the surface, however, is in a different situation (Fig. 5.11). If it is moving slowly, it cannot escape from the liquid because of the attractions of neighboring molecules. If, however, it is moving rapidly (a high-energy molecule) and is moving upward, it can escape from the liquid and enter the gaseous space above it.

If the container is open, this process continues until eventually all the molecules have left the liquid and it has dried up. However, if the liquid is in a closed container, as in Figure 5.12, this does not happen because the water vapor molecules in a closed container cannot diffuse away (as they do if the container is open). They remain in the air space above the liquid, where they behave, of course, like any other gas molecules, meaning that they move rapidly in straight lines until they collide with something. Some of these water vapor molecules move *downward* and thus can be recaptured by the liquid.

At this point we have an **equilibrium.** The amount of water in the vapor no longer increases or decreases, as long as the temperature does not change. (We will meet another type of equilibrium in Chapter 7.) After equilibrium is reached, the space above the liquid shown in Figure 5.12 contains air and water vapor, and it is possible to measure the partial pressure of the water vapor. We call this the **vapor pressure of the liquid.** Note that we measure the partial pressure *of a gas* but call it the vapor pressure *of the liquid.*

At equilibrium the rate of vaporization is equal to the rate of liquefaction.

**Figure 5.11**
Some molecules at the surface of a liquid are moving fast enough to escape into the gas phase.

**Figure 5.12**
In a closed container, molecules of liquid escape into the vapor phase, and vapor molecules are recaptured by the liquid.

Molecules from air ($O_2$, $N_2$)  Molecules of vapor

Molecules from air  Molecules of vapor

**Box 5H**

## Ethyl Chloride

Ethyl chloride spray is used to alleviate pain from minor muscular injuries and bruises. This compound evaporates very quickly so that the skin is cooled rapidly. The net result is a freezing effect. This minimizes swelling from bruises and reduces the pain suffered in the injury by deadening nerves. Athletes routinely use ethyl chloride spray.

The vapor pressure of a liquid is a physical property of the liquid at a given temperature. A liquid has the same vapor pressure no matter what gas is above it (air or any other gas) and no matter what the pressure of that gas is. In fact, the vapor pressure is the same even if a vacuum exists in the space above the gas. Vapor pressure is an intensive property: It is independent of the volume of the liquid.

The molecules that evaporate are the more energetic molecules. When they go into the gas phase, the molecules left behind are the less energetic ones. Therefore, the temperature of the liquid drops as a result of evaporation. This is what causes the cooling effect of perspiration. The liquid that comes out of the pores evaporates, so the skin temperature drops. The same cooling effect is felt when you come out of a swimming pool and the layer of water starts to evaporate from your skin.

The vapor pressure of a liquid depends on the temperature. Figure 5.13 shows that **vapor pressure increases as temperature increases.** Eventually

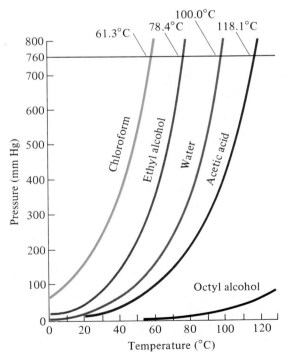

**Figure 5.13**
The change in vapor pressure with temperature for five liquids.

Boiling is a special form of evaporation during which bubbles form inside the liquid so that the surface of the liquid is greatly increased.

the liquid boils. The **boiling point** of a liquid is the **temperature at which its vapor pressure is equal to the pressure of the atmosphere in contact with the surface of the liquid.** The boiling point when the atmospheric pressure is 1 atm is called the **normal boiling point.** Thus, 100°C is the normal boiling point of water because that is the temperature at which water boils at 1 atm pressure.

The use of a pressure cooker is an example of boiling at higher temperatures. In this type of pot, food is cooked at, say, 2 atm, where the boiling point of water is 121°C. Since the food has been raised to such a high temperature, it cooks much faster than it would in an open pot, in which boiling water cannot get hotter than 100°C. Conversely, at low pressures water boils at lower temperatures. For example, at the top of a mountain, the boiling point might be 95°C.

Different liquids have different normal boiling points. There are two factors that cause liquids to have high or low boiling points:

**1.  Molecular weight and shape**    Where intermolecular forces are similar, boiling point increases with increasing molecular weight. For example, consider methane, $CH_4$, a gas whose normal boiling point is $-161°C$, and $n$-hexane, $C_6H_{14}$, a liquid with a normal boiling point of 69°C. Both compounds have fairly similar London dispersion forces and essentially no other intermolecular forces. The difference in boiling points is caused by the difference in molecular weights (16 for $CH_4$ and 86 for $C_6H_{14}$). But besides the effect of molecular weight, the shape of the molecule is also important, because the shape helps to determine the surface area. For example, compare $n$-pentane, a chain-like molecule

$$
\begin{array}{ccccccccc}
 & H & & H & & H & & H & & H \\
 & | & & | & & | & & | & & | \\
H-&C&-&C&-&C&-&C&-&C&-H \\
 & | & & | & & | & & | & & | \\
 & H & & H & & H & & H & & H \\
\end{array}
$$

Although all of the carbon bonds in dimethylpropane point to the corners of a tetrahedron, the 12 hydrogens on the outside of the molecule can be thought of as being on the surface of a sphere (Fig. 5.14).

with dimethylpropane, a spherically shaped molecule (Fig. 5.14).

$$
\begin{array}{c}
H \\
| \\
H-C-H \\
H \quad | \quad H \\
| \quad | \quad | \\
H-C-\!\!-C-\!\!-C-H \\
| \quad | \quad | \\
H \quad | \quad H \\
H-C-H \\
| \\
H
\end{array}
$$

These compounds have the same molecular weight (they are both $C_5H_{12}$), but the boiling point of $n$-pentane is 36.2°C, significantly higher than that of dimethylpropane, 9.5°C. In the chain-like molecule, the London dispersion forces act over a larger molecular surface than in the spherically shaped molecule of the same molecular weight. Thus, the $n$-pentane molecules need a higher temperature than the dimethylpropane molecules in order to move them from the liquid to the gas phase.

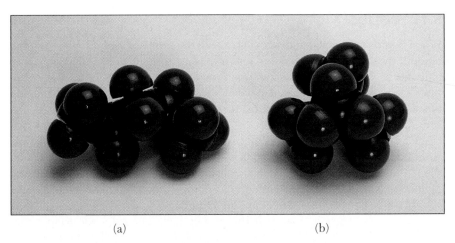

(a)                                                    (b)

**Figure 5.14**
Molecular models of (a) *n*-pentane and (b) dimethylpropane. (Photograph by Beverly March.)

**2.   Intermolecular forces**   Water (MW = 18) and methane (MW = 16) have about the same molecular weight. The normal boiling point of water is 100°C and that of methane is −161°C. The difference in boiling points is due to the fact that the $CH_4$ molecules must overcome only London dispersion forces in order to go into the vapor phase (low boiling point), whereas water molecules, being hydrogen-bonded to each other, need more kinetic energy (and a higher boiling temperature) to escape into the vapor phase.

## 5.10   SURFACE TENSION

When a drop of water is placed on a glass slide, the water does not spread out on the surface. Instead it maintains its spherical shape as if it were surrounded by a protective skin. Water bugs gliding on the surface of a pond appear to be walking on an elastic skin of water (Fig. 5.15). We can conclude from this that the surface of a liquid behaves differently from the rest of it. The surface has a certain tension, just as if a force were pulling it inward.

**Figure 5.15**
The water strider *Halobates* standing on the water.

This can be explained by the fact that molecules on the surface and those in the interior of the liquid experience different attractive forces. In the interior each and every molecule is surrounded by other molecules; therefore, the intermolecular attractions extend in every direction equally. Not so on the surface. A liquid molecule at the surface is attracted more to other liquid molecules beneath it than to the gas molecules above it. Therefore, there is a preferential pull toward the center of the liquid. This pull, called **surface tension,** crowds the molecules on the surface, thereby establishing a layer that is tough to penetrate. The surface tension of liquids depends on their intermolecular interactions. Water has a high surface tension because of the strong hydrogen bonding. A steel needle can easily be made to float on the surface of water, whereas the same needle pushed to the interior sinks to the bottom.

## 5.11   SOLIDS

When liquids are cooled and/or compressed, the intermolecular forces stop the random motion of the molecules, and a solid is formed. In the solid state, not only has each molecule or ion stopped moving through space, but the entire three-dimensional array of molecules or ions is usually orderly. The formation of solid from liquid is called **crystal formation** or **crystallization.**

All crystals have a regular shape that in many cases is obvious to the eye (Fig. 5.16). This regular shape often reflects the arrangement of the atoms within the crystal. For example, table salt is made of small crystalline cubes, and the $Na^+$ and $Cl^-$ ions in the crystal are also arranged in a cubic system. Because molecules or ions in a solid are usually closer together, solids almost always have a higher density than the corresponding liquids.

As can be seen in Figure 5.16, crystals have characteristic shapes and symmetries. We are familiar with the cubic nature of table salt and the hexagonal ice crystals in snow flakes. It is less well known, however, that **the same compound can have more than one type of solid state.** The best known example of this is the element carbon. Carbon has a crystalline form, diamond, that occurs when solidification takes place under very high pressure (thousands of atmospheres). Another crystalline form of carbon is the graphite in a pencil.

*Even in the solid state, molecules and ions do not stop moving completely. They vibrate around fixed points.*

*When two or more forms of a solid are possible, one of them is more stable than the others.*

**Figure 5.16**
Some crystals. (Photograph by Beverly March.)

(a) Ruby

(b) Garnet

(c) Sulfur

(d) Fluorite

The carbon atoms are packed differently in high-density, hard diamonds and in low-density, soft graphite (Fig. 5.17[a]).

Recently, a third form of carbon has been discovered in which each molecule contains 60 carbon atoms arranged in a structure having 12 pentagons and 20 hexagons as faces, resembling a soccer ball (Fig. 5.17[c]). Because the famous architect Buckminster Fuller (1895–1983) invented domes of a similar structure (he called them *geodesic domes*), the C-60 substance was named buckminsterfullerene, or bucky balls for short. This discovery has generated a whole new area of carbon chemistry. Similar cage-like carbon structures, containing 72, 80, and even larger numbers of carbon have been synthesized.

Soot is still a fourth form of solid carbon. This substance solidifies directly out of carbon vapor and is an *amorphous solid;* that is, its atoms have no set pattern and are arranged randomly (Fig. 5.17[d]). Another example of an amorphous solid is glass. In essence, glass is a frozen liquid.

As we have now seen, some crystalline solids consist of orderly arrays of ions (Fig. 3.2) and others of molecules (molecular solids). Ions are held in the crystal lattice by ionic bonds (Sec. 3.3). Molecules are held only by intermolecular forces, which are much weaker than ionic bonds. Therefore, molecular solids generally have much lower melting points than ionic solids.

**Figure 5.16** *continued*

(e)  Quartz

(f)  Muscovite

(g)  Amethystine quartz

(h)  Pyrite

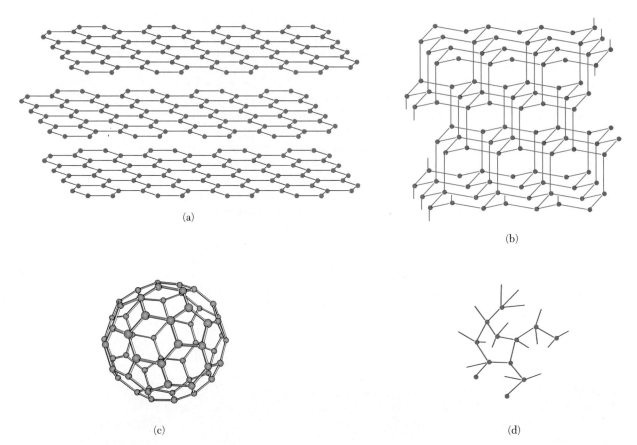

**Figure 5.17**
Solid carbon structures: (a) graphite, (b) diamond, (c) buckminsterfullerene, (d) soot.

Besides these, there are other types of solids. Some are extremely large molecules; each molecule may have $10^{22}$ or $10^{23}$ atoms, all connected by covalent bonds. In such a case, the entire crystal is one big molecule. We call such molecules **network solids** or **network crystals.** A good example is diamond (Fig. 5.17[b]). When you hold a diamond in your hand, you are holding a single gigantic molecule. Like ionic crystals, network crystals have very high melting points — if they can be liquefied at all. In many cases they cannot be. We can be fairly confident, then, that any solid with a relatively low melting point (less than 300°C) can only be a molecular solid and not an ionic crystal or a network crystal.

There are other very large molecules, called polymers and containing about $10^4$ to $10^7$ atoms, that do not form network crystals (Sec. 11.7).

## 5.12  FREEZING AND BOILING. PHASE CHANGES

The **melting point** is the temperature at which a solid changes to a liquid. The temperature at which the reverse process occurs is the **freezing point.** For any substance, the melting point is the same as the freezing point.

The old saying, "Diamonds are a girl's best friend," refers to the esthetic as well as to the economic value of diamonds in jewelry. A far more extensive use is found for diamonds that compete neither in size (carats) nor in reflective beauty with those used in rings and necklaces. Industrial diamonds, which are mostly manufactured from cheap graphite, have a variety of uses. Because of diamond's extreme hardness, they are used in drilling for oil, in the construction industry, in sawing and drilling granite and marble blocks, and in cutting grooves in concrete highways. Diamonds are also used in grinding lenses from glass blocks.

**Figure 5l**
Synthetic diamonds.

Imagine the following experiment: We heat a piece of ice that is initially at, let's say, −20°C. At first we don't see any change in the physical state of the ice. The temperature is increasing, but the appearance of the ice does not change. At 0°C, the ice melts and liquid water appears. While we continue the heating, more and more ice melts, but the *temperature stays constant* at 0°C until all the ice disappears. After all the ice has changed to liquid, the temperature of the water again begins to increase as more heat is added. These changes in the state of matter are called **phase changes. A phase** is any part of a system that looks uniform (homogeneous) throughout. Ice (solid) is one phase, and water (liquid) is another.

The diagram describing this experiment (Fig. 5.18) is called a heating curve. It shows that from −20 to 0°C added heat raises the temperature of the ice. After the ice reaches 0°C, **additional heat does not increase the temperature.** Instead, it melts the ice by breaking some of the hydrogen bonds that provide the rigid hexagonal structure (Fig. 5G.1) of solid ice. The heat necessary to melt 1 g of any solid is called its **heat of fusion.** For ice, this is a rather large value: 80 cal/g. During melting, the two phases (solid and liquid) coexist side by side. Only after the ice has completely melted does the temperature of the water rise, assuming that we continue to add heat. For every gram of water, it takes 100 cal to raise the temperature from 0 to 100°C, the normal boiling point. (Contrast this with the 80 cal that only melted the gram of ice but did not raise the temperature.)

The important aspect of this phase transition is that it is reversible. If we start with liquid water, let's say at room temperature, and cool it by immersing the container in a Dry Ice bath, the same process is observed, but in reverse.

The criterion of uniformity is the way it appears to our eyes and not as it is on the molecular level.

This was discussed in Section 1.8.

**Figure 5.18**
The heating curve of ice. The graph shows the effect of adding heat to 1 g of ice initially at −20°C.

The temperature drops until it reaches 0°C, and then the ice starts to crystallize. During the crystallization (phase transition) the temperature stays constant, but heat is given off. The amount of heat given off is exactly the same as the amount of heat absorbed during melting.

Similar considerations apply to the remaining part of the heating curve in Figure 5.18. When the temperature of the liquid reaches the boiling point, another phase change takes place: The liquid is vaporized. While vaporization is taking place, the temperature stays constant at the boiling point. The heat supplied at the boiling point does not increase the temperature; it changes the liquid to a vapor. The amount of heat necessary to vaporize 1 g of any liquid is called its **heat of vaporization.** For water this value is 540 cal/g (as with the heat of fusion, this is a very high value). As long as any liquid water remains, the temperature does not go above 100°C (at 1 atm pressure).

Transition from the solid state directly into the vapor state without going through the liquid state is called **sublimation** (Fig. 2.7[c]). Solids usually sublime only at reduced pressures (less than 1 atm). At high altitudes, where the atmospheric pressure is low, snow sublimes. Solid $CO_2$ (Dry Ice) sublimes at −78.5°C under 1 atm pressure. At 1 atm pressure, $CO_2$ can exist only as a solid or as a gas, never as a liquid.

## SUMMARY

Matter can exist in three different states: gaseous, liquid, and solid. Attractive forces between molecules tend to hold matter together, whereas kinetic energy tends to disorganize matter. At high temperatures (high kinetic energy), matter is in the gaseous state. Molecules in the gaseous state move randomly. Gases fill all the available space of their container. Gas molecules, in their random motion, collide with the walls of the container and thereby exert **pressure.** The **kinetic molecular theory** explains the behavior of gases. The pressure of the atmosphere is measured with a **barometer.** The units of pressure are 1 mm Hg = 1 torr and 760 mm Hg = 1 atm.

**Boyle's, Charles's,** and **Gay-Lussac's laws** describe the behavior of gases under different conditions. These laws are combined and expressed as the **combined gas law:**

$$\frac{P_1V_1}{T_1} = \frac{P_2V_2}{T_2}$$

**Avogadro's law** states that, under constant temperature and pressure, equal volumes of different gases contain the same number of molecules. The **ideal gas law,** $PV = nRT$, incorporates Avogadro's law into the combined gas law. Different gases in a mixture exert their own pressures independently. The total pressure is the sum of the partial pressures of the gases (**Dalton's law**).

**Intermolecular forces** are responsible for the condensation of gases into the liquid and solid states. The intermolecular forces discussed in the chapter are, in increasing order of magnitude, **London dispersion forces, dipole-dipole interactions,** and **hydrogen bonds.**

Liquids have **vapor pressures.** The most energetic molecules of liquids escape from the surface into the vapor phase and exert their own pressure. The vapor pressure of a liquid increases with increasing temperature. When the temperature of a liquid is raised so that the vapor pressure becomes equal to the atmospheric pressure, the liquid boils. The temperature at which this occurs is the **boiling point.** The boiling point of a liquid is determined by its molecular weight and shape and by intermolecular forces. Liquids have a **surface tension** that is dependent on the magnitude of the intermolecular forces.

Solids crystallize in well-formed geometrical shapes that often reflect the patterns in which the atoms are arranged within the crystals. Amorphous solids have in essence a frozen liquid structure. When a solid is heated, its temperature is raised until the melting point is reached; at the melting point there is a **phase change:** The solid becomes a liquid. After all the solid has melted, further heating raises the temperature of the liquid. At the boiling point there is a second phase change: The liquid becomes a gas; further heating can only increase the temperature of the gas. The heat necessary to convert 1 g of any liquid to vapor is the **heat of vaporization** of the liquid. The heat necessary to melt 1 g of any solid to liquid is its **heat of fusion.**

## KEY TERMS

Amorphous solid (Sec. 5.11)
Avogadro's law (Sec. 5.5)
Barometer (Sec. 5.3)
Boiling point (Sec. 5.9)
Boyle's law (Sec. 5.4)
Charles's law (Sec. 5.4)
Combined gas law (Sec. 5.4)
Dalton's law (Sec. 5.6)
Diffusion (Sec. 5.6)
Dipole-dipole attraction (Sec. 5.7)
Freezing point (Sec. 5.12)
Gay-Lussac's law (Sec. 5.4)
Graham's law (Sec. 5.6)
Heat of fusion (Sec. 5.12)
Heat of vaporization (Sec. 5.12)
Hydrogen bond (Sec. 5.7)
Ideal gas (Sec. 5.2)

Ideal gas law (Sec. 5.5)
Intermolecular forces (Secs. 5.1, 5.7)
Kinetic molecular theory (Sec. 5.2)
London dispersion forces (Sec. 5.7)
Manometer (Sec. 5.3)
Melting point (Sec. 5.12)
Network solid (Sec. 5.11)
Normal boiling point (Sec. 5.9)
Partial pressure (Sec. 5.6)
Phase (Sec. 5.12)
Pressure (Sec. 5.3)
STP (Sec. 5.5)
Sublimation (Sec. 5.12)
Surface tension (Sec. 5.10)
Vapor pressure (Sec. 5.9)
Volume (Sec. 5.2)

## PROBLEMS

**5.9**  What forces hold matter together?

### Gases and Pressure

**5.10**  In terms of the kinetic molecular theory, what causes (a) the pressure  (b) the temperature of gases?

**5.11**  A sphygmomanometer (Fig. 5C, page 134) reads a pressure of 120 mm Hg. Draw a diagram indicating the level of Hg in the evacuated reference arm and in the arm connected to the inflating cuff.

**5.12**  Express in atm:
(a)  20 mm Hg
(b)  22.7 in Hg
(c)  1760 mm Hg
(d)  standard pressure

**\*5.13**  Under certain weather conditions (just before rain) the air becomes less dense. How does this affect the barometric pressure reading?

### Gas Laws

**5.14**  If a gas undergoes a change in which both the pressure and the Kelvin temperature are doubled, does the volume of the gas increase, decrease, or stay the same? Explain.

**5.15**  At constant temperature, a sample of 3.0 L of Ne gas at 2.0 atm is compressed until the pressure is 3.524 atm. What is the new volume?

**5.16**  A sample of 23.0 L of $NH_3$ gas at 10.0°C is heated at constant pressure until it fills a volume of 50.0 L. What is the new temperature in °C?

**5.17**  If a sample of 4.17 L of $C_2H_6$ gas at 725°C is cooled to 175°C at constant pressure, what is the new volume?

**5.18**  A sample of $SO_2$ gas in a 3.71-L container is heated from 22 to 115°C. The initial pressure was 0.625 atm. If the volume does not change, what is the new pressure?

**5.19**  A sample of $B_2H_6$ gas in a 35-mL container is at a pressure of 450 mm Hg and a temperature of 625°C. If the gas is allowed to cool at constant volume until the pressure is 375 mm Hg, what is the new temperature in °C?

**\*5.20**  A certain quantity of helium gas is at a temperature of 27°C and a pressure of 1.00 atm. What will the new temperature be if its volume is doubled at the same time as its pressure is dropped to one-half of its original value?

**5.21**  A 480-mL balloon is filled with helium at 730 mm Hg barometric pressure. The balloon is released and climbs to an altitude where the barometric pressure is 395 mm Hg. What will the volume of the balloon be if, during the ascent, the temperature drops from 20 to 3°C?

**5.22**  An ideal gas occupies 56.44 L at 2.00 atm and 310.0 K. If the gas is compressed to 23.52 L and the temperature lowered to 281.3 K, what is the new pressure?

**5.23**  A sample of 30.0-mL Kr gas is at 756 mm Hg at 25.0°C. What is the new volume if the pressure is decreased to 325 mm Hg and the temperature to −12.5°C?

**5.24**  A sample of helium gas has a volume of 230.0 L at 5.50 atm and −203°C. What is its volume at STP?

**5.25**  A 26.4-mL sample of $C_2H_4$ gas has a pressure of 2.50 atm at 2.5°C. If the volume is increased to 36.2 mL and the temperature raised to 10°C, what is the new pressure?

**5.26**  The inside pressure of an automobile tire is 1.80 atm at a temperature of 20°C. What will the pressure in the tire be if, after 10 miles of driving, the temperature of the tire increases to 47°C?

**5.27**  In an autoclave, a constant amount of steam is generated at a constant volume. Under 1.00 atm pressure the steam temperature is 100°C. What pressure setting should be used to obtain a 165°C steam temperature for the sterilization of surgical instruments?

### Avogadro's Law and the Ideal Gas Law

**5.28**  A sample of a gas at 77°C and 1.33 atm occupies a volume of 50.3 L.
(a)  How many moles of the gas are present?
(b)  Does your answer depend on knowing what gas it is?

**5.29**  What volume is occupied by 0.56 mole of an ideal gas at a pressure of 0.85 atm and a temperature of 127°C?

**5.30**  What volume is occupied by 5.8 g of propane gas ($C_3H_8$) at 23°C and 1.15 atm pressure?

**5.31**  If 5.50 moles of BrCl gas occupies 20.2 L at 200.0°C, what is the pressure in mm Hg?

**5.32**  If 0.320 mole of $Cl_2$ gas occupies 11.2 L at 780 mm Hg, what is its absolute temperature?

**5.33** What volume, in milliliters, does 0.275 g of $UF_6$ gas occupy at 425°C and 365 torr?

**5.34** A hyperbaric chamber has a volume of 200 L. (a) How many moles of oxygen are needed to fill the chamber at room temperature (300 K) to 3.00 atm pressure? (b) How many grams of oxygen?

**5.35** What is the volume of 32 g of argon gas at 0°C and a pressure of 1.5 atm?

**\*5.36** At STP, 1 mole of every ideal gas occupies 22.4 L. What are the densities, in g/L, of $SO_2$, $CH_4$, and $F_2$ at STP?

**5.37** A sample of cyanogen gas weighing 6.802 g occupies 3.74 L at 25.0°C and 650.0 mm Hg. Calculate the molecular weight of cyanogen.

**5.38** How many molecules of CO are there in 100 L of CO at STP?

**5.39** Of the four properties $n$, $P$, $V$, $T$, which are extensive and which are intensive?

## Dalton's and Graham's Laws

**5.40** Three gases in open vessels are all at one end of a room. One is a perfume with the molecular formula $C_{10}H_{18}O_2$, the second is $NH_3$, and the third is HCl. If you sit in the corner farthest from the samples, which odor reaches you first? Which last?

**5.41** The pressure of $CH_4$ over water registered 342 mm Hg. If the water had a partial pressure of 18.2 mm Hg, what was the partial pressure of $CH_4$?

**\*5.42** The rate of diffusion of an unknown gas is three times lower than that of helium. If helium has a molecular weight of 4.0 and a diffusion rate of 60 cm/min, what is the molecular weight of the unknown gas?

## Intermolecular Forces

**5.43** Under which condition does water vapor behave most ideally:
(a) 0.5 atm, 400 K          (c) 0.01 atm, 500 K
(b) 4 atm, 500 K

**5.44** Which molecule has the strongest intermolecular interaction: (a) CO (b) HF (c) $Cl_2$?

**5.45** Explain how helium can become a liquid at a sufficiently low temperature even though the atoms of helium are nonpolar.

**5.46** What kind of intermolecular interactions are there in (a) $NH_3$ (b) $Cl_2$ (c) HBr

## Evaporation, Condensation, and Boiling Point

**5.47** Even though the $H_2S$ molecule has almost twice the mass of the $H_2O$ molecule, $H_2S$ is a gas at room temperature, while water is a liquid. Explain why this is so.

**5.48** Why does the temperature of a liquid drop as a result of evaporation?

**5.49** Two compounds, $n$-octane and tetramethylbutane, have the same molecular weight, and the London forces acting between the molecules are the same. Yet $n$-octane has a boiling point of 126°C, while tetramethylbutane boils at 107°C. Explain.

Octane

Tetramethylbutane

**5.50** Using Figure 5.13, estimate the vapor pressure of ethyl alcohol
(a) at 30°C          (c) at 70°C
(b) at 0°C

**5.51** Using Figure 5.13, find the boiling point of (a) water, (b) ethyl alcohol, and (c) chloroform at 400 mm Hg.

**5.52** The normal boiling point of a substance depends on both the mass of the molecule and the intermolecular interactions. Considering these, rank in order of increasing boiling point and explain your answer:
(a) HCl, HBr, HI          (b) $O_2$, HCl, $H_2O_2$

**5.53** Explain, in terms of the behavior of the molecules, why a puddle of water dries up after it stops raining.

**\*5.54** Kinetic energy increases with increasing temperature, whereas potential energy stays constant. On the basis of this simple principle, explain how a gas condenses to a liquid when the temperature is decreased.

**\*5.55** Considering the vapor pressure of ethyl alcohol and that of water at room temperature (Fig. 5.13), which would give a greater cooling effect if used to sponge off a patient?

## Surface Tension

**5.56** Define surface tension.

**5.57** Which has the higher surface tension: (a) water (b) $n$-hexane, $C_6H_{14}$?

## Solids and Phase Changes

**5.58** Define: (a) crystal  (b) amorphous solid.

**5.59** Considering their structures, explain why diamond is harder than buckminsterfullerene.

**\*5.60** Explain why buckminsterfullerene can accommodate a metal atom, such as lanthanum, within its structure, while neither graphite nor diamond can do this.

**5.61** Draw the heating curve for acetic acid if its melting point is 17°C and its normal boiling point 118°C.

**5.62** Define sublimation.

## Boxes

**5.63** (Box 5A) What is Dry Ice?

**5.64** (Box 5B) Why should you collect the mercury from a broken thermometer and store it under water?

**5.65** (Box 5C) A patient with hypertension has a blood pressure reading of 140/110. What do these numbers mean? If you took the reading on this patient with a sphygmomanometer, at what pressure would you hear the first tapping sound? At what pressure would you stop hearing the tapping sound?

**5.66** (Box 5D) What happens when we lower the diaphragm in our chest cavity?

**5.67** (Box 5E) In carbon monoxide poisoning the hemoglobin is incapable of transporting oxygen to the tissues. How does the oxygen get delivered to the cells when the patient is put into a hyperbaric chamber?

**5.68** (Box 5F) In circulating blood, where is the pressure of $CO_2$ the highest? The lowest?

**5.69** (Box 5G) If you fill a bottle with water, cap it, and freeze to −10°C, the bottle will crack. Explain.

**5.70** (Box 5G) Why is slow freezing of tissues more damaging than quick freezing?

**5.71** (Box 5I) What are industrial diamonds manufactured from?

## Additional Problems

**5.72** On the basis of what you have learned of intermolecular forces, predict which liquid has the highest surface tension:
   (a)  hexane, $C_6H_{14}$        (c)  water, $H_2O$
   (b)  chloroform, $HCCl_3$

**\*5.73** The density of a gas is 0.00300 g/cc at 100°C and 1.00 atm. What is the molecular weight of the gas?

**5.74** Which gas molecules diffuse faster, HCl or $C_3H_8$?

**5.75** An ideal gas occupies 387 mL at 275 mm Hg and 75°C. If the pressure is changed to 1.36 atm and the temperature increased to 105°C, what is the new volume?

**\*5.76** Because natural gas (methane) has no odor, gas companies add a warning gas so that leaks can be detected. Two candidates for this are $NH_3$ and $CH_3SH$, because of their strong odors. Which could be used to warn the household faster of an impending disaster? How much faster?

**5.77** If 60.0 g of $NH_3$ occupies 35.1 L under 2.58 atm pressure, what is the temperature of the gas, in °C?

**5.78** Differentiate between evaporation and sublimation.

**\*5.79** What is the density of $CH_4$ at 1 atm and at 35°C?

**\*5.80** Explain why gases are transparent.

**5.81** Given that 2.0 L of diborane gas weighs 2.48 g at STP, calculate its molecular weight.

**5.82** Distinguish among a molecular crystal, a network crystal, and an ionic crystal.

**5.83** The normal boiling point of hexane is 69°C, and that of pentane is 36°C. Predict which of these has a higher vapor pressure at 20°C.

**\*5.84** The CO and NO molecules have approximately the same mass. Which molecule has a greater intermolecular interaction? (*Hint:* Look at Table 3.2.)

# CHAPTER 6

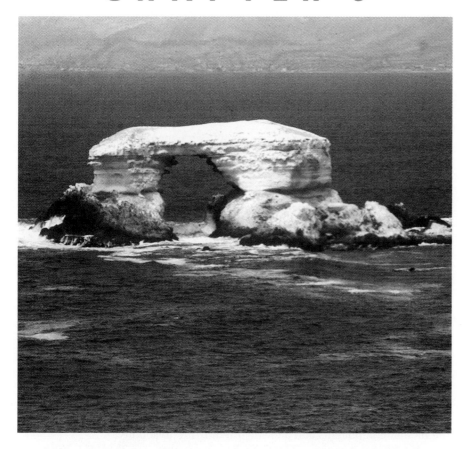

*The sea is the world's largest solution. (Courtesy of Viaggi Avventurosi.)*

# Solutions and Colloids

## 6.1  INTRODUCTION

In Chapter 5, we discussed pure substances—systems having only one component. Such systems are the easiest to study, so it was convenient to begin with them. In our daily lives, however, we more frequently encounter systems having more than one component. Such systems—for example, air, smoke, seawater, milk, blood, rocks—are **mixtures.**

If a mixture is uniform throughout, we call it a **homogeneous mixture.** In a homogeneous mixture the molecules (or ions) are thoroughly mixed. A homogeneous mixture is called a **solution** (Fig. 6.1).

In contrast, in most rocks we can see distinct regions separated from each other by well-defined boundaries. Such rocks are **heterogeneous mixtures.** For another example, a mixture of sand and sugar is heterogeneous. Even though sand and sugar are both present, we can easily distinguish between the two; the mixing is not at the molecular level (see also Fig. 2.2[c]).

**Figure 6.1**
Making a homogeneous solution. A green solid, nickel nitrate, is stirred into water, where it dissolves with stirring to form a homogeneous solution. (Photograph by Charles D. Winters.)

Many alloys are solid solutions; one example is stainless steel, which is mostly iron but also contains carbon, chromium, and other elements.

We normally do not use the terms "solute" and "solvent" in talking about solutions of gases in gases or solids in solids.

Mixtures are classified on the basis of how they look to the unaided eye. There are systems, however, that fall between homogeneous and heterogeneous mixtures. Cigarette smoke, muddy water, and blood plasma may look homogeneous, but they do not have the transparency of air or seawater. These mixtures are classified as **colloidal dispersions.** We will deal with such systems in Section 6.7.

Although mixtures can contain many components, we shall restrict our discussion mainly to two-component systems, with the understanding that everything we say can be extended to multicomponent systems.

## 6.2  TYPES OF SOLUTIONS

When we think of a solution, we normally think of a liquid. Liquid solutions are the most common kind, but there are also solutions that are gases or solids. In fact, all mixtures of gases are solutions. Because gas molecules are far apart from each other and there is much empty space between them, two or more gases can mix with each other in any proportions. Because the mixing is at the molecular level, a true solution always forms. That is, there are no heterogeneous mixtures of gases.

With solids, we are at the other extreme. Anytime we mix solids we get a heterogeneous mixture. Since even microscopic pieces of solid still contain many billions of particles (molecules, ions, or atoms), there is no way to obtain mixing at the molecular level. Homogeneous mixtures of solids (that is, true solutions) do exist, but we make them by first melting the solids, then mixing the molten components and allowing the mixture to solidify.

The five most common types of solutions are shown in Table 6.1. Examples of other types (gas in solid, liquid in gas, and so on) are also known but are much less important. In this chapter, we deal almost entirely with liquid solutions (the first three types listed in Table 6.1).

When a solution consists of a solid or a gas dissolved in a liquid, the liquid is called the **solvent** and the solid or gas the **solute.** A solvent may have several solutes dissolved in it, even of different types. A common example is beer, in which a liquid (alcohol), a gas (carbon dioxide), and a solid (malt) are dissolved in the solvent, water (Fig. 6.2).

When a liquid is dissolved in another liquid, there is a question about which is the solvent and which the solute. In most cases the one present in the greater amount is called the solvent, but there is no rigid rule about this.

| TABLE 6.1 | Types of Solutions | | |
|-----------|--------------------|--------------------------|--------------------------|
| Solute | Solvent | Appearance of Solution | Example |
| Gas      in | Liquid | Liquid | Carbonated water |
| Liquid   in | Liquid | Liquid | Wine |
| Solid    in | Liquid | Liquid | Salt water (saline solution) |
| Gas      in | Gas | Gas | Air |
| Solid    in | Solid | Solid | 14-carat gold |

The water vapor evaporated by the sun from oceans, lakes, and rivers condenses and forms clouds of water vapor that eventually fall as rain. The raindrops contain small amounts of $CO_2$, $O_2$, and $N_2$. Table 6A shows that, of these gases, $CO_2$ is the most soluble in water. When $CO_2$ dissolves in water, it forms carbonic acid, $H_2CO_3$.

**B o x   6A**

## Acid Rain

**TABLE 6A   Solubility of Some Gases in Water**

| Gas | Solubility (g per kg of $H_2O$ at 20°C and 1 atm) |
|---|---|
| $O_2$ | 0.0434 |
| $N_2$ | 0.0190 |
| $CO_2$ | 1.688 |
| $H_2S$ | 3.846 |
| $SO_2$ | 112.80 |
| $NO_2$ | 0.0617 |

The acidity caused by the $CO_2$ is not harmful. However, a serious acid rain problem is caused by contaminants that result from industrial pollution. Burning coal or oil that contains sulfur generates sulfur dioxide, which has a high solubility in water. Sulfur dioxide in the air is oxidized to sulfur trioxide. The combination of water and sulfur dioxide results in sulfurous acid, and sulfur trioxide and water give sulfuric acid. Smelting industries produce other soluble gases as well. The result is that in many parts of the world, especially downwind from heavily industrial areas, acid rain pours down on forests and lakes. This damages vegetation and kills fish. Such is the situation in the northeastern United States, where in the Adirondacks and New England areas, as well as in eastern Canada, acid rain has been observed with increasing frequency (Fig. 6A).

**Figure 6A**
Fir trees killed by acid rain, on Mount Mitchell in North Carolina. (Peter Arnold, Inc./© Ray Pfortner.)

## 6.3   CHARACTERISTICS OF SOLUTIONS

The following are some properties of true solutions:

**1.** The distribution of particles in a solution is uniform. Every part of the solution has exactly the same composition and properties as every other part. That, in fact, is the definition of homogeneous. This means that we cannot usually tell a solution from a pure solvent simply by looking at it. A glass of pure water looks the same as a glass of water containing dissolved salt or sugar. In some cases we can tell by looking, for example, if the solution is colored and we know that the solvent is colorless.

**2.** The components of a solution do not separate on standing. A solution of vinegar (acetic acid in water), for example, can remain in the closet for many years without separating out.

**3.** A solution cannot be separated into its components by filtration. Both the solvent and the solute pass through the filter paper.

**4.** For any given solute and solvent, it is possible to make solutions of many different compositions. For example, we can easily make a solution of 1 g of

**Figure 6.2**
Beer is a solution in which a liquid (alcohol), a solid (malt), and a gas ($CO_2$) are dissolved in the solvent water. (Photograph by Beverly March.)

We use the word "clear" to mean "transparent." A solution of copper sulfate in water is blue, but clear.

glucose in 100 g of water, or 2 g, or 6, or 8.7, or any amount at all, up to the solubility limit (Sec. 6.4).

**5.**    Solutions are almost always transparent. They may be colorless or colored, but you can usually see through them.

**6.**    Solutions can be separated into pure components, not by filtration but by such other methods as distillation (Sec. 2.2) and chromatography, which you may learn about in the laboratory portion of this course. The separation of a solution into its components is a physical, not a chemical, change.

## 6.4   SOLUBILITY

Suppose we wish to make a solution of table salt (NaCl) in water. We take some water, add a few grams of salt, and stir. At first we see the particles of salt suspended in the water, but soon all the salt dissolves. Now let us add more salt to the same solution and continue to stir. Again the salt all dissolves. Can we repeat this process indefinitely? The answer is no; there is a limit. The solubility of table salt at 25°C is 36.2 g per 100 g of water. If we add more salt than that, the excess solid does not dissolve but remains suspended as long as we keep stirring, and it sinks to the bottom after we stop.

Tables of solubility values are found in handbooks.

The **solubility** of a solid in a liquid is **the maximum amount of that solid that dissolves in a given amount of liquid at a given temperature.** Solubility is a *physical constant,* like melting point or boiling point. Each solid has a different solubility in every liquid. Some solids have a very low solubility in a particular solvent. We often call these solids insoluble. Others have a much higher solubility. We call these soluble, but even for soluble solids, there is always a solubility limit. The same is true for gases dissolved in liquids.

For solutions of liquids in liquids, however, the situation may be different. Some liquids are essentially insoluble in other liquids (gasoline in water), and others are soluble to a limit. For example, 100 g of water dissolves about 4 g of ethyl ether (another liquid). If we add more ether than that, we see two layers (Fig. 6.3). Some liquids, however, are *completely* soluble in other liquids, no matter how much is present. The most important example is ethyl alcohol and water, which form a solution no matter what quantities of each are mixed. We say that the water and ethyl alcohol are *miscible* in all proportions.

We use the word "miscible" to refer to a liquid dissolving in a liquid.

When a solvent contains all the solute it can hold at a given temperature, we call the solution **saturated.** Any solution containing a lesser amount of solute is **unsaturated.** It may seem surprising, but there are also solutions in which the solvent holds *more* solute than it can normally hold at a given temperature! Such solutions are called **supersaturated** and are fairly common (p. 167).

Whether a particular solute dissolves in a particular solvent depends on several factors.

### Nature of the Solute and Solvent

Here the rule is **"like dissolves like."** The more similar two compounds are, the more likely it is that one is soluble in the other. This is not an absolute rule, but it does apply in most cases.

**Figure 6.3**
Diethyl ether and water form two layers. A separatory funnel permits the bottom layer to be drawn off.

Diethyl ether

Water

When we say "like," we mostly mean alike in *polarity*. That is, polar compounds dissolve in polar compounds and nonpolar compounds dissolve in nonpolar compounds. For example, the liquids benzene and carbon tetrachloride are nonpolar compounds. They dissolve in each other, and other nonpolar materials, such as gasoline and camphor, dissolve in them. On the other hand, ionic compounds (such as NaCl) and polar compounds (such as table sugar) are insoluble in all of these solvents.

The most important polar solvent is water, and we have already seen that most ionic compounds are soluble in water, as are small covalent compounds that can form hydrogen bonds with water. It is worth noting that even polar molecules are usually insoluble in water if they cannot either react with the water or form hydrogen bonds with water molecules. Water as a solvent is discussed in Section 6.6.

Polar compounds dissolve in polar compounds because the positive end of one molecule attracts the negative end of the other.

## Temperature

For most solids and liquids that dissolve in liquids, the rule is **solubility increases with increasing temperature.** Sometimes the increase is great. For example, the solubility of glycine, a building block of proteins, is 52.8 g in 100 g of water at 80°C but only 33.2 g at 30°C.

If we prepare a saturated solution of glycine in 100 g of water at 80°C, it will hold 52.8 g of glycine. If we now allow it to cool to 30°C, where the solubility is 33.2 g, we might expect the excess glycine, 19.6 g, to come out of solution and form crystals. It often does, but on many occasions it does not. Even though the solution contains more glycine than the water can normally hold at 30°C, the excess may well stay in solution. This is an example of the supersaturated solution referred to earlier. The excess glycine stays in solution because the molecules need a surface on which to begin crystallizing, and no such surface is present.

Supersaturated solutions are not indefinitely stable, however. If we shake or stir the solution, we may find that all of the excess solid precipitates out at once. Another way to crystallize the excess solute is to add a grain of glycine

The difference between saturated, unsaturated, and supersaturated solutions is illustrated in Figure 6.4.

**Figure 6.4**
The difference in behavior of unsaturated, saturated, and supersaturated solutions when a crystal of the solute is added. (From Kotz and Purcell, *Chemistry and Chemical Reactivity.*)

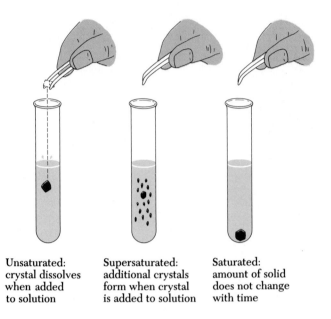

Unsaturated:
crystal dissolves
when added
to solution

Supersaturated:
additional crystals
form when crystal
is added to solution

Saturated:
amount of solid
does not change
with time

crystal. This is called **seeding.** The seed crystal provides the surface onto which the solute molecules can converge.

For gases, **solubility in liquids almost always decreases with increasing temperature.** This explains why a bottle of beer or a carbonated soft drink foams up when opened at room temperature. When the beverage is cold, the $CO_2$ it contains is more soluble. At room temperature it is less soluble, and more of it comes out of solution. This additional gaseous $CO_2$ increases the pressure. When the pressure is released, the gas pushes out some of the liquid.

The effect of temperature on the solubility of gases in water can have important consequences for fish. Oxygen is only slightly soluble in water, but fish need that oxygen to live. When the temperature of the water is increased, perhaps by the output from a power plant, the solubility of the oxygen decreases and may become so low that fish die. This is called *thermal pollution.*

### Pressure

Henry's law states that the solubility of a gas in a liquid is directly proportional to the pressure.

Pressure has little effect on the solubility of liquids or solids. For gases, however, **the higher the pressure, the greater the solubility of a gas in a liquid.** This is the basis of the hyperbaric medicine discussed in Box 5E. When the pressure is increased, more oxygen dissolves in the blood plasma and reaches the tissues at higher-than-normal pressures (2 to 3 atm).

### 6.5   CONCENTRATION  UNITS

The amount of a solute dissolved in a given quantity of solvent is called the **concentration.** This can be expressed in a number of ways. Some concentration units are better suited than others for some purposes. Sometimes qualitative terms are good enough. We say a solution is **dilute** or **concentrated.** This

## The Bends

Deep-sea divers encounter high pressures. In order for them to breathe properly under such conditions, oxygen must be supplied under pressure. At one time this was achieved with compressed air. As pressure increases, the solubility of the gases in the blood increases. This is especially true for nitrogen, which constitutes almost 80 percent of the air.

When a diver comes up and the pressure on the body decreases, the solubility of nitrogen in the blood decreases as well. As a consequence, the previously dissolved nitrogen in the blood and in the tissues starts to form small bubbles, especially in the veins. The formation of gas bubbles, called **the bends,** can prevent blood circulation. If allowed to develop uncontrolled, the resulting pulmonary embolism can be fatal.

If the ascent is gradual, the dissolved gases are removed by regular exhalation and diffusion through the skin. Divers use decompression chambers, where the high pressure is gradually reduced to normal pressure.

Because of the problem caused by nitrogen, divers often use a helium-oxygen mixture instead of air. The solubility of helium in blood is less affected by pressure than is that of nitrogen. The sudden decompression and ensuing bends are important not only in deep-sea diving but also in high-altitude flight, especially orbital flight.

tells us little about the concentration, but we know that a concentrated solution contains more solute than a dilute solution.

For most purposes, we need quantitative concentrations. For example, a nurse must know how much glucose to give to a patient. Many methods of expressing concentration exist, but in this chapter we deal only with the three most important: percent concentration, molarity, and parts per million (ppm).

### Percent Concentration

There are three different ways of representing percent concentration. The most common is *weight* of solute per *volume* of solution. If 10 g of sugar is dissolved in enough water so that the total volume is 100 mL, the concentration is 10 percent w/v. We need to know the total volume of the solution, not the volume of the solvent. For this purpose we use **volumetric flasks,** which come in different sizes. Volumetric flasks of different sizes are shown in Figures 6.5 and 6.6.

Percent concentration (% w/v) is the number of grams of solute in 100 mL of solution.

---

**EXAMPLE**

How would we make 500 mL of a 1.5 percent w/v solution of KCl in water?

**Answer**
We find out how many grams of KCl there are in this solution by taking the percentage:

$$\frac{1.5 \text{ g KCl}}{100 \text{ mL}} \times 500 \text{ mL} = 7.5 \text{ g KCl}$$

We then put 7.5 g of KCl into a beaker, add some water, and stir until the KCl is all dissolved. We then transfer the solution to a 500-mL volumetric flask and fill with water to the mark. We don't know exactly how much water we have added, but we do know that the total volume is 500 mL.

**PROBLEM 6.1**   How would you make 250 mL of a 4.4 percent w/v KBr solution in water? Assume that a 250-mL volumetric flask is available.

---

### EXAMPLE

If 3.7 g of sodium nitrate is dissolved in enough water to make 250.0 mL of solution, what is the percentage of $NaNO_3$ w/v?

### Answer

We divide the weight of the solute by the volume of solution and multiply by 100:

$$\frac{3.7 \text{ g}}{250.0 \text{ mL}} \times 100 = 1.5\% \text{ w/v}$$

The percentage is 1.5% w/v.

---

**PROBLEM 6.2**   If 7.7 g of lithium iodide is dissolved in enough water to make 400.0 mL of solution, what is the percentage of LiI w/v?

---

A second type of percentage unit is *weight* of solute per *weight* of solution (w/w). Calculations are essentially the same as with w/v except that the *weight* of solution is used instead of the volume. A volumetric flask is not used for these solutions. (Why not?)

Finally, there is *volume* of solute per *volume* of solution (v/v). This unit is used only for solutions of liquids in liquids, most notably alcoholic beverages. For example, 40 percent v/v ethyl alcohol in water means that 40 mL of ethyl alcohol has been added to enough water to make 100 mL of solution.*

A 40% v/v solution of ethyl alcohol in water is 80 proof. Proof in the United States is twice the percent concentration (v/v) of ethyl alcohol in water.

### Molarity

For many purposes it is easiest to express concentration by the weight or volume percentage methods just discussed, but when we want to focus on the *number of molecules* present we need another way. For example, a 5 percent solution of sugar in water does not contain the same concentration of solute molecules as a 5 percent solution of alcohol in water. That is why chemists often

---

* A 40 percent v/v ethyl alcohol solution contains not 60 mL of water for every 40 mL of alcohol, but somewhat less than that. *Volumes of liquids are not always additive.* When one liquid is dissolved in another, the new volume may be more or less than the sum of the volumes of the individual liquids because the molecules move either closer to each other (because of greater attraction) or farther apart.

use molarity. The **molarity** of a solution is defined as the **number of moles of solute dissolved in 1 L of solution.** Thus, in the same volume of solution, a 0.2-molar solution of sugar in water contains the same number of molecules of solute as a 0.2-molar solution of alcohol. In fact, this holds true for equal volumes of any solutions, as long as the molarities are the same. The units of molarity are moles per liter, and the symbol is $M$.

$$\text{Molarity } (M) = \frac{\text{moles } (n)}{\text{liters } (V)}$$

We can prepare a given volume of a given molarity in essentially the same way we prepare a solution of given w/v concentration, except that we use moles instead of grams in our calculations. We can always find out how many moles of solute there are in any solution of known molarity by the formula

Molarity × volume in liters = number of moles

$$M \times V = \text{moles}$$

The solution is then prepared as shown in Figure 6.5.

(a)

(b)

(c)

**Figure 6.5**
(a) A 0.0100 $M$ solution of $KMnO_4$ is made by adding enough water to 0.395 g of $KMnO_4$ to make 0.250 L of solution. (b) To ensure the correct solution volume, the $KMnO_4$ is placed in a volumetric flask and dissolved in a small amount of water. (c) After dissolving is complete, sufficient water is added to fill the flask to the mark. The flask contains 0.250 L of solution. (Photographs by Charles D. Winters.)

---

**EXAMPLE**

How do we prepare 2.0 L of a 0.15 $M$ solution of NaOH?

**Answer**

First we find out how many moles of NaOH there will be in this solution:

$$M \times V = \text{moles}$$

$$\frac{0.15 \text{ mole NaOH}}{1 \text{ L}} \times 2.0 \text{ L} = 0.30 \text{ mole NaOH}$$

We need 0.30 mole of NaOH to make up this solution. How many grams is this? To convert moles to grams, we multiply by the appropriate conversion factor (Sec. 4.3). The formula weight of NaOH = 40.0, so

$$0.30 \text{ mole NaOH} \times \frac{40.0 \text{ g NaOH}}{1 \text{ mole NaOH}} = 12 \text{ g NaOH}$$

We therefore weigh out 12 g of NaOH, put it into a beaker, add some water, stir until the solid has dissolved, and then transfer the solution to a 2-L volumetric flask and fill with water to the 2-L mark.

---

**PROBLEM 6.3**    How would you prepare 2.0 L of a 1.06 $M$ aqueous solution of KCl?

---

**EXAMPLE**

What is the molarity of a solution made by dissolving 75 g of glucose, $C_6H_{12}O_6$, in enough water to have 550 mL of solution?

**Answer**

We are looking for molarity, that is, moles per liter. First we find the number of moles of glucose, then the number of liters of solution, and then divide. To find moles of glucose, we multiply grams of glucose by the conversion factor 1 mole glucose/180.0 g glucose:

$$75 \text{ g glucose} \times \frac{1 \text{ mole glucose}}{180.0 \text{ g glucose}} = 0.42 \text{ mole glucose}$$

To find liters of solution, we divide mL by 1000:

$$550 \text{ mL} \times \frac{1 \text{ L}}{1000 \text{ mL}} = 0.55 \text{ L}$$

To find moles of glucose per liter of solution, we divide:

$$\frac{0.42 \text{ mole glucose}}{0.55 \text{ L solution}} = 0.76 \text{ } M$$

---

**PROBLEM 6.4**    If 2.3 g of KI is dissolved in enough solvent to make 250 mL of solution, what is the molarity of the solution?

---

**EXAMPLE**

The concentration of sodium chloride in blood serum is approximately 0.14 $M$. What volume of blood serum contains 2.0 g of NaCl?

**Answer**

Since we know the concentration in *moles* per liter, we must first find out how many moles there are in 2.0 g of NaCl (formula weight = 58.5) by multiplying by the proper conversion factor:

$$2.0 \text{ g NaCl} \times \frac{1 \text{ mole NaCl}}{58.5 \text{ g NaCl}} = 0.034 \text{ mole NaCl}$$

We next find liters:

$$0.034 \text{ mole NaCl} \times \frac{1 \text{ L}}{0.14 \text{ mole NaCl}} = 0.24 \text{ L} = 240 \text{ mL}$$

Blood serum is the liquid part of the blood that remains after the cellular particulates and the fibrinogen have been removed.

**PROBLEM 6.5** If 0.300 $M$ glucose ($C_6H_{12}O_6$) solution is available for intravenous infusion, how many milliliters are needed to deliver 10.0 g of glucose?

**EXAMPLE**

How many grams of HCl are there in 225 mL of concentrated HCl, whose concentration is 6.00 $M$?

**Answer**

First we find out how many moles of HCl there are:

$$M \times V = \text{moles}$$

$$\frac{6.00 \text{ moles}}{1 \text{ L}} \times 0.225 \text{ L} = 1.35 \text{ moles}$$

We get grams by multiplying by the conversion factor 1 mole = 36.5 g:

$$1.35 \text{ moles HCl} \times \frac{36.5 \text{ g HCl}}{1 \text{ mole HCl}} = 49.3 \text{ g HCl}$$

**PROBLEM 6.6** How many grams of $H_2SO_4$ are there in 50.0 mL of a 3.0 $M$ $H_2SO_4$ solution?

**Dilution**

We frequently prepare solutions by diluting more concentrated solutions rather than by weighing out pure solute (Fig. 6.6). Since we are adding only solvent, the number of moles of solute remains unchanged. Before we dilute, the equation that applies is

$$M_1V_1 = \text{moles}$$

(a)

(b)

(c)

**Figure 6.6**
Making a solution by dilution. (a) A 100-mL volumetric flask is filled to the mark with 0.100 $M$ $K_2Cr_2O_7$. (b) This is transferred to a 1.00-L volumetric flask. (c) The 1.00-L volumetric flask is filled to the mark with distilled water. The concentration of the now diluted $K_2Cr_2O_7$ is 0.0100 $M$. (Photographs by Charles D. Winters.)

After we dilute, the volume and molarity have both changed and we have

$$M_2V_2 = \text{moles}$$

but the number of moles of solute is the same before and after, so we can say that

$$M_1V_1 = M_2V_2$$

This is a handy equation that we can use for dilution problems.

---

### EXAMPLE

How do we prepare 200 mL of a 3.5 $M$ solution of acetic acid if we have a bottle of concentrated acetic acid (6.0 $M$)?

**Answer**
We use the equation

$$M_1V_1 = M_2V_2$$

$$\frac{6.0 \text{ moles}}{1 \text{ L}} \times V_1 = \frac{3.5 \text{ moles}}{1 \text{ L}} \times 0.200 \text{ L}$$

$$V_1 = \frac{(3.5 \text{ moles/L})(0.200 \text{ L})}{6.0 \text{ moles/L}} = 0.12 \text{ L}$$

We put 0.12 L, or 120 mL, of the concentrated acetic acid in a 200-mL volumetric flask, add some water and mix, and then fill to the mark with water.

---

**PROBLEM 6.7** How would you prepare 200.0 mL of 0.200 $M$ HCl from a 10.0 $M$ stock HCl solution?

---

A similar equation,

$$\%_1V_1 = \%_2V_2$$

can be used for dilution problems involving percent concentrations.

---

**EXAMPLE**

How do we prepare 500 mL of a 0.5 percent w/v solution of NaOH if we have a stock solution of 50 percent w/v NaOH on hand?

**Answer**

$$(50\%)V_1 = (0.5\%)(500 \text{ mL})$$

$$V_1 = \frac{(0.5\%)(500 \text{ mL})}{50\%} = 5 \text{ mL}$$

We add 5 mL of the concentrated solution to a 500-mL volumetric flask, add some water and mix, and then fill to the mark with water.

---

**PROBLEM 6.8**   A concentrated solution of 15 percent w/v KOH solution is available. How would you prepare 20.0 mL of a 0.10 percent w/v KOH solution?

---

### Parts per Million

Sometimes we need to deal with very dilute solutions, for example, 0.0001 percent. In such cases it is more convenient to use **parts per million (ppm).** For example, if drinking water is polluted with lead ions to the extent of 1 ppm, it means that there is 1 mg of lead ions in 1 kg (1 L) of the water. Some solutions are so dilute that we use **parts per billion (ppb).** Modern methods of analysis allow us to detect concentrations this small. Some substances are harmful even at concentrations measured in ppb. One such substance is dioxin, an impurity in the 2,4,5-T herbicide sprayed by the United States as a defoliant in Vietnam.

$$\text{ppm} = \frac{\text{g solute}}{\text{g solution}} \times 10^6$$

$$\text{ppb} = \frac{\text{g solute}}{\text{g solution}} \times 10^9$$

## 6.6   WATER AS A SOLVENT

Water has many unusual properties. Among these are very low vapor pressure, high heat of vaporization, high boiling point, and high surface tension. All of these are caused by the hydrogen-bonded structure of water. In Box 5G we noted that ice has a lower density than liquid water, a property that causes it to float on top of water.

Water covers about 75 percent of the earth's surface in the form of oceans, ice caps, glaciers, lakes, and rivers. Water vapor is always present in the atmosphere. Life evolved in water, and without it life as we know it could not exist. The human body is about 60 percent water. This water is located both inside the cells of the body (intracellular) and outside (extracellular). Most of the important chemical reactions in living tissue occur in aqueous solution; water serves as a solvent to transport reactants and products from one place in the body to another. Water is also itself a reactant or product in many biochemical reactions.

The properties that make water such a good solvent are its polarity and its hydrogen-bonding capacity. Water dissolves many compounds because it **solvates** ions and molecules. Solutions in which water is the solvent are called **aqueous solutions.**

**Figure 6.7**
Water molecules remove anions and cations from the surface of an ionic solid.

Surface of ionic solid

Solvated ions

**Figure 6.8**
Anions and cations solvated by water.

## How Water Dissolves Ionic Compounds

We learned in Section 3.3 that ionic compounds in the solid state are composed of a regular array of ions. The crystal is held together by ionic bonds, which are attractions between positive and negative ions. Water, of course, is a polar molecule. When a solid ionic compound is added to water, the ions at the surface of the crystal become surrounded by water molecules. The negative ions (anions) attract the positive ends of water molecules, and the positive ions (cations) attract the negative ends of water molecules (Fig. 6.7). Each ion attracts two or three water molecules, and the combined force of attraction is usually enough for the ion to be completely dislodged from its position in the crystal. Once the ion is removed from the crystal, it is *completely* surrounded by water molecules (Fig. 6.8). We call these ions **hydrated.** A more general term is **solvated.** The **solvation layer**—that is, the surrounding shell of water molecules—acts as a cushion. It prevents a solvated anion from colliding directly with a solvated cation and thus keeps the solvated ions in solution.

Not all ionic solids are soluble in water. Some—for example, AgCl and $Ca_3(PO_4)_2$—are insoluble in water. However, even "insoluble" solids have at least a tiny solubility. Some rules for predicting solubilities were given in Section 4.8.

Many buildings and monuments in urban areas throughout the world are decaying, ruined by air pollution. The main culprit in this process is gypsum. The most common stones used for buildings and stone monuments are limestone and marble, both of which are largely calcium carbonate. In the absence of polluted air these can last for thousands of years, and many statues and buildings from ancient times (Babylonian, Egyptian, Greek, and others) have survived until recently with little change, and indeed still do, in rural areas. But in urban areas the air is polluted with $SO_2$ and $SO_3$, which mostly come from the combustion of coal and petroleum products containing small amounts of sulfur compounds as impurities (Box 6A). They react with the calcium carbonate at the surface of the stones to form calcium sulfate. When calcium sulfate interacts with water in the rain it forms the dihydrate gypsum. The problem is that gypsum has a larger volume than the original marble or limestone, causing the surface of the stone to expand, resulting in flaking. This is how statues such as those in the Parthenon (in Athens, Greece) become noseless and later faceless.

## Box 6C

## Hydrates and Air Pollution: The Decay of Buildings and Monuments

**Figure 6C**
Effects of air pollution on statues. The photo at the left was taken at the Lincoln Cathedral in England in 1910; the one on the right in 1984. (Dean and Chapter of Lincoln.)

## Solid Hydrates

The attraction between ions and water molecules is so strong in some cases that water molecules are an integral part of the crystal structure of many solids. Water molecules in a crystal are called **water of hydration.** The substances that contain water in their crystals are themselves called **hydrates.** For example, gypsum and plaster of Paris are hydrates of calcium sulfate: Gypsum is $CaSO_4 \cdot 2H_2O$, calcium sulfate dihydrate, and plaster of Paris is $(CaSO_4)_2 \cdot H_2O$, calcium sulfate monohydrate.

Some of these crystals hold their water tenaciously. To remove it, the crystals must be heated for some time at a high temperature (Fig. 6.9). The

The dot in the formula indicates that the $H_2O$ is present in the crystal, but it is not covalently bonded to the ions of $Ca^{2+}$ or $SO_4^{2-}$.

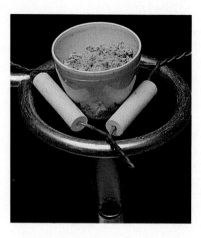

**Figure 6.9**
When blue hydrated copper sulfate is strongly heated in a crucible, it changes to the white anhydrous salt. (Photograph by Charles D. Winters.)

crystal without its water is called **anhydrous.** In many cases the anhydrous crystals are so strongly attracted to water that they take it from the water vapor in the air. That is, some anhydrous crystals become hydrated on standing in the open air. Crystals that do this are called **hygroscopic.**

If we want hygroscopic compounds to remain anhydrous, they must be placed in sealed containers that contain no water vapor.

Hydrated crystals often look different from the anhydrous forms. For example, copper sulfate pentahydrate, $CuSO_4 \cdot 5H_2O$, is blue, but the anhydrous form is white (see Fig. 6.9).

The difference between hydrated and anhydrous crystals can sometimes have an effect in the body. For example, the compound sodium urate in the anhydrous form exists as spherical crystals, but in the monohydrate the crystals are needle-shaped (Fig. 6.10). The deposition of sodium urate monohydrate in the joints (mostly in the big toe) causes gout.

(a)

**Figure 6.10**
Sodium urate crystals. (a) Spherical anhydrous crystals. (b) The needle-shaped monohydrate crystals that cause gout. (Courtesy of P. A. Dieppe, P. A. Bacon, A. N. Bamji and I. Watt and the Gower Medical Publishing Co., Ltd., London.)

(b)

### Electrolytes

Ions in water migrate from one place to another, maintaining their charge. Because of this, solutions of ions conduct electricity. They can do this because the ions in the solution migrate independently; as shown in Figure 6.11, the cations migrate to a negative electrode, called the **cathode,** and the anions migrate to a positive electrode, the **anode.** In doing so, they carry the electric current and complete the circuit initiated by the battery, and the electric bulb lights up.

Substances that conduct an electric current when dissolved in water or when in the molten state are called **electrolytes.** Sodium chloride is an electrolyte. Hydrated $Na^+$ ions carry positive charge and hydrated $Cl^-$ ions carry negative charge. Substances that do not conduct electricity are called **nonelectrolytes.** Distilled water is a nonelectrolyte. The light bulb shown in Figure 6.11 does not light up if only distilled water is placed in the container. However, when tap water is placed in the beaker, the bulb lights dimly. Tap water contains enough ions to carry electricity, but the concentration of these ions is so low that only a small amount of electricity is conducted.

This experiment shows that electric conductance depends on the concentration of ions. **The higher the ion concentration, the greater the electric conductance.** However, there are differences in electrolytes. If we take a 0.1 $M$ aqueous solution of NaCl and compare it with a 0.1 $M$ HF solution, we find that the NaCl solution lights a bulb brightly, but the HF solution lights it only dimly. We might have expected the two solutions to behave similarly. They have the same concentration, 0.1 $M$, and each "molecule" provides two ions, a cation and an anion ($Na^+$ and $Cl^-$; $H^+$ and $F^-$). The reason they behave differently is that, whereas $Na^+Cl^-$ dissociates completely to two ions (each hydrated and each moving independently), in the case of HF only a few molecules are dissociated into ions. Most of the HF molecules do not dissociate, and undisso-

**Figure 6.11**
Conductance by an electrolyte.

Battery

Anode
(positive electrode)

Cathode
(negative electrode)

**Box 6D**

**Hypoglycemia and Potassium Balance**

In the body, potassium ions are mainly concentrated inside the cells, where the concentration is about 0.158 $M$. The $K^+$ concentration in the blood and in the interstitial fluids is much lower (about 0.004 $M$). Among the functions of $K^+$ in the heart cells is maintenance of the proper contractions, and the $K^+$ ions in the extracellular fluids help to control nerve transmission to the muscles.

In a condition called **hypoglycemia,** the body output of insulin is elevated and blood sugar is depleted. This may suddenly shift the already small amount of $K^+$ from the extracellular media into the cells. The general result is lack of nerve impulses going to the muscles and to the extremities of the body. Muscular weakness and numbness of fingers and toes are symptoms of low $K^+$ content. Since the heartbeat is also influenced, later symptoms may include tachycardia (fast heartbeat) and, still later, weak pulse and falling blood pressure. Potassium may be obtained from food (bananas, orange and pineapple juice, veal, chicken, and pork) or, if necessary, intravenously as KCl solution. The latter is used to prevent severe lack of $K^+$ from causing cardiac arrest.

Sweating causes loss of $K^+$ ions. In hot weather, when athletes perspire heavily, they lose a large amount of $K^+$. That is why football and baseball players, long-distance runners, and other athletes often suffer serious muscle cramps in such weather.

ciated molecules do not conduct electricity. Compounds that dissociate completely are called **strong electrolytes,** and those that dissociate into ions only partially are called **weak electrolytes.**

Electrolytes are important components of the body because they help to maintain the acid-base balance and the water balance. The most important cations in body tissues are $Na^+$, $K^+$, $Ca^{2+}$, and $Mg^{2+}$. The most abundant anions in the body are bicarbonate, $HCO_3^-$; chloride, $Cl^-$; monohydrogen phosphate, $HPO_4^{2-}$; and dihydrogen phosphate, $H_2PO_4^-$.

### How Water Dissolves Covalent Compounds

Water is a good solvent not only for ionic compounds; it also dissolves many covalent compounds. In a few cases, the covalent compounds dissolve because they *react* with the water. An important example is the covalent compound HCl. HCl is a gas (with a penetrating, choking odor) that attacks the eyes. When dissolved in water, the HCl molecules react with water to give ions:

$$HCl + H_2O \longrightarrow Cl^- + \quad H_3O^+$$
$$\text{Hydronium ion}$$

Another example is the gas sulfur trioxide, which reacts as follows:

$$SO_3 + 2H_2O \longrightarrow HSO_4^- + H_3O^+$$

Because HCl and $SO_3$ are completely converted to ions in dilute aqueous solution, these solutions are ionic solutions and behave just as other electrolytes do (they conduct a current) even though HCl and $SO_3$ are themselves covalent compounds, unlike salts such as NaCl.

However, most covalent compounds that dissolve in water do not react with water. They dissolve because the water molecules surround the entire

In aqueous solution, $H^+$ does not exist; it combines with water and forms hydronium ion, $H_3O^+$.

covalent molecule and solvate it. For example, when methanol, $CH_3OH$, is dissolved in water, the methanol molecules are solvated by the water molecules (Fig. 6.12).

There is a simple way to predict which covalent compounds dissolve in water and which do not. **Covalent compounds dissolve in water if they can form hydrogen bonds with water, provided that the molecules are fairly small.** You may recall from Section 5.7 that hydrogen bonding is possible between two molecules if one of them contains an O, N, or F atom and the other contains an O—H, N—H, or F—H bond. Every water molecule contains an O atom and O—H bonds. Therefore, **water can form hydrogen bonds to any molecule that contains an O, N, or F atom or an O—H, N—H, or F—H bond.** If these molecules are small enough, they will be soluble in water. How small? In general, they should have no more than about three carbon atoms for each O, N, or F.

For example, acetic acid, $CH_3COOH$, is soluble in water, but benzoic acid, $C_6H_5COOH$, is not. Similarly, ethyl alcohol, $C_2H_5OH$, is soluble, but dipropyl ether, $C_6H_{14}O$, is not. Table sugar, $C_{12}H_{22}O_{11}$, is very soluble in water. Though it contains a large number of carbon atoms, it has so many oxygen atoms that it forms many hydrogen bonds with water molecules, and thus this molecule is very well solvated. **Covalent molecules that do not contain O, N, or F atoms are almost always insoluble in water,** except in the rare cases where they react with water. For example, methyl alcohol, $CH_3OH$, is infinitely soluble in water, but methyl chloride, $CH_3Cl$, is not.

## Water in the Body

Water is important in the body not only because it dissolves ionic substances as well as some covalent compounds, but also because it hydrates all polar molecules in the body. It thus serves as a vehicle to carry most of the organic compounds, nutrients, and fuels the body uses, as well as waste material. Blood and urine are but two examples of aqueous body fluids.

Beyond that, the hydration by water of macromolecules such as proteins, nucleic acids, and polysaccharides allows the proper motions within these molecules.

## 6.7 COLLOIDS

Up to now, we have been discussing solutions. In a true solution the maximum diameter of the solute particles is about 1 nm. If the diameter of the solute particles is larger than this, we no longer have a true solution—we have a colloid. In a **colloid** (or **colloidal dispersion**), the diameter of the solute particles ranges from about 1 to 1000 nm. Particles of this size usually have a very large surface area, and that accounts for the two basic characteristics of colloidal systems:

**1.** They scatter light and therefore appear turbid, cloudy, or milky.
**2.** Although colloidal particles are large, they form stable dispersions; that is, they do not form separate phases that settle out.

Like true solutions, colloids can exist in a variety of phases (Table 6.2).

**Figure 6.12**
Solvation of a covalent compound by water. The dotted lines represent hydrogen bonds.

**TABLE 6.2  Types of Colloidal Systems**

| Type | Example |
| --- | --- |
| Gas in gas | None |
| Gas in liquid | Whipped cream |
| Gas in solid | Marshmallows |
| Liquid in gas | Clouds, fog |
| Liquid in liquid | Milk, mayonnaise |
| Liquid in solid | Cheese, butter |
| Solid in gas | Smoke |
| Solid in liquid | Jelly, plasma extender |
| Solid in solid | Dried paint |

$1 \text{ nm} = 1 \times 10^{-9} \text{ m}$

**Figure 6.13**
The Tyndall effect. The test tube on the left contains a true solution; the one on the right, a colloidal solution of starch. A laser beam, generated by the apparatus at the far left, passes through both tubes. In the colloidal solution the beam leaves a trail of scattered light. In the true solution it is invisible. (Photograph by Beverly March.)

Figure 6.14 shows ordinary household dust particles magnified 2200 times.

All colloids exhibit the following characteristic effect. When we shine light through a colloidal system and look at the system from a 90° angle, we see the pathway of the light without seeing the individual colloidal particles (Fig. 6.13). This is called the **Tyndall effect** and is due to light scattering. Smoke, starch solutions, and fog, to name a few examples, all exhibit the Tyndall effect. We are all familiar with the sunbeams that can be seen when sunlight passes through dusty air. This, too, is an example of the Tyndall effect. Ordinary solutions do not show this effect.

Colloidal systems are stable. Mayonnaise stays emulsified and does not separate into oil and water. When the size of colloidal particles is larger than about 1000 nm, the system is unstable and separates into phases. Such systems are called **suspensions.** For example, if we take a lump of soil and disperse it in water, we get a muddy suspension. The soil particles are anywhere from $10^3$ to $10^6$ nm in diameter. The muddy mixture scatters light and therefore appears turbid, but it is not a stable system. If left alone, the soil particles soon settle to the bottom, with clear water above the sediment. Therefore, soil in water is a suspension, not a colloidal system.

The properties of true solutions, suspensions, and colloids are summarized in Table 6.3.

**Figure 6.14**
Ordinary household dust particles, magnified 2200 times. (Courtesy of David Scharf.)

**TABLE 6.3   Properties of Three Types of Mixtures**

| Property | Solutions | Colloids | Suspensions |
|---|---|---|---|
| Particle size (nm) | 0.1–1.0 | 1–1000 | >1000 |
| Filterable with ordinary paper | No | No | Yes |
| Homogeneous | Yes | Borderline | No |
| Settles on standing | No | No | Yes |
| Behavior to light | Transparent | Tyndall effect | Translucent or opaque |

Oil and water do not mix. Even when we stir them vigorously and the oil droplets are dispersed in the water, the two phases separate as soon as we stop stirring. However, there are a number of stable colloidal systems made of oil and water; these are **emulsions.** For example, the oil droplets in milk are dispersed in aqueous solution. This is possible because in milk there is a protective colloid, the milk protein called casein. The casein molecules surround the oil droplets, and since they are polar and carry a charge, they protect and stabilize the oil droplets. Casein is thus an **emulsifying agent.** A similar emulsifying agent is egg yolk, which is the ingredient in mayonnaise that coats the oil droplets and prevents them from separating.

What makes a colloidal dispersion stable? To answer this question, we must first realize that the colloidal particles are in constant motion. Just look at the dust particles dancing in a ray of sunlight that enters your room. Actually you are not seeing the dust particles themselves; they are too small to see. What you are seeing are flashes of scattered light. The motion of the dust particles dispersed in air is a random, chaotic motion. It is the motion of any colloidal particle suspended in a solvent. This is called **Brownian motion** (Fig. 6.15).

The cause of Brownian motion is the buffeting and collisions the colloidal particles suffer from the solvent molecules (in the case of the dust particles, the solvent is air). This constant Brownian motion creates favorable conditions for collisions between colloidal particles. When such large particles collide, they stick together, combine to give larger particles, and finally settle out of the solution.

So why do colloidal particles remain in solution? For two reasons:

**1.** Most colloidal particles carry a large *solvation layer.* If the solvent is water, as in the case of protein molecules in the blood, the colloidal particles are surrounded by a large number of water molecules, which move together with the colloidal particles and cushion them. When two colloidal particles collide as a result of Brownian motion, they do not actually touch each other; only the solvent layers collide. This way, the particles do not stick together and thus stay in solution.

**2.** The large surface area of colloidal particles acquires *charges* from the solution. All colloids in a particular solution acquire the same kind of charge, for

Brownian motion is named for Robert Brown (1773–1858), the Scottish botanist who first observed it in 1827.

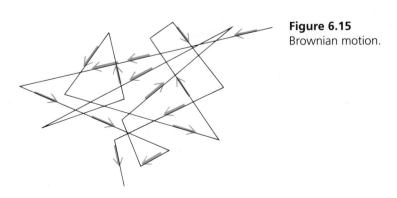

**Figure 6.15**
Brownian motion.

example, negative charge. This leaves a net positive charge in the dispersion medium. When a charged colloidal particle encounters another colloidal particle, the two repel each other because of their like charge.

Thus, the combined effect of solvation layer and surface charge keeps colloidal particles in a stable dispersion. By using these effects, chemists can either increase or decrease the stability of a colloidal system. If we want to get rid of a colloidal dispersion, we can remove either the solvation layer or the surface charge, or both. For example, proteins in the blood form a colloidal dispersion. If we want to isolate a protein from blood, we may want to precipitate it. We can accomplish this in two ways: either remove the hydration layer or remove the surface charges. If we add solvents such as ethanol or acetone, which have great affinity for water, the water will be removed from the solvation layer of the protein, and then the unprotected protein molecules will stick together when they collide and form a sediment. Similarly, by adding electrolytes such as NaCl to the solution, we can remove the charges from the surface of the proteins (by a mechanism that is too complicated to discuss here). Without their protective charges, two protein molecules no longer repel each other. Instead, when they collide they stick together and precipitate from the solution.

## 6.8   COLLIGATIVE PROPERTIES. FREEZING-POINT DEPRESSION

Certain properties of solutions depend only on the *number* of solute particles dissolved in a given amount of solvent and not on the *nature* of these particles. Such properties are called **colligative properties.** For example, one such property is **freezing-point depression.** One mole ($6.02 \times 10^{23}$ particles) of *any* solute dissolved in 1000 g of water lowers the freezing point of the water by 1.86°C. The *nature* of the solute does not matter, only the number of particles.

This principle is used in a number of practical ways. In the winter we use salts (sodium chloride and calcium chloride) to melt snow and ice on our streets. The salts dissolve in the melting snow and ice, and this lowers the freezing point of the water. Another application is the use of antifreeze in automobile radiators. Since water expands on freezing (Box 5G), the water in a car's cooling system can crack the engine block of a parked car when the outside temperature falls below 0°C. The addition of antifreeze prevents this because it makes the water freeze at a much lower temperature. The most common antifreeze is ethylene glycol, $C_2H_6O_2$.

*Note that in preparing a solution for this purpose we do not use molarity. That is, we do not need to measure the total volume of the solution.*

---

**EXAMPLE**

If we add 275 g of ethylene glycol per 1000 g of water in a car radiator, what will the freezing point of the solution be?

**Answer**

Since $C_2H_6O_2$ has a molecular weight of 62.0, 275 g contains

$$275 \text{ g } C_2H_6O_2 \times \frac{1 \text{ mole } C_2H_6O_2}{62.0 \text{ g } C_2H_6O_2} = 4.44 \text{ moles } C_2H_6O_2$$

Each mole lowers the freezing point by 1.86°C, so the freezing point is lowered by

$$\frac{1.86°C}{1 \text{ mole}} \times 4.44 \text{ moles} = 8.26°C$$

The freezing point of the water will be lowered from 0°C to −8.26°C, and the radiator will not crack if the outside temperature remains above −8.26°C (17.1°F).

**PROBLEM 6.9**   If we add 215 g of $CH_3OH$ to 1000 g of water, what will be the freezing point of the solution?

If a solute is ionic, then each mole of solute dissociates (breaks up) to more than 1 mole of particles. For example, if we dissolve 1 mole (58.5 g) of NaCl in 1000 g of water, the solution contains 2 moles of solute particles: 1 mole each of $Na^+$ and $Cl^-$. The freezing point of water will be lowered by *twice* 1.86°C, or 3.72°C.

**EXAMPLE**

What will be the freezing point of the solution if we dissolve 1 mole of $K_2SO_4$ in 1000 g of water?

**Answer**
One mole of $K_2SO_4$ dissociates to produce 3 moles of ions: 2 moles of $K^+$ and 1 mole of $SO_4^{2-}$. The freezing point will be lowered by $3 \times 1.86°C = 5.58°C$, and the solution will freeze at −5.58°C.

**PROBLEM 6.10**   What will be the freezing point of the solution if 2.5 moles of $Al(NO_3)_3$ is added to 1000 g of water?

Besides freezing-point depression, there are several other colligative properties, including vapor-pressure lowering, boiling-point elevation, and osmotic pressure. We discuss only the last of these because biologically it is the most important.

## 6.9  OSMOTIC PRESSURE

In order to understand osmotic pressure, let us consider the beaker shown in Figure 6.16(b). In this beaker two compartments are separated by an osmotic semipermeable membrane. A **semipermeable membrane** is a thin slice of some material, such as cellophane, that contains very tiny holes (far too small for us to see) that are big enough to let small solvent molecules pass through, but not big enough to let large solute molecules pass. In the right-hand compartment of the beaker is pure solvent—in this example, water; in the left-hand compartment is a solution of sugar in water.

An osmotic membrane is a very selective semipermeable membrane that allows only solvent molecules, and nothing else, to pass through.

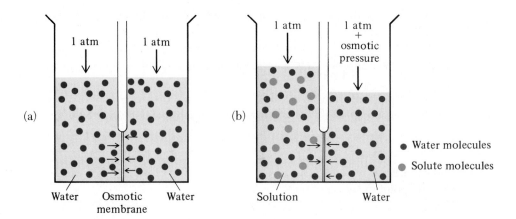

**Figure 6.16**
Osmotic pressure. (a) Two compartments separated by an osmotic semipermeable membrane both contain only solvent molecules that can pass through the membrane. (b) The compartment on the right contains only solvent, the one on the left both solute and solvent. Solute molecules cannot pass through the membrane. The solvent molecules move to the left compartment in an effort to dilute the solution, raising the liquid level on that side.

The sugar molecules are too big to go through the membrane and thus are forced to remain in the left-hand compartment. The water molecules, however, easily go back and forth; as far as they are concerned, there is no membrane. Therefore, water molecules go from the right to the left side because *they will try to dilute the sugar solution.* After all, if the membrane were not there, this is exactly what would happen. After a time, the two sides would have equal concentration, even without stirring. But the membrane is there, which means that the two sides cannot achieve equal concentration because none of the sugar molecules can get to the right side. Nevertheless, the water molecules keep trying to get to the left side. **This passage of solvent molecules from the dilute to the concentrated side of a semipermeable membrane** is called **osmosis.** When a significant number of water molecules have moved from the right to the left side, the liquid level on the right side goes down and that on the left side goes up, as shown in Figure 6.16(b).

The process cannot continue indefinitely because a liquid seeks its own level and because gravity prevents the difference in levels from becoming too great. Eventually the process stops with the levels unequal. The levels can be made equal again if we apply an external pressure to the higher side. **The amount of external pressure necessary to equalize the levels** is called the **osmotic pressure.**

Although this discussion assumes that one compartment contained pure solvent, the same principle applies if both compartments contain solutions, as long as the concentrations are different. The solution of higher concentration always has a higher osmotic pressure than the one of lower concentration, which means that *the flow of solvent is always from the more dilute to the more concentrated side.* Of course, it is the number of particles that matters, so we

Some ions are small but still do not go through the membrane because they are solvated by a shell of water molecules (see Fig. 6.8).

must remember that, in ionic solutions, each mole of solvent gives rise to more than 1 mole of particles. For convenience in calculation, we define a new term, **osmolarity (osmol)**, as **molarity multiplied by the number of particles produced by each molecule of solute.**

---

### EXAMPLE

A 0.89 percent w/v NaCl solution is referred to as a *physiological saline solution* because it has the same concentration of salts as normal human blood, though blood contains several salts and saline solution has only NaCl. What is the osmolarity of this solution?

### Answer

0.89% w/v NaCl = 8.9 g NaCl in 1 L of solution

$$\frac{8.9 \text{ g NaCl}}{1 \text{ L solution}} \times \frac{1 \text{ mole NaCl}}{58.5 \text{ g NaCl}} = \frac{0.15 \text{ mole NaCl}}{\text{L solution}} = 0.15 \, M$$

Osmolarity = $0.15 \times 2 = 0.30$ osmol

---

**PROBLEM 6.11**  What is the osmolarity of a 3.3 percent w/v $Na_3PO_4$ solution?

---

**Figure 6.17**
The effect of osmosis on a carrot. The carrot on the left is in a solution that contains a high concentration of NaCl. Water flowing out of the carrot has caused it to shrink. The carrot on the right, in pure water, is slightly swollen. (Photograph by Charles D. Winters.)

For the effect of the osmolarity of a solution on a carrot, see Figure 6.17.

Osmotic pressure is a colligative property. The osmotic pressure generated by a solution across the semipermeable membrane—the difference between the heights of the two columns in Figure 6.16(b)—depends on the osmolarity of the solution. If the osmolarity increases by a factor of 2, the osmotic pressure will also increase by a factor of 2.

Osmotic pressure is very important in biological organisms because cell membranes are semipermeable membranes. Therefore, biological fluids must have the proper osmolarity. For example, the red blood cells in the body are suspended in a medium called plasma that must have the same osmolarity as the red blood cells. **Two solutions with the same osmolarity** are called **isotonic,** and plasma is said to be isotonic with the red blood cells. This means that no osmotic pressure is generated across the cell membrane.

What would happen if we suspended red blood cells in distilled water instead of in plasma? Inside the red blood cells the osmolarity is about the same as in a physiological saline solution, 0.30 osmol. Distilled water has zero osmolarity. As a consequence, water flows into the red blood cells. The volume of the cells increases, and the cells swell, as shown in Figure 6.18(b). The membrane cannot resist the osmotic pressure, and the red blood cells eventually burst, spilling their contents into the water. We call this process **hemolysis.**

Solutions with an osmolarity (and hence osmotic pressure) lower than that of suspended cells are called **hypotonic solutions.** Obviously, it is very important that in intravenous feeding and blood transfusion we always use *isotonic* solutions and never *hypotonic* solutions. The latter would simply kill the red blood cells by hemolysis.

Iso-osmotic, hypo-osmotic, and hyper-osmotic are recent synonyms for isotonic, hypotonic, and hypertonic, respectively.

(a)

(b)

(c)

**Figure 6.18**
Red blood cells: (a) in an isotonic solution, (b) in a hypotonic solution, (c) in a hypertonic solution. (Peter Arnold, Inc./© Phillip A. Harrington.)

Equally important, we should not use *hypertonic* solutions either. A **hypertonic solution** has a greater osmolarity (and greater osmotic pressure) than the reference cells. If red blood cells are placed in a hypertonic solution (for example, 0.5 osmol glucose solution), water flows from the cells into the glucose solution through the semipermeable cell membrane. This process, called **crenation,** shrivels the cells, as shown in Figure 6.18(c).

As already mentioned, 0.89 percent NaCl (physiological saline) is isotonic with red blood cells and is used in intravenous injections.

A 5.5 percent glucose solution is also isotonic and is used in intravenous feeding.

## 6.10   DIALYSIS

In osmosis there is an osmotic semipermeable membrane that allows only solvent and not solute molecules to pass. If, however, the openings in the membrane are somewhat larger, then *small* solute molecules can also get through, but large solute molecules, such as macromolecular and colloidal particles, cannot. This process is called **dialysis.** For example, ribonucleic acid is an important biological molecule that we will study in Chapter 23. When biochemists prepare ribonucleic acid solutions, they must remove small molecules,

B o x 6F

## Magnesium Salts in Hypertonic Solution

In coastal areas, seawater can penetrate drinking-water wells if the water table becomes depleted. The water in such wells contains higher-than-normal concentrations of magnesium salts. These salts, especially magnesium sulfate and magnesium citrate, are incompletely absorbed in the intestine. The concentration of these salts in the intestine creates a hypertonic condition relative to neighboring tissues. As a consequence, water flows from the tissues into the intestine, diluting the stool and causing diarrhea. At the same time the tissues are dehydrated.

This is also the principle used to treat hemorrhoids in a sitz bath. When hemorrhoidal tissue is swollen, a hypertonic solution of magnesium sulfate draws out water and shrinks the tissue. Swollen feet respond to a hypertonic solution when soaked in a hot magnesium sulfate bath.

such as NaCl and ethyl alcohol, from the solution in order to get a pure nucleic acid preparation. This is done by placing the nucleic acid solution in a dialysis bag (cellophane) of sufficient pore size to allow all the small molecules to diffuse and retain only the large nucleic acid molecules. If the dialysis bag is suspended in flowing distilled water, all the NaCl and other small molecules leave the bag and, after a certain amount of time, it contains only the pure nucleic acids.

This is also the way our kidneys work. The millions of nephrons, or kidney cells, have very large surface areas in which the capillaries of the blood vessels are in contact with the nephrons. The kidneys serve as a gigantic filtering machine. The waste products of the blood dialyse out through semipermeable membranes in the glomeruli and enter collecting tubes that carry the urine to the ureter. Large protein molecules and cells are retained in the blood.

The glomeruli of the kidney are fine capillary blood vessels in which the body's waste products are removed from the blood.

B o x 6G

## Hemodialysis

The kidneys' main function is to remove toxic waste products from the blood. When the kidneys are not functioning properly, these waste products may threaten life. Hemodialysis is an artificial process that performs the same filtration function (see Figure 25.6).

The patient's blood is circulated through a long tube of cellophane membrane suspended in an isotonic solution, and then returned to the patient's vein. The cellophane membrane retains all the large molecules (for example, proteins) but allows the small ones, including the toxic wastes, to pass through. Thus, the wastes are removed from the blood by dialysis.

If the cellophane tube were suspended in distilled water, other small molecules and ions, such as $Na^+$, $Cl^-$, and glucose, would also be removed from the blood. The isotonic solution used in hemodialysis consists of 0.6 percent NaCl, 0.04 percent KCl, 0.2 percent $NaHCO_3$, and 0.72 percent glucose (all w/v). In this manner, no glucose or $Na^+$ is lost from the blood.

A patient is usually on an artificial kidney machine for 4 to 7 hours. During this time the isotonic bath is changed every 2 hours. Kidney machines allow people with kidney failure to lead a fairly normal life, though they must take these hemodialysis treatments regularly.

## SUMMARY

Systems containing more than one component are **mixtures. Homogeneous mixtures** are uniform throughout. **Heterogeneous mixtures** exhibit well-defined boundaries between phases. When a solution consists of a solid or gas dissolved in a liquid, the liquid is the **solvent** and the solid or gas is the **solute.** The **solubility** of a substance is the maximum amount of substance that dissolves in a given amount of solvent at a given temperature.

"Like dissolves like" means that polar molecules are soluble in polar solvents and nonpolar molecules in nonpolar solvents. The solubility of solids and liquids in liquids usually increases with temperature; the solubility of gases usually decreases with an increase in temperature. Percent concentration is given in either weight per unit volume **(w/v)** or weight per unit weight **(w/w). Molarity** is the number of moles of solute in 1 L of solution. Water is the most important solvent because it dissolves polar compounds and ions through hydrogen bonding and dipole interactions. Hydrated ions are surrounded by water molecules (hydration layer) that move together with the ion and cushion it from collisions with other ions. Aqueous solutions of ions and molten salts are **electrolytes** and conduct electricity.

**Colloids** exhibit a scattering of light called the **Tyndall effect.** Colloids are stable mixtures in spite of the large size of the colloidal particles (1 to 1000 nm). The stability is due to the solvation layer that cushions the colloids from direct collisions and to an electric charge on the surface of colloidal particles.

**Colligative properties** are properties that depend only on the number of solute particles present. **Osmotic pressure** is one of the most important colligative properties. Osmotic pressure operates across an osmotic semipermeable membrane that allows only solvent molecules to pass but screens out all other molecules. In osmotic pressure calculations, concentration is measured in **osmolarity,** which is the molarity of the solution multiplied by the number of particles produced by dissociation of the solute. Red blood cells in **hypotonic solution** swell and burst; this is called **hemolysis.** Red blood cells in **hypertonic solution** shrink; this is called **crenation.** Some semipermeable membranes allow small solute molecules to pass along with solvent molecules. In **dialysis,** such membranes are used to separate large molecules from smaller ones.

## KEY TERMS

Acid rain (Box 6A)
Anhydrous (Sec. 6.6)
Anode (Sec. 6.6)
Aqueous solution (Sec. 6.6)
Brownian motion (Sec. 6.7)
Cathode (Sec. 6.6)
Colligative property (Sec. 6.8)
Colloid (Sec. 6.7)
Concentrated (Sec. 6.5)
Crenation (Sec. 6.9)
Dialysis (Sec. 6.10)
Dilute (Sec. 6.5)
Electrolyte (Sec. 6.6)
Emulsion (Box 6E)
Freezing-point depression (Sec. 6.8)
Hemodialysis (Box 6G)
Hemolysis (Sec. 6.9)
Henry's law (Sec. 6.4)
Heterogeneous mixture (Sec. 6.1)

Homogeneous mixture (Sec. 6.1)
Hydrate (Sec. 6.6)
Hygroscopic (Sec. 6.6)
Hypertonic (Sec. 6.9)
Hypoglycemia (Box 6D)
Hypotonic (Sec. 6.9)
Isotonic (Sec. 6.9)
Miscible (Sec. 6.4)
Molarity (Sec. 6.5)
Osmolarity (Sec. 6.9)
Osmosis (Sec. 6.9)
Osmotic membrane (Sec. 6.9)
Osmotic pressure (Sec. 6.9)
Parts per million (Sec. 6.5)
Percent concentration (Sec. 6.5)
Saturated solution (Sec. 6.4)
Seeding (Sec. 6.4)
Semipermeable membrane (Sec. 6.9)
Solubility (Sec. 6.4)

Solute (Sec. 6.2)
Solution (Sec. 6.1)
Solvated ions (Sec. 6.6)
Solvent (Sec. 6.2)
Supersaturated solution (Sec. 6.4)

Suspension (Sec. 6.7)
Tyndall effect (Sec. 6.7)
Unsaturated solution (Sec. 6.4)
Water of hydration (Sec. 6.6)

## PROBLEMS

### Solutions and Their Characteristics

**6.12** What is the difference between a homogeneous and a heterogeneous mixture?

**6.13** A solution is made by dissolving glucose in water. Which is the solvent and which the solute?

**6.14** Suppose you are given a solution containing either salt or sugar dissolved in water. (a) Can you tell which it is by looking at it? (b) Give a simple way by which you could easily tell which it is.

**6.15** In each of the following tell whether the solutes and solvents are gases, liquids or solids:
  (a) bronze (see Box 2C)
  (b) cup of coffee
  (c) car exhaust
  (d) champagne

**6.16** Give a familiar example of solutions of each of these types:
  (a) liquid in liquid     (c) gas in liquid
  (b) solid in liquid      (d) gas in gas

**6.17** Are mixtures of gases true solutions or heterogeneous mixtures? Explain.

### Solubility

**6.18** If two liquids, such as benzene and toluene, are completely soluble in each other, can they form a supersaturated solution? Explain.

**6.19** The solubility of a compound is 2.5 g in 100 mL of aqueous solution at 25°C. If you put 1.12 g of the compound in a 50-mL volumetric flask at 25°C and add sufficient water to fill it to the 50-mL mark, what kind of solution do you get, saturated or unsaturated?

**6.20** Gasoline is a nonpolar solvent; water is a polar solvent. Decide whether each of these compounds is more likely to be soluble in gasoline or in water:
  (a) sugar         (d) $Na_2SO_4$
  (b) $CCl_4$        (e) diethyl ether
  (c) KCl

**6.21** On the basis of polarity and of hydrogen-bonding capacity, list the following in order of increasing solubility in water:
  (a) $CH_3-O-CH_3$
  (b) $CH_3-CH_2-CH_2-CH_3$
  (c) $(CH_3)_2S=O$
  (d) $HO-CH_2-CH_2-OH$

**6.22** Suppose you are trying to dissolve 2.0 g of a certain solid drug in water. You add it to 100 mL of water and part, but not all, of it dissolves. Give two procedures you might try to dissolve the entire 2.0 g.

**6.23** How can you convert a saturated solution at room temperature to a supersaturated one?

**6.24** The solubility of aspartic acid in water is 0.500 g in 100 mL at 25°C. If you dissolve 0.251 g aspartic acid in 50.0 mL water at 50°C and let the solution cool to 25°C without stirring, shaking, or otherwise disturbing the solution, would you get a saturated, unsaturated, or supersaturated solution?

**6.25** Near power plants, warm water is discharged into a river. Sometimes, dead fish are observed in the area. Why do fish die from the warm water?

**6.26** If a bottle of beer, after being opened, is allowed to stand for several hours, it becomes "flat" (it loses $CO_2$). Explain.

**6.27** Can you dissolve (a) more solid glucose in water at 2 atm pressure than at 1 atm? (b) Answer the same question for the gas $SO_2$.

### Concentration Units

**6.28** Describe exactly how you would make
  (a) 250.0 g of a 34.6 percent w/w solution of KCl in water
  (b) 156.00 g of a 4.00 percent w/w solution of NaOH in water
  (c) 300.00 g of a 0.32 percent w/w solution of formic acid (HCOOH) in water

**6.29** Describe exactly how you would make
(a) 410 mL of a 27 percent v/v solution of ethyl alcohol, $C_2H_5OH$, in water
(b) 330 mL of a 1.8 percent v/v solution of ethyl acetate, $C_4H_8O_2$, in water
(c) 3.85 L of an 8.00 percent v/v solution of benzene $C_6H_6$, in chloroform, $HCCl_3$

**6.30** Describe exactly how you would make
(a) 250 mL of a 3.6 percent w/v solution of NaCl in water
(b) 625 mL of a 4.9 percent w/v solution of glycine, $C_2H_5NO_2$, in water
(c) 43.5 mL of a 13.7 percent w/v solution of $Na_2SO_4$ in water
(d) 518 mL of a 2.1 percent w/v solution of acetone, $C_3H_6O$, in water

**6.31** Calculate the w/v percentage of each of these solutes:
(a) 15.3 g NaCl in 725 mL of solution
(b) 72 g sucrose in 1.23 L of solution
(c) 2.35 g $Ba(NO_3)_2$ in 325 mL of solution

**6.32** Describe how you would prepare 250 mL of 0.10 M NaOH from solid NaOH and water.

**6.33** Assuming that the appropriate volumetric flasks are available, describe exactly how you would make
(a) 175 mL of a 1.14 M solution of $NH_4Br$ in water
(b) 1.35 L of a 0.825 M solution of NaI in water
(c) 330 mL of a 0.16 M solution of ethanol, $C_2H_5OH$, in water

**6.34** What is the molarity of each of these solutions?
(a) one made by dissolving 37 g of KCl in enough water to give 375 mL of solution
(b) one made by dissolving 75.9 g of sucrose, $C_{12}H_{22}O_{11}$, in enough water to give 725 mL of solution
(c) one made by dissolving 7.1 g of $(NH_4)_2SO_4$ in enough water to give 2.35 L of solution

**6.35** The concentration of stomach acid, HCl, is approximately 0.10 M. What volume of stomach acid contains 0.25 g of HCl?

**6.36** What is the molarity of a 2.0 percent w/v HCl solution?

**6.37** If 79.49 g of $H_2SO_4$ is dissolved in enough solvent to make 2.00 L of solution, what is the molarity?

**6.38** If 3.88 g of $BaCl_2$ is dissolved in enough solvent to make 500.0 mL of solution, what is the molarity?

**6.39** In the reaction

$$CO_2 + H_2O \longrightarrow H_2CO_3$$

1.25 g of $CO_2$ was added to enough water to make 50.0 mL of solution. What is the molarity of the $H_2CO_3$ formed? Assume the reaction goes to completion.

**6.40** If 0.456 g of NaF is dissolved in enough water to make 100 mL of solution, what is the molarity of the solution?

**6.41** You have a concentrated HCl solution labeled 6.0 M. How would you prepare 750 mL of 0.25 M HCl solution?

**6.42** How many grams of solute are present in each of the following aqueous solutions?
(a) 625 mL of a 2.00 M solution of $HNO_3$
(b) 1.90 L of a 0.286 M solution of alanine, $C_3H_7NO_2$
(c) 180 mL of a 0.0081 M solution of $CaSO_4$

**6.43** A student has a stock solution of 30.0 percent w/v $H_2O_2$. Describe how she should prepare 250 mL of a 0.25 percent $H_2O_2$ (hydrogen peroxide) solution.

**6.44** You have a stock solution of 50.0 percent w/v NaOH. Describe how you would prepare 100 mL of 0.50 percent w/v NaOH.

**6.45** Describe exactly how you would prepare each of the following aqueous solutions, in each case by diluting a more concentrated solution:
(a) 550 mL of a 0.90 M solution of $HNO_3$, starting with 4.0 M $HNO_3$
(b) 1.75 L of a 2.5 M solution of KOH, starting with 12.0 M KOH
(c) 385 mL of a 0.0095 M solution of glycine, $C_2H_5NO_2$, starting with 2.0 M glycine

**6.46** An aspirin preparation contains 0.00002 g salicylic acid as a contaminant in 100 g of aspirin. What is this in parts per billion?

**\*6.47** Dioxin is considered to be poisonous in concentrations above 2 ppb. If a lake containing $1 \times 10^7$ L has been contaminated by 0.1 g of dioxin, did the concentration reach a dangerous level?

## Water as a Solvent

**6.48** How is a cation such as $K^+$ solvated by water?

**6.49** Define:
(a) hygroscopic          (c) electrolyte
(b) water of hydration   (d) anhydrous

**6.50** Considering polarities, electronegativities, and similar concepts learned in Chapter 3, classify the following as strong electrolyte, weak electrolyte, or nonelectrolyte:

(a) NaCl   (d) HF
(b) $C_2H_5OH$ (ethanol)   (e) $C_6H_{12}O_6$ (glucose)
(c) KOH

**6.51** Which would produce the brightest light in the conductance apparatus in Figure 6.11?

(a) 0.1 $M$ KCl   (c) 0.5 $M$ sucrose
(b) 0.1 $M$ $(NH_4)_3PO_4$

**6.52** Ethanol, $C_2H_6O$, is very soluble in water. Describe how water dissolves ethanol.

**6.53** Predict which of these covalent compounds is soluble in water:

(a) $C_2H_6$   (d) $NH_3$
(b) $CH_3OH$   (e) $CCl_4$
(c) HF

## Colloids

**6.54** If the dust in the air of a room shows the Tyndall effect, what is the average size of the dust particles?

**6.55** On the basis of Tables 6.1 and 6.2, classify the following systems as homogeneous, heterogeneous, or colloidal mixtures:

(a) physiological saline solution
(b) orange juice
(c) a cloud
(d) wet sand
(e) suds
(f) milk

**6.56** Table 6.2 shows no examples of a gas-in-gas colloidal system. Considering the definition of a colloid, explain why this is so.

**6.57** How do acquired surface charges on a colloidal particle make the colloidal dispersion stable?

**6.58** Define Brownian motion. On the basis of Brownian motion, explain why many colloidal systems become unstable upon heating.

## Colligative Properties

**6.59** Calculate the freezing points of solutions made by dissolving 1 mole of each of the following ionic solutes in 1000 g of $H_2O$:

(a) NaCl   (c) $(NH_4)_2CO_3$
(b) $MgCl_2$   (d) $Al(HCO_3)_3$

**6.60** If we add 245 g of ethylene glycol $(C_2H_6O_2)$ per 1000 g of water to a car radiator, what will be the freezing point of the solution?

**6.61** Methanol is used as an antifreeze. How many grams of methanol, $CH_3OH$, do you need per 1000 g of water for an aqueous solution to stay liquid at $-20°C$?

**6.62** In winter, after a snow storm, salt, NaCl, is spread to melt the ice on roads. How many grams of salt per 1000 g of ice is needed to make it liquid at $-5.0°C$?

**6.63** How many moles of glycerol, $C_3H_8O_3$, which does not dissociate in aqueous solution, must you add to 1000 g of water to make a solution that freezes at $-10°C$?

## Osmosis

**6.64** What is the difference between an osmotic semipermeable membrane and a semipermeable membrane?

**6.65** In each case, tell which side (if either) rises and why. The solvent is water.

| | A | B |
|---|---|---|
| (a) | 1% glucose | 5% glucose |
| (b) | 1 $M$ glucose | 5 $M$ glucose |
| (c) | 1 $M$ NaCl | 1 $M$ glucose |
| (d) | 1 $M$ NaCl | 1 $M$ $K_2SO_4$ |
| (e) | 3% NaCl | 3% NaI |
| (f) | 1 $M$ NaBr | 1 $M$ NaCl |

**6.66** An osmotic semipermeable membrane that allows only water to pass separates two compartments, A and B. Compartment A contains 0.9 percent NaCl and B contains 3 percent glycerol, $C_3H_8O_3$. (a) In which compartment will the level of solution rise?   (b) Which compartment has the higher osmotic pressure (if either)?

**6.67** Calculate the osmolarity of

(a) 5.2 $M$ $Na_2CO_3$   (c) 2.8 $M$ LiBr
(b) 0.33 $M$ $Al(NO_3)_3$   (d) 0.009 $M$ $K_3PO_4$

**6.68** Define:

(a) isotonic   (c) crenation
(b) hypertonic   (d) hemolysis

**\*6.69** A 0.9 percent NaCl solution is isotonic with blood plasma. Which solution would crenate red blood cells?

(a)  0.3% NaCl

(b)  0.9 $M$ glucose

(c)  0.9% glucose (MW = 180)

## Boxes

**6.70** (Box 6A) In 1992 a volcano in the Philippines emitted large quantities of $CO_2$ and $SO_2$ gases as high as 2 to 3 miles into the atmosphere. Would you expect an increase in acid rain locally or globally?

**6.71** (Box 6A) Oxides of nitrogen (NO, $NO_2$, $N_2O_3$) are also responsible for acid rain. Which acids do you suppose can be formed from these oxides?

**6.72** (Box 6A) What makes normal rain water slightly acidic?

**6.73** (Box 6A) How many grams of dissolved $SO_2$ can 7.50 kg water hold?

**6.74** (Box 6B) Why do deep-sea divers use a helium-oxygen mixture in the tank instead of air?

**6.75** (Box 6C) What is the chemical formula for the main component of limestone and marble?

**\*6.76** (Box 6C) Write balanced equations (two steps) for the conversion of marble to gypsum dihydrate.

**6.77** (Box 6D) Why do long-distance runners suffer muscle cramps in hot and humid weather?

**6.78** (Box 6D) Explain how hypoglycemia causes a lack of nerve impulses to the muscles.

**6.79** (Box 6E) What is the protective colloid in (a) milk   (b) mayonnaise?

**\*6.80** (Box 6F) Body tissue is isotonic with 0.9 percent NaCl solution. What would be a good w/v concentration of epsom salt ($MgSO_4 \cdot 7H_2O$) solution to use as a sitz bath for shrinking hemorrhoids? ($MgSO_4 \cdot 7H_2O$ dissociates into two particles.)

**\*6.81** (Box 6G) The artificial kidney machine uses a solution containing 0.6 percent NaCl, 0.04 percent KCl, 0.2 percent $NaHCO_3$, and 0.72 percent glucose (all w/v). Show that this is an isotonic solution.

## Additional Problems

**6.82** Government regulations permit a 6 ppb concentration of a certain pollutant. How many grams of pollutant are allowed in one ton (1016 kg) of water?

**\*6.83** Both methanol, $CH_3OH$, and ethylene glycol,

$C_2H_6O_2$, are used as antifreeze. Which is more efficient; that is, which produces a lower freezing point if equal weights of each are added to the same weight of water?

**6.84** Orange juice is a stable colloidal system. When you add vodka to orange juice, the colloidal particles settle to the bottom. The ethyl alcohol in the vodka has great attraction for the water. Explain how you destabilize the colloidal system of orange juice by mixing it with vodka.

**6.85** A solution of $As_2O_3$ has a molarity of $2 \times 10^{-5} M$. What is this concentration in ppm? (Assume the density of the solution is 1.00 g/mL.)

**6.86** If solid starch is stirred in cold water, the mixture is opaque, it does not produce a Tyndall effect, and it settles on standing. When the same starch is stirred in boiling water and then allowed to cool, the solution exhibits a Tyndall effect, and it does not settle on standing. What can you say about the sizes of the starch particles in these two mixtures?

**6.87** You have a 20 percent w/v stock solution of NaOH. Describe how you would prepare 250 mL of a 0.50 $M$ NaOH solution.

**\*6.88** The average osmolarity of sea water is 1.18. How much pure water would have to be added to 1.0 mL of sea water for it to achieve the osmolarity of blood (0.30 osmol)?

**6.89** If 10.0 g of ethyl alcohol and 30.0 g of water are mixed and the solution has a final volume of 38.0 mL, what is the concentration of ethyl alcohol (a) in % w/v and   (b) in % w/w?

**6.90** What effect does increasing the temperature have on the solubility of most substances?

**6.91** How can we make a solution of a solid in a solid?

**6.92** You know that an 0.89% saline (NaCl) solution is isotonic with blood. In a real-life emergency you run out of physiological saline solution and have only KCl as a salt, and distilled water. Would it be all right to make an 0.89% aqueous KCl solution and use it for intraveneous infusion?

**\*6.93** Carbon dioxide and sulfur dioxide are soluble in water because they react with the water. Write possible equations for these reactions.

**6.94** Which will have greater osmotic pressure: (a) a 0.9 percent w/v NaCl solution or   (b) a 25 percent w/v solution of a nondissociating dextran with a molecular weight of 15 000?

**\*6.95** If an 0.500 osmolar solution of glucose generates an osmotic pressure of 10 mm Hg, what are the osmolarity and the molarity of an NaCl solution that generates an osmotic pressure of 25 mm Hg?

# CHAPTER 7

*Iron powder burning in a Bunsen flame.*
*(Photograph by Charles D. Winters.)*

# Reaction Rates and Equilibrium

## 7.1 INTRODUCTION

Some chemical reactions take place rapidly; others are very slow. For example, hydrogen gas and oxygen gas react with each other to form water:

$$2H_2(g) + O_2(g) \longrightarrow 2H_2O(l)$$

This reaction is extremely slow. A tank containing a mixture of $H_2$ and $O_2$ shows no measurable change even after many years.

In contrast to such a slow reaction, consider what happens when you take one or two aspirin tablets for a slight headache. Very often, the pain disappears in half an hour or so. This shows that the aspirin must have reacted with compounds in the body within that time.

Many known reactions are much faster still. For example, if we add a solution of silver ions to a solution of chloride ions (Sec. 4.8)

$$Ag^+(aq) + Cl^-(aq) \longrightarrow AgCl(s)$$

a precipitate of silver chloride forms almost instantaneously. This reaction is essentially complete in considerably less than one second.

In the course of several years, a few molecules of $H_2$ and $O_2$ will react, but not enough for us to detect.

This is a net ionic reaction. Spectator ions are not shown.

**The study of reaction rates** is called **chemical kinetics.** The **rate** of a reaction is **the change in concentration of a reactant (or product) per unit time.** For example, consider the reaction

$$CH_3{-}Cl + I^- \xrightarrow[\text{acetone}]{} CH_3{-}I + Cl^-$$

carried out in the solvent acetone. To determine the reaction rate, we can measure the concentration of iodomethane, $CH_3I$, in the acetone at periodic time intervals, say every ten minutes. The rate of the reaction is the increase in the concentration of iodomethane divided by the time interval. For example, the concentration might increase from 0 to 0.12 mole/L over a period of 30 minutes. If so, the rate of the reaction during that time is

The rate could also be determined by following the decrease in concentration of $CH_3Cl$, if that is more convenient.

$$\frac{0.12 \text{ mole/L} - 0.0 \text{ mole/L}}{30 \text{ min}} = \frac{0.12 \text{ mole/L}}{30 \text{ min}} = 0.004 \frac{\text{mole/L}}{\text{min}}$$

This unit is read "0.004 mole per liter per minute." During each minute of the reaction, an average of 0.004 mole of chloromethane, $CH_3Cl$, has been converted to iodomethane, for each liter of solution. Every reaction has its own rate, which must be measured in the laboratory.

Chemical reactions are very important to us, both the ones that we carry out in the laboratory and the ones that take place inside our bodies. The rates of these reactions are also important. A reaction that goes more slowly than we need may be useless, whereas a reaction that goes too fast may be dangerous. We would like to have some idea as to what causes the enormous variety in reaction rates. In the next three sections we examine this question.

## 7.2   MOLECULAR COLLISIONS

In order for two molecules or ions to react with each other, they must first collide. As we saw in Chapter 5, molecules in gases and liquids are constantly in motion and frequently collide with each other. If we want a reaction to take place between two compounds, A and B, we generally mix them together, either by dissolving them in the same solvent or, if they are gases, by simply allowing them to mix. In either case, we can be sure that the constant motion of all the molecules will cause frequent collisions between molecules of A and B. In fact, we can even calculate how many such collisions will take place within a given period of time.

When we make such calculations, we generally get a surprising result. There are so many collisions between A and B molecules that most reactions should be over in considerably less than one second. Since the actual reactions are generally much slower, we must conclude that **most collisions do not result in a reaction.** In most cases, when a molecule of A collides with a molecule of B, they simply bounce apart without reacting. **A collision that results in a reaction** is called an **effective collision.**

Why are some collisions effective and others not? There are two main reasons.

**1.**   In most cases, for a reaction to take place between A and B, one or more covalent bonds must be broken in A or B or both, and energy is required for this to happen. This energy comes from the collision between A and B. If the energy

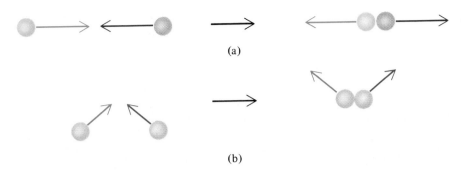

**Figure 7.1**
**Figure 7.1**
The energy of molecular collisions varies. Two fast-moving molecules colliding head-on (a) have a higher collision energy than two slower-moving molecules colliding at an angle (b).

of the collision is large enough, the bonds can break and the reaction will take place. If the collision energy is too low, the molecules will bounce apart without reacting. **The minimum energy necessary for the reaction to happen** is called the **activation energy.**

The energy of any collision depends on the relative speed of the colliding objects and on the angle of approach. Much greater damage is done in a head-on collision of two cars both going 40 miles per hour than in a collision in which a car going 20 miles per hour sideswipes one going 10 miles per hour. It is the same with molecules. There is a much greater collision energy when two fast-moving molecules collide head-on, for example, than when two slow-moving molecules collide at an angle (see Fig. 7.1).

**2.** Even if two molecules collide with an energy greater than the energy of activation, a reaction still may not take place if the molecules are not oriented properly when they collide. Consider, for example, the reaction between $H_2O$ and HCl:

$$H_2O + HCl \longrightarrow H_3O^+ + Cl^-$$

For this reaction to take place, the molecules must collide in such a way that the H of the HCl hits the O of the water, as shown in Figure 7.2(a). A collision in which the Cl of the HCl hits the O, as shown in Figure 7.2(b), cannot lead to reaction even if there is sufficient energy.

**Figure 7.2**
(a) The two molecules are properly oriented for a reaction to take place. (b) The orientation is such that a reaction cannot take place, because the Cl, not the H, collides with the O. The colored arrows show the path of the molecules.

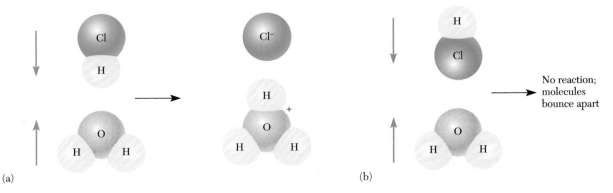

Coming back to the example at the beginning of this chapter, we can now see why the reaction between $H_2$ and $O_2$ is so slow. The $O_2$ and $H_2$ molecules are constantly colliding, but the percentage of effective collisions is extremely tiny at room temperature.

## 7.3   ACTIVATION ENERGY AND ENERGY DIAGRAMS

Figure 7.3 shows a typical energy diagram for a reaction. The products have a lower energy than the reactants; we might therefore expect the reaction to take place rapidly. As the curve shows, however, the reactants cannot be converted to products without the necessary activation energy. The activation energy is like a hill. If we are in a mountainous region, we may find that the only way to go from one point to another is to climb over a hill. It is the same in a chemical reaction. Even though the products may have a lower energy than the reactants ($H_2O$ has a lower energy than $H_2 + O_2$), they cannot be formed unless the reactants "go over the hill," that is, gain energy of activation.

Let us look into this more closely. In a typical reaction, bonds are broken *and* new bonds are formed. For example, when $H_2$ reacts with $N_2$ to give $NH_3$, six covalent bonds must break and six new covalent bonds must be formed (Fig. 7.4). Breaking a bond requires energy, but forming a bond releases energy. In a "downhill" reaction, of the type shown in Figure 7.3, the amount of energy released in making the new bonds is greater than that required to break the original bonds. That is, the reaction is exothermic. Yet it may well have a substantial activation energy, or energy barrier, because, in most cases, at least one bond must break *before* any new bonds can form. This means energy must be put in before we get any back. This is like the situation you might find yourself in if somebody offered to let you buy into a business from which, for an investment of $100 000, you could get an income of $40 000 per year, beginning in

> Molecules tend to go spontaneously from higher to lower energy states.

> "Downhill" reactions are exothermic. "Uphill" reactions are endothermic.

**Figure 7.3**
Energy diagram for a typical reaction.

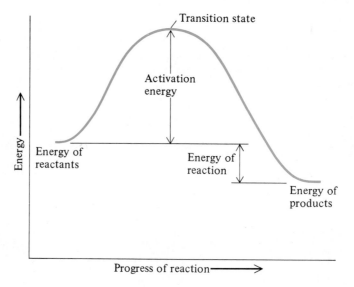

one year. In the long run you would do very well, but first you need to put up the $100 000 right now (the activation energy).

Every reaction has a different energy diagram. In some cases the energy of the products is higher than that of the reactants ("uphill" reactions, Fig. 7.5), but in almost all cases there is an energy hill—the activation energy. The activation energy is inversely related to the rate of the reaction. **The lower the activation energy, the faster the reaction; the higher the activation energy, the slower the reaction.**

The top of the energy hill is called the **transition state.** When the reacting molecules reach this point, one or more original bonds are partially broken and one or more new bonds may be in the process of formation. The entire system at this point—all the atoms in both molecules—is called the **activated complex.**

Let us look once more at the reaction between $H_2$ and $O_2$. The activation energy is so high that at room temperature no measurable reaction occurs even after many years. But if we light a match, the situation suddenly changes. The heat from the match causes the reaction not only to take place, but to take place so rapidly that the whole mixture explodes as soon as the match is lit. What has happened?

The heat of the match has supplied activation energy to some of the molecules in its vicinity, causing them to react. (It has done this by making the molecules move faster—see the next section.) Once they react, they give off additional energy (the heat of the reaction, which is 137 000 cal for each mole of $O_2$ used up). This additional energy becomes the activation energy for the neighboring molecules, which in turn give off still more. The result is that the small amount of energy from the lighted match is enough to start a process that becomes self-sustaining so rapidly that the mixture explodes as soon as the match is lit.

This is a dramatic example of the importance of activation energy. In most cases explosions do not take place, but activation energy is required for the large majority of reactions.

**Figure 7.4**
In the reaction $3H_2 + N_2 \rightarrow 2NH_3$, six covalent bonds (three single and one triple) break, and six new bonds form.

**Figure 7.5**
Energy diagram for an endothermic reaction, in which the energy of the products is greater than that of the reactants.

A number of gases are routinely used in hospitals. Some of these are flammable. Flammable gases do not burn when mixed with the oxygen in the air because there is no source of activation energy. A lighted cigarette or match, however, or a spark from faulty wiring in hospital equipment can provide the activation energy and cause a fire or an explosion. That is why it is absolutely necessary, when using such gases, to make sure that there is no source of fire or spark, and extreme precautions must be taken to ensure that this is so. Today, more and more hospitals prohibit smoking altogether, but other possible sources of sparks or flame must still be guarded against.

Hospitals also use pure or enriched oxygen for patients with breathing difficulties. Oxygen does not burn, but it does react with anything combustible. Pure oxygen, in which the $O_2$ concentration is 100%, compared with the 20% concentration in air, undergoes such reactions much faster. Therefore, precautions against fire and sparks are also necessary when using pure oxygen or oxygen-enriched mixtures.

Even when sparking from faulty wiring, x-ray machines, and other equipment is eliminated, there remains a danger of sparks from static electricity, which can build up—especially when the air is dry—on nonconducting materials such as nylon or rubber, or even from combing your hair. To prevent sparking in surgical rooms, the floors are made of antistatic ceramic tiles or antistatic polyvinyl chloride (PVC) or linoleum covers. Linen and cotton are used instead of wool or nylon, and the air is moistened to 55% relative humidity.

## 7.4 FACTORS AFFECTING RATES OF REACTION

In the previous section we saw that reactions occur as a result of collisions between fast-moving molecules possessing a certain minimum energy (the activation energy). In this section we examine some of the factors that affect activation energies and reaction rates.

### Nature of the Reactants

As we have already pointed out, every reaction has its own rate and its own activation energy. In general, reactions that take place between ions in aqueous solution (Sec. 4.8) are extremely rapid, almost instantaneous. Activation energies for these reactions are very low because usually no covalent bonds need to be broken. As we might expect, reactions between covalent molecules, whether in aqueous solution or not, are much slower. Many of these require anywhere from 15 minutes to 24 hours or more for most of the starting compounds to be converted to the products. There are, of course, reactions that take a good deal longer, but such reactions are seldom useful.

### Concentration

For reactions in the gas phase, an increase in pressure usually increases the rate.

Consider a reaction $A + B \rightarrow C + D$. In most cases, a reaction rate increases when the concentration of either or both reactants (A or B) is increased. For many reactions, though by no means all, there is a direct relationship between concentration and rate; that is, when the concentration of a reactant is doubled,

Box 1D points out that a sustained body temperature of 41.7°C (107°F) is invariably fatal. We can now see why a high fever is dangerous. Normal body temperature is 37°C (98.6°F), and all of the many reactions in the body—including respiration, digestion, and the synthesis of various compounds—take place at that temperature. If an increase of 10°C causes the rates of most reactions approximately to double, then an increase of even 1°C makes them go significantly faster.

Fever is a protective mechanism, and a small increase in temperature allows the body to kill germs faster by mobilizing the immune defense mechanism. But this increase must be small: An increase of 1°C brings the temperature to 100.4°F; an increase of 3°C brings it to 104°F. A temperature higher than 104°F increases reaction rates to the danger point.

One can easily detect the increase in reaction rates when a patient has a high fever. The pulse rate increases and breathing is faster in an effort to supply increased amounts of oxygen for the fast reactions.

the rate also doubles. All of this is easily understandable on the basis of the collision theory. If we double the concentration of A, there are twice as many molecules of A in the same volume, so the molecules of B in that volume now collide with twice as many A molecules per second as before. Since the reaction rate depends on the number of collisions per second, the rate is doubled (Fig. 7.6).

## Temperature

In virtually all cases, reaction rates increase with increasing temperature. An approximate rule for many reactions is that every time the temperature goes up by 10°C, the rate of reaction doubles. This rule is far from exact, but it is not far from the truth in many cases. As you can see, this is quite a large effect and says,

**Figure 7.6**
When heated in air, steel wool glows but does not burn rapidly, because the $O_2$ concentration of the air is only about 20%. When the glowing steel wool is put into 100% $O_2$ (in the photo) it burns vigorously. (Photograph by Leon Lewandowski.)

Box 7C

## The Effects of Lowering Body Temperature

As with a significant increase in body temperature, a substantial decrease below 37°C is also harmful because reaction rates are lower than they should be. However, it is sometimes possible to take advantage of this effect. In some heart operations, it is necessary to stop the flow of oxygen to the brain for a considerable time. At 37°C, the brain cannot survive without oxygen for more than about 5 minutes without permanent damage. When the patient's body temperature is deliberately lowered to about 28 to 30°C (82.4 to 86°F), however, the oxygen flow can be stopped for a considerable time without causing damage because reaction rates are lowered. At 25.6°C (78°F) the body's oxygen consumption is reduced by 50 percent.

There are seven 10° increments between 20° and 90°, and $2^7 = 128$.

for example, that if we run a reaction at 90°C instead of at room temperature (20°C), the reaction will go about 128 times faster. Putting it another way, if it takes 20 hours to convert 100 g of A to product C at 20°C, then it would take only about 10 minutes at 90°C. Temperature, therefore, is a powerful tool that lets us increase the rates of reactions that are inconveniently slow. It also lets us decrease the rates of reactions that are inconveniently fast. In some cases we choose to run reactions at low temperatures because explosions would result or the reactions would otherwise be out of control at room temperature.

What causes reaction rates to increase with increasing temperature? Once again, we turn to collision theory. Here temperature has two effects:

**1.** In Section 5.2 we learned that temperature is related to the average kinetic energy of molecules. When the temperature increases molecules move faster, which means they collide more frequently. More frequent collisions mean higher reaction rates. However, this factor is much less important than the second factor.

**2.** Recall from Section 7.2 that a reaction between two molecules takes place only if there is an *effective collision* — a collision with an energy greater than the activation energy. However, when the temperature increases, not only is the average speed of the molecules greater, but there is also a different *distribution* of speeds. The number of very fast molecules increases much more than the average speed (Fig. 7.7). This means that the number of effective collisions

**Figure 7.7**
Distribution of kinetic energies (molecular velocities) at two temperatures. The vertical line at the right indicates the energy (velocity) necessary to pass through the activation energy barrier. The shaded areas represent the fraction of molecules that have kinetic energies (velocities) greater than the activation energy.

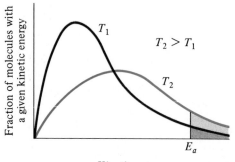

rises even more than the total number of collisions. Not only are there more collisions, but the percentage of collisions that have an energy greater than the activation energy also rises. It is this factor that is mainly responsible for the sharp increase in reaction rates with increasing temperature.

## Presence of a Catalyst

**Any substance that increases the rate of a reaction without itself being used up** is called a **catalyst.** Many catalysts are known, some that increase the rate of only one reaction, others that can affect several reactions. Although we have seen that reactions can be speeded up by increasing the temperature, in some cases they are still too slow even at the highest temperatures we can conveniently reach. In other cases it is not feasible to increase the temperature, for example, because other, unwanted, reactions would also be speeded up. In such cases a catalyst, if we can find the right one for a given reaction, can prove very valuable (Fig. 7.8). Many important industrial processes rely on catalysts (see Box 7E), and virtually all reactions that take place inside living organisms are catalyzed by enzymes (Chapter 19).

Catalysts work by allowing the reaction to take a different pathway, one with a lower activation energy (Fig. 7.9). Without the catalyst, the reactants would have to get over the energy hill. The catalyst provides a lower hill. As we have seen, a lower activation energy means a higher reaction rate. Each catalyst has its own way of providing an alternate pathway. Many catalysts do it by

**Figure 7.8**
In this dish $Cl^-$ ions act as a catalyst in the decomposition of $NH_4NO_3$. (Photograph by Charles D. Winters.)

**Figure 7.9**
Energy diagram for a catalyzed reaction. The dashed line shows the energy curve for the uncatalyzed process. The catalyst provides an alternate pathway whose activation energy is lower.

providing a surface on which the reactants can meet. Thus, the reaction between ethylene gas, $C_2H_4$, and hydrogen gas, $H_2$, to give ethane gas, $C_2H_6$,

$$C_2H_4 + H_2 \xrightarrow{\ Pt\ } C_2H_6$$

We often write the catalyst over or under the arrow.

goes so slowly without a catalyst that it is not practical even if we increase the temperature to any reasonable level. But if the mixture of gases is shaken with finely divided solid platinum, the reaction takes place at a convenient rate. The $C_2H_4$ molecules and the $H_2$ molecules meet each other on the surface of the platinum where the proper bonds can be broken and the reaction can proceed.

Catalysts can be homogeneous or heterogeneous. A **homogeneous catalyst** is in the same phase as the reactants, for example, enzymes in body tissues. A **heterogeneous catalyst** is in a separate phase from the reactants, for example, the solid platinum in the reaction between the gases $C_2H_4$ and $H_2$.

## 7.5   REVERSIBLE REACTIONS AND EQUILIBRIUM

Many reactions are irreversible. When a piece of paper is completely burned, the products are $CO_2$ and $H_2O$. Anybody who took pure $CO_2$ and $H_2O$ and tried to make them react to give paper and oxygen would not succeed. A tree, of course, turns $CO_2$ and $H_2O$ into wood and oxygen, and we, in sophisticated factories, make paper from the wood, but this is not the same as taking $CO_2$, $H_2O$, and energy and directly combining them in a single process to get paper and oxygen. We can certainly consider the burning of paper an irreversible reaction.

Other reactions, however, are reversible. A **reversible reaction is one that can be made to go in either direction.** For example, if carbon monoxide is mixed with water in the gas phase at a high temperature, hydrogen and carbon dioxide are produced:

$$CO + H_2O \longrightarrow CO_2 + H_2$$

But, if we desire, we can also make the reaction take place the other way: We can mix hydrogen and carbon dioxide to get carbon monoxide and water vapor:

$$CO_2 + H_2 \longrightarrow CO + H_2O$$

This reaction, like many others, is reversible. Let us see what happens when we run a reversible reaction. We will add some carbon monoxide to water vapor in the gas phase. The two compounds begin to react at a certain rate (the *forward* reaction):

$$CO + H_2O \longrightarrow CO_2 + H_2$$

As the reaction proceeds, the concentrations of CO and $H_2O$ gradually decrease, because they are being used up. This means that the *rate* of the reaction also gradually decreases because the rate of a reaction depends on the concentration of the reactants (Sec. 7.4).

The units of concentration in the gas phase are the same as in solution: moles per liter.

But what is happening in the other direction? Before we added the carbon monoxide, there was no carbon dioxide or hydrogen present. As soon as the forward reaction begins, however, it produces small amounts of these substances, and we now have some $CO_2$ and $H_2$. These two compounds will now, of course, begin reacting with each other (the *reverse* reaction):

$$CO_2 + H_2 \longrightarrow CO + H_2O$$

At first the reverse reaction is very slow, but as the concentrations of $H_2$ and $CO_2$ (produced by the forward reaction) gradually increase, the rate of the reverse reaction also gradually increases.

We have a situation, then, in which the rate of the forward reaction gradually decreases as time goes on while the rate of the reverse reaction (which began at zero) gradually increases. Eventually the two rates become equal. When this point is reached, we call the process a **dynamic equilibrium** or just an **equilibrium.**

What happens in the container once we reach equilibrium? If we measure the concentrations of the substances in the container, we find that **no change in concentration takes place after equilibrium is reached** (Fig. 7.10). Whatever the concentrations of all the substances, they will remain the same forever unless something is done to disturb the equilibrium (this is discussed in Sec. 7.8). This does not mean that all the concentrations must be equal—they can, in fact, be all different and usually are—but it does mean that, whatever they are, they no longer change once equilibrium has been reached, no matter how long we wait.

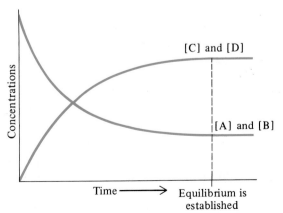

**Figure 7.10**
Changes in the concentrations of A, B, C, and D in the A + B $\rightleftharpoons$ C + D system, as equilibrium is approached beginning with A and B only. This equilibrium lies far to the right.

**Figure 7.11**
An equilibrium can be approached from either direction.

At the start:
A mixture of
CO and $H_2O$.

At equilibrium:
All four substances
present.

At the start:
A mixture of
$CO_2$ and $H_2$.

Given the fact that the concentrations of all the substances, both reactants and products, no longer change, can we say that nothing is happening? No, we know that both reactions are going on; all the molecules are constantly reacting —the CO and $H_2O$ are being changed to $CO_2$ and $H_2$, and the $CO_2$ and $H_2$ are being changed to CO and $H_2O$. But because the rates of the forward and reverse reactions are the same, none of the concentrations can change.

In the example just discussed, we approached equilibrium by adding carbon monoxide to water vapor. We could also have done it by adding carbon dioxide to hydrogen. In either case, we eventually get an equilibrium mixture containing the same four compounds (Fig. 7.11).

It is not necessary to begin with equal amounts. We could, for example, take 10 moles of carbon monoxide and 0.2 mole of water vapor. We would still arrive at an equilibrium mixture of all four compounds.

Reversible chemical reactions are not the only places in which we find dynamic equilibria. They are very common in the body and elsewhere in science.

*Another way to look at it is that the concentration of carbon monoxide (and of the other three compounds) does not change because it is being used up as fast as it is being formed.*

*We have already met an example of dynamic equilibrium in this book: the equilibrium between a liquid and its vapor (Sec. 5.9).*

## 7.6 EQUILIBRIUM CONSTANTS

Chemical equilibria can be treated by a simple mathematical expression. First let us write

$$aA + bB + \cdots \rightleftharpoons cC + dD + \cdots$$

as the general equation for all reversible reactions. In this equation, the capital letters stand for substances—$H_2O$ and $CO_2$, for instance—and the small letters are the coefficients of the balanced equation. The double arrow $\rightleftharpoons$ is used to show that the reaction is reversible. The dots mean that more than two substances can be present on either side.

Once equilibrium is reached, the following equation is valid:

*This expression was first derived in 1863 by Cato Guldberg (1836–1902) and Peter Waage (1833–1900), two Norwegian chemists.*

$$K = \frac{[C]^c[D]^d \cdots}{[A]^a[B]^b \cdots} \qquad \text{The equilibrium expression}$$

*It is understood that the concentration of the species within the brackets is always expressed in moles per liter.*

The brackets [ ] are a shorthand way of saying "concentration." Let us examine this equation. The $K$ is a constant called the **equilibrium constant.** What this expression tells us is that, when we multiply the concentrations of the substances on the right-hand side of the chemical equation and divide this product by the concentrations of the substances on the left-hand side (after raising each number to the appropriate power), *we get a number that does not change:* the equilibrium constant.

Let us look at examples of how to set up equilibrium expressions.

---

**EXAMPLE**

Write the equilibrium expression for the reaction

$$CO + H_2O \rightleftharpoons CO_2 + H_2$$

**Answer**

$$K = \frac{[CO_2][H_2]}{[CO][H_2O]}$$

This expression tells us that, at equilibrium, the concentration of carbon dioxide multiplied by the concentration of hydrogen and divided by the concentrations of water and carbon monoxide is a constant ($K$) and will not change. Mathematically, it would be just as correct to write the left-hand compounds on top, but the universal custom is to write them as shown here, with the products on top and the reactants on the bottom.

The concentration terms in brackets in the expression for $K$ are for species in solution, either liquid or gaseous.

---

**EXAMPLE**

Write the equilibrium expression for

$$PCl_5 \rightleftharpoons Cl_2 + PCl_3$$

**Answer**

$$K = \frac{[Cl_2][PCl_3]}{[PCl_5]}$$

---

**PROBLEM 7.1** Write the equilibrium expression for the reaction

$$SO_2 + H_2O \rightleftharpoons H_2SO_3$$

---

**EXAMPLE**

Write the equilibrium expression for

$$O_2 + 4ClO_2 \rightleftharpoons 2Cl_2O_5$$

**Answer**

$$K = \frac{[Cl_2O_5]^2}{[O_2][ClO_2]^4}$$

In this case the chemical equation has coefficients other than unity, so the equilibrium expression contains powers.

**PROBLEM 7.2**   Write the equilibrium expression for the reaction

$$2NH_3 \rightleftharpoons N_2 + 3H_2$$

Now let us see how $K$ is calculated.

---

**EXAMPLE**

Some $H_2$ is added to $I_2$ at 427°C and the reaction

$$I_2(g) + H_2(g) \rightleftharpoons 2HI(g)$$

is allowed to come to equilibrium. When this point is reached, the concentrations are found to be $[I_2] = 0.42$ mole/L, $[H_2] = 0.025$ mole/L, and $[HI] = 0.76$ mole/L. Calculate $K$ at 427°C.

**Answer**

The equilibrium expression is

$$K = \frac{[HI]^2}{[I_2][H_2]}$$

Substituting the concentrations, we get

$$K = \frac{(0.76)^2}{(0.42)(0.025)} = 55$$

---

**PROBLEM 7.3**   What is the equilibrium constant for the reaction

$$PCl_3 + Cl_2 \rightleftharpoons PCl_5$$

if the equilibrium concentrations are $[PCl_3] = 1.66\ M$, $[Cl_2] = 1.66\ M$, and $[PCl_5] = 1.66\ M$?

---

The last example shows us that the reaction between $I_2$ and $H_2$ to give HI has an equilibrium constant of 55. What does this mean? At constant temperature, equilibrium constants remain the same no matter what concentrations we have. That is, at 427°C, if we begin by adding, say, 5 moles of $H_2$ to 5 moles of $I_2$, the two reactions take place and equilibrium is reached. At that point, the concentration of HI, squared, divided by the concentrations of $I_2$ and $H_2$ equals 55. If we begin instead, at 427°C, with 7 moles of $H_2$ and 2 of $I_2$, or 0.4 mole of $H_2$ and 17 moles of $I_2$, or *any* number of moles of HI, once equilibrium is reached, the value of $[HI]^2/[I_2][H_2]$ is again 55. It makes no difference what concentrations of the three substances we begin with. At 427°C, as long as all three substances are present and equilibrium has been reached, the concentrations of the three substances so adjust themselves that the value of the equilibrium expression is 55.

The equilibrium constant is different for every reaction. Some reactions have a large $K$, others a small $K$. A reaction with a very large $K$ proceeds almost to completion (to the right). For example, $K$ for

$$HCl + H_2O \rightleftharpoons H_3O^+ + Cl^-$$

This symbol $\rightleftharpoons$ means that the equilibrium lies far to the right.

is about 10 000 000, or $10^7$, at 25°C, meaning that at equilibrium $[H_3O^+]$ and $[Cl^-]$ must be very large and $[HCl]$ and $[H_2O]$ very small, in order that $[H_3O^+][Cl^-]/[HCl][H_2O]$ be equal to $10^7$. This means that if we add HCl to $H_2O$ we can be certain that, when equilibrium has been reached, an essentially complete reaction has taken place.

On the other hand, a reaction with a very small $K$, say $10^{-7}$, hardly goes at all. Equilibrium effects are most obvious in reactions with $K$ values between $10^3$ and $10^{-3}$. In such cases, the reaction goes part way, and significant concentrations of all substances are present at equilibrium. An example is the reaction between carbon monoxide and water, discussed in Section 7.5, for which $K$ is equal to 10 at 600°C.

The equilibrium constant for a given reaction remains the same no matter what happens to the concentrations, but this is not true for changes in temperature. The value of $K$ does change when the temperature is changed.

## 7.7 EQUILIBRIUM AND REACTION RATES

As pointed out in the previous section, the equilibrium expression is valid only after equilibrium has been reached. Before that, there is no equilibrium and the equilibrium expression is not valid. But how long does it take for a reaction to reach equilibrium? For this there are no easy answers. Some reactions, if the reactants are well mixed, reach equilibrium in less than one second; others will still not get there even after millions of years.

There is no relationship between the rate of a reaction (how long it takes to reach equilibrium) and the value of $K$. It is possible to have a large $K$ and a low rate, as in the reaction between $H_2$ and $O_2$ to give $H_2O$, which does not reach equilibrium for millions of years (Sec. 7.1), or a small $K$ and a high rate, as well as reactions in which both the rate and $K$ are large or both are small.

## 7.8 LE CHATELIER'S PRINCIPLE

When a reaction has reached equilibrium, the forward and reverse reactions are taking place at the same rate, and the concentrations of all the components do not change as long as we don't do anything to the system. But what happens if we do? In 1888 Henri Le Chatelier (1850–1936) put forth the statement known as **Le Chatelier's principle: If an external stress is applied to a system in equilibrium, the system reacts in such a way as to partially relieve the stress.** Let us look at four types of stress that can be put on chemical equilibria.

### Addition of Reaction Components

Suppose that this reaction has reached equilibrium:

$$CH_3COOH + C_2H_5OH \underset{}{\overset{HCl}{\rightleftharpoons}} CH_3COOC_2H_5 + H_2O$$

| Acetic acid | Ethyl alcohol | Ethyl acetate | |

The HCl is a catalyst, but this does not affect the equilibrium.

This means that the reaction flask contains all of these substances and that the concentrations no longer change. We now disturb the system by adding some acetic acid from outside. The result is that the concentration of acetic acid suddenly increases, which increases the rate of the forward reaction. As a consequence, the concentrations of products (ethyl acetate and water) begin to increase. At the same time the concentrations of reactants are decreasing. Now, an increase in the concentrations of products causes the rate of the reverse reaction to increase, but the rate of the forward reaction is decreasing, so eventually the two rates will be equal again and a new equilibrium is established.

*The equilibrium constant K of course remains the same.*

When that happens, the concentrations once again remain constant, but they are not the same as they were before. The concentrations of ethyl acetate and water are higher now, and the concentration of ethyl alcohol is lower. The concentration of acetic acid is higher *because we added some,* but it is less than it was immediately after we made the addition.

This always happens when we add more of any component to a system in equilibrium. The addition constitutes a stress; the system relieves the stress by increasing the concentrations of the components on the other side. We say that the equilibrium shifts in the opposite direction:

Adding
CH$_3$COOH         $CH_3COOH + C_2H_5OH \xrightleftharpoons{HCl} CH_3COOC_2H_5 + H_2O$

The addition of acetic acid causes the reaction to move toward the right: More $CH_3COOC_2H_5$ and $H_2O$ are formed and some of the $CH_3COOH$ and $C_2H_5OH$ are used up. The same thing happens if we add $C_2H_5OH$.

On the other hand, if we add $H_2O$ or $CH_3COOC_2H_5$, the reaction shifts to the left:

$CH_3COOH + C_2H_5OH \xrightleftharpoons{HCl} CH_3COOC_2H_5 + H_2O$    Adding
CH$_3$COOC$_2$H$_5$

We can summarize by saying that **the addition of any component causes the equilibrium to shift to the opposite side** (see Fig. 7.12).

## Removal of a Reaction Component

It is not always as easy to remove a component from a reaction mixture as it is to add one, but there are often ways to do it. The removal of a component, or even a decrease in its concentration, lowers the corresponding reaction rate and changes the position of the equilibrium. In this case, the reaction is shifted toward the side from which the reactant was removed.

In the case of the acetic acid–ethyl alcohol equilibrium, ethyl acetate has the lowest boiling point of the four components and can be removed by distillation. If this is done, the equilibrium shifts to that side:

$CH_3COOH + C_2H_5OH \xrightleftharpoons{HCl} CH_3COOC_2H_5 + H_2O$    Removing
CH$_3$COOC$_2$H$_5$

**Figure 7.12**
The tube at the left contains a saturated solution of silver acetate ($Ag^+$ ions and $CH_3COO^-$ ions) in equilibrium with solid silver acetate. When silver ions, in the form of silver nitrate solution, are added the equilibrium

$$Ag^+(aq) + CH_3COO^-(aq) \rightleftharpoons CH_3COOAg(s)$$

shifts to the right, producing more solid silver acetate, as can be seen in the tube on the right. (Photograph by Charles D. Winters.)

The concentrations of $CH_3COOH$ and $C_2H_5OH$ decrease and that of $H_2O$ increases. The effect of removing a component is thus the opposite of adding one: **The removal of a component causes the equilibrium to shift to the side the component was removed from.**

No matter what happens to the individual concentrations, **the value of the equilibrium constant remains unchanged.**

### Change in Temperature

The effect of a change in temperature on a reaction that has reached equilibrium depends on whether the reaction is exothermic (gives off heat) or endothermic (requires heat). Let us look at an exothermic reaction:

$$2H_2(g) + O_2(g) \rightleftharpoons 2H_2O(l) + 137\,000 \text{ cal}$$

Heat is not the only form of energy that affects equilibria. The statements made in the text regarding endothermic or exothermic reactions can be generalized to endergonic and exergonic reactions. A practical illustration of this is the use of sunglasses with adjustable shading. The compound silver chloride, AgCl, is incorporated in the glasses. This compound, on exposure to sunlight, produces metallic silver, Ag.

$$\text{Light} + 2Ag^+ + 2Cl^- \rightleftharpoons 2Ag + Cl_2$$

The more silver produced, the darker the glasses. At night, or when the wearer goes indoors, the reaction is reversed according to Le Chatelier's principle. In this case addition of sunlight drives the equilibrium to the right; its removal drives it to the left.

This equation tells us that 2 moles of $H_2$ react with 1 mole of $O_2$ to give 2 moles of $H_2O$ and heat. If we look upon the heat as a product of this reaction, then we can use the same type of reasoning as we did before. An increase in temperature means that we are adding heat. Since heat is a product, its addition pushes the equilibrium to the opposite side. We can therefore say that, if the reaction is at equilibrium and we increase the temperature, the reaction goes to the left — the concentrations of $H_2$ and $O_2$ increase and that of $H_2O$ decreases. This is true of all exothermic reactions.

For an example of the effect of temperature on an equilibrium, see Figure 7.13.

> An **increase** in temperature drives an **exothermic** reaction toward the **reactants** (to the left).
>
> A **decrease** in temperature drives an **exothermic** reaction toward the **products** (to the right).

For an endothermic reaction, of course, the opposite is true:

> An **increase** in temperature drives an **endothermic** reaction toward the **products.**
>
> A **decrease** in temperature drives an **endothermic** reaction toward the **reactants.**

Remember that *a change in temperature changes not only the position of equilibrium but the value of* K *as well.*

**Figure 7.13**
Effect of temperature on the equilibrium:

$$2NO_2(g) \rightleftharpoons N_2O_4(g) + 13\ 700 \text{ cal}$$

$NO_2$ is a brown gas; $N_2O_4$ is colorless. The concentration of $NO_2$ is greater in the right-hand picture, where the temperature is 50°C, than in the left-hand picture, where the temperature is 0°C, because the decrease in temperature has driven this exothermic reaction to the right, favoring the production of $N_2O_4$. (Photographs by Charles D. Winters.)

Human beings and animals need proteins and other nitrogen-containing compounds in order to live. Ultimately, the nitrogen in these compounds comes from the plants that we eat. There is plenty of $N_2$ in the atmosphere, but the only way that nature converts this $N_2$ to compounds usable by biological organisms is by certain bacteria that have the ability to ''fix'' atmospheric $N_2$, that is, convert it to ammonia, $NH_3$. Most of these bacteria live in the roots of certain plants such as clover, alfalfa, peas, and beans. However, the amount of $N_2$ fixed by such bacteria each year is far less than the amount necessary to feed all the humans and animals in the world.

The world today can support its population only by artificial fixing, primarily by the **Haber process,** which converts $N_2$ to $NH_3$ by the reaction

$$N_2(g) + 3H_2(g) \rightleftharpoons 2NH_3(g) + 22\ 000\ cal$$

As indicated above, this is an exothermic equilibrium process. Early workers on the problem of fixing nitrogen were troubled by a conflict between equilibrium and rate. Since this is an exothermic reaction, an increase in temperature drives the equilibrium to the left, so the best results (largest possible yield of $NH_3$) should be obtained at low temperatures. But at low temperatures, the *rate* is too low to produce any meaningful amounts of $NH_3$. In 1908 Fritz Haber (1868–1934) solved this problem by discovering a catalyst that permits the reaction to take place at a convenient rate at 500°C.

The $NH_3$ produced by the Haber process is converted to fertilizers, which are used all over the world (Fig. 7E). Without these fertilizers, food production would diminish so much that widespread starvation would result.

**B o x  7E**

# The Haber Process

**Figure 7E**
An industrial plant for the manufacture of ammonia. (M. W. Kellogg Company.)

**Addition of a Catalyst**

As we saw in Section 7.4, a catalyst increases the rate of a reaction without itself being changed. For a reversible reaction, catalysts always increase the rates of both the forward and reverse reactions to the same extent. Therefore, **the addition of a catalyst has no effect on the position of equilibrium.** However, adding a catalyst to a system not yet at equilibrium causes it to reach equilibrium faster than it would without the catalyst.

## SUMMARY

Some reactions are fast; others are slow. The **rate of a reaction** is the change in concentration of a reactant or a product per unit time. Reactions take place when molecules or ions collide. The rate of a reaction increases with an increasing number of **effective collisions,** that is, ones that lead to a reaction. The energy necessary for a reaction to begin is the **activation energy.** Effective collisions are those that have (a) more than the activation energy and (b) the proper orientation in space.

The lower the activation energy, the faster the reaction. An energy diagram shows the progress of a reaction. The position at the top of the curve is the **transition state.**

Reaction rates generally increase with increasing concentration and temperature and also depend on the nature of the reactants. The rate of some reactions can be increased by the addition of a **catalyst,** which lowers the activation energy.

Many reactions are reversible and eventually reach equilibrium—some slowly, some quickly. At **equilibrium,** the forward and reverse reactions take place at equal rates, and concentrations do not change. Every equilibrium has an **equilibrium expression** and an **equilibrium constant,** $K$, which does not change when concentrations change but does change when the temperature changes. The equilibrium constant is large for some reactions and small for others. There is no relationship between $K$ and the rate of a reaction.

**Le Chatelier's principle** tells us what happens when we put stress on a system in equilibrium. Addition of a component causes the equilibrium to shift to the opposite side. Removal of a component causes the equilibrium to shift to the side the component is taken from. Increasing the temperature drives an exothermic equilibrium to the side of the reactants. Addition of a catalyst has no effect on the position of equilibrium.

## KEY TERMS

Activated complex (Sec. 7.3)
Activation energy (Sec. 7.2)
Catalyst (Sec. 7.4)
Effective collision (Sec. 7.2)
Equilibrium (Sec. 7.5)
Equilibrium constant (Sec. 7.6)
Equilibrium expression (Sec. 7.6)
Haber process (Box 7E)

Heterogeneous catalyst (Sec. 7.4)
Homogeneous catalyst (Sec. 7.4)
Kinetics (Sec. 7.1)
Le Chatelier's principle (Sec. 7.8)
Rate of a reaction (Sec. 7.1)
Reversible reaction (Sec. 7.5)
Transition state (Sec. 7.3)

## PROBLEMS

### Reaction Rates

**7.4** Define reaction rate. What might be a typical unit for reaction rate?

**7.5** If the concentration of a product increases from 0.320 mole/L to 0.750 mole/L over a period of 1 hour and 10 minutes, what is the average rate per minute?

**7.6** In the reaction

$$CH_3Br + I^- \longrightarrow CH_3I + Br^-$$

we start the reaction with an *initial* $I^-$ concentration of 0.200 $M$. After ten minutes the $I^-$ concentration is 0.176 $M$. What is the rate of reaction?

### Molecular Collisions and Activation Energy

**7.7** Define: (a) activation energy  (b) effective collision

**7.8** Give two reasons why most collisions between $N_2$ and $O_2$ molecules do not result in a reaction to give NO.

**7.9** In the reaction

$$H_2 + Cl_2 \longrightarrow 2HCl$$

how many covalent bonds are broken and how many new ones are formed?

**7.10** Is the breaking of chemical bonds an exothermic or an endothermic process, or is no energy released or absorbed? Explain.

**7.11** Why are reactions between ions in aqueous solution generally much faster than reactions between covalent molecules?

**7.12** A certain reaction is exothermic by 9 kcal/mole and has an activation energy of 14 kcal/mole. Draw an energy curve for this reaction, and label the transition state.

### Factors Affecting Rates of Reaction

**7.13** A quart of milk quickly spoils if left at room temperature but keeps for several days in a refrigerator. Explain.

**7.14** Explain why the rates of most reactions increase when the concentration of reactants is increased.

**7.15** If a certain reaction takes 16 hours to go to completion at 10°C, at about what temperature should we run it if we want it to go to completion in 1 hour?

**\*7.16** In most cases when we run a reaction by mixing a fixed quantity of A with a fixed quantity of B, the rate of the reaction begins at a maximum and then decreases as time goes by. Explain.

**7.17** If you were running a reaction and wished it to go faster, what three things might you try to accomplish this?

**7.18** Give two reasons why reaction rates increase with increasing temperatures.

**7.19** Explain how a catalyst increases the rate of a reaction.

**7.20** If you add a piece of marble ($CaCO_3$) to a 6 $M$ HCl solution at room temperature you will see some bubbles form around the marble as $H_2$ gas slowly rises. If you crush another piece of marble and add it to the same solution at the same temperature, you will see vigorous gas formation, so much so that the solution appears to be boiling. Explain.

### Reversible Reactions and Equilibrium

**7.21** Burning a piece of paper is an irreversible reaction. Can you give some other examples of irreversible reactions?

**7.22** If the reaction $PCl_3 + Cl_2 \rightleftharpoons PCl_5$ is at equilibrium, are the concentrations of $PCl_3$, $Cl_2$, and $PCl_5$ necessarily equal? Is the concentration of $PCl_3$ necessarily equal to that of $Cl_2$? Explain.

### Equilibrium Constants

**7.23** Write equilibrium expressions for
(a) $CO + Cl_2 \rightleftharpoons COCl_2$
(b) $N_2O_4 \rightleftharpoons 2NO_2$
(c) $2H_2S + 3O_2 \rightleftharpoons 2H_2O + 2SO_2$
(d) $2C_2H_6 + 7O_2 \rightleftharpoons 4CO_2 + 6H_2O$

**7.24** Write the chemical equations corresponding to the following equilibrium expressions:

(a) $K = \dfrac{[CO_2][H_2O]}{[H_2CO_3]}$

(b) $K = \dfrac{[CaO]^2}{[Ca]^2[O_2]}$

(c) $K = \dfrac{[P_4O_{10}]}{[P_4][O_2]^5}$

(d) $K = \dfrac{[HF]^3[PF_3]}{[F_2]^3[PH_3]}$

**7.25** When a mixture containing $H_2$, $CO_2$, $H_2O$, and CO at 987°C reached equilibrium, the concentrations were

$$[H_2] = 2.63\ M \qquad [CO_2] = 0.096\ M$$
$$[H_2O] = 0.81\ M \qquad [CO] = 0.473\ M$$

What is the equilibrium constant at 987°C for the reaction

$$H_2 + CO_2 \rightleftharpoons H_2O + CO$$

**7.26** When a mixture containing $N_2O_4$ and $NO_2$ at 25°C reached equilibrium, the concentrations were

$$[N_2O_4] = 0.877\ M \qquad [NO_2] = 0.0714\ M$$

What is the equilibrium constant at 25°C for the reaction

$$2NO_2 \rightleftharpoons N_2O_4$$

**7.27** When a mixture containing NOCl, NO, and $Cl_2$ at 25°C reached equilibrium, the concentrations were

$$[NO] = 2.6\ M \qquad [NOCl] = 3.7\ M$$
$$[Cl_2] = 0.20\ M$$

What is the equilibrium constant at 25°C for the reaction

$$2NOCl \rightleftharpoons 2NO + Cl_2$$

**7.28** The reaction

$$NH_3 + H_2O \rightleftharpoons NH_2^- + H_3O^+$$

is very fast but has an equilibrium constant of $10^{-34}$. Would you be wise to try to run it?

**7.29** Equilibrium constants for several reactions are
(a) $4.5 \times 10^8$      (d) $3 \times 10^{-7}$
(b) 32      (e) 0.0032
(c) 4.5
Which of them favor the formation of products and which favor the formation of reactants?

**7.30** A particular reaction has an equilibrium constant of 1.13 under one set of conditions and 1.72 under a different set of conditions. Which conditions would be more advantageous in an industrial process to obtain the maximum amount of products?

## Equilibrium and Reaction Rates

**7.31** (a) If a reaction is very exothermic—that is, the products have a much lower energy than the reactants—can we be reasonably certain that it will take place rapidly? (b) If a reaction is very endothermic—that is, the products have a much higher energy than the reactants—can we be reasonably certain that it will take place extremely slowly or not at all? (Think carefully about this one.)

## Le Chatelier's Principle

**7.32** The reaction

$$2HI(g) \rightleftharpoons H_2(g) + I_2(g)$$

is endothermic. If the reaction is at equilibrium, tell whether the equilibrium will be shifted to the right or the left if we
(a) remove some HI
(b) add some $H_2$
(c) remove some $I_2$
(d) increase the temperature
(e) add a catalyst

**7.33** The reaction

$$2O_3(g) \rightleftharpoons 3O_2(g)$$

is exothermic. If the reaction is at equilibrium, tell whether the equilibrium will be shifted to the right or the left if we
(a) remove some $O_3$
(b) remove some $O_2$
(c) add some $O_3$
(d) decrease the temperature
(e) add a catalyst

**7.34** The reaction

$$C(s) + H_2O(g) \rightleftharpoons CO(g) + H_2(g)$$

is exothermic. If the reaction is at equilibrium, tell whether the equilibrium will be shifted to the right or to the left if we
(a) add some $H_2$

(b) remove some CO
(c) remove some $H_2O$
(d) decrease the temperature
(e) add a catalyst

**7.35** The reaction

$$4NO_2 + 6H_2O \rightleftharpoons 7O_2 + 4NH_3$$

is an endothermic process. What happens to the position of equilibrium if we lower the reaction temperature from 450 to 420°C?

**7.36** Is there any change in conditions that changes the equilibrium constant $K$ of a given reaction?

**\*7.37** The equilibrium constant at 1127°C for the endothermic reaction

$$2H_2S \rightleftharpoons 2H_2 + S_2$$

is 571. If the mixture is at equilibrium, what happens to $K$ if we
(a) add some $H_2S$
(b) add some $H_2$
(c) lower the temperature to 1000°C?

**Boxes**

**7.38** (Box 7A) What precautions must be taken in hospitals when working with flammable gases?

**7.39** (Box 7A) Oxygen gas, which is frequently used in hospitals, does not burn. Why then is it a fire hazard?

**7.40** (Box 7B) In a bacterial infection body temperature may rise to 101°F. Does this body defense kill the bacteria directly by heat or by another mechanism? If so, which mechanism?

**7.41** (Boxes 7B and 7C) Why is a high fever dangerous? Why is a low body temperature dangerous?

**7.42** (Box 7C) Why do surgeons sometimes lower body temperatures during heart operations?

**7.43** (Box 7D) What reaction takes place when sunlight hits the compound silver chloride?

**7.44** (Box 7E) If the equilibrium for the Haber process is unfavorable at high temperatures, why do the factories nevertheless use high temperatures?

**Additional Problems**

**7.45** A mixture of hydrogen and oxygen gases can be kept for many years without reacting, but if we light a match the two react so rapidly that an ex-

plosion results. Explain how the lighted match causes such a large change in the reaction rate.

**7.46** The equilibrium constant for

$$COCl_2 \rightleftharpoons CO + Cl_2$$

is 25. A measurement made on the equilibrium mixture found that the concentrations of CO and $Cl_2$ were each 0.80 $M$. What is the concentration of $COCl_2$ at equilibrium?

**\*7.47** Assume that there are two different reactions taking place at the same temperature. In A, two different spherical molecules collide to yield a product. In B, the shape of the colliding molecules is rod-like. Each reaction has the same number of collisions per second and the same activation energy. Which reaction goes faster?

**\*7.48** How could you increase the rate of a gaseous reaction without adding more reactants or a catalyst and without changing the temperature?

**\*7.49** Is it possible for an endothermic reaction to have zero activation energy?

**7.50** Write equilibrium expressions for these ionic equilibria:
(a) $PO_4^{3-} + 2H_2O \rightleftharpoons 2OH^- + H_2PO_4^-$
(b) $H_3O^+ + NO_3^- \rightleftharpoons H_2O + HNO_3$
(c) $OH^- + H_3O^+ \rightleftharpoons 2H_2O$

**7.51** The reaction

$$C_2H_4 + Cl_2 \rightleftharpoons C_2H_4Cl_2$$

has an equilibrium constant of 10. What is the equilibrium concentration of $C_2H_4Cl_2$ if the equilibrium concentration of $C_2H_4$ is 0.2 $M$ and that of $Cl_2$ is 0.4 $M$?

**\*7.52** Draw an energy diagram for an exothermic reaction that yields 75 kcal/mole. The energy of activation is also 75 kcal/mole.

**\*7.53** In the reaction $2HI \rightarrow H_2 + I_2$ the rate of appearance of $I_2$ is measured. The following concentrations were found at the end of the times shown:

| Time (min) | $[I_2]$ |
|:---:|:---:|
| 0 | 0 |
| 10 | 0.30 |
| 20 | 0.60 |
| 30 | 0.90 |

What is the rate of reaction?

**7.54** For the reaction

$$H_2 + Br_2 \rightleftharpoons 2HBr$$

the equilibrium concentrations are $[H_2] = 0.37\ M$, $[Br_2] = 0.70\ M$, $[HBr] = 3.6\ M$. What is the equilibrium constant?

**7.55** What factors determine whether a reaction run at a given temperature will be fast or slow?

**\*7.56** In an endothermic reaction, $A \rightarrow B$, the activation energy was 100 kcal/mole. Would the activation energy of the reverse reaction, $B \rightarrow A$, also be 100 kcal/mole, or would it be more or less? Explain with the aid of a diagram.

**\*7.57** The equilibrium constant for the reaction $\alpha$-D-glucose $\rightleftharpoons$ $\beta$-D-glucose is 1.5 at 30°C. This reaction takes place only in aqueous solution. If you begin with a fresh solution of $1\ M$ $\alpha$-D-glucose in water, what will be its concentration when equilibrium is reached?

**7.58** In the reaction

$$H_2 + Cl_2 \longrightarrow 2HCl$$

a 10°C increase in temperature doubles the rate of reaction. If the rate of reaction at 15°C is 2.8 moles of HCl per liter per second, what is the rate at −5°C and at 45°C?

**7.59** In the reaction

$$N_2O_4 \longrightarrow 2NO_2$$

the concentration of $N_2O_4$ was measured at the end of the times shown:

| Time(s) | $[N_2O_4]$ |
|---------|-----------|
| 0 | 0.200 |
| 10 | 0.180 |
| 20 | 0.160 |
| 30 | 0.140 |

What is the rate of reaction?

# CHAPTER 8

*Some familiar acidic and basic products sold in supermarkets. (Photograph by Beverly March.)*

# Acids and Bases

## 8.1  INTRODUCTION

We frequently encounter acids and bases in our daily lives. Oranges, lemons, and vinegar are examples of acidic foods. Sulfuric acid is in our automobile batteries. As for bases, we take antacid tablets for heartburn and use household ammonia as a cleaning agent. What do these substances have in common, and why are acids and bases usually discussed together?

In 1884 a young Swedish chemist named Svante Arrhenius (1859–1927) proposed an answer to the first of these questions by giving what was then a new definition of acids and bases. In modern terms, his definitions are

> An **acid** is a **substance that produces $H_3O^+$ ions in aqueous solution.**
>
> A **base** is a **substance that produces $OH^-$ ions in aqueous solution.**

This definition of acid is a slight modification of what Arrhenius himself said, which was that an acid produces hydrogen ions, $H^+$. Today we know that $H^+$ ions cannot exist in water. A $H^+$ ion is a bare proton, and a charge of $+1$ is too concentrated for such a tiny particle (Sec. 3.2). Because of this, any $H^+$ ion in water immediately combines with an $H_2O$ molecule to give a **hydronium ion, $H_3O^+$**:

$$H^+ + H_2O \longrightarrow H_3O^+$$

The word "alkali" is an older term for "base"; it means the same thing.

Although all chemists know that acidic aqueous solutions do not contain $H^+$ ions, they frequently use the terms "$H^+$" and "proton" when they really mean "$H_3O^+$." The three terms are generally interchangeable.

219

Apart from this modification, the Arrhenius definitions of acid and base are still valid and useful today, after more than 100 years, as long as we are talking about aqueous solutions. When we dissolve an acid in water, it reacts with the water to produce $H_3O^+$. For example, hydrogen chloride, HCl, in its pure state is a poisonous, choking gas. When dissolved in water, it reacts as follows:

$$H-\ddot{\underset{..}{Cl}}: + H-\ddot{\underset{..}{O}}-H \longrightarrow \left[H-\underset{\overset{|}{H}}{\overset{..}{O}}-H\right]^+ + :\ddot{\underset{..}{Cl}}:^-$$

The bottle on the shelf labeled "HCl" is actually not HCl but a solution of $H_3O^+$ and $Cl^-$ ions in water.

With bases the situation is slightly different. Many bases are solid metallic hydroxides, such as KOH, NaOH, $Mg(OH)_2$, and $Ca(OH)_2$. These compounds are ionic even in the solid state. When dissolved in water, the ions merely separate, and each ion is solvated by water molecules (Sec. 6.6). For example,

$$NaOH(s) \xrightarrow{H_2O} Na^+(aq) + OH^-(aq)$$

Other bases, though, are not hydroxides. These bases produce $OH^-$ ions in water by reacting with the water. The most important example of this kind of base is ammonia, $NH_3$. Like HCl, ammonia is a poisonous, choking gas. When dissolved in water, it reacts as follows:

$$NH_3 + H_2O \rightleftharpoons NH_4^+ + OH^-$$

Ammonia produces $OH^-$ ions by taking $H^+$ from water molecules and leaving $OH^-$ behind. This equilibrium lies well to the left. In a 1 $M$ solution of $NH_3$ in water, only about 4 molecules of $NH_3$ out of every 1000 have reacted to form $NH_4^+$. Nevertheless, some $OH^-$ ions are produced, so $NH_3$ is therefore a base.

Bottles of $NH_3$ in water are sometimes labeled "ammonium hydroxide" or "$NH_4OH$" but this gives a false impression of what is really in the bottle. Most of the $NH_3$ molecules have not reacted with the water, so the bottle contains mostly $NH_3$ and $H_2O$ and only a little $NH_4^+$ and $OH^-$.

The boiling point of ammonia is $-33°C$.

We show the equation with equilibrium arrows because it is reversible.

Another reason not to use these labels is that $NH_4OH$ cannot be isolated away from water as a pure substance.

## 8.2   ACID AND BASE STRENGTH

All acids are not equally strong. According to the Arrhenius picture of acids, a **strong acid** is one that **reacts almost completely with water to form $H_3O^+$ ions.** A **weak acid** reacts much less completely. There are six common acids that react completely, or almost completely, when dissolved in water. They are shown in Table 8.1 (Fig. 8.1). Like HCl, the other five acids listed exist in water almost entirely as $H_3O^+$ and the corresponding anion. All other acids are weaker and produce a much smaller concentration of $H_3O^+$ ions. For example, acetic acid in water exists primarily as acetic acid molecules. The equilibrium

$$CH_3COOH + H_2O \rightleftharpoons CH_3COO^- + H_3O^+$$

Acetic                Acetate<br>
acid                 ion

lies well to the left, and only a few acetic acid molecules are converted to acetate ions. Many acids are even weaker than acetic acid.

**STRONG ACIDS**   Sulfuric acid, $H_2SO_4$, is used in a great many industrial processes. More tons of sulfuric acid are manufactured in the United States than any other chemical, organic or inorganic. It is a strong acid and a powerful oxidizing agent, as well as a dehydrating agent. The dissolving of $H_2SO_4$ in water gives off a lot of heat. The acid should always be added to water, and *not* water to it, because of the danger of splashing.

Hydrochloric acid, HCl, is an important acid in chemistry laboratories. Pure HCl is a gas, and the HCl in laboratories is an aqueous solution. HCl is the acid in the gastric fluid in your stomach, where it is secreted at a strength of about 5 percent.

Nitric acid, $HNO_3$, is a strong oxidizing agent (Fig. 8A). A drop of this acid causes the skin to turn yellow. This is due to a reaction with skin protein. A yellow color with nitric acid has long been a test for proteins.

**WEAK ACIDS**   Acetic acid, $CH_3COOH$, is present in vinegar (about 5 percent). Vinegar is made by allowing wine to oxidize in air, whereupon the ethyl alcohol in the wine is oxidized to acetic acid (Sec. 12.3). Pure acetic acid is called *glacial* because of its melting point of 17°C, which means that it freezes on a moderately cold day.

Boric acid, $H_3BO_3$, is a solid. Solutions of boric acid in water were once used as antiseptics, especially for eyes. Boric acid is toxic when swallowed.

Phosphoric acid, $H_3PO_4$, is one of the strongest of the weak acids. The ions produced from it—$H_2PO_4^-$, $HPO_4^{2-}$, and $PO_4^{3-}$—are important in biochemistry (see also Sec. 14.12). Phosphates are used in detergents.

**STRONG BASES**   Sodium hydroxide, NaOH, also called lye, is the most important of the strong bases. It is a solid whose aqueous solutions are used in many industrial processes, including the manufacture of glass and soap.

Potassium hydroxide, KOH, also a solid, is used for many of the same purposes as NaOH.

**WEAK BASES**   Ammonia, $NH_3$, the most important weak base, is a gas with many industrial uses. One of its chief uses is for fertilizers. A 5 percent solution is sold in supermarkets as a cleaning agent, and weaker solutions are used as ''spirits of ammonia'' to revive people who have fainted.

Magnesium hydroxide, $Mg(OH)_2$, is a solid that is insoluble in water. A suspension of about 8 percent $Mg(OH)_2$ in water is called milk of magnesia and is used as a laxative.

**B o x  8A**

# Some Important Acids and Bases

**Figure 8A**
Copper (*left*) and zinc (*center*) react with concentrated nitric acid. (From H. C. Metcalfe, J. E. Williams, and J. F. Castka, *Modern Chemistry*, Holt, Rinehart and Winston, 1986.)

**Figure 8.1**
Some strong acids and strong bases.
(Photograph by Beverly March.)

These bases are strong because they are soluble in water and dissociate completely to OH$^-$.

A similar situation exists with bases. There are only four common strong bases (Table 8.1), all of the metal hydroxide type (Fig. 8.1). As we saw in Section 8.1, ammonia is a weak base because the equilibrium

$$NH_3 + H_2O \rightleftharpoons NH_4^+ + OH^-$$

lies to the left.

It is important to understand that the *strength* of an acid or base is not related to its *concentration*. HCl is a strong acid, whether it is concentrated or dilute, because in its reaction with water the equilibrium lies well to the right:

$$HCl + H_2O \rightleftharpoons H_3O^+ + Cl^-$$

Acetic acid is a weak acid, whether it is concentrated or dilute, because in its reaction with water the equilibrium lies well to the left.

In Section 6.6 we saw that electrolytes (substances that produce ions in solution) can be strong or weak. The strong acids and bases in Table 8.1 are strong electrolytes. Almost all other acids and bases are weak electrolytes.

**TABLE 8.1   The Six Strong Acids and Four Strong Bases**

| Acids | | Bases | |
|-------|------|-------|------|
| *Formula* | *Name* | *Formula* | *Name* |
| $HClO_4$ | Perchloric acid | NaOH | Sodium hydroxide |
| HI | Hydroiodic acid | KOH | Potassium hydroxide |
| HBr | Hydrobromic acid | LiOH | Lithium hydroxide |
| $H_2SO_4$ | Sulfuric acid | $Ba(OH)_2$ | Barium hydroxide |
| $HNO_3$ | Nitric acid | | |
| HCl | Hydrochloric acid | | |

## 8.3 BRØNSTED-LOWRY ACIDS AND BASES

The Arrhenius definitions of acid and base are very useful in aqueous solutions. But what if water is not involved? In 1923 new definitions of acids and bases were proposed simultaneously by Brønsted and Lowry. These definitions are more general and do not require that the solvent be water. They are

Johannes Brønsted (1879–1947) was a Danish chemist, and Thomas Lowry (1874–1936) was an English chemist.

> An **acid** is a **proton donor**.
>
> A **base** is a **proton acceptor**.
>
> An **acid-base reaction** is the **transfer of a proton**.

As noted in Section 8.1, this is precisely how acids react with water. An acid gives a proton, $H^+$, to water:

$$HCl + H_2O \rightleftharpoons H_3O^+ + Cl^-$$

$$\text{Acid} \quad \text{Base} \qquad \text{Acid} \quad \text{Base}$$

We have, in this equation, two new ideas:

**1.** The water, because it accepts the proton, is a base by the Brønsted-Lowry definition.

**2.** The hydronium ion is an acid since it can now, at least in theory, give a proton back to the $Cl^-$. Similarly, the $Cl^-$ is a base because it can take the proton back.

To repeat, an acid, in the Brønsted-Lowry picture, is a proton donor. When it gives up its proton, it always becomes a base. This base is called the **conjugate base** of the acid. Some examples are given in Table 8.2.

**TABLE 8.2  Some Acids and Their Conjugate Bases, in Decreasing Order of Acid Strength**

| | Acid | | | Conjugate Base | |
|---|---|---|---|---|---|
| **Strong Acids** | $HI$ | Hydroiodic acid | $I^-$ | Iodide ion | **Weak Bases** |
| | $H_2SO_4$ | Sulfuric acid | $HSO_4^-$ | Hydrogen sulfate ion | |
| | $HCl$ | Hydrochloric acid | $Cl^-$ | Chloride ion | |
| | $HNO_3$ | Nitric acid | $NO_3^-$ | Nitrate ion | |
| | $H_3O^+$ | Hydronium ion | $H_2O$ | Water | |
| | $HSO_4^-$ | Hydrogen sulfate ion | $SO_4^{2-}$ | Sulfate ion | |
| | $H_3PO_4$ | Phosphoric acid | $H_2PO_4^-$ | Dihydrogen phosphate ion | |
| | $CH_3COOH$ | Acetic acid | $CH_3COO^-$ | Acetate ion | |
| | $H_2CO_3$ | Carbonic acid | $HCO_3^-$ | Bicarbonate ion | |
| | $H_2S$ | Hydrogen sulfide | $HS^-$ | Hydrogen sulfide ion | |
| | $H_2PO_4^-$ | Dihydrogen phosphate ion | $HPO_4^{2-}$ | Hydrogen phosphate ion | |
| | $NH_4^+$ | Ammonium ion | $NH_3$ | Ammonia | |
| | $C_6H_5OH$ | Phenol | $C_6H_5O^-$ | Phenylate ion | |
| **Weak Acids** | $HCO_3^-$ | Bicarbonate ion | $CO_3^{2-}$ | Carbonate ion | **Strong Bases** |
| | $HPO_4^{2-}$ | Hydrogen phosphate ion | $PO_4^{3-}$ | Phosphate ion | |
| | $H_2O$ | Water | $OH^-$ | Hydroxide ion | |
| | $C_2H_5OH$ | Ethyl alcohol | $C_2H_5O^-$ | Ethoxide ion | |
| | $NH_3$ | Ammonia | $NH_2^-$ | Amide ion | |

Similarly, a base is a proton acceptor. When it accepts a proton, it always becomes an acid called the **conjugate acid** of that base. The examples in Table 8.2 illustrate this also. Every acid and the species that remains when it loses its proton are called a **conjugate acid-base pair.** Thus HCl and Cl$^-$ are a conjugate acid-base pair, as are H$_3$O$^+$ and H$_2$O, and CH$_3$COOH and CH$_3$COO$^-$.

Note the following points from the examples in Table 8.2.

**1.** An acid can be positively or negatively charged, or it can be neutral. A base can be negatively charged, or it can be neutral.

**2. Some acids can give up two or three protons** (other acids, not shown in Table 8.2, can give up even more). Such acids are called **diprotic** or **triprotic.** For example, H$_2$SO$_4$ can give up two protons to become SO$_4^{2-}$ and H$_3$PO$_4$ can give up three to become PO$_4^{3-}$. In the Brønsted-Lowry picture, however, each acid is considered to give up only one proton **(monoprotic).** Thus H$_2$SO$_4$ is treated as if it gives up a single proton to become HSO$_4^-$, and HSO$_4^-$ is regarded as a separate acid in its own right (note that HSO$_4^-$ is listed separately in Table 8.2).

**3.** Some substances appear on *both* sides of Table 8.2. For example, HCO$_3^-$, the bicarbonate ion, can give up a proton to become CO$_3^{2-}$ (thus it is an acid) or take a proton to become H$_2$CO$_3$ (thus it is a base). A substance that can act as either an acid or a base is called **amphoteric.** The most important amphoteric substance in Table 8.2 is of course water, which takes a proton to become H$_3$O$^+$ or loses one to become OH$^-$.

**4.** A substance cannot be a Brønsted-Lowry acid unless it contains a hydrogen atom, but not all hydrogen atoms can in fact be given up. For example, acetic acid, CH$_3$COOH, has four hydrogens but is monoprotic. It gives up only one of them. Similarly, phenol, C$_6$H$_5$OH, gives up only one of its six hydrogens.

We have seen that HCl reacts with water as follows:

$$\text{HCl} + \text{H}_2\text{O} \rightleftharpoons \text{H}_3\text{O}^+ + \text{Cl}^-$$

Acid    Base        Acid     Base

The equilibrium lies far to the right. Out of every 1000 HCl molecules dissolved in water, more than 990 are converted to Cl$^-$ ions. However, H$_3$O$^+$ is an acid and Cl$^-$ a base. Why doesn't H$_3$O$^+$ give a proton to Cl$^-$ to produce HCl and H$_2$O? What we are really asking here is, why does the equilibrium lie far to the right in this case?

We can answer this question by using the concept of acid and base strength. In the Brønsted-Lowry picture, **acid strength** is the **willingness to give up the proton** and **base strength** is the **willingness to accept the proton.** Because HCl is more willing to give up its proton, it is a stronger acid than H$_3$O$^+$. Looking at it the other way, H$_2$O is a stronger base than Cl$^-$ and is therefore more willing to accept the proton. **In the battle for the proton, the stronger base always wins.** Since H$_2$O is a stronger base than Cl$^-$, it gets the proton.

It should be obvious that the stronger an acid is, the weaker must be its conjugate base, since the more willing an acid is to donate the proton, the less willing is the conjugate base to take it back.

The acids in Table 8.2 are listed in order of *decreasing* strength (strongest at the top, weakest at the bottom). It follows that the conjugate bases are listed in order of *increasing* strength (strongest at the bottom, weakest at the top). We can therefore use this table (or any other table that lists acids with the strongest at the top) to predict whether or not any acid will donate its proton to any base. The rule is that **any acid in Table 8.2 donates its proton to the conjugate base of any acid *below* it but does not donate its proton to the conjugate base of any acid above it.**

---

**EXAMPLE**

Which of these reactions proceed from left to right?
(a)  $H_3O^+ + CH_3COO^- \rightleftharpoons H_2O + CH_3COOH$
(b)  $H_3O^+ + I^- \rightleftharpoons HI + H_2O$
(c)  $H_2CO_3 + OH^- \rightleftharpoons HCO_3^- + H_2O$
(d)  $HPO_4^{2-} + NH_3 \rightleftharpoons PO_4^{3-} + NH_4^+$

These reactions are all equilibria, and what we are really doing is predicting in which direction the equilibria lie.

**Answer**

All we need to do is find the acids and bases in Table 8.2. Reactions (a) and (c) proceed because $H_3O^+$ is above $CH_3COOH$ and $H_2CO_3$ is above $H_2O$. Reactions (b) and (d) do not proceed because $H_3O^+$ is below HI and $HPO_4^{2-}$ is below $NH_4^+$.

---

**PROBLEM 8.1**   Which of these reactions proceed from left to right?
(a)  $CH_3COO^- + NH_4^+ \rightleftharpoons CH_3COOH + NH_3$
(b)  $H_3PO_4 + HS^- \rightleftharpoons H_2PO_4^- + H_2S$

---

## 8.4   ACID DISSOCIATION CONSTANTS

In Section 8.2 we learned that acids vary in the extent to which they produce $H_3O^+$ when added to water. The six strong acids (see Table 8.1) are completely converted to $H_3O^+$ and the corresponding anions, whereas weak acids undergo this reaction to a lesser extent.

Since these reactions are all equilibria, we can use equilibrium constants to tell us quantitatively just how strong any weak acid actually is. The reaction that takes place when a weak acid is added to water is

$$HA + H_2O \rightleftharpoons H_3O^+ + A^-$$

HA is the general formula for an acid.

The equilibrium expression (Sec. 7.6) is

$$K = \frac{[H_3O^+][A^-]}{[HA][H_2O]} \tag{8.1}$$

Notice that this expression contains the concentration of water. Since water is the solvent, its concentration changes very little when we add HA, so we can treat $[H_2O]$ as constant and rearrange Equation 8.1 to bring the constant terms together on the left side:

$$K[H_2O] = \frac{[H_3O^+][A^-]}{[HA]} \tag{8.2}$$

**TABLE 8.3   Some $K_a$ and p$K_a$ Values[a] for Weak Acids, in Decreasing Order of Strength**

| Name | Formula | $K_a$ | p$K_a$ |
|---|---|---|---|
| Phosphoric acid | $H_3PO_4$ | $7.5 \times 10^{-3}$ | 2.1 |
| Hydrofluoric acid | HF | $3.5 \times 10^{-4}$ | 3.45 |
| Formic acid | HCOOH | $1.8 \times 10^{-4}$ | 3.75 |
| Lactic acid | $CH_3CH(OH)COOH$ | $1.4 \times 10^{-4}$ | 3.85 |
| Acetic acid | $CH_3COOH$ | $1.8 \times 10^{-5}$ | 4.75 |
| Carbonic acid | $H_2CO_3$ | $4.3 \times 10^{-7}$ | 6.37 |
| Dihydrogen phosphate ion | $H_2PO_4^-$ | $6.2 \times 10^{-8}$ | 7.21 |
| Boric acid | $H_3BO_3$ | $7.3 \times 10^{-10}$ | 9.14 |
| Ammonium ion | $NH_4^+$ | $5.6 \times 10^{-10}$ | 9.25 |
| Phenol | $C_6H_5OH$ | $1.3 \times 10^{-10}$ | 9.9 |
| Bicarbonate ion | $HCO_3^-$ | $5.6 \times 10^{-11}$ | 10.25 |
| Hydrogen phosphate ion | $HPO_4^{2-}$ | $2.2 \times 10^{-13}$ | 12.66 |

[a] The p$K_a$ is explained in Section 8.8.

When we multiply one constant, $K$, by another, $[H_2O]$, we get a third constant, which we call $K_a$:

$$K_a = K[H_2O] = \frac{[H_3O^+][A^-]}{[HA]} \qquad (8.3)$$

$K_a$ is also called the **acidity constant** and the **ionization constant**.

The $K_a$ is called the **acid dissociation constant,** and every weak acid has one. Table 8.3 lists some weak acids and their dissociation constants. The larger the $K_a$, the stronger the acid. We can see this by looking at the equation

$$HA + H_2O \rightleftharpoons H_3O^+ + A^-$$

A stronger acid produces more $H_3O^+$ ions (and $A^-$ ions also) than a weaker acid. This increases the numerator of Equation 8.3 so that $K_a$ is larger for the stronger acid.

The importance of $K_a$ is that it immediately tells us how strong an acid is. For example, Table 8.3 shows us that although acetic acid, formic acid, and phenol are all weak acids, their strengths as acids are not the same. The table gives $K_a$ for acetic acid as $1.8 \times 10^{-5}$. Thus formic acid, with a $K_a$ of $1.8 \times 10^{-4}$, is ten times stronger than acetic acid (an increase of 1 in the exponent increases the acidity ten times), while phenol, whose $K_a$ is $1.3 \times 10^{-10}$, is about a hundred thousand times weaker than acetic acid.

## 8.5   SOME PROPERTIES OF ACIDS AND BASES

Chemists stopped tasting chemicals when some of them got sick or died from doing this.

Chemists of today do not taste the substances they work with, but 150 and 200 years ago it was routine to do so. That is how we know that acids taste sour and bases taste bitter. The sour taste of lemons, vinegar, and many other foods is due to the acids they contain.

Both strong acids and strong bases damage living tissue. Externally, they cause chemical burns. Internally, they are even more harmful and can cause

## Acid and Alkali Burns of the Cornea

In the laboratory every student must wear safety glasses to prevent the accidental splashing of chemicals into the eyes. The outermost part of the eye is the cornea. This transparent tissue is very sensitive to chemical burns, whether caused by acids or by bases. Acids and bases in small amounts and in low concentrations that would not seriously damage other tissues, such as skin, may cause severe corneal burns. If not treated promptly, these can lead to permanent loss of vision.

Acids are less dangerous than bases.

For acid in the eyes, first aid should be provided by immediately washing the eyes with a steady stream of cold water. The fast removal of the acid is of utmost importance. Time should not be wasted looking for some mild alkaline solution to neutralize the acid because the removal of the acid is more helpful than the neutralization. Even if it is sulfuric acid in the eyes, washing with copious quantities of water is *the* remedy. The heat generated by the interaction of sulfuric acid with water (lots of water) will be removed by the washing. The extent of the burn can be assessed after all the acid has been washed out.

Not so with alkali (base) burns. Bases are dangerous chemicals, not only in the laboratory and in industry but also in the home. Concentrated ammonia or sodium hydroxide of the "liquid plumber" type can cause severe damage when splattered in the eyes. Again, immediate washing is of utmost importance, but alkali burns can later develop ulcerations in which the healing wound deposits scar tissue that is not transparent, so that vision is impaired. Fortunately, the cornea is a tissue with no blood vessels and therefore can be transplanted without immunological rejection problems. Today, corneal grafts are common.

severe damage to the digestive tract. One way they cause internal damage is by breaking down the organism's protein molecules.

The most important property of acids and bases is that they react with each other in a process called **neutralization.** This is an appropriate name because, when a strong corrosive acid such as hydrochloric acid reacts with a strong corrosive base such as sodium hydroxide, the product (a solution of ordinary table salt in water) has neither acidic nor basic properties. We call such a solution **neutral.** Neutralization reactions are further discussed in the next section.

In the laboratory, you must be careful not to get any strong acids or bases on your skin or, even worse, in your eyes, where they can cause blindness.

## 8.6  REACTIONS OF ACIDS

Acids have several characteristic reactions. In all of them, they react as proton donors, giving their protons to a base. In this section we look at three types.

### Reactions with Metals, Metal Hydroxides, and Oxides

Acids react with certain metals (called *active metals*) to produce hydrogen ($H_2$) and a salt. An example is (Fig. 8.2):

$$Mg(s) + 2HCl(aq) \longrightarrow MgCl_2(aq) + H_2(g)$$

Active metal     Acid     A salt     Hydrogen

## Drugstore Antacids

**Figure 8C**
A commercial remedy for excess stomach acid. The bubbles are $CO_2$, from the reaction between an acid (citric acid) and the $HCO_3^-$ ion (from sodium bicarbonate). (Photograph by Charles D. Winters.)

Stomach fluid is normally quite acidic because of its HCl content, but sometimes we get "heartburn" that may be caused by excess stomach acidity. Then many of us take an antacid, which, as the name implies, is a substance that neutralizes acids—in other words, a base. The word "antacid" is a medical term, not one used by chemists. It is, however, found on the labels of many medications available in drugstores and supermarkets. Almost all of them use weak bases (hydroxides and/or carbonates) to decrease the acidity of the stomach (Fig. 8C). The active ingredients of some of these antacids are as follows (some brands are sold in more than one formulation, but we list only one in each case):

Alka-Seltzer: sodium bicarbonate, $NaHCO_3$; citric acid; and aspirin

Brioschi: sodium bicarbonate

Chooz: calcium carbonate, $CaCO_3$

Di-gel: $CaCO_3$; magnesium hydroxide, $Mg(OH)_2$; and simethicone (an antiflatulent)

Gaviscon: aluminum hydroxide, $Al(OH)_3$, and magnesium carbonate, $MgCO_3$

Gelusil: $Al(OH)_3$, $Mg(OH)_2$, and simethicone

Maalox: $Mg(OH)_2$ and $Al(OH)_3$

Mylanta: $Al(OH)_3$, $Mg(OH)_2$, and simethicone

Phillips' Milk of Magnesia: $Mg(OH)_2$

Riopan: magnesium aluminum hydrate (magaldrate)

Rolaids: aluminum sodium dihydroxy carbonate, $AlNa(OH)_2CO_3$

Tempo: $CaCO_3$, $Al(OH)_3$, $Mg(OH)_2$, and simethicone

Tums: $CaCO_3$

**Figure 8.2**
A ribbon of magnesium metal reacts with aqueous HCl to give $H_2$ gas and aqueous $MgCl_2$. (Photograph by Charles D. Winters.)

Acids react with metal hydroxides to give water and a salt:

$$HNO_3 + KOH \longrightarrow H_2O + KNO_3$$
$$\text{Acid} \qquad \text{Base} \qquad\qquad \text{A salt}$$

A **salt** is an **ionic compound.** We may consider all salts to be composed of a negative ion from an acid and a positive ion from a base. For example, potassium nitrate, $KNO_3$, is made of nitrate ions, $NO_3^-$, from nitric acid, $HNO_3$, and potassium ions, $K^+$, from potassium hydroxide, KOH. The *actual* equation for the reaction between $HNO_3$ and KOH is therefore (Sec. 4.8)

$$H_3O^+ + NO_3^- + K^+ + OH^- \rightleftharpoons H_2O + H_2O + NO_3^- + K^+$$

This can be simplified by omitting the spectator ions ($K^+$ and $NO_3^-$):

$$H_3O^+ + OH^- \rightleftharpoons H_2O + H_2O$$

This is the actual ionic equation when *any* strong acid reacts in water with *any* strong base. Note that even in this view an acid and a base are converted to a salt and water, and the salt ends up dissolved in the water; an example is $KNO_3$ in the case mentioned above.

Strong acids also react with metal oxides to give the same salt that is obtained from the corresponding hydroxide; for example,

$$2H_3O^+ + 2NO_3^- + CaO \longrightarrow 3H_2O + Ca^{2+} + 2NO_3^-$$

When an acid is added to the bicarbonate ion, $CO_2$ gas is given off:

$$HCO_3^- + H_3O^+ \rightleftharpoons H_2O + H_2CO_3 \rightleftharpoons CO_2(g) + H_2O$$

The acid protonates $HCO_3^-$, and the resulting $H_2CO_3$, being unstable, breaks down to $CO_2$ and water. This reaction is used in a number of ways, of which we discuss two.

**CAKE RISING**  Baking powder consists of sodium bicarbonate, $NaHCO_3$, and a solid acidic compound. When the powder is dry, little or no reaction takes place, but in the presence of water the acid and bicarbonate ion react to produce $CO_2$ gas, which is what causes a cake to rise. Potassium acid tartrate, $KC_4H_5O_6$, called cream of tartar, was the earliest acidic compound used for this purpose and is still one of the best, but because of its high cost it has been largely replaced in recent years by monocalcium phosphate or similar compounds. Because of its use in baking powders, $NaHCO_3$ is often called baking soda. Baking powder is seldom used for bread. Bread rising is accomplished by yeast, which produces $CO_2$ by fermenting the sugars present in the dough.

**FIRE EXTINGUISHERS**  There are several types of fire extinguishers, but one type uses the reaction between $NaHCO_3$ and an acid to produce $CO_2$. The pressure of the $CO_2$ forces water out of the fire extinguisher. In another type of extinguisher $CO_2$ itself, in the form of a foam, is used to extinguish the flames.

**Box 8D**

# Rising Cakes and Fire Extinguishers

## Reactions with Carbonates

When a strong acid is added to a carbonate such as sodium or potassium carbonate, bubbles of gas are rapidly given off. The gas, carbon dioxide, is formed in this manner:

$$2H_3O^+ + CO_3^{2-} \rightleftharpoons H_2CO_3 + 2H_2O$$
$$H_2CO_3 \rightleftharpoons CO_2(g) + H_2O$$

The direct product of the proton transfer is carbonic acid, $H_2CO_3$, but this compound is very unstable and breaks down to $CO_2$ and $H_2O$. As shown in Tables 8.2 and 8.3, carbonic acid is a weak acid. By the rule given in Section 8.3, not only the six strong acids but any acid stronger than $H_2CO_3$ reacts with carbonates to give $CO_2$ gas.

## Reactions with Ammonia and Amines

Any acid stronger than $NH_4^+$ (see Table 8.2) is strong enough to react with $NH_3$ in this way:

$$HCl + NH_3 \rightleftharpoons NH_4^+ + Cl^-$$

In Section 15.2 we will meet a family of compounds called amines, which are similar to ammonia except that one or more of the three hydrogen atoms are replaced by carbon. A typical example is methylamine, $CH_3-NH_2$. The base strength of most amines is similar to that of $NH_3$, which means that amines react with acids in the same way:

$$HCl + CH_3-NH_2 \rightleftharpoons CH_3-NH_3^+ + Cl^-$$

Neither $NH_3$ nor amines survive as such in the presence of an acid. They are converted to the corresponding positive ions (conjugate acids). This is very important in the chemistry of the body, as we shall see in later chapters.

## 8.7 SELF-IONIZATION OF WATER

We have seen that an acid produces $H_3O^+$ ions in water and a base produces $OH^-$ ions. Suppose we have absolutely pure water, with no added acid or base. Surprisingly enough, even pure water contains a very small number of $H_3O^+$ and $OH^-$ ions. They come from the reaction

This reaction is called the **self-ioniza-tion of water.**

$$H_2O + H_2O \rightleftharpoons H_3O^+ + OH^-$$

Acid  Base  Acid  Base

in which one water molecule acts as an acid and gives a proton to a second water molecule, which is therefore acting as a base.

What is the extent of this reaction; how far does it go? The answer is, very little. As shown by the arrows, it is an equilibrium lying far to the left. We shall soon see exactly how far, but first let us write the equilibrium expression:

$$K = \frac{[H_3O^+][OH^-]}{[H_2O]^2}$$

This expression contains the concentration of water, but since only a very few water molecules react, the concentration of water remains essentially constant. Since this is so, it is convenient to change the expression to

$$K[H_2O]^2 = [H_3O^+][OH^-]$$

As we did with the dissociation constant for acids (Sec. 8.4), we can replace $K[H_2O]^2$ by a new constant, $K_w$. Our equilibrium expression is now

$$K_w = [H_3O^+][OH^-]$$

The $K_w$ is called the **ion product of water.** At room temperature, $K_w$ has the value of $1.0 \times 10^{-14}$. Since in pure water $H_3O^+$ and $OH^-$ are formed by the same process (the self-ionization of water), their concentrations must be equal. This means that, in pure water,

$$[H_3O^+] = 1 \times 10^{-7} \text{ moles/L}$$
$$[OH^-] = 1 \times 10^{-7} \text{ moles/L}$$

Thus, in pure water, each liter of $H_2O$ contains only $1 \times 10^{-7}$ mole of $H_3O^+$ and $1 \times 10^{-7}$ mole of $OH^-$. These are very small concentrations, not enough to make pure water a conductor of electricity. Water is not an electrolyte.

The equation $K_w = [H_3O^+][OH^-]$ is important because it applies not only to pure water; **it is valid in any water solution.** It says that the product of $[H_3O^+]$ and $[OH^-]$ in any aqueous solution is equal to $1 \times 10^{-14}$. If, for example, we add 0.01 mole of HCl to 1 L of pure water, it reacts essentially completely to give $H_3O^+$ ions. The concentration of $H_3O^+$ will be 0.01 $M$, or $1 \times 10^{-2} M$. This means that the $[OH^-]$ *must be* $1 \times 10^{-12} M$ because the product *must be* $1 \times 10^{-14}$.

---

**EXAMPLE**

The [OH⁻] of an aqueous solution is $1 \times 10^{-4}$ $M$. What is the [H₃O⁺]?

**Answer**

We substitute into the equation:

$$K_w = [H_3O^+][OH^-] = 1 \times 10^{-14}$$
$$[H_3O^+](10^{-4}) = 1 \times 10^{-14}$$
$$[H_3O^+] = \frac{1 \times 10^{-14}}{1 \times 10^{-4}} = 1 \times 10^{-10} \ M$$

---

**PROBLEM 8.2**  The [OH⁻] of an aqueous solution is $1 \times 10^{-12}$ $M$. What is the [H₃O⁺]?

---

Aqueous solutions can have a very high [H₃O⁺], but the [OH⁻] must then be very low, and vice versa. Any solution with a [H₃O⁺] higher than $1 \times 10^{-7}$ $M$ is **acidic.** In such solutions, of course, [OH⁻] must be lower than $1 \times 10^{-7}$ $M$. The higher the [H₃O⁺], the more acidic the solution is. Similarly, any solution with an [OH⁻] higher than $1 \times 10^{-7}$ $M$ is **basic.**

Pure water, in which [H₃O⁺] and [OH⁻] are equal (they are both $1 \times 10^{-7}$ $M$) is **neutral,** that is, neither acidic nor basic.

Not only pure water, but *any* aqueous solution in which [H₃O⁺] = [OH⁻] = $1 \times 10^{-7}$ $M$ is a neutral solution.

## 8.8 pH

Because the [H₃O⁺] of any aqueous solution is such an important characteristic, chemists, workers in the health sciences, and other scientists must frequently talk about it. When they do so, it is inconvenient to keep saying, for example, "$6 \times 10^{-11}$ $M$, $2 \times 10^{-2}$ $M$," or other numbers of this form. Fortunately, a shorter way exists. Chemists have defined a number, called the pH, to express these concentrations in a more convenient way. The mathematical definition of **pH** is

$$\textbf{pH} = -\textbf{log[H}_3\textbf{O}^+\textbf{]}$$

When the [H₃O⁺] is of the form $1 \times 10^{-A}$ $M$, where the A stands for any number, we don't have to worry about the mathematical definition—the pH is simply A. Instead of saying that a [H₃O⁺] is $1 \times 10^{-4}$ $M$, we simply say that the pH is 4. As you can see, all we do is take the exponent of the [H₃O⁺] and drop the minus sign.

---

**EXAMPLE**

The [H₃O⁺] of a certain liquid soap is $1 \times 10^{-9}$ $M$. What is the pH?

**Answer**
The pH is 9.

---

**PROBLEM 8.3** The $[H_3O^+]$ of an acidic solution is $1 \times 10^{-5}$ $M$. What is the pH?

---

**EXAMPLE**

The pH of black coffee is 5. What is the $[H_3O^+]$?

**Answer**
The $[H_3O^+]$ is $1 \times 10^{-5}$ $M$.

---

**PROBLEM 8.4** The pH of tomato juice is 4. What is the $[H_3O^+]$?

---

**EXAMPLE**

The $[OH^-]$ in a strongly alkaline solution is $1 \times 10^{-1}$ $M$. What is the pH?

**Answer**
Here we are given $[OH^-]$, so we must first calculate $[H_3O^+]$ before we can find pH:

$$[H_3O^+][OH^-] = K_w$$
$$[H_3O^+](10^{-1}) = 10^{-14}$$
$$[H_3O^+] = \frac{10^{-14}}{10^{-1}} = 10^{-13}$$

Since $[H_3O^+] = 10^{-13}$, the pH is 13.

---

**PROBLEM 8.5** The $[OH^-]$ of an acidic solution is $1 \times 10^{-12}$. What is the pH?

| TABLE 8.4 | Two-place Logarithms |
|-----------|---------------------|
| Number | Logarithm |
| 1 | 0.00 |
| 2 | 0.30 |
| 3 | 0.48 |
| 4 | 0.60 |
| 5 | 0.70 |
| 6 | 0.78 |
| 7 | 0.85 |
| 8 | 0.90 |
| 9 | 0.95 |
| 10 | 1.00 |

The convenience of this system is readily apparent. It is much easier to say "the pH is 12" than to say "the $[H_3O^+] = 1 \times 10^{-12}$ $M$," and yet both mean the same thing.

Suppose that the $[H_3O^+]$ is $7 \times 10^{-4}$. What do we do then? That's when we need to use logarithms. Now that inexpensive scientific calculators are readily available, this is no problem at all. We simply take the logarithm of $7 \times 10^{-4}$ with the calculator and get the answer $-3.15$. Just drop the minus sign (because pH $= -\log[H_3O^+]$) and you have the pH: 3.15.

If you don't have ready access to a scientific calculator, you can calculate fractional pH values using a two-place logarithm table such as that shown as Table 8.4.

---

**EXAMPLE**

What is the pH of a solution whose $[H_3O^+]$ is $7 \times 10^{-4}$ $M$?

**Answer**

The number is 7 *times* $10^{-4}$. We must find the logarithm of 7 and of $10^{-4}$ and *add* the logs (because in order to multiply numbers we must add their logarithms):

$$\log 7 = 0.85 \text{ (Table 8.4)}$$
$$\log 10^{-4} = -4$$

Adding, we get

$$-4 + 0.85 = -3.15$$

Since pH = $-\log[H_3O^+]$, we drop the minus sign, and the pH is 3.15.

**PROBLEM 8.6**   What is the pH of a solution whose $[H_3O^+]$ is $8 \times 10^{-9}\, M$?

---

**EXAMPLE**

A solution has a pH of 11.70. What is the $[H_3O^+]$?

**Answer**

**CALCULATOR METHOD**   Since this is the reverse of the method just described, we first convert 11.70 to $-11.70$ and then find the antilogarithm on the calculator ("inverse log" on some calculators). When we do this we get $[H_3O^+] = 2.0 \times 10^{-12}\, M$.

**LOGARITHM TABLE METHOD**   Here, too, we must do the *reverse* of what we did in the last example. The first step is to add a minus sign, so 11.70 becomes $-11.70$. Then we convert this to the *next highest negative integer plus a positive decimal.*

This is the key step of this procedure.

$$-11.70 = -12 + 0.30$$

The last step is to take the antilogarithm of both terms. The antilogarithm of $-12$ is $10^{-12}$. The antilogarithm of 0.30 (Table 8.4) is 2. The $[H_3O^+]$ is therefore $2 \times 10^{-12}\, M$.

**PROBLEM 8.7**   The pH of blood is 7.4. What is the $[H_3O^+]$?

---

As already indicated, pH is universally used to express the acidity of aqueous solutions (you probably had heard of pH even before you took this course). **The lower the pH, the more acidic the solution; the higher the pH, the more basic the solution.** Pure water, with a $[H_3O^+]$ of $1 \times 10^{-7}\, M$, has a pH of 7 and is neutral.

Acidic solutions have a pH below 7.
Basic solutions have a pH above 7.

| TABLE 8.5 | pH Values for Common Materials |
|---|---|
| **Material** | **pH** |
| Battery acid | 0.5 |
| Gastric juice | 0.9–1.8 |
| Soft drinks | 2.0–4.0 |
| Lemon juice | 2.2–2.4 |
| Vinegar | 2.4–3.4 |
| Tomatoes | 4.0–4.4 |
| Black coffee | 5.0–5.1 |
| Urine | 5.5–7.5 |
| Rain (unpolluted) | 6.2 |
| Milk | 6.3–6.6 |
| Saliva | 6.5–7.5 |
| Pure water | 7.0 |
| Blood | 7.35–7.45 |
| Bile | 7.6–8.6 |
| Pancreatic fluid | 7.8–8.0 |
| Seawater | 8–9 |
| Soap | 8–10 |
| Milk of magnesia | 10.5 |
| Household ammonia | 11.7 |
| Lye (1 $M$ NaOH) | 14 |

**Figure 8.3**
The indicators show that vinegar has a pH of about 3, a carbonated beverage has a pH of about 4 to 5, and a household cleaner is strongly basic. (Photograph by Charles D. Winters.)

All fluids in the human body are aqueous; that is, the only solvent is water. Consequently, all body fluids have a pH value. Some of them have a narrow pH range, others a wide range. The pH of blood must be between 7.35 and 7.45 (slightly basic). If it goes outside these limits, illness and even death will result (see Box 8E). In contrast, the pH of urine can vary from 5.5 to 7.5. Table 8.5 gives the pH values of some body fluids, as well as of some other common materials (see Fig. 8.3).

One thing we must remember when we see pH values is that a difference of one pH unit means a multiplication by 10 of the $[H_3O^+]$. For example, a pH of 3 does not sound very different from a pH of 4, but the first means a $[H_3O^+]$ of $10^{-3}$ $M$ and the second means a $[H_3O^+]$ of $10^{-4}$ $M$. The $[H_3O^+]$ of the pH 3 solution is ten times the $[H_3O^+]$ of the pH 4 solution.

There are two ways to measure the pH of an aqueous solution. In one method pH paper is used (Fig. 8.4). When a drop of solution is placed on this paper, the paper turns a certain color. The pH is determined by comparing the color of the paper with the colors on a chart supplied with the paper.

pH paper is made by soaking plain paper with a mixture of indicators. An **indicator** is a **substance that changes color at a certain pH.** An example is the compound methyl orange. When a drop of methyl orange solution is added to any aqueous solution with a pH of 3.2 or lower it turns red (and turns the whole solution red). Added to an aqueous solution with a pH of 4.4 or higher, it turns yellow (and turns the whole solution yellow). These are the particular limits and colors for this indicator. Other indicators have other limits and colors.

Litmus paper turns red in acidic solution and blue in basic solution. This indicator is often used as an approximate guide to acidity or basicity.

**Figure 8.4**
Rolls of pH paper. (Photograph by Beverly March.)

Phenolphthalein, for example, is colorless below pH 8.2 but pink above pH 10.0. The colors and pH values of these and some other indicators are shown in Figure 8.5.

The other method for determining pH is more accurate. In this method a pH meter is used (Fig. 8.6). The electrode of the meter is dipped into the unknown solution and the pH, to the nearest tenth of a pH unit, is read on a dial.

In Section 8.4 we saw that $K_a$ values for weak acids are given in exponential notation. Because of this, $pK_a$ values are often used instead. The $pK_a$ is defined as $-\log K_a$ and is calculated from $K_a$ values in exactly the same way pH is calculated from $[H_3O^+]$. Table 8.3 lists $pK_a$ as well as $K_a$ values for some weak acids.

The higher the $pK_a$, the weaker the acid.

**Figure 8.5**
Some acid-base indicators. Note that some have two color changes.

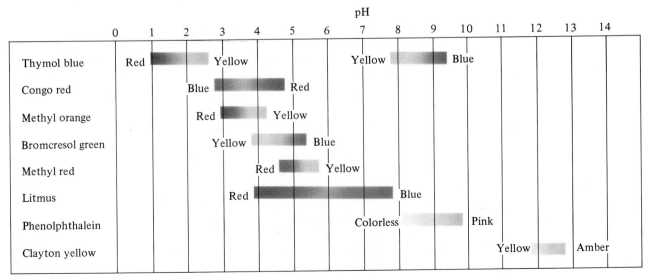

**Figure 8.6**
A pH meter. The blue solution is 0.1 $M$ CuSO$_4$. This is the salt of a strong acid and a weak base, so the pH is 4.5 (Sec. 8.9). (Photograph by Charles D. Winters.)

## 8.9 THE pH OF AQUEOUS SALT SOLUTIONS

When an acid is added to water, the solution becomes acidic (pH below 7); when a base is added to water, the solution becomes basic (pH above 7). Suppose we add a salt to water? What happens to the pH then? At first thought, you might suppose that all salts are neutral and that addition of any salt to pure water leaves the pH at 7. This is in fact true for some salts. They are neutral and do not change the pH of pure water. Many salts are acidic or basic, however, and cause the pH of pure water to change. In order to see why, let us look at sodium acetate, a typical basic salt.

Like almost all salts, sodium acetate is 100 percent ionized in water, so that when we add CH$_3$COONa to water we get a solution containing Na$^+$ and CH$_3$COO$^-$ ions. The Na$^+$ ions do not react with the water, but the CH$_3$COO$^-$ ions do:

$$CH_3COO^- + H_2O \rightleftharpoons CH_3COOH + OH^-$$

Because this is a reaction with water, it is sometimes called **hydrolysis.**

This is an equilibrium reaction lying well to the left. In a 0.1 $M$ solution of sodium acetate, only about one out of every 13 500 acetate ions is converted to acetic acid and OH$^-$ ions at equilibrium. Nevertheless, OH$^-$ ions *are* being produced, enough to cause the pH to rise to about 8.9. Because this is higher than 7, the solution is basic.

We can now understand why sodium acetate is a basic salt. Some of the acetate ions react with water to produce OH$^-$ ions, but none of the Na$^+$ ions react with water at all. Because OH$^-$ ions are being produced, the solution is basic. It is true that acetic acid is also being produced, but the pH of an aqueous solution depends only on the concentration of H$_3$O$^+$ (and of OH$^-$) and *not* on the concentration of any other acid or base.

How can we predict whether any particular salt will produce an acidic, basic, or neutral solution when added to water? This is easily done if we can remember the six strong acids and four strong bases listed in Table 8.1. Acetate ions, CH$_3$COO$^-$, react with water because the corresponding acid, acetic acid, is a weak acid. Sodium ions do not react with water because the corresponding base, NaOH, is a strong base.

The four strong bases are NaOH, KOH, LiOH, and Ba(OH)$_2$. All other bases are considered weak. The six strong acids are HI, HBr, HCl, HNO$_3$, H$_2$SO$_4$, and HClO$_4$.

All salts can be thought of as being combinations of an acid and a base. We can predict whether a salt will be acidic, neutral, or basic in water solution when we know something about the strength of the acid and base. We have four cases:

**1.** *A salt of a strong base and a weak acid.* As we see from the sodium acetate example, such salts are **basic salts** and raise the pH of pure water.

**2.** *A salt of a strong acid and a weak base.* We have the opposite behavior here. An example is ammonium chloride, $NH_4Cl$. The $Cl^-$ ion does not react with $H_2O$, but the $NH_4^+$ ion does:

$$NH_4^+ + H_2O \rightleftharpoons H_3O^+ + NH_3$$

The newly formed $H_3O^+$ lowers the pH, as it does for all salts in this category. Such salts are **acidic** (see Fig. 8.6).

**3.** *A salt of a strong acid and a strong base.* In this case, neither ion reacts with water, and the solution remains neutral. These are **neutral salts.** Examples are $NaCl$, $KNO_3$, and $Na_2SO_4$.

**4.** *A salt of a weak acid and a weak base.* Both ions react with water. Sometimes the effects cancel each other and sometimes not. We will not try to make predictions in such cases.

---

**B o x 8E**

## Acidosis and Alkalosis

The pH of blood is normally between 7.35 and 7.45. If the pH goes lower than this, the condition is called **acidosis.** A blood pH higher than 7.45 is called **alkalosis.** Both of these are abnormal conditions. Acidosis leads to depression of the acute nervous system. Mild acidosis can result in fainting; a more severe case can cause coma. Alkalosis leads to overstimulation of the nervous system, muscle cramps, and convulsions. If the acidosis or alkalosis persists for a sufficient period of time, or if the pH gets too far away from 7.35 to 7.45, the patient dies.

Acidosis has several causes. One type, called *respiratory acidosis,* is caused by difficulty in breathing **(hypoventilation).** An obstruction in the windpipe or diseases such as pneumonia, emphysema, asthma, or congestive heart failure diminish the amount of $CO_2$ that leaves the body through the lungs. (You can even produce mild acidosis by holding your breath.) The pH of the blood is decreased because the $CO_2$, unable to escape fast enough, remains in the blood, where it decreases the $[HCO_3^-]/[H_2CO_3]$ ratio.

*Respiratory alkalosis* is the reverse. It arises from rapid or heavy breathing, called **hyperventilation.** This can come about from fever, infection, the action of certain drugs, or even hysteria. Here the excessive loss of $CO_2$ raises the $[HCO_3^-]/[H_2CO_3]$ ratio and the pH.

Acidosis caused by other factors is called *metabolic acidosis.* Two causes of this condition are starvation or fasting and heavy exercise. When the body doesn't get enough food, it burns its own fat, and the products of this reaction are acidic compounds that get into the blood. This sometimes happens to people on fad diets. Heavy exercise causes the muscles to produce excessive amounts of lactic acid—this is what makes you feel tired and sore. Metabolic acidosis is also caused by a number of metabolic irregularities. For example, the disease diabetes mellitus produces acidic compounds called ketone bodies (Sec. 21.7).

*Metabolic alkalosis* can also result from various metabolic irregularities and in addition can be caused by excessive vomiting. The contents of the stomach are strongly acidic (pH $\approx$ 1), and loss of substantial amounts of this acidic material raises the pH of the blood.

## 8.10   BUFFERS

In Section 8.8 we saw that the body must keep the pH of blood between 7.35 and 7.45 (Box 8E). Yet we frequently eat acidic foods such as oranges, lemons, sauerkraut, and tomatoes, and this eventually adds considerable quantities of $H_3O^+$ to the blood. How does the body manage to keep the pH of blood so constant? It does it with buffers. A **buffer solution** is **a solution whose pH does not change very much when $H_3O^+$ or $OH^-$ ions are added to it.**

How do we make a buffer solution, and how does it work? All we need to do to make a buffer solution is to add to water approximately equal quantities of a weak acid and its conjugate base. For example, if we dissolve 1 mole of acetic acid (a weak acid) and 1 mole of its conjugate base (in the form of sodium acetate) in 1 L of water, we have a very good buffer solution.

How does it work? It works because we have substantial concentrations of *both* $CH_3COOH$ and $CH_3COO^-$. If $H_3O^+$ is added, the $CH_3COO^-$ ions remove it by reacting with it. If $OH^-$ is added, the $CH_3COOH$ molecules perform the same function:

$$H_3O^+ + CH_3COO^- \rightleftharpoons CH_3COOH + H_2O$$

$$OH^- + CH_3COOH \rightleftharpoons CH_3COO^- + H_2O$$

(Note from Table 8.2 that both equilibria lie well to the *right*.) The important thing here is that when the $CH_3COO^-$ ions remove $H_3O^+$, they are converted to $CH_3COOH$ molecules. Since a substantial amount of $CH_3COOH$ is already present, all that happens is that the $[CH_3COO^-]$ goes down a bit and the $[CH_3COOH]$ goes up a bit.

We saw in Section 8.4 that the following equilibrium expression holds for acetic acid (a similar one holds for any weak acid):

$$K_a = \frac{[H_3O^+][CH_3COO^-]}{[CH_3COOH]}$$

If we rearrange this, we get

$$[H_3O^+] = K_a \frac{[CH_3COOH]}{[CH_3COO^-]}$$

This equation tells us that the $[H_3O^+]$ is equal to the constant $K_a$ multiplied by the ratio of $[CH_3COOH]$ to $[CH_3COO^-]$. We've just seen, however, that $H_3O^+$ ions added to the buffer solution make only a slight change in this ratio: the $[CH_3COOH]$ goes up a bit and the $[CH_3COO^-]$ down a bit. Therefore the $[H_3O^+]$, and the pH, are only slightly changed.

The same thing happens if we add $OH^-$ to the buffer, but in the opposite direction. The $OH^-$ ions are removed by the $CH_3COOH$, which in the process is converted to $CH_3COO^-$. Thus the $[CH_3COO^-]$ goes up a bit and the $[CH_3COOH]$ down a bit, so that the $[H_3O^+]$, and the pH, are only slightly changed.

In essence, a buffer solution works because it is able to neutralize, and hence remove from the solution, most of any $OH^-$ or $H_3O^+$ ions that are added. This effect can be quite powerful. For example, pure water has a pH of 7. If we add 0.1 mole of HCl to 1 L of pure water, with no buffer, the pH drops to 1. If we add 0.1 mole of NaOH to 1 L of pure water, the pH rises to 13. A buffer solution containing 1 $M$ $H_2PO_4^-$ ion and 1 $M$ $HPO_4^{2-}$ ion (its conjugate base) has a pH of 7.21. If we add 0.1 mole of HCl to 1 L of this solution, the pH drops only to 7.12. If we add 0.1 mole of NaOH, the pH rises only to 7.30. In summary,

$$
\begin{array}{lll}
\text{Pure } H_2O + HCl & \text{pH 7} & \longrightarrow 1 \\
\text{Buffer } + HCl & \text{pH 7.21} & \longrightarrow 7.12 \\
\text{Pure } H_2O + NaOH & \text{pH 7} & \longrightarrow 13 \\
\text{Buffer } + NaOH & \text{pH 7.21} & \longrightarrow 7.30
\end{array}
$$

It is obvious from this example that a buffer really does the job: It holds the pH relatively constant (Fig. 8.7).

Every buffer solution has two characteristics, its pH and its capacity.

## Buffer pH

In the example just given, the pH of the $H_2PO_4^-/HPO_4^{2-}$ buffer solution is 7.21 as long as the concentrations of $H_2PO_4^-$ and $HPO_4^{2-}$ are equal. If you look at Table 8.3, you see that 7.21 is the $pK_a$ of the $H_2PO_4^-$ ion. This is not a coincidence. If we make a buffer solution by mixing equimolar concentrations of any weak acid and its conjugate base, the pH of the solution is always equal to the $pK_a$ of the acid.

This allows us to prepare buffer solutions to maintain almost any pH. For example, if we want to maintain a pH of 9.14, we could make a buffer solution from boric acid and its conjugate base (see Table 8.3).

**Figure 8.7**
Buffer solutions. (a) The pH meter shows the pH of water that contains a trace of acid (and bromcresol green indicator, see Fig. 8.5). The solution at the right is a buffer that has the same pH as human blood, 7.4. (b) When 5 mL of 0.10 $M$ HCl is added to each solution, the pH of the water drops several units, while the pH of the buffer stays constant. (Photograph by Charles D. Winters.)

(a)

(b)

**Box 8F**

## The Henderson-Hasselbalch Equation

Whenever we work with buffers, a useful equation is

$$pH = pK_a + \log \frac{[A^-]}{[HA]}$$

where [HA] is the concentration of the acid and [A⁻] that of its conjugate base. This equation, called the **Henderson-Hasselbalch equation,** is easily derived from Equation 8.3. Suppose we wish to make a buffer solution whose pH does not correspond exactly to the $pK_a$ of any acid we may have on hand. We can make up such a solution by choosing an acid with a $pK_a$ that is not too far from the desired pH and adjusting the concentrations of the weak acid (HA) and its conjugate base (A⁻) to obtain the desired pH. The Henderson-Hasselbalch equation allows us to calculate the proper ratio.

---

**EXAMPLE**
How can we make a buffer solution with a pH of 7.99, using an acid-base pair from Table 8.3?

**Answer**
The closest $pK_a$ value in Table 8.3 is 7.21, the value for $H_2PO_4^-$. We therefore use $H_2PO_4^-/HPO_4^{2-}$ as our buffer and adjust the $[HPO_4^{2-}]/[H_2PO_4^-]$ ratio to get a pH of 7.99:

$$pH = pK_a + \log \frac{[HPO_4^{2-}]}{[H_2PO_4^-]}$$

$$7.99 = 7.21 + \log \frac{[HPO_4^{2-}]}{[H_2PO_4^-]}$$

$$0.78 = \log \frac{[HPO_4^{2-}]}{[H_2PO_4^-]}$$

In Table 8.4 (or on a calculator) we find that the antilogarithm of 0.78 is 6. Taking the antilogarithm of both sides, we have

$$6 = \frac{[HPO_4^{2-}]}{[H_2PO_4^-]}$$

---

**EXAMPLE**

What is the pH of buffer systems containing equimolar quantities of (a) $H_3PO_4$ and $H_2PO_4^-$   (b) $H_2CO_3$ and $HCO_3^-$ ?

**Answer**
The pH is equal to the $pK_a$ of the acid, which we find in Table 8.3: (a) pH = 2.1   (b) pH = 6.37.

---

**PROBLEM 8.8**   What is the pH of buffer systems containing equimolar quantities of (a) $H_3BO_3$ and $H_2BO_3^-$   (b) $HCO_3^-$ and $CO_3^{2-}$ ?

Suppose the concentrations of the weak acid and its conjugate base are not equal. In that case, the pH of the buffer will be a bit higher or lower (but not by very much) as long as significant amounts of both acid and conjugate base are present (Box 8F). The most efficient buffers do contain equimolar concentra-

Thus, in order to get a buffer solution of pH 7.99, we must use a molar concentration of $HPO_4^{2-}$ that is six times that of $H_2PO_4^-$.

---

**PROBLEM 8.9**
How can we make a buffer solution of pH 4.15, using an acid-base pair from Table 8.3?

---

We can also use the Henderson-Hasselbalch equation to calculate the pH if we know the $[A^-]/[HA]$ ratio.

---

**EXAMPLE**
What is the pH of an $H_2PO_4^-/HPO_4^{2-}$ buffer system in which $[HPO_4^{2-}]$ is twice $[H_2PO_4^-]$?

**Answer**

$$pH = pK_a + \log \frac{[HPO_4^{2-}]}{[H_2PO_4^-]}$$

$$pH = 7.21 + \log \frac{2}{1}$$

The log of 2 (from Table 8.4 or a calculator) is 0.30. Thus

$$pH = 7.21 + 0.30 = 7.51$$

---

**PROBLEM 8.10**
What is the pH of a buffer solution in which $[CH_3COO^-] = 0.20\ M$ and $[CH_3COOH] = 0.60\ M$?

---

tions, and buffer efficiency decreases as the ratio becomes more unequal. If the acid-to-conjugate-base ratio is more than about 15 to 1 or less than about 1 to 15, the solution is essentially no longer a buffer.

### Buffer Capacity

We could make a buffer solution by dissolving 1 mole each of $CH_3COONa$ and $CH_3COOH$ in 1 L of $H_2O$, or we could use only 0.1 mole of each. Both solutions have the same pH: 4.75. However, the former has a **buffer capacity** ten times that of the latter. If we add 0.2 mole of HCl to the former solution, it performs the way we expect—the pH drops only a little (to 4.57). If we add 0.2 mole of HCl to the latter solution, however, the pH drops all the way to 1.0 because the buffer has been **swamped out.** The amount of $H_3O^+$ added has exceeded the buffer capacity. The first 0.1 mole of HCl completely neutralizes essentially all of the $CH_3COO^-$ present. After that, the solution contains only $CH_3COOH$ and is no longer a buffer, so that the second 0.1 mole of HCl is able to lower the pH to 1.0.

### Blood Buffers

To hold the pH of the blood at close to 7.4, the body uses three buffer systems: carbonate, phosphate, and proteins (proteins are discussed in Chapter 18). The most important of these is the carbonate system. The acid is carbonic acid, $H_2CO_3$; the base is the bicarbonate ion, $HCO_3^-$. The $pK_a$ of $H_2CO_3$ is 6.37 (see Table 8.3). Since the pH of an equimolar mixture of acid and conjugate base is equal to its $pK_a$, a buffer made of equal concentrations of $H_2CO_3$ and $HCO_3^-$ has a pH of 6.37.

Blood, however, has a pH of 7.4. The carbonate buffer can maintain this pH only if $[H_2CO_3]$ and $[HCO_3^-]$ are not equal. In fact, the necessary $[HCO_3^-]/[H_2CO_3]$ ratio is about 10 to 1. The normal concentrations in blood are about 0.025 $M$ $HCO_3^-$ and 0.0025 $M$ $H_2CO_3$. The buffer works, as all buffers do, because any added $H_3O^+$ is neutralized by the $HCO_3^-$ and any added $OH^-$ is neutralized by the $H_2CO_3$.

The fact that the $[HCO_3^-]/[H_2CO_3]$ ratio is 10 to 1 means that this is a better buffer for acids, which lower the ratio and thus improve buffer efficiency, than for bases, which raise the ratio. This is in harmony with the actual functioning of the body, since under normal conditions larger amounts of acidic than basic substances enter the blood. The 10-to-1 ratio is easily maintained under normal conditions because the body can very quickly increase or decrease the amount of $CO_2$ entering the blood.

The second buffering system of the blood is made up of hydrogen phosphate ion, $HPO_4^{2-}$, and dihydrogen phosphate ion, $H_2PO_4^-$. In this case, a 1.6-to-1 ratio of $[HPO_4^{2-}]$ to $[H_2PO_4^-]$ is necessary to maintain a pH of 7.4. This is well within the limits of good buffering action.

## 8.11   TITRATION, EQUIVALENTS, AND NORMALITY

In laboratories, the question frequently arises as to how much acid or base is present in a given amount of solution. To measure this, we use a method called titration. If a solution is acidic, **titration** consists of **adding base to it until all the acid is neutralized.** To do this, we need two things: (1) a means of measuring how much base is added and (2) a means of telling just when the acid is completely neutralized. Titration is shown in Figure 8.8. The **buret** is the means of measuring the amount of base. An **indicator** in the solution tells us when the acid is exactly neutralized (we could also use a pH meter) because at that point there is a sudden large change in pH. This is called the **end point.**

It is important to understand that *titration is not a method of determining the acidity (or basicity) of a solution.* If we want to do that, we measure the pH, which is the only measurement of solution acidity or basicity. Titration is a method for determining the total acid or base concentration of a solution, which is not the same as the acidity. For example, a 0.1 $M$ solution of HCl in water has a pH of 1, but a 0.1 $M$ solution of acetic acid has a pH of 2.9. The solutions have very different acidities, but the total concentration of acid is the same and each of these solutions neutralizes the same amount of NaOH solution.

When we have reached the end point in a titration, we can easily read, directly on the buret, exactly how many milliliters of NaOH solution were required to neutralize the acid. How do we then calculate the amount of acid? Before we answer that, we must first define equivalents and normality.

(a)

(b)

(c)

**Figure 8.8**
Titration. An acid of unknown concentration is in the Erlenmeyer flask. When base from the buret is added, the acid is neutralized. The end point is reached when the indicator color change becomes permanent. (Photographs by Charles D. Winters.)

We have seen that some acids are diprotic ($H_2SO_4$, $H_2CO_3$) or triprotic ($H_3PO_4$). Similarly, there are some bases that produce two or three $OH^-$ ions in solution—for example, $Ba(OH)_2$, $Mg(OH)_2$, and $Al(OH)_3$. For acids and bases, we define the **equivalent weight (EW)** as **the formula weight (FW) divided by the number of $H^+$ or $OH^-$ ions produced.**

---

**EXAMPLE**

What is the equivalent weight of these acids and bases: (a) $HNO_3$, FW = 63   (b) NaOH, FW = 40   (c) $H_2SO_4$, FW = 98   (d) $Mg(OH)_2$, FW = 58?

**Answer**
In each case we divide the formula weight by the number of $H^+$ or $OH^-$ ions produced by the acid or base:
(a)  $HNO_3$      EW = 63/1 = 63
(b)  NaOH        EW = 40/1 = 40
(c)  $H_2SO_4$      EW = 98/2 = 49
(d)  $Mg(OH)_2$    EW = 58/2 = 29

---

**PROBLEM 8.11**   What is the equivalent weight of these acids and bases: (a) $CH_3COOH$, FW = 60.0   (b) $Ba(OH)_2$, FW = 171.3   (c) $H_3PO_4$, FW = 98.0?

Because we often deal with very small quantities, we use **milliequivalents:** 1 meq = 1/1000 eq.

We define an **equivalent (eq)** of any substance as its **equivalent weight expressed in grams.** Thus 1 eq of $HNO_3$ is 63 g, and 49 g is 1 eq of $H_2SO_4$. This definition is similar to that of the mole, which is the *formula* weight expressed in grams, and we deal with equivalents in a manner similar to the way we deal with moles. That is, if 49 g is 1 eq of $H_2SO_4$, then 4.9 g = 0.1 eq, 98 g = 2 eq, and so forth.

---

**EXAMPLE**

How many equivalents are there in (a) 18 g of $H_2SO_4$  (b) 51 g of $Mg(OH)_2$?

**Answer**
We multiply the number of grams by the appropriate conversion factor:

(a) $18 \text{ g } H_2SO_4 \times \dfrac{1 \text{ eq } H_2SO_4}{49 \text{ g } H_2SO_4} = 0.37 \text{ eq } H_2SO_4$

(b) $51 \text{ g } Mg(OH)_2 \times \dfrac{1 \text{ eq } Mg(OH)_2}{29 \text{ g } Mg(OH)_2} = 1.8 \text{ eq } Mg(OH)_2$

---

**PROBLEM 8.12**   How many equivalents are there in (a) 73.4 g of $H_2SO_3$ (b) 1.74 g of $Ba(OH)_2$?

---

Why have chemists gone through the trouble of defining equivalent weight and equivalent? The answer to this is present in the word used: "equivalent." In an acid-base reaction, **one equivalent of any acid exactly neutralizes one equivalent of any base.** Since most acid-base reactions occur in aqueous solution, we can go one step further and define a new concentration unit, **normality (N),** as **equivalents of solute per liter of solution.**

---

**EXAMPLE**

What is the normality of a solution made by dissolving 8.5 g of $H_2SO_4$ in enough water to make 500 mL of solution?

**Answer**
First we need the number of equivalents of $H_2SO_4$ per liter of solution, and we know that the equivalent weight of $H_2SO_4$ is 49. Thus,

$$8.5 \text{ g} \times \dfrac{1 \text{ eq}}{49 \text{ g}} = 0.17 \text{ eq } H_2SO_4$$

To find normality we divide equivalents of $H_2SO_4$ by liters of solution:

$$\dfrac{0.17 \text{ eq}}{0.500 \text{ L}} = 0.34 \, \dfrac{\text{eq}}{\text{L}} = 0.34 \, N$$

---

**PROBLEM 8.13**   What is the normality of a solution made by dissolving 3.4 g of $Mg(OH)_2$ in enough water to make 1250 mL of solution?

---

Concentrations of acids and bases in many laboratories are often expressed in normalities.

There is a simple relationship between the normality and the molarity (Sec. 6.5) of an acidic or basic solution. Normality is simply the molarity multiplied by the number of $H^+$ or $OH^-$ ions produced by one molecule of acid or base. For example, in HCl only one $H^+$ ion is produced. This means that, for all HCl solutions, the normality is equal to the molarity. Sulfuric acid produces two $H^+$ ions, so for $H_2SO_4$ solutions the normality is always twice the molarity; thus a 6 $M$ solution of $H_2SO_4$ is 12 $N$. *Normality is always equal to or greater than molarity.*

Now that we have defined normality, we can come back to titration and see how an unknown acid concentration is calculated. We know that, at the end point, the number of equivalents of acid equals the number of equivalents of base:

$$Eq_{acid} = Eq_{base}$$

Since normality multiplied by volume gives the number of equivalents, the following equation is valid:

$$V_{acid}N_{acid} = V_{base}N_{base}$$

If we are titrating an acid with a base, we begin by putting the acid in an Erlenmeyer flask. In doing so, we must carefully measure the volume of acid added (because this is $V_{acid}$). At the end point, we read the buret to find the volume of base added ($V_{base}$). The normality of the base ($N_{base}$) must be known before we do the titration. It is then simple to calculate the unknown ($N_{acid}$).

A pipet is often used for this (Fig. 8.9).

**Figure 8.9**
Laboratory worker measuring a volume with a pipet. (FourByFive.)

---

**EXAMPLE**

If 50.0 mL of an acid solution of unknown concentration is titrated with 0.32 $N$ base and it takes 24.6 mL of base to reach the end point, what is the normality of the acid?

**Answer**

$$V_{acid}N_{acid} = V_{base}N_{base}$$
$$(50.0 \text{ mL})(N_{acid}) = (24.6 \text{ mL})(0.32 \ N)$$
$$N_{acid} = \frac{(24.6 \text{ mL})(0.32 \ N)}{50.0 \text{ mL}} = 0.16 \ N$$

The concentration of the acid is 0.16 $N$.

---

**PROBLEM 8.14**   A sample of 50.00 mL of an acid solution of unknown concentration was titrated with 0.152 $N$ base, and it took 32.5 mL of base to reach the end point. What was the normality of the acid?

Note that this answer is correct no matter which acid is in the flask and no matter which base is in the buret. The advantage of using normalities is that the simple equation $V_{acid}N_{acid} = V_{base}N_{base}$ works for any acid and any base, no matter how many $H^+$ or $OH^-$ ions they produce.

The same procedure and calculations can be used to determine an unknown base concentration. In this case, a known amount of the base is put in the flask, and the buret is filled with an acid solution of known concentration.

## SUMMARY

By the **Arrhenius** definitions, **acids** are substances that produce $H_3O^+$ ions in water and **bases** are substances that produce $OH^-$ ions in water. A **strong acid or base** produces high concentrations of these ions. The **Brønsted-Lowry** definitions expand this concept beyond water: An acid is a proton donor, and a base is a proton acceptor. Every acid has a **conjugate base.** An **amphoteric** substance, such as water, can act as either an acid or a base. Table 8.2 shows which Brønsted-Lowry acid-base reactions will proceed; the favored direction is always toward the weaker acid and base.

The strengths of weak acids are expressed by $K_a$ **values;** the higher the $K_a$, the stronger the acid. Strong acids and bases are corrosive. Acids neutralize metals, metal hydroxides, and oxides to give **salts,** which are made up of positive and negative ions. Acids also react with carbonates, ammonia, and amines.

In pure water, a small percentage of molecules undergo the reaction $2H_2O \rightleftharpoons H_3O^+ + OH^-$, enough to produce a concentration of $10^{-7}$ $M$ each of $H_3O^+$ and $OH^-$. The **ion product,** $K_a = [H_3O^+][OH^-]$, is equal to $10^{-14}$ in any aqueous solution. Hydronium ion concentrations are generally expressed in **pH units,** with $pH = -\log[H_3O^+]$. Solutions with pH less than 7 are **acidic;** those with pH higher than 7 are **basic. Neutral** solutions have a pH of 7. The pH is measured with indicators or with a pH meter.

Many salts hydrolyze in water. Salts of strong acids and weak bases are acidic; salts of weak acids and strong bases are basic; salts of strong acids and strong bases are neutral.

A **buffer solution** does not change its pH very much when a strong acid or base is added. Buffers are made up of approximately equal concentrations of a weak acid and its conjugate base. Every buffer solution has a pH and a capacity. The most important buffers for blood are carbonate and phosphate.

The concentration of aqueous solutions of acids and bases can be measured by **titration,** in which a base of known concentration is added to an acid of unknown concentration (or vice versa) until an end point is reached, at which point the solution is completely neutralized. By the use of **normalities,** the unknown concentration can be calculated.

## KEY TERMS

Acid (Sec. 8.1)
Acid dissociation constant (Sec. 8.4)
Acidic salt (Sec. 8.9)
Acidic solution (Sec. 8.8)
Acidity constant (Sec. 8.4)
Acidosis (Box 8E)
Alkali (Sec. 8.1)
Alkalosis (Box 8E)
Amphoteric (Sec. 8.3)
Base (Sec. 8.1)
Basic salt (Sec. 8.9)
Basic solution (Sec. 8.8)
Brønsted-Lowry acid and base (Sec. 8.3)
Buffer (Sec. 8.10)
Buffer capacity (Sec. 8.10)
Conjugate acid and base (Sec. 8.3)

Diprotic acid (Sec. 8.3)
End point (Sec. 8.11)
Equivalent (Sec. 8.11)
Equivalent weight (Sec. 8.11)
Henderson-Hasselbalch equation (Box 8F)
Hyperventilation (Box 8E)
Hypoventilation (Box 8E)
Indicator (Sec. 8.8)
Ion product (Sec. 8.7)
Ionization constant (Sec. 8.4)
$K_a$ (Sec. 8.4)
$K_w$ (Sec. 8.7)
Milliequivalent (Sec. 8.11)
Monoprotic acid (Sec. 8.3)
Neutral salt (Sec. 8.9)
Neutral solution (Secs. 8.5, 8.8)

Neutralization (Sec. 8.5)
Normality (Sec. 8.11)
pH (Sec. 8.8)
Salt (Sec. 8.6)
Self-ionization of water (Sec. 8.7)

Strong acid or base (Sec. 8.2)
Titration (Sec. 8.11)
Triprotic acid (Sec. 8.3)
Weak acid or base (Sec. 8.2)

## PROBLEMS

### Arrhenius Acids and Bases

**8.15** Write equations for the reactions that take place when each of the following acids is added to water:

(a) HF
(b) HBr
(c) $H_2SO_3$
(d) $H_2SO_4$
(e) $HCO_3^-$

Which of these reactions proceed nearly to completion?

**8.16** Write equations for the reactions that take place when the bases LiOH and $(CH_3)_2NH$ are each added to water.

### Brønsted-Lowry Theory

**8.17** (a) What are the Arrhenius definitions of acid and base? (b) What are the Brønsted-Lowry definitions?

**8.18** Write formulas for the conjugate bases of these acids:

(a) HBr
(b) $H_3PO_4$
(c) $H_2SO_4$
(d) $NH_4^+$
(e) $H_2O$
(f) $H_3O^+$
(g) $HS^-$
(h) $HPO_4^{2-}$
(i) $HCO_3^-$
(j) $H_3BO_3$

**8.19** Write formulas for the conjugate acids of these bases:

(a) $OH^-$
(b) $NH_2^-$
(c) $NH_3$
(d) $CO_3^{2-}$
(e) $HCO_3^-$
(f) $PO_4^{3-}$
(g) $H_2O$
(h) $Br^-$
(i) $CH_3-NH_2$
(j) $HCOO^-$

**8.20** Tell whether each of these acids is monoprotic, diprotic, or triprotic:

(a) $H_2SO_4$
(b) $HNO_3$
(c) $H_3PO_4$
(d) $CH_3COOH$
(e) $H_2CO_3$

**8.21** Is there a contradiction between the Brønsted-Lowry and Arrhenius definitions of acid strength? Explain.

**8.22** Which of these reactions proceed from left to right?

(a) $HNO_3 + NH_3 \rightleftharpoons NO_3^- + NH_4^+$
(b) $H_2O + SO_4^{2-} \rightleftharpoons OH^- + HSO_4^-$
(c) $H_3O^+ + PO_4^{3-} \rightleftharpoons H_2O + HPO_4^{2-}$
(d) $H_2S + CO_3^{2-} \rightleftharpoons HS^- + HCO_3^-$
(e) $HCO_3^- + H_2O \rightleftharpoons CO_3^{2-} + H_3O^+$
(f) $CH_3COOH + H_2PO_4^- \rightleftharpoons$
$$CH_3COO^- + H_3PO_4$$

### Acid Dissociation Constants

**8.23** Looking at the values in Table 8.3, indicate which is the stronger acid in each of the following pairs:

(a) $H_2CO_3$ and HF
(b) $H_3BO_3$ and $CH_3COOH$
(c) HCOOH and $HCO_3^-$

### Acid and Base Properties

**8.24** What is the most important chemical property of acids and bases?

**8.25** Write equations for the reaction of HCl with

(a) $Na_2CO_3$
(b) Mg
(c) NaOH
(d) $NH_3$
(e) $CH_3-NH_2$

### Self-ionization of Water

**8.26** Calculate the $[OH^-]$ of each of the following aqueous solutions, given that $[H_3O^+] =$

(a) $10^{-9} M$
(b) $10^{-2} M$
(c) $10^{-7} M$
(d) $10^1 M$

**8.27** Determine the $[H_3O^+]$ of each of the following aqueous solutions, given that $[OH^-] =$

(a) $10^{-8} M$
(b) $10^{-4} M$
(c) $10^{-7} M$
(d) $10^0 M$

## pH and p$K_a$

**8.28** What is the pH of a solution having a $[H_3O^+]$ of
(a) $10^{-8}$ $M$  (d) $10^0$ $M$
(b) $10^{-10}$ $M$  (e) $10^{-7}$ $M$
(c) $10^{-2}$ $M$
Which of these solutions are acidic, which basic, and which neutral?

**8.29** What is the pH of a solution having a $[OH^-]$ of
(a) $10^{-11}$ $M$  (d) $10^{-4}$ $M$
(b) $10^{-3}$ $M$  (e) $10^{-7}$ $M$
(c) $10^{-13}$ $M$
Which of these solutions are acidic, which basic, and which neutral?

**8.30** What is the pH of a solution having a $[H_3O^+]$ of
(a) $3 \times 10^{-9}$  (c) $8 \times 10^{-12}$
(b) $6 \times 10^{-2}$  (d) $5 \times 10^{-7}$
Which of these solutions are acidic, which basic, and which neutral?

**8.31** What is the $[H_3O^+]$ of an aqueous solution having a pH of
(a) 11  (e) 2.40
(b) 3  (f) 13.10
(c) 7  (g) 6.52
(d) 1
Which of these solutions are acidic, which basic, and which neutral?

**8.32** What is the $[OH^-]$ of an aqueous solution having a pH of
(a) 12  (d) 14
(b) 1  (e) 7
(c) 4
Which of these solutions are acidic, which basic, and which neutral?

**8.33** Three acids have the following p$K_a$ values: butyric acid, 4.82; barbituric acid, 5.00; lactic acid, 3.85. Which is the strongest of the three and which the weakest?

## The pH of Aqueous Salt Solutions

**8.34** Give three examples of acidic salts and three of basic salts.

**8.35** If we dissolve some solid sodium carbonate, $Na_2CO_3$, in pure water, the resulting solution is basic, that is, the pH rises above 7. Explain how this happens.

**8.36** Tell whether each of these salts is acidic, basic, or neutral:
(a) $K_3PO_4$  (e) $(NH_4)_2SO_4$
(b) $Li_2CO_3$  (f) $NaNO_3$
(c) $BaBr_2$  (g) $Al(ClO_4)_3$
(d) $MgCl_2$  (h) $LiI$

## Buffers

**\*8.37** What is the connection between buffer action and Le Chatelier's principle?

**8.38** Write equations to show what happens when, to a buffer solution containing equimolar amounts of $CH_3COOH$ and $CH_3COO^-$, we add (a) $H_3O^+$ (b) $OH^-$.

**8.39** Write equations to show what happens when, to a buffer solution containing equimolar amounts of $HPO_4^{2-}$ and $H_2PO_4^-$, we add (a) $H_3O^+$ (b) $OH^-$.

**8.40** What is the pH of a buffer solution made by dissolving 0.1 mole of hydrofluoric acid, HF, and 0.1 mole of sodium fluoride, NaF, in 1 L of water?

**8.41** What is meant by buffer capacity?

**8.42** The pH of a solution made by dissolving 1 mole of propionic acid and 1 mole of sodium propionate in 1 L of water is 4.85. (a) What would the pH be if we used 0.1 mole of each (in 1 L of water) instead of 1 mole? (b) With respect to buffer action, how would the two solutions differ?

**8.43** What are the two main buffers in blood? Describe how each works.

## Titration, Equivalents, and Normality

**8.44** What is the purpose of titration?

**8.45** What is the equivalent weight of
(a) HI  (e) $NH_3$
(b) $Ba(OH)_2$  (f) $H_2SO_3$
(c) $HClO_4$  (g) RbOH
(d) $Sc(OH)_3$  (h) $H_3AsO_4$

**8.46** How many equivalents are there in
(a) 72.5 g of KOH
(b) 37 g of oxalic acid $H_2C_2O_4$ (a diprotic acid)
(c) 124 g of $H_3PO_4$
(d) 2.45 g of $Al(OH)_3$

**8.47** What is the normality of a solution made by dissolving 12.7 g of HCl in enough water to make 1 L of solution?

**8.48** What is the normality of a solution made by dissolving 3.4 g of $Ba(OH)_2$ in enough water to make 450 mL of solution?

**8.49** What is the normality of a solution made by dissolving 73.2 g of $H_3PO_4$ in enough water to make 400 mL of solution?

**8.50** Describe exactly how you would prepare each of the following solutions (in each case assume you have the solid bases):
(a) 400.0 mL of a 0.75 N solution of NaOH
(b) 1.0 L of a 0.071 N solution of $Ba(OH)_2$

**8.51** What is the normality of
(a) 3.0 M $HNO_3$        (d) 2.7 M $Ca(OH)_2$
(b) 4.5 M $H_2SO_4$      (e) 2.05 M $H_3PO_4$
(c) 0.34 M KOH

**8.52** If 25.0 mL of an aqueous solution of $H_2SO_4$ requires 19.7 mL of 0.72 M NaOH to reach the end point, what is the normality of the $H_2SO_4$ solution?

**8.53** A sample of 40.0 mL of a 0.24 N solution of NaOH is titrated with 0.74 N $H_2SO_4$ solution. How many mL of the $H_2SO_4$ solution are required to reach the end point?

**8.54** A 0.300 M solution of $H_2SO_4$ was used to titrate 10.00 mL of an unknown base; 15.00 mL of acid was required to neutralize the basic solution. What was the normality of the base?

**8.55** A solution of an unknown base was titrated with 0.150 N HCl, and 22.0 mL of acid was needed to reach the end point of the titration. How many equivalents of the unknown base were in the solution?

**8.56** What is the molarity of
(a) 3.2 N HCl              (d) 0.73 N $NH_3$
(b) 18 N $H_2SO_4$        (e) 0.069 N $Ba(OH)_2$
(c) 0.066 N KOH

**\*8.57** The usual concentration of $HCO_3^-$ ions in blood plasma is approximately 24 meq/L. How would you make up 1 L of a solution containing this concentration of $HCO_3^-$ ions?

## Boxes

**8.58** (Box 8A) Which weak base is used as a laxative?

**8.59** (Box 8A) Which strong acid turns your skin yellow?

**8.60** (Box 8B) With respect to corneal burns, which are more dangerous, strong acids or strong bases?

**8.61** (Box 8B) What is the most important *immediate* first aid in any chemical burn of the eyes?

**8.62** (Box 8C) Which of the antacids listed in Box 8C contain no hydroxides?

**8.63** (Box 8D) What compounds generate the $CO_2$ in fire extinguishers?

**8.64** (Box 8D) What compound in baking powder produces the sponginess of a cake?

**8.65** (Box 8E) Define: (a) hypoventilation  (b) respiratory alkalosis  (c) metabolic alkalosis.

**8.66** (Box 8E) What causes (a) hypoventilation  (b) respiratory alkalosis  (c) metabolic alkalosis?

**8.67** (Box 8F) Calculate the pH of an aqueous solution containing
(a) 0.8 M lactic acid and 0.4 M lactate ion
(b) 0.3 M $NH_3$ and 1.50 M $NH_4^+$
(c) 0.42 M HF and 1.38 M $F^-$

**\*8.68** (Box 8F) You wish to make a buffer whose pH is 8.21. You have available a stock solution of 1.00 L of 0.100 M $NaH_2PO_4$, as well as some solid $Na_2HPO_4$. How many grams of the solid $Na_2HPO_4$ must be added to the stock solution to accomplish this task? (Assume the volume remains at 1.00 L.)

**8.69** (Box 8F) Describe how you would make a buffer solution of pH 4.55, using $CH_3COOH$/ $CH_3COO^-$ as the buffer system.

**\*8.70** (Box 8F) You want to make a $CH_3COOH$/ $CH_3COO^-$ buffer solution with a pH of 5.60. The acetic acid concentration is to be 0.10 M. What should the acetate ion concentration be?

## Additional Problems

**8.71** If pure water self-ionizes to give $H_3O^+$ and $OH^-$ ions, why doesn't pure water conduct an electric current?

**8.72** Write equations to show what happens when, to a buffer solution containing equal amounts of HF and $F^-$, we add (a) $H_3O^+$  (b) $OH^-$.

**\*8.73** Assume that you have a dilute solution of HCl (0.1 M) and a concentrated solution of acetic acid (5 M). Which solution is more acidic? Explain.

**\*8.74** A scale of $K_b$ values for bases could be set up in a manner similar to that for the $K_a$ scale for acids. However, this is generally considered unnecessary. Can you explain why?

**\*8.75** 4.00 grams of an unknown acid was dissolved in 1.00 L of solution. This was titrated with 0.600 $N$ NaOH, and 38.7 mL of the NaOH solution was needed to neutralize the acid. What is the equivalent weight of the acid?

**8.76** If we add 0.1 mole of $NH_3$ to 0.5 mole of HCl dissolved in enough water to make 1 L of solution, what happens to the $NH_3$? Will any of it remain? Explain.

**8.77** Three acids have the following $K_a$ values: caproic acid, $1.4 \times 10^{-5}$; acrylic acid, $5.6 \times 10^{-5}$; uric acid, $1.3 \times 10^{-4}$. Which is the strongest acid of the three and which the weakest?

**8.78** What is the $[H_3O^+]$ of a solution that has a pH of 1.1?

**8.79** Do a 1 $M$ $CH_3COOH$ solution and a 1 $M$ HF solution have the same pH? Explain.

**8.80** If the $[OH^-]$ of a solution is $1 \times 10^{-14}$, (a) what is the pH of the solution? (b) What is the $[H_3O^+]$?

**8.81** What is the normality of a solution made by dissolving 0.492 g of the diprotic acid oxalic acid, $H_2C_2O_4$, in enough water to make 1.75 L of solution?

**8.82** What is the pH of a solution that has a $[H_3O^+]$ of $6 \times 10^{-12}$?

**\*8.83** If a solution of $NH_3$ in water (aqueous ammonia) is added to a solution of sodium ethoxide (containing ethoxide ions $C_2H_5O^-$), will an acid-base reaction take place? If so, write an ionic equation for the reaction.

**8.84** The $pK_a$ value of barbituric acid is 5.00. If the $H_3O^+$ and the barbiturate ion concentrations are each 0.00300 $M$, what is the concentration of the undissociated barbituric acid?

# CHAPTER 9

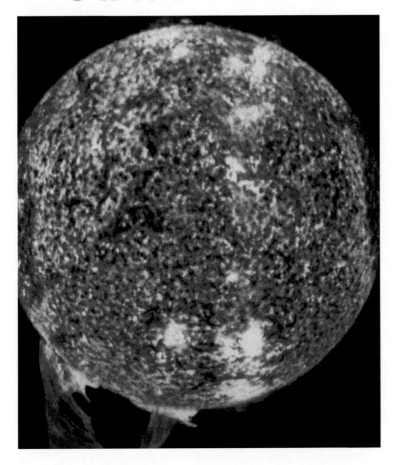

*The energy given out by the sun is produced by nuclear fusion. (Courtesy of NASA.)*

# Nuclear Chemistry

## 9.1 INTRODUCTION

Every so often, a scientist makes the kind of discovery that changes the future of the world in some significant way. In 1896 a French scientist, Henri Becquerel (1852–1908), made one of the most important of these discoveries. Completely by accident he found that a certain rock was giving off mysterious radiation. The discovery was totally unexpected; nobody had ever had the faintest idea that anything of the kind was happening, and yet that rock, and others like it, had been giving off radiation since long before there were humans on the earth. It was simply that, before 1896, nobody knew it.

Becquerel's rock contained uranium. The radiation given off by the uranium caused photographic plates to "fog" even though they were wrapped in black paper. This meant that the radiation coming from the rock must be more energetic than visible light because it could pass through the paper.

Becquerel and the Curies were awarded the Nobel Prize in 1903 for this discovery. Rutherford won the Nobel Prize in 1908.

Shortly after Becquerel's discovery, other scientists, most notably Pierre (1859–1906) and Marie (1867–1934) Curie and Ernest Rutherford (1871–1937), began to study this radiation, which Marie Curie called *radioactivity*. They soon learned that there were three types, which they named alpha, beta, and gamma, after the first three letters of the Greek alphabet.

> **Alpha particles** contain two protons and two neutrons and are thus the same as helium nuclei. They have an atomic number of 2.
>
> **Beta particles** are electrons. Beta "radiation" is just a stream of electrons.
>
> **Gamma rays** are not particles; they are a high-energy form of electromagnetic radiation.

Gamma rays are only one form of **electromagnetic radiation.** There are many others, including visible light. All consist of waves (Fig. 9.1). The only difference between one form of electromagnetic radiation and another is the **wavelength** ($\lambda$), which is the distance from one wave crest to the next. All electromagnetic radiation travels at the same speed in vacuum (the speed of light, which is $3.0 \times 10^{10}$ cm/s). Because of this, the different forms of electromagnetic radiation have different frequencies. The **frequency** ($\nu$) of a radiation is the number of crests that pass a given point in one second. **The longer the wavelength, the lower the frequency.** Mathematically they are related by the equation

$$\lambda = c/\nu$$

where $c$ is the speed of light. There is also a direct relationship between the frequency of electromagnetic radiation and its energy: **The higher the frequency, the higher the energy.** Electromagnetic radiation comes in packets; the smallest units are called **photons.**

Gamma rays are electromagnetic radiation of very high frequency. We cannot see them because our eyes are not sensitive to waves of this frequency, but they can be detected by instruments (Sec. 9.4).

The only radiation we know of with even higher frequency (and energy) than gamma rays are cosmic rays.

Another kind of radiation, called **x-rays,** has frequencies (and energies) higher than those of visible light but less than those of gamma rays. Figure 9.2 shows the position of all these waves on the **electromagnetic spectrum.**

**Figure 9.1**
Two electromagnetic waves with different wavelengths.

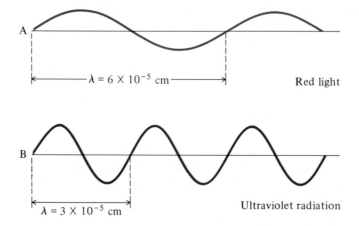

$\lambda = 6 \times 10^{-5}$ cm — Red light

$\lambda = 3 \times 10^{-5}$ cm — Ultraviolet radiation

Materials that emit radiation (alpha, beta, or gamma) are called **radioactive. Radioactivity comes from the atomic nucleus** and not from the electron cloud. Table 9.1 summarizes the properties of the particles and rays that come out of radioactive nuclei, along with the properties of some other particles and rays. Note that x-rays are not considered to be a form of radioactivity. They do not come out of the nucleus but are generated in other ways.

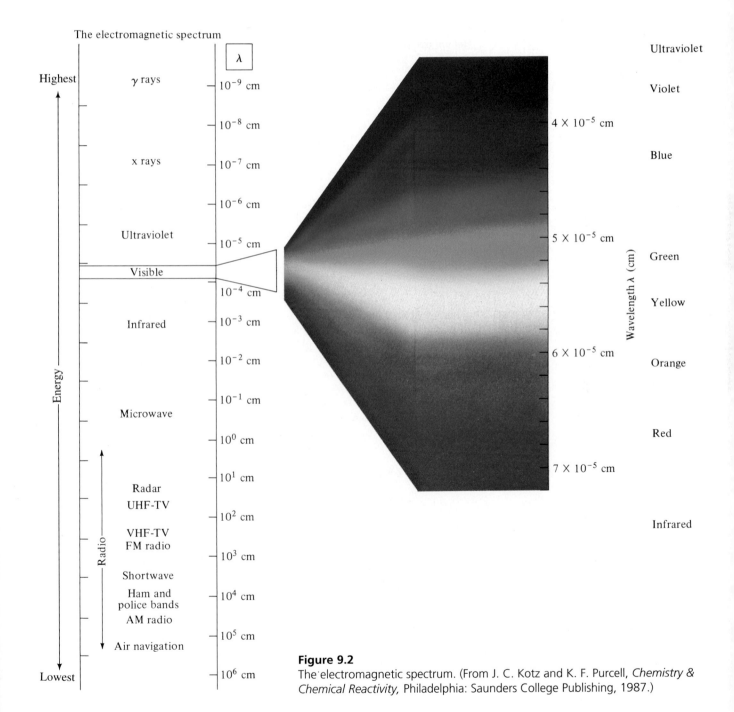

**Figure 9.2**
The electromagnetic spectrum. (From J. C. Kotz and K. F. Purcell, *Chemistry & Chemical Reactivity*, Philadelphia: Saunders College Publishing, 1987.)

**TABLE 9.1   Particles and Rays Frequently Encountered in Radiation**

| Particle or Ray | Common Name of Radiation | Symbol | Charge | Atomic Mass Units |
|---|---|---|---|---|
| Proton | Proton beam | $_1^1H$ | $+1$ | 1 |
| Electron | Beta particle | $_{-1}^0e$ or $\beta$ | $-1$ | $0.00055 \left( \dfrac{1}{1835} \right)$ |
| Neutron | Neutron beam | $_0^1n$ | 0 | 1 |
| Helium nucleus | Alpha particle | $_2^4He$ or $\alpha$ | $+2$ | 4 |
| Energetic radiation $\begin{cases} \\ \\ \end{cases}$ | Gamma ray | $\gamma$ | 0 | 0 |
|  | X-ray |  | 0 | 0 |

## Characteristics of Radioactivity

We have said that humans cannot see gamma rays. This is also true for the other kinds of radioactivity. We cannot see alpha or beta particles either. Not only that, but we also cannot hear them, smell them, or feel them. They are undetectable by our senses, which is why they were undiscovered for most of human history. Radioactivity can be detected only by instruments. We will discuss this subject in Section 9.4.

## Rutherford's Experiment

The names alpha, beta, and gamma were assigned by Ernest Rutherford. Rutherford placed a piece of radium in a lead container that had a small opening. The radioactive material emitted radiation that passed through the opening and then between two poles of a magnet (Fig. 9A). One ray was deflected to the right, as negatively charged particles would be, and another to the left, as expected from positively charged particles. A third ray passed through without deflection, indicating no charge. The three rays were detected when they darkened a photographic plate. Rutherford named the positively charged rays alpha rays, the negatively charged rays beta rays, and the undeflected rays gamma rays.

**Figure 9A**
Rutherford's experiment.

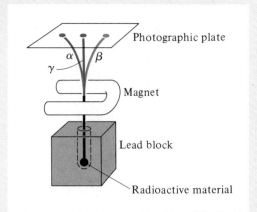

## 9.2 WHAT HAPPENS WHEN A NUCLEUS EMITS RADIOACTIVITY? NATURAL TRANSMUTATION

As mentioned in Section 2.4, different nuclei are made up of different numbers of protons and neutrons. It is customary to indicate these numbers with subscripts and superscripts placed to the left of the atomic symbol. The number of protons is shown as a subscript and the mass number (total number of protons and neutrons) as a superscript. For example, $_1^1H$, $_1^2H$, and $_1^3H$ are the symbols for the three isotopes of hydrogen. We call them hydrogen-1, hydrogen-2, and hydrogen-3, respectively.

*Hydrogen-2 is also called deuterium, and hydrogen-3 is called tritium.*

### Radioactive and Stable Nuclei

Not all nuclei are radioactive. Some isotopes (Sec. 2.4) are radioactive and others are stable. There are more than 300 naturally occurring isotopes. Of these, 264 are stable, meaning that these nuclei never give off any radioactivity. As far as we can tell, they will last forever. The remainder are radioactive—they do give off radioactivity. Furthermore, more than 1000 artificial isotopes have been made in laboratories.

*A radioactive isotope is often called a* **radioisotope.**

*All artificial isotopes are radioactive.*

Isotopes in which the numbers of protons and neutrons are balanced seem to be stable. In the lighter elements, this occurs when the numbers of protons and neutrons are approximately equal. For example, $_6^{12}C$ is a stable nucleus (six protons and six neutrons) and so are $_8^{16}O$ (eight protons and eight neutrons) and $_{10}^{20}Ne$ (ten protons and ten neutrons). Among the heavier elements, stability requires more neutrons than protons. Lead-206, one of the most stable isotopes, contains 82 protons and 124 neutrons.

*The role of neutrons seems to be to provide binding energy to overcome the repulsion between protons.*

### Beta Emission

If there is a serious imbalance in the proton-to-neutron ratio, a nucleus will emit an alpha or beta particle and thereby make the ratio more favorable and the nucleus more stable. For example, if a nucleus has an excess number of neutrons, it can stabilize itself by converting a neutron to a proton and an electron:

$$_0^1n \longrightarrow {_1^1}H + {_{-1}^0}e$$

This process is called **beta emission** because the electrons (beta particles) that are produced are emitted from the nucleus. For example, phosphorus-32 is a beta emitter:

*The electrons of the beta emission come from the nucleus and not from the surrounding electron cloud.*

$$_{15}^{32}P \longrightarrow {_{16}^{32}}S + {_{-1}^0}e$$

When the unstable phosphorus-32 is converted to sulfur-32, stability is achieved (16 protons and 16 neutrons).

**Every time a nucleus emits a beta particle, it is transformed into another nucleus with the same mass number but an atomic number one unit greater.**

In balancing nuclear equations, **the sum of the masses (superscripts) must be equal on both sides of the equation. The same is true for the atomic numbers (subscripts).** To find the proper symbol for a nucleus corresponding to a certain atomic number, use the periodic table on the inside cover of this book.

*Note that for balancing purposes the electron is given an atomic number of $-1$.*

---

**EXAMPLE**

What nucleus is $^{14}_{6}C$, a beta emitter, converted to?

$$^{14}_{6}C \longrightarrow ? + _{-1}^{0}e$$

**Answer**

The $^{14}_{6}C$ nucleus has six protons and eight neutrons. One of its neutrons is converted to a proton and an electron. The electron is emitted (this is the beta particle), but the proton remains. The nucleus now has seven protons (therefore it must be nitrogen) and seven neutrons. The equation is

$$^{14}_{6}C \longrightarrow ^{14}_{7}N + _{-1}^{0}e$$

Note that this equation is balanced: the sum of the masses is 14 on both sides; and the sum of the atomic numbers is 6 on both sides.

---

**PROBLEM 9.1**    Iodine-139 is a beta emitter. What is the reaction product?

### Alpha Emission

The changing of one element into another is called **transmutation.** It happens naturally every time an element gives off beta particles. What about alpha emission? Transmutation takes place here also. For heavy elements, the loss of alpha particles is an especially important stabilization process. For example,

$$^{238}_{92}U \longrightarrow ^{234}_{90}Th + ^{4}_{2}He$$
$$^{210}_{84}Po \longrightarrow ^{206}_{82}Pb + ^{4}_{2}He + \gamma$$

The symbol $\gamma$, the Greek letter gamma, represents a gamma ray.

Both alpha and beta emission can be either "pure" or mixed with gamma rays.

In the second example, stable lead-206 is obtained after alpha emission. Note that the polonium-210 emits both alpha particles and gamma rays.

Can you deduce, from the examples given, the general rule for alpha emission? **The new nucleus always has a mass number four units lower and an atomic number two units lower than the original.** Like beta emission, alpha emission always means that natural transmutation has taken place.

### Positron Emission

A positron is a particle that has the same mass as an electron, but a charge of $+1$, rather than $-1$. Its symbol is $\beta^+$. Positron emission is much less frequent than beta emission, but some nuclei do emit positrons. When it happens the nucleus is transformed into another nucleus with the same mass number, but an atomic number one unit less. An example is

$$^{11}_{6}C \longrightarrow _{+1}^{0}e + ^{11}_{5}B$$

### Gamma Emission

Besides alpha, beta, and positron emitters, one occasionally encounters pure gamma emitters:

$$^{11}_{5}B* \longrightarrow ^{11}_{5}B + \gamma$$

In this equation, the $^{11}_5$B* symbolizes the boron nucleus in a high-energy (excited) state. In this case, there is no transmutation. The boron is still boron, but it is in a lower-energy form (more stable) after the emission of excess energy in the form of gamma rays.

## 9.3 HALF-LIFE

Suppose we have 40 g of a radioactive isotope, say $^{90}_{38}$Sr. It contains many atoms, of course (in this case, about $2.7 \times 10^{23}$). These atoms are unstable and decay. In this case they give off beta particles and thereby become $^{90}_{39}$Y atoms. But do they all decay at once? No, they do not. They decay one at a time, at a fixed rate. For strontium-90, the decay rate is such that one half of our sample (about $1.35 \times 10^{23}$ atoms) will have decayed by the end of 28.1 years. **The time it takes for one half of any sample of radioactive material to decay** is called the **half-life** ($t_{\frac{1}{2}}$).

When a nucleus gives off radiation, it is said to decay.

It does not matter how big or small a sample is. For example, in the case of our 40 g of strontium-90, 20 g will be left at the end of 28.1 years (the rest will have been converted to yttrium-90). It will then take another 28.1 years for half of the remainder to decay, so that at the end of that time we will have 10 g. If we wait still a third period of 28.1 years, then 5 g will be left. If we had begun with 100 g, then 50 g would be left after the first 28.1-year period.

Figure 9.3 shows the radioactive decay curve of $^{131}_{53}$I. Inspection of this graph shows that, at the end of eight days, half of the original $^{131}_{53}$I has disappeared. Thus, the half-life of iodine-131 is eight days. It would take a total of 16 days, or two half-lives, for three fourths of the original iodine-131 to decay.

---

### EXAMPLE

If 10 mg of fresh $^{131}_{53}$I is administered to a patient, how much is left in the body after 32 days?

### Answer

We know that 32 days corresponds to four half-lives because $t_{\frac{1}{2}}$ for this isotope is eight days. After one half-life, 5 mg is left; after two half-lives, 2.5 mg; after three half-lives, 1.25 mg; and after four half-lives, 0.62 mg.

---

**PROBLEM 9.2** Barium-122 has a half-life of 2 minutes. You just obtained a fresh sample weighing 10 g. It takes 10 minutes to set up the experiment in which barium-122 will be used. How many grams of barium-122 will be left when you begin the experiment?

---

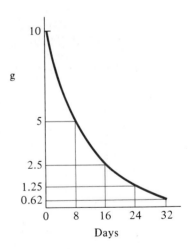

**Figure 9.3**
The decay curve of iodine-131.

It must be noted that, in theory, it would take an infinite amount of time for all of a radioactive sample to disappear. However, most of the radioactivity disappears after five half-lives; by that time, only 3 percent of the original is left. After ten half-lives less than 0.1 percent of the activity remains.

It is important to note that the half-life of an isotope is independent of temperature and pressure, and indeed of all other conditions, and is a property of the particular isotope only. It does not depend on what kind of other atoms surround the particular nucleus, that is, what kind of molecule the nucleus is

# Radioactive Dating

Carbon-14, with a half-life of 5730 years, can be used to date archeological objects as old as 60 000 years. The dating is based on the principle that an organism, plant or animal, as long as it lives, maintains a constant carbon-12/carbon-14 ratio. When the organism dies, the carbon-12 level remains constant but the carbon-14 decays through the reaction

$$^{14}_{6}C \longrightarrow \,^{14}_{7}N + \,^{0}_{-1}e$$

Thus, the changing carbon-12/carbon-14 ratio can be used to determine the date of an artifact.

For example, in charcoal made from a tree that has recently died, the carbon-14 gives a radioactive count of 13.70 disintegrations per minute per gram of carbon. In a piece of charcoal found in a cave in France near some ancient cave paintings (Fig. 9B.1), the carbon-14 count was 1.71 disintegrations/min for each gram of carbon. From this information the cave paintings can be dated. After one half-life, the number of disintegrations per minute per gram is 6.85; after two half-lives it is 3.42, and after three half-lives it is 1.71. Therefore, three half-lives have passed since the paintings were done. Since carbon-14 has a half-life of 5730 years, the paintings are about 3 × 5730 = 17 190 years old.

The famous Shroud of Turin, a piece of linen cloth with the image of a man's head on it (Fig. 9B.2), was believed by many to be the original cloth that was wrapped around Jesus Christ. However, radioactive dating showed, with 95 percent certainty, that the plants from which the linen was obtained were alive sometime between AD 1260 and 1380, proving that it could not have been the shroud of Christ. It was not necessary to destroy the shroud in order to make the tests. Scientists used just a few square centimeters of cloth from its edge.

Rock samples can be dated on the basis of their lead-206 and uranium-238 content. The assumption is that all of the lead-206 has come from the decay of uranium-238,* which has a half-life of 4.5 billion years. The oldest rock found on the earth is a granite outcrop in Greenland, dated to be 3.7 × 10⁹ years old.

On the basis of dating meteorites, the estimated age of the solar system is 4.6 × 10⁹ years.

---

* Uranium-238 does not decay *directly* to lead-206. It first goes to thorium-234, which then also decays to radium-230, and so on. Lead-206 is the final result of a series of such transmutations.

**Figure 9B.1**
A prehistoric cave painting (Hall of Bulls) from Lascaux, Dordogne, France.

**Figure 9B.2**
The Shroud of Turin: (a) as it appears to the eye, and (b) photographic negative, which brings out the image more clearly. (Santa Visalli/The IMAGE BANK.)

(a)                    (b)

**TABLE 9.2  Half-lives of Some Radioactive Nuclei**

| Name | Symbol | Half-life | Radiation |
|------|--------|-----------|-----------|
| Hydrogen-3 (tritium) | $^3_1H$ | 12.26 years | Beta |
| Carbon-14 | $^{14}_6C$ | 5730 years | Beta |
| Phosphorus-32 | $^{32}_{15}P$ | 14.3 days | Beta |
| Potassium-40 | $^{40}_{19}K$ | $1.28 \times 10^9$ years | Beta + gamma |
| Cobalt-60 | $^{60}_{27}Co$ | 5.2 years | Gamma |
| Strontium-90 | $^{90}_{38}Sr$ | 28.1 years | Beta |
| Technetium-99m | $^{99m}_{43}Tc$ | 6.0 hours | Gamma |
| Iodine-131 | $^{131}_{53}I$ | 8 days | Beta + gamma |
| Mercury-197 | $^{197}_{80}Hg$ | 65 hours | Gamma |
| Polonium-210 | $^{210}_{84}Po$ | 138 days | Alpha |
| Radon-222 | $^{222}_{86}Rn$ | 3.8 days | Alpha |
| Uranium-238 | $^{238}_{92}U$ | $4 \times 10^9$ years | Alpha |

part of. **We do not know any way to speed up radioactive decay or slow it down.**

Some half-lives are given in Table 9.2. Even this brief sampling indicates that there are tremendous differences in half-lives. Some isotopes, such as technetium-99m, decay and disappear in a day; others, such as uranium-238, will be radioactive for billions of years. Very short-lived isotopes, especially the artificial heavy elements (Sec. 9.7) with atomic numbers greater than 100, have half-lives of the order of seconds. The usefulness, or the inherent danger, in some of these radioactive isotopes is related to their half-lives.

The "m" in technetium-99m means *metastable*.

In assessing long-range health effects of atomic-bomb damage or of nuclear power plant accidents, such as what happened at Three Mile Island, Pennsylvania, in 1979 or Chernobyl (in the former Soviet Union) in 1986 (see Box 9H), we can see that radioactive isotopes with long half-lives, such as $^{85}_{36}Kr$ ($t_{\frac{1}{2}} = 10$ years) or $^{60}_{27}Co$ ($t_{\frac{1}{2}} = 5.2$ years), are more important than the short-lived ones. On the other hand, when a radioactive isotope is used in medical diagnosis or therapy, short-lived isotopes are more useful because they disappear faster from the body. Among the isotopes listed in Table 9.2, those with half-lives of only hours or days, such as $^{99m}_{43}Tc$, $^{32}_{15}P$, $^{131}_{53}I$, and $^{197}_{80}Hg$, find much use in medicine.

## 9.4  CHARACTERISTICS OF RADIATION

As already noted, radioactivity is not detectable by our senses. We cannot see it, hear it, feel it, or smell it. Then how do we know it is there? Alpha, beta, gamma, and x-rays, as well as proton beams, all have a property we can use to detect them: When they interact with matter, they usually knock electrons out of the electron cloud surrounding the atomic nucleus and thereby create positively charged ions from neutral atoms. Because of this, we call all of them **ionizing radiation.**

Radiation is characterized by two physical measurements: (a) its **intensity,** which is the number of particles or photons emerging per unit time, and (b) the **energy** of each particle or photon.

### Intensity

To measure intensity we take advantage of the ionizing property of radiation. Devices such as the **Geiger-Müller counter** (Fig. 9.4) or the **proportional counter** contain a gas such as helium or argon. When a radioactive nucleus emits beta particles (electrons), these particles ionize the gas, and the instrument registers this by indicating that an electric current is passing between two electrodes. In this way, the instrument counts particle after particle.

Other measuring devices, such as **scintillation counters,** have a material called a phosphor that emits a unit of light when an alpha or beta particle or gamma ray strikes it. Again the particles are counted one by one. The quantitative measure of radioactivity can be reported in counts/minute or counts/second.

Alpha and beta particles and gamma rays all have different energies as they leave the nucleus. For example, cobalt-60 emits both beta particles and gamma rays when it decays in a number of distinct steps:

**1.**   $^{60}_{27}\text{Co} \longrightarrow \, ^{60}_{28}\text{Ni*} + \, ^{0}_{-1}\text{e}$

The electron emitted in this step has an energy of $7.15 \times 10^6$ kcal/mole. The resultant new nucleus, nickel-60, is in a high-energy state (indicated by the asterisk) and releases some of its excess energy by emitting a gamma ray:

**2.**   $^{60}_{28}\text{Ni*} \longrightarrow \, ^{60}_{28}\text{Ni*} + \gamma$

This gamma ray has an energy of $26.90 \times 10^6$ kcal/mole. The resultant nickel-60 is still in an excited state, and another gamma ray is emitted before the nickel settles down to its most stable, lowest-energy state, which we call the **ground state:**

**3.**   $^{60}_{28}\text{Ni*} \longrightarrow \, ^{60}_{28}\text{Ni} + \gamma$

This second gamma ray has an energy of $30.67 \times 10^6$ kcal/mole.

A television picture tube works on a similar principle.

**Figure 9.4**
The Geiger-Müller counter. The counter is made of a glass cylinder. The inner surface is coated with a material that conducts electricity. In the center is the collector electrode, made of tungsten wire. This is positively charged to about 1200 volts. The space between is filled with helium or argon gas at a low pressure. A thin Mylar plate (window) at the end keeps the gas in and at the same time allows the ionizing radiation to pass through. When the radiation enters the tube, it converts an argon or helium atom to a positive ion, which goes to the negative electrode and is counted.

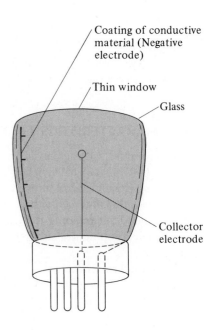

Coating of conductive material (Negative electrode)

Thin window

Glass

Collector electrode

Therefore, a person exposed to cobalt-60 treatment in cancer therapy would in essence get three different kinds of radiation, though in practice the least energetic, the beta particles, are screened out and the patient gets only the $26.90 \times 10^6$ and $30.67 \times 10^6$ kcal/mole gamma rays.

These three different energies constitute the **spectrum of radiation** of cobalt-60. Each isotope has its own spectrum of radiation, which may contain one or more different energetic particles and rays.

Even penetrating gamma radiations can be stopped by lead. Lead-shielded enclosures and leaded aprons minimize the exposure to radiation of x-ray technicians and other health scientists.

## 9.5  RADIATION DOSIMETRY AND THE EFFECTS OF RADIATION ON HUMAN HEALTH

A common unit of radiation activity or intensity is the **curie** (Ci), named in honor of Marie Curie, whose lifelong work with radioactive materials greatly helped our understanding of nuclear phenomena. One curie is defined as $3.7 \times 10^{10}$ disintegrations per second (or counts per second). This is radiation of very high intensity, the amount a person would get from 1 g of pure $^{286}_{88}$Ra. This activity is too high for regular medical use, and the most common units used in the health sciences are small fractions of this:

> 1 millicurie (mCi) = $3.7 \times 10^7$ counts/s
>
> 1 microcurie ($\mu$Ci) = $3.7 \times 10^4$ counts/s

---

**EXAMPLE**

A radioactive isotope with an activity of 100 mCi per vial is sold to a hospital. The vial contains 10 mL of liquid. The instruction is to administer 2.5 mCi intravenously. How much of the liquid should one use?

**Answer**
The activity of a sample is directly proportional to the amount present, and so

$$2.5 \; \cancel{mCi} \times \frac{10 \text{ mL}}{100 \; \cancel{mCi}} = 0.25 \text{ mL}$$

---

**PROBLEM 9.3**   A radioactive isotope in a 9.0-mL vial has an activity of 300 mCi. A patient is required to take 50 mCi intravenously. How much liquid should be used for the injection?

---

Another unit of radiation activity or intensity is the **becquerel** (Bq). One becquerel is one disintegration per second.

In studying the effect of radiation on the body, neither the energy of the radiation (in kcal/mole) nor its activity (in Ci) alone or in combination is of particular importance. The critical question is what kind of effects such radiation produces in the body. Three different units are used to describe the **effects of radiation:** roentgens, rads, and rems. One **roentgen** is the amount of radiation that produces ions having $2.58 \times 10^{-4}$ coulombs per kilogram (a coulomb

Wilhelm Roentgen (1845–1923) discovered x-rays in 1895.

is a unit of electrical charge). The number of roentgens is the **effective dose delivered** by the radiation and is therefore a measure of exposure to a particular form of radiation.

However, exposure, or delivered energy, does not take into account the effect on tissue. Radiation damages body tissue by causing ionization. In order for ionization to occur, the delivered energy must be absorbed by the tissue. The measure of the amount of **radiation absorbed** is the **rad,** which stands for **r**adiation **a**bsorbed **d**ose. A newer unit than the rad is the **gray** (Gy). One Gy equals 100 rads.

The relationship between delivered dose in roentgens and absorbed dose in rads is that exposure to 1 roentgen yields 0.97 rad of absorbed radiation in water, 0.96 rad in muscle, and 0.93 rad in bone. This relationship holds for photons with energies between $2.3 \times 10^5$ and $70 \times 10^6$ kcal/mole of photons. For lower-energy photons, such as x-rays between $2.3 \times 10^5$ and $46 \times 10^5$ kcal/mole, each roentgen yields 3 rads of absorbed dose in bone. This latter effect exists because bone contains heavier elements (such as calcium, with atomic number 20) than other tissues, and these elements absorb low-energy radiation in proportion to their atomic numbers. This is the principle of diagnostic x-rays, wherein soft tissue lets the radiation through to strike a photographic plate but the bones absorb the radiation and cast a shadow on the plate.

The most important biological measure of radiation is the **rem,** which stands for **r**oentgen **e**quivalent **m**an. One rem is a measure of the effect of the radiation when 1 roentgen is absorbed by a person. Other units are the **millirem** (mrem) (1 mrem = $1 \times 10^{-3}$ rem) and the **sievert** (Sv) (1 Sv = 100 rem). The need for such units arises because 1 rad of absorbed energy from different sources can cause different damage to the tissues. One rad from alpha rays or neutrons causes ten times more damage than 1 rad from x-rays or gamma rays.

The various radiation units are summarized in Table 9.3.

Although alpha particles cause more damage than x-rays or gamma rays, they have a very low penetrating power and cannot pass through the skin. Consequently, they are not harmful to humans or animals as long as they do not get into the body some other way. If they do get in, however, they are quite harmful. They can get inside, for example, if a person swallows or inhales a small particle of a substance that emits alpha particles.

Beta particles are less damaging to tissue than alpha particles but penetrate farther and so are generally more harmful. But gamma rays, which can completely penetrate the skin, are by far the most dangerous and harmful form of radiation.

**TABLE 9.3   Radiation Dosimetry**

| Measurement | International Unit | Conventional Unit |
|---|---|---|
| **Activity** | 1 becquerel (Bq) = 1 disintegration/s | 1 curie (Ci) = $3.7 \times 10^{10}$ disintegrations/s |
| **Production of ions** | 1 roentgen (R) = $2.58 \times 10^{-4}$ coulomb/kg | 1 roentgen (R) = $2.58 \times 10^{-4}$ coulomb/kg |
| **Absorbed dose** | 1 gray (Gy) = 1 joule/kg | 1 rad = 0.01 Gy |
| **Absorbed dose by humans** | 1 sievert (Sv) = 1 joule/kg | 1 rem = 0.01 Sv |

Most of our exposure to ionizing radiation comes from natural sources (Table 9.4). Radon gas is the main cause of this exposure. Radon has more than 20 isotopes, all radioactive, but the most important is radon-222, an alpha particle emitter. Radon-222 is a natural decay product of uranium-238, which is widely distributed in the earth's crust. Radon poses a particular health hazard among radioactive elements because it is a gas at normal temperatures and pressures. This means that radon enters our lungs with the air we breath and gets trapped in the mucous lining of the lung. Radon-222 has a half-life of 3.8 days. It decays naturally and produces, among other isotopes, two harmful alpha emitters: polonium-218 and polonium-214. Radon-222 and its decay products can cause lung cancer over a long exposure.

The Environmental Protection Agency sets a standard of 4 pCi/L (one pico-curie, pCi, is $10^{-12}$ curie) as a safe exposure level. A survey of single-family homes in the United States showed that 7 percent exceeded this level. Most radon seeps into dwellings through cracks in cement foundations and around pipes. It accumulates mostly in basements. The remedy is to ventilate basements as well as houses enough to reduce the radiation levels. In a notorious case, in Grand Junction, Colorado, houses were built from bricks made from uranium tailings. Obviously the radiation levels in these buildings were unacceptably high. They could not be controlled and the buildings had to be destroyed. In our modern radiation-conscious age more and more home buyers request a certification of radon levels before buying a house.

## Box 9C

## The Indoor Radon Problem

**Figure 9C**
A number of testing devices are available to determine whether radon is building up in a home. (Photograph by Larry Cameron.)

Therefore, for comparison purposes and for determining exposure from all kinds of sources, the **equivalent dose** is an important measure. If an organ receives radiation from different sources, the total effect can be summed up in rems (or mrem or Sv). For example, 10 mrem of alpha particles and 15 mrem of gamma radiation give a total of 25 mrem absorbed equivalent dose. Table 9.4 shows the amount of radiation exposure an average person obtains yearly from both natural and artificial sources. People who work in nuclear medicine are of

**TABLE 9.4 Average Exposure to Radiation from Common Sources[a]**

| | Dose (mrem/year)[b] |
|---|---|
| *Naturally Occurring Radiation* | |
| Cosmic rays | 27 |
| Terrestrial radiation (rocks, buildings) | 28 |
| Inside human body (K-40 and Ra-226 in the bones) | 39 |
| Radon in the air | 200 |
| Total | 294 |
| *Man-made (Artificial) Radiation* | |
| Medical x-rays[c] | 39 |
| Nuclear medicine | 14 |
| Consumer products | 10 |
| Nuclear power plants | 0.5 |
| All others | 1.5 |
| Total | 65.0 |
| Grand total | 359[d] |

[a] Source: *National Council on Radiation Protection and Measurements* NCRP Report No. 93 (1987).
[b] 100 mrem = 1 mSv.
[c] Individual medical procedures may expose certain parts of the body to much higher doses. For instance, one chest x-ray gives 27 mrem, and a diagnostic GI series gives 1970 mrem.
[d] The federal safety standard for allowable occupational exposure is about 5000 mrem/year. However, it has been suggested that this be lowered to 4000 mrem/yr or even lower, to reduce the risk of cancer stemming from low levels of radiation.

Symptoms of radiation sickness are nausea, vomiting, a decrease in the white blood cell count, and loss of hair.

**Figure 9.5**
Worker with a radiation badge.
(Phototake/© Yoav Levy.)

course exposed to greater amounts. In order to assure that exposures do not get too high, they wear radiation badges (Fig. 9.5). A single whole-body irradiation of 25 rem is noticeable in the blood count, and 100 rem causes the typical symptoms of radiation sickness. A dose of 400 rem causes death within one month in 50 percent of exposed persons, and 600 rem is almost invariably lethal within a short time. It should be noted that up to 50 000 rem is needed to kill bacteria and up to $10^6$ rem to inactivate viruses.

Most of us, fortunately, never get a single dose of more than a few rem and so never suffer from any form of radiation sickness. This does not mean, however, that small doses are totally harmless. The harm comes in two ways:

1. Small doses of radioactivity over a period of years can cause cancer, especially blood cancers such as leukemia. Nobody knows how many cancers have been caused this way because the doses are so small and continue for so many years that they cannot generally be measured. Also, because there seem to be so many other causes of cancer, it is difficult or impossible to decide if any particular case is caused by radiation.

2. If any form of radiation strikes an egg or sperm cell, it can cause a change in the genes (Sec. 23.13). Such changes are called **mutations.** If this particular egg or sperm cell mates, grows, and becomes a new individual, that individual may have mutated characteristics, which are usually harmful and frequently lethal.

Since radiation carries so much potential for harm, it would be nice if we could totally escape it. But can we? Table 9.4 shows that this is impossible. Naturally occurring radiation, which we call **background radiation,** is present everywhere on earth. As Table 9.4 shows, this background radiation is much more than the average radiation level from artificial sources (which are mostly diagnostic x-rays). Therefore, if we stopped all forms of artificial radiation, including medical uses, we would still be exposed to the background radiation.

## 9.6 MEDICAL USES OF RADIOACTIVE MATERIALS

Radioactive isotopes have two main uses in medicine: diagnosis and therapy (Table 9.5).

**TABLE 9.5  Some Medically Useful Radioactive Isotopes**

| | Isotope | Use |
|---|---|---|
| $^{3}_{1}H$ | Tritium (hydrogen-3) | Measure water content of body |
| $^{11}_{6}C$ | Carbon-11 | Brain scan with positron emission transverse tomography (PET) to trace glucose pathway |
| $^{14}_{6}C$ | Carbon-14 | Radioimmunoassay |
| $^{32}_{15}P$ | Phosphorus-32 | Detection of eye tumors |
| $^{51}_{24}Cr$ | Chromium-51 | Albumism diagnosis, size and shape of spleen, gastrointestinal disorders |
| $^{59}_{26}Fe$ | Iron-59 | Bone marrow function, diagnosis of anemias |
| $^{60}_{27}Co$ | Cobalt-60 | Treatment of cancer |
| $^{67}_{31}Ga$ | Gallium-67 | Whole-body scan for tumors |
| $^{75}_{34}Se$ | Selenium-75 | Pancreas scan |
| $^{81m}_{36}Kr$ | Krypton-81m | Lung ventilation scan |
| $^{85}_{38}Sr$ | Strontium-85 | Bone scan for bone diseases, including cancer |
| $^{99m}_{43}Tc$ | Technetium-99m | Brain, liver, kidney, bone marrow scans; diagnosis of damaged heart muscles |
| $^{131}_{53}I$ | Iodine-131 | Diagnosis of thyroid malfunction; treatment of hyperthyroidism and thyroid cancer |
| $^{197}_{80}Hg$ | Mercury-197 | Kidney scan |

## Box 9E

# CAT and Nmr Scans

One particular x-ray diagnostic technique, computer assisted tomography **(CAT scan)** has seen a phenomenal growth in the more than twenty years since its invention. Allan M. Cormack of Tufts University and Godfrey N. Hounsfield, an English scientist, shared the 1979 Nobel prize in medicine for this invention.

CAT scans can reveal small differences in the density of tissues because different tissues slow down x-rays to different extents. A modern CAT scan needs one rotation of the x-ray source around the body. It has a large number of detectors to measure simultaneously the x-rays passing through the body. A computer then clearly reconstructs the internal structure of the body.

CAT scans are used mainly to detect brain tumors because no other technique is so accurate in locating tissues within the skull. Not only are internal structures displayed (Fig. 9E.1), but one can even distinguish among the various fluid-filled cavities in the brain and between gray and white matter. The use of CAT scans is not restricted to the brain—any part of the body can be examined with this tool.

**Figure 9E.1**
A CAT scan of a human brain.
(The IMAGE BANK/© Bill Varie.)

Other body-scanning techniques, which operate on different principles, most notably nuclear magnetic resonance (nmr) scans (Fig. 9E.2), can be used for the same diagnostic purposes. The nmr technique has the advantage that it does not use ionizing radiation (Fig. 9E.3). In hospitals nmr is usually called magnetic resonance imaging (mri).

**Figure 9E.2**
An nmr scan. (Phototake/
© Debbie Breen.)

**Figure 9E.3**
A medical nmr imager. (Photograph by Beverly March; courtesy of Long Island Jewish Hospital.)

## Diagnosis

This is the more widely used aspect of nuclear medicine.

Chemically and metabolically, a radioactive isotope in the body behaves exactly the same way as the nonradioactive isotopes of the same element. In the simplest form of diagnosis, a radioactive isotope is injected intravenously and a technician uses the various types of detectors to monitor how the radiation is distributed.

The use of iodine-131 to diagnose malfunctioning thyroid glands is a good example. The thyroid glands in the neck produce a hormone, thyroxine, that controls the overall rate of metabolism (use of food) in the body. Thyroxine contains iodine atoms. When radioactive iodine-131 is administered into the blood stream, the thyroid glands take it up and incorporate it into the thyroxine (see Box 12J). A normally functioning thyroid absorbs about 12 percent of the administered iodine within a few hours. An overactive thyroid (hyperthyroidism) absorbs and localizes iodine-131 in the glands faster, and an underactive thyroid (hypothyroidism) does this much more slowly than normal. Thus, by counting the radioactive particles emitted from the neck, one can determine the rate of uptake of iodine-131 into the glands and diagnose hyper- or hypothyroidism.

Most organ scans are similarly based on the preferential uptake of some radioactive isotopes by a particular organ (Fig. 9.6).

Another important use of radioactive isotopes is to learn what happens to ingested material. The foods and drugs swallowed or otherwise taken in by the body are transformed, decomposed, and excreted. To understand the pharmacology of a drug, it is important to know how these processes occur. For example, a certain drug may be effective in treating certain bacterial infections. Before this drug can be used in a clinical trial, the manufacturer must prove that the drug is not harmful to humans. In a typical case this is done first in animal studies. The drug in question is synthesized, and some radioactive isotope, such as hydrogen-3, carbon-14, or phosphorus-32, is incorporated into its structure. The drug is administered to the animals, and after a certain period the animals are sacrificed. The fate of the drug is then followed by isolating from the body any radioactive compounds formed.

*This is called tagging a drug.*

A typical pharmacological experiment was the study of the effects of tetracycline. This powerful antibiotic tends to accumulate in bones and is not given

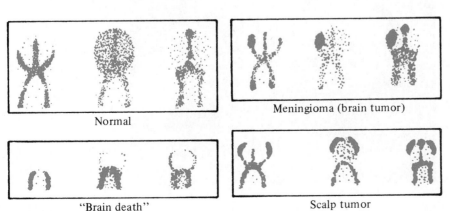

Normal

Meningioma (brain tumor)

"Brain death"

Scalp tumor

**Figure 9.6**
A comparison of dynamic scan patterns for normal and pathological brains. The studies were performed by injecting technetium-99m into blood vessels. (From *CRC Handbook in Clinical Laboratory Science*, Vol. 1, Nuclear Medicine. CRC Press, Inc.)

to pregnant women because it is transferred to the bones of the fetus. A particular tetracycline was tagged with radioactive hydrogen-3, and its uptake in rat bones was monitored in the presence and absence of a sulfa drug. With the aid of a scintillation counter, the activity of the maternal and fetal bones was determined, and it was found that the sulfa drug helped to minimize this undesirable side effect of tetracycline, that is, its accumulation in the fetal bones.

The metabolic fate of essential chemicals in the body can also be followed with radioactive tracers. A number of normal and pathological body functions have been elucidated by the use of radioactive isotopes.

### Therapy

The main use of radioactive isotopes in therapy is the selective destruction of pathological cells and tissues. Remember that radiation, whether from gamma rays or other sources, is detrimental to cells. Ionizing radiation damages cells, especially those that divide rapidly. The damage may be great enough to destroy the cells or to sufficiently alter the genes so that multiplication of the cells slows down.

Cancerous cells are the main target for such ionizing radiation. Radiation is usually used when the cancer is not well localized and therefore surgical removal is not the complete answer, and also when the cancerous cells are in a metastatic state. It is also used for preventive purposes when one wants to eliminate any possible remaining cancerous cells after surgery has been performed. The idea, of course, is to kill cancerous but not normal cells. Therefore, radiation such as x-rays or gamma rays from cobalt-60 is highly *collimated*, that is, aimed at a small part of the body where the cancerous cells are suspected. Despite this pinpointing, the radiation kills normal cells along with the cancerous cells. Because the radiation is most effective against rapidly dividing cancer cells, rather than normal cells, and because the radiation is aimed at a specific location, the damage to healthy tissues is minimized. Figure 9.7 shows a patient receiving radiation treatment.

Another way to localize radiation damage in therapy is to use specific radioactive isotopes. In the case of thyroid cancer, large doses of iodine-131 are

A metastatic state exists when the cancerous cells have broken off from their primary site(s) and are moving to other parts of the body.

Because of this, radiation sickness is a side effect of radiation therapy.

**Figure 9.7**
Patient receiving radiation treatment. (Photograph by Beverly March; courtesy of Long Island Jewish Hospital.)

In Section 9.3 we saw that the half-lives of some radioactive isotopes are very short. What do we do when a medically useful isotope is one with a short half-life? For example, $^{99m}_{43}$Tc is very useful for brain and other scans but has a half-life of only six hours. If it were produced in a laboratory and then shipped by normal channels to a hospital, most of it would decay by the time it got there. The solution to this problem is an isotope generator. In this case, what is shipped to the hospital is a container of $^{99}_{42}$Mo. This isotope decays by beta emission, with a half-life of 66.7 hours, to $^{99m}_{43}$Tc, which can be removed as it is formed and used before it decays.

**Box 9F**

## Isotope Generators for Medical Use

given, which are taken up by the glands. The isotope, which has high radioactivity, kills all the cells of the gland, the cancerous as well as the healthy ones, but does not appreciably damage other organs. In a similar manner, localization can be achieved by implanting a tiny grain of yttrium-90 in the pituitary gland. This beta emitter destroys the gland but nothing much beyond that. The hope is that this treatment will slow the growth of tumors elsewhere in the body because the pituitary gland stimulates cell reproduction, and without the growth hormone of the pituitary, the cell proliferation of tumors will cease.

## 9.7 NUCLEAR FUSION. ARTIFICIAL TRANSMUTATION

It is estimated that 98 percent of all matter in the universe is made up of hydrogen and helium. This is explained by the "big bang" theory of the formation of the universe. This theory postulates that our universe started with an explosion (big bang) in which matter was formed out of energy, and, at the beginning, only the lightest element, hydrogen, was in existence. Later, as the universe expanded, stars were born when the hydrogen clouds collapsed under gravitational forces. In the cores of these stars hydrogen nuclei fused together and formed helium.

The transformation of hydrogen nuclei into a helium nucleus liberates a very large amount of energy in the form of photons (Box 9G), largely by this equation:

$$^{2}_{1}H + ^{3}_{1}H \longrightarrow ^{4}_{2}He + ^{1}_{0}n + energy$$

This process, called **fusion,** is how the sun makes its energy. If we can ever achieve a controlled version of this fusion reaction, we shall be able to solve our energy problems. However, this may not occur for a long time.

All the **transuranium elements** (elements with atomic numbers greater than 92) from 93 to 109 are artificial and have been prepared by a process in which heavy nuclei are bombarded with light ones. Many, as their names indicate, were first prepared at the Lawrence Laboratory of the University of California, Berkeley, by Glenn Seaborg (1912–    ) (Nobel laureate in chemistry, 1951) and his colleagues:

The reactions occurring in the sun are essentially the same as those in the hydrogen bomb.

$$^{244}_{96}Cm + ^{4}_{2}He \longrightarrow ^{245}_{97}Bk + ^{1}_{1}H + 2\,^{1}_{0}n \qquad \text{Berkelium}$$

$$^{238}_{92}U + ^{12}_{6}C \longrightarrow ^{246}_{98}Cf + 4\,^{1}_{0}n \qquad \text{Californium}$$

$$^{252}_{98}Cf + ^{10}_{5}B \longrightarrow ^{257}_{103}Lr + 5\,^{1}_{0}n \qquad \text{Lawrencium}$$

---

**Box 9G**

**Energy from Mass**

$c = 3.0 \times 10^8$ m/s

When a deuterium and a tritium nucleus are transformed into a helium nucleus and a neutron, a large amount of energy is given off (see the text). Where does this energy come from? If we weighed the nuclei of the two hydrogen isotopes and compared this with the weight of the helium nucleus plus that of a neutron, we would find that the combined mass of the former is *greater* than the combined mass of the latter. When the deuterium and tritium nuclei are converted to helium and a neutron, the extra mass has to go *somewhere*. Where does it go? **It is converted to energy.** We even know, from the equation of Albert Einstein (1879–1955), how much energy we can get from the conversion of any amount of mass:

$$E = mc^2$$

This says that the mass ($m$) that was lost, in kilograms, multiplied by the square of the velocity of light ($c^2$), in m²/s², is equal to the amount of energy created ($E$), in joules. For example, 1 g of matter completely converted to energy would produce $8.8 \times 10^{13}$ J, which is enough energy to boil 34 000 000 L of water initially at 20°C. As you can see, we get a lot of energy from a little bit of mass.

---

These transuranium elements are unstable, and most have very short half-lives. For example, that of $^{257}_{103}$Lr is 0.65 second.

## 9.8 NUCLEAR FISSION. ATOMIC ENERGY

In the 1930s Enrico Fermi (1901–1954) and his colleagues in Rome, and Otto Hahn (1879–1968), Lisa Meitner (1878–1968), and Fritz Strassman (1902–    ) in Germany, tried to produce new transuranium elements by bombarding uranium-235 with neutrons. To their surprise, Hahn and his co-workers found that, rather than fusion, they obtained **nuclear fission** (fragmentation of large nuclei into smaller pieces):

$$^{235}_{92}U + {}^1_0n \longrightarrow {}^{139}_{56}Ba + {}^{94}_{36}Kr + 3\,{}^1_0n + \gamma$$

In this reaction the uranium-235 nucleus is broken into smaller particles, but the most important product is energy, which is produced because the products have less mass than the starting materials (Box 9G). This form of energy is called **atomic energy** and has been used for both war (in the atomic bomb) and peace.

In the fission of uranium-235, each fission produces three neutrons, which in turn can generate more fission by colliding with other uranium-235 nuclei. If even one of these neutrons produces new fission, the process becomes a self-propagating **chain reaction** (Fig. 9.8) that continues at a constant rate. If all three neutrons are allowed to produce new fission, the rate of the reaction increases constantly and eventually culminates in a nuclear explosion. The rate of reaction can be controlled by inserting boron control rods to absorb neutrons.

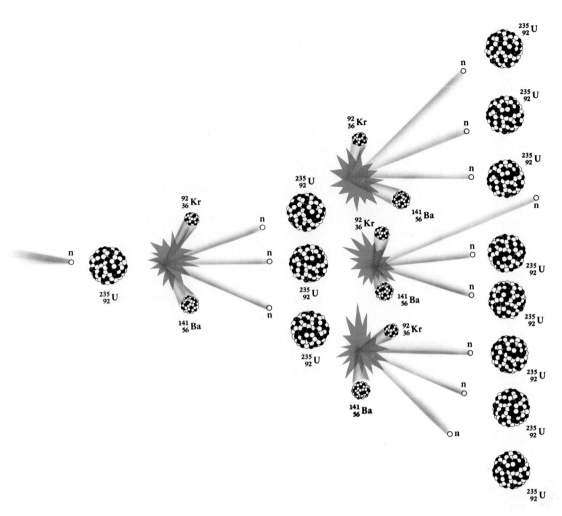

**Figure 9.8**
A chain reaction is begun when a neutron collides with a nucleus of $^{235}_{92}$U. (From J. C. Kotz and K. F. Purcell, *Chemistry & Chemical Reactivity,* Philadelphia: Saunders College Publishing, 1987.)

In nuclear plants, the energy produced by fission is sent to heat exchangers and used to generate steam, which drives a turbine to produce electricity (Fig. 9.9). Today about 10 percent of the electrical energy in the United States is supplied by nuclear plants. The opposition to nuclear plants is based on safety considerations and on the unsolved problems of waste disposal. Nuclear plants in general have had good safety records, but accidents such as those at Chernobyl (Box 9H; Fig. 9.10) and Three Mile Island have caused concern. The location of reactors in highly populated areas is definitely objectionable.

Waste disposal is a long-range problem. The fission products of nuclear reactors are highly radioactive themselves, with long half-lives. Thus, storing

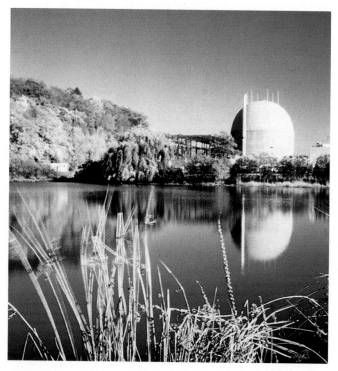

**Figure 9.9**
Haddam Neck nuclear reactor (NUS). (Courtesy of Atomic Industrial Forum.)

**Box 9H**

## Radioactive Fallout from Nuclear Accidents

On April 26, 1986, an accident occurred at the nuclear reactor in the town of Chernobyl in the former Soviet Union (Fig. 9.10). It was a clear reminder of the dangers involved in this industry and of the far-reaching contamination that such accidents create. In Sweden, more than 500 miles away from the accident, the radioactive cloud increased the background radiation from four to 15 times normal. The radioactive cloud reached England, about 1300 miles away, a week later. As a consequence the natural background radiation was increased by 15 percent.

Radioactivity from iodine-131 was 400 Bq/L in milk and 200 Bq/kg in leafy vegetables. Even some 4000 miles away in Spokane, Washington, iodine-131 activity of 242 Bq/L was found in rainwater, and smaller activities—1.03 Bq/L of ruthenium-103 and 0.66 Bq/L of cesium-137—were also recorded. These levels are not harmful.

Closer to the source of the nuclear accident, however, in neighboring Poland, potassium iodide pills were given to children. This was done to prevent radioactive iodine-131 (which might come from contaminated food) from concentrating in their thyroid glands, which could lead to cancer. Soon after the accident about 115 000 people were evacuated from the immediate area and nearby fallout zones. But even that was not sufficient, and more than three years later many additional villages in the fallout path were evacuated.

**Figure 9.10**
The nuclear power plant at Chernobyl after the accident. (Photograph by
V. Zufarov; courtesy of Fotokhronika Tass.)

them underwater or burying them underground (Fig. 9.11) just postpones the
disposal for generations to come. The fear is that corrosion and leakage from
such storage tanks may contaminate water supplies. Remember that, up to now,
no one has ever discovered a way to speed up or slow down radioactive decay.

It is true that most other ways of generating large amounts of electrical
power also have their environmental problems. For example, burning coal or oil
contributes to the accumulation of $CO_2$ in the atmosphere (Box 10E) and to
acid rain (see Box 6A).

**Figure 9.11**
Storing nuclear wastes in a storage room carved out of an underground salt bed.
(Courtesy U.S. Department of Energy.)

## SUMMARY

Certain isotopes are radioactive; others are stable. The three major types of radioactivity are **alpha particles** (helium nuclei), **beta particles** (electrons), and **gamma rays** (high-energy radiation). **Radioactivity** comes from the nucleus, not from the electron cloud. When a nucleus emits an alpha or beta particle, it is changed to the nucleus of another element. When an alpha particle is emitted, the new element has an atomic number two units lower and a mass number four units lower. When a beta particle is emitted, the new element has the same mass number and an atomic number one unit higher. **Positrons** (positive electrons) are emitted by fewer isotopes, but in this case the new element has an atomic number one unit lower.

Each radioactive isotope decays at a fixed rate described by its **half-life,** which is the time required for half the sample to decay. Radiation is detected and counted by devices such as Geiger-Müller counters.

The main unit of intensity of radiation is the **curie,** which is $3.7 \times 10^{10}$ disintegrations per second. Radiation can be harmful in several ways. For medical purposes and to measure potential radiation damage, the absorbed dose, measured in **rads,** is an important factor. Different particles damage body tissues differently. The **rem** is a measure of the damage. Radioactive nuclei are used in medicine for both diagnosis and therapy.

Helium is synthesized in the interiors of stars by **fusion** of hydrogen nuclei. The energy released in this process is the energy of our sun. Artificial elements are created by bombarding atomic nuclei with high-energy nuclei. With certain nuclei, such bombardment may cause the nucleus to split **(fission).** Nuclear fission releases large amounts of energy, which can be either controlled (nuclear reactors) or uncontrolled (nuclear weapons).

## KEY TERMS

Alpha emission (Sec. 9.2)
Alpha particle (Sec. 9.1)
Atomic energy (Sec. 9.8)
Background radiation (Sec. 9.5)
Becquerel (Sec. 9.5)
Beta emission (Sec. 9.2)
Beta particle (Sec. 9.1)
CAT scan (Box 9E)
Chain reaction (Sec. 9.8)
Curie (Sec. 9.5)
Electromagnetic radiation (Sec. 9.1)
Frequency (Sec. 9.1)
Gamma emission (Sec. 9.2)
Gamma rays (Sec. 9.1)
Gray (Sec. 9.5)
Half-life (Sec. 9.3)
Ionizing radiation (Sec. 9.4)

Magnetic resonance imaging (mri) (Box 9E)
Nmr scan (Box 9E)
Nuclear fission (Sec. 9.8)
Nuclear fusion (Sec. 9.7)
Photon (Sec. 9.1)
Positron (Sec. 9.2)
Rad (Sec. 9.5)
Radioactive dating (Box 9B)
Radioactive isotope (Sec. 9.2)
Radioactivity (Sec. 9.1)
Radioisotope (Sec. 9.2)
Rem (Sec. 9.5)
Sievert (Box 9.5)
Transmutation (Sec. 9.2)
Transuranium element (Sec. 9.7)
Wavelength (Sec. 9.1)
X-ray (Sec. 9.1)

## PROBLEMS

### Radioactivity and Electromagnetic Radiation

**9.4** How did Becquerel discover radioactivity?

**9.5** What is the difference between an alpha particle and a proton?

**\*9.6** In each case given the frequency, give the wavelength in cm or nm ($10^{-9}$ m) and tell what kind of radiation it is:
(a) $7.5 \times 10^{14}$/s
(b) $1.0 \times 10^{10}$/s
(c) $1.1 \times 10^{15}$/s
(d) $1.5 \times 10^{18}$/s

**\*9.7** Microwaves are a form of electromagnetic radiation that is used for the rapid heating of foods. What is the frequency of a microwave with a wavelength of 5.8 cm?

**9.8** What are the differences between alpha particles, beta particles, and gamma rays?

**9.9** How do x-rays differ from gamma rays? In what ways are they similar?

**9.10** Which has the longest wavelength: (a) infrared (b) ultraviolet (c) x-rays? Which has the highest energy?

## Radioactive and Stable Isotopes

**9.11** Write the symbol for a nucleus with
(a) 9 protons and 10 neutrons
(b) 15 protons and 17 neutrons
(c) 37 protons and 50 neutrons

**9.12** In each pair, tell which isotope is more stable:
(a) $^1_1H$ and $^3_1H$     (c) $^{40}_{20}Ca$ and $^{43}_{20}Ca$
(b) $^{18}_8O$ and $^{16}_8O$     (d) $^{26}_{14}Si$ and $^{28}_{14}Si$

**9.13** How many artificial stable (nonradioactive) nuclei are known?

## Natural Transmutation and Nuclear Reactions

**9.14** Samarium-151 is a beta emitter. What is the transmutation product of the nuclear reaction?

**9.15** The following nuclei turn into new nuclei by emitting beta particles. In each case write the formula for the new nucleus:
(a) $^{159}_{63}Eu$     (c) $^{242}_{95}Am$
(b) $^{141}_{56}Ba$

**\*9.16** Chromium-51 is used in diagnosing pathology of the spleen. The nucleus of this isotope captures an electron according to the equation

$$^{51}_{24}Cr + {}^{0}_{-1}e \longrightarrow ?$$

What is the transmutation product?

**9.17** The following nuclei turn into new nuclei by emitting alpha particles. In each case write the formula for the new nucleus:
(a) $^{210}_{83}Bi$     (c) $^{174}_{72}Hf$
(b) $^{238}_{94}Pu$

**9.18** Copper-60 is a positron emitter. What is the transmutation product of this nuclear reaction?

**9.19** Phosphorus-29 is a positron emitter. Write the balanced nuclear equation.

**9.20** When aluminum-27 is bombarded with alpha particles, one of the products is $^{30}_{15}P$. What is the other product? Write the balanced nuclear equation.

**\*9.21** When nitrogen-14 is bombarded with alpha particles, one of the reaction products is a proton. What is the other product? Write the balanced nuclear equation.

**9.22** For each of the following, write a balanced nuclear equation and identify the radiation emitted:
(a) beryllium-10 changes to boron-10
(b) europium-151\* changes to europium-151
(c) thallium-195 changes to mercury-195
(d) plutonium-239 changes to uranium-235

**9.23** In the radioactive decay of uranium-238, the first two steps are (1) alpha emission and (2) beta emission. Write the balanced nuclear equations for these two steps.

**9.24** What kind of emission does not result in transmutation?

**9.25** Complete the following nuclear reactions:
(a) $^{16}_8O + {}^{16}_8O \longrightarrow ? + {}^4_2He$
(b) $^{235}_{92}U + {}^1_0n \longrightarrow {}^{90}_{38}Sr + ? + 3\,{}^1_0n$
(c) $^{13}_6C + {}^4_2He \longrightarrow {}^{16}_8O + ?$
(d) $^{210}_{83}Bi \longrightarrow {}^{0}_{-1}e + ?$
(e) $^{12}_6C + {}^1_1H \longrightarrow ? + \gamma$

## Half-life

**9.26** What is a decay curve?

**9.27** The half-life of uranium-232 is 70 years. How long will it take 5.0 g of this element to be reduced to 1.25 g?

**\*9.28** A rock containing 1 mg plutonium-239/kg rock is found in a glacier. The half life of plutonium-239 is 25,000 years. If this rock was deposited 100,000 years ago during an Ice Age, how much plutonium-239, per kg of rock, was in the rock at that time?

**9.29** The element radium is extremely radioactive. If you converted a piece of radium metal to radium chloride (the weight of radium remaining the same), would it become less radioactive?

**9.30** In what ways can we increase the rate of radioactive decay? Decrease it?

**\*9.31** A scientist prepared 78.0 g of pure $^{135}_{57}La$. At the end of 58.5 hours, only 9.75 g was left. What is the half-life of this isotope?

**\*9.32** A patient receives 200 mCi of iodine-131, which has a half-life of eight days. (a) If 12 percent of this is taken up by the thyroid gland after two hours, what will be the activity of the thyroid after two hours, in mCi and in counts per minute? (b) After 24 days how much activity will remain in the thyroid gland?

## Characteristics of Radiation

**9.33** Why is it necessary to use instruments to detect radioactivity?

**9.34** What is the difference between a Geiger-Müller counter and a scintillation counter? Do they measure the same properties?

**9.35** What do Geiger-Müller counters measure: (a) intensity or (b) energy of radiation?

## Radiation Dosimetry

**9.36** A radioactive isotope with an activity of 80.0 mCi per vial is delivered to a hospital. The vial contains 7.0 cc of liquid. The instruction is to administer 7.2 mCi intravenously. How many cubic centimeters of liquid should be used for one injection?

**9.37** State clearly the properties that are measured in
(a) rads      (e) grays
(b) rems      (f) becquerels
(c) roentgens    (g) sieverts
(d) curies

**9.38** What is the source of the largest amount of radiation our bodies are exposed to? Can we control this exposure?

**9.39** Which radiation has the greatest penetrating power: alpha particles, beta particles, or gamma rays?

**9.40** Assuming the same amount of effective radiation, in rads, from three sources, which would be the most damaging to the tissues: alpha particles, beta particles, or gamma rays?

**9.41** How many sieverts are there in 1 mrem?

## Medical Uses of Radioactive Materials

**9.42** Which radioactive isotopes are used for
(a) cancer treatment    (d) brain scans
(b) pancreas scans     (e) kidney scans
(c) bone scans

**\*9.43** If a radioactive isotope is used for diagnosis (for example, $^{99m}_{43}$Tc for a brain scan), it should emit only particles that have high penetrating power so that it can be detected by a scintillation counter outside the skull. Write the nuclear decay equation for $^{99m}_{43}$Tc.

**9.44** Selenium-75 is used in the diagnosis of pancreatic diseases. It decays by beta emission and also emits gamma rays. Of these gamma rays, 94% have an energy of 10.6 kcal/mole, 2% have 14.3 kcal/mole, and 3% have 13.8 kcal/mole. Which of the three has the greatest penetrating power?

**9.45** Cobalt-60, with a half-life of 5.26 years, is used in cancer therapy. The energy of the radiation from cobalt-62 (half-life, 14 minutes) is even higher. Why isn't this isotope also used for cancer therapy?

**9.46** Radioactive isotopes can be used for cancer therapy, especially if they are going to be concentrated in the diseased organ. Name two radioactive isotopes that are used in cancer therapy because of this property.

**9.47** Iodine-131 is used both for diagnosis and for therapy. What is the difference in the way iodine-131 is administered for diagnosis and for therapy?

**9.48** Match the radioactive isotope with its proper use:
_____ (a) $^{14}_{6}$C      1. Cancer therapy
_____ (b) $^{60}_{27}$Co     2. Brain scan
_____ (c) $^{99m}_{43}$Tc    3. Thyroid scan
_____ (d) $^{131}_{53}$I      4. Artifact dating

## Fusion and Fission

**9.49** What is the most abundant nucleus in the universe?

**9.50** What is the product of the fusion of hydrogen nuclei?

**9.51** Look at the three equations on p. 269 describing the production of transuranium elements. What product is common to all these nuclear reactions?

**9.52** Element 109 was first prepared in 1982. A single atom of this element, with a mass number of 266 ($^{266}_{109}$Une), was made by bombarding a bismuth-209 nucleus with an iron-58 nucleus. What other products, if any, must have been formed besides $^{266}_{109}$Une?

**9.53** What is a chain reaction?

**9.54** Why are some people opposed to nuclear reactors?

## Boxes

**9.55** (Box 9A) If a naturally occurring radioactive nucleus emits neutrons, alpha particles, and gamma rays, how many spots would you see on the photographic plate in Rutherford's experiment?

**9.56** (Box 9A) Explain why, in Rutherford's experiment, the three rays were deflected in different directions.

*__9.57__ (Box 9B) Carbon-14 dating of the Shroud of Turin indicated that the plant from which the shroud was made was alive around AD 1350. How many half-lives does this correspond to?

**9.58** (Box 9B) The half-life of carbon-14 is 5730 years. The wrapping of an Egyptian mummy gave off 7.5 counts per minute per gram of carbon. A piece of linen purchased today would give an activity of 15 counts per minute per gram of carbon. How old is the mummy?

**9.59** (Box 9B) If a tree dies and the trunk remains undisturbed for 22 920 years, what percentage of the original $^{14}$C content is still present?

**9.60** (Box 9C) How does radon-222 produce polonium-218?

**9.61** (Box 9D) How did the painters of radioactive watch dials develop cancer?

**9.62** (Box 9E) What kind of tissues can we see with CAT scans that cannot be seen with regular x-rays?

**9.63** (Box 9E) What are the advantages of an nmr scan over a CAT scan?

**9.64** (Box 9F) What is meant by the term "isotope generator"?

**9.65** (Box 9G) Explain how the fusion of hydrogen nuclei to produce a helium nucleus gives off energy.

**9.66** (Box 9H) In the nuclear accident, one of the radioactive nuclei that people are concerned about is iodine-131. This is so because iodine is easily vaporized, can be carried by the winds, and can cause radioactive fallout hundreds, even thousands, of miles away. Why is iodine-131 especially harmful?

## Additional Problems

**9.67** Use Table 9.4 to determine the percentage of annual radiation we receive from
(a) naturally occurring sources
(b) diagnostic medical sources
(c) nuclear power plants

**9.68** You have an old wristwatch that still has radium paint on its dial. Measurement of the radioactivity of the watch shows a beta-ray count of 0.50 counts/s. 1.0 microcurie radiation of this sort produces 1000 mrem/year. How much radiation, in mrem, do you expect from the wristwatch if you wear it for a year?

**9.69** Where does the energy of the atomic bomb come from?

*__9.70__ What is the ground state of a nucleus?

**9.71** Neon-19 and sodium-20 are positron emitters. What are the products in each case?

**9.72** A beta emitter sends an electron out of the nucleus, yet there are no electrons in the nucleus. Explain where the electron comes from.

**9.73** Explain why (a) it is impossible to have a completely pure sample of any radioactive isotope, (b) beta emission of a radioactive isotope creates a new isotope with an atomic number one unit higher than that of the radioactive isotope.

**9.74** When uranium-238 decays it emits alpha particles sequentially. How many alpha particles are emitted during the decay of uranium-238 to radon-222?

**9.75** Californium-244 was first produced in 1950 by bombarding curium-242 with alpha particles. What were the other reaction products?

**9.76** Which of these is the most stable isotope?
(a) $^{14}_{6}C$        (c) $^{10}_{6}C$
(b) $^{12}_{6}C$

*__9.77__ The half-lives of some oxygen isotopes are: oxygen-14 = 71 s; oxygen-15 = 124 s; oxygen-19 = 29 s; and oxygen-20 = 14 s. Oxygen-16 is the stable nonradioactive isotope. Do the half-lives indicate anything about the stability of the other oxygen isotopes?

*__9.78__ Selenium-75 has a half-life of 120.4 days. This means that it would take 602 days (five half-lives) to diminish to 3 percent of the original quantity. Yet this isotope is used for pancreatic scans without any fear that the radioactivity will cause undue harm to the patient. Suggest a possible explanation.

**9.79** Yttrium-90, which emits beta particles, is used in radiotherapy. What is the decay product of yttrium-90?

**9.80** The half-life of nitrogen-16 is 7 seconds. How long does it take for 100 mg of nitrogen-16 to be reduced to 6.25 mg?

**9.81** Which radiation will cause more ionization, x-rays or radar?

# PART II

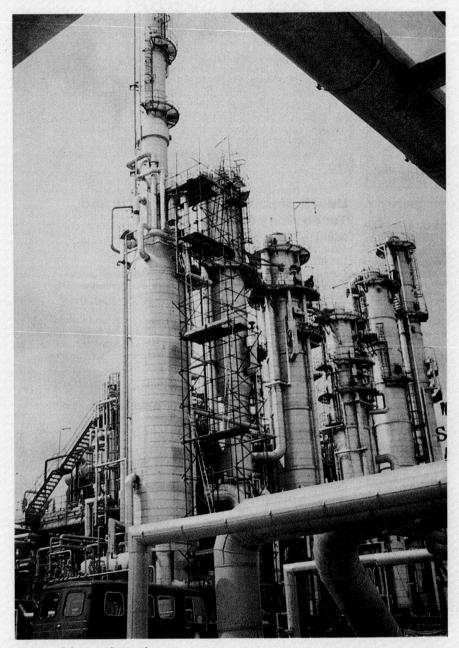

*Photograph by Beverly March.*

# Organic Chemistry

# INTERVIEW

# Carl Djerassi

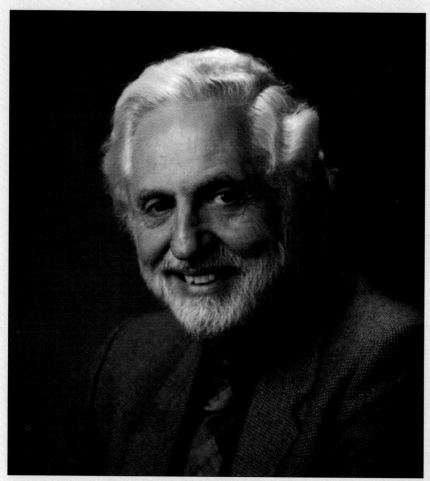

(*Photograph by Beverly March.*)

Carl Djerassi, Professor of Chemistry at Stanford University, is a man of many achievements, in science and in letters. Born in Vienna in 1923, he was forced to leave Europe in 1939 in the face of the Nazi occupation. Although he had not completed high school, he was allowed to enroll at Newark Junior College. Following this he went to Tarkio College and then to Kenyon College, where he graduated summa cum laude in 1942.

Still short of his nineteenth birthday, he joined the CIBA Corporation, where he and Charles Huttrer synthesized pyribenzamine, one of the first two antihistamines—drugs that act to control allergies. He left CIBA for the University of Wisconsin, where he received his Ph.D. in 1945, at the age of 22. For the next seven years he worked first at CIBA (until 1949) and then at Syntex, S. A., a fledgling pharmaceutical corporation then located in Mexico City. Here he was part of the research team that produced the first synthesis of cortisone, an important hormone, and later synthesized norethindrone, the first oral contraceptive and one that is still widely used all over the world.

In 1952 he moved to Wayne University (now Wayne State) in Detroit, as Associate Professor, and in 1959 accepted a Professorship at Stanford University, where he has remained ever since. Both at Wayne and at Stanford he maintained an association with Syntex and served as president of Syntex Research from 1968 to 1972. In 1968 he helped to found the Zoecon Corporation, of which he also became president and which pioneered in developing hormonal methods of insect control that do not rely on conventional insecticides.

Dr. Djerassi has always been involved in steroid chemistry, but his early interest in synthesis turned to the equally important area of the structural determination of naturally occurring steroids, of which there are many thousands. When he began his career, determining the structure of a single

280

naturally occurring steroid was a difficult process that would often take years. As an illustration of the difficulty, it may be mentioned that two German chemists, Heinrich Wieland and Adolph Windaus, were awarded Nobel Prizes in 1927 and 1928, partly for determining the structures of a few steroids, and these structures were later shown (by Wieland) to be wrong!

Beginning approximately in the 1950s instrumental methods of structure determination slowly began to replace the laborious older chemical methods, and Dr. Djerassi was one of the pioneers in developing and using techniques such as optical rotatory dispersion, circular dichroism, and mass spectrometry for steroids and other organic compounds. Owing to the work of Dr. Djerassi and other pioneers, most unknown structures now can be determined in a few days or less.

Dr. Djerassi has had a prolific career, publishing more than 1000 research papers and seven books dealing with the topics mentioned above as well as with alkaloids, antibiotics, lipids, and terpenoids, and with chemical applications of computer artificial intelligence. For this work he has received many awards, including the National Medal of Science (1973), the National Medal of Technology (1991), the first Wolf Prize in Chemistry (1978), the Perkin Medal of the Society for Chemical Industry (1975), The American Chemical Society's Award in Pure Chemistry (1958), the Roussel Prize (1988), the Priestley medal, the highest award given by the American Chemical Society (1992), and many others. In 1978 he was inducted into the National Inventor's Hall of Fame. He has also received a dozen honorary degrees and is a member of the U.S. National Academy of Sciences as well as the Academies of Science of several other countries.

In recent years Dr. Djerassi, without abandoning his interest in chemistry, has turned to literature. He has published a collection of short stories, two novels (Cantor's Dilemma and The Bourbaki Gambit), a collection of poems, and two books of autobiography, one entitled Steroids Made it Possible and the other, intended for a more general audience, called The Pill, Pygmy Chimps, and Degas's Horse. Under the auspices of the Djerassi Foundation, he has established an artist's colony near Woodside, California, that provides residences and studio space for visual, literary, choreographic, and musical artists.

## Why did you choose a career in organic chemistry?

I really didn't choose it; it sort of happened to me. I was the child of two physicians and it was always assumed that I would go to medical school and become a practicing physician. When I arrived in this country at the age of 16 I had taken no chemistry and had not even graduated from high school. Fortunately, I got straight into a Junior College in New Jersey and began a pre-med curriculum that included first-year chemistry. I had a first-rate teacher there, named Nathan Washton, who got me interested in chemistry. After one semester at Tarkio College in Missouri (the alma mater of Wallace Carothers, the inventor of nylon), I went to Kenyon College, a small (at that time) college in Ohio, that had only two faculty members in chemistry. Chemistry classes were very small, but I got a first-class education, and that is where I decided to become a chemist, rather than go to medical school.

## Could you tell us about your work in developing antihistamines?

When I graduated from Kenyon I needed to earn some money, so I got a job as a junior chemist at CIBA Pharmaceutical Corporation, in New Jersey. Just at that time the company became interested in antihistamines, and I was one of only two chemists working on this project. So despite my youth, I was involved in the synthesis of one of the first two antihistamines produced in this country, pyribenzamine. This compound, which was synthesized during the first year after I graduated from college, turned out to make a significant contribution in the treatment of allergies.

## What got you interested in the chemistry of steroids?

While working at CIBA, I began taking classes at NYU and Brooklyn Polytechnic Institute at night, so as to get an advanced degree, but commuting from New Jersey after a day of work was murderous, and, after a year of night school, I decided I would go to graduate school full time. CIBA was involved in steroid projects, and even though I was working on antihistamines, I started reading books on steroids, especially Louis Fieser's Natural Products Related to Phenanthrene, a superb book. This turned me on to steroid chemistry, so when I went full time to the University of Wisconsin, I was prepared to work in this field. Fortunately, Wisconsin had two young Assistant Professors in this area, and I did my Ph.D. with one of them, A. L. Wilds, on the conversion of androgens to estrogens, which at that time was a tough problem.

## How did you get the idea to try to synthesize an oral contraceptive?

We did not set out with that objective. We set out with the objective of developing an orally active progestational hormone; in other words, a compound that would mimic the biological properties of progesterone. At that time progesterone was clinically used for menstrual disorders and infertility, but there were ideas about using it as a contraceptive, because it is progesterone that naturally stops further ovulation after an ovum is fertilized. However, progesterone itself is not active by mouth, and daily injections would be needed. By this time I was Associate Director of Chemical Research at

Syntex Corporation, located in Mexico City. By combining ideas discovered by previous investigators, we set out to synthesize a steroid that would not only be active by mouth, but would also have enhanced progestational activity. This compound was 19-nor-17α-ethynyltestosterone (norethindrone), whose synthesis we completed on 15 October, 1951. It was first tested for menstrual disorders and fertility problems and then as an oral contraceptive. Forty years after its synthesis it is still the active ingredient of about a third of all the oral contraceptives used throughout the world.

## How do you feel about the social impact of your contraceptive work?

On the whole, it was enormously positive, and if I could do it over again, there is no question that I would proceed to do so. By now about 13–14 million women in the United States and 50–60 million in the world use the pill, making it the most widely used method of reversible birth control. Population growth is probably the biggest problem facing us in the world, assuming that the possibility of nuclear warfare is now greatly diminished, and the widespread use of oral contraceptives helps in controlling that. I think the development of these contraceptives is one of the most important contributions that chemistry has made to society. New drugs cure diseases of individuals during their lifetimes, but the use of contraceptives has implications for generations, because if you do not control the production of offspring, you do not control future generations. Furthermore, these compounds have had an enormous impact on women, empowering them to be in control of their own fertility.

## Do you think there will be further research in oral contraception, for example, the development of a pill for men?

I think there will be very little fundamental new research in that area over the next couple of decades, and a pill for men is completely out of the question for the next 15 to 20 years. The reason is that the pharmaceutical industry has turned its back on this field for a number of very complex reasons, not only in the United States, but in other industrialized countries as well. The only really new approach in chemical birth control in the last 25 years is the French "morning-after pill," RU486 —at present used almost exclusively as an early abortifacient, although it can also be employed as a method of birth control.

## Tell us about your work on insect control.

Conceptually there was a relationship to our work on oral contraceptives, because in a way you could say that steroid oral contraceptives were true biorational methods of human birth control, since progesterone—our conceptional lead compound—is really nature's contraceptive. That was a model on which insect control could be based. At this time (the late 1960s) I was in charge of research at Syntex in addition to being Professor at Stanford University. Governments and the public realized that conventional methods of insect control—largely spraying with chlorinated hydrocarbons such as DDT—were damaging the environment. DDT and similar compounds were now being banned, and a new approach was needed. We formed a new company called Zoecon. Two insect hormones—the moulting hormone ecdysone, which is a complicated steroid, and the juvenile hormone, based on a much simpler sesquiterpene skeleton—had been discovered in the 1960s, and we decided to focus on the latter. Insects pass through a juvenile stage controlled by the juvenile hormone, whose production is later shut off by another hormone so that the insect

can then mature. Our biorational approach was to synthesize an artificial juvenile hormone which would continue to be applied to immature insects, so that the insect would never reach the stage at which it could reproduce. This turned out to be a new biorational approach to controlling mosquitoes, fleas, cockroaches, and other insects that do their damage as adults, and it was approved by the Environmental Protection Agency for public use.

## You have been a leader in both the synthetic and structure determination areas in organic chemistry. Which of these activities gave you more satisfaction?

In my early days—the 1950s and 1960s—undoubtedly structure determination was more satisfying, especially because I was involved in the development of new physical methods for structure determination. However, these methods have now made structural determination much more routine and automatic. For example, mass spectrometry enables us to identify compounds on a micro scale, which was unheard of before.

## When you embarked on a research project, what proportion of your driving force was provided by (a) pure intellectual curiosity, (b) consideration of potential benefits to society, and (c) material rewards?

In my own work, material rewards never played any initial role, because they always came much later. To a large extent it was initially pure intellectual curiosity, but very soon I started to look for the potential benefits to society. My interest has always been in the biological areas, and I was never interested in war-related research. I never had the slightest question about the societal appropriateness of the work I was doing.

**What is your opinion about the interplay between academia and industry in developing new inventions, including chemical inventions?**

I am a great believer in this, and have always been connected with a University and with chemical companies. This made me a much better professor, because I became aware of the many steps needed to take a laboratory discovery up to practical realization. Conversely, I was a much better industrial research director because of what I learned in the university. For example, should scientists be concerned with societal implications of their work? Obviously yes, but it is not easy for a scientist who stays only in the ivory tower. You ought to have a responsibility for taking your work a step further. There should, however, be guidelines to prevent actual or potential conflicts of interest.

**Do you feel that scientists should or should not be advocates for social or political positions?**

Scientists should be advocates, in that every intelligent person should be, provided he or she is well informed. Just because you are a scientist does not mean that you are well informed on any particular issue, but when scientists deal with their own specialties, of course they should. Also, scientists should be interested in the societal implications of their work, just as non-scientists should be informed about certain areas of science, simply because they have such an enormous impact on their daily lives.

**What advice would you give to a student who wants to become a medical research worker?**

Learn chemistry, especially organic chemistry. You don't have to become a chemist, but in medical areas a knowledge of organic chemistry is indispensable for any area of medicine, because we are learning more and more how the body operates on a molecular level. A parallel would be that a person wanting to be a physicist must study mathematics.

**You have trained many students in chemical research. On the whole, have they made many contributions to society after they left your laboratories?**

Yes. I've trained somewhere between three and four hundred graduate students and postdoctoral fellows in universities and many others in industry as well. Training these students is probably the biggest professional contribution that one makes. These are very intimate relationships, comparable to those between parents and children. The mentor becomes a role model for the student.

**If you had to do it all over again, would you become a scientist?**

Yes, but not necessarily the same scientist. If I were starting in the 1990s, I would not become an organic chemist. I would still learn a lot of organic chemistry, but I would now go into a more biologically oriented discipline.

**Do you think that the adverse conditions of your youth—for example, the Nazi upheaval and the necessity for taking your schooling in several languages— led you to become an overachiever?**

It's likely, though I could never repeat that experiment. I had been an average student when I had to leave Vienna because of the Nazi takeover. I knew that I would be going to the United States, but had to wait a year and a half for a visa, so I went to an American school in Bulgaria to learn English. There I became a superb student, because I knew that grades would count in continuing my education in America. So it seems that the adverse conditions did have an impact.

---

Interview conducted by Dr. Jerry March.

# CHAPTER 10

*A petroleum refinery. (SuperStock, Inc.)*

# Organic Chemistry.
# Alkanes

## 10.1 INTRODUCTION

**Organic chemistry** is the **study of compounds that contain carbon.** What is there about the element carbon that justifies our making it the basis of a whole branch of chemistry, whereas compounds that may contain any of the other elements, but not carbon, are all classified as inorganic compounds? One answer is that organic and inorganic compounds, in general, differ in many properties, some of which are shown in Table 10.1. Most of these differences stem from the fact that all the chemical bonds in most organic compounds are covalent, while many inorganic compounds have ionic bonding. Figure 10.1 shows some organic and inorganic compounds.

Organic and inorganic compounds differ in properties because they differ in structure, not because they obey different natural laws. There is only one set of natural laws for all compounds.

**Figure 10.1**
(a) Some organic compounds. Solids: on the left, *meta*-nitroaniline; on the right, aspartic acid. Liquids: cyclohexane and methanol. (b) Some inorganic compounds. Clockwise from the top: copper(II) nitrate, ammonium sulfate, potassium dichromate, and iron(III) sulfate. (Photographs by Beverly March.)

(b)

(a)

It is important to note that there are many exceptions to the statements in Table 10.1—some organic compounds behave like inorganic compounds, and vice versa—but still, the generalizations in Table 10.1 are true for the vast majority of compounds in both groups.

But how did the name "organic" come about? In the early days of chemistry, scientists thought that there were two classes of compounds. One class was produced by living organisms—these compounds they called "organic." The other class, found in minerals or rocks, they called "inorganic." They also believed that chemists working in laboratories could not synthesize any organic compound starting only from inorganic compounds. They thought that a "vital force," possessed only by living organisms, was necessary to produce organic compounds. This kind of theory is very easy to disprove if indeed it is wrong. All it takes is one experiment in which a chemist makes an organic compound

| TABLE 10.1   A Comparison of Properties of Organic and Inorganic Compounds | |
|---|---|
| **Organic Compounds** | **Inorganic Compounds** |
| Bonding is almost entirely covalent. | Many compounds have ionic bonds. |
| Compounds may be gases, liquids, or solids with low melting points (less than 360°C). | Mostly high-melting solids (above 360°C). |
| Mostly insoluble in water. | Many soluble in water. |
| Mostly soluble in organic solvents such as gasoline, benzene, carbon tetrachloride. | Almost entirely insoluble in organic solvents. |
| Solutions in water or any other solvent do not conduct electricity. | Water solutions conduct electricity. |
| Almost all burn. | Very few burn. |
| Reactions are usually slow. | Reactions are often very fast. |

starting only from inorganic ones. Such an experiment was carried out by Friedrich Wöhler (1800–1882) in 1828. He heated an aqueous solution of two inorganic compounds, ammonium chloride and silver cyanate, and to his surprise obtained urea,

$$NH_4Cl + AgNCO \xrightarrow{heat} NH_2CONH_2 + AgCl$$
$$\text{Urea}$$

which definitely fit the old definition of "organic" because it had been isolated from human urine.

This single experiment of Wöhler's was enough to disprove the "doctrine of vital force," though it took several years and a number of additional experiments for the entire chemical world to accept the fact. This meant that the words "organic" and "inorganic" no longer had real definitions, since, for example, urea could be obtained from both sources. A few years later Friedrich Kekulé (1829–1896) assigned the modern definition—**organic compounds are those containing carbon**—and this has been accepted ever since. The chemistry of living organisms is no longer called organic chemistry; today we call that branch of science **biochemistry.**

In this experiment, Wöhler was trying to prepare ammonium cyanate, then an unknown compound. Shortly afterward he did prepare ammonium cyanate, which yielded urea when heated:

$$NH_4NCO \xrightarrow{heat} NH_2CONH_2$$

Kekulé put forth the structural theory of organic chemistry (Sec. 10.3) and also solved the difficult question of the structure of benzene (Sec. 11.9).

## 10.2 SOURCES OF ORGANIC COMPOUNDS

You might suppose that, with 108 elements other than carbon to choose from, there would be many more inorganic compounds known than organic, but in fact the opposite is true. There are more than five million known organic compounds, but only about 200 000 to 300 000 known inorganic compounds. Furthermore, more than 100 000 new organic compounds are becoming known every year. This is another reason a whole branch of chemistry is devoted to carbon compounds.

What is there about carbon that permits the formation of so many compounds? The answer is

**1.** Carbon atoms form stable bonds with other carbon atoms, so that both long and short chains of carbon atoms, and even whole networks, are possible.
**2.** Carbon atoms also form stable bonds with certain other atoms, including hydrogen, oxygen, nitrogen, the halogens, and sulfur.
**3.** A carbon atom forms four bonds, and these can be made up in many ways, so that many combinations and arrangements of atoms are possible.

We have used the expression "known compound." What does this mean? A compound becomes "known" when a chemist obtains that compound, determines its formula, and measures some of its properties. He or she then publishes the results of the work in a chemical journal, describing the properties and how the compound was obtained. The compound is then "known." In many cases, the chemist then uses the compound to make some other compound, or it decomposes or is discarded. In any event, it no longer exists. Still, it remains "known" because any chemist can read that journal and obtain the compound once more by following the original procedure.

There are two principal ways of obtaining known as well as new organic compounds: isolation from nature and synthesis.

It is likely that only a few hundred thousand of the millions of known compounds actually exist at any one time.

## Isolation from Nature

Living organisms are chemical factories. Every plant and animal makes hundreds of organic compounds; even microorganisms such as bacteria synthesize organic compounds. This process is called **biosynthesis.** One way to get organic compounds is by extraction from a biological source. In this book we will meet many compounds that have been obtained in this way. Some important examples are ethyl alcohol, acetic acid, cholesterol, cane sugar (sucrose), nicotine, quinine, and ATP (adenosine triphosphate). Many thousands of organic compounds have been and still are being extracted from nature. The process is a long way from being finished. The number of animal and especially plant species is so large that in many cases nobody has even tried to extract the compounds they contain. Besides this, nature also supplies us with two other important sources of organic compounds: petroleum (see Box 10C) and coal.

Coal and petroleum were formed from the remains of organisms that lived millions of years ago.

## Synthesis

Ever since Wöhler synthesized urea, organic chemists have been constantly making the same compounds that nature also makes. In recent years the methods for doing this have been so greatly improved that there are few natural organic compounds, no matter how complicated, that cannot be synthesized by chemists (Fig. 10.2). It must be emphasized that *the compounds made by chemists are identical to those in nature.* The ethyl alcohol made by a chemist is exactly the same as the ethyl alcohol we get by distilling wine. The molecules are the same and so are all the properties; thus there is no way that anyone can tell whether a given sample of ethyl alcohol was made by a chemist or obtained from nature.

Organic chemists have not rested with duplicating nature's compounds, however. They also synthesize compounds not found in nature. In fact, the majority of the more than 5 000 000 known organic compounds are purely synthetic and do not exist in living organisms. Many of them are used only for

**Figure 10.2**
A modern laboratory for synthetic organic chemistry. (Photograph by Beverly March.)

research, but a large number find more practical use. A few examples are acetylene, carbon tetrachloride, diethyl ether, DDT, and TNT. A very important area in which synthetic organic compounds find much use is drugs. Many modern drugs—for example, aspirin, Valium, Vasotec, Prozac, Zantac, Cardizem, Inderal, Lasix, sulfadiazine, and Enovid—are synthetic organic compounds not found in nature.

## 10.3 STRUCTURES OF ORGANIC COMPOUNDS

The molecular formula for ethyl alcohol is $C_2H_6O$. As we learned in Section 3.4, a *molecular formula* shows which atoms, and how many of each, are present in a molecule. A *structural formula* shows more than this: It shows not only all the atoms present in the molecule, but also all the bonds that connect the atoms to each other. The structural formula for ethyl alcohol is

$$\begin{array}{ccc} & H & H \\ & | & | \\ H- & C- & C-O-H \\ & | & | \\ & H & H \end{array}$$

Remember that a straight line indicates a pair of electrons in a covalent bond.

However, ethyl alcohol is not the only compound whose molecular formula is $C_2H_6O$. An entirely different compound, dimethyl ether, is also $C_2H_6O$. The structural formula for this compound is

$$\begin{array}{ccc} H & & H \\ | & & | \\ H-C- & O- & C-H \\ | & & | \\ H & & H \end{array}$$

**Compounds that have the same molecular formula but different structural formulas** are called **isomers.** Ethyl alcohol and dimethyl ether, both being $C_2H_6O$, are therefore isomers. Isomers are not at all unusual; in fact, they are very common. There are three isomers of $C_5H_{12}$, five of $C_4H_8$, and in many cases dozens and even hundreds of different isomers for a given molecular formula. The various isomers sharing a given molecular formula all have different properties because **the properties of a compound depend on its structural formula and not on its molecular formula.** That is why in organic chemistry we almost always use structural rather than molecular formulas. For example, the isomers we just mentioned, dimethyl ether and ethyl alcohol, have entirely different properties (Table 10.2). If we just said "$C_2H_6O$" no one would know which compound we meant.

This kind of isomerism is called **structural isomerism.** We will meet other kinds in Sections 10.9 and 16.2.

Although organic chemists use structural formulas all the time, there are times when even these are not enough to let us understand the behavior of organic compounds. We need pictures, or better yet, three-dimensional models, that show the shapes of organic molecules. Such models for ethyl alcohol and dimethyl ether are shown in Figure 10.3. Models of this type are very useful in the study of organic chemistry and biochemistry. Of course, the proper angles (approximately) must be used in these models. In Section 3.6 we saw that carbon points its four bonds to the corners of a tetrahedron and that the bond angles are about 110°.

Model sets are often sold in college book stores, but you can make your own by using marshmallows or gumdrops for atoms and toothpicks or matchsticks for bonds.

| TABLE 10.2 Some Properties of Ethyl Alcohol and Dimethyl Ether | | |
|---|---|---|
| | $\begin{array}{c} H \quad H \\ | \quad | \\ H-C-C-O-H \\ | \quad | \\ H \quad H \end{array}$ | $\begin{array}{c} H \quad\quad H \\ | \quad\quad | \\ H-C-O-C-H \\ | \quad\quad | \\ H \quad\quad H \end{array}$ |
| Property | Ethyl Alcohol | Dimethyl Ether |
| Physical state at room temperature | Liquid | Gas |
| Boiling point (°C) | 78 | −23 |
| Melting point (°C) | −117 | −138 |
| Reacts with sodium | Yes | No |
| Poisonous (in moderate amounts) | No | Yes |
| Anesthetic (in small amounts) | No | Yes |

Nearly all of the millions of known organic compounds are made up of only a few elements. More than 95 percent of them contain no elements other than C, H, O, N, S, and the four halogens. These elements form the following numbers of bonds (see Sec. 3.4):

| | | | | |
|---|---|---|---|---|
| H  1 | C  4 | N  3 | O  2 | F  1 |
| | | | S  2 | Cl  1 |
| | | | | Br  1 |
| | | | | I  1 |

This statement does not hold for atoms with charges. For example, as we already saw, the nitrogen in the ammonium ion, $NH_4^+$, forms four bonds.

These numbers are very important, so you should memorize them as soon as possible. One reason that so many organic compounds can be made from so few elements is that stable compounds can be formed from almost any combination of these elements, as long as each atom has the correct number of bonds. This means that if we can draw the structural formula of a molecule, it is usually possible to go to the laboratory and make it (chances are it has been made already).

**Figure 10.3**
Models of (a) ethyl alcohol and (b) dimethyl ether.

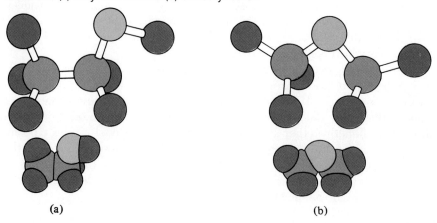

(a)                    (b)

**TABLE 10.3    Possible Double and Triple Bonds**

$$-\overset{|}{C}=\overset{|}{C}- \qquad\qquad -C\equiv C-$$

$$-\overset{|}{C}=N- \qquad\qquad -C\equiv N$$

$$-\overset{|}{C}=O$$

$$\left.\begin{array}{l} -\overset{|}{C}=S \\[4pt] -N=N- \end{array}\right\} \text{Less important than the others}$$

Remember (Sec. 3.4) that carbon and nitrogen can form double and triple bonds as well as single bonds and that oxygen can form both double and single bonds (Table 10.3). All these possibilities give rise to a vast number of possible organic compounds, for example,

Methane    Ethene    Formaldehyde    Hydrogen cyanide

Carbon dioxide    Propyne    Methylamine    Methanol

## 10.4  HYBRID ORBITALS

The concept of Lewis structures and the VSEPR model (Sec. 3.6) give us an insight into chemical bonding and allow us to predict the shapes of molecules. But these models are insufficient to predict certain important characteristics of organic molecules. For example, the Lewis model cannot explain why a carbon-carbon double bond is more reactive than a carbon-carbon single bond. Neither is it clear why a carbon atom forms four equal bonds with four hydrogens in methane ($CH_4$) when its valence electrons are in different atomic orbitals ($2s^2\, 2p^2$).

In order to explain these and other properties, theoretical chemists came up with the idea of **hybrid orbitals.** Let's take the case of methane, which is made of four C—H bonds, all of which have the same properties. The Lewis structure is simple:

$$\text{H}:\overset{\displaystyle ..}{\underset{\displaystyle ..}{\text{C}}}:\text{H}$$

The four covalent bonds are formed by overlap between the atomic orbitals of the carbon atom and the atomic orbitals ($1s$) of four hydrogen atoms. But the valence electrons in a carbon atom do not occupy four orbitals. Two electrons are in an $s$ orbital and the other two are in $p$ orbitals that are 90° to each other (see Fig. 2.11).

We can schematically show it like this:

$$\boxed{\uparrow\downarrow} \qquad \boxed{\uparrow} \quad \boxed{\uparrow} \quad \boxed{\phantom{\uparrow}}$$
$$\;2s \qquad\quad 2p_x \;\; 2p_y \;\; 2p_z$$

From this picture we should expect carbon to form only two bonds. The two bonds should form by overlap of the two $2p$ electrons with orbitals from outside atoms. The $2s$ orbital, being filled, should not form any bonds at all. But we know that a carbon atom forms four bonds, and that the formula of methane is $CH_4$.

What happens is that the carbon atom promotes one of its electrons from the $2s$ orbital into the empty $2p_z$ orbital:

$$\boxed{\uparrow} \qquad \boxed{\uparrow} \quad \boxed{\uparrow} \quad \boxed{\uparrow}$$
$$\;2s \qquad\; 2p_x \;\; 2p_y \;\; 2p_z$$

This promotion requires the expense of a certain amount of energy, but the ability to form four rather than two bonds returns the energy investment many-fold.

We can now explain why all four bonds in methane are equal and also why the bond angles are 109.5°. We say that the valence orbitals of a carbon atom are neither $s$ nor $p$ but a *hybrid* or *mixture* of orbitals. Since three $p$ orbitals are mixed with one $s$ orbital, we call the hybrid orbitals **$sp^3$,** meaning that each of them has one-fourth $s$ character and three-fourths $p$ character. An $sp^3$ orbital is neither spherical nor dumbbell-shaped, but has a mixture of the two shapes (Fig. 10.4).

Each of the four hybrid $sp^3$ orbitals contains one electron, so that four equivalent C—H bonds are formed simply by the overlap between the four carbon $sp^3$ orbitals and four hydrogen $s$ orbitals. This overlap allows the sharing of two electrons between a carbon and a hydrogen atom. The bond formed between them is a $\sigma$ bond.

The $sp^3$ orbitals of a carbon atom can similarly overlap with other orbitals such as the $p$ orbital of F or Cl or indeed any $s$ or $p$ orbital of any atom, or even the $sp^3$ orbital of another carbon atom.

We previously encountered diagrams of this type in Section 2.6.

No stable compound with the formula $CH_2$ has ever been found.

**Figure 10.4**
The four $sp^3$ orbitals.

The four equivalent $sp^3$ orbitals get as far away from each other as they can. As the VSEPR theory showed us, this means that the bond angles will be 109.5°.

---

**EXAMPLE**

The compound chloroethane can be represented as follows:

```
   H   H
   |   |
H— C — C —Cl
   |   |
   H   H
```
Structural formula

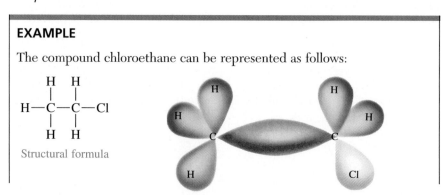

> Each of the molecular orbitals shown in color is made up by the overlap of an $sp^3$ orbital from a carbon atom with an $s$, $p$, or $sp^3$ orbital of another atom.

In fact we can assume that in any case where a carbon atom bonds with four other atoms, it uses $sp^3$ hybrid orbitals.

**PROBLEM 10.1** Describe the orbitals that make up all the bonds in (a) $CH_3F$ (b) $CH_3$—$CH_2$—F.

## 10.5 HYDROCARBONS

We begin our study of organic compounds with the **hydrocarbons,** which are **compounds that contain only carbon and hydrogen.** Even with only two elements, there are many thousands of known hydrocarbons and a much larger number of possible ones that have not yet been synthesized.

There are several families of hydrocarbons:

**Alkanes** have only single bonds.

**Alkenes** have C=C double bonds.

**Alkynes** have C≡C triple bonds.

**Aromatic hydrocarbons** have benzene rings (Sec. 11.9).

We discuss alkanes in this chapter and reserve the other types for Chapter 11.

## 10.6 ALKANES

**Alkanes** are **hydrocarbons with only single bonds.** The simplest one, *methane*, has only one carbon atom:

$$-\overset{\displaystyle |}{\underset{\displaystyle |}{C}}-$$

Since carbon forms four bonds and hydrogen forms one, methane must have this formula:

$$H-\overset{\displaystyle H}{\underset{\displaystyle H}{C}}-H \qquad CH_4$$

Methane

In a similar way, we can construct the next two alkanes, ethane and propane:

$$H-\overset{\displaystyle H}{\underset{\displaystyle H}{C}}-\overset{\displaystyle H}{\underset{\displaystyle H}{C}}-H \qquad C_2H_6 \qquad\qquad H-\overset{\displaystyle H}{\underset{\displaystyle H}{C}}-\overset{\displaystyle H}{\underset{\displaystyle H}{C}}-\overset{\displaystyle H}{\underset{\displaystyle H}{C}}-H \qquad C_3H_8$$

Ethane                                   Propane

**Figure 10.5**
This camper will use the propane tank for fuel. (FourByFive.)

All three of these compounds are gases. Methane, also called *marsh gas,* is the main constituent of *natural gas,* which is used as a fuel in many homes and factories. Propane is also used as a fuel (Fig. 10.5).

Propane is the fuel sold as "bottled gas."

When written out as just shown, these formulas occupy a lot of space. Therefore, chemists generally write them in a shorthand way:

Methane   $CH_4$

Ethane    $CH_3{-}CH_3$    or    $CH_3CH_3$

Propane   $CH_3{-}CH_2{-}CH_3$    or    $CH_3CH_2CH_3$

Occasionally the hydrogen atoms are written before the carbons, $H_3C{-}CH_3$, to emphasize the $C{-}C$ bond.

These shorthand formulas are called **condensed structural formulas.** Though they do not explicitly show all the bonds (or even any of the bonds), we must not forget that the bonds are there. Of course, as we learned in Section 10.3, none of these formulas show the real shape of the molecules. For that we need models.

Figure 10.6 shows models for the first three alkanes, of two different types. As we could predict from the fact that carbon atoms use $sp^3$ hybrid orbitals to form their bonds (see Fig. 10.4), the carbon atoms in all alkanes point their four bonds to the corners of a tetrahedron, so that all angles are approximately 110°. This means that when we write, for example,

$$CH_3{-}CH_2{-}CH_3$$

for the structure of propane (as chemists often do), we must remember that, in the real molecule, the three carbons are not really in a straight line, as the formula suggests. They are actually connected by an angle of about 110°:

$$H_3C \overset{\overset{\textstyle H_2}{\textstyle C}}{\diagdown \underset{110°}{\phantom{x}} \diagup} CH_3$$

We write the condensed formula in a straight line because it is convenient to do so, in spite of the fact that it is geometrically inaccurate.

**Figure 10.6**
Models of methane, ethane, and propane.

As you might suppose, it is possible to make carbon chains containing many more than three carbon atoms. In fact, alkanes of this type can contain any number of carbon atoms. Table 10.4 lists the name and condensed formula for all those up to ten carbons. It is important that you memorize these names because they are the basis for naming most of the other organic compounds. Note that the names of all alkanes end in "-ane."

In all the alkanes in Table 10.4 the carbons form a continuous chain; in other words, there is no **branching.** Such alkanes are called *normal* alkanes,

In Section 11.7 we shall see that such chains can be many thousands of carbon atoms long and still be perfectly stable.

**TABLE 10.4 Molecular and Condensed Structural Formulas, and Boiling and Melting Points, for the First Ten Normal Alkanes**

| Molecular Formula[a] | Condensed Structural Formula | Name | Boiling Point (°C) | Melting Point (°C) |
|---|---|---|---|---|
| $CH_4$ | $CH_4$ | Methane | −164 | −182 |
| $C_2H_6$ | $CH_3CH_3$ | Ethane | −89 | −183 |
| $C_3H_8$ | $CH_3CH_2CH_3$ | Propane | −42 | −190 |
| $C_4H_{10}$ | $CH_3CH_2CH_2CH_3$ | Butane | 0 | −138 |
| $C_5H_{12}$ | $CH_3CH_2CH_2CH_2CH_3$ | Pentane | 36 | −130 |
| $C_6H_{14}$ | $CH_3CH_2CH_2CH_2CH_2CH_3$ | Hexane | 69 | −95 |
| $C_7H_{16}$ | $CH_3CH_2CH_2CH_2CH_2CH_2CH_3$ | Heptane | 98 | −91 |
| $C_8H_{18}$ | $CH_3CH_2CH_2CH_2CH_2CH_2CH_2CH_3$ | Octane | 126 | −57 |
| $C_9H_{20}$ | $CH_3CH_2CH_2CH_2CH_2CH_2CH_2CH_2CH_3$ | Nonane | 151 | −51 |
| $C_{10}H_{22}$ | $CH_3CH_2CH_2CH_2CH_2CH_2CH_2CH_2CH_2CH_3$ | Decane | 174 | −30 |

[a] The general formula for all alkanes, normal or branched, is $C_nH_{2n+2}$, where $n$ = the number of carbons.

abbreviated *n*-alkanes. Branched alkanes are also possible. For example, consider this compound:

$$CH_3-CH-CH_3 \qquad \text{Isobutane}$$
$$\qquad\quad | $$
$$\qquad\quad CH_3$$

There are four carbons and ten hydrogens in this molecule, so the molecular formula is $C_4H_{10}$. This is the same as that of butane (Table 10.4), which means, of course, that butane and this new compound are isomers. Because of this, we call it *isobutane*. As in other cases of isomerism, butane and isobutane have different structures and different properties.

Notice that isobutane has a carbon atom (the one in the middle) that is connected to *three* other carbon atoms, but butane has no such carbon. We define four types of carbon atoms:

> A **quaternary carbon** is attached to four other carbons.
>
> A **tertiary carbon** is attached to three other carbons.
>
> A **secondary carbon** is attached to two other carbons.
>
> A **primary carbon** is attached to one other carbon or to no other carbons.

Why can't there be a quintary carbon?

Note that normal alkanes have only primary or secondary carbons and branched alkanes have at least one tertiary or quaternary carbon.

We have seen that there are two alkanes containing four carbons: butane and isobutane. What about five-carbon alkanes? Of course, there is one normal compound, pentane (there can never be more than one normal alkane of any particular chain length), but there are *two* branched compounds:

$$CH_3-CH-CH_2-CH_3 \quad \text{Tertiary carbon} \qquad CH_3-C-CH_3 \quad \text{Quaternary carbon}$$

Isopentane · Neopentane

If we go to six carbons, we find that, besides *normal*-hexane, there are four branched alkanes:

A · B · C · D

for a total of five isomers (all having the molecular formula $C_6H_{14}$). If we were to continue, we would find the number of isomers growing steadily. There are

75 alkanes with 10 carbons (all $C_{10}H_{22}$), 366 319 with 20 carbons (all $C_{20}H_{42}$), and more than four billion with 30 carbons (all $C_{30}H_{62}$). Only a few of this vast number of alkanes are known compounds, but almost all of them could be made fairly easily, if we really wanted to. Why haven't we done it? Simply because there are more important things to do. We don't generally synthesize new compounds without a good reason.

There are some important points to notice about the formulas we just gave. Take a look at the formula marked **A.** First we note that **A** is not the same as **B.** Both have a continuous chain of five carbons (the **main chain**) and a —$CH_3$ attached to the main chain (this —$CH_3$ constitutes a **side chain**), but in **A** the side chain is attached to the *second* carbon of the main chain, while in **B** it is attached to the *third* carbon of the main chain. This means that **A** and **B** are different compounds. Another way we can see the difference is to look at the tertiary carbons (both **A** and **B** have one tertiary carbon). In **A** the tertiary carbon is connected to *two* —$CH_3$ groups and one 3-carbon chain. In **B** the tertiary carbon is connected to *one* —$CH_3$ and two 2-carbon chains.

Now let us look once more at compound **A.** We wrote it like this:

$$CH_3—\underset{\underset{\displaystyle CH_3}{|}}{CH}—CH_2—CH_2—CH_3$$

A

that is, a five-carbon main chain with an extra —$CH_3$ on the second carbon of the chain, pointing down. But we could have shown it pointing up, or on the fourth carbon (up or down), or we could even have twisted the chain:

$$\underset{A}{\overset{\overset{\displaystyle CH_3}{|}}{CH_3—CH—CH_2—CH_2—CH_3}} \qquad \underset{A}{\overset{\overset{\displaystyle CH_3}{|}}{CH_3—CH_2—CH_2—CH—CH_3}}$$

$$\underset{A}{\underset{\underset{\displaystyle CH_3 \quad CH_2—CH_3}{| \quad\quad |}}{CH_3—CH—CH_2}} \qquad \underset{A}{\underset{\underset{\displaystyle CH_3}{|}}{CH_3—CH_2—CH_2—CH—CH_3}}$$

$$\underset{A}{\overset{\overset{\displaystyle CH_3 \qquad CH_3}{| \qquad\quad |}}{CH_2—CH_2—CH—CH_3}} \qquad \underset{A}{\underset{\underset{\displaystyle CH_3—CH_2}{|}}{\underset{\underset{\displaystyle CH_2}{|}}{\overset{\overset{\displaystyle CH_3}{|}}{CH—CH_3}}}}$$

All of these are equally correct ways of writing the *same compound* (2-methylpentane) because they all show the *same connections*. In all these formulas there is one tertiary carbon connected to two —$CH_3$ groups and one 3-carbon chain. Looking at it another way, in all of them there is a main chain of five carbons and a —$CH_3$ side chain connected to the second carbon. Of course, the real molecule does not look like any of these. The model does get much closer, though; at least the angles are correct (Fig. 10.7).

**Figure 10.7**
Some of the shapes of the 2-methylpentane molecule (**A** in the text). The curved arrows show the rotations that change one shape to another.

As Figure 10.7 shows, the atoms in **A** are free to rotate around the bonds, so that the molecule may assume any of a number of shapes (these shapes are called **conformations**), though the 110° angles are present in all of them. There is *free rotation* about the C—C bonds of all the alkanes, as shown in Figure 10.8.

## 10.7   IUPAC NOMENCLATURE OF ALKANES AND ALKYL HALIDES

The IUPAC system for naming organic compounds was made official in 1892.

How do we name all the compounds we just met in Section 10.6? The International Union of Pure and Applied Chemistry (IUPAC) is an organization that, among other things, issues rules for naming chemical compounds, so that all the chemists of the world can use the same names. Most people do use the IUPAC names for most compounds, but unfortunately some compounds are still called by their "common" names, which are not part of the IUPAC system. Since it is important for students to know the names that people actually use, we will be using some of these common names in this book. A major advantage of the

**Figure 10.8**
Free rotation in the propane molecule. Three configurations are shown.

|  |  |  |
|---|---|---|
| (a) | (b) | (c) |

Back to position (a)

IUPAC names is that they are *systematic*. This means that a person can derive the name by looking at the structural formula, and vice versa. It is not necessary to memorize a different name for every structure. Common names, which are often shorter, do not have this advantage.

An important part of the IUPAC naming system is the idea of the **group.** If we take a hydrogen away from methane, we are left with a $CH_3$—.

$$ \underset{\begin{array}{c} | \\ H \end{array}}{\overset{\begin{array}{c} H \\ | \end{array}}{H-\underset{|}{\overset{|}{C}}-H}} \xrightarrow{-H} \underset{\begin{array}{c} | \\ H \end{array}}{\overset{\begin{array}{c} H \\ | \end{array}}{H-\underset{|}{\overset{|}{C}}-}} $$

Methyl group

This is called the **methyl group.** Note that we changed the ending of methane from "-ane" to "-yl." In a similar way we can have the **ethyl group,** the **propyl group,** and so on:

$$ CH_3-CH_2- \qquad \text{Ethyl group} $$
$$ CH_3-CH_2-CH_2- \qquad \text{Propyl group} $$

The propyl group is also called *n*-propyl.

All of these are called **alkyl groups** because they are derived from **alk**anes. Alkyl groups are not independent compounds. They do not exist by themselves. They are *parts* of molecules, not whole molecules themselves. We are now ready to give the IUPAC rules for naming alkanes and alkyl halides.

> **1.** Find the longest continuous chain of carbon atoms. Assign a parent name from Table 10.4 based on this number.

As an example, let us try this on our compound **A:**

$$ CH_3-\underset{\underset{CH_3}{|}}{CH}-CH_2-CH_2-CH_3 $$

The longest continuous chain has *five* carbons, so the parent name (from Table 10.4) is *pentane.*

> **2.** Find whatever groups are *not* part of the longest continuous chain. Name these as prefixes.

If we look at compound **A,**

$$ CH_3-\underset{\underset{CH_3}{|}}{CH}-CH_2-CH_2-CH_3 $$

we see that the only group that is not part of the chain is $CH_3$—, the methyl group. We therefore put *methyl* in front of pentane to get methylpentane.

We are almost at the end of the naming process, but not quite. If we apply the above two rules to compound **B** on page 296, we also get methylpentane (try it). Since we can't have the same name for two different compounds, we need another rule:

**3.** Assign numbers to groups by counting from one end of the chain. A chain has two ends, and the end we start from is the one that gives the lowest possible numbers to the groups.

$$\overset{1}{CH_3}-\overset{2}{CH}-\overset{3}{CH_2}-\overset{4}{CH_2}-\overset{5}{CH_3} \qquad \text{2-Methylpentane}$$
$$| \atop CH_3$$

A

If we count from left to right, we arrive at the number 2 for the position of the methyl group. Our final name for compound **A** is 2-methylpentane. If we had begun at the other end we would have said 4-methylpentane, but this is not a correct name because 2 is lower than 4. It doesn't matter whether we begin at the left or the right. *All that matters is that we come out with the lowest numbers.* If we use these rules to name the other formulas on page 297 (try it), we find that the name 2-methylpentane applies to all of them, which is another indication that they are all different ways of drawing the same compound. As for compound **B**,

$$\overset{1}{CH_3}-\overset{2}{CH_2}-\overset{3}{CH}-\overset{4}{CH_2}-\overset{5}{CH_3} \qquad \text{3-Methylpentane}$$
$$| \atop CH_3$$

B

this is 3-methylpentane no matter which end we start from.

---

**PROBLEM 10.2**   Name the following compounds by the IUPAC system:

(a)  $CH_3-CH_2-CH_2-CH_3$

(b)  $CH_3-CH_2-CH-CH_3$
$\phantom{(b)\ CH_3-CH_2-C}| \atop \phantom{(b)\ CH_3-CH_2-}CH_3$

(c)  $CH_3-CH_2-CH_2-CH_2-CH-CH_2-CH_3$
$\phantom{(c)\ CH_3-CH_2-CH_2-CH_2-C}| \atop \phantom{(c)\ CH_3-CH_2-CH_2-CH_2-}CH_2$
$\phantom{(c)\ CH_3-CH_2-CH_2-CH_2-C}| \atop \phantom{(c)\ CH_3-CH_2-CH_2-CH_2-}CH_3$

---

Let us now look at some other cases.

**4.** If there are two or more identical groups, we use these prefixes:

| Number of Identical Groups | Prefix | Number of Identical Groups | Prefix |
|---|---|---|---|
| 2 | di- | 5 | penta- |
| 3 | tri- | 6 | hexa- |
| 4 | tetra- | | |

and so on. For example, let's name compounds **C** and **D**:

$$CH_3-CH_2-\underset{\underset{\displaystyle CH_3}{|}}{\overset{\overset{\displaystyle CH_3}{|}}{C}}-CH_3 \qquad CH_3-\underset{\underset{\displaystyle CH_3}{|}}{CH}-\underset{\underset{\displaystyle CH_3}{|}}{CH}-CH_3$$

**C**             **D**

The longest continuous chain in each has four carbons, so the parent name in both cases is butane. In both cases there are two methyl groups, so both compounds are dimethylbutane. As required by rule 3, we use numbers to tell them apart:

Compound **C** is 2,2-dimethylbutane.

Compound **D** is 2,3-dimethylbutane.

Note that we must use a separate number for every prefix, even if two of the same groups are on the same carbon. Compound **C** is 2,2-dimethylbutane, and it would be incorrect to call it 2-dimethylbutane. It is also incorrect to say 2,2-methylbutane. If there are two identical groups we must say "di-." Here are examples showing the use of "tri-" and "tetra-":

Why isn't compound **C** called 3,3-dimethylbutane?

The comma is used to separate position numbers; the hyphen connects the number to the name.

$$CH_3-\underset{\underset{\displaystyle CH_2}{|}}{\overset{\overset{\displaystyle CH_3}{|}}{CH}}-\underset{}{\overset{\overset{\displaystyle CH_3}{|}}{CH}}-CH-CH_3$$

2,3,4-Trimethylhexane      2,2,4,4-Tetramethylhexane

---

**PROBLEM 10.3**   Name:

(a) $CH_3-\underset{\underset{\displaystyle CH_3}{|}}{CH}-\underset{\underset{\displaystyle CH_3}{|}}{CH}-CH_2-CH_3$

(b) $CH_3-\underset{\underset{\displaystyle CH_3}{|}}{CH}-CH_2-\underset{\underset{\displaystyle CH_3}{|}}{\overset{\overset{\displaystyle CH_3}{|}}{C}}-CH_3$

---

**5.** If there are two or more different groups, we put all of them into the prefix in alphabetical order.

$$CH_3-CH_2-CH_2-\underset{\underset{\displaystyle CH_3}{|}}{CH}-\underset{\underset{\displaystyle CH_2}{|}}{CH}-\underset{\underset{\displaystyle CH_3}{|}}{CH}-CH_3$$

The longest continuous chain in this molecule has seven carbons, so the parent name is heptane. When we look for the groups, we find two methyl groups and one ethyl group. Since ethyl comes before methyl alphabetically, the prefixes will be ethyl- followed by dimethyl-. Numbering to give the lowest numbers, we come out with

3-ethyl-2,4-dimethylheptane

Note that we must put the methyl groups together. It is wrong to say 2-methyl-3-ethyl-4-methylheptane.

---

**PROBLEM 10.4**   Name:

$$\text{CH}_3\text{—CH—}\overset{\displaystyle \text{CH}_3}{\underset{\displaystyle \text{CH}_2}{\overset{|}{\underset{|}{\text{C}}}}\text{—CH}_2\text{—CH}_2\text{—CH}_3}$$

$$\underset{\text{CH}_3}{\overset{|}{\text{CH}_2}} \quad \underset{\text{CH}_3}{\overset{|}{\text{CH}_2}}$$

---

In rule 6 we extend this system to **alkyl halides,** which are compounds containing carbon and halogens and only single bonds (most also contain hydrogen):

> **6.**   The four halogens are named by prefixes:
>   **F**   fluoro-        **Br**   bromo-
>   **Cl**   chloro-       **I**   iodo-

All other rules are the same. For example,

Br       Cl

$\text{CH}_3\text{—CH—CH}_3$    $\text{Cl—CH}_2\text{—CH—CH}_2\text{—CH}_3$

2-Bromopropane     1,2-Dichlorobutane

F

$\text{CH}_3\text{—CH—CH}_2\text{—C—CH}_3$

CH$_2$—Br  F

1-Bromo-4,4-difluoro-2-methylpentane

The prefixes bromo-, fluoro-, and methyl- are listed in alphabetical order. Let us look at the last example. There are two ways of taking the longest continuous chain, but we choose the one shown in color

F

$\text{CH}_3\text{—CH—CH}_2\text{—C—CH}_3$

CH$_2$—Br  F

because that way the bromine is a substituent on the parent chain instead of on a side chain.

Note that we ignore "di-" in determining alphabetical order. Ethyl- comes before *di*methyl-.

**PROBLEM 10.5**  Name:

$$\begin{array}{ccc} & \text{Cl} & \text{Cl} \\ & | & | \end{array}$$

(a)  $CH_3—CH_2—CH—CH—CH_3$

$$\begin{array}{ccc} & \text{Br} & \text{I} \\ & | & | \end{array}$$

(b)  $CH_3—CH_2—CH_2—\underset{\underset{\text{I}}{|}}{C}—\underset{\underset{\text{Cl}}{|}}{C}—CH_3$

On page 299 we met a series of alkyl groups, none of which are branched. In some compounds, however, we need to use branched alkyl groups as well. The most important of these are shown in Table 10.5. Isopropyl is a **secondary** group because the **bond by which it is connected to the parent is on a secondary carbon.** The unbranched alkyl groups we have already met are all primary groups. Except for methyl and ethyl, the unbranched ones are often called *normal* alkyl groups. Here is an example of the use of a branched group in the naming of an alkane:

$$CH_3—CH_2—CH_2—\underset{\underset{\underset{\underset{\text{CH}_3}{|}}{\text{CH—CH}_3}}{|}}{CH}—CH_2—CH_2—CH_3$$

4-Isopropylheptane

**TABLE 10.5  The Most Important Branched Alkyl Groups**

| Name | Structure | Type of Group |
|------|-----------|---------------|
| Isopropyl | $\underset{\overset{|}{CH_3—CH—}}{\overset{CH_3}{|}}$ | Secondary |
| Isobutyl | $\underset{\overset{|}{CH_3—CH—CH_2—}}{\overset{CH_3}{|}}$ | Primary |
| *sec*-Butyl | $\underset{\overset{|}{CH_3—CH_2—CH—}}{\overset{CH_3}{|}}$ | Secondary |
| *tert*-Butyl | $\underset{\overset{|}{\underset{CH_3}{|}CH_3—C—}}{\overset{CH_3}{|}}$ | Tertiary |
| Neopentyl | $\underset{\overset{|}{\underset{CH_3}{|}CH_3—C—CH_2—}}{\overset{CH_3}{|}}$ | Primary |

Branched groups are also often used in the naming of alkyl halides:

$$CH_3$$
$$|$$
$$CH_3—CH—CH_2—Cl$$

IUPAC name: 1-chloro-2-methylpropane

Common name: isobutyl chloride

---

**PROBLEM 10.6**   Give the IUPAC and common name for

$$CH_3$$
$$|$$
(a) $CH_3—CH—CH_3$      (b) $CH_3—C—CH_2—I$
$$|$$                            $$|$$
$$Cl$$                           $$CH_3$$

---

## 10.8   CYCLOALKANES

Carbon atoms can be connected to each other in rings as well as in chains, giving rise to a series of hydrocarbons called **cycloalkanes.** The rings may be of any size from three carbons on up, but the most common and most important are five- and six-membered rings. Some examples are shown in Table 10.6.

For cyclic compounds it is customary to use the very condensed formulas shown on the right in Table 10.6. These formulas show only the C—C bonds. Although the carbon and hydrogen atoms are omitted from the formulas, we must not forget that they are there.

Cycloalkanes are named by the same IUPAC rules we discussed in Section 10.7, except that we need a new rule for numbering because unlike a chain, a ring has *no* ends. The rule is

*Note the way these compounds are named.*

> **7.**   We may begin at any position and may proceed either clockwise or counterclockwise, but we must end up with the lowest possible numbers.

**TABLE 10.6   Three Common Cycloalkanes**

| Name | Structural Formula | Very Condensed Structural Formula | Boiling Point (°C) | Melting Point (°C) |
|------|--------------------|-----------------------------------|--------------------|--------------------|
| Cyclopropane | | | −33 | −128 |
| Cyclopentane | | | 49 | −94 |
| Cyclohexane | | | 81 | 7 |

**EXAMPLE**

Name these compounds:

(a) [structure: cyclopentane with CH₃]  (b) H₃C—[cyclohexane ring]—CH₃  (c) [cyclooctane with Br substituents]

**Answer**

(a)  This compound is methylcyclopentane. No number is needed.

(b)  To get the lowest numbers, we can begin at either methyl group and count in the direction of the other. Either way, we get 1,3-dimethylcyclohexane:

H₃C—[cyclohexane ring, positions 1, 2, 3]—CH₃

(c)  We get the lowest numbers only by starting at the position with two bromines. The correct name is 1,1,3,6-tetrabromocyclooctane. If we went clockwise instead of counterclockwise, we would get 1,1,4,7. This would be wrong because it is a higher set of numbers. The lowest is always correct.

**PROBLEM 10.7**   Name:

(a) [cyclopentane with Cl]   (b) [cyclobutane with CH₂CH₃ and CH₂CH₃]   (c) [cyclohexane with CH₃ groups]

*cis*-1,2-Dimethylcyclopentane

*trans*-1,2-Dimethylcyclopentane

**Figure 10.9**
The two stereoisomers of 1,2-dimethylcyclopentane.

## 10.9   STEREOISOMERISM IN CYCLIC COMPOUNDS

There is a major difference between cycloalkane structures and the structures of the branched and unbranched alkanes we considered before (we may call those *acyclic* alkanes). In the acyclic compounds there is free rotation about all the C—C bonds. This free rotation cannot exist in the cyclic compounds because the ring prevents it. Because of this we have a new kind of isomerism here, which can be illustrated by the models shown in Figure 10.9. These two compounds not only have the same molecular formula, $C_7H_{14}$ (and are therefore isomers), but also the same structural formula:

[cyclopentane with CH₃ and CH₃ substituents]

## The Actual Shape of the Cyclohexane Ring

In the text, rings are drawn as if they are flat (planar). Most rings, however, are not planar. Since six-membered rings are the most important, let us look at them. The cyclohexane ring has this shape:

We call it the **chair** conformation. (Can you see a chair in it?) The ring adopts this shape because doing so allows all the carbon angles to be about 110° (the normal carbon angle). If this molecule were a planar hexagon, the angles would have to be 120°.

Let us look at the 12 hydrogens in cyclohexane:

The lines show the C—H bonds.

Six of the hydrogens (marked *a*) are pointing up or down (three up, three down), and the other six (marked *e*) are pointing approximately in the "plane" of the ring. The six up-and-down hydrogens are called **axial** hydrogens; the other six are **equatorial.** You might conclude that there should be two stereoisomers of, say, bromocyclohexane: one with the bromine axial and the other with it equatorial. There is only one, however, because the two molecules rapidly interconvert by going through an intermediate stage called the **boat** conformation:

Equatorial bromine (more stable)    Boat conformation    Axial bromine

It is true that the two molecules have different structures (nonsuperimposable), but there is no stereoisomerism here because we cannot separate the two forms. When we try to make only one of them, we get an equilibrium mixture of both.

In general, however, molecules with groups other than hydrogen in equatorial positions are more stable than those with groups in axial positions.

By our naming system, both are called 1,2-dimethylcyclopentane. Yet they are different compounds, with different structures and different properties. How can we tell that two molecules have different structures? There is a very simple test. Mentally try to superimpose the two structures—slide one on top of the other and see if it fits. If it does, the molecules are identical; if not, they are different. When we try this with the models in Figure 10.9, we find that they *don't* fit. One molecule has both methyl groups on the same side of the ring, and the other has one methyl above and the other below. If we could twist the second molecule so as to put both methyl groups on the same side, we might be able to superimpose it on the first molecule, but the lack of rotation about the C—C bonds prevents us from doing this without breaking the ring.

**Compounds that have the same structural formulas but different three-dimensional shapes (cannot be superimposed)** are called **stereoisomers.** Since they are different compounds we cannot give them exactly the same name. Both compounds in Figure 10.9 are called 1,2-dimethylcyclopentane, but we use the prefixes *cis* and *trans* to tell them apart. **Cis** means the **two groups are on the same side; trans** means **they are on opposite sides:**

*cis*-1,2-Dimethylcyclopentane

*trans*-1,2-Dimethylcyclopentane

*trans*-1-Chloro-3-methylcyclohexane

*cis*-1-Chloro-3-methylcyclohexane

Note that it is not difficult to draw these isomers.

Cis and trans isomers are found in rings of all sizes, from three-membered on up. All that is necessary is that at least two groups, the same or different, be substituted for H atoms on different carbons of a cycloalkane ring. In general, the properties of cis and trans isomers, both physical and chemical, are different. They may be similar, but they are not identical.

Cis-trans isomers are sometimes called **geometrical isomers.**

## 10.10 PHYSICAL PROPERTIES

The most important physical properties of organic compounds are physical state (including boiling and melting points), solubility, and density.

The properties of cycloalkanes are very similar to those of the acyclic alkanes.

### Physical State

In general, both the melting and the boiling points of hydrocarbons increase with increasing molecular weight (see Tables 10.4 and 10.6). The low-molecular-weight alkanes, including cycloalkanes, up to about 5 carbons, are gases at room temperature. From about 5 to 17 carbons, they are liquids. Larger alkanes are white, waxy solids. Solid paraffin is a mixture of high-molecular-weight alkanes.

In any state, alkanes are colorless, odorless, and tasteless.

## The Biological Effects of Alkanes

Because they are so unreactive, alkanes in general are not particularly poisonous. Gasoline, which is a mixture of alkanes, should not be swallowed because it generally contains additives, some of which can be quite harmful. Liquid alkanes can cause damage if they get into the lungs, though by a physical rather than a chemical process. They dissolve the lipid molecules in the cell membranes and cause pneumonia-like symptoms. For this reason, anyone who swallows liquid alkanes should *not* be induced to vomit them up because some might be forced into the lungs in that manner.

Liquid alkanes can also harm the skin by a similar physical process. Human skin is kept moist by natural body oils. Liquid alkanes dissolve these oils and cause the skin to dry out. However, mixtures of high-molecular-weight liquid alkanes, sold in drugstores under the name *mineral oil*, soften and moisten skin. *Petroleum jelly* ("Vaseline" is one brand) is a mixture of solid and liquid hydrocarbons also used to protect skin. Because alkanes are not soluble in water, a coating of petroleum jelly protects skin from too much contact with water. For example, it can protect babies from diaper rash caused by the skin's contact with urine.

### Solubility

Even though carbon has a slightly higher electronegativity than hydrogen (Sec. 3.7), alkanes are nonpolar compounds. Therefore *they are not soluble in water,* which dissolves only ions and polar compounds. Alkanes are soluble in each other, however—an example of "like dissolves like"—as well as in other nonpolar organic compounds, such as benzene, diethyl ether, and carbon tetrachloride.

### Density

Each liquid alkane has a slightly different density, but all are less dense than water. Since none of them are soluble in water, this means that alkanes always float on water.

## 10.11   CHEMICAL PROPERTIES

### Combustion

The most important chemical property of alkanes is that they burn. Much of our current civilization depends on this property, called *combustion*. Petroleum (Box 10C) is a mixture of organic compounds consisting mostly of alkanes. From petroleum we get gasoline, diesel fuel, and fuel oil, which we use to power our cars, trucks, airplanes, and ships and to heat our houses. Natural gas, another important fuel, is chiefly methane, with a small amount of other low-molecular-weight alkanes. Other sources of energy, such as coal, hydroelectric, and nuclear power, provide us with less than half our energy needs.

The complete combustion of any alkane gives only two products: carbon dioxide and water. For example, this is the equation for the combustion of propane (bottled gas):

Combustion reactions are always exothermic.

$$C_3H_8 + 5O_2 \longrightarrow 3CO_2 + 4H_2O + heat$$

Petroleum, a thick black liquid found underground, is a mixture of hundreds, probably thousands, of different alkanes, ranging from methane up to alkanes with 40 or more carbons. It also contains varying amounts of sulfur-, oxygen-, and nitrogen-containing compounds. Because it is such a complex mixture, petroleum must be refined before use. The most important refinery process is distillation (Fig. 10C), in which the crude petroleum is separated into fractions, each with a different boiling range (hence a different average molecular weight). The major fractions, each a mixture of many compounds, are

Gasoline (about $C_4$ to $C_{12}$, boiling range 20 to 200°C)

Kerosene (about $C_{10}$ to $C_{14}$, boiling range 200 to 275°C)

Fuel oil and diesel fuel (about $C_{14}$ to $C_{18}$, boiling range 275 to 350°C)

Lubricating oil (about $C_{16}$ to $C_{20}$, boiling higher than 350°C)

Residue, used for greases and asphalt (more than $C_{20}$)

The most important fraction is gasoline, and it would be convenient if a very large proportion of the alkanes in petroleum had molecular weights that put them in this fraction. However, nature has not made petroleum like that; more than half of it consists of $C_{13}$ and higher alkanes.

Another disadvantage of petroleum is that a high percentage of the molecules are normal (unbranched) alkanes. This is a disadvantage because unbranched alkanes give rise to much more "knocking" when used in automobile engines. Knocking takes place when the gasoline-air mixture in the cylinder explodes before the spark plug ignites it. The tendency for a fuel to cause knocking can be measured; the *octane number* is used for this. The higher the octane number of a gasoline, the less knocking it will cause. Branched alkanes have higher octane numbers than unbranched alkanes. Another way to increase octane number is to add the compound lead tetraethyl, but this practice has been mostly phased out in the United States because, when burned, the lead fouls up the catalytic converters designed to reduce air pollution. (Also, lead itself is a poisonous pollutant.)

Though most petroleum is eventually used for fuel, about 2 percent is used to synthesize organic compounds, as mentioned in Section 10.2. This sounds like a small percentage, but so much petroleum is refined that even 2 percent is a very large volume indeed. More than 50 percent of all industrial synthetic organic compounds are made from this source. These chemicals are eventually turned into dyes, explosives, drugs, plastics, artificial fibers, detergents, insecticides, and other materials.

Gases

Gasoline

Kerosene

Fuel oil

Crude oil and vapor inlet

Lubricating oil

Residue

**Figure 10C**
Distillation of petroleum. The tower may be about 100 feet tall.

## Carbon Dioxide and Carbon Monoxide

In August 1986 a huge volume of $CO_2$ was released from beneath Lake Nyos (in Cameroon, Africa), a lake situated in the crater of a volcano, killing more than 1700 local inhabitants.

The complete combustion of fuels produces carbon dioxide. This occurs when enough oxygen is present. When the combustion is incomplete, carbon monoxide may also be produced. Oxidation of carbon compounds takes place not only in furnaces and in our cars but also in our bodies. In this case the product is entirely $CO_2$, which we exhale.

Carbon dioxide is a minor component of our atmosphere, although it is produced in large quantities by the action of volcanos (Fig. 10D.1) as well as by the burning of fossil fuels. However, the $CO_2$ in the air is used in photosynthesis of carbohydrates by plants. It is also converted to solid carbonates, first by dissolving in water to form carbonic acid and later by interacting with cations, for example, with $Ca^{2+}$ to form insoluble calcium carbonate (Fig. 10D.2).

Carbon monoxide is produced in incomplete combustion. For example, 4 to 7 percent of automobile exhaust gas and of chimney gases is CO. It is highly poisonous because it strongly combines with hemoglobin (Box 3C) and prevents the transport of oxygen to the tissues. Carbon monoxide binds to hemoglobin 200 times more strongly than oxygen does and is therefore very difficult to remove once it is absorbed in the blood.

Carbon monoxide poisoning is particularly lethal because it has no warning signals. Carbon monoxide is odorless and colorless, and its action is rapid.

Carbon monoxide itself is also produced in the body by the breakdown of heme, but this amount of CO blocks only about 1 percent of the hemoglobin activity and produces no symptoms. When 10 to 30 percent of the hemoglobin molecules are combined with carbon monoxide, headache and dizziness develop, combined with dimmed vision and nausea. A greater amount of blocked hemoglobin results in unconsciousness, depressed heart action, coma, and death. Treatment involves administration of a mixture of oxygen and carbon dioxide (see also Box 5E) and blood transfusion if necessary.

**Figure 10D.1**
Erupting volcanos produce $CO_2$, among other gases. (J. Stewart Lowther, University of Puget Sound.)

**Figure 10D.2**
These stalactites, in Mammoth Cave, Kentucky, are made of $CaCO_3$ that has precipitated out of water solution. (Photograph by Beverly March.)

## The Greenhouse Effect

The world gets most of its energy by burning fossil fuels—not only petroleum but also coal and natural gas. If the combustion is complete, all the carbon in these materials is converted to carbon dioxide. The carbon in the fossil fuels has been stored in the earth for hundreds of millions of years. In the last century or two, large amounts of it have been converted to carbon dioxide in this way. Not all the carbon dioxide remains in the atmosphere (some of it dissolves in the oceans and lakes, and some is converted to rocks in the form of calcium and magnesium carbonates), but measurements show that the $CO_2$ content of the atmosphere is slowly increasing every year and has been doing so for decades.

This increase in the $CO_2$ content of the atmosphere poses a potential problem for the following reason. Carbon dioxide has the property of allowing visible light to pass through it but absorbing infrared light (see Sec. 9.1). In order for the earth to maintain a more or less constant average temperature, it must give off about as much energy as it receives. The energy it receives, from the sun, is mostly in the form of visible light. Since the $CO_2$ in the atmosphere allows this to pass through, the energy goes right through to the earth's surface. But the energy reemitted by the earth is mostly in the form of infrared light, not visible light, so it is absorbed by the atmospheric $CO_2$. Once a $CO_2$ molecule absorbs this energy it does not keep it, but reemits it in all directions, sending some of it back down to the surface of the earth. The net consequence is that the atmospheric $CO_2$ does nothing to stop the sun's energy from reaching the surface of the earth, but it does stop some energy from going back into space. This is called the **greenhouse effect.**

We might therefore expect a gradual increase in the average temperature of the earth's surface, as the amount of $CO_2$ in the atmosphere increases every year. It would not take a very large increase in the average temperature to cause substantial effects on the earth's climate. A rise of about 4°C would probably cause enough Antarctic ice to melt to flood most of the world's coastal cities. But has the average temperature really been rising? Measurements show that earth's average temperature did rise, by a total of about 0.6°C, from 1880 to 1940, but then fell, by about 0.3°C, from 1940 to about 1975, even though the concentration of atmospheric $CO_2$ was still increasing all that time. Since about 1975 the average temperature has been slowly rising again. Consequently, we cannot be certain at this time that the greenhouse effect will really have major effects on the climate. There are several other factors involved, and much research must still be carried out.

However, complete combustion takes place only under ideal conditions. Most combustion of gasoline and other fuels is not complete, and a variety of other products are formed, including soot and carbon monoxide. It is because of the production of the very poisonous carbon monoxide that one must not run an automobile engine in a closed garage.

### Other Reactions

Apart from combustion, alkanes undergo almost no reactions. They do not react with strong acids, bases, or oxidizing or reducing agents. To an organic chemist, **the chief chemical property of alkanes (including cycloalkanes) is that they are chemically inert.** There are a few exceptions (for example, alkanes react with $Cl_2$ at high temperatures or under the influence of ultraviolet light), but these are unimportant for our purposes. Actually, the inertness of alkanes is quite useful, as we shall see in the next section.

## 10.12   FUNCTIONAL GROUPS

Let us look at the structural formula of the compound 1-butanol:

$$CH_3-CH_2-CH_2-CH_2-OH$$

The only difference between this molecule and the alkane butane is the —OH group (called a **hydroxy group**). The molecule may be thought of as consisting of two parts: an alkane-like portion and the —OH group. We use the term **functional group** to refer to an atom or group of atoms that substitutes for a hydrogen atom. In 1-butanol, the —OH is a functional group. Functional groups can consist of any combination of N, O, S, H, halogen, and even C atoms, but some of them are more important than others. The most important functional groups are shown in Table 10.7.

Just as there are many alkanes, so are there a large number of compounds containing any given functional group, because in general each functional group can replace hydrogen at any position of any alkane. For example, see how many compounds can be made from alkanes of up to four carbons if we substitute —OH for —H (note that we do this *mentally*):

Since alkanes are inert, we cannot make —OH compounds in this way in practice.

| | | |
|---|---|---|
| $CH_4$ <br> Methane | $CH_3OH$ | |
| $CH_3CH_3$ <br> Ethane | $CH_3CH_2OH$ | |
| $CH_3CH_2CH_3$ <br> Propane | $CH_3CH_2CH_2OH$ | $CH_3-\overset{\displaystyle |}{\underset{\displaystyle OH}{CH}}-CH_3$ |
| $CH_3CH_2CH_2CH_3$ <br> Butane | $CH_3CH_2CH_2CH_2OH$ | $CH_2-\overset{\displaystyle |}{\underset{\displaystyle OH}{CH}}-CH_2CH_3$ |

$$CH_3-\underset{\underset{\textstyle CH_3}{|}}{CH}-CH_3 \qquad CH_3-\overset{\overset{\textstyle OH}{|}}{\underset{\underset{\textstyle CH_3}{|}}{C}}-CH_3 \qquad CH_3-\underset{\underset{\textstyle CH_3}{|}}{CH}-CH_2OH$$

Isobutane

The five alkanes give rise to eight hydroxy-containing compounds. All of these compounds are called alcohols, an **alcohol being a carbon compound containing an —OH group.** The general formula for an alcohol is

$$R-OH$$

**TABLE 10.7  The Most Important Functional Groups**

| Group | Type of Compound | Typical Example | |
|---|---|---|---|
| $\diagdown$C=C$\diagup$ | Alkene | $CH_2$=$CH_2$ | Ethylene |
| —C≡C— | Alkyne | HC≡CH | Acetylene |
| R—F (or Cl, Br, I) | Alkyl halide | $CH_3$—Cl | Chloromethane |
| R—OH | Alcohol | $CH_3$—OH | Methanol |
| R—O—R′ | Ether | $CH_3CH_2$—O—$CH_2CH_3$ | Diethyl ether |
| R—SH | Thiol (mercaptan) | $CH_3CH_2$—SH | Ethanethiol |
| R—NH$_2$ | Amines | $CH_3$—NH$_2$ | Methylamine |
| R$_2$NH | | $CH_3$—NH—$CH_3$ | Dimethylamine |
| R$_3$N | | $CH_3$—N—$CH_3$ $\quad$ $CH_3$ | Trimethylamine |
| R—C(=O)—H | Aldehyde | $CH_3$—C(=O)—H | Acetaldehyde |
| R—C(=O)—R′ | Ketone | $CH_3$—C(=O)—$CH_3$ | Acetone |
| R—C(=O)—OH | Carboxylic acid | H—C(=O)—OH | Formic acid |
| R—C(=O)—OR′ | Carboxylic ester | $CH_3$—C(=O)—O—$CH_2CH_3$ | Ethyl acetate |
| R—C(=O)—NH$_2$ | Amides | $CH_3$—C(=O)—NH$_2$ | Acetamide |
| R—C(=O)—NHR′ | | $CH_3$—C(=O)—NH—$CH_3$ | N-Methylacetamide |
| R—C(=O)—NR′R″ | | $CH_3$—C(=O)—N—$CH_3$ $\quad$ $CH_3$ | N,N-Dimethylacetamide |

where the **R** is a symbol that stands for **any alkyl group.** The alcohols shown above contain eight of these R groups, and we have already learned their names: methyl, ethyl, n-propyl, isopropyl, and so forth. These are all the possible non-cyclic alkyl groups with up to four carbons, but there are many others with five or more carbons. Note that R groups may be branched or unbranched.

The family of alcohols, symbolized R—OH, have many properties in common. We can see why by looking at their structures. All of them have an alkyl group connected to an —OH, for example:

$$CH_3-\underset{\underset{CH_3}{|}}{CH}-CH_2-CH_2-OH$$

## Taxol, a New Anticancer Drug and the Environment

Many naturally occurring organic compounds have a multiplicity of functional groups. An example is **taxol,** which can be extracted from the bark of the Pacific Yew tree found in the old-growth forests of the Pacific Northwest. As can be seen in the formula, this one compound belongs to six of the families in Table 10.7: alkene, alcohol, ether, ketone, carboxylic ester, and amide.

Taxol

The reason taxol is important is that it shows promise in the treatment of ovarian and breast cancer. Taxol acts by preventing the division of cancer cells. It does this by blocking the disassembly of microtubules within the cells. Before dividing, the cell must disassemble these units, and taxol prevents this.

Although taxol is still being clinically tested as an anticancer drug, there is a problem. Until recently the entire supply of this drug came from the Pacific Yew tree, but to get 10 grams of the drug it is necessary to cut down 25 trees, and it has been estimated that 25 000 trees will be necessary to supply enough taxol just for the clinical trials. If the drug passes the clinical trials and is approved for medical use the situation would be even worse for the yew trees, and indeed, the species could be completely wiped out, which of course would eliminate the trees as a source of the drug anyway. The obvious answer is to obtain taxol some other way. What about synthesis? Despite its complicated structure, early in 1994 chemists working in laboratories did succeed in synthesizing taxol, but their methods are too expensive to supply the amount needed. At present the most promising approach seems to be a method of extracting taxol from the needles and twigs of the yew tree itself, and from closely related plants. This is a renewable source and does not kill the tree, as stripping the bark does. Some of the closely related plants are much easier to grow than the yew trees, which are very old, and some of them contain, not taxol itself, but similar compounds that can be converted to taxol in chemical laboratories.

---

We saw in the previous section that alkanes are inert. *Therefore, so are the alkane-like portions of other molecules (alkyl groups).* Any compound that reacts with an alcohol reacts not with the alkyl part but only with the —OH. Since the —OH group is the same in all alcohols, it follows that all alcohols have many properties in common. This does not mean that the properties of alcohols are *identical.* Each alcohol has its own set of specific properties, and anyone who works with any particular alcohol must be aware of the properties of that alcohol. For example, methanol, $CH_3OH$, is poisonous, but ethanol,

Methanol is also called methyl alcohol, and ethanol is also called ethyl alcohol.

$CH_3CH_2OH$, is present in wine, beer, and whisky. Still, the properties of an entire family are similar enough to make it convenient to study organic compounds by family. We will do just that in the next five chapters.

## SUMMARY

**Organic chemistry** is the study of carbon compounds. The properties of organic compounds are generally very different from those of inorganic compounds. Many organic compounds are obtained from natural sources, but many others, which do not exist in nature, are synthesized in laboratories. Organic chemists use structural formulas and models to show three-dimensional shapes of molecules. The number of bonds formed by C, N, O, H, and the halogens are four, three, two, one, and one, respectively.

In order to explain the differing reactivities of double and single covalent bonds, chemists developed the concept of **hybrid orbitals.** A carbon atom that bonds to four other atoms uses orbitals that are neither $s$ nor $p$ but a mixture of these called $sp^3$ hybrid orbitals. An **$sp^3$ orbital** is neither spherical nor dumbbell-shaped but a mixture of the two. In forming a single covalent bond, an $sp^3$ orbital of the carbon atom overlaps with an $s$ or $p$ orbital of any atom or with an $sp^3$ orbital of another carbon atom.

**Hydrocarbons** contain only carbon and hydrogen. **Alkanes,** which are hydrocarbons containing only single bonds, are named by the **IUPAC system,** in which (1) the longest continuous chain is the parent, (2) other groups are named as prefixes, and (3) chains are numbered to produce the lowest numbers for the groups. There is free rotation about the C—C bonds of alkanes. **Alkyl halides** are named the same way as alkanes. **Cycloalkanes** have the carbon atoms connected in a ring. Cyclic compounds with two substituents on different carbons can exist as **cis and trans isomers. Stereoisomers** are compounds with the same structural formula but different three-dimensional shapes.

The melting and boiling points of alkanes increase with increasing molecular weight. Alkanes are insoluble in water and soluble in each other. They are less dense than water. Though alkanes burn, they are otherwise chemically inert and undergo few reactions.

Most organic molecules that are not alkanes contain **functional groups.** These groups, of which about a dozen are very important, are used to divide organic compounds into families. The letter **R** stands for any alkyl group.

## KEY TERMS

Alkane (Sec. 10.6)
Alkyl group (Sec. 10.7)
Alkyl halide (Sec. 10.7)
Axial hydrogen (Box 10)
Boat conformation (Box 10A)
Branched alkane (Sec. 10.6)
Chair conformation (Box 10A)
Cis isomer (Sec. 10.9)
Condensed structural formula (Sec. 10.6)
Conformation (Sec. 10.6)
Cycloalkane (Sec. 10.8)
Equatorial hydrogen (Box 10A)
Functional group (Sec. 10.12)
Geometrical isomers (Sec. 10.9)
Greenhouse effect (Box 10E)
Group (Sec. 10.7)
Hybrid orbital (Sec. 10.4)

Hydrocarbon (Sec. 10.5)
Isomers (Sec. 10.3)
IUPAC system (Sec. 10.7)
Main chain (Sec. 10.6)
Normal alkane (Sec. 10.6)
Octane number (Box 10C)
Organic chemistry (Sec. 10.1)
Parent name (Sec. 10.7)
Primary carbon (Sec. 10.6)
Quaternary carbon (Sec. 10.6)
Secondary carbon (Sec. 10.6)
Side chain (Sec. 10.6)
$sp^3$ orbital (Sec. 10.4)
Stereoisomerism (Sec. 10.9)
Structural isomerism (Sec. 10.3)
Tertiary carbon (Sec. 10.6)
Trans isomer (Sec. 10.9)

## PROBLEMS

### Organic and Inorganic Compounds

**10.8**  Which are organic compounds?
(a)  LiCl
(b)  $C_3H_7Br$
(c)  $C_2Cl_4$
(d)  $C_7H_{10}$
(e)  $KMnO_4$
(f)  $C_{12}H_{22}O_{11}$

**10.9**  Do liquid organic compounds conduct electricity?

**10.10**  In each case, pick out the compound with the higher melting point:
(a)  NaCl vs. $CH_3Cl$
(b)  $C_3H_7OH$ vs. KOH
(c)  $CH_3(CH_2)_5CH_3$ vs. $CH_3CH_3$

**10.11**  An unknown liquid has a melting point of 87°C, is insoluble in water, and burns. Is it likely to be an inorganic or an organic compound?

### Sources of Organic Compounds

**10.12**  (a) What is meant by the expression "known compound"?   (b) What are the two ways in which compounds become known?

**10.13**  Is there any difference between vitamin C made synthetically and vitamin C isolated from oranges, assuming that both are chemically pure?

### Isomers

**10.14**  There are two different compounds, butene and cyclobutane, that have entirely different properties, though both have the molecular formula $C_4H_8$. How is this possible?

**10.15**  For each of the following pairs, tell whether the structures shown are (1) different formulas for the same compound,   (2) isomers, or   (3) different compounds that are not isomers:

(a)  $CH_3-CH_2-\underset{\underset{CH_3}{|}}{CH}-CH_2-CH_3$   and

$CH_3-\underset{\underset{CH_3}{|}}{CH}-\overset{\overset{CH_3}{|}}{CH}-CH_3$

(b)  $CH_3-CH-CH_3$

$\underset{\underset{O\;-\!-\;CH_2}{|}}{CH_2}\quad \underset{}{CH_2}-CH_3$   and

$CH_3-CH_2-CH_2-O-CH_2-\underset{\underset{CH_3}{|}}{CH}-CH_3$

(c)  $CH_3-\underset{\underset{CH_3}{|}}{CH}-CH_2-\underset{\underset{CH_3}{|}}{CH}-CH_3$   and

$\underset{\underset{CH_3}{|}\;\underset{\underset{CH_3}{|}}{CH}-CH_3}{\overset{\overset{CH_3}{|}}{CH}-CH_2}$

(d)  $\underset{\underset{CH_2-CH_2}{|}}{CH_2-CH_2}$   and   $CH_3-\underset{\underset{CH_3}{|}}{CH}-CH_3$

(e)  $\underset{\underset{\underset{O}{}}{H_2C\qquad CH_2}}{H_2C-CH_2}$   and

$CH_3-\underset{\underset{O}{\|}}{C}-CH_2-CH_3$

(f)  $CH_2{=}\underset{\underset{CH_3}{|}}{C}-CH{=}CH_2$   and

$HC{\equiv}C-\underset{\underset{CH_3}{|}}{CH}-CH_3$

(g)  $\underset{\underset{CH_2}{}}{\underset{H_2C\qquad CH_2}{H_2C-CH}}^{NH_2}$   and   $\underset{\underset{NH}{}}{\underset{H_2C\qquad CH_2}{H_2C-CH}}^{CH_3}$

(h)  $\underset{\underset{O}{}}{\underset{H_2C\qquad CH_2}{H_2C-CH_2}}$   and

$CH_3-CH_2-O-CH_2-CH_3$

(i)  $CH_3-CH_2-CH_2-\overset{\overset{CH_3}{|}}{NH}$   and

$CH_3-\underset{\underset{CH_3}{|}}{CH}-CH_2-NH_2$

**10.16** Draw structural formulas for
  (a) two different compounds of molecular formula $C_2H_4FCl$
  (b) three different compounds of molecular formula $C_3H_6Cl_2$

**10.17** Draw structural formulas for all compounds whose molecular formula is $C_3H_8O$.

**\*10.18** (a) From among the following eight structural formulas, pick out the isomers.  (b) Draw structural formulas for two other isomers of the same molecular formula as the ones in this problem:

  (a) $CH_2{=}CH{-}CH_2{-}O{-}CH_2{-}CH_3$

  (b) $CH_3{-}\underset{\underset{O}{\|}}{C}{-}CH_2{-}\underset{\underset{CH_3}{|}}{CH}{-}CH_3$

  (c) $CH_3{-}\underset{\underset{CH_3}{|}}{CH}{-}CH_2{-}\underset{\overset{O}{\|}}{C}{-}H$

  (d) ⬠—OH

  (e) $CH_3{-}C{\equiv}C{-}CH_2{-}CH_2{-}CH_2{-}OH$

  (f) (furan ring)—$CH_3$

  (g) $CH_3{-}CH_2{-}\underset{\underset{CH_2}{\|}}{C}{-}CH_2{-}OH$

  (h) $CH_3{-}O{-}CH{=}\underset{\underset{CH_3}{|}}{C}{-}CH_3$

### Bonding in Organic Compounds

**10.19** List the principal elements that make up organic compounds and the number of bonds they form.

**10.20** Some of the following formulas are incorrect (that is, they do not represent any real compound) because they have atoms with an incorrect number of bonds. Which are they, and which atoms have an incorrect number of bonds?

  (a) $H{-}\underset{\underset{H}{|}}{N}{-}\underset{\underset{H}{|}}{C}{-}H$

  (b) $H{-}\underset{\underset{H}{|}}{C}{=}\underset{\underset{Cl}{|}}{C}{-}\underset{\underset{H}{|}}{C}{=}\underset{\underset{H}{|}}{C}{=}\underset{\underset{Br}{|}}{C}{-}Br$

  (c) $O{-}\underset{\underset{H}{|}}{\overset{\overset{H}{|}}{C}}{-}H{-}\underset{\underset{H}{|}}{C}{=}\underset{\underset{H}{|}}{C}{-}H$

  (d) $H{-}\underset{\underset{H}{|}}{C}{-}\underset{\underset{Cl}{\|}}{C}{-}C{-}O{-}\underset{\underset{H}{|}}{N}{-}H$

  (e) $H{-}C{\equiv}C{-}\underset{\underset{Br}{|}}{C}{-}\underset{\underset{H}{|}}{C}{-}O{-}\underset{\underset{H}{|}}{C}{=}N{-}\underset{\underset{H}{|}}{\overset{\overset{H}{|}}{C}}{-}H$

  (f) $H{-}C{\equiv}C{-}\underset{\underset{H}{|}}{C}{-}\underset{\underset{N}{|}}{C}{-}C{=}H$

  (g) $H{-}\underset{\underset{H}{|}}{C}{-}O{-}\underset{\underset{Br}{|}}{C}{-}\underset{\underset{H}{|}}{C}{-}\underset{\underset{H}{|}}{N}{-}H$

  (h) $H{-}O{-}\underset{\underset{H}{|}}{C}{=}\underset{\underset{Br}{|}}{\overset{\overset{Cl}{|}}{C}}{-}N$

**10.21** Complete these structural formulas by supplying the missing hydrogens:

  (a) $C{-}C{=}C{-}\underset{\underset{C}{|}}{C}{-}C$

  (b) $N{-}C{-}C{-}O{-}C{-}\underset{\overset{C}{|}}{C}{-}Br$

  (c) $Cl{-}C{=}C{-}C{\equiv}C{-}\underset{\underset{C}{|}}{N}{-}C$

### Hybrid Orbitals

**10.22** What do we mean by the term hybrid orbital?

**10.23** What is the shape of (a) an $s$ orbital (b) a $p$ orbital (c) an $sp^3$ orbital?

**10.24** The promotion of an electron from a $2s$ to a $2p_z$ orbital
  (a) requires energy input
  (b) goes without energy gain or loss
  (c) produces energy

**10.25** Show the overlap of orbitals in (a) $CBr_4$   (b) $CH_3CH_2Br$

## Structure of Alkanes

**10.26** The condensed formula for *n*-butane is $CH_3CH_2CH_2CH_3$. Explain why this formula does not show the geometry of the real molecule.

**10.27** Draw condensed or very condensed structural formulas for

(a)

(b)

(c)

**10.28** Pick out all the tertiary carbons and all the quaternary carbons:

(a) $ClCH_2CH_2CH_2CH_3$

(b)

(c)

## Nomenclature of Alkanes and Alkyl Halides

**10.29** (a) Draw structural formulas for all compounds whose molecular formula is $C_6H_{14}$.   (b) Give the IUPAC name of each.

**10.30** Draw and name all compounds whose molecular formula is (a) $C_3H_7Cl$   (b) $C_5H_{11}Br$

**10.31** Name each of the following by the IUPAC system:

(a) $CH_3CH_2CH_2CH_2CH_2CH_2CH_3$

(b)

(c)

(d)

(e)

(f)

(g)

**10.32** Draw structural formulas for
(a) 3-methylhexane
(b) 2,3-dimethylbutane
(c) 2,4,4-trimethylheptane
(d) 2,2,4-trimethylheptane
(e) octabromopropane
(f) 5-isopropylnonane

**10.33** Name each of the following halides by the IUPAC system:

(a)

(b) $CH_3-CH-CH_2-CH-I$
 $\quad\quad\;\;|\quad\quad\quad\;\;|$
 $\quad\quad CH_2I\quad\quad\;\; CH_3$

(c) $Br-CH_2-CH_2-\overset{\displaystyle Br}{\underset{\displaystyle Br}{\overset{|}{\underset{|}{C}}}}-Br$

(d) $CH_3-CH-\overset{\displaystyle I}{\overset{|}{CH}}-CH-CH_3$
 $\quad\quad\;\;|\quad\quad\quad\quad\;|$
 $\quad\quad Cl\quad\quad\quad CH_3$

(e) $CH_2-CH-CH_2-CH-CH_3$
 $\quad\;|\quad\quad\;\;|\quad\quad\quad\;|$
 $\quad Cl\quad\; Cl\quad\quad\; Br$

(f) $CH_3-\overset{\displaystyle I}{\underset{\displaystyle I}{\overset{|}{\underset{|}{C}}}}-\overset{|}{\underset{\displaystyle CH_2I}{CH}}-CH_3$

**10.34** Name these groups:

(a) $CH_3CH_2-$

(b) $CH_3-\overset{\displaystyle CH_3}{\underset{\displaystyle CH_3}{\overset{|}{\underset{|}{C}}}}-$

(c) $CH_3-\overset{\displaystyle CH_3}{\overset{|}{CH}}-CH_2-$

(d) $Cl-$

(e) $CH_3-\overset{\displaystyle CH_3}{\overset{|}{CH}}-$

(f) $CH_3CH_2CH_2CH_2CH_2-$

**10.35** Give the IUPAC name for
(a) isobutane
(b) isopropyl bromide
(c) *sec*-butyl fluoride
(d) neopentyl chloride
(e) *tert*-butyl iodide

## Cycloalkanes

**10.36** Name each of the following by the IUPAC system (ignore stereoisomerism):

(a)

(b)

(c)

(d)

(e)

**10.37** Name the following cyclic halogen-containing compounds by the IUPAC system (ignore stereoisomerism):

(a)

(b)

(c)

(d)

**10.38** There are six *cycloalkanes* with the molecular formula $C_5H_{10}$. Draw all of them and give the IUPAC names.

**10.39** Draw the condensed or very condensed structural formulas for

(a)

(b) 

(c)

**10.40** Draw structural formulas for
(a)  *trans*-1,3-dimethylcyclobutane
(b)  *cis*-1,4-dichlorocyclohexane
(c)  1,3-diethyl-4,4-dimethylcyclopentane

**10.41** Give IUPAC names:

(a) 

(b) 

(c) 

## Properties of Alkanes

**10.42** The compound *n*-decane, $C_{10}H_{22}$, is an unbranched alkane. Predict the following:
(a)  Does it dissolve in water?
(b)  Does it dissolve in *n*-hexane?
(c)  Will it burn when ignited?
(d)  Is it a liquid, solid, or gas at room temperature?
(e)  Is it more or less dense than water?

**10.43** What are the chief chemical properties of the alkanes and cycloalkanes?

**10.44** Balance the complete combustion reaction of
(a)  cyclobutane
(b)  3,3-dimethylpentane
(c)  *n*-decane

## Functional Groups

**10.45** State which family each belongs to:

(a)  $CH_3-CH-Br$
         $|$
        $CH_3$

(b)  $CH_3-\overset{\displaystyle CH_3}{\underset{\displaystyle CH_3}{C}}-OH$

(c)  $CH_3-C\equiv CH$

(d) 

(e) 

(f)  $CH_3-CH-CH_2-CH-CH_3$
              $|$               $|$
            $CH_3$          $CH_3$

(g) 

(h)  $CH_3-\underset{\displaystyle \overset{\|}{O}}{C}-CH_2CH_3$

(i)  $CH_3-O-\underset{\displaystyle \overset{\|}{O}}{C}-CH_2CH_3$

**10.46** Draw the structural formula for one member of each of the following classes (other than the one shown in Table 10.7):
(a)  alkene                    (f)   ether
(b)  carboxylic acid    (g)  thiol
(c)  amine                    (h)  aldehyde
(d)  amide                    (i)   carboxylic ester
(e)  alcohol                  (j)   ketone

**10.47** List all the functional groups in Table 10.7 that contain oxygen atoms.

**10.48** Each of the following compounds has two or three functional groups. Put each into as many families as is appropriate:

(a) $CH_3-O-CH_2CH_2-NH-CH_3$

(b) $CH_2=CH-CH_2-Cl$

(c) $CH_2-CH-CH_2$
$\quad\ |\quad\ \ |\quad\ \ |$
$\quad OH\ \ OH\ \ OH$

(d) $CH_3-CH-CH_2CH_2-OH$
$\qquad\ \ |$
$\qquad\ \ F$

(e) $CH_3-\underset{\underset{O}{\|}}{C}-CH_2-\underset{\underset{O}{\|}}{C}-O-CH_3$

(f) 

**10.49** Tell which functional groups are present in

(a) 

Aspirin

(b) 

2,4-D (a herbicide)

(c) $HO-\underset{\underset{O}{\|}}{C}-CH_2-CH_2-\underset{\underset{NH_2}{|}}{CH}-COOH$

Glutamic acid

(d) 

*para*-Aminobenzoic acid

**Boxes**

**10.50** (Box 10A) Explain why it is not possible to isolate separate isomers of axial and equatorial chlorocyclohexane.

**10.51** (Box 10B) The solubility of body oils in alkanes decreases with increasing molecular weight of the alkanes. On the basis of this principle, explain how petroleum jelly protects the skin but gasoline causes chapping.

**10.52** (Box 10C) Besides its use as fuel, what other uses of petroleum can you name?

**10.53** (Box 10C) Why is it necessary to distill petroleum?

**10.54** (Box 10D) Why is blood transfusion and not just administration of $O_2$ necessary in severe carbon monoxide poisoning?

**10.55** (Box 10D) Carbon dioxide is a product of the burning of fossil fuels, as well as of the breathing of humans and animals. Explain why carbon dioxide is still a minor component of the atmosphere.

**10.56** (Box 10E) Certain gases, among them $CO_2$, generate the greenhouse effect. What property of these gases promotes an increase in the temperature of the atmosphere?

**10.57** (Box 10F) What kind of functional groups does taxol have?

**10.58** (Box 10F) How many carboxylic ester groups are present in the taxol molecule?

**\*10.59** (Box 10F) How many OH groups in taxol are located at primary, secondary, and tertiary carbons?

**10.60** (Box 10F) What is the mode of anticancer activity of taxol?

**Additional Problems**

**10.61** The following names are all incorrect. State what is wrong with each (do not try to give correct names)

(a) 4-methylpentane
(b) 1-methylbutane
(c) 2-dimethylbutane
(d) 2,3-methylhexane
(e) 2-methylcyclohexane
(f) 2-ethyl-4,4-dimethylpentane
(g) 2-ethylpentane
(h) cycloethane

**10.62** Write structural formulas for

(a) An alcohol that has five carbons

(b) A carboxylic acid that has three carbons
(c) A ketone that has three carbons
(d) An ether that has four carbons
*(e) A cyclic alcohol that has four carbons

**10.63** Describe the orbitals that overlap, making up all bonds in the following compounds:

(a)
$$F-\underset{\underset{H}{|}}{\overset{\overset{H}{|}}{C}}-\underset{\underset{F}{|}}{\overset{\overset{F}{|}}{C}}-F$$

(b)
$$H-\underset{\underset{H}{|}}{\overset{\overset{Cl}{|}}{C}}-H$$
$$H-\underset{\underset{H}{|}}{\overset{\overset{H}{|}}{C}}-H$$
$$H-\underset{\underset{H}{|}}{\overset{\overset{H}{|}}{C}}-Cl$$

**10.64** Classify all the carbons as primary, secondary, tertiary, or quaternary:

(a) $CH_3-\underset{\overset{|}{CH_3}}{\overset{}{CH}}-CH_2-CH_3$

(b) $CH_3-\underset{\underset{CH_3}{|}}{\overset{\overset{CH_3}{|}}{C}}-CH_2Cl$

(c)

Br

CH₃

Br

(d) $CH_3-CH_2-CH_3$

**10.65** Draw the structural formula for
(a) 2,3-dichlorohexane
(b) 1,2,4-trichlorocyclohexane
(c) 1,1-dimethylcyclobutane
(d) 2-sec-butyl-1-isopropyl-3-neopentylcyclo-heptane

**10.66** Define:
(a) organic chemistry
(b) hydrocarbon
(c) alkane
(d) functional group

**10.67** What important experiment was carried out by Wöhler in 1828?

**10.68** Draw structural formulas and give IUPAC names for all compounds whose molecular formula is $C_7H_{16}$.

**\*10.69** Write structural formulas for two different amines, each of which has two carbons.

**\*10.70** As can be seen in Table 10.4, an increase in the carbon chain by one carbon increases both the melting point and the boiling point. This increase is greater going from $CH_4$ to $C_2H_6$ and from $C_2H_6$ to $C_3H_8$ than it is in going from $C_8H_{18}$ to $C_9H_{20}$ or from $C_9H_{20}$ to $C_{10}H_{22}$. What do you think is the reason for this?

**10.71** Identify the functional groups:

(a) $\underset{\underset{CH_3}{|}}{\overset{\overset{CH_2OH}{|}}{C}}=O$

(b) $\overset{O}{\overset{||}{C}}-OH$
$\underset{\underset{CH_3}{|}}{\underset{\overset{|}{O}}{\overset{|}{C}}}=O$

(c) $\underset{\underset{CH_3}{|}}{H-\underset{}{C}}-OH$ with $\overset{\overset{H}{|}}{C}=O$

**\*10.72** Draw structural formulas and give IUPAC names for all the dibromopropanes ($C_3H_6Br_2$).

**10.73** Suppose you have a sample of 2-methylheptane and a sample of 2-methyloctane. Could you tell the difference by looking at them? What color would they be? How could you tell which is which?

# CHAPTER 11

*These floor tiles are made of an addition polymer, polyvinyl chloride. (Photograph by Beverly March.)*

# Alkenes, Alkynes, and Aromatic Compounds

## 11.1 INTRODUCTION

In Chapter 10 we discussed the alkanes, including the cycloalkanes. In this chapter we deal with the other families of hydrocarbons—alkenes, alkynes, and aromatic hydrocarbons.

**Alkenes** contain **carbon-carbon double bonds; alkynes** contain **carbon-carbon triple bonds.** Because compounds with double and triple bonds can add more hydrogen (see Sec. 11.6), alkenes and alkynes are called **unsaturated hydrocarbons.** Alkanes, which cannot add any more hydrogen, are called **saturated hydrocarbons.**

Alkenes are also called **olefins,** an older name that is still widely used.

## 11.2 NOMENCLATURE OF ALKENES

Alkenes are named by the same IUPAC rules we met in Section 10.7, but with several additional rules to handle the double bonds.

1. **The suffix "-ene" is used when double bonds are present.**

**Box 11A**

**Terpenes**

**Terpenes** are a family of compounds found widely distributed throughout the plant world. Many are alkenes, but some contain other functional groups as well. Many are cyclic or bicyclic (two rings). Hundreds of different ones are known. Some have important biological uses; others we use for various commercial purposes. Among the latter are compounds called *essential oils,* which are what give plants their pleasant odors. For thousands of years, essential oils have been extracted from flowers and other parts of plants and used in perfumes and cosmetics. All terpenes have two things in common: They are made by plants (some are made by animals as well) and they follow the **isoprene rule.** Isoprene is a common name for 2-methyl-1,3-butadiene:

$$CH_2\!=\!\overset{\overset{\textstyle CH_3}{\textstyle |}}{C}\!-\!CH\!=\!CH_2$$

Isoprene

Isoprene itself is not a terpene, but the isoprene rule says that all terpenes are made up of isoprene units—that is, the carbon atoms of any terpene molecule can be divided into pieces, each having the $C\!-\!C\!-\!C\!-\!C$ skeleton. Because of this, the

$$\overset{\textstyle |}{C}$$

number of carbon atoms in a terpene is always a multiple of five. Terpenes can have 10 carbon atoms, or 15 (sesquiterpenes), or 20, or 30, or 40. They are found in nature in a bewildering array of structures, of which we can look at only a few.

We will see how the isoprene rule works by using dashed lines to divide the molecules into isoprene units. One of the simplest terpenes is menthol.

Menthol

Because an alkene must have a C=C bond, the simplest alkene has two carbon atoms:

$$H_2C\!=\!CH_2$$

Ethene or ethylene

In this molecule, the longest chain has two carbon atoms. If this were an alkane, it would be ethane. Because it has a double bond, the suffix "-ane" is changed to "-ene" and we call it **ethene.** This compound is more frequently called by its common name, ethylene. Similarly, $CH_3\!-\!CH\!=\!CH_2$ is called propene or propylene.

When the chain has four or more carbons, there is structural isomerism because the double bond can be in more than one position. Consider these compounds:

$$CH_2\!=\!CH\!-\!CH_2\!-\!CH_3 \qquad CH_3\!-\!CH\!=\!CH\!-\!CH_3$$

The rules we know so far say that both these compounds are butene, but they are obviously different compounds, with different structural formulas (though they are isomers because both are $C_4H_8$). We need a new rule.

$CH_3\!-\!CH\!=\!CH_2$ is the condensed formula for

$$H\!-\!\overset{\overset{\textstyle H}{\textstyle |}}{\underset{\textstyle |}{C}}\!-\!\overset{\textstyle |}{\underset{\textstyle H}{C}}\!=\!\overset{\textstyle |}{\underset{\textstyle H}{C}}\!-\!H.$$

2.   When a chain has four or more carbons, **we number the position of the double bond, using only the lower of the possible numbers.**

Menthol is one of the chief constituents of mint, especially Japanese peppermint oil.

Two larger terpenes are

Vitamin A

β-Carotene

β-Carotene, a tetraterpene, has 40 carbons (and is therefore made up of eight isoprene units) and 11 double bonds. It is the compound that gives the orange color to carrots. As you can see, β-carotene consists of two vitamin A units joined together. Because the body can break the carotene down, carrots are a good source of vitamin A.

Thus these two compounds are

$$\overset{1}{C}H_2=\overset{2}{C}H-\overset{3}{C}H_2-\overset{4}{C}H_3 \qquad \overset{1}{C}H_3-\overset{2}{C}H=\overset{3}{C}H-\overset{4}{C}H_3$$

1-Butene                 2-Butene

As in all the cases of isomerism we have seen so far, 1-butene and 2-butene have different physical and chemical properties.

In 1-butene the double bond is between carbons 1 and 2, but in the name we use only the 1. It is understood that the double bond goes from the digit shown to the next higher digit.

### 3. Branches and halogen substituents are named as in the rules for alkanes.

**EXAMPLE**

Name this compound:

$$CH_3-\underset{\underset{Cl}{|}}{\overset{\overset{CH_3}{|}}{C}}-CH_2-\overset{\overset{CH_3}{|}}{C}H-\overset{\overset{CH_3}{|}}{C}=CH-CH_3$$

**Answer**

The longest chain has seven carbons and includes a double bond. Therefore the parent name is heptene. The double bond goes from the second to the third carbon, so the parent is 2-heptene. (If we counted from the other end, it would be 5-heptene, but 2 is lower than 5.) When we add the other groups, we get

$$\underset{7}{CH_3}-\underset{6|}{\overset{\overset{\displaystyle CH_3}{|}}{C}}-\underset{5}{CH_2}-\underset{4}{\overset{\overset{\displaystyle CH_3}{|}}{CH}}-\underset{3}{\overset{\overset{\displaystyle CH_3}{|}}{C}}=\underset{2}{CH}-\underset{1}{CH_3}$$
$$\underset{Cl}{|}$$

6-Chloro-3,4,6-trimethyl-2-heptene

As this example shows, the double bond takes precedence over alkyl and halo groups in determining which end to start numbering from. We begin numbering at the end that gives the double bond the lowest number, no matter where the alkyl or halo groups are.

---

**PROBLEM 11.1** Name:

$$Cl-\overset{\overset{\displaystyle CH_3}{|}}{CH}-CH_2-CH=\overset{\overset{\displaystyle Cl}{|}}{C}-CH_3$$

---

**4. The "longest chain" must include the double bond.**

$$CH_3-CH_2-\underset{1\|}{\overset{2}{C}}-\underset{}{\overset{3}{CH_2}}-\overset{4}{CH_3}$$
$$\underset{CH_2}{}$$

2-Ethyl-1-butene

This compound contains a five-carbon chain, but that chain does not contain the double bond. The longest chain with the double bond contains four carbons, so the parent is *but* and not *pent*.

---

**PROBLEM 11.2** Name:

$$CH_3-CH_2-CH_2-\overset{\overset{}{|}}{C}=CH-CH_3$$
$$\underset{}{\overset{}{|}}$$
$$CH_2$$
$$\overset{}{|}$$
$$CH_3$$

---

**5. Cyclic compounds containing double bonds are called cycloalkenes.** When there is only one double bond in a ring, it always goes from the 1 to the 2 position, so it is not necessary to show the number in the name.

Cyclobutene

1-Chlorocyclopentene
(*not* 1-chloro-1-cyclopentene)

3,4-Dimethylcyclohexene

**PROBLEM 11.3**  Name:

(a)

(b)

6. **Compounds with two or more double bonds are given the suffix "-diene," "-triene," and so on.**

$$CH_2\!\!=\!\!CH\!-\!CH\!\!=\!\!CH_2$$

1,3-Butadiene

$$CH_2\!\!=\!\!\overset{\overset{\displaystyle CH_3}{\displaystyle |}}{C}\!-\!CH\!\!=\!\!CH\!-\!CH_2\!-\!CH\!\!=\!\!CH\!-\!CH_3$$

2-Methyl-1,3,6-octatriene

## 11.3  *sp²* HYBRIDIZATION AND π ORBITALS

In Section 10.4 we learned that a carbon atom uses *sp³* orbitals when it is connected to four other atoms. The carbon atom of a double bond, however, is connected to only three other atoms, not four. It hybridizes its orbitals in a different way. The hybridization takes place only among three orbitals; the fourth orbital remains unchanged.

**Figure 11.1**
The three *sp²* orbitals and the unchanged *p* orbital.

Three *sp²* orbitals are formed.

Since the mixing or hybridization of three orbitals involves one *s* and two *p* orbitals, we call the hybrid orbitals ***sp²* orbitals.** Each of them has one-third *s* character and two-thirds *p* character.

The three *sp²* orbitals all lie in one plane, and the angle between them is 120°. The fourth orbital, the unchanged *p* orbital, forms a 90° angle with the *sp²* orbitals (Fig. 11.1).

When two carbon atoms, each having *sp²* orbitals, approach each other, they form a sigma bond by sharing the two electrons from the overlapping *sp²* orbitals (Fig. 11.2).

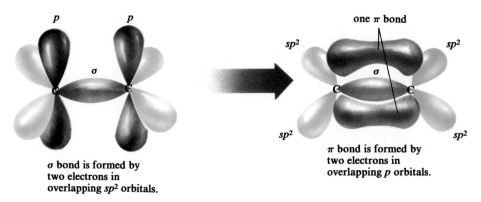

σ bond is formed by
two electrons in
overlapping *sp²* orbitals.

π bond is formed by
two electrons in
overlapping *p* orbitals.

**Figure 11.2**
Two *sp²* orbitals overlap to form a σ bond and two *p* orbitals overlap sideways to form a π bond.

The two *p* orbitals also form a bond. This bond is not the result of direct overlap, as in the sigma bond, but of the overlap of two parallel *p* orbitals (sideways overlap). An orbital formed by the overlap of two parallel *p* orbitals is called a **π** (Greek letter pi) **orbital,** and the bond is called a **π bond.** A carbon-carbon double bond is therefore made of one σ bond and one π bond. Since the π bond is weaker than the σ bond, it is more reactive and easier to break.

*There is less overlap here, so the π bond is weaker than the σ bond.*

Pi bonds can exist only in double or triple (Sec. 11.8) bonds. The parallel alignment and the sideways overlap of two *p* orbitals in the absence of a σ bond is simply not strong enough to form a stable single bond.

Pi bonds and the double bonds that contain them are not restricted to carbon-carbon double bonds. Carbon-oxygen and carbon-nitrogen double bonds are also made of a σ bond and a π bond.

## 11.4 GEOMETRY AND STEREOISOMERISM OF ALKENES

The fact that a double bond is made of one σ bond and one π bond has an important consequence for the structure of alkenes. The π bond can only exist if the two *p* orbitals are parallel to each other. If one of the *p* orbitals were to rotate, so that it was no longer parallel to the other (Fig. 11.3), the bond would be broken. This does not normally happen, because the molecule is more stable with two bonds (a σ and a π) rather than with just a σ bond.

*The bond can be broken by the addition of heat or light.*

**Figure 11.3**
The *p* orbitals of a C═C double bond remain parallel because if one rotated, it would break the π bond.

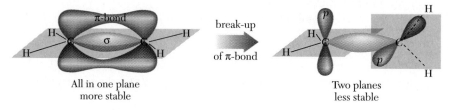

All in one plane
more stable

break-up

of π-bond

Two planes
less stable

The fact that the $p$ orbitals remain parallel (to produce the $\pi$ bond by sideways parallel overlap) also affects the $sp^2$ orbitals. We saw in Section 11.3 that the $sp^2$ orbitals are all perpendicular to the $p$ orbitals. Therefore, a molecule like ethylene is planar, with angles of 120° (the angles between $sp^2$ orbitals):

Other alkenes also have this geometry. In each case, six atoms—the two C=C atoms and the four atoms directly attached to them—are all in one plane. Furthermore, **there is no rotation about the double bond.** This is in sharp contrast to the case of single bonds, where free rotation takes place (Sec. 10.6, Fig. 10.8).

---

**B o x   11B**

**Pheromones**

Insects emit compounds called **pheromones** to transmit messages to each other. These compounds are remarkably species-specific; that is, each insect is sensitive only to the pheromones of its own species. One of the main uses for these compounds is as sex attractants. Normally, the female emits the pheromone and the male is attracted by its scent. Only a very small quantity is necessary, in some cases as little as $10^{-12}$ g.

Pheromones are often very simple compounds, though usually with long chains. Two examples are 9-tricosene, the sex attractant of the common housefly, and bombykol, secreted by the female silkworm moth:

The 9-tricosene secreted by the female housefly is the cis isomer. The trans isomer does not attract male houseflies and is not secreted by the female. Similarly, the bombykol of the silkworm moth is the trans-cis compound, and none of the three other stereoisomers work. (This is a general rule in nature: In most cases where more than one stereoisomer is possible, nature uses only one.)

The discovery of pheromones has given us a possible way to control harmful insects without polluting the environment (as happens with chlorinated insecticides; see Box 12I). Because pheromones have simple structures, they are easily synthesized in laboratories. A small quantity of a synthetic pheromone can be used to lure male insects into a trap, where they can be killed. Without males, the females cannot propagate. Some of this has already been done—for example, with Japanese beetles—and it is likely that its use will increase.

Because there is no rotation about the double bond, we can have the same type of stereoisomerism we saw for cycloalkanes in Section 10.9. There are two different compounds called 2-butene. This is easily seen if we draw the formulas this way, remembering that the molecules are planar, with 120° angles:

$$H_3C-C=C-H \qquad H_3C-C=C-CH_3$$

trans-2-Butene          cis-2-Butene

As in the case of cycloalkanes, we use the prefix *cis* for the isomer in which the two similar or identical groups are on the same side and *trans* for the one in which they are on opposite sides. As usual, the boiling points, freezing points,

---

**Box 11C**

## Cis-Trans Isomerism in Vision

(a)

(b)

**Figure 11C**
Molecular models of (a) 11-*cis*-retinal and (b) *all-trans*-retinal show the different shapes of the two isomers. (Photographs by Beverly March.)

The retina, the light-detector layer in the back of our eyes, contains colored compounds called *visual pigments*. They are insoluble in water and can be extracted from the retina with aqueous detergents. In the dark these pigments are reddish (their name, rhodopsin, comes from a Greek word meaning rose-colored), but the color fades upon exposure to light.

The rhodopsin molecules contain a protein called opsin plus a derivative of vitamin A (see Box 11A) called 11-*cis*-retinal:

11-*cis*-Retinal

*all-trans*-Retinal

In the dark, 11-*cis*-retinal is stable and fits nicely into the folds of the surrounding opsin. When light hits the rhodopsin, the 11-*cis*-retinal becomes *all-trans*-retinal and no longer fits into the cavity of opsin. The opsin and the *all-trans*-retinal separate. The change in rhodopsin conformation is eventually transmitted to the nerve cells and then the brain. The stereoisomerism of retinal is thus an important part of the vision process.

Note that only one of the five double bonds is affected in this transformation, but when this one changes from cis to trans, the shape of the entire molecule is changed (Fig. 11C). An enzyme later catalyzes the change of *all-trans*-retinal back to 11-*cis*-retinal so that it can once again bind the opsin and wait for the next exposure to light.

reactivities, and all other physical and chemical properties of the two isomers are different. Because their structures are similar, most properties are usually similar (see Table 11.1) *but not identical.* The fact that there is no free rotation about the C=C bond prevents them from interconverting. A bottle containing, say, the trans isomer will continue to hold only the trans isomer no matter how long it remains on the shelf. (When the compounds are heated to very high temperatures or treated with certain catalysts, rotation can take place, and then cis-trans conversion does occur.)

Not all double-bond compounds show cis-trans stereoisomerism. The test for its presence is very simple. **In order for there to be cis-trans stereoisomers, both double-bonded carbons must be connected to two different groups.** If *either* carbon is connected to two identical groups, there are no stereoisomers. It is easy to see why:

1,1-Dichloropropene

In this case, one of the two C=C carbons is connected to two chlorines. Thus, if we interchange the groups on either carbon, we get the same compound and not a different one. On the other hand, when both carbons of the double bond are attached to two different groups, we do have cis-trans isomers:

*cis*-2-Butene          *trans*-2-Butene

---

**EXAMPLE**

In each case decide whether cis-trans isomers exist: (a) 1-butene (b) 2-methyl-2-butene (c) 1-chloro-2-methyl-2-buten

**Answer**

(a)
1-Butene

Because one double-bonded carbon is connected to two hydrogens, there are no cis-trans isomers.

(b)
2-Methyl-2-butene

Here one double-bonded carbon is connected to two methyl groups, so again there are no cis-trans isomers.

(c)

1-Chloro-2-methyl-2-butene

This case is very similar to (b); but the presence of the Cl makes the two groups on the left-hand carbon different. Since the carbon on the right is also connected to two different groups, cis-trans isomers do exist.

**PROBLEM 11.4**  In each case, decide whether cis-trans isomers exist:
(a) 2-pentene  (b) 1-pentene  (c) 3-ethyl-2-pentene.

## 11.5  PHYSICAL PROPERTIES OF ALKENES

The physical properties of alkenes are very similar to those of the corresponding alkanes (Sec. 10.10). Up to about 5 carbons, alkenes are gases at room temperature. From about 5 to 17 carbons they are liquids, and then solids above that. Table 11.1 shows boiling and melting points for some typical compounds. Note that virtually all alkenes are insoluble in water and have densities much lower than that of water (Table 11.1).

Following the "like dissolves like" principle, alkenes are generally soluble in alkanes and in each other.

**TABLE 11.1  Physical Properties of Some Alkenes**

| Name | Structural Formula | BP(°C) | MP(°C) | Density (g/mL) |
|---|---|---|---|---|
| Ethene (ethylene) | $CH_2=CH_2$ | −104 | −169 | — |
| Propene | $CH_2=CHCH_3$ | −47 | −185 | — |
| 1-Butene | $CH_2=CHCH_2CH_3$ | −6 | −185 | 0.595 |
| cis-2-Butene | | 4 | −139 | 0.621 |
| trans-2-Butene | | 1 | −105 | 0.604 |
| Methylpropene (isobutylene) | | −7 | −140 | 0.594 |
| 1-Pentene | $CH_2=CHCH_2CH_2CH_3$ | 30 | −138 | 0.641 |
| 1-Hexene | $CH_2=CHCH_2CH_2CH_2CH_3$ | 63 | −140 | 0.673 |
| Cyclohexene | | 83 | −104 | 0.810 |

## 11.6 CHEMICAL PROPERTIES OF ALKENES: ADDITION REACTIONS

Unlike alkanes, which are inert to almost all chemical reagents (Sec. 10.11), alkenes undergo many reactions. Since the only difference between an alkane and an alkene is the double bond, we are not surprised to learn that most of the reactions of alkenes take place at the double bond. These reactions follow the pattern

$$-C{=}C- \ + \ AB \ \longrightarrow \ -\overset{A}{\underset{|}{C}}-\overset{B}{\underset{|}{C}}-$$

*AB is a general symbol to show something adding to a double bond.*

They are called **addition reactions** because something is added to the double bond. One of the two bonds of the double bond (the one shown in color) is broken, and two new bonds are formed. The original double bond is converted to a single bond. We will discuss only a few of the many known reactions of this type.

### Addition of H₂

General reaction
$$-C{=}C- \ + \ H_2 \ \xrightarrow{\text{catalyst}} \ -\overset{H}{\underset{|}{C}}-\overset{H}{\underset{|}{C}}-$$

Specific example
$$CH_3CH{=}CH_2 \ + \ H_2 \ \xrightarrow{\text{Pt}} \ CH_3CH_2CH_3$$
Propene            Propane

Any alkene can be converted to the corresponding alkane by reaction with hydrogen gas and a catalyst, which may be finely divided platinum, nickel, rhodium, or some other transition metal or transition metal compound. The alkene and the catalyst are shaken in a bottle containing hydrogen gas at a pressure slightly above 1 atm. In this quick and convenient reaction, called **catalytic hydrogenation,** the unsaturated hydrocarbon (alkene) is *saturated* with hydrogen to become a saturated hydrocarbon (alkane).

*Ethene is converted to ethane, cyclohexene to cyclohexane, and so forth.*

*In Section 17.3 we shall see how catalytic hydrogenation is used to solidify liquid oils.*

---

**PROBLEM 11.5** Write the structural formula of the product:

$$CH_3CH_2CH_2CH{=}CH_2 \ + \ H_2 \ \xrightarrow{\text{Pt}}$$

---

### Addition of Br₂ and Cl₂

*The letter X is frequently used to stand for a halogen atom.*

General reaction
$$-C{=}C- \ + \ X_2 \ \longrightarrow \ -\overset{X}{\underset{|}{C}}-\overset{X}{\underset{|}{C}}-$$

Specific example
$$CH_3CH{=}CHCH_3 \ + \ Br_2 \ \longrightarrow \ CH_3\overset{}{\underset{\underset{Br}{|}}{CH}}-\overset{}{\underset{\underset{Br}{|}}{CH}}CH_3$$

2-Butene                2,3-Dibromobutane
(cis or trans)

(a)          (b)          (c)

**Figure 11.4**
(a) A drop of $Br_2$ dissolved in $CCl_4$ is added to an unknown liquid. If the color remains after stirring (b), it indicates the absence of unsaturation. If the color disappears (c), the unknown is unsaturated. (Photographs by Beverly March.)

Bromine adds to almost all alkenes to give a compound that has two bromines on adjacent carbons. Chlorine reacts in the same way. The reaction is very similar to the one with $H_2$, and again the double bond of an alkene is converted to a single bond. The reaction is not useful for $F_2$ or $I_2$.

---

**PROBLEM 11.6**  Write the structural formula of the product:

$$\text{(cyclopentene with } CH_3) + Cl_2 \longrightarrow$$

---

Because almost all alkenes add bromine, this reaction is often used as a test for the presence of unsaturation (a double or triple bond). Organic chemists have at their disposal many such simple laboratory tests. What we want in a laboratory test is something that will give a quick answer (less than five minutes) and is simple to run. This test for unsaturation is an excellent example. All we do is add a drop of bromine (usually dissolved in carbon tetrachloride) to our unknown liquid and we get our answer at once. If the deep red color of $Br_2$ disappears, a double or triple bond is present (the color disappears because the $Br_2$ is reacting with the double or triple bond)(Fig. 11.4). Chlorine gives the same reaction but is not used as a test for unsaturation because it is difficult to see the pale green chlorine disappearing.

A quantitative version of this test is used in lipid chemistry to determine the degree of unsaturation (see Box 17A).

**Addition of HX**

General reaction

$$-C=C- + HX \longrightarrow -\overset{\underset{\displaystyle |}{H}}{C}-\overset{\underset{\displaystyle |}{X}}{C}-$$

Specific example

$$\text{(cyclohexene)} + HBr \longrightarrow \text{(bromocyclohexane)}$$

Cyclohexene          Bromocyclohexane

Any of the four hydrogen halides—HF, HCl, HBr, and HI—add to the double bond of an alkene to give the corresponding alkyl halide. The reaction is very similar to $H_2$ and $X_2$ addition. Once again, a double bond is converted to a single bond. However, we do encounter one new factor in this reaction. In $H_2$ and $X_2$ addition, the *same* group adds to each carbon atom (H and H or X and X). Here a *different* group is adding to each carbon (H and X). This means that, for most alkenes, two different products are possible. For example, when HCl adds to propene we can get either 1-chloropropane or 2-chloropropane:

$$CH_3CH{=}CH_2 + HCl \longrightarrow CH_3{-}\overset{\displaystyle Cl}{\underset{\displaystyle |}{C}}H{-}\overset{\displaystyle H}{\underset{\displaystyle |}{C}}H_2 \quad or \quad CH_3{-}\overset{\displaystyle H}{\underset{\displaystyle |}{C}}H{-}\overset{\displaystyle Cl}{\underset{\displaystyle |}{C}}H_2$$

Propene         2-Chloropropane       1-Chloropropane
                     (main product)

More than 100 years ago, the Russian chemist Vladimir Markovnikov (1838–1904), after studying many such cases, formulated a rule that predicts which product will be exclusively or predominantly formed. **Markovnikov's rule** states that, **in the addition of HX, the hydrogen goes to the carbon that already has more hydrogens.** The X goes to the carbon with fewer hydrogens. The only hydrogens we count are those *directly* attached to the carbons of the double bond. Applying this rule to propene, we find

*Markovnikov's rule is frequently stated as "the rich get richer."*

$$\overset{3}{C}H_3{-}\overset{2}{C}H{=}\overset{1}{C}H_2 \quad \text{Two hydrogens}$$

One hydrogen

Therefore, the hydrogen of HX goes to the end carbon, and the main product is 2-chloropropane. The other compound, 1-chloropropane, is formed only in small quantities or not at all.

---

**PROBLEM 11.7** Write the structural formula of the principal product:

$$CH_3CH_2{-}\underset{\displaystyle \underset{\displaystyle CH_3}{|}}{C}{=}CH_2 + HBr \longrightarrow$$

---

## Addition of $H_2O$

General reaction
$$-\overset{|}{C}{=}\overset{|}{C}- + H_2O \xrightarrow{H^+} -\overset{\displaystyle H}{\underset{\displaystyle |}{C}}-\overset{\displaystyle OH}{\underset{\displaystyle |}{C}}-$$

Specific example
$$CH_3{-}\underset{\displaystyle \underset{\displaystyle CH_3}{|}}{C}{=}CH_2 + H_2O \xrightarrow{H_2SO_4} CH_3{-}\overset{\displaystyle OH}{\underset{\displaystyle \underset{\displaystyle CH_3}{|}}{C}}{-}\overset{\displaystyle H}{\underset{\displaystyle |}{C}}H_2$$

Isobutylene                    *tert*-Butyl alcohol

**Box 11D**

## The Mechanism of HX Addition

Organic chemists spend a great deal of time investigating the mechanisms of reactions. A **mechanism** describes exactly how a reaction takes place: which bonds break, in what order, and so forth. Additions of HBr, HCl, HF, HI, and $H_2O$ to double bonds all have essentially the same mechanism. The mechanism begins when an $H^+$ ion approaches the double bond and forms a bond by using one of the electron pairs of the double bond:

$$-\underset{|}{C}=\underset{|}{C}- + H^+ \longrightarrow -\underset{|}{\overset{H}{\underset{|}{C}}}-\overset{+}{\underset{|}{C}}- \qquad \text{First step of the mechanism}$$

Carbocation

It is only natural that a double bond, which has a high electron density (because there are four electrons), should be attractive to a positive ion like $H^+$. One of the ways we know that it is $H^+$ attacking (and not, say, HBr) is that $H_2O$ alone does not add to an alkene. An acid catalyst is necessary to supply $H^+$ ions.

Let us now look at the ion formed from the alkene and the $H^+$. This ion has a positive charge on a carbon atom. Such ions are called **carbocations** (some chemists call them **carbonium ions**). A carbocation is extremely reactive. It cannot last for more than a tiny fraction of a second, and so of course it is impossible to isolate it in a stable salt, as we can isolate the $Na^+$ ion as NaCl, for example. As soon as it is formed, the carbocation reacts with any species in its vicinity that can supply a pair of electrons. In the case of HX this is a halide ion, so the second step of the mechanism is

$$-\underset{|}{\overset{H}{\underset{|}{C}}}-\overset{+}{\underset{|}{C}}- + Br^- \longrightarrow -\underset{|}{\overset{H}{\underset{|}{C}}}-\underset{|}{\overset{Br}{\underset{|}{C}}}-$$

In the case of water, there is no negative ion available, but the O of $H_2O$ has two unshared electron pairs, one of which is used:

In the absence of a catalyst, water does not react with alkenes. But if an acid catalyst—usually sulfuric acid—is added, water does add to C=C double bonds to give alcohols. The reaction is called **hydration.** As shown in the isobutylene reaction, Markovnikov's rule is followed: The H of the $H_2O$ goes to the carbon that already has more hydrogens.

---

**PROBLEM 11.8** Write the structural formula of the principal product:

$$CH_3-CH=CH_2 + H_2O \xrightarrow{H_2SO_4}$$

---

Most industrial ethyl alcohol, $CH_3CH_2OH$, is made by the hydration of ethylene, $CH_2=CH_2$.

The conversion of alkenes to alcohols is important not only in the organic chemist's laboratory and industrially but also in the body, where the hydration of double bonds is catalyzed by enzymes rather than by acids. An important example is the hydration of fumaric acid to give malic acid, catalyzed by the enzyme fumarase:

$$-\overset{H}{\underset{|}{\overset{|}{C}}}-\overset{+}{\underset{|}{\overset{|}{C}}}- + \ddot{O}H_2 \xrightarrow[\text{step}]{\text{second}} -\overset{H}{\underset{|}{\overset{|}{C}}}-\overset{\overset{+}{O}H_2}{\underset{|}{\overset{|}{C}}}- \xrightarrow[\text{step}]{\text{third}} -\overset{H}{\underset{|}{\overset{|}{C}}}-\overset{OH}{\underset{|}{\overset{|}{C}}}- + H^+$$

In this case a third step follows: loss of $H^+$ to give the alcohol. Note that the catalyst, $H^+$, that started the reaction is recovered unchanged.

A knowledge of mechanisms helps us to understand the facts about a reaction: why some go but others don't, and why those that go proceed the way they do. For example, understanding the mechanism of HX addition allows us to understand why Markovnikov's rule works (which is something Markovnikov himself didn't understand because the mechanism was unknown in his time). If we add HX to propene, the first step is attack by $H^+$. This could lead to two different carbocations:

$$\underset{\text{Propene}}{\overset{3\quad 2\quad 1}{CH_3CH{=}CH_2}} + H^+ \begin{array}{l} \xrightarrow[\text{at C-1}]{\text{attack}} CH_3-\overset{+}{CH}-CH_3 \quad \text{Secondary carbocation} \\[2mm] \xrightarrow[\text{at C-2}]{\text{attack}} CH_3-CH_2-\overset{+}{CH_2} \quad \text{Primary carbocation} \end{array}$$

We know from much other evidence that secondary carbocations (where the positive charge is on a secondary carbon) are more stable than primary carbocations (tertiary carbocations are even more stable). Therefore, the $H^+$ is much more likely to attach to C-1 rather than C-2 because that gives a secondary rather than a primary carbocation. Once the secondary carbocation is formed, it must give the secondary halide or alcohol:

$$CH_3-\overset{+}{CH}-CH_3 + X^- \longrightarrow CH_3-\overset{X}{\underset{|}{CH}}-CH_3 \quad \text{Second step of the mechanism}$$

A carbocation is an example of an **intermediate,** which is a species that is formed in a reaction but then is used up so quickly that it cannot be isolated.

$$\underset{\text{Fumaric acid}}{\overset{\displaystyle H}{\underset{\displaystyle HOOC}{}}{C}{=}{C}{\overset{\displaystyle COOH}{\underset{\displaystyle H}{}}}} + H_2O \xrightarrow{\text{fumarase}} \underset{\text{Malic acid}}{HOOC-\overset{OH}{\underset{|}{CH}}-CH_2-COOH}$$

This is one of the steps in the metabolism of carbohydrates, fats, and proteins (see Sec. 20.4).

Besides the addition reactions just discussed, two other chemical properties of double-bond compounds are worthy of mention. First, like alkanes and all other hydrocarbons, alkenes burn; the products of complete combustion are $CO_2$ and $H_2O$:

$$CH_2{=}CH_2 + 3O_2 \longrightarrow 2CO_2 + 2H_2O$$

Second, double-bond compounds can be polymerized to give long chains. This topic is treated in the following section.

## 11.7 ADDITION POLYMERS

Possibly the most important reaction of alkenes and other compounds that contain C=C bonds is that, when treated with appropriate catalysts, they add to each other to give long chains. For example, the simplest alkene, ethylene, polymerizes as follows:

$$
\begin{array}{cccccccccc}
H & H & H & H & H & H & H & H & H & H \\
| & | & | & | & | & | & | & | & | & | \\
C{=}C & + & C{=}C & + & C{=}C & + & C{=}C & + & C{=}C \\
| & | & | & | & | & | & | & | & | & | \\
H & H & H & H & H & H & H & H & H & H
\end{array}
$$

catalyst ↓

$$
\begin{array}{cccccccccc}
H & H & H & H & H & H & H & H & H & H \\
| & | & | & | & | & | & | & | & | & | \\
-C & -C & -C & -C & -C & -C & -C & -C & -C & -C- \\
| & | & | & | & | & | & | & | & | & | \\
H & H & H & H & H & H & H & H & H & H
\end{array}
$$

Newly formed bonds are shown in color

Polyethylene

Under the influence of the catalyst, each double bond becomes a single bond, and the extra electrons form new single bonds that tie the molecules together into long chains. **A long chain made up of repeating units** is called a **polymer,** and the process of making it is **polymerization.** Later in the book we will meet many biological polymers. The compounds made by polymerizing alkenes do not exist in nature; they are synthetic polymers. The simple compound that is polymerized (in the above case, ethylene) is called a **monomer.**

There are two main types of synthetic polymers. The ones we are discussing in this section are called **addition polymers** because they are made by adding double-bonded molecules to each other. (The other kind, **condensation polymers,** are discussed in Boxes 14D and 15E.) The polymer made by polymerizing ethylene is called polyethylene. The molecules of this polymer are essentially unbranched alkanes thousands of units long. When another group is present in the monomer, the polymer will have this group on every second carbon atom. For example, polymerization of chloroethene (common name, vinyl chloride) gives the polymer polyvinyl chloride (PVC):

"Poly," from the Greek *polys,* means many.

$$
\underset{\text{Vinyl chloride}}{CH_2{=}\overset{\overset{\displaystyle Cl}{|}}{CH} + CH_2{=}\overset{\overset{\displaystyle Cl}{|}}{CH} + CH_2{=}\overset{\overset{\displaystyle Cl}{|}}{CH}} \longrightarrow
$$

The repeating unit in PVC is
—CH$_2$—CH—.
     |
     Cl

$$
\underset{\text{Polyvinyl chloride}}{-CH_2{-}\overset{\overset{\displaystyle Cl}{|}}{CH}{-}CH_2{-}\overset{\overset{\displaystyle Cl}{|}}{CH}{-}CH_2{-}\overset{\overset{\displaystyle Cl}{|}}{CH}-}
$$

---

**PROBLEM 11.9** Draw the structural formula of polyacrylonitrile, made by polymerizing cyanoethene, $CH_2{=}CH{-}CN$.

---

An important example of a naturally-occurring addition polymer is rubber. Natural rubber is obtained upon coagulation of the sap (called *latex*) of the *Hevea* tree. Rubber differs from the other polymers discussed in this section in that the monomer has *two* double bonds, so the polymer still contains double bonds:

Newly formed bonds are shown in color

The monomer is 2-methyl-1,3-butadiene, which has the common name **isoprene.** All the double bonds in natural rubber have the cis configuration, so the chemical name is *all-cis*-polyisoprene. The most important property of rubber is its elasticity; it can be extended a considerable amount, and then, when released, it snaps back to its original dimensions. However, raw rubber is too soft and tacky to be of much use, and it must be hardened by heating it with sulfur, a process called **vulcanization.** This process creates sulfur *cross links,* which hold the polymer chains together:

Two to three percent sulfur gives rubber that is used for such things as shoe soles, erasers (which is where rubber got its name; it was used to rub out pencil marks), and rubber bands, while more sulfur (5 to 8 percent) gives rubber that can be used for tires.

Although natural rubber is a highly useful material, it is nevertheless not the best material for all applications where an elastic substance is needed. Chemists have been able to polymerize dienes other than isoprene to obtain **synthetic rubbers,** some of which are copolymers. Worldwide, nearly two thirds of all rubber used today is synthetic. The largest use of rubber is in tires, and most tires contain both natural rubber (which withstands the heat better) and one or more kinds of synthetic rubber.

Some of the more important commercial addition polymers are shown in Table 11.2. Saran is an example of a **copolymer,** which is a polymer made up of two or more different monomers.

As you can tell by looking at some of the familiar names in Table 11.2, synthetic polymers are very important in modern life, and we would find it hard

**TABLE 11.2  Some Important Addition Polymers**

| Monomer | | Polymer | |
|---------|---|---------|---|
| *Name* | *Structure* | *Structure* | *Name* |
| Ethylene | $CH_2{=}CH_2$ | $-CH_2-CH_2-CH_2-CH_2-$ | Polyethylene |
| Propylene | $CH_2{=}\overset{\displaystyle CH_3}{\underset{\displaystyle \vert}{CH}}$ | $-CH_2-\overset{\displaystyle CH_3}{\underset{\vert}{CH}}-CH_2-\overset{\displaystyle CH_3}{\underset{\vert}{CH}}-$ | Polypropylene |
| Styrene | $CH_2{=}CH$ (phenyl) | $-CH_2-CH-CH_2-CH-$ (phenyl, phenyl) | Polystyrene |
| Vinyl chloride (chloroethene) | $CH_2{=}\overset{\displaystyle Cl}{\underset{\vert}{CH}}$ | $-CH_2-\overset{\displaystyle Cl}{\underset{\vert}{CH}}-CH_2-\overset{\displaystyle Cl}{\underset{\vert}{CH}}-$ | Polyvinyl chloride (PVC) |
| Acrylonitrile (cyanoethene) | $CH_2{=}\overset{\displaystyle CN}{\underset{\vert}{CH}}$ | $-CH_2-\overset{\displaystyle CN}{\underset{\vert}{CH}}-CH_2-\overset{\displaystyle CN}{\underset{\vert}{CH}}-$ | Polyacrylonitrile |
| Methyl methacrylate | $CH_2{=}\overset{\displaystyle CH_3}{\underset{\displaystyle COOCH_3}{C}}$ | $-CH_2-\overset{\displaystyle CH_3}{\underset{\displaystyle COOCH_3}{C}}-CH_2-\overset{\displaystyle CH_3}{\underset{\displaystyle COOCH_3}{C}}-$ | Acrylic plastics |
| Tetrafluoroethene | $CF_2{=}CF_2$ | $-\overset{F}{\underset{F}{C}}-\overset{F}{\underset{F}{C}}-\overset{F}{\underset{F}{C}}-\overset{F}{\underset{F}{C}}-$ | Teflon |
| Vinylidine chloride  +  Vinyl chloride | $CH_2{=}\overset{Cl}{\underset{Cl}{C}}$  +  $CH_2{=}\overset{}{\underset{Cl}{CH}}$ | $-\overset{H}{\underset{H}{C}}-\overset{Cl}{\underset{Cl}{C}}-\overset{H}{\underset{H}{C}}-\overset{H}{\underset{Cl}{C}}-$ | Saran (copolymer) |

to get along without them (Fig. 11.5). Among the synthetic polymers are plastics. Some of these are rigid and are used for such things as combs, ballpoint pens, toys, and bottles for soft drinks and other liquids. Other plastics are produced as thin sheets and are used to wrap foods and other merchandise. Others are used as artificial fibers (nylon and Dacron, Boxes 14D and 15E), synthetic rubbers and leathers, adhesives, and coatings (for example, the glazed surface of refrigerators), and other uses. Figure 11.6 shows a plastic garment used by surgeons as protection when treating patients with AIDS.

In polymers the chemical structure is less important in defining physical properties than are other attributes such as molecular weight or the orientation of the molecules. The polymer polyethylene supplies an example. Low-molecular-weight polyethylene (MW 500 to 600) has the consistency of grease and is used for lubrication. The ordinary polyethylene bag (low-density polyethylene) used in packaging owes its flexibility to short branches along the chain (the

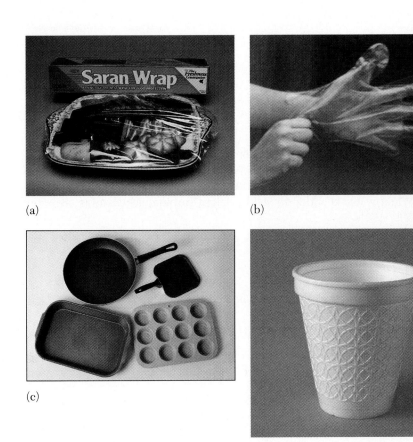

(a)

(b)

(c)

(d)

**Figure 11.5**
Some articles made from synthetic addition polymers. (a) Saran wrap, a copolymer of vinyl chloride and vinylidene chloride. (b) A polyethylene glove. (c) Teflon-coated kitchenware. (d) A cup made from polystyrene. (Photographs by Beverly March.)

branches prevent the molecules from getting very close to each other, so that crystals cannot be formed), while high-density linear polyethylene (without branches) is more rigid and is used in manufacturing pipes, bottles, toys, and a host of houseware items. Both have molecular weights up to a few hundred thousand. Polyethylene that is solidified (crystallized) while being extended exhibits even more interesting properties. Here the molecular weights are up to millions, and the molecules are stretched out, unfolded, and oriented in one direction. Such polyethylene films and fibers (manufactured under the name of Spectra) are so strong that they are used in manufacturing bullet-proof vests, riot helmets, and cut-resistant gloves for surgeons to wear under surgical latex gloves.

The importance of polymers can be judged by the fact that they make up more than half the output of our chemical factories.

## 11.8 ALKYNES

Alkynes are named by the rules used for alkenes except that the suffix "-yne" is used for the triple bond:

$$HC{\equiv}CH \qquad CH_3{-}C{\equiv}CH \qquad CH_3{-}\underset{\underset{\displaystyle CH_3}{|}}{CH}{-}C{\equiv}C{-}CH_3$$

Ethyne          Propyne                 4-Methyl-2-pentyne
(acetylene)

**Figure 11.6**
A plastic garment and other materials used as protection when treating people with AIDS. (Courtesy of the National Museum of Health and Medicine, Armed Forces Institute of Pathology, Washington, D.C.)

---

**PROBLEM 11.10** Name:

(a) $CH_3CH_2CH_2—C{\equiv}C—CH_3$

(b) $CH_3—\underset{\underset{Cl}{|}}{CH}—\underset{\underset{CH_3}{|}}{CH}—C{\equiv}CH$

---

The high temperature of an acetylene flame makes it useful for torches that can cut and weld steel.

Ethyne, the simplest and by far the most important alkyne, is much more often called by its common name, acetylene. It is commercially produced by two methods. The older method (discovered in the nineteenth century) is very simple. When water is added to solid calcium carbide, acetylene is given off (Fig. 11.7):

$$CaC_2 + 2H_2O \longrightarrow HC{\equiv}CH + Ca(OH)_2$$

Calcium carbide

In the other method, methane is heated to a very high temperature:

$$2CH_4 \xrightarrow{1500°C} HC{\equiv}CH + 3H_2$$

We can see the geometry of triple bonds by examining the hybridization. A triple-bond carbon is attached to only two other atoms, so it hybridizes as follows:

$$\boxed{\uparrow}\ \boxed{\uparrow}\ \boxed{\uparrow}\ \boxed{\uparrow} \longrightarrow \boxed{\uparrow}\ \boxed{\uparrow}\ \boxed{\uparrow}\ \boxed{\uparrow}$$
$$s\quad p_x\quad p_y\quad p_z \qquad sp\quad sp\quad p_y\quad p_z$$

We now have two hybrid orbitals, each called $sp$. The **$sp$ orbitals** have 50 percent $s$ and 50 percent $p$ character, and the angle between the two is 180°. Both of the remaining $p$ orbitals are perpendicular to the $sp$ orbitals and to each other (Fig. 11.8). Thus two carbon atoms that have $sp$ hybrid orbitals form a

**Figure 11.7**
Calcium carbide is a brown solid (right). When water is added (left) bubbles of acetylene gas and solid $Ca(OH)_2$ are produced. The fact that the reaction gives off much heat is indicated by a wisp of steam rising from the reaction. (Photograph by Charles D. Winters.)

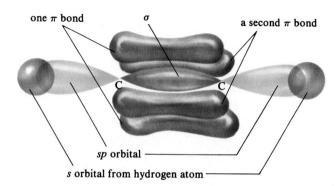

one $\pi$ bond     $\sigma$     a second $\pi$ bond

*sp* orbital

*s* orbital from hydrogen atom

**Figure 11.8**
A triple bond is made of one $\sigma$ and two $\pi$ bonds.

triple bond: one $\sigma$ bond formed by the direct overlap of *sp* hybrid orbitals, and two $\pi$ bonds formed by the parallel sideways overlap of the $p_y$ and $p_z$ orbitals of both carbons (Fig. 11.8). The remaining *sp* orbitals (one from each carbon) form $\sigma$ bonds with the *s* orbitals of hydrogen atoms. This causes acetylene to be a linear molecule—all four atoms are in a straight line:

$$H—C≡C—H$$

The same geometry is found for all other alkynes: The two carbon atoms and the two atoms connected to them are all in a straight line.

Alkynes also undergo addition reactions. The same substances that add to double bonds ($H_2$, $Br_2$, HCl, and so on) also add to triple bonds, though somewhat less readily. Since the product of addition to a triple bond still has a double bond, it is often possible to get a second addition reaction:

$$HC≡CH + Br_2 \longrightarrow \underset{\underset{Br}{|}}{\overset{\overset{Br}{|}}{HC}}=\underset{\underset{Br}{|}}{\overset{\overset{Br}{|}}{CH}} + Br_2 \longrightarrow H—\underset{\underset{Br}{|}}{\overset{\overset{Br}{|}}{C}}—\underset{\underset{Br}{|}}{\overset{\overset{Br}{|}}{C}}—H$$

$$CH_3—C≡CH + HBr \longrightarrow CH_3—\underset{}{\overset{\overset{Br}{|}}{C}}=CH_2 + HBr \longrightarrow CH_3—\underset{\underset{Br}{|}}{\overset{\overset{Br}{|}}{C}}—CH_3$$

$$CH_3—C≡C—CH_3 + H_2 \xrightarrow{Pt} CH_3—CH=CH—CH_3 + H_2 \xrightarrow{Pt} CH_3CH_2CH_2CH_3$$

---

**PROBLEM 11.11**  Draw the structural formula for the principal product after 1 mole of reagent adds and after 2 moles of reagent add:
(a)  $CH_3—C≡C—CH_2CH_2CH_3 + Cl_2 \longrightarrow$
(b)  $HC≡C—CH_2CH_2CH_3 + HBr \longrightarrow$

---

Because alkynes react with $Br_2$, they also give positive tests for unsaturation. It is therefore not possible to tell an alkene from an alkyne by this test alone. Note from the reactions just shown that the addition of HBr obeys Markovnikov's rule.

## 11.9  AROMATIC HYDROCARBONS

All of the hydrocarbons discussed so far—alkanes, alkenes, and alkynes (including the cyclic ones)—are called **aliphatic hydrocarbons.** More than 150 years ago, it became apparent to the early organic chemists that there was another kind of compound, one whose chemical properties are quite different from those of the aliphatic compounds. Because some of them have pleasant odors, they were named **aromatic compounds.** Today we know that most aromatic compounds do not have pleasant odors (in fact, some are downright unpleasant).

The most common aromatic hydrocarbon is benzene. Between 1860 and 1880, benzene and the other aromatic compounds known at that time presented a major challenge to the theories of organic structure. The problem was that benzene has a molecular formula of $C_6H_6$, and a compound with so few hydrogens for its six carbons (compare hexane, $C_6H_{14}$) should be quite unsaturated, with several double and/or triple bonds. But benzene does not *behave* like an unsaturated compound. It does not give a positive $Br_2$ test; in fact, it does not react with $Br_2$ at all except in the presence of a catalyst (Sec. 11.11). When benzene does undergo reactions, they are *substitution* reactions, not *addition* reactions:

Ethylene (addition)

$$CH_2{=}CH_2 + Br_2 \xrightarrow[\text{fast}]{\text{no catalyst}} \underset{\underset{}{\overset{\overset{Br \quad Br}{|\quad\ |}}{CH_2{-}CH_2}}}{}$$

Benzene (substitution)

$$C_6H_6 + Br_2 \xrightarrow[\text{slow}]{\text{Fe}} C_6H_5Br + HBr$$

This is a major difference in chemical properties and shows that aromatic compounds, despite their low hydrogen content, cannot be regarded as unsaturated molecules. The full story of how the dilemma was finally resolved is too long to tell here. We merely say that the man who did the most to resolve it was Kekulé (Fig. 11.9), who proposed that the six carbons of benzene are connected in a ring. Today we know that benzene consists of a ring of six carbon atoms, each connected to a hydrogen atom (structure **A**):

Incomplete structure of benzene

**A**

This gives each carbon three bonds. Carbon, of course, forms four bonds, and so each carbon has one remaining electron. These six electrons are located in a cloud (made up of three molecular orbitals) that looks like two donuts (Fig. 11.10). All twelve atoms—six carbons and six hydrogens—are in the same

The physical properties of aromatic hydrocarbons—boiling points, solubilities, densities, etc.—are similar to those of the aliphatic hydrocarbons.

Benzene, the simplest and most important aromatic hydrocarbon, was discovered by Michael Faraday (1791–1867) in 1825.

If aromatic compounds were unsaturated, they would add $Br_2$ the same way alkenes and alkynes do.

Benzene is an important industrial compound, but it must be handled carefully because it is toxic. Not only is it poisonous if ingested in the liquid form, but the vapor form is also toxic and can be absorbed either by breathing or through the skin. Long-term inhalation can cause liver damage and cancer.

These three orbitals are also called $\pi$ orbitals, because they result from the sideways overlap of $p$ orbitals.

**Figure 11.9**
Friedrich August Kekulé. (Courtesy of
E. F. Smith Memorial Collection,
Center for History of Chemistry,
University of Pennsylvania.)

plane (benzene is a *flat* molecule); one donut lies just above the plane, the other just below it.

The picture of benzene shown as structure **A** does not violate the rule that carbon forms four bonds. Each carbon is connected to three other atoms by ordinary covalent bonds. The fourth bond can be thought of as being shared by all six carbons to create the donut-shaped clouds (the **aromatic sextet, or ring,** of electrons). It is true that Figure 11.10 is not a Lewis structure (Sec. 3.4), but benzene and other aromatic compounds simply cannot be represented by a single Lewis structure.

However, there is a way to represent benzene by Lewis structures. In this method we must draw not one, but two, Lewis structures:

**Figure 11.10**
Benzene. The two donut-shaped clouds hold a total of six electrons.

The double-headed arrow is used to indicate resonance. It has no other use.

We start with the three ordinary covalent bonds for each carbon (structure **A**) and put in the extra six electrons as three double bonds. Because there are two ways to do this, we get two Lewis structures, **B** and **C.** Both are valid Lewis structures. However, neither of these structures is the correct formula for benzene. (As we have already said, benzene cannot be represented by a single Lewis structure.) We know that neither can be the correct structure because if either one were, then benzene would have three double bonds and would undergo addition reactions. **When we write these structures, we understand that the true benzene is neither of them but has a structure halfway between them.** In other words, *real* benzene is an *average* of **B** and **C**.

The use of fictitious structures to represent a real structure that is the average of them is called **resonance.** It must always be understood that structures like **B** and **C** are fictitious—they don't exist. The real molecule is an average, or **hybrid,** of the two fictitious ones.

Wherever we find resonance, we find stability. The real structure is always more stable than the fictitious structures. (If any fictitious structure were more stable, it would exist.) The benzene ring is greatly stabilized by resonance. That is why benzene does not undergo addition reactions. If, say, $Br_2$ added to benzene, it would have to pull two electrons out of the aromatic sextet, resulting in a molecule without an aromatic sextet.

*Chemists often use fictitious resonance structures because they obey the rules for Lewis structures, and chemists are comfortable with them.*

A reaction that doesn't happen

This does not happen because molecules with aromatic sextets are more stable than similar molecules (such as **D**) without aromatic sextets.

Both the resonance picture and the aromatic sextet picture shown in Figure 11.10 predict that all six C—C bonds in benzene should be the same length. Bond distances can be measured by x-ray crystallography, and such measurements show that the six bonds are indeed all the same length.

We can now define an **aromatic compound** as **one that contains an aromatic loop (sextet) of electrons,** just as benzene itself does. Because of the aromatic sextet, which is donut-shaped, we use formulas like **E**, with a circle, to represent benzene and most other aromatic compounds. However, many chemists use the Kekulé type formulas **F** or **G**:

*When you see formulas like these, don't forget that each carbon is connected to a hydrogen. We don't usually show the six hydrogens, but they are there just the same.*

E      F      G

This is strictly a matter of preference. Whichever you choose, you should realize that they all mean the same thing. The real structure of benzene is that shown in Figure 11.10.

## 11.10 NOMENCLATURE OF BENZENE DERIVATIVES

A benzene ring may have one or more of its six hydrogens replaced by other groups. They are named as follows.

## One Group

Some monosubstituted benzenes are named by giving the name of the group followed by the parent "benzene":

| Chlorobenzene | Nitrobenzene | Ethylbenzene | Bromobenzene |

Others, however, are always named by common names:

| Toluene | Phenol | Aniline | Benzoic acid |

These four compounds are so important that their names should be memorized.

Note that all six benzene hydrogens are equivalent and so it makes no difference whether we write the group up, down, right, or left.

## Two Groups

When there are two groups on a benzene ring (disubstituted benzenes), three isomers are possible, whether the groups are the same or different:

| 1,2 *(ortho)* | 1,3 *(meta)* | 1,4 *(para)* |

All of these are dichlorobenzene, and, as usual, we distinguish the *positions* of the groups by prefixes. There are two ways to do this. We can use numbers, following the rule for cyclohexanes (rule 7 in Section 10.8). However, a much more common way is to use prefixes: *ortho* for 1,2; *meta* for 1,3; and *para* for 1,4.

The letters *o*, *m*, and *p* are often used as abbreviations for *ortho, meta,* and *para.*

When one of the two groups produces a compound that has a common name, we use that name as the parent:

*o*-Bromotoluene    *p*-Nitrophenol

---

**PROBLEM 11.12**   Name:

(a)    (b)    (c)

---

## Three or More Groups

The rules do not permit the use of *ortho, meta,* and *para* when more than two groups are present. For these cases we must use numbers, following rule 7 in Section 10.8:

1,2,4-Tribromobenzene    2,3-Dinitroaniline    1,4-Dibromo-2,6-dichlorobenzene

4-Iodo-3,5-dimethylphenol    2,4,6-Trinitrotoluene (TNT)

---

**PROBLEM 11.13**   Name:

(a)

(b)

---

## The Phenyl Group

In many cases it is necessary to name the benzene ring as a group. Although you might suppose that this is called the benzyl group, the name used is not that, but **phenyl.** For example:

Diphenylmethane

As you can see, compounds like this would be difficult to name using the benzene ring as a parent. Remember that the phenyl group is $C_6H_5$—; that is, one of the six benzene hydrogens has been removed to make it a group (just as the methyl group, $CH_3$—, has one less hydrogen than methane).

One reason you shouldn't make a mistake and call the phenyl group "benzyl" is that there *is* a benzyl group. It is $C_6H_5CH_2$—. Here are some examples of both:

Phenyl chloride    Benzyl chloride      1-Benzyl-3-phenylcyclopentane
(chlorobenzene)

## 11.11   REACTIONS OF AROMATIC COMPOUNDS

Most of the known reactions of aromatic compounds are **aromatic substitutions.** We will discuss three of them.

*We have stressed that aromatic compounds undergo substitution reactions rather than additions.*

### Nitration

A nitro compound

When an aromatic compound is heated with a mixture of concentrated nitric and sulfuric acids, one of the hydrogens of the ring is replaced by a nitro group, —$NO_2$. Because this is such a convenient reaction, aromatic nitro compounds are very common.

### Sulfonation

A sulfonic acid

Heating an aromatic compound with concentrated sulfuric acid alone gives a sulfonic acid. The one shown above is called benzenesulfonic acid. Sulfonic acids are strong acids, about as strong as sulfuric acid.

### Halogenation

Aromatic rings react with bromine, in the presence of an iron catalyst, to give brominated aromatic compounds. The bromine replaces a hydrogen of the ring. Chlorine behaves in the same way and produces a chlorinated aromatic compound.

## 11.12 FUSED AROMATIC RINGS

All of the aromatic compounds we have mentioned up to now possess a single intact benzene ring. We have devoted so much space to these compounds because they are so common, both in laboratories and in our bodies. There are also compounds with two or more benzene rings fused together. The three simplest of these are

Note that naphthalene does not have two *complete* benzene rings (nor do anthracene or phenanthrene have three complete benzene rings). Two complete benzene rings would require 12 carbon atoms, but naphthalene has only 10. Because it does not have two complete aromatic sextets, we do not draw circles inside the rings; instead, we represent these compounds by single resonance forms (other resonance forms can also be drawn).

Naphthalene      Anthracene      Phenanthrene

Unlike benzene, which is a liquid, these compounds are all solids at room temperature. Naphthalene is used to make moth balls.

Fused aromatic compounds undergo aromatic substitutions, just as single-ring benzenes do.

## 11.13 HETEROCYCLIC COMPOUNDS

**A heterocyclic compound** is **one that contains a ring in which at least one atom is not carbon.** Heterocyclic compounds are very important in biochemistry, and we will meet many of them in later chapters. Some heterocyclic compounds are aromatic; others are not. Some examples of nonaromatic heterocyclic compounds are

Tetrahydropyran

Pyrrolidine

Piperidine

These compounds are nonaromatic because they do not have an aromatic sextet.

# Carcinogenic Fused Aromatics and Smoking

Some years ago it was discovered that certain organic compounds, when rubbed on the skin of rats, caused skin cancers. **A compound that causes cancer** is called a **carcinogen.** Although many compounds have now been found to be carcinogenic, the first to be discovered were a group of fused aromatic hydrocarbons, all of which have at least four rings and at least one angular junction (that is, structures resembling phenanthrene rather than anthracene). Some of these are

Dibenz[*a,h*]anthracene          Benzo[*a*]pyrene          Benzo[*e*]pyrene

These carcinogens are produced by automobile exhausts and are found in cigarette smoke. If you smoke, you greatly increase your chances of getting cancer. A sizable proportion of cancers, especially lung cancers (Fig. 11F), but other cancers also, are caused by cigarette smoking.

About 200 years ago a London surgeon, Percivall Pott, found that chimney sweeps (boys employed to clean chimneys) were especially prone to cancer of the scrotum and other parts of the body. Today it is known that these cancers were caused by fused aromatic hydrocarbons present in the chimney soot.

**Figure 11F**
A smoker's lung showing carcinoma. (Courtesy of Anatomical Collections, National Museum of Health and Medicine, Armed Forces Institute of Pathology, Washington, D.C.)

Some of the most important aromatic heterocyclic compounds are

In pyrimidine, two nitrogens (in the 1 and 3 positions) replace CH groups of benzene. As we shall see in Section 23.2, pyrimidine rings are important constituents of nucleic acids.

Pyrrole is an example of a five-membered heterocyclic aromatic compound. Although there are only five atoms in the ring, there are six electrons (including the unshared pair on the nitrogen atom), so the ring is aromatic. Note that pyridine has the same formula as benzene except that one CH has been replaced by an N. There is no hydrogen on the N because nitrogen forms three bonds and not four.

## SUMMARY

**Alkenes** are hydrocarbons containing double bonds. They are named the same way alkanes are except that the suffix "-ene" is used and the double bond is given the lower of the two carbon numbers.

Carbons forming double bonds use $sp^2$ orbitals. One of the two bonds in a double bond is $\sigma$, the other, $\pi$. All this causes alkenes to be planar molecules. There is no rotation about a double bond, so that **cis and trans isomers** occur in suitable cases.

The chief chemical property of alkenes is that they undergo **addition reactions,** for example with $H_2$ (in the presence of a catalyst), $Br_2$, $Cl_2$, HBr, HCl, and $H_2O$. Unsymmetrical reagents like HBr add according to **Markovnikov's rule:** The hydrogen goes to the carbon that already has more hydrogens. Compounds containing double bonds can be **polymerized** to give **addition polymers,** which are used for plastics, artificial fibers, synthetic rubbers, and coatings. The **polymers** consist of long chains of repeating units.

**Alkynes** are hydrocarbons containing triple bonds, which consist of one $\sigma$ and two $\pi$ bonds. They are linear molecules. The most important is acetylene (ethyne), which is used in acetylene torches. Alkynes also undergo addition reactions.

All the preceding hydrocarbons are **aliphatic. Aromatic hydrocarbons** and their derivatives have an aromatic ring of electrons like the one in benzene. The benzene ring is especially stable and is found in many naturally occurring compounds. Aromatic compounds undergo **substitution** (rather than addition) reactions, including nitration, sulfonation, and halogenation. The prefixes *ortho, meta,* and *para* are usually used to name disubstituted aromatic compounds.

Many aromatic compounds (such as naphthalene or anthracene) have two or more **fused rings.** Others are **heterocyclic**—they have one or more atoms in the ring that are not carbon. Some aromatic heterocyclic rings are six-membered, others are five-membered.

## Summary of Reactions

1. Addition of $H_2$ to alkenes (Sec. 11.6):

$$-\overset{|}{C}=\overset{|}{C}- + H_2 \xrightarrow{\text{catalyst}} -\overset{|}{\underset{H}{C}}-\overset{|}{\underset{H}{C}}-$$

2. Addition of $Br_2$ or $Cl_2$ to alkenes (Sec. 11.6):

$$-\overset{|}{C}=\overset{|}{C}- + X_2 \longrightarrow -\overset{|}{\underset{X}{C}}-\overset{|}{\underset{X}{C}}- \qquad X = \text{Br or Cl}$$

3. Addition of HX to alkenes (Sec. 11.6):

$$-\overset{|}{C}=\overset{|}{C}- + HX \longrightarrow$$

$$-\overset{|}{\underset{H}{C}}-\overset{|}{\underset{X}{C}}- \qquad X = \text{F, Cl, Br, or I}$$

Reactions 1 through 3 can also be done with alkynes.

4. Addition of $H_2O$ to alkenes (Sec. 11.6):

$$-\overset{|}{C}=\overset{|}{C}- + H_2O \xrightarrow{H^+} -\overset{|}{\underset{H}{C}}-\overset{|}{\underset{OH}{C}}-$$

5. Nitration of aromatic rings (Sec. 11.11):

$$\bigcirc\!\!\!\!\hexagon + HNO_3 \xrightarrow{H_2SO_4} \hexagon\!\!\!\!\bigcirc\text{—}NO_2$$

6. Sulfonation of aromatic rings (Sec. 11.11):

$$\bigcirc\!\!\!\!\hexagon + H_2SO_4 \longrightarrow \hexagon\!\!\!\!\bigcirc\text{—}SO_2OH$$

7. Halogenation of aromatic rings (Sec. 11.11):

$$\bigcirc\!\!\!\!\hexagon + X_2 \xrightarrow{Fe} \hexagon\!\!\!\!\bigcirc\text{—}X \qquad X = \text{Br or Cl}$$

## KEY TERMS

Addition polymer (Sec. 11.7)
Addition reaction (Sec. 11.6)
Aliphatic hydrocarbon (Sec. 11.9)
Alkene (Sec. 11.1)
Alkyne (Sec. 11.8)
Aromatic compound (Sec. 11.9)
Aromatic sextet (Sec. 11.9)
Aromatic substitution (Sec. 11.11)
Carbocation (Box 11D)
Carbonium ion (Box 11D)
Carcinogen (Box 11F)
Catalytic hydrogenation (Sec. 11.6)
Copolymer (Sec. 11.7)
Fused aromatic ring (Sec. 11.12)
Heterocyclic compound (Sec. 11.13)
Hybrid (Sec. 11.9)

Hydration (Sec. 11.6)
Intermediate (Box 11D)
Isoprene rule (Box 11A)
Markovnikov's rule (Sec. 11.6)
Mechanism (Box 11D)
Monomer (Sec. 11.7)
Olefin (Sec. 11.1)
Pheromone (Box 11B)
Pi bond (Sec. 11.3)
Pi orbital (Sec. 11.3)
Polymer (Sec. 11.7)
Polymerization (Sec. 11.7)
Resonance (Sec. 11.9)
Saturated hydrocarbon (Sec. 11.1)
$sp$ orbital (Sec. 11.8)
$sp^2$ orbital (Sec. 11.3)
Unsaturated hydrocarbon (Sec. 11.1)

## PROBLEMS

**11.14** Pick out the unsaturated compounds and classify each as an alkene or alkyne:

(a) $CH_3-\overset{|}{\underset{CH_3}{CH}}-CH_2CH_3$

(b) $CH_2{=}CH-CH_2CH_3$

(c)

(d) $CH_3-C\equiv C-CH_3$

(e) $CH_3-\overset{\overset{\displaystyle Br}{|}}{C}H-CH_2CH_3$

(f) [benzene ring with $CH_3$ substituent]

## Nomenclature of Alkenes

**11.15** Give the IUPAC name (ignore cis-trans isomerism):

(a) $CH_3-CH=CH_2$

(b) $CH_3-\overset{\overset{\displaystyle Br}{|}}{C}=\overset{\overset{\displaystyle Br}{|}}{C}-CH_3$

(c) $Cl-\overset{\overset{\displaystyle Cl}{|}}{\underset{\underset{\displaystyle Cl}{|}}{C}}-CH_2-CH=CH_2$

(d) $CH_3CH_2-\overset{\overset{\displaystyle CH_2}{||}}{C}-CH_3$

(e) $CH_2=\overset{\overset{\displaystyle CH_3}{|}}{\underset{\underset{\underset{\underset{\displaystyle CH_3}{|}}{CH_2}}{|}}{C}}-CH_2-\overset{\overset{\displaystyle CH_3}{|}}{C}H-CH_3$

(f) $H_3C$ [cyclopentene ring]

(g) [cyclohexene ring with Cl substituent]

**11.16** Draw the structural formula for
(a) propylene
(b) 1-pentene
(c) 2,4-dimethyl-2-pentene
(d) 1,3-dibromo-5-methyl-3-octene
(e) 1,3-dimethylcyclohexene
(f) 2,4-heptadiene
(g) 2-methyl-1,3-cyclohexadiene

**11.17** Give the IUPAC name:
(a) $CH_2=CH-CH=CH_2$
(b) $CH_3CH_2-CH=CH-CH_2-CH=CH_2$

(c) $CH_3-CH=C=\overset{\overset{\phantom{x}}{}}{\underset{\underset{\displaystyle Cl}{|}}{C}}-CH_2CH_3$

(d) [cyclohexadiene ring with $CH_3$ substituent]

## sp² Hybridization

**11.18** If a carbon atom uses $sp^2$ orbitals to form bonds, how many $sp^2$ orbitals does it use?

**11.19** The angle between $sp^2$ orbitals is 120°. What is the angle between an $sp^2$ orbital and the remaining $p$ orbital?

## Stereoisomerism of Alkenes

**11.20** In each case tell whether cis-trans isomers exist. If they do, draw the two and label them cis and trans:
(a) 1-butene
(b) 2-butene
(c) 2-methyl-2-butene
(d) 2-hexene
(e) 3-methyl-3-hexene
(f) 3-ethyl-3-hexene
(g) 1-chloro-3-ethyl-3-hexene

**11.21** The compound 1,1-dichloropropene is shown on page 331. (a) Why isn't this compound called 3,3-dichloropropene? (b) Is there a compound with this name? If so, draw its structural formula. (c) Is there a compound 2,2-dichloropropene?

**11.22** Name the following cis or trans isomers:

(a) $\underset{H}{\overset{CH_3}{\diagdown}}C=C\underset{H}{\overset{CH_2CH_3}{\diagup}}$

(b) $\underset{H}{\overset{Cl-CH_2}{\diagdown}}C=C\underset{CH_2-Cl}{\overset{CH_3}{\diagup}}$

(c) $\underset{CH_3-\overset{\overset{\phantom{x}}{}}{\underset{\underset{\displaystyle Cl}{|}}{C}}H}{\overset{H}{\diagdown}}C=C\underset{CH_2CH_2CH_3}{\overset{H}{\diagup}}$

(d) $\underset{CH_3CH_2}{\overset{CH_3}{\diagdown}}C=C\underset{CH_2CH_3}{\overset{CH_3}{\diagup}}$

## Addition Reactions

**11.23** Define addition reaction. Write the equation for any addition reaction to propene.

**11.24** Write structural formulas for the principal reaction products when ethylene, $CH_2{=}CH_2$, reacts with

(a) $Br_2$  (d) $H_2O + H_2SO_4$
(b) $H_2 + Pt$  (e) $Cl_2$
(c) HBr

**11.25** Write structural formulas for the principal reaction products.

(a) $CH_3{-}CH{=}CH{-}CH_3 + H_2 \xrightarrow{Pt}$

(b) $CH_3{-}\underset{\underset{CH_3}{|}}{C}{=}CH{-}CH_3 + HBr \longrightarrow$

(c) + $Cl_2 \longrightarrow$

(d) $CH_2{=}CH{-}\underset{\underset{CH_3}{|}}{CH}{-}CH_3 + H_2O \xrightarrow{H_2SO_4}$

(e) $CH_3{-}\underset{\underset{CH_3}{|}}{C}{=}CH{-}CH_2CH_3 + HCl \longrightarrow$

(f) + HBr $\longrightarrow$

**11.26** Show how to make the following compounds starting from propene, $CH_3{-}CH{=}CH_2$ (write equations):

(a) $CH_3CH_2CH_3$  (c) $CH_3\underset{\underset{Br}{|}}{CH}\underset{\underset{Br}{|}}{CH_2}$

(b) $CH_3\underset{\underset{Br}{|}}{CH}CH_3$  (d) $CH_3\underset{\underset{OH}{|}}{CH}CH_3$

**11.27** Write the structural formula for the principal reaction product, if a reaction takes place. Where there is no reaction, say so.

(a) $CH_3{-}CH{=}CH{-}CH_3 + Cl_2 \longrightarrow$

(b) $CH_3{-}CH{=}CH{-}CH_3 + I_2 \longrightarrow$

(c) $CH_3{-}\underset{\underset{CH_3}{|}}{C}{=}CH_2 + H_2 \xrightarrow{no\ catalyst}$

(d) + HI $\longrightarrow$

**11.28** Write structural formulas for all possible addition products when HBr is added to each of the following. In each case predict which of the possible products is actually formed, in accord with Markovnikov's rule:

(a) $CH_3{-}CH{=}CH_2$

(b) $CH_3{-}\underset{\underset{CH_3}{|}}{C}{=}CH{-}CH_3$

(c) $CH_3CH_2{-}\underset{\underset{CH_3}{|}}{C}{=}CH_2$

(d) $CH_3CH_2{-}CH{=}CH{-}CH_2CH_3$

(e)

**11.29** What reagents and/or catalysts are necessary to make each of the following conversions in the laboratory?

(a) $CH_3{-}\underset{\underset{CH_3}{|}}{C}{=}CH_2 \longrightarrow CH_3{-}\overset{\overset{Br}{|}}{\underset{\underset{CH_3}{|}}{C}}{-}\overset{\overset{Br}{|}}{CH_2}$

(b)

(c) $CH_2{=}CH{-}CH_2{-}CH{=}CH_2 \longrightarrow$
$CH_3CH_2CH_2CH_2CH_3$

(d) $CH_3{-}\underset{\underset{CH_3}{|}}{C}{=}CH{-}CH_3 \longrightarrow CH_3{-}\overset{\overset{CH_3}{|}}{\underset{\underset{OH}{|}}{C}}{-}CH_2CH_3$

(e) $CH_3{-}CH{=}CH_2 \longrightarrow CH_3{-}\underset{\underset{I}{|}}{CH}{-}CH_3$

**11.30** Could you make 1-chloropropane from propene? Explain.

**11.31** Assume that you are given a liquid that could be either cyclohexane or cyclohexene. What simple test could you perform to tell which it is? Describe exactly what you would do and what you would see in each case.

## Addition Polymers

**11.32** Draw structural formulas for the polymers made from

(a) $CH_2{=}CH{-}CH_3$
(b) $Cl{-}CH{=}CH{-}Cl$

**\*11.33** Which monomers would you use to make the following polymers?

$$
\begin{array}{cccccc}
\text{Cl} & \text{CH}_3 & \text{Cl} & \text{CH}_3 & \text{Cl} & \text{CH}_3 \\
| & | & | & | & | & | \\
\end{array}
$$
(a)  $-\text{CH}-\text{CH}-\text{CH}-\text{CH}-\text{CH}-\text{CH}-$

$$
\begin{array}{ccc}
\text{COOCH}_3 & \text{COOCH}_3 & \text{COOCH}_3 \\
| & | & | \\
\end{array}
$$
(b)  $-\text{CH}_2-\text{CH}-\text{CH}_2-\text{CH}-\text{CH}_2-\text{CH}-$

## Alkynes

**11.34** Name the three orbitals that make up a C≡C triple bond.

**11.35** Give the IUPAC name:

(a)  $\text{CH}_3-\text{C}\equiv\text{C}-\text{CH}_2\text{CH}_3$

(b)  $\text{HC}\equiv\text{C}-\underset{\underset{\text{Br}}{|}}{\text{CH}}-\underset{\underset{\text{Br}}{|}}{\text{CH}}-\text{Br}$

(c)  $\text{Cl}-\text{CH}_2\text{CH}_2-\underset{\underset{\text{Cl}}{|}}{\text{CH}}-\text{C}\equiv\text{C}-\text{Cl}$

**11.36** Draw structural formulas for (a) propyne (b) 1,4-dibromo-2-butyne (c) 3-bromo-1-iodo-4,4-dimethyl-1-pentyne

**11.37** Why is cis-trans isomerism not possible for alkynes?

**11.38** It is possible to add 1 mole of $Br_2$ to 1 mole of propyne, $\text{CH}_3-\text{C}\equiv\text{CH}$, and then to add 1 mole of HBr to the product. Write equations for these two reactions.

## Aromatic Compounds

**11.39** Define aromatic compound.

**11.40** Draw the two resonance forms of benzene. Is the real benzene represented by either of them? Explain.

**11.41** What kind of damage is caused when benzene is ingested in the liquid or vapor form?

**11.42** The compound *para*-dichlorobenzene,

has a rigid geometry that allows no free rotation. Yet there are no cis-trans isomers for this structure (or any other benzene ring). Explain why.

**11.43** Why do aromatic compounds undergo substitution rather than addition reactions?

**11.44** Draw structural formulas for
(a)  *meta*-dichlorobenzene
(b)  *ortho*-ethylaniline
(c)  *para*-nitrophenol
(d)  phenylcyclopentane
(e)  benzyl fluoride
(f)  *cis*-1,3-dibenzylcyclohexane
(g)  *ortho*-diphenylbenzene
(h)  *meta*-isopropylbenzoic acid

**11.45** Name:

(a)

(b)

(c)

(d)

(e)

(f)

(g)

**11.46** Name:

(a)

(b)

(c)

(d)

(e) $CH_3-CH-CH-CH_2CH_3$

(f)

(g)

**11.47** Write structural formulas for the principal reaction products:

(a)

$+ HNO_3 \xrightarrow{H_2SO_4}$

(b)

$+ Br_2 \xrightarrow{Fe}$

(c)

$+ H_2SO_4 \longrightarrow$

**11.48** What reagents and/or catalysts are necessary to make each of the following conversions in the laboratory?

(a)

(b)

*(c)

*(d)

**11.49** Three possible products can be formed in the chlorination of bromobenzene. Draw the three structures.

**11.50** Draw structural formulas for the principal reaction products, if a reaction takes place. Where there is no reaction, say so.

(a)

$+ Br_2 \xrightarrow[\text{catalyst}]{\text{no}}$

(b)

$+ Cl_2 \xrightarrow{Fe}$

### Fused and Heterocyclic Compounds

**11.51** The structure we have given for naphthalene,

, is only one of three resonance forms. Draw the other two (note that each carbon atom must have one double bond in each form).

**11.52** Aromatic substitution can also be done on naphthalene. When naphthalene is sulfonated with concentrated $H_2SO_4$, two (and only two) different naphthalenesulfonic acids are produced. Draw the two structures.

**11.53** Define: (a) heterocyclic compound (b) heterocyclic aromatic compound

**11.54** There are three different methylpyridines. Draw the three structural formulas.

### Boxes

**11.55** (Box 11A) Limonene is the chief terpene in orange peel. Show where the isoprene units are in this molecule:

Limonene

**11.56** (Box 11A) Phytol is a terpene that is part of the chlorophyll molecule. Show where the isoprene units are in this molecule:

Phytol

**11.57** (Box 11A) How many carbon atoms are there in the $\beta$-carotene molecule? In vitamin A?

**11.58** (Box 11B) Assume that $1 \times 10^{-12}$ g of *cis*-9-tricosene is the amount secreted by a single housefly. How many molecules is this? (The molecular formula is $C_{23}H_{46}$.)

**11.59** (Box 11C) (a) In which isomer is the end-to-end distance longer, *all-trans-* or 11-*cis*-retinal? (b) Which does not fit into the opsin cavity?

**11.60** (Box 11D) Show the mechanism of the reaction between HCl and propene.

**11.61** (Box 11D) Show the mechanism of the reaction between $H_2O$ and isobutene, $CH_3\!-\!\overset{\displaystyle |}{\underset{\displaystyle CH_3}{C}}\!=\!CH_2$, catalyzed by $H_2SO_4$.

**\*11.62** (Box 11D) Polymerization reactions can occur through a mechanism involving carbocation intermediates, for example:

$$CH_2\!=\!CH_2 + H^+ \longrightarrow CH_3\!-\!CH_2{}^+$$
$$CH_3\!-\!CH_2{}^+ + CH_2\!=\!CH_2 \longrightarrow$$
$$CH_3CH_2CH_2\!-\!CH_2{}^+$$

Complete the mechanism for the preparation of polyisobutylene:

**11.63** (Box 11E) Why is it necessary to vulcanize rubber?

**11.64** (Box 11E) Why are synthetic rubbers manufactured?

**\*11.65** (Box 11E) *Neoprene* is an important synthetic rubber that is made by polymerizing the monomer 2-chloro-1,3-butadiene (called chloroprene). Draw the structural formula of neoprene.

**11.66** (Box 11F) What is a carcinogen? What kind of carcinogens are found in cigarette smoke?

**11.67** (Box 11F) What compounds caused the scrotum cancers found in London chimney sweeps 200 years ago?

## Additional Problems

**11.68** (a) Why is it incorrect to call this compound 1,2- or 2,3- dimethylcyclohexene?

(b) What is the correct numbering?

**11.69** Write the structural formula and name for the simplest (a) alkene  (b) alkyne  (c) aromatic hydrocarbon

**11.70** The following names are all incorrect. State what is wrong with each (do not try to give correct names):
(a) *cis*-2-methyl-2-pentene
(b) 3-bromo-*meta*-dichlorobenzene
(c) 2,2-dimethyl-1-pentene
(d) 3-pentene
(e) 1-bromo-3-butene
(f) *para*-methylbromobenzene
(g) 2-bromopropyne
(h) 3-bromo-2,3-pentadiene
(i) 1-bromobutene

**11.71** There are two different bromonaphthalenes. Draw the two structural formulas.

**\*11.72** For each carbon atom in this compound,
$$\overset{1}{C}H_3-\overset{2}{C}H=\overset{3}{C}H-\overset{4}{C}\equiv\overset{5}{C}H,$$ tell whether the hybridization is *sp*, *sp*², or *sp*³.

**11.73** What is the difference between a saturated and an unsaturated compound?

**11.74** Explain why a $\pi$ bond is weaker than a $\sigma$ bond.

**11.75** What is the product of this reaction:

$$CaC_2 + H_2O \longrightarrow$$

**\*11.76** Cyclohexyne, , is not a stable compound (it is unknown), but cyclohexene, , is a perfectly stable, known compound. Suggest a reason for the instability of cyclohexyne.

**11.77** Do aromatic rings have double bonds? Are they unsaturated? Explain.

**11.78** Write structural formulas for all the different mononitro products that are possible in the nitration of
(a) *meta*-bromotoluene
(b) naphthalene

**11.79** Show how tetrafluoroethene, $F_2C=CF_2$, polymerizes to give the polymer Teflon.

**11.80** Draw the structures of and name all the isomers of dibromopropene (there are seven, including cis and trans isomers).

# CHAPTER 12

*Wine contains 10 to 13 percent ethanol.*
*(Photograph by Beverly March.)*

# Alcohols, Phenols, Ethers, and Halides

## 12.1 INTRODUCTION

In Chapters 10 and 11 we studied the hydrocarbons: alkanes, alkenes, alkynes, and aromatic compounds. Section 10.12 introduced the concept of **functional group.** A particular functional group, such as the hydroxyl group, —OH, can replace a hydrogen at almost any position of any hydrocarbon of any of the classes we have studied.

In this and the next three chapters, we will study the most important types of organic compounds—those listed in Table 10.7. All of these functional groups contain either O, N, S, or halogen atoms. In this chapter we study those compounds that are simple derivatives of $H_2O$, $H_2S$, and the hydrogen halides, HX (Table 12.1).

**TABLE 12.1 Types of Compounds Derived from $H_2O$, $H_2S$, and HX**

| Formula | Name |
|---------|------|
| H—O—H | Water |
| R—O—H | Alcohols |
| R—O—R′ | Ethers[a] |
| H—S—H | Hydrogen sulfide |
| R—S—H | Thiols (mercaptans) |
| R—S—R′ | Thioethers[a] |
| H—X | Hydrogen halide |
| R—X | Alkyl halides |

[a] The R′ can be the same as R or different.

361

## 12.2 NOMENCLATURE OF ALCOHOLS

As already noted, **alcohols** are **compounds containing —OH groups connected to an alkyl carbon.** If the —OH is connected directly to an aromatic ring, the compound is not an alcohol but a **phenol** (see Sec. 12.5).

Alcohols are classified as **primary, secondary, or tertiary** depending on whether the —OH is connected to a primary, a secondary, or a tertiary carbon atom (Table 12.2). Recall that a primary carbon is connected to one (or no) other carbon, a secondary to two other carbons, and a tertiary to three other carbons.

---

**PROBLEM 12.1**   Classify these alcohols as primary, secondary, or tertiary:

(a)   $CH_3CH_2$—CH—OH
                     |
                    $CH_3$

(b)   $CH_3CH_2$—CH—$CH_2$—OH
                     |
                    $CH_3$

              $CH_3$
               |
(c)   $CH_3$—C—OH
               |
              $CH_3$

---

There are two ways to name alcohols. Simple alcohols are often named by giving the name of the alkyl group (Sec. 10.7) followed by the word "alcohol";

**TABLE 12.2   Classification, Common Names, and IUPAC Names for Some Alcohols**

| Structural Formula | Classification | Common Name | IUPAC Name |
|---|---|---|---|
| $CH_3$—OH | Primary | Methyl alcohol | Methanol |
| $CH_3CH_2$—OH | Primary | Ethyl alcohol | Ethanol |
| $CH_3CH_2CH_2$—OH | Primary | *n*-Propyl alcohol | 1-Propanol |
| $CH_3$—CH—$CH_3$   OH | Secondary | Isopropyl alcohol | 2-Propanol |
| $CH_3CH_2$—CH—$CH_3$   OH | Secondary | *sec*-Butyl alcohol | 2-Butanol |
| $CH_3$   $CH_3$—C—OH   $CH_3$ | Tertiary | *tert*-Butyl alcohol | 2-Methyl-2-propanol |
| ⬡—$CH_2$—OH | Primary | Benzyl alcohol | Phenylmethanol |

Table 12.2 gives examples. The other way to name alcohols is by the IUPAC rules we use for alkanes, alkyl halides (Sec. 10.7), and alkenes (Sec. 11.2), except that we need two additional rules:

1. The "longest" chain must include the carbon bearing the —OH group.
2. The suffix "-ol" is used to represent the —OH group, and the lowest possible number is assigned to the carbon bearing this group.

*When we assign the lowest number, the —OH takes preference over all alkyl and halo groups.*

Table 12.2 shows how the IUPAC system applies to some simple alcohols. Note that the prefixes iso, normal, tertiary, and so on are not used in the IUPAC names.

*For many complicated molecules there are no simple names; IUPAC names must be used.*

**EXAMPLE**

Give the IUPAC name for

$$\overset{\displaystyle CH_3}{\underset{\displaystyle |}{}}$$
$$\overset{\displaystyle CH_2}{\underset{\displaystyle |}{}}$$
$$\underset{4}{CH_3}-\underset{3}{CH_2}-\underset{2}{CH}-\underset{1}{CH_2}-OH \qquad \text{2-Ethyl-1-butanol}$$

**Answer**

This molecule has a five-carbon chain (shown in color), but this chain does not contain the carbon bearing the —OH group. The longest chain that contains this carbon has four carbons, and so the parent is butane. Because there is an —OH group, the suffix is "-ol." Thus the name is 2-ethyl-1-butanol, with the OH-bearing carbon given the lowest possible number.

**EXAMPLE**

Give the IUPAC name for

$$\underset{1}{CH_3}-\underset{2}{CH_2}-\underset{3}{CH}-\underset{4}{CH_2}-\underset{5}{C}-\underset{6}{CH_3} \qquad \text{5-Bromo-5-methyl-3-hexanol}$$

with Br on carbon 5, OH on carbon 3, and $CH_3$ on carbon 5.

**Answer**

The longest chain has six carbons, and it contains the —OH group; thus the parent is hexanol. The chain is numbered as shown, because the —OH carbon becomes 3. If the chain were numbered the other way, this would be 4, and the rule is that the —OH group must be given the lowest number, even though this means that the methyl and the bromo groups must be numbered 5 and 5, respectively.

---

**EXAMPLE**

Give the IUPAC name for

3-Methylcyclohexanol

**Answer**

Since the —OH group is given preference, the carbon to which it is attached is C-1, meaning that the —CH$_3$ group is on C-3 (since 3 is lower than 5, which we would get by counting the other way).

---

Because the —OH is always at the 1 position in a ring, we do not show the 1 in the name.

**PROBLEM 12.2**    Give the IUPAC name for

Compounds with two —OH groups on the same carbon are usually not stable.

Many compounds contain more than one —OH group in a molecule (though not generally two on the same carbon atom). Molecules of this type are called diols, triols, and so forth. Two of the most important are (common names) ethylene glycol and glycerol (also called glycerin):

Ethylene glycol
(1,2-ethanediol)

Glycerol
(1,2,3-propanetriol)

## 12.3  CHEMICAL PROPERTIES OF ALCOHOLS

This was mentioned in Section 8.1.

Metallic hydroxides, such as NaOH, Mg(OH)$_2$, and Al(OH)$_3$, are bases, some strong and some weak, because they produce OH$^-$ ions in water. Alcohols do not behave this way. The —OH in an alcohol is connected to the carbon by a covalent bond and not by an ionic bond (as in the metal hydroxides). Therefore, alcohols have no basic properties. Nor are they acidic, despite the hydrogen connected to the oxygen (they are weaker acids than water). When an alcohol is dissolved in pure water, the pH remains 7.

Alcohols are neutral compounds.

We give three reactions of alcohols in this section (we will see some others in Chapters 13 to 15).

## Dehydration to Give an Alkene

General reaction

$$-\overset{|}{\underset{H}{C}}-\overset{|}{\underset{OH}{C}}- \xrightarrow[\text{heat}]{H_2SO_4} -\overset{|}{C}=\overset{|}{C}- + H_2O$$

Specific example

$$CH_3-\overset{|}{\underset{OH}{CH}}-CH_3 \xrightarrow[\text{heat}]{H_2SO_4} CH_3-CH=CH_2 + H_2O$$

2-Propanol                                                          Propene

When heated with sulfuric or phosphoric acid as catalyst, an alcohol loses an OH group and an H atom on the adjacent carbon to give an alkene. This reaction can be performed only on alcohols that have at least one hydrogen on the adjacent carbon. (Methanol, $CH_3OH$, for example, cannot undergo this reaction.) Note that this is the reverse of the addition of water to a double bond (Sec. 11.6). That is, we can heat 2-propanol with sulfuric acid and get propene. If desired, we can then take the propene and convert it back to 2-propanol, as shown in Section 11.6. In this case the catalyst is the same, and we can shift the equilibrium either way by using Le Chatelier's principle (Sec. 7.8).

> When two groups are removed from adjacent carbons to give a double bond, the reaction is called **elimination.**

> Many reactions go both forward and backward, but in most cases different reagents are required for each direction.

---

**PROBLEM 12.3**   Draw the structural formula of the product:

$$CH_3-CH_2-\overset{|}{\underset{OH}{CH}}-CH_2-CH_3 + H_2SO_4 \longrightarrow$$

---

> Here the catalyst is not written over the arrow. In organic chemistry we often write catalysts as if they were reagents, and vice versa.

In Section 11.6 we saw that the hydration reaction is also carried out in the body, which uses enzymes as catalysts rather than sulfuric or phosphoric acid. The same is true for the dehydration reaction; for example, the dehydration of citric acid,

$$H-\overset{H}{\underset{HOOC}{\overset{|}{C}}}-\overset{OH}{\underset{COOH}{\overset{|}{C}}}-\overset{H}{\underset{H}{\overset{|}{C}}}-COOH \xrightarrow{\text{enzyme}} H-\overset{}{C}=\overset{}{\underset{COOH}{\overset{|}{C}}}-\overset{H}{\underset{H}{\overset{|}{C}}}-COOH$$

Citric acid                                                *cis*-Aconitic acid

is part of the citric acid cycle (Sec. 20.4). Note that the dehydration of citric acid involves exactly the same changes in bonding as the dehydration of 2-propanol (that is why we classify them as the same type of reaction). In both cases an H and an OH are removed from adjacent carbons, leaving a C=C double bond. The rest of the bonds in both molecules do not change. It is important that you be able to recognize that the same thing is happening in both cases, despite the considerable differences in the structures of citric acid and 2-propanol. In fact, almost every molecule with OH and H on adjacent carbon atoms is capable of giving this reaction under the proper conditions.

> There are only a limited number of reactions possible for any class of compounds, and it is not surprising that the same reactions are found in the body and in the laboratory.

## Dehydration to Give an Ether

Remember that R stands for any alkyl group.

General reaction

$$2R\text{—}O\text{—}H \xrightarrow{H_2SO_4} R\text{—}O\text{—}R + H_2O$$

Specific example

$$2CH_3CH_2\text{—}OH \xrightarrow[140°]{H_2SO_4} CH_3CH_2\text{—}O\text{—}CH_2CH_3 + H_2O$$

Ethanol      Diethyl ether

An alkene is not the only possible product when an alcohol is heated with sulfuric acid. It is also possible for water to be lost from two molecules of alcohol—a hydrogen atom from one molecule and an —OH group from the other—to give an ether, R—O—R:

$$R\text{—}O\text{—}H \quad H\text{—}O\text{—}R \longrightarrow R\text{—}O\text{—}R$$

Ether

This reaction was discovered around AD 1200 by alchemists, who were astonished to get a liquid (diethyl ether) that would quickly evaporate in their hand.

For most alcohols this reaction gives a very low yield.

This is one of the oldest known organic reactions. Since ethanol can react with sulfuric acid to give ethylene, $CH_3CH_2OH \rightarrow CH_2{=}CH_2$, how can we control the reaction to get diethyl ether instead? The answer for this alcohol is that we use temperature as the controlling factor. Reaction at 140°C produces mostly the ether, whereas at 180°C the product is mostly ethylene. For other alcohols, the ether-forming reaction is largely useful only when there is no hydrogen on an adjacent carbon. We have seen that methanol cannot give an alkene, but it can give dimethyl ether, $CH_3\text{—}O\text{—}CH_3$.

---

**PROBLEM 12.4** Draw the structural formula of the product:

$$+ H_2SO_4 \longrightarrow$$

---

## Oxidation

The three classes of alcohols behave differently toward oxidizing agents.

### Primary Alcohols

The [O] represents an unspecified oxidizing agent.

General reaction

$$R\text{—}CH_2\text{—}OH \xrightarrow{[O]} R\overset{\|}{\underset{O}{C}}\text{—}H \xrightarrow{[O]} R\overset{\|}{\underset{O}{C}}\text{—}OH$$

Primary    Aldehyde     Carboxylic
alcohol                acid

Specific example

$$CH_3CH_2\text{—}OH \xrightarrow[H_2SO_4]{K_2Cr_2O_7} CH_3\overset{\|}{\underset{O}{C}}\text{—}H \xrightarrow[H_2SO_4]{K_2Cr_2O_7} CH_3\overset{\|}{\underset{O}{C}}\text{—}OH$$

Ethanol     Acetaldehyde     Acetic
acid

Primary alcohols are oxidized by many oxidizing agents. The immediate product is an aldehyde, which results from the removal of two hydrogens:

$$
\underset{\underset{\text{H}}{|}}{\overset{\overset{\text{H}}{|}}{\text{R}-\text{C}-\text{O}}}\underset{\text{H}}{}\longrightarrow \text{R}-\overset{\overset{\text{H}}{|}}{\text{C}}=\text{O}
$$

Aldehyde

During the oxidation water is produced (not shown in the equation). The two hydrogens lost by the alcohol combine with an oxygen from the oxidizing agent.

Aldehydes are themselves readily oxidized by the same oxidizing agents (Sec. 13.5) to give carboxylic acids. Therefore, the usual product of oxidation of a primary alcohol is the corresponding carboxylic acid. It is often possible to stop the reaction at the aldehyde stage by distilling the mixture so that the aldehyde (which usually has a lower boiling point than either the alcohol or the carboxylic acid) is removed before it can be oxidized further. A common oxidizing agent in this reaction is a mixture of potassium dichromate, $K_2Cr_2O_7$, and sulfuric acid, but potassium permanganate, $KMnO_4$, and bromine are also commonly used.

An important biological example of this reaction is oxidation of ethanol to acetic acid, which is catalyzed by enzymes. This is what happens when wine is oxidized to vinegar. The oxidizing agent in this case is the $O_2$ in the air.

---

**PROBLEM 12.5**  Draw the structural formula of the first and second products:

$$
\underset{\underset{|}{\overset{|}{\text{CH}_3}}}{\text{CH}_3-\text{CH}-\text{CH}_2-\text{OH}} \xrightarrow[\text{H}_2\text{SO}_4]{\text{K}_2\text{Cr}_2\text{O}_7}
$$

---

## Secondary Alcohols

General reaction

$$
\underset{\underset{\text{OH}}{|}}{\text{R}-\text{CH}-\text{R}'} \xrightarrow{[\text{O}]} \underset{\underset{\text{O}}{\|}}{\text{R}-\text{C}-\text{R}'}
$$

Secondary alcohol    Ketone

Specific example

$$
\underset{\underset{\text{OH}}{|}}{\text{CH}_3-\text{CH}-\text{CH}_3} \xrightarrow[\text{H}_2\text{SO}_4]{\text{K}_2\text{Cr}_2\text{O}_7} \underset{\underset{\text{O}}{\|}}{\text{CH}_3-\text{C}-\text{CH}_3}
$$

2-Propanol    Acetone

Secondary alcohols are oxidized to ketones in exactly the same way primary alcohols are oxidized to aldehydes, and by the same oxidizing agents. The difference is that, unlike aldehydes, ketones are resistant to further oxidation, so this reaction is an excellent method for the preparation of ketones.

---

**PROBLEM 12.6**  Draw the structural formula of the product:

$$
\underset{\underset{\text{CH}_3\ \ \text{OH}}{|\quad\ \ |}}{\text{CH}_3-\text{CH}-\text{CH}-\text{CH}_2\text{CH}_3} \xrightarrow[\text{H}_2\text{SO}_4]{\text{K}_2\text{Cr}_2\text{O}_7}
$$

---

**Tertiary Alcohols**

$$R-\underset{\underset{R''}{|}}{\overset{\overset{R'}{|}}{C}}-OH \xrightarrow{[O]} \text{no reaction}$$

Tertiary alcohols, not having a hydrogen on the —OH carbon, do not react with oxidizing agents.

Note that some of the equations we have written in this section are not balanced; some are not even complete. It is difficult to balance these oxidation reactions, though it could be done if necessary. Most of the time, organic chemists do not find it necessary. Their focus is on the starting compound, reagent, and product and not on the balanced equation. This is the custom among organic chemists, and we shall follow that custom. This does not mean that the reactions are not balanced by nature. All reactions that actually take place follow the conservation of mass law and are balanced.

## 12.4   SOME IMPORTANT ALCOHOLS

### Ethanol

When people say "alcohol" and mean a specific compound, ethanol, $CH_3CH_2OH$, is almost always the one that is meant. The production of ethanol by fermentation is as old as civilization. For the entire span of recorded history, human beings have been making ethanol-containing beverages by this process. Ethanol is sometimes called *grain alcohol* because large quantities are made by fermentation of such grains as corn, wheat, and rye, but fruits (especially grapes), vegetables (e.g., potatoes), and molasses (from sugar cane) are also used (Fig. 12.1).

*Organic reactions rarely produce only one product. Most often, side reactions yield a variety of products in small quantities. In contrast, most biochemical reactions are very specific and yield only one product.*

**Figure 12.1**
Fermentation of molasses. This mixture will be distilled to give rum. (FourByFive.)

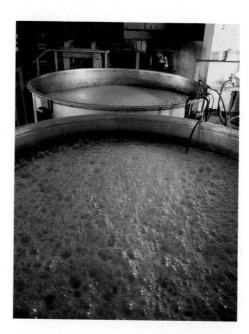

## Alcoholic Beverages

Despite the great differences in alcoholic beverages, virtually all of them consist of at least 99 percent water and ethanol. They differ from one another in two ways: the relative percentages of water and ethanol, and the composition of the other 1 percent. Beer contains 3 to 4 percent ethanol, and table wine contains 10 to 13 percent. These beverages are made by direct fermentation (Fig. 12.1). The yeast cells that catalyze the fermentation cannot live when the ethanol concentration gets higher than about 13 percent. The only way to get a higher ethanol concentration is to distill the wine. This process was not discovered until about AD 1200. Thus in all human history up to that time, people had nothing stronger to drink than table wine.

After 1200, distillation became a common process, and today we have spirits and fortified wines. Spirits are the direct products of distillation, often aged in bottles or casks, or given various other treatments. The ethanol content of spirits varies from 35 to 50 percent. Fortified wines (such as sherry and port) are produced in the normal way, but extra ethanol is added to bring the concentration to about 20 percent. In the United States, the ethanol concentration of beverages is expressed as **proof,** which is twice the percent ethanol (v/v). Thus 86-proof whiskey contains 43 percent ethanol.

The difference between one bottle of spirits and another of the same proof is in the composition of the last 1 percent. The compounds that make up this final 1 percent, called **congeners,** may be higher alcohols (such as 1-butanol), aldehydes, ketones, carboxylic esters, or other compounds. They get into the spirits either by being carried over in the original distillation or by absorption during the aging process (for example, from wooden casks). Alcohol itself is tasteless; the congeners provide the taste. Vodka is pure ethanol and water and has no congeners. Gin is made by adding juniper berries to vodka and has only a small amount of congeners. Rye, scotch, rum, bourbon, and brandy have greater amounts of congeners. Some alcoholic beverages also contain carbon dioxide (beer, champagne), which causes the alcohol to get into the blood faster, increasing the speed of intoxication.

The United States government puts a high tax on beverage ethanol. The tax on a 1-L bottle of 86-proof spirits is about $3.00. Laboratories and hospitals are permitted to use ethanol without paying the tax, but the government rigidly controls such usage with frequent inspections. The tax can also be avoided by using denatured alcohol (see text), but this can be done only when the denaturants are not harmful in the intended use.

In fermentation, sugars (represented here as $C_6H_{12}O_6$) are converted to ethanol and carbon dioxide by enzymes present in yeast cells:

$$C_6H_{12}O_6 \xrightarrow{\text{zymase}} 2CH_3CH_2OH + 2CO_2$$

The sugar content of grains and vegetables is very low, but they do contain starch, which is hydrolyzed to sugars during the fermentation process.

Of all alcohols, ethanol is the least toxic (except for glycerol), and it is the only alcohol present in alcoholic beverages (Box 12A), except perhaps for tiny amounts of others.

By law, the ethanol used in all alcoholic beverages and drugs must be made by fermentation, but industrial ethanol is mostly made by the hydration of ethylene (Sec. 11.6),

$$CH_2{=}CH_2 + H_2O \xrightarrow{H_2SO_4} CH_3CH_2{-}OH$$

Sugars and starches are discussed in Chapter 16.

## Box 12B

## Alcoholism and Disulfuram

In the United States and many other countries, alcoholism is a major social problem. Alcohol is a psychologically addictive drug, and many alcoholics remain intoxicated for weeks at a time. Alcohol causes deterioration of the liver (cirrhosis of the liver, caused almost entirely by alcohol intake, is a major cause of death) and loss of memory. People (not necessarily alcoholics) who drive while intoxicated are responsible for an estimated 40 percent of traffic deaths in the United States. Ethanol is not a stimulant, but a depressant. The individual may feel stimulated, but sensory perception is impaired and the reflexes are slowed.

The first step in normal ethanol metabolism is oxidation to acetaldehyde,

$$CH_3CH_2OH \xrightarrow{[O]} CH_3-\overset{\displaystyle O}{\underset{\displaystyle \|}{C}}-H \xrightarrow{[O]} \text{other products}$$

Ethanol              Acetaldehyde

which is then oxidized further. One treatment for alcohol addiction involves use of the drug disulfuram (trade name Antabuse),

$$\underset{CH_3CH_2}{\overset{CH_3CH_2}{>}}N-\overset{\overset{\displaystyle S}{\|}}{C}-S-S-\overset{\overset{\displaystyle S}{\|}}{C}-N\overset{CH_2CH_3}{\underset{CH_2CH_3}{<}}$$

Disulfuram

which interferes with the second step of the metabolism process. The ethanol is oxidized to acetaldehyde, but the acetaldehyde, which can no longer be oxidized, builds up in the blood, causing nausea, sweating, and vomiting. Knowing that this will happen if he or she drinks after taking disulfuram often keeps the person from drinking, though this is obviously not a permanent cure for alcoholism in all cases.

because this process is cheaper. The ethanol made by the two processes is of course identical.

Apart from its beverage use, ethanol is used as a solvent for medicines, perfumes, and varnishes, as a body rub (isopropyl alcohol is also used for this purpose—see below), and, in a 70 percent solution, as an antiseptic (Box 18H) used to prepare patients for surgery and to clean surgical instruments.

Pure ethanol is called **absolute alcohol.** It is often found in laboratories in this form and in the form of a solution containing 5 percent water (95% alcohol). **Denatured alcohol** is ethanol that has been made unfit to drink by the addition of small quantities of a poison such as methanol (see Box 12C) or benzene. Denatured alcohol is used in some chemical laboratories and factories (where the addition of the poison does not affect its use) because no tax has to be paid on it.

*Solutions in which ethanol is a solvent are called* **tinctures.**

### Methanol

Sometimes called *wood alcohol* because at one time it was made from wood, methanol, $CH_3OH$, is now made by subjecting hydrogen and carbon monoxide to high temperatures and pressures in the presence of a catalyst. Methanol is

## The Poisonous Effects of Methanol

Methanol is highly toxic. Ingestion of 15 mL causes blindness and 30 mL, death. Methanol in the body is oxidized to formaldehyde and formic acid. We are not certain whether formic acid or formaldehyde poisons the cells of the retina and causes blindness, but we do know that, as with methanol, the ingestion of formic acid also causes acidosis in the blood, which can lead to death. The antidote for methanol and formic acid poisoning is to drink a sodium bicarbonate solution to neutralize the acid.

very toxic (Box 12C). It is used as a solvent for paints, shellacs, and varnishes as well as to make formaldehyde:

$$CH_3OH \xrightarrow{[O]} H-\overset{\displaystyle \underset{\|}{O}}{C}-H$$

Formaldehyde

### Isopropyl Alcohol (2-Propanol)

Often called *rubbing alcohol*, isopropyl alcohol, $CH_3-\overset{OH}{\underset{|}{CH}}-CH_3$, is useful for this purpose because it cools the skin by evaporation. It thus helps to lower fever. It is also an astringent—it hardens the skin and decreases the size of the pores, limiting secretions. It is used as a solvent for cosmetics, perfumes, and skin creams.

Isopropyl alcohol has a large heat of vaporization (though not as large as that of water), so when it evaporates it removes a lot of heat (see Sec. 5.12).

### Menthol, Ethylene Glycol, and Glycerol

Menthol is a terpene alcohol (Box 11A):

It has a pleasant, minty odor and is used in shaving creams, cough drops ("mentholated"), cigarettes, and toothpastes.

Ethylene glycol, $HO-CH_2CH_2-OH$, is a viscous (thick), toxic liquid. Its most important use is as antifreeze for automobile radiators, where it lowers the freezing point of the water (Fig. 12.2). Although any solute will lower the freezing point, ethylene glycol is preferred because it is relatively cheap, infinitely soluble in water, noncorrosive to metal, and has a higher boiling point than water (thus it won't evaporate in summer).

Menthol in throat sprays and lozenges soothes the respiratory tract.

See Section 6.8 for a discussion of freezing-point lowering.

Ethylene glycol is colorless; the color of most antifreezes comes from additives.

**Figure 12.2**
Antifreeze being poured into an automobile radiator. (Photograph by Beverly March.)

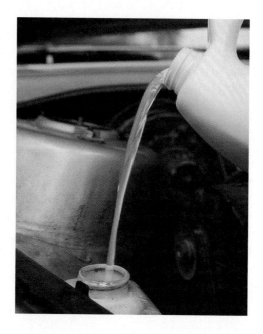

Glycerol (also called glycerin),

$$CH_2—CH—CH_2$$
$$|\quad\ \ |\quad\ \ |$$
$$OH\quad OH\quad OH$$

Glycerol is relatively inexpensive because it is a by-product in the production of soap from fats (Sec. 17.3).

is an even more viscous liquid than ethylene glycol. It is sweet, nontoxic, and infinitely soluble in water. These properties make it a good sweetening agent, solvent for medicines, and lubricant for suppositories and in chemical laboratories (for putting glass tubes through rubber stoppers). It is also used as a moistening agent (humectant) in cosmetics, skin creams, lotions, tobacco, and food. Another important use of glycerol is in the manufacture of nitroglycerine (Box 12D).

## 12.5 PHENOLS

The same name, *phenol*, is used for a single compound and for a whole class of compounds. There are, unfortunately, several examples of this practice in the nomenclature of organic compounds.

As we have said, phenols are compounds containing an —OH connected directly to an aromatic ring. The simplest phenol is phenol itself. Other phenols are generally named as derivatives of phenol (Sec. 11.10), though some have common names. As is evident, the suffix "-ol" is used for phenols as well as for alcohols (with a few exceptions, such as hydroquinone):

Phenol      *o*-Chlorophenol      *m*-Cresol      Hydroquinone

In 1847 an Italian chemist named Ascanio Sobrero (1812–1888) discovered that glycerol reacts with a mixture of nitric and sulfuric acids to give a liquid called nitroglycerine (or glyceryl trinitrate):

Box 12D

# Nitroglycerine, an Explosive and a Drug

$$\begin{array}{ccc}
CH_2-OH & & CH_2-O-NO_2 \\
| & & | \\
CH-OH + HNO_3 & \xrightarrow{H_2SO_4} & CH-O-NO_2 \\
| & & | \\
CH_2-OH & & CH_2-O-NO_2
\end{array}$$

Glycerol · Nitroglycerine

When Sobrero heated a small quantity of this liquid, it exploded. At that time there were few chemical explosives known, so nitroglycerine factories sprang up in many countries to make a compound that has many uses both for war and for peace (in digging canals and tunnels, blasting hills for road building, and mining, for example).

A problem soon arose because nitroglycerine is so explosive and difficult to handle that these factories would occasionally blow up. This problem was essentially solved by a Swedish chemist, Alfred Nobel (1833–1896), whose brother had been killed when Nobel's nitroglycerine factory exploded in 1864. In 1866 Nobel found that a clay-like substance called diatomaceous earth would absorb liquid nitroglycerine so that it would not explode without a fuse. He called the new material **dynamite,** and this is still one of our most important explosives (Fig. 12D).

Surprising as it may seem, nitroglycerine is also used in medicine, to treat angina pectoris. This condition, the symptoms of which are sharp chest pains, is caused by a reduced flow of blood to the heart. Nitroglycerine, available either in liquid form (diluted with alcohol to render it nonexplosive), or as tablets, or in paste form, relaxes the smooth muscles of the blood vessels and dilates the arteries, allowing more blood to reach the heart and reducing or eliminating the chest pains.

Another compound used for the same purpose is isoamyl nitrite (also called amyl nitrite):

$$\begin{array}{c}
CH_3 \\
| \\
CH_3-CH-CH_2-CH_2-O-NO
\end{array}$$

However, this compound has an unpleasant odor.

**Figure 12D**
Dynamite explosion in an open-pit mine. (FourByFive.)

Some phenols are important natural compounds. Examples are vanillin, which gives flavor to vanilla; tetrahydrourushiol, a principal irritant of poison ivy; and thymol, which is used in medicine to kill fungi and hookworms:

Vanillin contains three functional groups: It is not only a phenol but also an ether and an aldehyde. Its IUPAC name is 4-hydroxy-3-methoxybenzaldehyde.

Vanillin          Tetrahydrourushiol          Thymol
(2-isopropyl-5-methylphenol)

Phenol itself is a solid that is fairly soluble in water. An aqueous solution of phenol is called *carbolic acid*. This solution was one of the earliest antiseptics because it is toxic to bacteria (Box 12E). Today it is no longer used directly on patients because it burns the skin, but it is still used to clean surgical and medical instruments.

Because phenols are acidic, they dissolve in aqueous NaOH solution. This enables us to distinguish them from other, nonacidic, organic compounds.

The most important chemical property of phenols is that, unlike alcohols, they are acidic. As shown in Table 8.3 (p. 226), phenols have $K_a$ values around $10^{-10}$ (p$K_a$ = about 10), which makes them weak acids, considerably stronger than water but much weaker than carboxylic acids (p$K_a$ about 5). One phenol that is an especially strong acid is picric acid, 2,4,6-trinitrophenol (p$K_a$ = 0.4):

---

**B o x   12E**

**Phenols as Antiseptics**

Before the middle of the eighteenth century, even the simplest surgery was very dangerous because wounds very often became infected by bacteria. Because physicians of the time had no idea that bacteria existed, they had no way to counteract them. In 1867 Joseph Lister (1827–1912), who had learned of Louis Pasteur's research on bacteria, discovered that solutions of phenol kill bacteria. Lister applied this substance to surgical instruments and to the patients themselves, and the results were dramatic. A much higher percentage of patients survived surgery.

Phenol burns healthy tissue, however, and a search was made for substitutes. Today, a number of phenols are used as antiseptics. Among these is 4-*n*-hexylresorcinol, which has a much stronger antibacterial action than phenol and is so nontoxic that it is used in throat lozenges and mouthwashes. Other phenolic antiseptics are *o*-phenylphenol, methyl salicylate (oil of wintergreen), and thymol (mentioned in the text).

4-*n*-Hexylresorcinol          *o*-Phenylphenol          Methyl salicylate
(oil of wintergreen)

Picric acid dissociating when added to water

Picric acid is used to treat burns.

## 12.6 ETHERS

Ethers are compounds whose formula is R—O—R′, where R and R′ can be the same or different. Ethers are generally named by giving the names of the groups and adding the word "ether":

$$CH_3—O—CH_3 \qquad CH_3—O—CH_2CH_3$$
Dimethyl ether \qquad Ethyl methyl ether

$$CH_3CH_2—O—CH_2CH_3$$
Diethyl ether

Isopropyl phenyl ether

If the groups are the same, we use the prefix di-, as in diethyl ether, but many people omit the di- and simply say "ethyl ether."

---

**PROBLEM 12.7** Name:

(a) $CH_3CH_2—\underset{\underset{CH_3}{|}}{CH}—O—\underset{\underset{CH_3}{|}}{CH}—CH_2CH_3$  (b)

---

Certain heterocyclic ethers (Sec. 11.13) are important in the laboratory, among them

Ethylene oxide \qquad Tetrahydrofuran \qquad Dioxane

Cyclic ethers don't have "ether" in their names.

By far the most important ether is diethyl ether, which is often simply called "ether." This was formerly the most important general anesthetic, though it has now been largely replaced by other compounds (Box 12F). Diethyl ether is still an important laboratory solvent, as are tetrahydrofuran and dioxane. Ethylene oxide, a gas, is used in the manufacture of plastics and other compounds.

## Box 12F

## Anesthetics

Before the mid-nineteenth century, not only did many patients die from surgery because of infection (Box 12E), but there were no anesthetics, so surgery was extremely painful; many patients died from shock. Physicians did use alcohol and such drugs as opium, but they were usually reluctant to operate except in extreme circumstances because the risks to the patient were so great. These problems were resolved by the discovery of antiseptics (Box 12E) and anesthetics. Diethyl ether was first used as a general anesthetic in 1842 (Figure 12F).

Diethyl ether has many advantages as an anesthetic. It is safe to the patient and easy to administer. However, it also has disadvantages: It is relatively slow acting, irritates the respiratory tract, and has a nauseating aftereffect. It is also highly flammable, and its vapor mixed with air is explosive. Because of these disadvantages, a number of other anesthetics are now used. Some of these are also ethers, the most important being enflurane and isoflurane.

$$F_2CH-O-CF_2-\underset{\underset{Cl}{|}}{CH}-F \qquad F_3C-O-\underset{\underset{Cl}{|}}{CH}-CF_3$$

Enflurane (Ethrane)                    Isoflurane

Other anesthetics are not ethers at all. One of the most important of today's anesthetics is halothane, a simple halogen derivative of ethane. This compound is nonflammable and nonexplosive.

$$H-\underset{\underset{Cl}{|}}{\overset{\overset{Br}{|}}{C}}-\underset{\underset{F}{|}}{\overset{\overset{F}{|}}{C}}-F$$

Halothane (1-bromo-1-chloro-2,2,2-trifluoroethane)

It does not cause discomfort and is considered safe, though it does depress respiratory and cardiovascular action. Another important anesthetic is the simple inorganic compound nitrous oxide, $N_2O$ (also called laughing gas), which is used by dentists because its effects wear off quickly.

**Figure 12F**
Early demonstration of the use of ether as an anesthetic. (Courtesy of the National Library of Medicine.)

B o x 12G

## Deodorants

Body odor is caused by the action of certain skin bacteria on natural secretions that are part of perspiration. The secretions themselves have no odor, but the bacteria react with compounds in the secretion to produce compounds with unpleasant odors. Throughout most of history people dealt with this by using perfumes and/or soap and water, or else many simply chose to smell bad. Today in the United States and other developed countries the use of personal deodorants is widespread. These deodorants operate by killing the bacteria that react with the secretions. The most common antibacterial agent in deodorants today is called *triclosan* (chemical name 2,4,4'-trichloro-2'-hydroxydiphenyl ether).

Triclosan          Hexachlorophene

This compound is both a phenol and an ether, as well as a halide. A typical solid deodorant has less than one percent triclosan, the rest being inactive ingredients (that is, they don't kill bacteria), mostly ethanol (or 1,2-propanediol) and water. A fragrance is also usually added. At one time most deodorants contained the compound hexachlorophene (this was also used in soap and even in baby lotion), but this compound proved too toxic and is no longer used in any cosmetic products.

---

The chief chemical property of ethers is that, like the alkanes, they are virtually inert and do not react with most reagents. It is this property that makes some of them such valuable solvents. There are, however, two chemical properties of ethers that laboratory workers must be aware of:

**1.** Like many organic compounds, ethers are flammable. Diethyl ether, the most common, is especially dangerous to work with because of its low boiling point (35°C), only a few degrees above room temperature. This means that the air in a room where ether is being used contains substantial amounts of ether vapor, which can quickly travel to a flame elsewhere in the room. It is therefore important that no source of flame (including pilot lights) be present in any room where diethyl ether is being used.

**2.** On standing for long periods of time, diethyl ether (and some other ethers) is oxidized by air to give high-boiling explosive compounds called peroxides. To prevent this, diethyl ether must be stored in the dark in sealed containers away from any source of heat.

Since 1992 the Clean Air Act has mandated the use of "oxygenated" gasoline in metropolitan areas of the United States during the winter months because of carbon monoxide pollution. By law, the oxygen content of this new gasoline must be at least 2.7% by weight. Since gasoline ordinarily contains mostly hydrocarbons, with only very small amounts of oxygen-containing compounds (Box 10C), gasoline companies achieve this goal by adding certain alcohols or ethers, mostly either methyl *tert*-butyl ether (MTBE) or ethanol, to the gasoline.

$$CH_3-O-\underset{\underset{CH_3}{|}}{\overset{\overset{CH_3}{|}}{C}}-CH_3 \qquad CH_3CH_2-OH$$

Methyl *tert*-butyl ether          Ethanol

## 12.7 PHYSICAL PROPERTIES OF ALCOHOLS, PHENOLS, AND ETHERS

All the hydrocarbons considered in Chapters 10 and 11 have similar physical properties (this also includes the halides). Basically, the properties we are most interested in are boiling points (and melting points), which increase fairly regularly with increasing molecular weight, and solubilities: They are all insoluble in water but generally are soluble in each other.

When we come to alcohols, phenols, and ethers, we find a big difference in physical properties, as shown in Table 12.3.

### Boiling Point

Ethers have about the same boiling points as alkanes of similar molecular weight. Thus, with respect to boiling points, ethers are essentially no different from hydrocarbons or halides.

This is not true for alcohols. The boiling points of alcohols are much higher than those of the corresponding alkane and ether, and those of the diols are much higher still. This is easily explained by hydrogen bonding. Each alcohol molecule can form a hydrogen bond with another one, but an ether molecule cannot:

Review Section 6.6 for a discussion of which molecules can form hydrogen bonds.

$$R-O\cdots H-O \qquad\qquad R-O \quad R-O$$

| Alcohol | Alcohol | | Ether | Ether |

(no hydrogen bond possible)

Alcohols have higher boiling points because hydrogen bonds must be broken during boiling, and this requires energy, available only at higher temperatures. Diols, with two —OH groups, have hydrogen bonds on two sides, so their boiling points are higher still. Hydrogen bonding also accounts for the viscosity of such liquids as ethylene glycol and glycerol.

### Solubility in Water

A look at Table 12.3 shows that both alcohols *and* ethers are soluble in water, up to about three or four carbons. This behavior is, of course, completely different from that of the hydrocarbons and halides. Again, the reason is hydrogen bonding. Both alcohols and ethers have an oxygen atom, so both can form hydrogen bonds with water and so can dissolve.

Ethers can form hydrogen bonds with water but not with other ether molecules, because water has the group —OH but an ether molecule does not.

$$R-O\cdots H-O \qquad\qquad R-O\cdots H-O$$

| Alcohol | Water | Ether | Water |

**TABLE 12.3**   Molecular Weights, Boiling Points, and Solubilities of Some Simple Alkanes, Alcohols, and Ethers

| Name | Structural Formula | MW | BP (°C) | Solubility in Water |
|------|--------------------|------|------|---------------------|
| Ethane | $CH_3—CH_3$ | 30 | −88 | Insoluble |
| Methanol | $CH_3—OH$ | 32 | 65 | Soluble |
| Propane | $CH_3—CH_2—CH_3$ | 44 | −42 | Insoluble |
| Dimethyl ether | $CH_3—O—CH_3$ | 46 | −23 | Soluble |
| Ethanol | $CH_3—CH_2—OH$ | 46 | 78 | Soluble |
| Butane | $CH_3—CH_2—CH_2—CH_3$ | 58 | 0 | Insoluble |
| Ethyl methyl ether | $CH_3—CH_2—O—CH_3$ | 60 | 11 | Soluble |
| 1-Propanol | $CH_3—CH_2—CH_2—OH$ | 60 | 97 | Soluble |
| Ethylene glycol | $HO—CH_2—CH_2—OH$ | 62 | 198 | Soluble |
| Pentane | $CH_3—CH_2—CH_2—CH_2—CH_3$ | 72 | 36 | Insoluble |
| Diethyl ether | $CH_3—CH_2—O—CH_2—CH_3$ | 74 | 35 | Slightly soluble |
| Methyl propyl ether | $CH_3—CH_2—CH_2—O—CH_3$ | 74 | 39 | Slightly soluble |
| 1-Butanol | $CH_3—CH_2—CH_2—CH_2—OH$ | 74 | 117 | Slightly soluble |
| 1,3-Propanediol | $HO—CH_2—CH_2—CH_2—OH$ | 76 | 214 | Soluble |

One of the most widespread illegal drugs in the United States is *marijuana*, obtained from the flowering tops of the hemp plant (*Cannabis sativa*, Figure 12H). A stronger form of this drug, obtained from a resin extruded by the same plant, is called *hashish*. The active ingredient in both marijuana and hashish is a compound that is both a phenol and a cyclic ether, *tetrahydrocannabinol* (THC):

Tetrahydrocannabinol

The typical marijuana smoked in the United States contains 2 to 3 percent THC; hashish contains about twice as much. THC produces immediate psychological effects, which the users call getting "high" or "stoned." There may be an intensification of sights and sounds. Effects that can be measured in the laboratory include increased pulse rate and bloodshot eyes.

There has been much debate about whether marijuana is harmful and whether it should be legalized. The Food and Drug Administration has approved a synthetic form of THC under the name of Dronabinol as a treatment for the wasting condition associated with AIDS (Box 24I). The drug is used to overcome severe nausea and stimulate the appetite of AIDS patients who often suffer from anorexia and weight loss. Synthetic THC has also been approved for cancer patients undergoing chemotherapy.

**Figure 12H**
Cannabis plant. (Courtesy of the National Library of Medicine.)

**Figure 12.3**
Water molecules surround the —OH end of 1-heptanol but do not come near the alkyl end. Because the alkyl end is unsolvated, 1-heptanol is insoluble in water.

Alcohols and ethers of higher molecular weight, however, do not dissolve in water because the water molecules cannot solvate the large R groups. For example, consider 1-heptanol. This molecule consists of an alkyl chain of seven carbons and an —OH group. As shown in Figure 12.3, the —OH forms hydrogen bonds with and is surrounded by water molecules, but the alkyl portion of the molecule has *no* attraction for water molecules. Because this part of the molecule cannot be surrounded by water molecules, 1-heptanol is insoluble in water.

*We call this part of the molecule hydrophobic, meaning water-hating.*

As you might expect, the physical properties of phenols are much like those of alcohols of similar molecular weight.

## 12.8 THIOLS, THIOETHERS, AND DISULFIDES

*Thiols are also called mercaptans, and thioethers are also called sulfides.*

Thiols and thioethers are sulfur analogs of the corresponding alcohols, phenols, and ethers (Table 12.1),

$$R—SH \qquad R—S—R'$$

Thiols        Thioethers

but are much less important chemically. Three properties of thiols are worth mentioning here.

*Gas companies take advantage of the odor of thiols. Natural gas (methane) has no odor, so the companies add a tiny amount of methanethiol, $CH_3SH$, so that gas leaks can be detected before a spark or a match sets off an explosion.*

**1.** Thiols have foul odors. You generally don't expect organic compounds to have pleasant odors (although some do), but thiols smell so bad that chemical companies put the word "stench" on the labels. The liquid squirted by skunks is a mixture of thiols and closely related compounds.
**2.** Thiols are weak acids, somewhat weaker than phenols.
**3.** Thiols are easily oxidizable to disulfides:

$$2RSH \xrightarrow{[O]} R—S—S—R$$

This reaction can be accomplished by many oxidizing agents. Note that the disulfides have an S—S bond. Disulfides are, in turn, easily reducible to thiols by several different reducing agents:

*The [H] represents an unspecified reducing agent.*

$$R—S—S—R \xrightarrow{[H]} 2R—SH$$

The easy conversion between thiols and disulfides is very important in protein chemistry, as we shall see in Section 18.9.

Thioethers are not very important, and we shall not discuss them further.

## Chlorinated Hydrocarbon Insecticides

Certain chlorinated hydrocarbons are widely used and effective insecticides. Two of the most important ones are DDT and chlordane:

1,1,1-Trichloro-2,2-bis(*p*-chlorophenyl)ethane (DDT)

Chlordane

In many places chlorinated hydrocarbons have been so effective in killing crop-eating insects that crop yields have increased enormously (Fig. 12I). Chlorinated hydrocarbons have also wiped out, in large areas of the world, a number of insects that carry such diseases as malaria and typhus.

However, chlorinated hydrocarbons are a two-edged sword. Despite their enormous benefits, they also have enormous disadvantages. They are toxic not just to insects but to all animal life, including human beings. They remain in the environment (soil and water) for years because they are not *biodegradable* (that is, there are no natural processes by which they decompose). Because of this, it is estimated that the tissues of the average person contain about 5 to 10 ppm of DDT. There is DDT in mother's milk. As a result of all these problems, some governments have banned some or all chlorinated hydrocarbons, and their worldwide use is declining.

**Figure 12I**
An airplane spraying insecticide on a field of crops. (Courtesy of U.S. Department of Agriculture.)

## 12.9 ALKYL AND ARYL HALIDES

We have already discussed the naming of simple alkyl halides (Sec. 10.7) and aryl halides (Sec. 11.10). These compounds are not important in living tissues, but they are important in chemical laboratories and in industry because they undergo many reactions and so can be converted to other compounds. We will be chiefly interested in looking at a few of the simpler ones.

An aryl halide is an aromatic halide.

Chloromethane, $CH_3Cl$, also called methyl chloride, is a gas at room temperature but can be liquefied under pressure. The pressurized liquid is used as a local anesthetic, whose effect is derived from its rapid evaporation. Ethyl chloride (chloroethane), $CH_3CH_2Cl$, is used for the same purpose (Box 5H).

Dichloromethane, $CH_2Cl_2$, also called methylene chloride, is a common laboratory solvent.

Trichloromethane, $CHCl_3$, more commonly known as chloroform, was one of the earliest anesthetics but is no longer used for this purpose because it is too toxic. It is still used (carefully) as a laboratory solvent.

Tetrachloromethane, $CCl_4$, better known as carbon tetrachloride, was once the most important dry-cleaning solvent. Because of its toxicity, it is no longer used for that purpose. Today most dry cleaning is done with perchloroethylene (tetrachloroethene), $Cl_2C{=}CCl_2$, which is as good a solvent but has a much lower toxicity.

Triiodomethane, $CHI_3$, also called iodoform, is a yellow solid. In the past it was used in hospitals as a general disinfectant, so much so that its characteristic odor has become known as "hospital odor."

Dichlorodifluoromethane, $CCl_2F_2$, known by the trade name Freon-12, is the most important member of a group of compounds called chlorofluorocarbons (CFCs), compounds that contain carbon, chlorine, and fluorine. Freon-12, a nontoxic, noncorrosive, nonreactive gas, has been used for many years as the refrigerant in virtually all refrigerators and air conditioners. Other major uses of Freon-12 and other CFCs have been as the propellant gas in aerosol sprays and in the manufacture of plastic foams used as insulators in refrigerators and other appliances. However, it is now known that Freon-12 and the other CFCs are a threat to the ozone layer in the earth's upper atmosphere, which shields us from the harmful rays of the sun (Fig. 12.4). The use of these compounds in aerosol sprays was discontinued several years ago, and use of them for other purposes, especially as refrigerants, is now also being discontinued, by international agreement. The compound that is largely replacing Freon-12 as a refrigerant is 1,1,1,2-tetrafluoroethane, $F_3CCH_2F$, which is considered to be much less of a threat to the ozone layer.

The compound *para*-dichlorobenzene,

$$Cl-\langle\bigcirc\rangle-Cl$$

sold under the trade name Para, is a moth repellent.

Polychlorinated biphenyls (PCBs),

$$Cl-\langle\bigcirc\rangle-\langle\bigcirc\rangle-Cl$$

have been used as insulators and heat exchangers, but because they are poisonous and not biodegradable, their manufacture has been banned in the United States since 1979. However, so much was used, that a great deal is still in the environment, and its removal presents a problem.

**Figure 12.4**
A chart showing the density of the earth's ozone layer in September 1992. As the scheme at the bottom shows, a blue color indicates a low density of ozone. The area around the south pole is entirely blue, showing very low ozone concentration. (The colorless circle directly at the pole indicates an area where no measurements were made.) (Courtesy of NASA.)

This is just one member of the family of PCBs.

# Iodine and Goiter

In the disease called goiter, the thyroid gland in the neck becomes enlarged (Fig. 12J). The relationship between this disease and iodine intake from food was established 130 years ago. Centuries before, goiter was treated by feeding patients such sources of iodine as burnt sea sponges. The iodine interacts with a protein called thyroglobulin, and the aromatic rings of the protein become iodinated. Two iodinated molecules interact, forming a thyroxine unit that is bound to a protein. The aromatic unit is then released to become the powerful thyroid hormone thyroxine.

Young mammals require thyroid hormones for normal growth and development. Lack of thyroid hormones during fetal development results in mental retardation. Low thyroid hormone levels in adults results in hypothyroidism, commonly called goiter, the symptoms of which are lethargy, obesity, and dry skin.

The iodine we get from our food and water is in the form of iodide ion, and the amount of it depends on the iodide content of the soil. Low-iodine soils the world over are responsible for what are called "goiter belts," large areas of a country where the average diet is very low in iodine. In the United States, approximately half of the table salt used is enriched with NaI (100 mg of iodide per gram of table salt). It is this iodized salt that has substantially reduced the incidence of goiter in this country, although it has not eliminated it completely.

**Figure 12J**
A patient suffering from goiter.
(Phototake.)

## SUMMARY

**Alcohols** have the formula **ROH,** where R is an alkyl group. Alcohols are classified as **primary, secondary,** or **tertiary** depending on whether the —OH is connected to a primary, secondary, or tertiary carbon. They can be named either by the IUPAC system or by common names. Alcohols are neither acidic nor basic. When treated with sulfuric acid they undergo dehydration to give alkenes (or, in some cases, ethers). Primary alcohols can be oxidized to aldehydes or carboxylic acids, and secondary alcohols to ketones. Tertiary alcohols resist oxidation. Some important alcohols are ethanol (the alcohol of alcoholic beverages), methanol (wood alcohol), isopropyl alcohol (rubbing alcohol), menthol, ethylene glycol (antifreeze), and glycerol.

Compounds in which an —OH group is directly connected to an aromatic ring are called **phenols.** Phenols are weak acids. Several of them are used as antiseptics and disinfectants.

**Ethers** have the formula **ROR′.** They are rather inert chemically and are used as laboratory solvents. Diethyl ether was formerly an important anesthetic.

Alcohols and phenols have high boiling points as a result of hydrogen bonding. Alcohols, phenols, and ethers are soluble in water if the molecules are small, also because of hydrogen bonding.

**Thiols, RSH,** have bad odors, are weak acids, and are easily oxidized to **disulfides, RSSR.** Some important alkyl halides are $CH_3Cl$, $CH_2Cl_2$, $CHCl_3$, and $CCl_4$, used mostly as solvents.

### Summary of Reactions

**1.** Dehydration of alcohols to give alkenes (Sec. 12.3):

**2.** Dehydration of alcohols to give ethers (Sec. 12.3):

$$2ROH \xrightarrow[\text{heat}]{H_2SO_4} R-O-R + H_2O$$

**3.** Oxidation of alcohols (Sec. 12.3):

Primary, Secondary, Tertiary

**4.** Oxidation of thiols (Sec. 12.8):

$$2RSH \xrightarrow{[O]} R-S-S-R$$

**5.** Reduction of disulfides (Sec. 12.8):

$$R-S-S-R \xrightarrow{[H]} 2RSH$$

## KEY TERMS

Alcohol (Sec. 12.2)
CFCs (Sec. 12.9)
Congener (Box 12A)
Dehydration (Sec. 12.3)
Denatured alcohol (Sec. 12.4)
Disulfide (Sec. 12.8)
Elimination (Sec. 12.3)
Ether (Sec. 12.6)
Mercaptan (Sec. 12.8)

PCBs (Sec. 12.9)
Phenol (Sec. 12.5)
Primary alcohol (Sec. 12.2)
Proof (Box 12A)
Secondary alcohol (Sec. 12.2)
Tertiary alcohol (Sec. 12.2)
Thioether (Sec. 12.8)
Thiol (Sec. 12.8)

## PROBLEMS

### Nomenclature of Alcohols

**12.8** What is the difference between an alcohol and a phenol?

**12.9** Give the IUPAC name and, where possible, the common name for these alcohols:

(a) $CH_3CH_2CH_2CH_2-OH$

(b) $CH_3CH_2-CH-CH_3$
$\qquad\qquad\quad |$
$\qquad\qquad\quad OH$

(c) $CH_3-\underset{\underset{CH_3}{|}}{\overset{\overset{CH_3}{|}}{C}}-OH$

(d) $\underset{\text{(benzene ring)}}{} CH_2-OH$

(e) $CH_3-\underset{\underset{F}{|}}{\overset{\overset{F}{|}}{C}}-\underset{\underset{H}{|}}{\overset{\overset{Cl}{|}}{C}}-CH_2-OH$

(f) $CH_3-\underset{}{CH}-\underset{\underset{OH}{|}}{CH}-\overset{\overset{CH_3}{|}}{CH}-CH_2CH_3$

(g) $\overset{\overset{Br}{|}}{CH_2}-CH-CH_2-\underset{\underset{OH}{|}}{\overset{\overset{CH_3}{|}}{CH}}-CH_3$

(h) $CH_3CH_2CH_2-\underset{\underset{CH_2CH_2-Cl}{}}{\overset{\overset{CH_2OH}{|}}{CH}}$

(i) $CH_3-CH_2-\underset{\underset{OH}{|}}{\overset{\overset{CH_3}{|}\\ \overset{CH_2}{|}}{C}}-CH_3$

(j) (cyclohexane ring with Br, $H_3C$, $CH_3$, OH substituents)

(k) (cyclopentane ring with $H_3C$, $-OH$, $CH_3$ substituents)

(l) $CH_2-CH-CH_2$
$\quad\ |\qquad |\qquad |$
$\quad OH\ \ OH\ \ OH$

**12.10** Tell whether each alcohol in Problem 12.9 is primary, secondary, or tertiary.

**12.11** What is the difference between a primary, a secondary, and a tertiary alcohol?

**12.12** Draw the structural formula for (a) any primary alcohol, (b) any secondary alcohol, (c) any tertiary alcohol

**12.13** Draw the structural formula for
(a) 2-propanol
(b) isobutyl alcohol
(c) cyclohexanol
(d) *meta*-bromobenzyl alcohol
(e) glycerol
(f) wood alcohol

**12.14** Draw the structural formula for
(a) 3-methyl-1-pentanol
(b) 2-methyl-3-hexanol
(c) 3-methyl-2-hexanol
(d) 2-chloro-3-heptanol
(e) 3,3-dibromocyclobutanol
(f) *para*-ethylbenzyl alcohol
(g) 1,2,3-pentanetriol

### Chemical Properties of Alcohols

**12.15** Write equations to show how 1-propanol, when treated with sulfuric acid, can give either an alkene or an ether.

**12.16** Show the structural formula of the principal organic product when ethanol is treated with
(a) $H_2SO_4$ at 140°C
(b) $H_2SO_4$ at 180°C
(c) excess $K_2Cr_2O_7$ and $H_2SO_4$

**12.17** Write the structural formula for the principal organic product (if no reaction, say so):
(a) $CH_3OH + H_2SO_4 \longrightarrow$

(b) (cyclohexane ring with Br and OH) $+ H_2SO_4 \longrightarrow$

(c) $CH_3CH_2$—CH—$CH_2CH_3$ + $K_2Cr_2O_7$ $\xrightarrow{H_2SO_4}$
         |
        OH

(d) $CH_3CH_2CH_2CH_2$—OH + $K_2Cr_2O_7$ $\xrightarrow[\text{excess}]{H_2SO_4}$

          $CH_3$
          |
(e) $CH_3$—C—OH + $KMnO_4$ $\longrightarrow$
          |
          $CH_3$

(f) ⬡—$CH_2OH$ + $KMnO_4$ $\xrightarrow{\text{excess}}$

## Some Important Alcohols

**12.18** Define: (a) denatured alcohol, (b) absolute alcohol, (c) tincture

**12.19** Which alcohol is
(a) oxidized to formaldehyde
(b) used for hospital rubdowns
(c) used to make nitroglycerine
(d) used as an antifreeze
(e) found in fats
(f) the alcohol of alcoholic beverages
(g) a cause of blindness and death when ingested in moderate quantities

## Phenols

**12.20** Name these phenols:

(a) ⬡—OH

(b) $NO_2$—⬡—OH

(c) Br—⬡(Br)—OH

(d) Cl—⬡(Cl)(—$NO_2$)—OH

**12.21** Complete these reactions:

(a) ⬡(OH)($CH_3$) + KOH $\longrightarrow$

(b) ⬡(OH)($NO_2$)($NO_2$) + NaOH $\longrightarrow$

(c) ⬡($NO_2$)(OH)($NO_2$)($NO_2$) + $H_2O$ $\longrightarrow$

**12.22** Which phenol mentioned in this chapter is
(a) used to treat burns
(b) a former antiseptic that is now considered too strong for use on patients
(c) a flavoring agent

## Ethers

**12.23** Name these ethers:
(a) $CH_3$—O—$CH_2CH_3$
(b) $CH_3CH_2$—O—$CH_2CH_3$

       $CH_3$       $CH_3$
       |           |
(c) $CH_3$—CH—O—CH—$CH_3$

(d) $CH_3$—O—CH—$CH_3$
            |
            $CH_3$

(e) ⬡—O—$CH_2CH_2CH_2CH_2CH_3$

(f) ⬠O (tetrahydrofuran ring)

**12.24** Draw the structural formulas for
(a) dimethyl ether
(b) ethyl propyl ether
(c) diisopropyl ether
(d) phenyl propyl ether
(e) benzyl ethyl ether

**12.25** What is the chief chemical property of ethers?

**12.26** Classify each of the following as (1) primary alcohol, (2) secondary alcohol, (3) tertiary alcohol, (4) phenol, (5) ether, or (6) none of these:

(a) $CH_3CH_2-\underset{\underset{OH}{|}}{CH}-CH_3$

(b) $CH_3-\underset{\underset{O}{\|}}{C}-CH_3$

(c) $CH_3CH_2-O-CH_2CH_3$

(d)

(e) $CH_3-\underset{\underset{CH_3}{|}}{CH}-CH_2-OH$

(f)

(g)

(h)

(i)

(j)

(k)

(l)

(m)

**12.27** Each of these compounds belongs to two or more of the first five classes mentioned in Problem 12.26. Tell which classes:

(a) $\underset{\underset{OH}{|}}{CH_2}-\underset{\underset{OH}{|}}{CH}-\underset{\underset{OH}{|}}{CH_2}$

(b) $CH_3-O-CH_2CH_2CH_2-OH$

(c)

(d)

(e)

(f)

(g)

(h)

**12.28** What is the chief chemical property of phenols?

**12.29** Draw the structural formulas for all seven possible compounds whose molecular formula is $C_4H_{10}O$. Some are alcohols and some are ethers.

## Physical Properties of Alcohols, Phenols, and Ethers

**12.30** Which of these molecules can form a hydrogen bond with another molecule of itself?

(a) $CH_3CH_2OH$

(b) $CH_3CH_2Cl$

(c) $CH_3CH_2-O-CH_3$

(d)

(e) benzene ring with CH$_2$—OH and Br substituents

(f) $CH_3CH_2$—O—$CH_2CH_2$—F

**12.31** Rank in order of increasing boiling point:

(a) $CH_3CH_2OH$

(b) cyclohexane ring with —OH

(c) $CH_3CH_2$—O—$CH_2CH_3$

(d) $CH_2$—CH—$CH_2$ with OH, OH, OH

**12.32** Predict whether each of the following is soluble in water:

(a) $CH_3OH$

(b) $CH_3CH_2CH_2CH_2CH_2CH_2OH$

(c) $CH_3OCH_2CH_3$

(d) $CH_3Cl$

(e) $CH_3CH_2CH_2CH_2$—O—$CH_2CH_3$

(f) benzene ring with OH, CH$_3$, CH$_3$

**12.33** 1-Pentanol is only slightly soluble in water. Predict the solubility in water of 1,5-pentanediol and 1,3,6-hexanetriol. Explain.

**12.34** Explain why the boiling point of ethylene glycol (198°C) is so much higher than that of 1-propanol (97°C) even though their molecular weights are about the same.

## Thiols and Disulfides

**12.35** If $CH_3SH$ is called methanethiol, name these compounds:

(a) $CH_3CH_2SH$

(b) $CH_3$—CH—$CH_3$ with SH

**12.36** Draw the structural formula for the principal organic product:

(a) $2CH_3CH_2$—SH $\xrightarrow{[O]}$

(b) $CH_3$—S—S—$CH_3$ $\xrightarrow{[H]}$

**12.37** Write an equation for

(a) the oxidation of ethanethiol, $CH_3CH_2SH$

(b) the reduction of the disulfide $CH_3CH_2CH_2$—S—S—$CH_2CH_2CH_3$

**12.38** The amino acid cysteine,

$$HS—CH_2—CH—COOH$$
with $NH_2$

is oxidized in the body to a disulfide. Draw its structure.

## Halides

**12.39** Which halogenated compound

(a) was formerly an anesthetic but is no longer used because it is too toxic

(b) was formerly a major dry-cleaning agent but is no longer used because it is too toxic

(c) is used as a local anesthetic

(d) is the cause of "hospital odor"

(e) has been an important refrigerant

**12.40** What are PCBs, and why are they harmful?

**12.41** Which halogenated compound (a) is used as a dry-cleaning agent today (b) is likely to become an important refrigerant?

**12.42** Why is Freon-12 no longer being used as a refrigerant?

## Boxes

**12.43** (Box 12A) What is the proof in each of the following?

(a) a cordial that contains 17 percent alcohol

(b) a vodka that contains 41 percent alcohol

(c) a table wine that contains 11 percent alcohol

**12.44** (Box 12A) Why does the average person get drunk faster from champagne than from table wine even if both contain the same percentage of alcohol?

**12.45** (Box 12A) Define congener.

**12.46** (Box 12A) How do we get an 86-proof whiskey from a fermentation broth that contains only 13 percent alcohol?

**12.47** (Box 12B) How does disulfuram remedy alcohol addiction?

**12.48** (Box 12B) What physiological effects of alcohol result in so many fatal accidents involving drunken drivers?

**12.49** (Box 12C) Distillation of wood yields wood alcohol (methanol). Why should this product never be used for drinking?

**12.50** (Box 12D) How does nitroglycerine help angina patients?

**12.51** (Box 12D) Do nitroglycerine and isoamyl nitrite have the same functional group? If not, how do they differ?

**12.52** (Box 12E) What phenolic compound is present in mouthwashes and throat lozenges?

**\*12.53** (Box 12F) Name the compound enflurane according to the method given in the text for naming ethers.

**12.54** (Box 12F) What are the advantages and disadvantages of diethyl ether as an anesthetic?

**12.55** (Box 12G) What causes body odor?

**12.56** (Box 12G) Why has triclosan replaced hexachlorophene in deodorants?

**12.57** (Box 12H) Find the ether linkage and the phenol group in tetrahydrocannabinol.

**12.58** (Box 12H) What is the medical use of THC?

**12.59** (Box 12I) What is meant by the term "biodegradable"?

**12.60** (Box 12J) In the absence of iodine in the diet, goiter develops. Explain why goiter is a regional disease.

## Additional Problems

**12.61** Write a balanced equation for the reaction in which glucose, $C_6H_{12}O_6$, is fermented to beverage alcohol.

**12.62** What is the difference in structure between thymol and menthol? Is either one acidic? Explain.

**12.63** Draw the structural formulas for
  (a) tetrahydrofuran
  (b) picric acid
  \*(c) *para*-cresol
  (d) ethylene oxide

**\*12.64** 2-Propanol is commonly used as a rubbing alcohol, to cool the skin. 2-Hexanol, also a liquid, is not suitable for this purpose. Why not?

**\*12.65** The $pK_a$ values of some phenols are: phenol = 9.89; *o*-chlorophenol = 8.49; *o*-methylphenol = 10.20. What is the effect of Cl and $CH_3$ substitution on the benzene ring on the acidity of the OH group?

**12.66** Decide which family of organic compounds is referred to in each case:
  (a) weak acids; very bad odors; contain sulfur
  (b) neutral compounds; contain oxygen; chemically almost inert
  (c) weak acids; many have been used as antiseptics
  (d) neutral compounds; oxidizable, low-molecular-weight compounds soluble in water

**12.67** Explain why glycerol is much thicker (more viscous) than ethylene glycol, which in turn is much thicker than ethanol.

**12.68** Explain why methanethiol, $CH_3SH$, has a lower boiling point (6°C) than methanol, $CH_3OH$ (65°C), even though it has a higher molecular weight.

**12.69** Draw the structural formulas for the principal organic products (if no reaction, say so):
  (a) $CH_3CH_2-SH + NaOH \longrightarrow$
  (b) $2 \langle\!\!\!\!\bigcirc\!\!\!\!\rangle - OH \xrightarrow{H_2SO_4}$
  (c) $CH_3CH_2-O-CH_2CH_3 + NaOH \longrightarrow$

**12.70** List in order of increasing solubility in water:
  (a) $CH_3CH_2CH_2OH$
  (b) $CH_3CH_2CH_2SH$
  (c) $CH_3CH_2-O-CH_2CH_3$

**12.71** Explain why the boiling point of 1-butanol (117°C) is so much higher than that of methyl propyl ether (39°C), even though they have the same molecular weight (they are isomers).

**12.72** Draw the structural formulas for the principal organic products when each of these compounds is treated with $K_2Cr_2O_7$ and $H_2SO_4$ (if no reaction, say so):
  (a) $CH_3CH_2CH_2-OH$

(b)

$$\text{C}_6\text{H}_5-\underset{\underset{\text{CH}_3}{|}}{\overset{\overset{\text{CH}_3}{|}}{\text{C}}}-\text{OH}$$

(c)  $CH_3CH_2-O-CH_3$

(d)  $CH_3CH_2-\underset{\underset{\text{OH}}{|}}{\text{CH}}-CH_2CH_3$

**12.73**  Of the chemical families studied in this chapter, which are acidic and which are neutral?

**12.74**  (a) Draw the structural formula for the two possible products when 1-butanol is oxidized. (b) Which of the two would you expect to form in greater amount?  (c) How could you get the other one, if that was what you wanted?

**12.75**  A large number of $NO_2$ groups on a compound makes it explosive (for example, nitroglycerine, trinitrotoluene-TNT). How would you handle a bottle of picric acid (Sec. 12.5)?

# CHAPTER 13

*Cinnamon, almonds, and vanilla beans all contain aldehydes. (Photograph by Beverly March.)*

# Aldehydes and Ketones

## 13.1 INTRODUCTION

One of the most important functional groups in organic chemistry is the **carbonyl group:**

$$\begin{array}{c} \phantom{} \\ -\!\!\overset{|}{\underset{\|}{C}}\!\!- \\ O \end{array}$$    The carbonyl group

This group is present in aldehydes and ketones, which we study in this chapter, as well as in other compounds, which will be discussed in Chapters 14 and 15. The carbonyl group is also very important in biochemistry.

The carbon of the carbonyl group must be connected to two other atoms or groups. When it is connected to a hydrogen and an alkyl group or aromatic ring (or to two hydrogens) the compound is called an **aldehyde;** when it is connected to two alkyl groups or aromatic rings (or one of each) the compound is a **ketone:**

Fats, carbohydrates, proteins, nucleic acids, urea, and many other biological compounds contain the carbonyl group.

$$H\!-\!\underset{\underset{O}{\|}}{C}\!-\!H \qquad R\!-\!\underset{\underset{O}{\|}}{C}\!-\!H \qquad \underset{\underset{O}{\|}}{\bigcirc}\!\!\overset{}{C}\!-\!H$$

Aldehydes

391

$$R-\overset{\overset{\displaystyle}{\underset{\displaystyle O}{\|}}}{C}-R'$$

Ketones

Do not confuse RCHO with alcohols, which are ROH.

Because they always contain at least one H connected to the C=O, aldehydes are often written RCHO to save space. Similarly, ketones may be written as RCOR'.

## 13.2 NOMENCLATURE

Both aldehydes and ketones can be named by the IUPAC rules. The system is the same as we have seen before, except that the suffix "-al" is used for aldehydes and "-one" for ketones.

---

### EXAMPLE

Name the following compounds.

(a) $CH_3-CH_2-\overset{\overset{\displaystyle}{\underset{\displaystyle O}{\|}}}{C}-H$     Propanal

The longest chain has three carbons; and the suffix for aldehydes is "-al." Therefore the name is propanal. Note that no number is needed because the aldehyde group is always at the end of the chain.

(b) $\overset{1}{CH_3}-\overset{2}{CH_2}-\overset{3}{\underset{\displaystyle O}{\overset{\displaystyle \|}{C}}}-\overset{4}{CH_2}-\overset{5}{CH_2}-\overset{6}{CH_3}$     3-Hexanone

There are six carbons in the chain; and the suffix is "-one." The compound is therefore 3-hexanone. In this case we need a number to specify the position of the ketone group, remembering, of course, that we start numbering at the end that gives the lowest possible number to the functional group.

(c) $Cl-\overset{5}{CH_2}-\overset{4}{\underset{\displaystyle OH}{CH}}-\overset{3}{CH_2}-\overset{2}{CH_2}-\overset{1}{\underset{\displaystyle O}{\overset{\displaystyle \|}{C}}}-H$     5-Chloro-4-hydroxy-pentanal

(d) $\overset{5}{CH_2}=\overset{4}{CH}-\overset{3}{CH_2}-\overset{2}{\underset{\displaystyle O}{\overset{\displaystyle \|}{C}}}-\overset{1}{CH_2}-Cl$     1-Chloro-4-penten-2-one

In this case both "-en-" and "-one" are placed in the suffix, and both are given numbers. The 2 goes just before the "-one," because if it were put anywhere else it would be confusing.

**PROBLEM 13.1** Name these compounds by the IUPAC system:

(a) $CH_3CH_2-CH-C-H$
　　　　　　　　|　　||
　　　　　　　Cl　　O

(b) $CH_3-C-CH_2CH_2-Cl$
　　　　　　||
　　　　　　O

As these examples show, the aldehyde or ketone carbon takes precedence over any previously considered group in numbering the chain. That is, we number the chain so as to give the lowest possible number to the carbon of the aldehyde or ketone group, even if such groups as C=C or —OH must get higher numbers. See Table 13.1 for the names of some other aldehydes and ketones.

Simple aldehydes often have common names, some of which are shown in Table 13.1. These names, which end in "aldehyde," are derived from the names of the corresponding carboxylic acids, which we will consider in Section 14.2. The compounds HCHO and $CH_3CHO$ are almost always called formaldehyde and acetaldehyde, respectively; the names methanal and ethanal are almost never used, perhaps because of possible confusion with methanol and ethanol.

Ketones also have common names, which simply consist of the word "ketone" preceded by the two group names (Table 13.1), though the simplest ketone, $CH_3COCH_3$, is almost always called *acetone*. There is a special system

This means that aldehyde and ketone groups take precedence over OH, halogen, and alkyl groups, as well as double and triple bonds.

**TABLE 13.1　Names and Boiling Points of Some Aldehydes and Ketones**

| Formula | IUPAC Name | Common Name | Boiling Point (°C) |
|---|---|---|---|
| H—CHO | Methanal | Formaldehyde | −21 |
| $CH_3$—CHO | Ethanal | Acetaldehyde | 21 |
| $CH_3CH_2$—CHO | Propanal | Propionaldehyde | 49 |
| $CH_3CH_2CH_2$—CHO | Butanal | Butyraldehyde | 76 |
| ⬡—CHO | Benzaldehyde | Benzaldehyde | 178 |
| $CH_3-C-CH_3$, O | Propanone | Acetone | 56 |
| $CH_3CH_2-C-CH_3$, O | Butanone | Ethyl methyl ketone | 80 |
| $CH_3CH_2-C-CH_2CH_3$, O | 3-Pentanone | Diethyl ketone | 102 |
| ⬡—C—$CH_3$, O | 1-Phenylethanone | Acetophenone | 203 |
| ⬡=O | Cyclohexanone | Cyclohexanone | 156 |

for aryl ketones. Thus, the compound that we might call methyl phenyl ketone is more often called acetophenone (Table 13.1). Cyclic ketones are very simply named (see cyclohexanone in Table 13.1).

## 13.3  PHYSICAL PROPERTIES

Remembering what we learned in Section 12.7 about the boiling points and solubilities of alcohols and ethers, what would we expect for aldehydes and ketones? From their structures we can see these things:

**1.**  Two molecules of an aldehyde or ketone cannot form hydrogen bonds with each other because neither molecule contains an O—H, N—H, or F—H bond (Sec. 6.6):

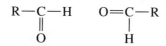

No hydrogen bond possible

We therefore expect their boiling points to be lower than those of alcohols, which can form hydrogen bonds with each other. Table 13.2 shows that both propanal and acetone have much lower boiling points than 1-propanol, which has about the same molecular weight.

**2.**  Aldehydes and ketones are polar molecules because the C=O bond is a polar bond:

The electrons in the C=O bond are closer to the more electronegative oxygen than to the carbon. This causes aldehyde and ketone molecules to behave as dipoles that attract each other. These attractions are not as strong as hydrogen bonds, but they do cause aldehydes and ketones to boil at higher temperatures than the nonpolar alkanes or the less polar ethers (Table 13.2).

*Electronegativity and dipoles are discussed in Section 3.7.*

| TABLE 13.2  Molecular Weights, Boiling Points, and Solubilities of Some Compounds | | | | |
|---|---|---|---|---|
| Name | Formula | MW | BP (°C) | Solubility in Water |
| Butane | $CH_3$—$CH_2$—$CH_2$—$CH_3$ | 58 | 0 | Insoluble |
| Ethyl methyl ether | $CH_3$—$CH_2$—O—$CH_3$ | 60 | 11 | Soluble |
| Propanal | $CH_3$—$CH_2$—$\overset{\text{O}}{\underset{\|}{\text{C}}}$—H | 58 | 49 | Soluble |
| Acetone | $CH_3$—$\overset{\text{O}}{\underset{\|}{\text{C}}}$—$CH_3$ | 58 | 56 | Soluble |
| 1-Propanol | $CH_3$—$CH_2$—$CH_2$—OH | 60 | 97 | Soluble |

**3.** Because aldehydes and ketones contain oxygen atoms, they do form hydrogen bonds with water:

$$R-\overset{\|}{\underset{O\cdots H-O}{C}}-H \qquad R-\overset{\|}{\underset{O\cdots H-O}{C}}-R'$$

Aldehyde  Water        Ketone  Water

Therefore, their solubility in water is about the same as that of alcohols and ethers: The small molecules are soluble (Table 13.2), but the larger ones are not, for the reasons discussed in Section 12.7 for alcohols and ethers (see Fig. 12.3).

Most aldehydes and ketones have strong odors. The odors of ketones are generally pleasant, and many of them are used in perfumes and as flavoring agents (see Box 13D). The odors of aldehydes vary. You may be familiar with the smell of formaldehyde. If so, you know that it is not pleasant. However, many higher aldehydes are also used in perfumes.

Formaldehyde is a gas at room temperature, but we often encounter it as an aqueous solution.

## 13.4 PREPARATION OF ALDEHYDES AND KETONES BY OXIDATION

Aldehydes and ketones are most often made by the oxidation of primary and secondary alcohols, respectively:

$$RCH_2OH \xrightarrow{[O]} R-\overset{\|}{\underset{O}{C}}-H$$

Primary alcohol        Aldehyde

$$R-\underset{OH}{\overset{|}{C}H}-R' \xrightarrow{[O]} R-\overset{\|}{\underset{O}{C}}-R'$$

Secondary alcohol        Ketone

We discussed these reactions in Section 12.3, and pointed out that it is not always easy to stop the oxidation of a primary alcohol at the aldehyde stage because aldehydes are easily oxidizable to carboxylic acids. We will discuss this reaction in the next section.

## 13.5 OXIDATION OF ALDEHYDES

We have already mentioned that aldehydes are easily oxidized to the corresponding carboxylic acids.

$$R-\overset{\|}{\underset{O}{C}}-H \xrightarrow{[O]} R-\overset{\|}{\underset{O}{C}}-OH$$

Many oxidizing agents will do this, even the oxygen in the air. Unless a bottle containing an aldehyde is kept tightly stoppered, the air will oxidize the aldehyde, and the bottle will soon contain the carboxylic acid as well as the aldehyde.

Ketones, on the other hand, resist oxidation:

$$R—\underset{\underset{O}{\|}}{C}—R' \xrightarrow[\substack{\text{mild} \\ \text{conditions}}]{[O]} \text{no reaction}$$

They can be oxidized by heating with very strong oxidizing agents, but no useful products are obtained.

When ketones are oxidized, C—C bonds are broken, leading to complex mixtures.

---

**PROBLEM 13.2**  Draw the structural formula for each product (if there is no reaction, say so):

(a) $CH_3CH_2CH_2—\underset{\underset{O}{\|}}{C}—H \xrightarrow[\text{mild conditions}]{[O]}$

(b) $CH_3—\underset{\underset{O}{\|}}{C}—CH_2CH_3 \xrightarrow[\text{mild conditions}]{[O]}$

---

The fact that aldehydes are easy to oxidize and ketones are not allows us to use simple tests to tell these compounds apart in the laboratory. Suppose we have a compound that we know is either an aldehyde or a ketone. All we have to do is treat it with a mild oxidizing agent. If it can be oxidized it is an aldehyde; otherwise it is a ketone. There are two tests commonly used for this: Tollens' and Benedict's tests.

### Tollens' Test

Tollens' reagent contains silver nitrate and ammonia in water. When these reagents are mixed, the silver ion combines with $NH_3$ to form the ion $Ag(NH_3)_2^+$. When this solution is added to an aldehyde, the complexed silver ion is reduced to silver metal. If the test tube is clean, the silver metal deposits on the inside of the test tube, forming a mirror (Fig. 13.1). The equation is

The $Ag(NH_3)_2^+$ ion is a coordination compound (Sec. 3.5), also called a complex ion.

$$R—\underset{\underset{O}{\|}}{C}—H + 2[Ag(NH_3)_2]^+ + 3OH^- \longrightarrow$$

$\text{Aldehyde}$

$$R—\underset{\underset{O}{\|}}{C}—O^- + 2Ag(s) + 4NH_3 + 2H_2O$$

Carboxylic    Mirror
acid
(as the ion)

Ketones are not oxidized under these conditions, so that no silver mirror forms.

### Benedict's Test

This is similar to Tollens' test in that a metal ion is reduced, and we know that a positive test has taken place when we see the product come out of solution. But here the oxidizing agent is copper(II) ion, which is reduced to copper(I) oxide by an aldehyde:

**Figure 13.1**
A positive Tollens' test (silver mirror). (Photograph by Beverly March.)

$$R-\underset{\underset{O}{\|}}{C}-H + 2Cu^{2+} + 5OH^- \longrightarrow R-\underset{\underset{O}{\|}}{C}-O^- + Cu_2O + 3H_2O$$

Aldehyde    Carboxylic    Red
acid (as    precipitate
the ion)

The formation of a red precipitate of copper(I) oxide and the disappearance of the blue color of $Cu^{2+}$ are the positive test. Benedict's solution is made by dissolving copper sulfate, sodium citrate, and sodium carbonate in water.

There is one type of ketone that does give positive tests with Tollens' and Benedict's reagents. This type of ketone has a hydroxy group on the carbon next to the C=O group:

General formula    $$R-\underset{\underset{OH}{|}}{CH}-\underset{\underset{O}{\|}}{C}-R'$$

Specific example    $$CH_3-\underset{\underset{OH}{|}}{CH}-\underset{\underset{O}{\|}}{C}-\bigcirc$$

These compounds are called α-hydroxy ketones.

It must therefore be borne in mind that an unknown compound that gives a positive Tollens' or Benedict's test might not be an aldehyde, but an α-hydroxy ketone instead. These tests are especially important in carbohydrate chemistry (Sec. 16.8).

## 13.6  OTHER CHEMICAL PROPERTIES OF ALDEHYDES AND KETONES

In Section 11.6 we saw that compounds containing C=C bonds undergo addition reactions. The same is true for C=O bonds. Unlike the C=C bond, the C=O bond is polar (Sec. 13.3):

$$\underset{O\ \delta-}{\overset{\delta+}{\underset{\|}{C}}}$$

so that negatively charged species are attracted to the C and positively charged species to the O. There are many known reactions of this type, but we look at only two (see also Box 13A).

### Addition of ROH. Formation of Acetals

General reaction:

$$R-\underset{\underset{O}{\|}}{C}-H + R'OH \longrightarrow R-\underset{\underset{OH}{|}}{\overset{\overset{OR'}{|}}{C}}-H \xrightarrow[R'OH]{H^+} R-\underset{\underset{OR'}{|}}{\overset{\overset{OR'}{|}}{C}}-H + H_2O$$

Aldehyde    Alcohol    Hemiacetal    Acetal
(not isolated)

## Box 13A

## Chloral Hydrate

When an aldehyde or ketone is dissolved in water, the water can add to the $C=O$ bond the same way an alcohol would:

$$R-\underset{\underset{O}{\|}}{C}-H + HOH \rightleftharpoons R-\underset{\underset{OH}{|}}{\overset{\overset{OH}{|}}{C}}-H$$

Hydrate
(unstable)

The compound formed is called a *hydrate*. Like simple hemiacetals, hydrates are very unstable and cannot be isolated. Therefore, this is not a general reaction of aldehydes and ketones.

There are some exceptions, however. One of these is chloral hydrate, made from the aldehyde chloral:

$$Cl-\underset{\underset{Cl}{|}}{\overset{\overset{Cl}{|}}{C}}-\underset{\underset{O}{\|}}{C}-H + H_2O \longrightarrow Cl-\underset{\underset{Cl}{|}}{\overset{\overset{Cl}{|}}{C}}-\underset{\underset{OH}{|}}{\overset{\overset{OH}{|}}{C}}-H$$

Chloral                    Chloral hydrate
(a stable solid)

Chloral hydrate, a stable solid (mp 57°C), is used as a sleeping tablet. Because it works rapidly, it is called "knockout drops." Another name, used more in detective stories and movies than in real life, is "Mickey Finn." When taken together with alcohol (for example, if added to a drink), chloral hydrate is dangerous and can easily be fatal.

Specific example:

$$CH_3-\underset{\underset{O}{\|}}{C}-H + CH_3CH_2OH \longrightarrow CH_3-\underset{\underset{OH}{|}}{\overset{\overset{OCH_2CH_3}{|}}{C}}-H \xrightarrow[CH_3CH_2OH]{H^+} CH_3-\underset{\underset{OCH_2CH_3}{|}}{\overset{\overset{OCH_2CH_3}{|}}{C}}-H + H_2O$$

Acetaldehyde        Ethanol                    (not isolated)

When an alcohol (R′OH) is added to an aldehyde, the positive part of the alcohol (H$^+$) adds to the O of the carbonyl group, and the negative part of the alcohol (R′O$^-$) adds to the C:

$$\underset{\underset{\delta-}{\overset{\delta+}{\underset{O}{\|}}}}{R-C}-H \quad R'O-H \rightleftharpoons R-\underset{\underset{OH}{|}}{\overset{\overset{OR'}{|}}{C}}-H \qquad \text{An addition reaction}$$

Hemiacetal
(not stable)

The compound produced in this way is both an alcohol and an ether. **A compound with an —OH, an —OR, and an —H on the same carbon** is called a **hemiacetal.** Simple hemiacetals are not stable and cannot be isolated. However, if a small amount of a strong acid (such as HCl) is present in the mixture, the hemiacetal reacts with a second molecule of alcohol (the —OR′ replaces the —OH) to produce a compound that has two ether groups on the same carbon:

$$R'-OH + R-\underset{\underset{\displaystyle OH}{|}}{\overset{\overset{\displaystyle OR'}{|}}{C}}-H \underset{}{\overset{H^+}{\rightleftharpoons}} R-\underset{\underset{\displaystyle OR'}{|}}{\overset{\overset{\displaystyle OR'}{|}}{C}}-H + H_2O \qquad \text{A substitution reaction}$$

Hemiacetal        Acetal

**A compound containing two —OR groups and an —H on the same carbon** is called an **acetal.**

Although in simple cases we cannot isolate hemiacetals, simple acetals are quite stable and are easily prepared by treating an aldehyde with an alcohol in the presence of HCl. This reaction is an equilibrium, and acetals can be cleaved to produce the alcohol and aldehyde by treating with aqueous HCl:

$$R-\underset{\underset{\displaystyle OR'}{|}}{\overset{\overset{\displaystyle OR'}{|}}{C}}-H + H_2O \underset{}{\overset{H^+}{\rightleftharpoons}} R-\underset{\underset{\displaystyle O}{\|}}{C}-H + 2R'OH$$

*A reaction with water is called* **hydrolysis.**

As with any other equilibrium, we can make this one go in the direction we want by using Le Chatelier's principle. If we want to hydrolyze an acetal, there should be a large amount of water present; this drives the equilibrium to the side of the aldehyde and alcohol. If we want to form an acetal, there should be very little water present (the best would be none at all). Acetals are generally made by adding the pure alcohol to the pure aldehyde and passing dry HCl gas through the mixture. The HCl is a catalyst for both the forward and the reverse reaction.

*Bases do not catalyze either reaction. Acetals are quite stable in the presence of a base.*

---

**PROBLEM 13.3** Draw the structural formula for the hemiacetal intermediate and for the acetal product:

$$CH_3-\underset{\underset{\displaystyle CH_3}{|}}{CH}-CH_2-\underset{\underset{\displaystyle O}{\|}}{C}-H + CH_3OH \xrightarrow{H^+}$$

---

Some ketones also react with alcohols to give the same type of product; however, it is usually called a **ketal** and the intermediate a **hemiketal:**

$$R-\underset{\underset{\displaystyle O}{\|}}{C}-R'' + R'OH \rightleftharpoons R-\underset{\underset{\displaystyle OH}{|}}{\overset{\overset{\displaystyle OR'}{|}}{C}}-R'' + R'OH \overset{H^+}{\rightleftharpoons} R-\underset{\underset{\displaystyle OR'}{|}}{\overset{\overset{\displaystyle OR'}{|}}{C}}-R'' + H_2O$$

Ketone        Hemiketal        Ketal<br>(unstable)

The reaction with ketones is more difficult than that with aldehydes, and many ketones do not form ketals.

We have said that hemiacetals (and hemiketals) are unstable and cannot in general be isolated. There are some exceptions. An important example is found when the —OH group of the alcohol and the CHO group of the aldehyde are in the same molecule:

*This is an addition reaction.*

Cyclic hemiacetal

In this case, the H of the —OH group adds to the C=O *in the same molecule* (you can see that it is not very far away), and a six-membered ring is produced. The most stable rings contain five or six atoms, and we are not surprised to learn that aldehydes that contain an —OH group on the fourth or fifth carbon of the chain spontaneously produce cyclic hemiacetals. These cyclic hemiacetals react with alcohols in the presence of $H^+$ to produce cyclic acetals:

*This is a substitution reaction.*

Cyclic hemiacetal                    Cyclic acetal

We shall see examples of cyclic hemiacetals and acetals when we study carbohydrates in Chapter 16. Because these groups are so important in carbohydrate chemistry, it is important that you be able to recognize them.

---

### EXAMPLE

In the following structures, identify the acetals, ketals, hemiacetals, and hemiketals:

(a) $CH_3CH_2$—O—$CH_2CH_2$—OH

(b) $CH_3$—CH—OH
$\qquad\quad$ |
$\qquad\quad$ $OCH_2CH_3$

(c) $CH_3CH_2$—O—$CH_2CH_2$—O—$CH_3$

(d) $CH_3CH_2$—C—$CH_3$
$\qquad\qquad$ |
$\qquad\quad$ $OCH_2CH_3$
with $OCH_3$ above the C

(e)

(f)

$$CH_3-O-CH-CH_2$$ (with ring: $H_2C$, $C$ ($H_2$), $O$)

(g)

(with ring: $H_2C$, $C$ ($H_2$), $O$, $CH-OCH_3$, $C$ ($H_2$), $H_2C$)

(h)

(with ring: $H_2C$, $C$ ($H_2$), $O$, $C$ with $OH$ and $CH_3$, $C$ ($H_2$), $H_2C$)

## Answer

Acetals, ketals, hemiacetals, and hemiketals all have two oxygen atoms connected to the same carbon. In compounds a, c, and f, there is no carbon connected to two oxygen atoms. Compounds c and f are ethers; compound a is an ether and an alcohol. None of them is an acetal, ketal, hemiacetal, or hemiketal. Compounds b and e are hemiacetals. Each of them has one carbon atom connected to an —OH group and an —OR group (shown below in color). Compound h is a hemiketal.

$$CH_3-CH-OH$$
$$\quad\quad |$$
$$\quad\quad O-CH_2CH_3$$

(b)
Hemiacetal

(ring structure with $H_2C$, $C$ ($H_2$), $O$, $CH-OH$, $H_2C$)

(e)
Hemiacetal

(ring structure with $H_2C$, $C$ ($H_2$), $O$, $C$ with $OH$ and $CH_3$, $C$ ($H_2$), $H_2C$)

(h)
Hemiketal

Compounds d and g have two —OR groups on the same carbon. Because compound g also has an H on this carbon, it is an acetal. Compound d, which has no H on this carbon, is a ketal.

(ring structure with $H_2C$, $C$ ($H_2$), $O$, $CH-OCH_3$, $C$ ($H_2$), $H_2C$)

(g)
Acetal

$$CH_3CH_2-\overset{\displaystyle OCH_3}{\underset{\displaystyle OCH_2CH_3}{C}}-CH_3$$

(d)
Ketal

**PROBLEM 13.4** Identify the acetals, ketals, hemiacetals, and hemiketals:

(a) (phenyl ring)$-CH-OCH_3$ with $OH$ below

(b) $CH_3CH_2-O-CH-$(cyclopentyl) with $OCH_3$ below

(c)  $CH_3CH_2CH_2$—O—CH—CH$_3$        (d)

$\qquad\qquad\qquad\qquad\qquad$ OH

---

**EXAMPLE**

Draw the structures of the aldehyde (or ketone) and alcohol(s) produced when compounds b, e, d, g, and h in the previous example break down.

**Answer**

Pick out the carbon connected to the two oxygen atoms (the one shown in color). This becomes the aldehyde or ketone carbon, the one connected to the $=O$:

(b) $CH_3$—C—OH $\longrightarrow$ $CH_3$—C$=$O + $CH_3CH_2$—OH

Becomes alcohol oxygen

Becomes alcohol oxygen

(e)

Becomes alcohol oxygen

(d) $CH_3CH_2$—C—CH$_3$ $\longrightarrow$

Becomes alcohol oxygen

$CH_3CH_2$—C—CH$_3$ + CH$_3$—OH + $CH_3CH_2$—OH

Becomes alcohol oxygen

(g)

Becomes alcohol oxygen

Becomes alcohol oxygen

(h)

**PROBLEM 13.5** Draw the structure of the aldehyde or ketone and alcohol produced upon breakdown of the four compounds in Problem 13.4.

## Addition of $H_2$. Reduction to Alcohols

General reactions:

$$R-\underset{\underset{O}{\|}}{C}-H \xrightarrow{\text{[H]}} R-CH_2-OH$$

Aldehyde                Primary alcohol

$$R-\underset{\underset{O}{\|}}{C}-R' \xrightarrow{\text{[H]}} R-\underset{\underset{OH}{|}}{CH}-R'$$

Ketone                Secondary alcohol

Specific examples:

$$CH_3-\underset{\underset{O}{\|}}{C}-H \xrightarrow[\text{catalyst}]{H_2} CH_3-CH_2OH$$

Acetaldehyde                Ethanol

Cyclohexanone        Cyclohexanol

In Section 11.6 we saw that C=C can be reduced to CH—CH. The C=O bonds of aldehydes and ketones undergo the same reaction, though the reducing agents are not always the same. Hydrogen gas and a catalyst reduce $\overset{\diagdown}{\underset{\diagup}{C}}$=O to $\overset{\diagdown}{\underset{\underset{H}{|}}{C}}$—OH and are frequently used for this purpose, but the reaction is slower than the one with C=C bonds. Thus, if there is a C=C bond in the same molecule, it is reduced first.

No matter what the reducing agent, *aldehydes are always reduced to primary alcohols and ketones to secondary alcohols.* (Can you see why?) These reactions are the opposite of the oxidation reactions discussed in Section 13.4.

The reduction of C=O to CH—OH is important in a number of biochemical processes. Since the body doesn't have $H_2$ available, it uses the reducing agent NADH instead (Box 13B).

**Box 13B**

**Biological Reduction of Aldehydes and Ketones**

Reduction of aldehydes and ketones also takes place in biological organisms, in which case organic reducing agents are used. One of the most important is the reduced form of **nicotinamide adenine dinucleotide, NADH.** An example is reduction of pyruvate ion to lactate ion, which takes place in the muscles:

$$CH_3-\overset{\delta+}{\underset{\underset{\delta-O}{\parallel}}{C}}-\overset{O}{\underset{\parallel}{C}}-O^- + NADH + H^+ \longrightarrow CH_3-\overset{H}{\underset{\underset{OH}{|}}{C}}-\overset{O}{\underset{\parallel}{C}}-O^- + NAD^+$$

Pyruvate ion                                      Lactate ion

The NADH itself becomes oxidized. In the oxidized form, it is known as $NAD^+$. After an NADH molecule performs its reducing function and is oxidized to $NAD^+$ in the process, the $NAD^+$ is now ready to oxidize some other molecule somewhere else. The body reuses its NADH molecules in this way over and over again.

---

**PROBLEM 13.6**  Draw the structural formula for each product:

(a)

$\xrightarrow{[H]}$

(b) $CH_3CH_2-\overset{O}{\underset{\parallel}{C}}-CH_2CH_3 \xrightarrow{[H]}$

---

### Keto–Enol Tautomerism

One other property of many aldehydes and ketones is worth noting. If the aldehyde or ketone possesses at least one hydrogen on the atom adjacent to the C=O group (called an $\alpha$-hydrogen), this hydrogen can migrate to the oxygen. An example is

$$H-\overset{H}{\underset{\underset{H\quad O}{|}}{C}}-\overset{O}{\underset{\parallel}{C}}-H \rightleftharpoons H-\overset{H}{\underset{\underset{O-H}{|}}{C}}=C-H$$

Keto form                         Enol form
More stable                       Less stable

This is an equilibrium, and in any such case both forms are present. The phenomenon is called **tautomerism.** For most simple aldehydes and ketones the **keto form** is much more stable than the **enol form,** and much less than 1 percent of the molecules are in the enol form at any time. But there are some cases in which the enol is more stable than the keto form, one such being the diketone 2,4-pentandione:

$$CH_3-\overset{O}{\underset{\parallel}{C}}-CH_2-\overset{O}{\underset{\parallel}{C}}-CH_3 \qquad CH_3-\overset{OH}{\underset{|}{C}}=CH-\overset{O}{\underset{\parallel}{C}}-CH_3$$

Keto form                         Enol form
Less stable                       More stable

Simple enols (compounds with a double bond and an OH group on the same carbon) are usually not stable enough to be isolated, because they tautomerize to the keto form.

Another compound in which the enol form is more stable is vitamin C (Table 26.2).

Before leaving this section, we should mention that **aldehydes and ketones are neither basic nor acidic;** they are neutral compounds like alcohols and ethers.

## 13.7 SOME IMPORTANT ALDEHYDES AND KETONES

### Formaldehyde, H—CHO

Although formaldehyde is a gas at room temperature (bp $-21°C$), it is soluble in water and so is normally found in laboratories as a 37 percent aqueous solution called *formalin*. This solution kills bacteria and is used to sterilize surgical instruments.

You may be familiar with formalin as a preservative for biological specimens. It does this by denaturing proteins (Sec. 18.10), hardening them. It is also used to embalm cadavers. Its use as a preservative and as an embalming liquid has declined, however, because it has been determined that formaldehyde is carcinogenic.

Industrially, formaldehyde is the most important aldehyde. Its principal use is in the production of polymers (Box 13C).

### Acetaldehyde, CH₃—CHO

Because its boiling point, $21°C$, is so close to room temperature, acetaldehyde is not easy to handle, either as a liquid or as a gas. For this reason, it is often converted to a cyclic trimer (a molecule made of three units) called *paraldehyde:*

$$3CH_3—CHO \longrightarrow$$

Paraldehyde

Paraldehyde was once used as a hypnotic or sleep-producer, but because of its disagreeable taste and odor it is seldom used today.

### Acetone, CH₃—CO—CH₃

A liquid with a moderately low boiling point ($56°C$), acetone is one of the few organic compounds that is infinitely soluble in water and also dissolves a great many organic compounds. For this reason it is one of the most important industrial solvents. It is used in paints, varnishes, resins, coatings, and nail polish, among other things.

Acetone is produced in the body as a product of one pathway of lipid metabolism (Sec. 21.7). Normally it does not accumulate because the small amount produced is oxidized to $CO_2$ and $H_2O$. However, in diabetic patients a larger quantity of acetone is produced than the body can oxidize completely, so

**Box 13C**

## Formaldehyde in Plastics and Plywood

In 1909 Leo Baekeland (1863–1944) announced the invention of the first completely synthetic plastic, made by combining formaldehyde with phenol under heat and pressure. These two compounds react to form a giant molecule (polymer) with a three-dimensional structure.

$$n\ H-\underset{\underset{O}{\|}}{C}-H + n\ \text{(phenol, OH)} \longrightarrow$$

Formaldehyde    Phenol

$+\ n\,H_2O$

Baekeland called his material *Bakelite.* More than 80 years later it is still an important plastic of the thermosetting type, meaning that it cannot be melted after it sets.

Formaldehyde also forms similar polymers with several other compounds, the most important being urea and melamine. Phenol-formaldehyde, urea-formaldehyde, and melamine-formaldehyde polymers are used not only as plastics but even more importantly as adhesives and coatings. Plywood consists of thin sheets of wood glued together by one of these polymers. They are also combined with wood chips or sawdust to make fiberboard. Besides Bakelite, the trade names Formica and Melmac are used for some of these materials. Some of these polymers are produced as porous foams and are used for building insulation. In recent years many of these products have been removed from the market because they contain tiny amounts of free formaldehyde, which can get into the air and possibly cause cancer.

much of it is excreted in the urine. A positive test for acetone in the urine is one way in which diabetes is diagnosed. In severe cases of diabetes acetone is exhaled, and its sweet odor can be detected in the patient's breath.

**α-Chloroacetophenone,** (phenyl)—CO—CH$_2$—Cl

A number of chlorine- and bromine-containing organic compounds are **lachrymators,** compounds that cause the eyes to tear. α-Chloroacetophenone has long been used as a tear gas, by police and other government officials to control

| Name | Formula | Source and Uses |
|------|---------|-----------------|
| Benzaldehyde | CHO (benzene ring) | Oil of almonds; used as flavoring |
| Cinnamaldehyde | CH=CH—CHO (benzene ring) | Oil of cinnamon; used as flavoring |
| Vanillin | CHO, OCH₃, OH (benzene ring) | From vanilla bean; used as flavoring |
| Muscone | (large ring with C=O and CH₃) | From musk deer; used in perfumes |
| Biacetyl (butanedione) | $CH_3-C-C-CH_3$ with two O (both double-bonded) | Flavoring for margarine |
| β-Ionone | (ring with H₃C, CH₃ groups) CH=CH—C—CH₃, O, CH₃ | From violets; used in perfumes |

Among other naturally occurring aldehydes and ketones are *cis*-retinal (Box 11C) and the sex hormones (Sec. 17.10).

**B o x 13D**

## Some Important Naturally Occurring Aldehydes and Ketones

unruly mobs. More recently, this compound has been used as the active ingredient in Mace, a handheld aerosol tear gas projector, used by individuals to protect themselves from an attacker. Mace has a range of about 15 feet.

Many aldehydes and ketones are found in nature. Some of these are shown in Box 13D.

## SUMMARY

Aldehydes and ketones contain the **carbonyl group,**
$-\overset{|}{C}=O$. **Aldehydes** are **RCHO; ketones** are **RCOR'.** Both types of compounds can be named by IUPAC rules or by common names. Aldehydes and ketones are polar and hence have higher boiling points than hydrocarbons, but their molecules cannot form hydrogen bonds with each other, so their boiling points are lower than those of the alcohols. Their solubility in water is about the same as that of alcohols.

Aldehydes are easily oxidized to carboxylic acids; ketones resist oxidation. Several tests for aldehydes depend on this difference. Aldehydes react with alcohols, in the presence of an acid catalyst, to produce **acetals. Hemiacetals** are intermediates. Ketones undergo this reaction (to produce **ketals**) less readily. These reactions are reversible, and acetals and ketals can be easily hydrolyzed to the aldehyde or ketone and the alcohol. Aldehydes can be reduced to primary alcohols, ketones to secondary alcohols.

Aldehydes and ketones that possess an $\alpha$-hydrogen exist in equilibrium with small amounts of their enol forms, a phenomenon called **tautomerism.** Important aldehydes are formaldehyde and acetaldehyde. The most important ketone is acetone.

## Summary of Reactions

**1.** Oxidation (Sec. 13.5):

$$R-\underset{\underset{O}{\|}}{C}-H \xrightarrow{[O]} R-\underset{\underset{O}{\|}}{C}-OH$$

Aldehyde      Carboxylic acid

$$R-\underset{\underset{O}{\|}}{C}-R' \xrightarrow[\substack{\text{Mild} \\ \text{conditions}}]{[O]} \text{no reaction}$$

Ketone

**2.** Formation of acetals and ketals (Sec. 13.6):

$$R-\underset{\underset{O}{\|}}{C}-H + R'OH \longrightarrow R-\underset{\underset{OH}{|}}{\overset{OR'}{|}}{C}-H \xrightarrow[R'OH]{H^+} R-\underset{\underset{OR'}{|}}{\overset{OR'}{|}}{C}-H$$

Aldehyde   Alcohol    Hemiacetal     Acetal

$$R-\underset{\underset{O}{\|}}{C}-R'' + R'OH \longrightarrow R-\underset{\underset{OH}{|}}{\overset{OR'}{|}}{C}-R'' \xrightarrow[R'OH]{H^+} R-\underset{\underset{OR'}{|}}{\overset{OR'}{|}}{C}-R''$$

Ketone        Hemiketal       Ketal

**3.** Hydrolysis of acetals (Sec. 13.6):

$$R-\underset{\underset{OR'}{|}}{\overset{OR'}{|}}{C}-H + H_2O \xrightarrow{H^+} R-\underset{\underset{O}{\|}}{C}-H + 2R'OH$$

Acetal        Aldehyde    Alcohol

**4.** Reduction (Sec. 13.6):

$$R-\underset{\underset{O}{\|}}{C}-H \xrightarrow{[H]} R-CH_2-OH$$

Aldehyde          Primary alcohol

$$R-\underset{\underset{O}{\|}}{C}-R' \xrightarrow{[H]} R-\underset{\underset{OH}{|}}{\overset{H}{|}}{C}-R'$$

Ketone       Secondary alcohol

## KEY TERMS

Acetal (Sec. 13.6)
Aldehyde (Sec. 13.1)
Benedict's test (Sec. 13.5)
Carbonyl group (Sec. 13.1)
Enol form (Sec. 13.6)
Hemiacetal (Sec. 13.6)
Hemiketal (Sec. 13.6)

Hydrolysis (Sec. 13.6)
Ketal (Sec. 13.6)
Keto form (Sec. 13.6)
Ketone (Sec. 13.1)
Lachrymator (Sec. 13.7)
Tautomerism (Sec. 13.6)
Tollens' test (Sec. 13.5)

## PROBLEMS

### Aldehydes and Ketones

**13.7** Which of these compounds have a carbonyl group?

    (a) $CH_3-\underset{\underset{O}{\|}}{C}-H$

(b) $CH_3CH_2-OH$

(c)

(d)

(e)

(f)  $CH_3CH_2-O-CH_3$

**13.8**  What is the difference in structure between an aldehyde and a ketone?

**13.9**  Which compounds are aldehydes, ketones, both, or neither?

(a)  $CH_3CH_2-\overset{\displaystyle O}{\underset{\displaystyle ||}{C}}-CH_3$

(b)  $CH_3CH_2-\overset{\displaystyle O}{\underset{\displaystyle ||}{C}}-OCH_3$

(c)

(d)  $CH_3-\overset{\displaystyle O}{\underset{\displaystyle ||}{C}}-CH_2CH_2-\overset{\displaystyle O}{\underset{\displaystyle ||}{C}}-H$

(e)

(f)  $CH_3-\overset{\displaystyle O}{\underset{\displaystyle ||}{C}}-OH$

**13.10**  Indicate the aldehyde and keto groups in the following compounds:

(a)  $H-\overset{\displaystyle O}{\underset{\displaystyle ||}{C}}-CH_2CH_2-\overset{\displaystyle O}{\underset{\displaystyle ||}{C}}-\overset{\displaystyle O}{\underset{\displaystyle ||}{C}}-CH_3$

(b)

(c)

**Nomenclature of Aldehydes and Ketones**

**13.11**  Name these compounds by the IUPAC system:

(a)  $CH_3-\overset{\displaystyle Br}{\underset{\displaystyle |}{CH}}-CH_2-\overset{\displaystyle O}{\underset{\displaystyle ||}{C}}-H$

(b)  $CH_3-\overset{\displaystyle OH}{\underset{\displaystyle |}{CH}}-\overset{\displaystyle O}{\underset{\displaystyle ||}{C}}-CH_2CH_2CH_3$

(c)

(d)  $CH_2{=}CH-CH_2CH_2-\overset{\displaystyle O}{\underset{\displaystyle ||}{C}}-H$

**13.12**  Give common names:

(a)

(b)  $CH_3-\overset{\displaystyle O}{\underset{\displaystyle ||}{C}}-CH_3$

(c)  $CH_3CH_2CH_2-\overset{\displaystyle O}{\underset{\displaystyle ||}{C}}-CH_3$

(d)  HCHO

(e)  $CH_3-CHO$

(f)

**13.13**  Draw the structural formulas for the following aldehydes:
(a)  butanal
(b)  formaldehyde
(c)  2,4-dimethylpentanal
(d)  propionaldehyde

**13.14** Draw structural formulas for the following ketones:
   (a)  acetophenone
   (b)  isopropyl methyl ketone
   (c)  2-hexanone
   (d)  hept-5-en-2-one
   (e)  diisopropyl ketone

## Physical Properties of Aldehydes and Ketones

**13.15** In each pair, pick the compound you would expect to have the higher boiling point:

   (a)  $CH_3CHO$ vs. $CH_3CH_2OH$

   (b)  (1) $CH_3-\overset{\overset{\displaystyle O}{\|}}{C}-CH_3$ vs.

   (2) $CH_3CH_2-\overset{\overset{\displaystyle O}{\|}}{C}-CH_3$

   (c)  $CH_3CH_2CH_2CHO$ vs. $CH_3CH_2CH_2CH_3$

   (d)  (1) $CH_3CH_2-\overset{\overset{\displaystyle O}{\|}}{C}-CH_3$ vs.

   (2) $CH_3CH_2-\overset{\underset{\displaystyle OH}{|}}{CH}-CH_3$

   (e)  (1) $HO-CH_2CH_2CH_2-OH$ vs.
   (2) $H-\overset{\overset{\displaystyle O}{\|}}{C}-CH_2-\overset{\overset{\displaystyle O}{\|}}{C}-H$

**13.16** Acetone, $CH_3COCH_3$, is completely soluble in water, but 4-heptanone,

$$CH_3CH_2CH_2COCH_2CH_2CH_3,$$

is almost completely insoluble in water. Explain.

**13.17** Why does acetone have a higher boiling point (56°C) than ethyl methyl ether (11°C), though their molecular weights are almost the same?

**13.18** Why does acetone have a lower boiling point (56°C) than 2-propanol (82°C), though their molecular weights are almost the same?

**13.19** Show how acetaldehyde and acetone can form hydrogen bonds with water molecules.

## Oxidation of Aldehydes

**13.20** Draw the structural formulas for the principal organic products. If no reaction, say so:

   (a)  $CH_3CH_2-\overset{\overset{\displaystyle O}{\|}}{C}-H$ + Tollens' solution $\longrightarrow$

   (b)  $CH_3-\overset{\overset{\displaystyle O}{\|}}{C}-CH_3$ + Tollens' solution $\longrightarrow$

   (c)  $+ H_2 \xrightarrow{\text{catalyst}}$

   (d)  $CH_3CH_2CH_2-\overset{\overset{\displaystyle O}{\|}}{C}-H + H_2 \xrightarrow{\text{catalyst}}$

**13.21** If you were given a compound that could be either $CH_3CH_2CH_2CH_2CHO$ or

$$CH_3CH_2-\overset{\overset{\displaystyle O}{\|}}{C}-CH_2CH_3,$$

what simple laboratory tests could you use to tell which it is?

**13.22** Show by means of an equation how Benedict's solution reacts with an aldehyde.

**13.23** What is the difference between the Benedict's and the Tollens' tests?

**13.24** Which of the following compounds will give positive tests with Tollens' or Benedict's reagents?

   (a)  $CH_3CH_2CH_2-\overset{\overset{\displaystyle O}{\|}}{C}-CH_2CH_2CH_3$

   (b)  $CH_2-\overset{\overset{\displaystyle O}{\|}}{C}-H$

   (c)  $CH_3-\overset{\underset{\displaystyle OH}{|}}{CH}-\overset{\overset{\displaystyle O}{\|}}{C}-CH_2CH_3$

   (d)  $CH_3-\overset{\underset{\displaystyle OH}{|}}{CH}-CH_2-\overset{\overset{\displaystyle O}{\|}}{C}-CH_3$

   (e)

## Other Chemical Properties of Aldehydes and Ketones

**13.25** Which is the positive end of the $C{=}O$ group and which is the negative end?

**13.26** Write the structural formulas for the principal organic products. If no reaction, say so:

   (a)  $CH_3-\overset{\overset{\displaystyle O}{\|}}{C}-H + CH_3CH_2CH_2OH \xrightarrow{H^+}$

(b) + CH$_3$OH $\xrightarrow{H^+}$

(c) + CH$_3$OH $\xrightarrow{H^+}$

*(d) + H$_2$O $\xrightarrow{H^+}$

**13.27** Write the structural formulas for the principal organic products. If no reaction, say so:

(a) CH$_3$—C—CH$_3$ + H$_2$ $\xrightarrow{catalyst}$
(with O double bond)

(b) CH$_3$CH$_2$—C—H + H$_2$ $\xrightarrow{catalyst}$
(with O double bond)

(c) + H$_2$ $\xrightarrow{catalyst}$

(d) + H$_2$O $\xrightarrow{H^+}$

(e) + H$_2$O $\xrightarrow{OH^-}$

**13.28** Aldehydes undergo addition reactions with HCN. Show the product when HCN adds to acetaldehyde. (The positive part of the HCN is H$^+$; the negative part is CN$^-$.)

## Acetals, Ketals, Hemiacetals, and Hemiketals

**13.29** Tell which are acetals, hemiacetals, or neither:

(a)

(b) CH$_3$CH$_2$—CH—OCH$_3$ with OH

(c) CH$_3$O—CH$_2$CH$_2$—OCH$_3$

(d)

(e)

(f)

**13.30** Tell which are ketals, hemiketals, or neither:

(a) CH$_3$—C(OCH$_3$)(OH)—CH$_2$CH$_2$CH$_3$

(b)

(c)

(d)

(e)

(f)

(g)

**13.31** Tell which are acetals, hemiacetals, ketals, hemiketals, or none of these:

(a)

$$\begin{array}{c} H_2 \\ C \\ H_2C \qquad O \\ | \qquad\quad | \\ H_2C\text{—}CH\text{—}OH \end{array}$$

(b)

$$\begin{array}{c} CH_2OH \\ | \\ C\text{—}O \\ /H \qquad\quad \backslash \\ HO\text{—}HC \qquad CH\text{—}OH \\ | \qquad\qquad | \\ HC\text{—}CH \\ | \qquad\quad | \\ HO \qquad\quad OH \end{array}$$

(c)

(d) $CH_3CH_2\text{—}O\text{—}CH_2\text{—}OH$

**13.32** Show the hemiacetal and acetal (or hemiketal and ketal) that would form in each case:

(a) $CH_3CH_2\text{—}\overset{\displaystyle O}{\underset{\displaystyle \|}{C}}\text{—}H + CH_3CH_2OH \xrightarrow{H^+}$

(b) $+ CH_3OH \xrightarrow{H^+}$

(c) $+ CH_3OH \xrightarrow{H^+}$

(d) (hemiacetal only)

$$\underset{\displaystyle OH}{\underset{\displaystyle |}{CH_2}}CH_2CH_2CH_2\text{—}\overset{\displaystyle O}{\underset{\displaystyle \|}{C}}\text{—}H \xrightarrow{H^+}$$

**13.33** Draw the structures of the aldehyde or ketone and alcohol produced when these acetals, hemiacetals, or ketals break down:

(a) $CH_3CH_2\text{—}\overset{\displaystyle OCH_2CH_3}{\underset{\displaystyle |}{CH}}\text{—}OCH_2CH_3$

(b) $CH_3CH_2\text{—}\overset{\displaystyle OCH_3}{\underset{\displaystyle \underset{\displaystyle OH}{|}}{\overset{\displaystyle |}{C}}}\text{—}CH_3$

(c)

$$\begin{array}{c} OCH_3 \\ | \\ C \\ / \quad \backslash \end{array}$$

(d)

$$\begin{array}{c} CH_2\text{—}O \\ / \qquad\qquad \backslash \\ HO\text{—}HC \qquad\quad CH\text{—}OCH_2CH_3 \\ \backslash \qquad\qquad / \\ CH_2\text{—}CH_2 \end{array}$$

(e)

$$\begin{array}{c} H_2 \\ C \\ H_2C \qquad CH\text{—}CH_3 \\ \backslash \qquad\quad | \\ O\text{—}C\text{—}OH \\ | \\ CH_2CH_3 \end{array}$$

(f)

$$\begin{array}{c} O \\ H_2C \qquad CH\text{—}OH \\ | \qquad\qquad | \\ H_2C \qquad CH_2 \\ \backslash \; H \; / \\ C \\ | \\ CH_3 \end{array}$$

## Reduction of Aldehydes and Ketones

**13.34** Draw the structure of the aldehyde or ketone that can be reduced to produce each of these alcohols. If none exists, say so.

(a) $CH_3\text{—}\overset{\displaystyle OH}{\underset{\displaystyle |}{CH}}\text{—}CH_3$

(b) $CH_2OH$

(c) $CH_3OH$

(d) $CH_3\text{—}\overset{\displaystyle CH_3}{\underset{\displaystyle \underset{\displaystyle OH}{|}}{\overset{\displaystyle |}{C}}}\text{—}CH_3$

(e) OH

**13.35** Explain why reduction of an aldehyde always gives a primary alcohol and reduction of a ketone always gives a secondary alcohol.

## Tautomerism

**13.36** Which of these compounds undergo tautomerism?

(a) $CH_3-\underset{\underset{O}{\|}}{C}-H$

(b) $CH_3-\underset{\underset{O}{\|}}{C}-CH_3$

(c) $\underset{\underset{O}{\|}}{\overset{C-H}{\bigcirc}}$

(d) $\underset{\underset{O}{\|}}{\overset{C-CH_3}{\bigcirc}}$

(e) $CH_3-\underset{\underset{Cl}{|}}{\overset{Cl}{C}}-\underset{\underset{O}{\|}}{C}-H$

(f) ⬡=O

(g) $H-\underset{\underset{O}{\|}}{C}-H$

**13.37** Draw the enol forms of these aldehydes and ketones:

(a) $CH_3CH_2CHO$

(b) $CH_3CH_2-\underset{\underset{O}{\|}}{C}-CH_3$

(c) $CH_3CH_2-\underset{\underset{O}{\|}}{C}-\bigcirc-CH_3$

(d) ⬠=O

**13.38** Draw the structural formula for the keto form of each of these enols:

(a) $CH_2{=}CH-OH$

(b) $CH_3-CH{=}\underset{\underset{OH}{|}}{C}-CH_3$

(c) $HO-CH{=}CH-CH{=}CH-OH$

(d) ⬠–OH

## Some Important Aldehydes and Ketones

**13.39** What is the most important industrial use for formaldehyde?

**\*13.40** Show how three molecules of acetaldehyde can add to each other to give paraldehyde (draw the bonds and show how they change in the course of the reaction).

**13.41** Why is the sweet odor of acetone on the breath of a patient a bad clinical symptom?

**13.42** What is the active ingredient in Mace?

**13.43** How should you handle a compound known to be a lachrymator?

## Boxes

**13.44** (Box 13A) Chloral is one of the few aldehydes that form stable hemiacetals. Draw the structure of the hemiacetal that forms when chloral reacts with ethanol:

$$Cl-\underset{\underset{Cl}{|}}{\overset{\overset{Cl}{|}}{C}}-\underset{\underset{O}{\|}}{C}-H + CH_3CH_2OH \longrightarrow$$

**13.45** (Box 13A) What is the medical use of chloral hydrate?

**13.46** (Box 13B) Show how NADH + $H^+$ reduces acetaldehyde to ethanol.

**\*13.47** (Box 13B) Write the name and structure of the compound that is formed by the reduction of dihydroxyacetone, $HOCH_2-\underset{\underset{O}{\|}}{C}-CH_2OH$, using NADH + $H^+$.

**13.48** (Box 13C) What are some uses for Bakelite-type polymers?

**13.49** (Box 13C) What is a thermosetting resin?

**13.50** (Box 13C) Could a thermosetting resin be made by treating HCHO with m-cresol instead of phenol?

*m*-Cresol

**13.51** (Box 13D) What naturally occurring ketones are used in the manufacture of perfumes?

**13.52** (Box 13D) Classify β-ionone into as many chemical families as is appropriate.

**13.53** (Box 13D) Which naturally occurring compound is used (a) as an almond flavor   (b) as a flavor for margarine   (c) as a flavor for cinnamon?

## Additional Problems

**13.54** What kind of odors do aldehydes and ketones have?

**13.55** Propanal, $CH_3CH_2CHO$, and 1-propanol, $CH_3CH_2CH_2OH$, have about the same molecular weight (Table 13.2), yet their boiling points differ by almost 40°C. Explain this fact.

**13.56** Name these compounds by IUPAC or common names:

(a) $CH_3CH_2CH_2$—$\underset{\displaystyle O}{\overset{\displaystyle \|}{C}}$—$\underset{\displaystyle Br}{CH}$—$CH_2$—$Cl$

(b) 

(c) 

(d) $CH_3CH_2$—$\underset{\displaystyle O}{\overset{\displaystyle \|}{C}}$—$CH$=$CH_2$

(e) 

**\*13.57** Write the structural formula for
(a) any cyclic hemiacetal that has a six-membered ring
(b) any cyclic ketal that has a five-membered ring
(c) any noncyclic hemiketal

**\*13.58** Which, if any, of the three compounds in Problem 13.57 can be isolated?

**13.59** Draw the structural formulas for the principal organic products. If no reaction, say so.

(a) $CH_3$—$\underset{\displaystyle OCH_3}{\overset{\displaystyle OCH_3}{\underset{\displaystyle |}{\overset{\displaystyle |}{C}}}}$—$CH_2CH_2CH_3$ + $H_2O$ $\xrightarrow{H^+}$

(b)  + Tollens' solution $\longrightarrow$

(c) $CH_3$—$\underset{\displaystyle O}{\overset{\displaystyle \|}{C}}$—$H$ + $CH_3CH_2$—$OH$ $\xrightarrow{H^+}$

(d) One mole

$$CH_2=CH—CH_2CH_2—\underset{\displaystyle O}{\overset{\displaystyle \|}{C}}—CH_3$$

+ one mole $H_2$ $\xrightarrow{catalyst}$

**\*13.60** Predict the product of this reaction:

$$CH_3CH_2CH_2—\underset{\displaystyle O}{\overset{\displaystyle \|}{C}}—H + \underset{\displaystyle OH}{\underset{\displaystyle |}{CH_2}}—\underset{\displaystyle OH}{\underset{\displaystyle |}{CH_2}} \xrightarrow{H^+} H_2O + ?$$

What kind of compound is this?

**\*13.61** A student took a bottle of butanal from the shelf and discovered that the contents turned litmus red (that is, it was acidic). Yet aldehydes are neutral compounds. Could you explain to the student what caused the acidic behavior?

**\*13.62** Acetone is an important laboratory solvent. For similar compounds, solvation ability increases as molecular weight decreases. Can you give a reason why formaldehyde and acetaldehyde are not important laboratory solvents?

**13.63** Why can't two molecules of acetone form a hydrogen bond with each other?

**13.64** Explain why the compound prop-1-en-2-ol has never been isolated.

$$CH_2=\underset{\displaystyle OH}{\underset{\displaystyle |}{C}}—CH_3$$

**13.65** What is formalin and what is it used for?

**\*13.66** Can you give a reason why 2,4-pentandione (p. 404) is more stable in the enol than in the keto form?

**\*13.67** Write the structures of cyclic hemiacetals formed by ring closure of (a) 6-hydroxyhexanal (b) 4-hydroxy-5-iodoheptanal   (c) 4,5-dihydroxypentanal (two different ones in this case).

## Tautomerism

**13.36** Which of these compounds undergo tautomerism?

(a) $CH_3-\overset{\displaystyle O}{\underset{\displaystyle \|}{C}}-H$

(b) $CH_3-\overset{\displaystyle O}{\underset{\displaystyle \|}{C}}-CH_3$

(c) benzene ring with $\overset{\displaystyle C-H}{\underset{\displaystyle \|}{O}}$

(d) benzene ring with $\overset{\displaystyle C-CH_3}{\underset{\displaystyle \|}{O}}$

(e) $CH_3-\overset{\displaystyle Cl}{\underset{\displaystyle Cl}{\overset{\displaystyle |}{\underset{\displaystyle |}{C}}}}-\overset{\displaystyle O}{\underset{\displaystyle \|}{C}}-H$

(f) cyclohexanone ring with $O$

(g) $H-\overset{\displaystyle O}{\underset{\displaystyle \|}{C}}-H$

**13.37** Draw the enol forms of these aldehydes and ketones:

(a) $CH_3CH_2CHO$

(b) $CH_3CH_2-\overset{\displaystyle O}{\underset{\displaystyle \|}{C}}-CH_3$

(c) $CH_3CH_2-\overset{\displaystyle O}{\underset{\displaystyle \|}{C}}-$ benzene ring $-CH_3$

(d) cyclopentanone ring with $O$

**13.38** Draw the structural formula for the keto form of each of these enols:

(a) $CH_2{=}CH-OH$

(b) $CH_3-CH{=}\overset{\displaystyle C}{\underset{\displaystyle OH}{\overset{\displaystyle |}{C}}}-CH_3$

(c) $HO-CH{=}CH-CH{=}CH-OH$

(d) cyclopentene ring with $OH$

## Some Important Aldehydes and Ketones

**13.39** What is the most important industrial use for formaldehyde?

**\*13.40** Show how three molecules of acetaldehyde can add to each other to give paraldehyde (draw the bonds and show how they change in the course of the reaction).

**13.41** Why is the sweet odor of acetone on the breath of a patient a bad clinical symptom?

**13.42** What is the active ingredient in Mace?

**13.43** How should you handle a compound known to be a lachrymator?

## Boxes

**13.44** (Box 13A) Chloral is one of the few aldehydes that form stable hemiacetals. Draw the structure of the hemiacetal that forms when chloral reacts with ethanol:

$$Cl-\overset{\displaystyle Cl}{\underset{\displaystyle Cl}{\overset{\displaystyle |}{\underset{\displaystyle |}{C}}}}-\overset{\displaystyle O}{\underset{\displaystyle \|}{C}}-H + CH_3CH_2OH \longrightarrow$$

**13.45** (Box 13A) What is the medical use of chloral hydrate?

**13.46** (Box 13B) Show how $NADH + H^+$ reduces acetaldehyde to ethanol.

**\*13.47** (Box 13B) Write the name and structure of the compound that is formed by the reduction of dihydroxyacetone, $HOCH_2-\overset{\displaystyle O}{\underset{\displaystyle \|}{C}}-CH_2OH$, using $NADH + H^+$.

**13.48** (Box 13C) What are some uses for Bakelite-type polymers?

**13.49** (Box 13C) What is a thermosetting resin?

**13.50** (Box 13C) Could a thermosetting resin be made by treating HCHO with *m*-cresol instead of phenol?

*m*-Cresol

**13.51** (Box 13D) What naturally occurring ketones are used in the manufacture of perfumes?

**13.52** (Box 13D) Classify β-ionone into as many chemical families as is appropriate.

**13.53** (Box 13D) Which naturally occurring compound is used (a) as an almond flavor   (b) as a flavor for margarine   (c) as a flavor for cinnamon?

## Additional Problems

**13.54** What kind of odors do aldehydes and ketones have?

**13.55** Propanal, $CH_3CH_2CHO$, and 1-propanol, $CH_3CH_2CH_2OH$, have about the same molecular weight (Table 13.2), yet their boiling points differ by almost 40°C. Explain this fact.

**13.56** Name these compounds by IUPAC or common names:

(a) $CH_3CH_2CH_2$—$\overset{\displaystyle O}{\underset{\displaystyle \phantom{O}}{C}}$—$\overset{}{\underset{\displaystyle Br}{CH}}$—$CH_2$—$Cl$

(b) 

(c) 

(d) $CH_3CH_2$—$\overset{\displaystyle O}{C}$—$CH$=$CH_2$

(e) 

**\*13.57** Write the structural formula for

  (a) any cyclic hemiacetal that has a six-membered ring

  (b) any cyclic ketal that has a five-membered ring

  (c) any noncyclic hemiketal

**\*13.58** Which, if any, of the three compounds in Problem 13.57 can be isolated?

**13.59** Draw the structural formulas for the principal organic products. If no reaction, say so.

(a) $CH_3$—$\overset{\displaystyle OCH_3}{\underset{\displaystyle OCH_3}{C}}$—$CH_2CH_2CH_3$ + $H_2O$ $\xrightarrow{H^+}$

(b) 
   + Tollens' solution $\longrightarrow$

(c) $CH_3$—$\overset{}{\underset{\displaystyle O}{C}}$—$H$ + $CH_3CH_2$—$OH$ $\xrightarrow{H^+}$

(d) One mole

$CH_2$=$CH$—$CH_2CH_2$—$\overset{}{\underset{\displaystyle O}{C}}$—$CH_3$

   + one mole $H_2$ $\xrightarrow{catalyst}$

**\*13.60** Predict the product of this reaction:

$CH_3CH_2CH_2$—$\overset{}{\underset{\displaystyle O}{C}}$—$H$ + $\overset{}{\underset{\displaystyle OH}{CH_2}}$—$\overset{}{\underset{\displaystyle OH}{CH_2}}$ $\xrightarrow{H^+}$ $H_2O$ + ?

What kind of compound is this?

**\*13.61** A student took a bottle of butanal from the shelf and discovered that the contents turned litmus red (that is, it was acidic). Yet aldehydes are neutral compounds. Could you explain to the student what caused the acidic behavior?

**\*13.62** Acetone is an important laboratory solvent. For similar compounds, solvation ability increases as molecular weight decreases. Can you give a reason why formaldehyde and acetaldehyde are not important laboratory solvents?

**13.63** Why can't two molecules of acetone form a hydrogen bond with each other?

**13.64** Explain why the compound prop-1-en-2-ol has never been isolated.

$CH_2$=$\overset{}{\underset{\displaystyle OH}{C}}$—$CH_3$

**13.65** What is formalin and what is it used for?

**\*13.66** Can you give a reason why 2,4-pentandione (p. 404) is more stable in the enol than in the keto form?

**\*13.67** Write the structures of cyclic hemiacetals formed by ring closure of (a) 6-hydroxyhexanal (b) 4-hydroxy-5-iodoheptanal   (c) 4,5-dihydroxypentanal (two different ones in this case).

## 14.2   NOMENCLATURE OF CARBOXYLIC ACIDS

Carboxylic acids are easily named by the IUPAC system. The suffix is "-oic acid"; otherwise the rules are the same as before. Because it is always at the end of the chain, the —COOH carbon is always number 1, although this number is understood and is not stated as part of the name.

In numbering chains, the —COOH group takes precedence over all groups we have met up to now.

---

**EXAMPLE**

$$\overset{5}{C}H_3-\overset{4}{C}H-\overset{3}{C}H_2-\overset{2}{C}H_2-\overset{1}{C}-OH \qquad \text{4-Bromopentanoic acid}$$

with Br on carbon 4 and O double-bonded to carbon 1.

The longest chain has five carbons, so the parent is pentanoic acid. The —COOH carbon is number 1, so the bromo is in the 4 position.

$$\overset{4}{C}H_3-\overset{3}{C}H-\overset{2}{C}-\overset{1}{C}-OH \qquad \text{3-Hydroxy-2,2-dimethylbutanoic acid}$$

with CH$_3$ on carbon 2, OH on carbon 3, CH$_3$ on carbon 2, and O double-bonded to carbon 1.

Here the parent is *buta-* because the longest chain has four carbons.

---

**PROBLEM 14.1**   Give the IUPAC name for

(a)  $CH_3(CH_2)_5 COOH$

(b)  $CH_3-CH-CH_2-CH-CH-COOH$
     with CH$_3$ on the second carbon, OH and CH$_2$CH$_3$ on the later carbons.

---

Although the IUPAC system is, as usual, easy to apply, it is generally used only for relatively complicated carboxylic acids. Those with simple structures almost always have common names, and these are almost always used in preference to the IUPAC names. Table 14.1 shows the most important of these, but there are hundreds of other carboxylic acids with common names.

Why do common names exist? The reason is that these acids were discovered more than 100 years ago, before the IUPAC system existed and, in most cases, before their structures were known. Since these compounds needed names, they were generally named according to their sources, using Latin word roots, as was customary. For example, formic acid was first isolated from red ants, and the Latin word for ant is *formica*. Similarly, acetic acid comes from vinegar (Latin, *acetum*) and butyric acid from rancid butter (Latin, *butyrum*). Three acids shown in Table 14.1 (caproic, caprylic, capric) have similar names. All of these were first isolated from goat fat (the Latin word for goat is *caper*).

Note that in Table 14.1 we have given common names for unbranched carboxylic acids up to 18 carbons long (stearic acid). These long-chain acids are important because esters of them constitute the fats, as well as some other lipids.

As Table 14.1 shows, **dicarboxylic acids** also have common names. These acids, from oxalic to adipic, are also important in biochemistry and in organic chemistry.

Formic acid is the compound chiefly responsible for the pain and blistering caused by the sting of red ants and bees.

The Formica used in kitchen counter tops is made from a formaldehyde resin (Box 13C). Formaldehyde got its name from formic acid, its oxidation product.

The unbranched acids having between 3 and 20 carbons are known as **fatty acids.** We will study them in Chapter 17.

# CHAPTER 14

*Aspirin is both a carboxylic acid and a carboxylic ester. (Photograph by Beverly March.)*

# Carboxylic Acids and Esters

## 14.1 INTRODUCTION

Anytime we eat foods with a sour or tart taste, it is likely that the taste is due to one or more carboxylic acids. Vinegar contains acetic acid, lemons contain citric acid, and the tart taste of apples is caused by malic acid. Carboxylic acids are very important in biochemistry and in organic chemistry, as we shall see in later chapters. The free acids themselves are present in biological systems less often than their salts and derivatives. **Derivatives of carboxylic acids** include carboxylic esters, studied in this chapter, and amides, discussed in Chapter 15.

Like aldehydes and ketones, carboxylic acids and their derivatives all contain the carbonyl group, C=O, but in carboxylic acids, the carbonyl carbon is always connected to a hydroxyl group, so all carboxylic acids have the formula

$$R-\underset{\underset{O}{\|}}{C}-OH$$

The $-\underset{\underset{O}{\|}}{C}-OH$ group is called the **carboxyl group** (because it is a combination of **carb**onyl and hyd**roxyl**). As we shall see, the properties of carboxylic acids are mostly quite different from those of aldehydes and alcohols. Just as R groups, derived from alkanes, RH, are called alkyl groups, groups derived from carboxylic acids, $R-\underset{\underset{O}{\|}}{C}-$, are called **acyl groups.**

Acyl chlorides and anhydrides, discussed in Section 14.10, are also acid derivatives.

An abbreviated form is either —COOH or —CO$_2$H, but we must not forget that the carbonyl group is there.

415

## 14.2 NOMENCLATURE OF CARBOXYLIC ACIDS

Carboxylic acids are easily named by the IUPAC system. The suffix is "-oic acid"; otherwise the rules are the same as before. Because it is always at the end of the chain, the —COOH carbon is always number 1, although this number is understood and is not stated as part of the name.

In numbering chains, the —COOH group takes precedence over all groups we have met up to now.

---

**EXAMPLE**

$$\overset{5}{CH_3}-\overset{4}{CH}-\overset{3}{CH_2}-\overset{2}{CH_2}-\overset{1}{C}-OH$$
with Br below carbon 4 and O double-bonded to carbon 1.

4-Bromopentanoic acid

The longest chain has five carbons, so the parent is pentanoic acid. The —COOH carbon is number 1, so the bromo is in the 4 position.

$$\overset{4}{CH_3}-\overset{3}{CH}-\overset{2}{C}-\overset{1}{C}-OH$$
with OH below carbon 3, CH₃ below and above carbon 2, and O double-bonded to carbon 1.

3-Hydroxy-2,2-dimethylbutanoic acid

Here the parent is *buta-* because the longest chain has four carbons.

---

**PROBLEM 14.1** Give the IUPAC name for

(a) $CH_3(CH_2)_5COOH$

(b) $CH_3-CH-CH_2-CH-CH-COOH$ with $CH_3$ below second carbon, $OH$ below fourth carbon, and $CH_2CH_3$ below fifth carbon.

---

Although the IUPAC system is, as usual, easy to apply, it is generally used only for relatively complicated carboxylic acids. Those with simple structures almost always have common names, and these are almost always used in preference to the IUPAC names. Table 14.1 shows the most important of these, but there are hundreds of other carboxylic acids with common names.

Why do common names exist? The reason is that these acids were discovered more than 100 years ago, before the IUPAC system existed and, in most cases, before their structures were known. Since these compounds needed names, they were generally named according to their sources, using Latin word roots, as was customary. For example, formic acid was first isolated from red ants, and the Latin word for ant is *formica*. Similarly, acetic acid comes from vinegar (Latin, *acetum*) and butyric acid from rancid butter (Latin, *butyrum*). Three acids shown in Table 14.1 (caproic, caprylic, capric) have similar names. All of these were first isolated from goat fat (the Latin word for goat is *caper*).

Note that in Table 14.1 we have given common names for unbranched carboxylic acids up to 18 carbons long (stearic acid). These long-chain acids are important because esters of them constitute the fats, as well as some other lipids.

As Table 14.1 shows, **dicarboxylic acids** also have common names. These acids, from oxalic to adipic, are also important in biochemistry and in organic chemistry.

Formic acid is the compound chiefly responsible for the pain and blistering caused by the sting of red ants and bees.

The Formica used in kitchen counter tops is made from a formaldehyde resin (Box 13C). Formaldehyde got its name from formic acid, its oxidation product.

The unbranched acids having between 3 and 20 carbons are known as **fatty acids.** We will study them in Chapter 17.

*Aspirin is both a carboxylic acid and a carboxylic ester. (Photograph by Beverly March.)*

# Carboxylic Acids and Esters

## 14.1 INTRODUCTION

Anytime we eat foods with a sour or tart taste, it is likely that the taste is due to one or more carboxylic acids. Vinegar contains acetic acid, lemons contain citric acid, and the tart taste of apples is caused by malic acid. Carboxylic acids are very important in biochemistry and in organic chemistry, as we shall see in later chapters. The free acids themselves are present in biological systems less often than their salts and derivatives. **Derivatives of carboxylic acids** include carboxylic esters, studied in this chapter, and amides, discussed in Chapter 15.

Like aldehydes and ketones, carboxylic acids and their derivatives all contain the carbonyl group, $C=O$, but in carboxylic acids, the carbonyl carbon is always connected to a hydroxyl group, so all carboxylic acids have the formula

$$R-\overset{\displaystyle \|}{\underset{\displaystyle O}{C}}-OH$$

The $-\overset{\|}{\underset{O}{C}}-OH$ group is called the **carboxyl group** (because it is a combination of **carb**onyl and hyd**roxyl**). As we shall see, the properties of carboxylic acids are mostly quite different from those of aldehydes and alcohols. Just as R groups, derived from alkanes, RH, are called alkyl groups, groups derived from carboxylic acids, $R-\overset{\|}{\underset{O}{C}}-$, are called **acyl groups.**

Acyl chlorides and anhydrides, discussed in Section 14.10, are also acid derivatives.

An abbreviated form is either $-COOH$ or $-CO_2H$, but we must not forget that the carbonyl group is there.

415

**TABLE 14.1   Names and Physical Properties of Some Carboxylic Acids[a]**

| Formula | Common Name | BP(°C) | MP(°C) | Solubility in Water (g/100 g of $H_2O$) |
|---|---|---|---|---|
| H—COOH | Formic acid | 101 | 8 | Infinite |
| $CH_3$—COOH | Acetic acid | 118 | 17 | Infinite |
| $CH_3$—$CH_2$—COOH | Propionic acid | 141 | −21 | Infinite |
| $CH_3$—$(CH_2)_2$—COOH | Butyric acid | 162 | −8 | Infinite |
| $CH_3$—$(CH_2)_3$—COOH | Valeric acid | 184 | −35 | 5 |
| $CH_3$—$(CH_2)_4$—COOH | Caproic acid | 205 | −2 | 1 |
| $CH_3$—$(CH_2)_6$—COOH | Caprylic acid | 240 | 16 | Insoluble |
| $CH_3$—$(CH_2)_8$—COOH | Capric acid | 268 | 31 | Insoluble |
| $CH_3$—$(CH_2)_{10}$—COOH | Lauric acid | — | 44 | Insoluble |
| $CH_3$—$(CH_2)_{12}$—COOH | Myristic acid | — | 54 | Insoluble |
| $CH_3$—$(CH_2)_{14}$—COOH | Palmitic acid | — | 63 | Insoluble |
| $CH_3$—$(CH_2)_{16}$—COOH | Stearic acid | — | 70 | Insoluble |
| HOOC—COOH | Oxalic acid | — | 190 | 9 |
| HOOC—$CH_2$—COOH | Malonic acid | — | 136 | 73 |
| HOOC—$(CH_2)_2$—COOH | Succinic acid | — | 190 | 8 |
| HOOC—$(CH_2)_3$—COOH | Glutaric acid | — | 99 | 65 |
| HOOC—$(CH_2)_4$—COOH | Adipic acid | — | 154 | 2 |
| HOOC \ C=C / COOH (cis, H, H) | Maleic acid | — | 130 | 4 |
| HOOC \ C=C / H (trans, H, COOH) | Fumaric acid | — | 286 | 55 |
| (benzene ring)—COOH | Benzoic acid | — | 122 | Insoluble |
| (benzene ring)—OH, COOH | Salicylic acid | — | 159 | Insoluble |

[a] See also Table 17.1.

The names of all the acid derivatives, as well as those of the aldehydes, are based on the names of the acids. Therefore, if you know the name of any carboxylic acid, you can name its derivatives as well.

## 14.3  PHYSICAL PROPERTIES OF CARBOXYLIC ACIDS

When compared with other compounds of about the same molecular weights, carboxylic acids have the highest boiling points of any compounds we have met so far. The simplest one, formic acid, HCOOH (MW = 46), has a boiling point of 101°C, considerably higher than that of ethanol (78°C), which has the same molecular weight.

As with alcohols (Sec. 12.7), the high boiling points of carboxylic acids are caused by hydrogen bonding, but the hydrogen bonding is stronger here because each molecule forms two hydrogen bonds with another molecule:

A **dimer** is a species made up of two identical units.

$$R-C\overset{O\cdots H-O}{\underset{O-H\cdots O}{}}C-R$$

Carboxylic acid dimer

From the discussion in Section 12.7 we would expect the smaller carboxylic acids to be quite soluble in water (why?), and Table 14.1 shows that this is so. The first four are infinitely soluble in water, but then the solubility drops rapidly and becomes insignificant after caprylic acid ($C_8$).

One other physical property of carboxylic acids must be mentioned. The liquid fatty acids from propionic ($C_3$) to capric ($C_{10}$) have extremely foul odors, about as bad as (though different from) those of the thiols (Sec. 12.8). Butyric acid is found in stale perspiration and is an important component of "locker room odor." Valeric acid smells even worse, and goats, which excrete the $C_6$, $C_8$, and $C_{10}$ acids, are not famous for their pleasant odors.

The higher fatty acids might also have foul odors, but their volatility is very low. You cannot smell a compound unless molecules of it reach your nose.

## 14.4  SOME IMPORTANT CARBOXYLIC ACIDS

### Acetic Acid, $CH_3COOH$

This acid is discussed in Box 8A.

### Oxalic Acid, HOOCCOOH

The simplest dicarboxylic acid, oxalic acid is present as the monopotassium salt, $HOOCCOO^- K^+$, in certain leafy vegetables, including spinach and rhubarb.

### Hydroxy Acids

Many hydroxy acids are present in various foods. Among the most important are lactic acid, present in sour milk, yogurt, sauerkraut, and pickles, and citric acid, found in citrus fruits:

$$CH_3-\underset{\underset{OH}{|}}{CH}-COOH \qquad HOOC-CH_2-\underset{\underset{COOH}{|}}{\overset{\overset{OH}{|}}{C}}-CH_2-COOH$$

Lactic acid                    Citric acid

Both of these acids are important in body processes. Lactic acid is produced in the muscles in the breakdown of glucose (Box 21B). Citric acid, in its salt form, is part of the Krebs cycle (Sec. 20.4). Both of these acids are normally present in blood. The intestines of human infants contain high concentrations of lactic acid due to the fermentation of mother's milk by the bacteria *Lactobacillus lipidus*. The lactic acid helps to prevent infections by suppressing the growth of harmful bacteria.

## Trichloroacetic Acid, $CCl_3COOH$

A 50 percent aqueous solution of trichloroacetic acid is used by dentists to cauterize gums. This strong acid denatures the bleeding or diseased tissue and allows the growth of healthy gum. Trichloroacetic acid solution is also used to cauterize canker sores.

## 14.5  PREPARATION OF CARBOXYLIC ACIDS

There are several ways to prepare carboxylic acids in the laboratory. We have already discussed the oxidation of primary alcohols and aldehydes (Secs. 12.3 and 13.5):

$$R-CH_2-OH \xrightarrow{[O]} R-\underset{\underset{O}{\|}}{C}-OH$$

$$R-\underset{\underset{O}{\|}}{C}-H \xrightarrow{[O]} R-\underset{\underset{O}{\|}}{C}-OH$$

The oxidation of a primary alcohol to a carboxylic acid is one of the oldest chemical reactions known, since even prehistoric humans found that their wine often turned to vinegar. The ethanol in the wine is oxidized to acetic acid by the oxygen in the air.

Other important methods for carboxylic acid preparation are the hydrolysis of carboxylic esters (Sec. 14.11) and of amides (Sec. 15.8).

## 14.6  THE ACIDITY OF CARBOXYLIC ACIDS

By far the most important property of carboxylic acids is the one implied by their name: They are acidic. Apart from the sulfonic acids (Sec. 11.11), carboxylic acids are the strongest organic acids. Still, compared with the strong inorganic acids (Table 8.1, p. 222), almost all carboxylic acids are weak. In Section 8.2 we saw that a strong acid reacts almost completely with water to form $H_3O^+$. The extent of this reaction for a typical unsubstituted carboxylic acid is only about 1 percent or less. That is, the equilibrium

$$R-\underset{\underset{O}{\|}}{C}-OH + H_2O \rightleftharpoons R-\underset{\underset{O}{\|}}{C}-O^- + H_3O^+$$

about 99%                                  about 1%

lies well to the left.

Although carboxylic acids are weak, they are strong enough to react with many bases stronger than water. Among these are hydroxide ion, $OH^-$, carbonate ion, $CO_3^{2-}$, bicarbonate ion, $HCO_3^-$, and ammonia (Sec. 8.6):

$$R-\underset{\underset{O}{\|}}{C}-OH + OH^- \longrightarrow R-\underset{\underset{O}{\|}}{C}-O^- + H_2O$$

$$R-\underset{\underset{O}{\|}}{C}-OH + CO_3^{2-} \longrightarrow R-\underset{\underset{O}{\|}}{C}-O^- + HCO_3^-$$

$$R—\underset{\underset{O}{\|}}{C}—OH + HCO_3^- \longrightarrow R—\underset{\underset{O}{\|}}{C}—O^- + H_2CO_3 \longrightarrow [H_2O + CO_2]$$

$$R—\underset{\underset{O}{\|}}{C}—OH + NH_3 \longrightarrow R—\underset{\underset{O}{\|}}{C}—O^- + NH_4^+$$

In all these reactions, the conversion of the carboxylic acid to its conjugate base (called a *carboxylate ion*) is essentially complete.

---

**PROBLEM 14.2** Complete the following reaction:

$$CH_3CH_2CH_2CH_2—\underset{\underset{O}{\|}}{C}—OH + OH^- \longrightarrow$$

---

At this point it is useful to compare the acidity of carboxylic acids with that of other oxygen-containing organic compounds. In Section 8.4 we learned that the acidity of an acid can be expressed by its equilibrium constant, $K_a$. Table 14.2 shows $K_a$ and $pK_a$ values for some organic compounds. As can be seen, carboxylic acids are much stronger acids than phenols, which in turn are somewhat stronger than thiols. Alcohols and aldehydes, on the other hand, are so weak (they are weaker acids than water) that we generally consider them to be neutral compounds, with no acidic properties at all.

As discussed in Section 8.8, $pK_a = -\log K_a$.

## 14.7 PROPERTIES OF CARBOXYLIC ACID SALTS

When a carboxylic acid reacts with a base, the acid is converted to the carboxylate ion, $RCOO^-$. Negative ions, of course, do not exist by themselves—there must also be positive ions present. As we learned in Chapter 8, the combination of a negative and a positive ion is called a salt, and the reaction of a carboxylic

**TABLE 14.2 The $K_a$ and $pK_a$ Values for Some Organic Compounds**

| Compound | Formula | $K_a$ | $pK_a$ |
|---|---|---|---|
| Formic acid | HCOOH | $1.8 \times 10^{-4}$ | 3.74 |
| Acetic acid | $CH_3COOH$ | $1.8 \times 10^{-5}$ | 4.74 |
| Benzoic acid | ⬡—COOH | $6.5 \times 10^{-5}$ | 4.19 |
| Phenol | ⬡—OH | $1.3 \times 10^{-10}$ | 9.89 |
| Ethanethiol | $CH_3CH_2—SH$ | $2.5 \times 10^{-11}$ | 10.6 |
| Ethanol | $CH_3CH_2—OH$ | Neutral | |
| Acetaldehyde | $CH_3—\underset{\underset{O}{\|}}{C}—H$ | Neutral | |

**SODIUM BENZOATE**   is an important preservative. It is added to many foods, especially acidic foods (those with a pH of less than about 4.5), to inhibit the growth of mold and bacteria.

**CALCIUM AND SODIUM PROPIONATE**   are also used as preservatives, chiefly in cheese, bread, and cakes. Some cheeses, especially hard ones of the Swiss cheese type, contain natural propionic acid. Propionate salts also inhibit the growth of mold, and thereby prevent the food from spoiling.

**ZINC 10-UNDECYLENATE,**   the zinc salt of 10-undecylenic acid,

$$CH_2{=}CH{-}(CH_2)_8{-}COOH$$

is used to treat fungal infections of the skin, especially *tinea pedis* (athlete's foot).

**ALUMINUM ACETATE**   is blended into creams and lotions used for minor skin irritations, such as those caused by diaper rash, insect bites, acne, or exposure to alkali.

acid with a base always produces a salt. These salts can be isolated as pure compounds by evaporating the water from the solution. Salts of carboxylic acids are ionic compounds, for example,

$$R{-}\underset{\underset{O}{\|}}{C}{-}O^- \ Na^+$$

and we are not surprised to find that they are always solids and are usually soluble in water (Sec. 6.6).

To name a carboxylate salt, always begin by naming the corresponding acid. Then change "-ic acid" to "-ate" and place the name of the positive ion before this word.

Note that the carboxylic acids themselves are often liquids, and many are insoluble in water.

---

**EXAMPLE**

Name the salts shown.

(a) $H{-}\underset{\underset{O}{\|}}{C}{-}O^- \ K^+$    Potassium formate

The corresponding acid, HCOOH, is formic *acid* (Table 14.1), and we change this to form*ate*. The positive ion is potassium, so the name is potassium formate.

(b) $CH_3{-}\underset{\underset{O}{\|}}{C}{-}O^- \ Na^+$    Sodium acetate

(c)

Lithium benzoate

The process is the same whether a common name or an IUPAC name is used for the acid.

(d) $\left( \begin{array}{c} \text{CH}_3 \\ | \\ \text{CH}_3-\text{C}-\text{CH}-\text{C}-\text{O}^- \\ | \quad | \quad \| \\ \text{Br} \quad \text{Br} \quad \text{O} \end{array} \right)_2 \text{Ca}^{2+}$     Calcium 2,3-dibromo-3-methylbutanoate (IUPAC)

**PROBLEM 14.3**   Name these salts:

(a) $\text{CH}_3(\text{CH}_2)_{10}-\text{C}-\text{O}^- \text{K}^+$
$\qquad\qquad\qquad\quad \|$
$\qquad\qquad\qquad\quad \text{O}$

(b) 
$\begin{array}{c} \text{COO}^- \text{Na}^+ \\ \\ \text{OH} \end{array}$

---

Because carboxylic acids are acidic, their structure in water depends on the pH, unlike the case for neutral compounds such as alcohols. When an alcohol, ROH, is dissolved in water, its structure remains ROH no matter what the pH. This is not the case for carboxylic acids. When a carboxylic acid is dissolved in water, the pH is always below 7. If we add enough base to the solution, (1) it will neutralize the acid (convert it to its salt), and (2) it will raise the pH. These two things always go together, and we can make the following useful statement:

A carboxylic acid can exist in water only at a low pH; a carboxylate ion (salt) can exist in water only at a moderate or high pH.

The latter part of this statement is true because the carboxylate ion is always converted to the acid when the pH is lowered:

$$\text{R}-\text{C}-\text{OH} \underset{\text{H}_3\text{O}^+}{\overset{\text{OH}^-}{\rightleftharpoons}} \text{R}-\text{C}-\text{O}^-$$
$$\qquad \| \qquad\qquad\qquad\qquad \|$$
$$\qquad \text{O} \qquad\qquad\qquad\qquad \text{O}$$
$$\text{Low pH} \qquad\qquad\qquad \text{High pH}$$

These interconversions are so easy and common that organic chemists consider carboxylic acids and their salts to be essentially interchangeable.

All of this allows us to go back and forth easily. If we have a carboxylic acid, all we have to do is add NaOH and we get the salt. If we have the salt, we just add HCl and get the acid.

The acidity of carboxylic acids often allows us to separate them from other organic compounds. As we have seen, most organic compounds, including most carboxylic acids, are insoluble in water, but almost all sodium carboxylates are soluble. To separate a carboxylic acid from a mixture that contains other insoluble compounds, we just shake the mixture with an aqueous solution of NaOH. The carboxylic acid is neutralized by the OH$^-$ and converted to its sodium salt, which dissolves in the water. The rest of the mixture is insoluble in water, and thus two layers form, which may easily be separated.

## 14.8   NOMENCLATURE OF CARBOXYLIC ESTERS

Acids are proton donors. In most acids, whether organic or inorganic, the proton that is donated comes from an O—H bond (HCl is an exception, of course). **If we replace the H from such an acid with an R,** the resulting compound is called an **ester.** If the acid is a carboxylic acid, the ester is called a **carboxylic ester** (we discuss other esters in Section 14.12). Carboxylic esters therefore have the structure

$$R-\overset{\overset{\textstyle O}{\|}}{C}-O-R' \quad \text{(abbreviated RCOOR')}$$

As usual, we use R' to show that the two R groups can be the same or different.

It is useful to think of a carboxylic ester as being derived from a carboxylic acid (the RCO part) and an alcohol (the OR' part):

Alcohol part

$$R-\overset{\overset{\textstyle O}{\|}}{C}-O-R'$$

Acid part

We will see in Section 14.10 that carboxylic esters can in fact be made from a carboxylic acid and an alcohol. In naming a carboxylic ester, it is not wise to start by trying to name the ester as a whole. Esters are named as derivatives of the corresponding carboxylic acids, and the best way to begin is by naming the acid, using either the IUPAC name or the common name. After that, the procedure is the same as for naming salts. The "-ic acid" is changed to "-ate," and this word is the second word in the ester name. The first word is the name of the R' group.

When a molecule contains both a carboxyl and a hydroxyl group, a cyclic ester, called a **lactone,** can be formed:

$$CH_3-\overset{\overset{\textstyle |}{OH}}{CH}-CH_2CH_2-COOH$$

$$\downarrow$$

---

**EXAMPLE**

Name the esters shown.

(a) $CH_3-\overset{\overset{\textstyle O}{\|}}{C}-O-CH_2CH_3$    Ethyl acetate

The corresponding acid, $CH_3COOH$, is ace*tic acid*. We change this to ace*tate*. The R group is ethyl, so the ester is called ethyl acetate. (The IUPAC name is ethyl ethanoate.)

(b) $CH_3CH_2CH_2-\overset{\overset{\textstyle O}{\|}}{C}-O-CH_2CH_2CH_2CH_3$    *n*-Butyl butyrate

The acid is butyric acid, and the alcohol is *n*-butyl alcohol. Thus the ester is *n*-butyl butyrate.

(c) 

$$\text{(structure)} \quad \overset{\overset{\textstyle CH_3}{|}}{\underset{\underset{\textstyle CH_3}{|}}{\overset{\overset{\textstyle}{}}{C}-O-C-CH_3}}$$

*tert*-Butyl salicylate

(d) $CH_3-\underset{\underset{\textstyle OH}{|}}{CH}-\overset{\overset{\textstyle O}{\|}}{C}-O-\text{(phenyl)}$    Phenyl lactate

The last example shows that the R′ of a carboxylic ester can be an aromatic ring. In such a case the "alcohol" portion of the ester is a phenol. This type of ester is called a *phenolic ester*.

---

**PROBLEM 14.4** Name these compounds:

(a) $CH_3CH_2CH_2CH_2\!-\!\overset{\displaystyle O}{\underset{\displaystyle \|}{C}}\!-\!OCH_2CH_3$   (b)

---

## 14.9 PHYSICAL PROPERTIES OF CARBOXYLIC ESTERS

Because ester molecules cannot form hydrogen bonds with each other, esters have lower boiling points than compounds of comparable molecular weight that do form hydrogen bonds (Table 14.3):

The boiling points of carboxylic esters are not only lower than those of the carboxylic acids, but even lower than those of the alcohols.

Carboxylic ester   Carboxylic ester   No hydrogen bond possible

Since carboxylic esters can form hydrogen bonds with water:

Carboxylic ester   Water

The low-molecular-weight esters, such as methyl acetate, are completely soluble in water. Solubility decreases rapidly, however, as molecular weight (and hence size of the nonpolar portion of the molecule) increases, and larger esters are insoluble in water.

we are not surprised to find that their solubility in water is about the same as that of carboxylic acids of the same molecular weight.

In Section 14.3 we learned that carboxylic acids have very disagreeable odors. In sharp contrast, the odors of the simple carboxylic esters are generally

**TABLE 14.3   Boiling Points of Some Esters Compared with Compounds of Similar Molecular Weight**

| Compound | | MW | BP(°C) |
|---|---|---|---|
| Butane | $CH_3CH_2CH_2CH_3$ | 58 | 0 |
| 1-Propanol | $CH_3CH_2CH_2\!-\!OH$ | 60 | 97 |
| Acetic acid | $CH_3COOH$ | 60 | 118 |
| Methyl formate | $HCOOCH_3$ | 60 | 32 |
| Hexane | $CH_3(CH_2)_4CH_3$ | 86 | 69 |
| 1-Pentanol | $CH_3(CH_2)_4\!-\!OH$ | 88 | 137 |
| Butyric acid | $CH_3(CH_2)_2\!-\!COOH$ | 88 | 165 |
| Ethyl acetate | $CH_3COOCH_2CH_3$ | 88 | 77 |

**B o x  14B**

## Carboxylic Esters as Flavoring Agents

The sweet and pleasant odors and tastes of many foods are due to complex mixtures of organic compounds, of which carboxylic esters generally are the most prevalent. Manufacturers of such products as soft drinks, ice cream, and maraschino cherries would like these products to taste and smell as much like the natural flavors as possible. In some cases they use extracts from the natural foods, but these are often too expensive, and artificial flavoring agents are very common. However, to reproduce synthetically the exact mixture found in a pineapple, for example, would be even more costly than using natural pineapple extract, and manufacturers have found that this is usually not necessary. Often the addition of only one or a few compounds, usually esters, makes the ice cream or soft drink taste natural.

The following are some carboxylic esters used as flavoring agents.

| Structure | Name | Flavor |
|---|---|---|
| HCOOCH$_2$CH$_3$ | Ethyl formate | Rum |
| CH$_3$COOCH$_2$CH$_2$CH(CH$_3$)$_2$ | Isopentyl acetate | Banana |
| CH$_3$COO(CH$_2$)$_7$CH$_3$ | Octyl acetate | Orange |
| CH$_3$(CH$_2$)$_2$COOCH$_2$CH$_3$ | Ethyl butyrate | Pineapple |
| CH$_3$(CH$_2$)$_2$COOCH$_3$ | Methyl butyrate | Apple |
| HCOOCH$_2$CH(CH$_3$)$_2$ | Isobutyl formate | Raspberry |
| | Methyl anthranilate | Grape |
| | Methyl salicylate | Wintergreen |

regarded as very pleasant. As a class, the esters have the most agreeable odors of all organic compounds. Many of the odors of fruits and flowers result from mixtures of carboxylic esters, and many of them (either naturally derived or synthetic) are used in perfumes and food flavorings (Box 14B).

## 14.10 PREPARATION OF CARBOXYLIC ESTERS

We have learned that a carboxylic ester can be considered as a combination of a carboxylic acid and an alcohol (or a phenol). In fact, carboxylic esters can be synthesized by heating a mixture of a carboxylic acid and an alcohol, provided that an acid catalyst (usually H$_2$SO$_4$) is present.

General reaction:

$$R-\underset{\underset{O}{\|}}{C}-OH + R'OH \xrightleftharpoons{H^+} R-\underset{\underset{O}{\|}}{C}-OR' + H_2O$$

Specific example:

$$CH_3-\overset{\underset{\|}{O}}{C}-OH + CH_3CH_2OH \overset{H^+}{\rightleftharpoons} CH_3-\overset{\underset{\|}{O}}{C}-OCH_2CH_3 + H_2O$$

Acetic acid        Ethanol        Ethyl acetate

In this reaction, called **esterification,** a molecule of water comes out, and the carboxylic acid and alcohol combine to form the ester:

It is known that in most cases, the O that comes out comes from the acid and not from the alcohol (note that the same ester would be formed even if it happened the other way).

$$R-\overset{\underset{\|}{O}}{C}\overset{O-H}{\phantom{x}} \quad \overset{H}{O}-R' \rightleftharpoons R-\overset{\underset{\|}{O}}{C}-O-R' + H-O-H$$

---

**EXAMPLE**

Draw the structural formula of the carboxylic ester obtained from each combination:

(a) $CH_3COOH + CH_3OH$

(b) $CH_3CH_2COOH +$ (phenol with OH)

(c) $HOOC-CH_2CH_2-COOH + CH_3CH_2OH$

**Answer**

(a) $CH_3-\overset{\underset{\|}{O}}{C}\overset{O-H}{\phantom{x}} \quad \overset{H}{O}-CH_3 \overset{H^+}{\rightleftharpoons} CH_3-\overset{\underset{\|}{O}}{C}-O-CH_3$

Acetic acid        Methanol        Methyl acetate

(b) $CH_3CH_2-\overset{\underset{\|}{O}}{C}\overset{O-H}{\phantom{x}} \quad \overset{H}{O}-$ $\overset{H^+}{\rightleftharpoons} CH_3CH_2-\overset{\underset{\|}{O}}{C}-O-$

Propionic acid        Phenol        Phenyl propionate

(c) **Monoester:**

$$HO-\overset{\underset{\|}{O}}{C}-CH_2CH_2-\overset{\underset{\|}{O}}{C}-OH + H-O-CH_2CH_3 \overset{H^+}{\rightleftharpoons}$$

Succinic acid        Ethanol

$$HO-\overset{\underset{\|}{O}}{C}-CH_2CH_2-\overset{\underset{\|}{O}}{C}-OCH_2CH_3$$

Monoethyl succinate

Diester:

$$CH_3CH_2-O-H + H-O-\underset{\underset{O}{\|}}{C}-CH_2CH_2-\underset{\underset{O}{\|}}{C}-OH + H-O-CH_2CH_3$$

$$\Updownarrow \; H^+$$

$$CH_3CH_2-O-\underset{\underset{O}{\|}}{C}-CH_2CH_2-\underset{\underset{O}{\|}}{C}-OCH_2CH_3$$

Diethyl succinate

**PROBLEM 14.5** Show the structural formulas of the carboxylic esters obtained from a combination of these carboxylic acids and alcohols:

(a)  $HCOOH + CH_3OH$

(b)  $+ CH_3-CH-OH \longrightarrow$
       with $CH_3$ below the CH

(c)  $+ CH_3CH_2OH \longrightarrow$

---

**EXAMPLE**

Identify the acid and alcohol (or phenol) portions of these carboxylic esters:

(a)  $CH_3CH_2-\underset{\underset{O}{\|}}{C}-O-CH_2-\underset{\underset{CH_3}{|}}{CH}-CH_3$

(b)  $CH_3-O-\underset{\underset{O}{\|}}{C}-CH_2$

(c)        Oil of wintergreen

(d)        Aspirin

**Answer**

The alcohol or phenol portions are shown in color:

(a) CH$_3$CH$_2$—C—O—CH$_2$—CH—CH$_3$
           ‖                 |
           O               CH$_3$

(b) CH$_3$—O—C—CH$_2$—C$_6$H$_5$
           ‖
           O

(c) salicylic OH, C—O—CH$_3$, ‖O

(d) O—C—CH$_3$ (‖O), C—OH (‖O)

**PROBLEM 14.6** Identify the acid and the alcohol or phenol portions of these carboxylic esters:

(a) CH$_3$CH$_2$CH$_2$—C—O—C$_6$H$_5$ (‖O)

(b) C$_5$H$_9$—O—C—CH$_2$—CH—CH$_3$ (‖O, |CH$_3$)

Although a great many carboxylic esters have been synthesized simply by heating a carboxylic acid with an alcohol in the presence of a strong acid catalyst, the reaction does have its problems: It is an equilibrium. To make it go to completion, the equilibrium must be shifted to the right. In most cases this is done by boiling the mixture. When the mixture of four components is boiled, the one with the lowest boiling point evaporates first and can be removed by distillation. If this is either the carboxylic ester or the water, the equilibrium shifts to the right (Le Chatelier's principle) and the reaction goes to completion. Since carboxylic esters often have lower boiling points than either acids or alcohols, this is usually a feasible procedure. However, if the alcohol or acid has the lowest boiling point, boiling will not produce the ester, and some other way must be found.

For centuries, many people knew that they could relieve pain by chewing willow bark (Hippocrates recommended it around 400 BC). In 1860 an organic chemist extracted the compound salicylic acid from willow bark and soon discovered that it was this compound that was responsible for pain relief. Physicians immediately began to prescribe salicylic acid because it worked well, but it did have serious drawbacks: It had a bad (sour) taste and irritated the mouth and stomach lining.

Salicylic acid is both a carboxylic acid and a phenol. Because it is a phenol, it forms an ester if treated with acetic anhydride:

Salicylic acid     Acetic anhydride     Aspirin (acetylsalicylic acid)

In 1893 a chemist at the Bayer Company in Germany did just that and converted salicylic acid to an ester that Bayer named "aspirin" (the chemical name is acetylsalicylic acid).

Aspirin turned out to be as good a pain-reliever as salicylic acid, or even better, and it did not have the same drawbacks. Its use grew very quickly, and it soon became the most widely used pain-reliever in the world, in fact, the most widely used drug of any kind.

Aspirin also reduces fever and the swelling caused by injuries and rheumatism. It is the most important drug in the treatment of arthritis, though it only relieves symptoms and does not cure the disease. It is one of the safest known drugs, and side effects are few.

However, as with most drugs, aspirin does have some side effects. It can prevent synthesis of stomach mucus, leading to ulceration by stomach acid of the unprotected stomach walls. It also prevents the aggregation of platelets whose function is to form a gel-like plug when a blood vessel is cut (see Box 25B). This side effect is now being turned to beneficial preventative use. Hundreds of thousands of patients at risk take small daily doses of aspirin to reduce or prevent the occurrence of strokes and heart attacks.

The drug is sold not only under the name aspirin, but is also the pain-relieving ingredient in such products as Anacin and Bufferin. Since all aspirin-containing products reduce pain equally well, the wise consumer buys the cheapest. See Box 17J for a discussion of how aspirin acts as an anti-inflammatory agent. For other nonprescription pain-relievers, see Box 15D.

Because of the equilibrium problem, carboxylic esters are more often prepared in laboratories not from the carboxylic acid but from two other types of acid derivatives: acyl chlorides and anhydrides. An **acyl chloride** is a compound in which a carbonyl group is connected to a chlorine:

An acyl chloride

Both acyl chlorides and anhydrides can be made from carboxylic acids.

# Polyesters

There are two major types of polymers. In Section 11.7 we discussed one of these: addition polymers. The other type, **condensation polymers,** are constructed from molecules that have two (or more) functional groups. For example, we saw in the text that an alcohol can be connected to a carboxylic acid to form an ester. The ester group "ties" the two molecules together (this is why it is often called an ester *linkage*). If instead of using an ordinary carboxylic acid and alcohol we run the same reaction with a dicarboxylic acid and a diol, the molecules link at both ends, and a long chain (a polymer) is formed.

In the most important polyester, the two monomers are terephthalic acid and ethylene glycol:

$$-CH_2-OH + HO-\underset{O}{\overset{}{C}}-\bigcirc-\underset{O}{\overset{}{C}}-OH + HO-CH_2CH_2-OH + HO-\underset{O}{\overset{}{C}}-\bigcirc-\underset{O}{\overset{}{C}}-OH + HO-CH_2CH_2OH$$

Terephthalic acid     Ethylene glycol

$$\downarrow -H_2O$$

$$-CH_2-O-\underset{O}{\overset{}{C}}-\bigcirc-\underset{O}{\overset{}{C}}-O-CH_2CH_2-O-\underset{O}{\overset{}{C}}-\bigcirc-\underset{O}{\overset{}{C}}-O-CH_2CH_2-O-$$

Polyester (Dacron)

**Figure 14D.1**
A garment made of Dacron. (Courtesy of E. I. Dupont de Nemours & Company, Inc.)

As you can see from the equation, a molecule of water is removed at each place a linkage is formed, and a chain in which all the links are ester groups has been formed. This particular polyester is called Dacron. It is widely used in clothing of all kinds (often just called polyester), as well as in tire cords, carpets, and many other products (Fig. 14D.1). The same polyester can also be produced in thin sheets, in which case it is called Mylar. Mylar is used to make recording and computer tape, among other things.

Dacron is also used to replace human arteries. In coronary bypass surgery, physicians prefer to use a section of the saphenous vein from the patient's leg to replace the diseased coronary artery. However, in nearly one third of patients this is unavailable because of previous surgery or previous disease (for example, phlebitis). In these cases synthetic arterial replacement is called for. Dacron and also Teflon (Table 11.2) have been used to make synthetic arteries. Dacron is also used to patch defective hearts (Fig. 14D.2).

**Figure 14D.2**
A Dacron patch is used to close an atrial septal defect in a heart patient. (Courtesy of Drs. James L. Monro and Gerald Shore and the Wolfe Medical Publications, London, England.)

An **anhydride** can be thought of as a molecule formed when two carboxylic acids combine and lose water:

$$R-\overset{\overset{\textstyle O-H}{|}}{\underset{\underset{\textstyle O}{\|}}{C}} \quad \overset{\overset{\textstyle H}{|}}{\underset{\underset{\textstyle O}{}}{O}}-\underset{\underset{\textstyle O}{\|}}{C}-R \longrightarrow R-\underset{\underset{\textstyle O}{\|}}{C}-O-\underset{\underset{\textstyle O}{\|}}{C}-R$$

An anhydride

Both acyl chlorides and anhydrides react easily and quickly with alcohols or phenols to give carboxylic esters. These reactions are not reversible.

Since these reactions are not reversible, yields are higher than with carboxylic acids.

General reactions:

$$R-\underset{\underset{\textstyle O}{\|}}{C}-Cl + R'OH \longrightarrow R-\underset{\underset{\textstyle O}{\|}}{C}-OR' + HCl$$

$$R-\underset{\underset{\textstyle O}{\|}}{C}-O-\underset{\underset{\textstyle O}{\|}}{C}-R + R'OH \longrightarrow R-\underset{\underset{\textstyle O}{\|}}{C}-OR' + RCOOH$$

Specific examples:

$$CH_3CH_2-\underset{\underset{\textstyle O}{\|}}{C}-Cl + CH_3OH \longrightarrow CH_3CH_2-\underset{\underset{\textstyle O}{\|}}{C}-OCH_3 + HCl$$

Propionyl chloride        Methyl propionate

$$CH_3-\underset{\underset{\textstyle O}{\|}}{C}-O-\underset{\underset{\textstyle O}{\|}}{C}-CH_3 + CH_3CH_2OH \longrightarrow CH_3-\underset{\underset{\textstyle O}{\|}}{C}-OCH_2CH_3 + CH_3COOH$$

Acetic anhydride        Ethyl acetate

---

**PROBLEM 14.7**  Draw the structural formulas for the products:

(a) $CH_3CH_2CH_2-\underset{\underset{\textstyle O}{\|}}{C}-Cl +$ $\longrightarrow$

(b) $CH_3CH_2-\underset{\underset{\textstyle O}{\|}}{C}-O-\underset{\underset{\textstyle O}{\|}}{C}-CH_2CH_3 + CH_3OH \longrightarrow$

---

## 14.11  HYDROLYSIS OF CARBOXYLIC ESTERS

Carboxylic esters undergo a number of reactions, but we consider only one of them in this chapter, the one that is the most important, both in the laboratory and in the body. This reaction, called **hydrolysis,** is the reverse of the esterification reaction discussed in Section 14.10. In hydrolysis, a carboxylic ester reacts with water to produce the carboxylic acid and the alcohol or phenol from which the ester was originally formed.

Carboxylic esters are neutral compounds because the —COOH proton has been replaced by an R group.

To hydrolyze a carboxylic ester we usually need a catalyst.

## Box 14E

## The Use of Absorbable Staples in Hysterectomy

**Figure 14E**
Staples used to close the vaginal vault in resection of the uterus. (Courtesy of United States Surgical Corporation.)

Another type of polyester (see Box 14D), a copolymer of lactic acid and glycolic acid, provides an improved method for surgical closures of the vagina. Staples (Fig. 14E) are manufactured from this polyester, which goes under the trade name Lactomer. Both lactic acid and glycolic acid have an OH group, so they form a copolymer as follows:

$$\text{HO}-\text{CH}-\text{COOH} + \text{HO}-\text{CH}_2-\text{COOH} \longrightarrow$$
$$\quad\quad\quad |$$
$$\quad\quad\quad \text{CH}_3$$

Lactic acid        Glycolic acid

$$\text{HO}-\text{CH}-\overset{\text{O}}{\underset{\text{||}}{\text{C}}}-\text{O}\!\!\left[\!\!-\text{CH}_2-\overset{\text{O}}{\underset{\text{||}}{\text{C}}}-\text{O}-\text{CH}-\overset{\text{O}}{\underset{\text{||}}{\text{C}}}-\text{O}\!\!-\right]_n\!\!\text{CH}_2-\overset{\text{O}}{\underset{\text{||}}{\text{C}}}-\text{OH} + n\text{H}_2\text{O}$$

The polyester starts to hydrolyze in the body after six to eight weeks, forming a mixture of the two acids, and the staples dissolve. Both acids are normal compounds in the body, so the staples do not produce a foreign material that would have to be removed by the phagocytes in the blood. The staples retain 50 percent of their initial strength for at least two months, which is more than adequate to heal the wounds in the vagina or uterine ligaments. They are also used in operations on the bladder and on the intestines.

The advantage of absorbable staples over sutures (absorbable or otherwise) is threefold: Stapling is a fast process and thus reduces the operating time and the ensuing trauma. It also greatly reduces the blood flow and consequent loss of blood during surgery. Finally, in suturing, each time the needle hits the tissue it produces a trauma. In stapling, the multiple trauma is reduced to one single shock, and healing is more rapid.

### Acid-Catalyzed Hydrolysis

Strong acids are frequently used as catalysts to hydrolyze esters.

General reaction:

$$\text{R}-\overset{\text{O}}{\underset{\text{||}}{\text{C}}}-\text{OR}' + \text{H}_2\text{O} \overset{\text{H}^+}{\rightleftharpoons} \text{R}-\overset{\text{O}}{\underset{\text{||}}{\text{C}}}-\text{OH} + \text{R}'\text{OH}$$

Specific example:

$$\text{CH}_2-\overset{\text{O}}{\underset{\text{||}}{\text{C}}}-\text{OCH}_2\text{CH}_3 + \text{H}_2\text{O} \overset{\text{H}^+}{\rightleftharpoons}$$

Ethyl phenylacetate

$$\text{CH}_2\text{COOH} + \text{CH}_3\text{CH}_2\text{OH}$$

Phenylacetic acid       Ethanol

This, of course, is precisely the reverse of the formation of esters, discussed in Section 14.10, which means that it is also an equilibrium reaction (the same equilibrium, approached from the other direction). This time we desire to shift the equilibrium the other way, and that isn't always easy. One way is to use a large excess of water.

---

**PROBLEM 14.8**  Draw the structural formulas for the products:

$$CH_3-\overset{O}{\underset{\|}{C}}-OCH_2CH_3 + H_2O \overset{H^+}{\rightleftharpoons}$$

---

### Base-Catalyzed Hydrolysis

Esters can also be hydrolyzed with a basic catalyst (typically, NaOH):

General reaction:

$$R-\overset{O}{\underset{\|}{C}}-O-R' + H_2O \xrightarrow{OH^-} R-\overset{O}{\underset{\|}{C}}-O^- + R'OH$$

Salt of acid
(carboxylate ion)

Specific example:

$$CH_3-\overset{O}{\underset{\|}{C}}-O-\overset{CH_3}{\underset{|}{CH}}-CH_3 + H_2O \xrightarrow{OH^-} CH_3-\overset{O}{\underset{\|}{C}}-O^- + CH_3-\overset{CH_3}{\underset{|}{CH}}-OH$$

Isopropyl acetate                                          Acetate          Isopropyl
                                                           ion              alcohol

---

**PROBLEM 14.9**  Draw the structural formulas for the products:

$$CH_3-\overset{CH_3}{\underset{|}{CH}}-\overset{O}{\underset{\|}{C}}-OCH_2CH_2CH_3 + H_2O \xrightarrow{OH^-}$$

---

As with acid catalysis, the alcohol R'OH is a product. In this case, however, the other product is not the carboxylic acid itself, but its salt, the carboxylate ion. However, this certainly does not present any problem, since, once the hydrolysis is over, the carboxylate ion can easily be converted to the free acid (if that is what is wanted) simply by adding HCl (Sec. 14.7).

Because we get the salt rather than the free acid, base-catalyzed hydrolysis of carboxylic esters is *not* reversible. Consequently, we needn't worry about shifting any equilibrium. Therefore, chemists usually prefer the base-catalyzed to the acid-catalyzed method.

The **base-catalyzed hydrolysis of carboxylic esters** has a special name, **saponification.** This word is derived from the soapmaking process, and, as we shall see in Section 17.3, soap is indeed made by saponification of fats, which are a type of carboxylic ester.

Since carboxylic acids cannot exist in basic solutions, it is only natural that in the presence of OH⁻ the product should be the carboxylate ion instead of the free acid.

Salts of carboxylic acids do not react with alcohols.

### Enzyme-Catalyzed Hydrolysis

For example, the enzyme **lipase** hydrolyzes fats, which are esters of fatty acids and glycerol.

Biological organisms contain many carboxylic esters, and the body is constantly forming and hydrolyzing them. When the body hydrolyzes carboxylic esters, it uses neither acids nor bases as catalysts, but enzymes, which are much more efficient. We will discuss the nature and action of enzymes in Chapter 19.

## 14.12  ESTERS AND ANHYDRIDES OF PHOSPHORIC ACID

We have defined an ester as the compound resulting when the H of an —OH acid is replaced by an R group. Most inorganic acids are —OH acids and can be converted to esters. The most important of these in biochemistry are the esters of phosphoric acid.

Phosphoric acid, $H_3PO_4$, is a triprotic acid and gives up its protons one at a time to sufficiently strong bases:

Phosphoric acid    Dihydrogen phosphate ion    Hydrogen phosphate ion    Phosphate ion

Because $H_3PO_4$ has three —OH groups, it can form mono-, di-, and triesters:

Monoester    Diester    Triester

Specific example:

Most formations of phosphate esters are catalyzed by enzymes.

Monomethyl phosphate

Note that the mono- and diesters still have —OH groups. Therefore they are still acidic and can give their protons to a base.

Esters of phosphoric acid, especially the mono- and diesters, are important biological compounds. We shall see many examples of them in later chapters, including phospholipids (Sec. 17.4), glucose 6-phosphate (Sec. 21.2), and nucleic acids (Chap. 23). A typical example is glycerol 1-phosphate:

At body pH values, mono- and diphosphate esters usually exist in their ionic forms, as shown for glycerol 1-phosphate.

Glycerol 1-phosphate

Like carboxylic acids, phosphoric acid can also be converted to an anhydride, called pyrophosphoric acid, but unlike the anhydrides formed from carboxylic acids, pyrophosphoric acid still has —OH groups and so can be converted to another anhydride, triphosphoric acid:

$$
\underset{\displaystyle \overset{\displaystyle O}{\|}}{HO-P-OH} + H-O-\underset{\displaystyle \overset{\displaystyle O}{\|}}{P}-OH \longrightarrow HO-\underset{\displaystyle \overset{\displaystyle O}{\|}}{P}-O-\underset{\displaystyle \overset{\displaystyle O}{\|}}{P}-OH
$$

Pyrophosphoric acid, $H_4P_2O_7$

$$
HO-\underset{OH}{\overset{O}{\|}}{P}-O-\underset{OH}{\overset{O}{\|}}{P}-OH + H-O-\underset{OH}{\overset{O}{\|}}{P}-OH \longrightarrow HO-\underset{OH}{\overset{O}{\|}}{P}-O-\underset{OH}{\overset{O}{\|}}{P}-O-\underset{OH}{\overset{O}{\|}}{P}-OH
$$

Triphosphoric acid, $H_5P_3O_{10}$

*At physiological pH values, pyrophosphoric acid exists as the pyrophosphate ion and triphosphoric acid exists as the triphosphate ion.*

These two anhydrides of phosphoric acid are extremely important in biochemistry. Or rather, it is their esters that are important. Both pyrophosphoric acid and triphosphoric acid still have —OH groups (which is why they are acids as well as anhydrides) and so form salts and esters. In the body these are generally the monoesters, which we show in the salt form because that is how they exist at body pH values:

$$
{}^-O-\underset{O_-}{\overset{O}{\|}}{P}-O-\underset{O_-}{\overset{O}{\|}}{P}-OR
$$

Alkyl diphosphate

$$
{}^-O-\underset{O_-}{\overset{O}{\|}}{P}-O-\underset{O_-}{\overset{O}{\|}}{P}-O-\underset{O_-}{\overset{O}{\|}}{P}-OR
$$

Alkyl triphosphate

The most important diphosphate and triphosphate in the body are called ADP and ATP, respectively. The R group is rather complicated, and we need not look at it now. The complete structures of ADP and ATP are given in Section 20.3.

Anhydrides of phosphoric acid are important in the body because the anhydride bonds are used to transfer energy. The P—O—P linkage is a **high-energy linkage,** which is just another way of saying that a lot of energy is given off when it is hydrolyzed. The body stores energy in part by forming P—O—P bonds. When it needs to use the energy, it hydrolyzes the linkage, so that the energy is given off, for example:

$$
{}^-O-\underset{O_-}{\overset{O}{\|}}{P}-O-\underset{O_-}{\overset{O}{\|}}{P}-O-\underset{O_-}{\overset{O}{\|}}{P}-OR \xrightarrow[\text{enzyme}]{H_2O}
$$

High-energy linkage

*These phosphoric anhydrides do not react with water in the absence of a catalyst. In the body specific enzymes catalyze the reaction.*

$$
{}^-O-\underset{O_-}{\overset{O}{\|}}{P}-OH + H-O-\underset{{}_-O}{\overset{O}{\|}}{P}-O-\underset{{}_-O}{\overset{O}{\|}}{P}-OR + \text{energy}
$$

Like the anhydrides of carboxylic acids (Sec. 14.10), anhydrides of phosphoric acid react with alcohols to give phosphate esters, for example:

$$
\underset{\text{Pyrophosphoric acid}}{\text{HO}-\underset{\underset{\text{OH}}{|}}{\overset{\overset{\text{O}}{\|}}{\text{P}}}-\text{O}-\underset{\underset{\text{OH}}{|}}{\overset{\overset{\text{O}}{\|}}{\text{P}}}-\text{OH}} + \text{CH}_3-\text{O}-\text{H} \longrightarrow
$$

$$
\underset{\substack{\text{Monomethyl} \\ \text{phosphate}}}{\text{HO}-\underset{\underset{\text{OH}}{|}}{\overset{\overset{\text{O}}{\|}}{\text{P}}}-\text{OCH}_3} + \underset{\substack{\text{Phosphoric} \\ \text{acid}}}{\text{HO}-\underset{\underset{\text{OH}}{|}}{\overset{\overset{\text{O}}{\|}}{\text{P}}}-\text{OH}}
$$

This is an important reaction in biochemistry. The body typically makes phosphate esters by this process. Because a phosphate group (shown in color) is transferred from the anhydride to the alcohol, this process is called **phosphorylation.** We shall see specific examples of phosphorylation in Chapters 20, 21, and 22.

### Other Inorganic Esters

Although this section has discussed only esters of phosphoric acid, all other —OH acids also form esters (the general term is **inorganic ester**). In Box 12D we met nitroglycerine, an ester of nitric acid, and isoamyl nitrite, an ester of nitrous acid. Esters of sulfuric acid are also important in biochemistry, especially in acidic polysaccharides (see Sec. 16.12). Many detergents are esters of sulfuric acid (see Box 17C). Esters made from thiols, RSH, rather than alcohols, ROH, are important in biochemistry (see Secs. 20.3 and 22.3).

## SUMMARY

Compounds that contain the **carboxyl group, —COOH,** are called **carboxylic acids.** Carboxylic acids can be named by the IUPAC system but are more often given common names. Because they form **dimers** by hydrogen bonding, they have high boiling points. They also have unpleasant odors. Their solubilities in water are about the same as those of other oxygen-containing organic compounds.

The most important chemical property of carboxylic acids is their acidity. They form salts when treated with strong bases like $OH^-$ and even with weaker bases such as $HCO_3^-$ and $NH_3$. These salts, called **carboxylates,** are soluble in water even when the acids themselves are not. A carboxylic acid can exist in water only at a low pH; a carboxylate ion can exist in water only at a moderate or high pH. Thus a carboxylic acid can always be converted to its salt by adding NaOH, and a carboxylate ion can always be converted to the acid by adding HCl.

An **ester** is made by replacing the H of an —OH acid with an R group. **Carboxylic esters** derived from carboxylic acids therefore have the formula **RCOOR'.** They are named by changing the "-ic acid" suffix to "-ate" and placing this new word after the name of the R' group. **Carboxylate salts** are named the same way, except that the name of the metal ion becomes the first word. Because carboxylic ester molecules cannot form hydrogen bonds with each other, their boiling points are low, but their solubilities are similar to those of the carboxylic acids. They generally have very pleasant odors.

Carboxylic esters, RCOOR′, are prepared by the reaction between an alcohol (R′OH) and either an acyl halide (RCOCl), an anhydride (R—CO—O—CO—R), or a carboxylic acid (RCOOH). In the last case, a strong acid catalyst (such as $H_2SO_4$) is needed, and an equilibrium mixture of reactants and products is formed, which must be driven to completion. The most important reaction of carboxylic esters is their **hydrolysis** (cleavage by water), which may be catalyzed by either acids or bases. For synthetic purposes, base catalysis is often preferred because the reaction is not reversible. Base-catalyzed ester hydrolysis is called **saponification.** In biological systems, ester hydrolysis is catalyzed by enzymes.

Esters and anhydrides of phosphoric acid, $H_3PO_4$, are very important in biochemistry. Phosphoric acid forms three series of esters: $(HO)_2PO(OR)$, $(HO)PO(OR)_2$, and $PO(OR)_3$. The P—O—P bond is a **high-energy bond,** and conversions between **alkyl diphosphates** (monoesters of pyrophosphoric acid, $H_4P_2O_7$) and **alkyl triphosphates** (monoesters of triphosphoric acid, $H_5P_3O_{10}$) are the major way the body transfers energy from one molecule to another. **Phosphorylation,** the transfer of a phosphate group from the —OH of one molecule to the —OH of another, is an important biological process.

## Summary of Reactions

**1.** Reaction of carboxylic acids with bases (Sec. 14.6):

**2.** Neutralization of carboxylate salts (Sec. 14.7):

**3.** Preparation of carboxylic esters (Sec. 14.10):
 a. From carboxylic acids and alcohols:
 b. From acyl chlorides and alcohols:
 c. From anhydrides and alcohols:

**4.** Hydrolysis of carboxylic esters (Sec. 14.11):
 a. Acid-catalyzed:
 b. Base-catalyzed (saponification):

## KEY TERMS

Acyl chloride (Sec. 14.10)
Acyl group (Sec. 14.1)
Anhydride (Sec. 14.10)
Carboxyl group (Sec. 14.1)
Carboxylate ion (Sec. 14.6)
Carboxylic acid (Sec. 14.1)
Carboxylic acid salt (Sec. 14.7)
Carboxylic ester (Sec. 14.8)
Condensation polymer (Box 14D)
Dimer (Sec. 14.3)
Ester (Sec. 14.8)
Ester hydrolysis (Sec. 14.11)

Esterification (Sec. 14.10)
Fatty acid (Sec. 14.2)
High-energy linkage (Sec. 14.12)
Hydrolysis (Sec. 14.11)
Inorganic ester (Sec. 14.12)
Lactone (Sec. 14.8)
Phenolic ester (Sec. 14.8)
Phosphate ester (Sec. 14.12)
Phosphoric anhydride (Sec. 14.12)
Phosphorylation (Sec. 14.12)
Polyester (Boxes 14D, 14E)
Saponification (Sec. 14.11)

## PROBLEMS

### Nomenclature of Carboxylic Acids

**14.10** Give IUPAC names for

(a) $CH_2CH_2CH_2CH_2—COOH$

(b) $CH_3—CH—CH—CH_2—CH—COOH$
         |      |              |
        $CH_3$  $CH_3$        $Br$

(c) $CH_3—CH——CH—COOH$ (with two phenyl groups attached)

(d) $Br—CH_2—CH—CH—CH—CH—COOH$
              |    |    |    |
             $Cl$  $Br$ $Cl$ $CH_3$

**14.11** Draw structural formulas for

(a) 2-hydroxypentanoic acid

(b) 3-bromo-2-methylbutanoic acid

(c) 5-cyclohexylhexanoic acid

*(d) 3-methyl-3-heptenoic acid

**14.12** Give common names for

(a) $CH_3—COOH$

(b) $CH_3(CH_2)_6COOH$

(c) $HOOCCH_2CH_2CH_2COOH$

(d) (benzene ring)—$COOH$

(e) $HOOC$        $COOH$
          \       /
           $C{=}C$
          /       \
         $H$       $H$

(f) $CH_3(CH_2)_{14}COOH$

**14.13** Draw structural formulas for

(a) formic acid          (d) malonic acid

(b) stearic acid         (e) salicylic acid

(c) succinic acid        (f) caproic acid

**14.14** Give the IUPAC names (common names given) for

(a) caproic acid, $CH_3(CH_2)_4COOH$

(b) capric acid, $CH_3(CH_2)_8COOH$

(c) pivalic acid, $CH_3—C—COOH$
                         |
with $CH_3$ above and $CH_3$ below the central C

**14.15** Why was butyric acid given that name?

**14.16** (a) Define fatty acid.

(b) Why are the fatty acids given that name?

### Physical Properties of Carboxylic Acids

**14.17** Propanoic acid has a boiling point of 141°C, considerably higher than that of 1-butanol (117°C), although they have the same molecular weight. Explain why.

**14.18** Draw the structure of the dimer formed when two molecules of acetic acid form hydrogen bonds with each other.

*14.19** Malonic acid forms an internal hydrogen bond (the H of one —COOH group forms a hydrogen bond with an O of the other —COOH group). Draw a structural formula that shows this.

**14.20** Rank the following compounds in decreasing order of solubility in water:

(a) $CH_3CH_2CH_2CH_2CH_2CH_2COOH$

(b) $CH_3CH_2Br$

(c) $CH_3COOH$

**14.21** Of the organic compounds studied so far, which have particularly bad odors? Which have particularly pleasant odors?

**14.22** Caproic acid has a solubility in water of 1 g per 100 g of $H_2O$ (Table 14.1). Which part of the molecule contributes to the solubility in water, and which part prevents greater solubility?

### Some Important Carboxylic Acids

**14.23** Draw structural formulas for (a) lactic acid (b) citric acid.

**14.24** Lactic acid is a monocarboxylic acid. What would you call citric acid?

**14.25** What is a medical use for $Cl_3CCOOH$?

**14.26** Which carboxylic acid suppresses the growth of harmful intestinal bacteria?

### Preparation of Carboxylic Acids

*14.27** Write an equation to show how wine turns to vinegar if allowed to stand in the open air.

## Acidity of Carboxylic Acids; Carboxylate Salts

**14.28** Name these salts:

(a) $CH_3CH_2-\overset{\overset{\displaystyle O}{\|}}{C}-O^-\ K^+$

(b) $[^-O-\overset{\overset{\displaystyle O}{\|}}{C}-CH_2-\overset{\overset{\displaystyle O}{\|}}{C}-O^-]\ Ca^{2+}$

(c) $CH_3(CH_2)_{10}-\overset{\overset{\displaystyle O}{\|}}{C}-O^-\ K^+$

(d) $[^-O-\overset{\overset{\displaystyle O}{\|}}{C}-\!\!\bigcirc\!\!-\overset{\overset{\displaystyle O}{\|}}{C}-O^-]2Na^+$

(see Box 14D)

**14.29** Draw structural formulas for
  (a)  potassium formate
  (b)  lithium butanoate
  (c)  monosodium oxalate
  (d)  calcium benzoate
  (e)  magnesium caproate
  (f)  dipotassium fumarate

**14.30** Draw the structural formulas of oxalic acid (a) at a very low pH,  (b) at a medium pH, and  (c) at a very high pH.

**14.31** Complete these equations:
  (a)  $CH_3COOH + OH^- \longrightarrow$
  (b)  $HCOOH + NH_3 \longrightarrow$
  (c)  $HOOC-CH_2-COOH +$
         $excess\ OH^- \longrightarrow$
  (d)  $CH_3-\underset{\underset{\displaystyle OH}{|}}{CH}-COOH + HCO_3^- \longrightarrow$
  (e)  $CH_3COO^- + HCl \longrightarrow$

  (f)  $\bigcirc\!\!\!-\!COO^-\quad + H_2SO_4 \longrightarrow$

**14.32** A small quantity of sodium acetate, $CH_3COO^-\ Na^+$, is added to an aqueous solution whose pH is 2. What happens to the salt? Draw its formula at this pH.

**14.33** (a) Butyric acid is dissolved in water. Draw its structural formula.  (b) Enough NaOH is added to raise the pH to 12. What happens to the butyric acid? Draw the formula now.  (c) Enough HCl is added to lower the pH to 2. Draw the formula now.

**14.34** You are given a mixture of capric acid and 3-decanone, neither of which is soluble in water. Based on the information in this chapter, give a simple method of separating them. Describe in detail exactly what you would do.

## Structure and Nomenclature of Carboxylic Esters

**14.35** Define ester.

**14.36** Tell whether each compound is a carboxylic ester and, if so, identify the alcohol (or phenol) and acid portions:

(a) $CH_3CH_2-\overset{\overset{\displaystyle O}{\|}}{C}-OCH_3$

(b) $CH_3-\underset{\underset{\displaystyle OCH_2CH_3}{|}}{CH}-OCH_2CH_3$

(c) $\bigcirc\!\!-O-\overset{\overset{\displaystyle O}{\|}}{C}-H$

(d) $\bigcirc\!\!-CH_2-\overset{\overset{\displaystyle O}{\|}}{C}-CH_2CH_3$

(e) $\bigcirc\!\!-O-\overset{\overset{\displaystyle O}{\|}}{C}-CH_2CH_3$

(f) $HO-\overset{\overset{\displaystyle O}{\|}}{C}-\!\!\bigcirc\!\!-\overset{\overset{\displaystyle O}{\|}}{C}-O-\bigcirc$

(g) $CH_3-\overset{\overset{\displaystyle O}{\|}}{C}-CH_2-\overset{\overset{\displaystyle O}{\|}}{C}-CH_3$

(h) $CH_3O-\overset{\overset{\displaystyle O}{\|}}{C}-CH_2-\overset{\overset{\displaystyle O}{\|}}{C}-OCH_3$

(i) 

**14.37** Name the following carboxylic esters:

(a) $H-\overset{\overset{\displaystyle O}{\|}}{C}-O-CH_2CH_3$

(b) $CH_3CH_2-\overset{\overset{\displaystyle O}{\|}}{C}-O-CH_3$

(c) $CH_3CH_2\!-\!O\!-\!\underset{\underset{O}{\|}}{C}\!-\!CH_3$

(d) $CH_3(CH_2)_6\!-\!\underset{\underset{O}{\|}}{C}\!-\!O\!-\!\underset{\underset{CH_3}{|}}{CH}\!-\!CH_3$

(e) $CH_3CH_2\!-\!O\!-\!\underset{\underset{O}{\|}}{C}\!-\!CH_2\!-\!\underset{\underset{O}{\|}}{C}\!-\!OCH_2CH_3$

(f)

(g) $CH_3(CH_2)_{14}\!-\!\underset{\underset{O}{\|}}{C}\!-\!O\!-\!\underset{\underset{CH_3}{|}}{\overset{\overset{CH_3}{|}}{C}}\!-\!CH_3$

(h) $CH_3(CH_2)_5\!-\!\underset{\underset{O}{\|}}{C}\!-\!OCH_3$

**14.38** Draw structural formulas for
  (a)  methyl formate
  (b)  ethyl butyrate
  (c)  *n*-propyl propionate
  (d)  *n*-pentyl valerate
  (e)  diethyl succinate
  (f)  benzyl acetate
  (g)  phenyl 2-methylheptanoate

## Physical Properties of Carboxylic Esters

**14.39** Arrange in order of increasing boiling point:
  (a)  $CH_3CH_2COOH$
  (b)  $CH_3CH_2CH_2CH_2CH_3$
  (c)  $CH_3CH_2COO^-\ Na^+$
  (d)  $CH_3CH_2COOCH_3$
  (e)  $CH_3CH_2CH_2CH_2OH$
**14.40** Would you predict $CH_3CH_2CH_2COOH$ to be more soluble in water than $CH_3CH_2COOCH_3$, less soluble, or to have about the same solubility? Explain.

## Preparation of Carboxylic Esters; Acyl Halides; Anhydrides

**14.41** Show the structural formulas of the carboxylic esters obtained from a combination of
  (a)  $CH_3COOH + CH_3CH_2OH$

(b) $+ CH_3OH$

(c)  $HOCH_2CH_2CH_3 + CH_3\!-\!\underset{\underset{CH_3}{|}}{CH}\!-\!CH_2COOH$

(d)

(e)  $HOOC(CH_2)_3COOH +$ excess $CH_3CH_2OH$

(f)  $Br\!-\!CH_2CH_2COOH +$

**14.42** Tell whether each of the following compounds is a carboxylic acid, a carboxylic ester, an acyl chloride, an anhydride, or none of these:

(a)  $CH_3COOH$

(b)  $CH_3\!-\!\underset{\underset{O}{\|}}{C}\!-\!CH_2\!-\!Cl$

(c)  $CH_3CH_2\!-\!\underset{\underset{O}{\|}}{C}\!-\!Cl$

(d)

(e)  $CH_3\!-\!\underset{\underset{O}{\|}}{C}\!-\!O\!-\!\underset{\underset{O}{\|}}{C}\!-\!CH_3$

(f)  $CH_3\!-\!\underset{\underset{OCH_3}{|}}{CH}\!-\!CH_2\!-\!OCH_3$

(g)  $CH_3\!-\!\underset{\underset{Cl}{|}}{CH}\!-\!COOH$

(h)

(i)

$$H_2C \overset{\overset{\displaystyle H_2}{\underset{|}{C}}}{\underset{H_2C}{\overset{|}{\phantom{.}}}} \begin{array}{c} CH_2 \\ | \\ CH_2 \\ O \end{array}$$

(j) 2-hydroxybenzene with $\overset{OH}{\phantom{.}}$ and $\overset{}{\underset{O}{C}}-O-CH_2CH_3$

(k) $HOOC-CH_2CH_2-COOCH_3$

**14.43** Complete the following equations:

(a) benzene-COOH $+ CH_3CH_2OH \overset{H^+}{\longrightarrow}$

(b) $CH_3(CH_2)_6-\overset{\overset{\displaystyle }{\underset{\displaystyle O}{C}}}{\phantom{.}}-Cl + CH_3OH \longrightarrow$

(c) $CH_3-\overset{\overset{\displaystyle }{\underset{\displaystyle O}{C}}}{\phantom{.}}-O-\overset{\overset{\displaystyle }{\underset{\displaystyle O}{C}}}{\phantom{.}}-CH_3 +$

$$CH_3-\underset{\underset{OH}{|}}{CH}-CH_3 \longrightarrow$$

(d) cyclopentane-$\overset{\overset{\displaystyle }{\underset{\displaystyle O}{C}}}{\phantom{.}}-Cl$ $+$ benzene with Br and OH $\longrightarrow$

(e) $CH_3COOH + CH_3CH_2OH \overset{\text{no catalyst}}{\longrightarrow}$

(f) $2CH_3-\overset{\overset{\displaystyle }{\underset{\displaystyle O}{C}}}{\phantom{.}}-O-\overset{\overset{\displaystyle }{\underset{\displaystyle O}{C}}}{\phantom{.}}-CH_3 +$

$$HO-CH_2CH_2CH_2-OH \longrightarrow$$

## Hydrolysis of Carboxylic Esters

**14.44** Write equations for the hydrolysis of ethyl butyrate, $CH_3CH_2CH_2CH_2COOCH_2CH_3$, as catalyzed by (a) acid  (b) base

**14.45** Define saponification.

**14.46** Draw the structural formula for each product:

(a) $CH_3-\overset{\overset{\displaystyle }{\underset{\displaystyle O}{C}}}{\phantom{.}}-OCH_3 + H_2O \overset{OH^-}{\longrightarrow}$

(b) $CH_3-\overset{\overset{\displaystyle }{\underset{\displaystyle O}{C}}}{\phantom{.}}-OCH_3 + H_2O \overset{H^+}{\longrightarrow}$

(c) benzene-$O-\overset{\overset{\displaystyle }{\underset{\displaystyle O}{C}}}{\phantom{.}}-CH_2CH_3 + H_2O \overset{OH^-}{\longrightarrow}$

(d) benzene-$\overset{\overset{\displaystyle }{\underset{\displaystyle O}{C}}}{\phantom{.}}-O-CH_2CH_3 + H_2O \overset{H^+}{\longrightarrow}$

*(e) $CH_3-O-\overset{\overset{\displaystyle }{\underset{\displaystyle O}{C}}}{\phantom{.}}-CH_2-\overset{\overset{\displaystyle }{\underset{\displaystyle O}{C}}}{\phantom{.}}-OCH_3$

$$+ H_2O \overset{OH^-}{\longrightarrow}$$

*(f) $CH_3CH_2-\overset{\overset{\displaystyle }{\underset{\displaystyle O}{C}}}{\phantom{.}}-O-CH_2CH_2-O-\overset{\overset{\displaystyle }{\underset{\displaystyle O}{C}}}{\phantom{.}}-CH_2CH_3$

$$+ H_2O \overset{OH^-}{\longrightarrow}$$

(g) $CH_3(CH_2)_8-\overset{\overset{\displaystyle }{\underset{\displaystyle O}{C}}}{\phantom{.}}-O-CH_2-\underset{\underset{CH_3}{|}}{CH}-CH_3$

$$+ H_2O \overset{OH^-}{\longrightarrow}$$

*14.47 A lactone is an internal ester. Show the product when the following lactones are hydrolyzed:

(a)

$$H_2C \overset{\overset{\displaystyle H_2}{\underset{|}{C}}}{\underset{H_2C}{\overset{|}{\phantom{.}}}} \begin{array}{c} CH_2 \\ | \\ C=O \\ O \end{array} + H_2O \overset{H^+}{\longrightarrow}$$

(b)

$$H_2C \overset{\overset{\displaystyle H_2}{\underset{|}{C}}}{\underset{O}{\overset{|}{\phantom{.}}}} \begin{array}{c} CH_3 \\ | \\ CH \\ | \\ C=O \end{array} + H_2O \overset{OH^-}{\longrightarrow}$$

## Esters and Anhydrides of Phosphoric Acid

**14.48** (a) Draw the structural formulas for the mono-, di-, and triethyl esters of phosphoric acid. (b) How many protons can each of these esters give up?

*14.49 Dihydroxyacetone, $CH_2-\overset{\overset{\displaystyle }{\underset{\displaystyle O}{C}}}{\phantom{.}}-CH_2$, and phosphoric acid form an important monoester called dihydroxyacetone phosphate. Write the equation for the formation of dihydroxyacetone phosphate.

**14.50** Show how triphosphoric acid is formed from three molecules of phosphoric acid. How many $H_2O$ molecules are split out?

**14.51** Draw the structural formulas of (a) pyrophosphoric acid, (b) triphosphoric acid.

**14.52** Write equations showing the hydrolysis of (a) methyl triphosphate to methyl diphosphate, (b) methyl diphosphate to methyl monophosphate.

**14.53** Draw the formulas for the acid form and the salt form of the monoethyl ester of triphosphoric acid.

**14.54** Write an equation to show how pyrophosphoric acid transfers a phosphate group to ethanol.

## Boxes

**14.55** (Box 14A) Draw the structural formula for calcium propionate.

**14.56** (Box 14B) Show how to prepare the carboxylic esters that give orange flavor and pineapple flavor.

**14.57** (Box 14C) Why is aspirin, a synthetic compound, a more satisfactory pain-reliever than salicylic acid, a natural product?

**14.58** (Box 14C) Write an equation to show how aspirin is prepared from acetyl chloride rather than acetic anhydride.

**14.59** (Box 14C) Why do people who do not have headaches and pains regularly take small doses of aspirin?

**14.60** (Box 14D) How does a condensation polymer differ from an addition polymer?

**14.61** (Box 14D) Draw the structure of a polyester obtained from the condensation of fumaric acid and 1,6-hexanediol:

Fumaric acid

$$HO—CH_2CH_2CH_2CH_2CH_2CH_2—OH$$

1,6-Hexanediol

**14.62** (Box 14E) Why do Lactomer staples disappear from the body after three to four months?

**14.63** (Box 14E) (a) Could you form an internal ester (a lactone) from lactic acid or glycolic acid?

(b) What would be their structures? (c) If such compounds were formed, how could they be separated from the polymers?

## Additional Problems

**14.64** Draw the structural formula for each of the following derivatives of acetic acid: (a) the ethyl ester (b) the anhydride (c) the acyl chloride.

**14.65** Of the organic compounds we have studied so far, which classes are acidic? Rank them in order of increasing acidity.

**14.66** What is the most important property of carboxylic acids?

**\*14.67** Draw the structural formula for dimethyl sulfate.

**14.68** If someone wanted to hydrolyze the ester methyl propionate, $CH_3CH_2\overset{\text{O}}{\underset{\|}{C}}—O—CH_3$, would it be better to use acid or base catalysis?

**14.69** Write the full equations for preparing the following esters from alcohols and anhydrides or acyl chlorides:

(a)

(b)

**\*14.70** Would you expect a molecule of an anhydride (Sec. 14.10) to react with two molecules of an alcohol to produce two carboxylic ester molecules?

**\*14.71** Write the structural formula for any compound that is both an ester and an anhydride of phosphoric acid.

**14.72** You are given a solid that could be either a carboxylic acid or a carboxylic ester. Describe a simple test (less than 5 minutes) for telling which of these classes it belongs to.

**14.73** Which of these types of compounds will produce $CO_2$ bubbles when added to a solution of sodium bicarbonate in water? (a) a carboxylic acid (b) a carboxylate salt (c) a carboxylic ester.

# CHAPTER 15

Der Apotheker. *A pharmacist in the 16th century mixing herbs and other medicinal ingredients. (Courtesy of the National Library of Medicine.)*

# Amines and Amides

## 15.1 INTRODUCTION

In this chapter we deal with organic compounds containing nitrogen. The two most important classes of nitrogen-containing organic compounds are amines and amides.

## 15.2 NOMENCLATURE OF AMINES

**Amines** are **derivatives of ammonia.** Ammonia has three hydrogens, and we classify amines according to the number of hydrogens replaced by alkyl or aryl groups:

| | |
|---|---|
| $NH_3$ | Ammonia |
| R—$NH_2$ | **Primary amines** have one hydrogen replaced by R (*two* H's remain) |
| R—NH—R′ | **Secondary amines** have two hydrogens replaced by R (*one* H remains) |
| R—N—R′<br>    \|<br>    R″ | **Tertiary amines** have all three hydrogens replaced by R (*no* H remains) |

The R groups in secondary and tertiary amines may be the same or different.

Note that the use of primary, secondary, and tertiary for amines is not quite the same as for alcohols. For alcohols we look at how many R groups are on a *carbon;* for amines we look at how many are on the *nitrogen.* Thus, for example, *tert*-butylamine is a *primary* amine because there is only one R group on the nitrogen, even though the R group is a tertiary alkyl group:

$$CH_3-\underset{\underset{CH_3}{|}}{\overset{\overset{CH_3}{|}}{C}}-NH_2 \qquad \textit{tert-}\text{Butylamine (a primary amine)}$$

---

**PROBLEM 15.1** Classify each as a primary, secondary, or tertiary amine:

(a) $CH_3-\underset{\underset{CH_3}{|}}{N}-CH_3$ 　　(b) ⬡—$NHCH_3$ 　　(c) $CH_3-\underset{\underset{CH_3}{|}}{CH}-NH_2$

---

Simple primary amines are named by adding the suffix "-amine" to the name of the group:

$$CH_3NH_2 \qquad \text{Methylamine}$$

$$CH_3-\underset{\underset{CH_3}{|}}{CH}-NH_2 \qquad \text{Isopropylamine}$$

Recall (Sec. 11.10) that the simplest aromatic amine is aniline:

⬡—$NH_2$ 　　Aniline

Simple secondary and tertiary amines are named the same way as the primary amines. We merely place all the R groups before the suffix "-amine" and use "di-" and "tri-" as necessary:

Secondary amines

$$CH_3-NH-CH_3 \qquad \text{Dimethylamine}$$

$$CH_3CH_2-NH-CH\underset{\diagdown CH_3}{\overset{\diagup CH_3}{}} \qquad \text{Ethylisopropylamine}$$

⬡—$NH$—⬡ 　　Diphenylamine

Tertiary amines

$$CH_3CH_2-\underset{\underset{\underset{CH_3}{|}}{\underset{CH_2}{|}}}{N}-CH_2CH_3 \qquad \text{Triethylamine}$$

⬡—$CH_2-\underset{\underset{CH_3}{|}}{N}-CH_2CH_3$ 　　Benzylethylmethylamine

**PROBLEM 15.2**   Name:

(a)  $CH_3CH_2NH_2$

(b)  $CH_3—NH—CH_2CH_3$

(c)  $CH_3CH_2—\underset{\underset{CH_3}{|}}{N}—CH_2CH_3$

---

When one R is aromatic, the amine is generally named as a derivative of aniline:

*N*-Methylaniline

*meta*-Bromo-*N*,*N*-diethylaniline

The capital letter *N* is used to show that the group is attached to a nitrogen.

Do not confuse the capital letter *N* with the small *n*, which stands for *normal*.

— *N* - ethyl aniline

---

**PROBLEM 15.3**   Name as a derivative of aniline:

para - nitro

---

When it is necessary to name $—NH_2$ as a group, it is called the *amino group*.

3-Amino-2-bromocyclopentanol

The $—OH$ group takes precedence over the $—NH_2$ group in numbering the ring.

---

**PROBLEM 15.4**   Name:

1,4 di amin cyclohexane

Box 15A

## Amphetamines (Pep Pills)

Amphetamine (benzedrine) and methamphetamine (methedrine) are synthetic amines that are powerful stimulants of the central nervous system. They reduce fatigue and diminish hunger by raising the glucose level of the blood. Because of these properties, they are widely used medicinally to counter mild cases of depression, to reduce hyperactivity in children, and as an appetite depressant for people who are trying to lose weight.

Amphetamine (benzedrine)

Epinephrine (adrenalin)

Methamphetamine (methedrine)

The action of amphetamines is similar to that of epinephrine (adrenalin), which has a very similar structure. Epinephrine is a hormone (Sec. 24.6) secreted by a gland when an individual feels excitement or fear. It helps to make glucose available to tissues that need it immediately, for example, when leg muscles must be used to run away from danger.

Amphetamines are used both legally (as prescribed by a physician) and illegally to elevate moods or reduce fatigue. They are called "uppers" or "pep pills." Methamphetamine is known as "speed," "ice," or "crank." Amphetamines are sometimes used by truck drivers on long-distance runs to stay awake all night and by athletes who want more strength in a game or a race. Abuse of these drugs can have severe effects on both the body and the mind. They are addictive, concentrate in the brain and nervous system, and can lead to long periods of sleeplessness, loss of weight, and paranoia.

As with ethers, secondary and tertiary amines can be heterocyclic. Section 11.13 discusses aromatic (pyridine, pyrimidine, pyrrole) as well as nonaromatic heterocyclic amines (pyrrolidine, piperidine), and we shall meet others later in this chapter. Many of these compounds have important biological functions.

## 15.3   PHYSICAL PROPERTIES OF AMINES

The boiling points of amines are between those of the alcohols on the one hand and the hydrocarbons and halides on the other. For example, the boiling point of $n$-propylamine, $CH_3CH_2CH_2NH_2$, is 48°C (Table 15.1). This is about halfway between the boiling point of butane (0°C) and that of 1-propanol (97°C), two compounds of about the same molecular weight. The reason is that a molecule of a primary or a secondary amine forms a hydrogen bond with another molecule of itself, but these hydrogen bonds are weaker than those formed by alcohols. As can be seen from the boiling points listed in Table 15.1,

**TABLE 15.1  Boiling Points and Solubilities of Ammonia and Some Primary, Secondary, and Tertiary Amines**

| Name | Formula | Boiling Point (°C) | Solubility in Water |
|---|---|---|---|
| Ammonia | $NH_3$ | −33 | Soluble |
| Methylamine | $CH_3NH_2$ | −6 | Soluble |
| Ethylamine | $CH_3CH_2NH_2$ | 17 | Soluble |
| n-Propylamine | $CH_3CH_2CH_2NH_2$ | 48 | Soluble |
| n-Butylamine | $CH_3CH_2CH_2CH_2NH_2$ | 78 | Soluble |
| Aniline | [benzene ring]—$NH_2$ | 184 | Soluble |
| Dimethylamine | $CH_3$—$NH$—$CH_3$ | 7 | Soluble |
| Ethylmethylamine | $CH_3CH_2$—$NH$—$CH_3$ | 36 | Soluble |
| Diethylamine | $CH_3CH_2$—$NH$—$CH_2CH_3$ | 56 | Soluble |
| Piperidine | [six-membered ring with N—H] | 106 | Soluble |
| Trimethylamine | $CH_3$—$\overset{\overset{\displaystyle CH_3}{\mid}}{N}$—$CH_3$ | 3 | Soluble |
| Triethylamine | $CH_3CH_2$—$\overset{\overset{\displaystyle CH_2CH_3}{\mid}}{N}$—$CH_2CH_3$ | 89 | Soluble |
| N,N-Dimethylaniline | [benzene ring]—$\overset{\overset{\displaystyle CH_3}{\mid}}{N}$—$CH_3$ | 194 | InSoluble |
| Pyridine | [six-membered ring with N] | 115 | Soluble |

amines of up to two or three carbons are gases at room temperature. Amines are somewhat more soluble in water than alcohols of comparable molecular weight.

The simple amines have strong odors resembling that of ammonia but not quite as sharp and pungent. The odor of amines also resembles that of raw fish, which is not surprising, since raw fish contains low-molecular-weight amines. Some diamines have especially bad odors. Two of these are

$$\underset{NH_2}{CH_2}-CH_2-CH_2-\underset{NH_2}{CH_2} \qquad \underset{NH_2}{CH_2}-CH_2-CH_2-CH_2-\underset{NH_2}{CH_2}$$

Putrescine (1,4-diaminobutane)      Cadaverine (1,5-diaminopentane)      The names give a hint of the odor.

These compounds, which are among the end products in the decomposition of proteins, are found in decaying flesh.

## 15.4 THE BASICITY OF AMINES. QUATERNARY SALTS

**Amines are bases**—this is their most important chemical property. They are the only important organic bases, and we can be fairly certain that any organic base we meet is an amine.

We saw in Chapter 8 that ammonia is a weak base. Since the amines are derivatives of ammonia, they share this property. When amines act as bases, they react with acids by accepting a proton (as do all Brønsted-Lowry bases). In Section 8.6 we saw that methylamine reacts with an acid such as HCl to give its conjugate acid, the methylammonium ion. Most other amines, whether primary, secondary, or tertiary, behave exactly the same way; they too are converted to positive ions:

Aliphatic amines are somewhat stronger bases than ammonia; aromatic amines are weaker.

$$CH_3NH_2 + HCl \longrightarrow CH_3\overset{+}{N}H_3 \quad Cl^-$$

Methylammonium chloride

$$CH_3NHCH_3 + HNO_3 \longrightarrow CH_3\overset{+}{N}H_2CH_3 \quad NO_3^-$$

Dimethylammonium nitrate

$N,N$-Dimethylanilinium fluoride

Carboxylic acids are strong enough to react with amines (primary, secondary, or tertiary) in the same way, for example,

$$CH_3CH_2-NH_2 + CH_3COOH \longrightarrow CH_3CH_2-\overset{+}{N}H_3 \quad CH_3COO^-$$

Ethylamine          Acetic acid          Ethylammonium acetate

---

**PROBLEM 15.5** Draw the structural formulas for each salt formed:

(a) $CH_3CH_2CH_2NH_2 + HCOOH \longrightarrow$

(b) [benzene ring]—NHCH$_3$  + HBr $\longrightarrow$

---

The ionic compounds produced when amines react with acids are called **amine salts.** Like all salts, they are solids at room temperature. As can be seen in the examples just given, they are named by changing "-amine" to "-ammonium" (or "-aniline" to "-anilinium") and adding the name of the negative ion.

There is an older way to represent and name amine salts. Since these methods are still occasionally used, we will look at an example.

Modern way:

$$CH_3CH_2-\overset{CH_3}{\underset{|}{N}}H + HCl \longrightarrow CH_3CH_2-\overset{CH_3}{\underset{|}{\overset{+}{N}}}H_2 \quad Cl^-$$

Ethylmethylammonium chloride

**Figure 15.1**
Two drugs that are amine salts labeled as hydrochlorides. (Photograph by Beverly March.)

Drugs that are amine salts are commonly named this way (Fig. 15.1).

Older way:

$$\underset{\substack{| \\ CH_3CH_2-NH}}{\overset{CH_3}{|}} + HCl \longrightarrow \underset{\substack{| \\ CH_3CH_2-NH \cdot HCl}}{\overset{CH_3}{|}}$$

Ethylmethylamine hydrochloride

The older method does not show the correct structural formula.

In Section 14.6 we saw that the structure of a carboxylic acid depends on the pH. The same is true for amines, but when an amine (primary, secondary, or tertiary) is added to water, the pH is always above 7 because amines are basic. If we now add a strong acid like HCl, it neutralizes the amine (converts it to its salt) and lowers the pH. Thus,

an amine can exist in water only at a high pH; an amine salt can exist in water only at a low pH:

$$RNH_2 \underset{OH^-}{\overset{H_3O^+}{\rightleftharpoons}} \overset{+}{R}NH_3$$

High pH          Low pH

Secondary and tertiary amines behave the same way.

As with carboxylic acids, we can easily go back and forth between amines and their salts. If we want to convert an amine to the salt, we just add a strong acid. If we want to convert an amine salt to the free amine, we just add a strong base.

This property of amines and their salts is quite useful in laboratories. Low-molecular-weight amines are soluble in water, but the higher ones are insoluble. However, virtually all amine *salts* are soluble in water. This provides a simple method for separating an insoluble amine from other organic compounds that are insoluble in water but are not basic. We simply add the mixture to a dilute aqueous solution of an acid such as HCl. The amine dissolves because it is converted to its (soluble) salt by the acid. The rest of the mixture is insoluble and thus easily removed.

Recall that ammonium salts are also soluble in water (Sec. 4.8).

Another important use of this property is in administering drugs (see Box 15B). Because drugs that are amines are often given in the form of their salts (Fig. 15.1), the amine itself is often called the **free base.**

As mentioned in the text, many drugs show the designation ·HCl in their chemical formulas and occasionally also in the generic names. Drugs with such different pharmacological properties as Novocain·HCl (an anesthetic) (Box 15I), Neo-Synephrine·HCl or phenylephrine·HCl (a decongestant), oxytetracycline·HCl (an antibiotic), morphine sulfate, cocaine·HCl, and methadone·HCl (narcotics) (see Sec. 15.9) are all ammonium salts. The drugs themselves are high-molecular-weight amines that are insoluble in aqueous body fluids such as blood plasma, urine, and cerebrospinal fluid. In order for the administered drugs to be effective in the target organs or cells, they must be absorbed and carried by the body fluids. When HCl or $H_2SO_4$ is added to these amines, they are converted to their respective salts, which are water-soluble.

**B o x   15B**

# The Solubility of Drugs in Body Fluids

Secondary amines react with nitrous acid ($HNO_2$) to give *N*-nitrosamines,

$$RNH + HONO \longrightarrow R-N-NO$$

with $R'$ below the nitrogen on each side.

An *N*-nitrosamine

*N*-Nitrosamines are carcinogens. This has raised an important question about the use of nitrites as additives in meats. Small amounts of sodium nitrite are added to such meats as cold cuts and frankfurters because it preserves the pink color associated with fresh meat. Nitrite ion adds to the myoglobin in the meat and binds to the iron center. This prevents oxidation of the iron with its accompanying brownish color. When the nitrite ion reaches the stomach, the high concentration of HCl there converts the $NO_2^-$ to HONO, which can react with secondary amines in the digestive tract to produce *N*-nitrosamines. So far, no one has proved that this is harmful to humans or found a better way to preserve the pink color of the meat.

This is an example of a potentially important social problem arising from the use of a chemical, and it is not easily solved. If we cannot find another economical way to preserve the color, do we risk the possibility of increased cancer (which may or may not be real)? There are strong arguments on both sides.

### Quaternary Ammonium Salts

Earlier in this section we saw examples of salts formed from primary, secondary, and tertiary amines that have three, two, and one hydrogen on the nitrogen, respectively:

$$\overset{+}{R}NH_3 \qquad R_2\overset{+}{N}H_2 \qquad R_3\overset{+}{N}H$$

It is also possible to have ions with *four* alkyl or aryl groups on the nitrogen, for example,

$$\left[ CH_3CH_2-\overset{\overset{\displaystyle CH_3}{|}}{\underset{\underset{\displaystyle CH_3}{|}}{N^+}}-CH_2CH_3 \right] Cl^-$$

Diethyldimethylammonium chloride

Certain quaternary salts are important biological compounds. In later chapters we will meet choline (Sec. 17.5), acetylcholine (Sec. 24.3), and other quaternary ammonium ions.

Such salts are called **quaternary ammonium salts** and are named the same way as the other ammonium salts. A major difference between them and the other three types is that quaternary salts cannot be made by acidifying any amine, nor can a proton be removed from them by a base. Thus they have the same structure in water at all pH values. Otherwise, their properties are the same as those of the other three types of amine salts. Quaternary ammonium salts are usually made by treatment of a tertiary amine with an alkyl halide.

General reaction:

$$R-\overset{\overset{\displaystyle R}{|}}{\underset{\underset{\displaystyle R}{|}}{N}} + RCl \longrightarrow \left[ R-\overset{\overset{\displaystyle R}{|}}{\underset{\underset{\displaystyle R}{|}}{N^+}}-R \right] Cl^-$$

Specific example:

Dimethylaniline     Chloroethane          Ethyldimethylanilinium chloride

All four R groups may be the same or they may be different. Aryl halides do not give this reaction.

---

**PROBLEM 15.6**   Draw the structural formula for the product:

$$CH_3CH_2 - \overset{\overset{\displaystyle CH_3}{|}}{N} - CH_2CH_3 + CH_3CH_2CH_2Cl \longrightarrow$$

---

## 15.5   AMIDES: NOMENCLATURE AND PHYSICAL PROPERTIES

**Amides** are **compounds in which a carbonyl group is connected to a nitrogen atom.** The two remaining bonds of the nitrogen can be attached to hydrogens, alkyl groups, or aromatic rings:

Amides are carboxylic acid derivatives. Just as a carboxylic ester can be thought of as being composed of a carboxylic acid and an alcohol, so can an amide be thought of as being composed of a carboxylic acid and ammonia or a primary or secondary amine (why not a tertiary amine?):

*In Chapter 18 we will discuss proteins, which are polyamides.*

Like other carboxylic acid derivatives, amides are named after the corresponding acids. The ending "-ic acid" or "-oic acid" is changed to "-amide."

Aspirin (Box 14C) is a safe and effective nonprescription pain-reliever, but it does cause side effects, such as bleeding ulcers, stomach upset, or allergic reactions in a few people. For these people there are two other nonprescription pain-relieving drugs available. One of these is acetaminophen, which is sold under such brand names as Tylenol (the chemical name is *para*-hydroxyacetanilide). Acetaminophen has about the same effectiveness as aspirin and also reduces fever, but it does not reduce swelling.

Acetaminophen         Ibuprofen

The other nonprescription pain reliever, ibuprofen, is more recent. For some years it was available only by prescription, under such brand names as Motrin and Rufen. However, the Food and Drug Administration now permits its sale without a prescription, and it is available as Advil and Nuprin. Like aspirin, ibuprofen not only relieves pain, but also reduces joint inflammation associated with arthritis and athletic injuries, menstrual cramps, and fever. Both aspirin and ibuprofen act by inhibiting the synthesis of prostaglandins (Sec. 17.12). Ibuprofen has fewer side effects than aspirin.

**EXAMPLE**

Acetamide     derived from     Acetic acid

Benzamide     derived from     Benzoic acid

**PROBLEM 15.7**   Name:

$$CH_3CH_2CH_2-\underset{\underset{O}{\|}}{C}-NH_2$$

When the nitrogen of an amide is connected to an alkyl or aryl group, the group is named as a prefix preceded by the letter *N*.

---

**EXAMPLE**

$$H-\underset{\underset{O}{\|}}{C}-NH-CH_3 \qquad \text{derived from} \qquad H-\underset{\underset{O}{\|}}{C}-OH$$

*N*-Methylformamide                                    Formic acid

$$CH_3CH_2-\underset{\underset{O}{\|}}{C}-\underset{\underset{CH_2CH_3}{|}}{N}-CH_2CH_3 \qquad \text{derived from} \qquad CH_3CH_2-\underset{\underset{O}{\|}}{C}-OH$$

*N,N*-Diethylpropionamide                                Propionic acid

$$CH_3CH_2CH_2-\underset{\underset{O}{\|}}{C}-\underset{\underset{CH_3}{|}}{N}-\bigcirc \qquad \text{derived from} \qquad CH_3CH_2CH_2-\underset{\underset{O}{\|}}{C}-OH$$

*N*-Methyl-*N*-phenylbutyramide                            Butyric acid

The IUPAC name is *N*-methyl-*N*-phenylbutanamide, derived from butanoic acid.

---

**PROBLEM 15.8**  Name:

(a) $CH_3CH_2CH_2CH_2-\underset{\underset{O}{\|}}{C}-N\bigcirc$   with phenyl and benzene substituents

(b) $Br-\bigcirc-\underset{\underset{O}{\|}}{C}-NH-CH_2CH_3$

---

Amides (except for those with no hydrogen atom attached to the nitrogen) form strong hydrogen bonds. The smallest amide, formamide, and some of its *N*-substituted derivatives are liquids, but all other amides are solids at room temperature.

## 15.6  ACID–BASE PROPERTIES OF ORGANIC COMPOUNDS. THE NEUTRALITY OF AMIDES

Because basicity is the most important property of amines, it is only natural to wonder if amides are also basic. The answer is no. Amides are neither acidic nor basic; they are neutral.

At this point, it is good to review the acid–base properties of the organic compounds we have studied. Table 15.2 lists the oxygen-, nitrogen-, and sulfur-containing compounds we have learned about. We can assume that any organic compound that contains only carbon, hydrogen, and/or halogens is neither acidic nor basic.

These listings are according to the Arrhenius definitions of acid and base (Sec. 8.1) and not the Brønsted–Lowry definitions (Sec. 8.3).

**TABLE 15.2  Acid-Base Properties of Organic Compounds**

| Acidic | Basic | Neutral |
|---|---|---|
| Sulfonic acids (strong) | Amines (weak) | Alcohols |
| Carboxylic acids (weak) | | Ethers |
| Phenols (very weak) | | Thioethers |
| Thiols (very weak) | | Aldehydes |
| | | Ketones |
| | | Carboxylic esters |
| | | Acyl chlorides |
| | | Anhydrides |
| | | Amides |

## 15.7  PREPARATION OF AMIDES

In Section 15.5 we saw that amides can be thought of as having a carboxylic acid portion and an ammonia or amine portion. However, amides cannot be made simply by adding ammonia or an amine to a carboxylic acid. Why not? Because, as we saw in Section 15.4, when we add ammonia or an amine to a carboxylic acid, we get a salt:

$$R-\underset{\underset{O}{\|}}{C}-OH + NH_3 \longrightarrow R-\underset{\underset{O}{\|}}{C}-O^- NH_4^+$$

We can get around this problem by starting with the acyl chloride or anhydride instead of the carboxylic acid (similar to the preparation of carboxylic esters, Sec. 14.10). Because acyl chlorides and anhydrides are neutral compounds, salts do not form.

General reactions:

$$R-\underset{\underset{O}{\|}}{C}-Cl \quad + \quad R'NH_2 \quad \longrightarrow \quad R-\underset{\underset{O}{\|}}{C}-NH-R' + HCl$$

Acyl chloride          Primary amine

$$R-\underset{\underset{O}{\|}}{C}-O-\underset{\underset{O}{\|}}{C}-R'' \quad + \quad R_2'NH \quad \longrightarrow \quad R-\underset{\underset{O}{\|}}{C}-NR_2' + R''COOH$$

Anhydride                Secondary amine

Specific examples:

Benzamide

N-Phenylacetamide
(acetanilide)

In Box 14D we met the polyesters, which are important synthetic fibers. Synthetic polyamides, especially nylon, are even more important.

Nylon is used to make shirts, dresses, stockings, underwear, and other kinds of clothing, as well as carpets, tire cord, rope, and parachutes. It also has nonfabric uses; for example, it is used in paint brushes, electrical parts, valves, gears, clips, and fasteners (Fig. 15E.1). It is very tough and strong, nontoxic, nonflammable, and resistant to chemicals.

**Figure 15E.1**
Articles made of nylon. (Courtesy of E. I. Dupont de Nemours & Company, Inc.)

**Figure 15E.2**
The formation of nylon 66. Hexamethylenediamine is dissolved in water (bottom layer), and adipoyl chloride is dissolved in hexane (top layer). The two compounds meet at the interface between the two layers to form nylon, which is being wound onto a stirring rod. (Photograph by Charles D. Winters.)

There are several types of nylon, but all are polyamides. The most important is made from hexamethylenediamine and adipoyl chloride (Fig. 15E.2). Like polyesters, polyamides are condensation polymers; each molecule has two ends that are "tied" together by amide linkages:

$$
-\overset{\underset{|}{H}}{\underset{H}{N}}-H \quad Cl-\overset{\underset{\|}{O}}{C}-CH_2CH_2CH_2CH_2-\overset{\underset{\|}{O}}{C}-Cl \quad H-\overset{\underset{|}{H}}{N}-CH_2CH_2CH_2CH_2CH_2CH_2-\overset{\underset{|}{H}}{N}-H \quad Cl-\overset{\underset{\|}{O}}{C}-CH_2CH_2-
$$

Adipoyl chloride        Hexamethylenediamine
(1,6-diaminohexane)

$\downarrow$ $-HCl$

$$
-\overset{\underset{|}{H}}{\underset{}{N}}-\overset{\underset{\|}{O}}{C}-CH_2CH_2CH_2CH_2-\overset{\underset{\|}{O}}{C}-\overset{\underset{|}{H}}{N}-CH_2CH_2CH_2CH_2CH_2CH_2-\overset{\underset{|}{H}}{N}-\overset{\underset{\|}{O}}{C}-CH_2CH_2-
$$

Polyamide (nylon 66)

An HCl molecule comes out every time an amide linkage is created. The nylon formed from adipoyl chloride and hexamethylenediamine is called nylon 66 because the acid and amine each have six carbons. Other nylons can be made from other dicarboxylic acid chlorides and diamines.

| Box 15F | Urea is a *diamide* of carbonic acid: |
|---|---|

$$HO-\underset{\underset{O}{\|}}{C}-OH \qquad NH_2-\underset{\underset{O}{\|}}{C}-NH_2$$

Carbonic acid           Urea

### Urea

It is an important biological compound that is soluble in water. Humans and other higher animals get rid of their waste nitrogen by converting it to urea and excreting it in the urine (Sec. 21.8).

Urea has important industrial uses, the most important being as a fertilizer. When added to the soil, it is slowly hydrolyzed to $CO_2$ and $NH_3$.

Tertiary amines do not undergo this reaction because they have no N—H bond.

Both acyl chlorides and anhydrides react readily with ammonia and with primary and secondary amines.

---

**PROBLEM 15.9**   Draw the structural formula for each product:

(a) $CH_3CH_2-\underset{\underset{O}{\|}}{C}-Cl + NH_3 \longrightarrow$

(b) $CH_3CH_2-\underset{\underset{O}{\|}}{C}-O-\underset{\underset{O}{\|}}{C}-CH_2CH_3 + CH_3-\underset{\underset{CH_3}{|}}{CH}-NH-CH_3 \longrightarrow$

---

## 15.8   HYDROLYSIS OF AMIDES

The most important reaction of amides is hydrolysis. This reaction is analogous to the hydrolysis of carboxylic esters; that is, the amide is cleaved into two pieces: the carboxylic acid part and the amine or ammonia part:

$$R-\underset{\underset{O}{\|}}{C}-NR_2' \xrightarrow{H_2O} R-\underset{\underset{O}{\|}}{C}-OH + R_2'NH$$

As with carboxylic ester hydrolysis, amide hydrolysis requires the presence of a strong acid or base, but there is a difference. In Section 14.11 we saw that base-catalyzed ester hydrolysis is not reversible (because the carboxylate *ion* is formed) but that acid-catalyzed ester hydrolysis is reversible because the carboxylic acid and alcohol that are produced can recombine to give the ester. With amides neither reaction is reversible because ions are formed in *both* cases. Let us look at the hydrolysis of a mono-*N*-substituted amide with both acid and base catalysis.

General reactions:

The H$^+$, of course, is added in the form of an acid, such as HCl, and the R'NH$_3^+$ will actually be the salt R'NH$_3$Cl. Similarly, in the basic reaction, the OH$^-$ may be in the form of NaOH, and the product will be the salt RCOONa.

$$R-\underset{\underset{O}{\|}}{C}-NHR' + H_2O \xrightarrow{H^+} R-\underset{\underset{O}{\|}}{C}-OH + R'\overset{+}{N}H_3$$

$$R-\underset{\underset{O}{\|}}{C}-NHR' + H_2O \xrightarrow{OH^-} R-\underset{\underset{O}{\|}}{C}-O^- + R'NH_2$$

In 1864 Adolph von Baeyer (1835–1917) added urea to diethyl malonate (the diethyl ester of malonic acid) in the presence of ethoxide ion ($CH_3CH_2O^-$). The product was a cyclic compound that Baeyer called barbituric acid:

Diethyl malonate          Urea                          Barbituric acid

As you can see, this compound is a cyclic amide. It was later discovered that certain derivatives of barbituric acid had powerful physiological effects. They could put people to sleep. Barbituric acid itself is not used for this purpose, but a number of derivatives, substituted at the $CH_2$ group, are important drugs. Some of them are

Pentothal                    Pentobarbital (Nembutal)

Secobarbital (Seconal)

To increase solubility in water, these drugs are often given in the form of their salts. Technically, only the salts should be called **barbiturates,** but in practice, all of these compounds are given that name, whether in the free form or the salt form.

Barbiturates have two principal effects. In small doses they are sedatives (tranquilizers); in larger doses they induce sleep. The duration of the effects varies. Pentothal acts within seconds, and its effect wears off quickly. It is used as a general anesthetic by dentists. Pentobarbital and secobarbital are longer-acting drugs (3 to 4 hours). They are often used to prepare patients for surgery.

Barbiturates are addictive. Because they act as tranquilizers, they are called "downers" (the opposite of amphetamines, Box 15A) or "goof balls." Barbiturates are responsible for many deaths. Addicts often find they need larger and larger doses to achieve the same effect, and sometimes they take too much. Barbiturates are especially dangerous when combined with alcohol, as they often are. The combined effect (called a **synergistic effect**) is usually greater than the sum of the effects from the two drugs taken separately.

Specific examples:

$$CH_3CH_2-\underset{\underset{O}{\|}}{C}-NH-CH_3 + H_2O \xrightarrow{H^+} CH_3CH_2-\underset{\underset{O}{\|}}{C}-OH + CH_3\overset{+}{N}H_3$$

*N*-Methylpropionamide        Propionic acid    Methylammonium ion

$$CH_3CH_2-\underset{\underset{O}{\|}}{C}-NH-CH_3 + H_2O \xrightarrow{OH^-} CH_3CH_2-\underset{\underset{O}{\|}}{C}-O^- + CH_3NH_2$$

*N*-Methylpropionamide        Propionate ion    Methylamine

Since the products of the hydrolysis are a carboxylic acid and an amine (or ammonia), *one of them must always be a salt.*

---

**EXAMPLE**

Show the products of acid-catalyzed and base-catalyzed hydrolysis of

(a)

Benzamide

(b) $CH_3-\underset{\underset{O}{\|}}{C}-\underset{\underset{CH_3}{|}}{N}-CH_3$    *N,N*-Dimethylacetamide

**Answer**

In both types of hydrolysis, cleavage takes place between the C=O carbon and the N:

(a)

(b) $CH_3-\underset{\underset{O}{\|}}{C}-\underset{\underset{CH_3}{|}}{N}-CH_3 + H_2O \xrightarrow{H^+} CH_3-\underset{\underset{O}{\|}}{C}-OH + CH_3-\underset{\underset{CH_3}{|}}{\overset{+}{N}}H_2$

$CH_3-\underset{\underset{O}{\|}}{C}-\underset{\underset{CH_3}{|}}{N}-CH_3 + H_2O \xrightarrow{OH^-} CH_3-\underset{\underset{O}{\|}}{C}-O^- + CH_3-\underset{\underset{CH_3}{|}}{N}H$

In our modern society the pressures and conflicts of daily life are much greater than they were in earlier times, when everything moved at a slower tempo. Drugs called **tranquilizers** are prescribed by physicians for the relief of anxiety disorders or for the short-term relief of the symptoms of anxiety or tension.

Most of these drugs belong to a family called the benzodiazepines, all of which have similar structures. The most popular is diazepam (Valium), but lorazepam, clorazepate, and chlordiazepoxide are also frequently prescribed. These drugs all have a seven-membered ring containing two nitrogens:

Diazepam (Valium)

Lorazepam (Ativan)

Clorazepate dipotassium
(Tranxene)

Chlordiazepoxide hydrochloride
(Librium)

These compounds are useful for relieving anxiety symptoms, but they are all habit-forming and must be used with care.

**PROBLEM 15.10** Show the products of acid-catalyzed and base-catalyzed hydrolysis of $CH_3CH_2CH_2-\overset{\displaystyle O}{\underset{\displaystyle \|}{C}}-NH-CH_3$.

The hydrolysis of amides is a very important reaction in biochemistry. Most amide hydrolysis in the body is catalyzed by enzymes rather than by acids or bases. Enzymes are much better catalysts, and biological amide hydrolysis is a very rapid reaction.

## 15.9 ALKALOIDS

For thousands of years, primitive tribes in various parts of the world have known that physiological effects can be obtained by eating or chewing the

Chewing the bark of the cinchona tree, found in the Andes, is an effective treatment for malaria (it cured the Spanish viceroy to Peru in 1638).

leaves, roots, or bark of certain plants. The effects vary with the plant. Some cure disease; others, such as opium, are addictive drugs. Still others are deadly poisons, such as the leaves of the belladonna plant.

The nature of the compounds that cause these effects was an interesting problem for chemists, and beginning around 1800, when analytical methods became advanced enough for the task, they began to extract the compounds responsible. They discovered that all of these compounds are weak bases and called them alkaloids (meaning alkali-like). Today we define an **alkaloid** as **a basic compound obtained from a plant.** Many of them have physiological activity of one kind or another. They are very widespread in nature, and their structures vary widely. We have already seen that virtually all organic bases are amines, and so we are not surprised to learn that all alkaloids are amines.

Thousands of different alkaloids have been extracted from the leaves, bark, roots, flowers, and fruits of plants.

We will look at a few of the more important alkaloids. Note that they generally have common names, usually taken from the name of the plant from which they are extracted, and, like other amines, their names end in "-ine."

## Coniine

One of the simplest alkaloids, coniine,

Coniine (2-propylpiperidine)

is very poisonous. It is found in the hemlock plant. The great philosopher Socrates (469–399 BC) committed suicide by drinking a cup of hemlock extract after he was found guilty of corrupting the youth of Athens (Fig. 15.2).

**Figure 15.2**
*The Death of Socrates*, as painted by the French artist Jacques David (1748–1825) in 1787. Socrates was sentenced to death but could have escaped. He chose to drink the cup of hemlock. (Courtesy of the Metropolitan Museum of Art, New York.)

## Nicotine

The active ingredient in tobacco is the alkaloid nicotine:

Nicotine

Nicotine has two heterocyclic rings— one aromatic (a pyridine ring) and one aliphatic (a pyrrolidine ring).

Many people ingest it by smoking or chewing. In small doses nicotine is a stimulant. It is not especially harmful in itself, but it is habit-forming and thus exposes smokers to the harmful effects of the other components of cigarette smoke (tars, carbon monoxide, the carcinogens mentioned in Box 11F, and other poisons).

In large doses nicotine is a poison. Solutions of it are used as an insecticide.

## Caffeine

As you may already know, caffeine is the compound that is mainly responsible for the stimulating action of coffee and tea. It is also present in chocolate, cocoa, and colas, and "stay-awake" pills (No-Doz). It is not only an alkaloid but also an amide (can you see why?) and a purine because it has the same ring system as the compound purine:

Caffeine                Purine

We shall meet other purines in Chapter 23.

## Atropine

Several poisonous plants, including henbane and belladonna, contain atropine:

Atropine

# Cocaine

Cocaine has a formula very similar to that of atropine:

Cocaine

The leaves of the coca bush, found in Peru and Bolivia, are rich in cocaine, and the local inhabitants have chewed them for centuries to obtain its stimulating action. Cocaine reduces fatigue, permits greater physical endurance, and gives a feeling of tremendous confidence and power. In some of the Sherlock Holmes stories, the great detective injects himself with a 7 percent solution of cocaine to overcome boredom. When the addictive property of cocaine became apparent, it was made illegal but is still in widespread use. In recent years a cheap, smokable form of cocaine, known as "crack," has created many new addicts and caused enormous social and criminal problems. A particularly unfortunate example of this is that the addiction of pregnant women to crack is often transferred to their babies.

In the nineteenth century cocaine was used as a local anesthetic for such things as tooth extractions (Fig. 15I). Because of its toxicity, however, it was replaced by synthetic compounds. The first of these was procaine hydrochloride (Novocain), but this has now largely been replaced for dental use by lidocaine (Xylocaine).

Procaine hydrochloride
(Novocain)

Lidocaine
(Xylocaine)

**Figure 15I**
Nineteenth-century advertisement for cocaine. (National Library of Medicine.)

Dilute solutions of atropine have been used for thousands of years to dilate the pupils of the eyes (make them larger). The name "belladonna" (which means beautiful lady) probably comes from the fact that Roman women used extracts from this plant to make themselves more attractive.

The American writer Ambrose Bierce defined "belladonna" in his satirical *Devil's Dictionary* as follows: "In Italian a beautiful lady; in English a deadly poison. A striking example of the essential identity of the two tongues."

## Morphine

Opium, the dried latex of the poppy plant (Fig. 15.3), has been used as a drug for centuries. It contains a number of alkaloids, the most important of which are morphine and its methyl ether, codeine.

Morphine            Codeine            Heroin

Morphine is a valuable drug in medicine, being one of the most effective pain-killers known. Drugs of this type are called *narcotics*. The major drawback of these drugs is that they are addictive. Heroin is even more harmful than morphine. Heroin is not found in nature, but is made from morphine by esterification of the OH groups with acetic anhydride (Sec. 14.10). The ester groups make the heroin less polar, and as a consequence it enters the brain more rapidly than morphine. Once there, however, it is converted back to morphine by ester hydrolysis and is trapped in the brain cells. The use of heroin is illegal, and morphine is restricted to hospital use. Codeine is sometimes used in cough syrups to decrease throat discomfort.

Chemists today are still searching for safe, effective substitutes for these compounds. A purely synthetic substitute for the pain-killing effect of morphine is Demerol, and this drug is widely prescribed as a pain-killer. Methadone, another synthetic compound, is used as a substitute for heroin to aid in withdrawal from that drug. However, methadone is itself addictive and cannot represent a complete solution to the problem of heroin addiction.

In general, nonpolar molecules pass from blood vessels into the brain more easily than polar molecules, because they have to cross the nonpolar membranes of the endothelial cells surrounding the blood vessels. (See the blood-brain barrier, Sec. 25.1.)

Demerol            Methadone

**Figure 15.3**
A poppy plant. (FourByFive.)

## Other Alkaloids

Other alkaloids worth mentioning are quinine, strychnine, and reserpine, all of which have complicated formulas with several rings:

Quinine

Strychnine

Reserpine

We mentioned at the beginning of this section that the bark of the cinchona tree is used to treat malaria. *Quinine,* which is extracted from this bark, is the actual compound responsible and is still used today for this purpose. *Strychnine,* which comes from the seeds of a plant called *Strychnos nux vomica,* has been used for many years as a poison for rats (sometimes for humans, too). It is also used medicinally, in low dosage, to counteract poisoning by central nervous system depressants. *Reserpine,* extracted from a shrub called Indian snake root, is used as a tranquilizer for patients in mental hospitals and as a drug that lowers blood pressure.

## SUMMARY

**Amines** are derivatives of ammonia, $NH_3$. They are classified as **primary ($RNH_2$), secondary ($RNHR'$), and tertiary ($RR'NR''$)**, depending on how many R groups are attached to the nitrogen. They are named by adding the suffix "-amine" to the name of the R groups. Their physical properties are similar to those of alcohols and ethers. The most important chemical property of amines is their basicity. They form salts with acids. Since these salts are soluble in water, even insoluble amines can be dissolved by treatment with acids. **Quaternary ammonium ions, $R_4N^+$,** have four groups attached to the nitrogen and a positive charge. Tertiary amines react with alkyl halides to give quaternary ammonium salts.

**Amides** are carboxylic acid derivatives that are combinations of a carboxylic acid and ammonia or a primary or secondary amine. They are named by changing "-ic acid" to "-amide," with an "N-" prefix to show any groups attached to the nitrogen. All amides are neutral compounds, and almost all are solids. They are made by treating ammonia or a primary or secondary amine with an acyl halide or an anhydride. They cannot be made by simply adding a carboxylic acid to ammonia or an amine (a salt is formed instead). Amides can be hydrolyzed in either acidic or basic solution.

**Alkaloids** are amines isolated from plants. Thousands are known, with widely varying structures ranging from simple to very complex. Many of them have physiological activity of various sorts. For example, quinine cures malaria, strychnine is a poison, morphine reduces severe pain, and nicotine is a stimulant.

### Summary of Reactions

1. Reaction of amines with acids (Sec. 15.4):

   Primary $\quad RNH_2 + HCl \longrightarrow R\overset{+}{N}H_3\ Cl^-$

   Secondary $\quad R_2NH + R'COOH \longrightarrow$
   $$R_2\overset{+}{N}H_2\ R'COO^-$$

   Tertiary $\quad R_3N + HNO_3 \longrightarrow R_3\overset{+}{N}H\ NO_3^-$

2. Reaction of tertiary amines with alkyl halides (Sec. 15.4):

   $$R_3N + R'Cl \longrightarrow R-\overset{\overset{\displaystyle R}{|}}{\underset{\underset{\displaystyle R}{|}}{N^+}}-R'\ Cl^-$$

3. Preparation of amides (Sec. 15.7):
   a. From acyl chlorides and ammonia or amines:

   $$R-\underset{\underset{\displaystyle O}{\|}}{C}-Cl + R_2'NH \longrightarrow R-\underset{\underset{\displaystyle O}{\|}}{C}-NR_2' + HCl$$

   b. From anhydrides and ammonia or amines:

   $$R-\underset{\underset{\displaystyle O}{\|}}{C}-O-\underset{\underset{\displaystyle O}{\|}}{C}-R + NH_3 \longrightarrow$$
   $$R-\underset{\underset{\displaystyle O}{\|}}{C}-NH_2 + R-\underset{\underset{\displaystyle O}{\|}}{C}-OH$$

4. Hydrolysis of amides (Sec. 15.8):
   a. Acid-catalyzed:

   $$R-\underset{\underset{\displaystyle O}{\|}}{C}-NH_2 + H_2O \xoverset{H^+}{\longrightarrow}$$
   $$R-\underset{\underset{\displaystyle O}{\|}}{C}-OH + NH_4^+$$

   b. Base-catalyzed:

   $$R-\underset{\underset{\displaystyle O}{\|}}{C}-NHR' + H_2O \xoverset{OH^-}{\longrightarrow}$$
   $$R-\underset{\underset{\displaystyle O}{\|}}{C}-O^- + R'NH_2$$

## KEY TERMS

Alkaloid (Sec. 15.9)
Amide (Sec. 15.5)
Amine (Sec. 15.2)
Amine salt (Sec. 15.4)
Amphetamine (Box 15A)
Barbiturate (Box 15G)
Free base (Sec. 15.4)
Narcotic (Sec. 15.9)

Nitrosamine (Box 15C)
Polyamide (Box 15E)
Primary amine (Sec. 15.2)
Quaternary ammonium salt (Sec. 15.4)
Secondary amine (Sec. 15.2)
Synergistic effect (Box 15G)
Tertiary amine (Sec. 15.2)
Tranquilizer (Box 15H)

## PROBLEMS

**15.11** What inorganic compound are amines derivatives of?

### Nomenclature of Amines

**15.12** What is the difference between primary, secondary, and tertiary amines? Compare with the difference between primary, secondary, and tertiary alcohols.

**15.13** In a secondary amine, how many hydrogens are attached to the nitrogen atom?

**15.14** Classify each of the following as primary, secondary, or tertiary amines:

(a) $CH_3-CH-NH_2$
  $|$
  $CH_3$

(b) $CH_3-CH-N-CH_2CH_3$
  $|$    $|$
  $CH_3$   $CH-CH_3$
       $|$
       $CH_3$

(c)
  $CH_3$
  $|$
  (ring)—N—(ring)

(d) $CH_3-\overset{\displaystyle CH_3}{\underset{\displaystyle CH_3}{C}}-CH_2-NH_2$

(e) $CH_3-\overset{\displaystyle CH_3}{\underset{\displaystyle CH_3}{C}}-NH-CH_3$

(f)
  (pyrrolidine ring)
  N
  $|$
  H

(g)
  (pyrrolidine ring)
  N
  $|$
  $CH_3$

**15.15** Name

(a) $CH_3CH_2NH_2$

(b) $CH_3-\overset{\displaystyle CH_3}{\underset{\displaystyle CH_3}{C}}-NH_2$

(c) $CH_3CH_2CH_2-NH-CH_3$

(d) $CH_3CH-NH-CH_2CH_2CH_3$
       $|$
       $CH_3$

(e)
  (triphenylamine: three benzene rings attached to N)

(f)
  (benzene ring)—$NH_2$

(g)
  (benzene ring with $NH_2$ and Br)

(h)
  (benzene ring)—$NH-CH_2CH_3$

(i)
  $NO_2$—(benzene ring)—$\overset{\displaystyle N-CH_3}{\underset{\displaystyle CH_3}{}}$

(j)
  (cyclohexane ring)—$NH_2$

(k) $CH_3$—(cyclohexane ring with $NH_2$)—$CH_3$

**15.16** Tell whether each amine in Problem 15.15 is primary, secondary, or tertiary.

**15.17** Which of the following contain (1) heterocyclic aromatic rings, (2) heterocyclic nonaromatic rings, (3) no heterocyclic rings at all:
  (a) nicotine          (f) methadone
  (b) codeine           (g) coniine
  (c) Novocain          (h) secobarbital
  (d) caffeine          (i) urea
  (e) quinine           (j) Demerol

**15.18** Draw the structural formula for
  (a) methylamine
  (b) di-*n*-butylamine
  (c) triethylamine
  (d) *m*-ethylaniline
  (e) *N*-ethylaniline
  (f) *sec*-butylisobutylamine
  (g) putrescine
  (h) 1,3,5-triaminopentane

## Physical Properties of Amines

**15.19** Without consulting any tables, arrange the following in order of increasing boiling point:
  (a) $CH_3CH_2CH_2CH_2NH_2$
  (b) $CH_3CH_2—O—CH_2CH_3$
  (c) $CH_3CH_2CH_2CH_2OH$
  (d) $CH_3CH_2—O—CH_3$

**15.20** Explain why amines have about the same solubility in water as alcohols of the same molecular weights.

**15.21** *n*-Propylamine $(CH_3CH_2CH_2NH_2)$, ethylmethylamine $(CH_3CH_2NHCH_3)$, and trimethylamine $\left( CH_3—\overset{\displaystyle CH_3}{\underset{}{N}}—CH_3 \right)$ all have the same molecular weight (they are isomers). Yet trimethylamine has a much lower boiling point than the other two (Table 15.1). Explain why.

## Basicity of Amines. Quaternary Salts

**15.22** Write the structural formulas for the principal products (if no reaction, say so).
  (a) $CH_3CH_2NH_2 + HCl \longrightarrow$
  (b) [benzene ring]$—NH_2$ $+ CH_3COOH \longrightarrow$

  (c) $CH_3\overset{\displaystyle CH_3}{\underset{}{CH}}—NH—CH_3 + H_2SO_4 \longrightarrow$

  (d) $CH_3—\overset{\displaystyle CH_3}{\underset{}{N}}—CH_3 + H_3PO_4 \longrightarrow$

  (e) $CH_3—\overset{\displaystyle CH_3}{\underset{\displaystyle CH_3}{N}}—CH_3 +$ [benzene ring]$—COOH \longrightarrow$

**15.23** Name these salts:
  (a) $CH_3CH_2—NH_3^+ \quad Br^-$
  (b) $CH_3CH_2—\overset{+}{N}H_2—CH_3 \quad CH_3COO^-$
  (c) [benzene ring]$\overset{+}{N}H_2—CH_2CH_3 \quad Cl^-$
  (d) $CH_3—\overset{\displaystyle CH_3}{\underset{\displaystyle CH_3}{\overset{+}{N}}}—CH_3 \quad OH^-$

**15.24** Draw the structural formulas for
  (a) dimethylammonium iodide
  (b) ethyltrimethylammonium hydroxide
  (c) tetraethylammonium chloride
  (d) anilinium nitrate

**15.25** If you dissolved $CH_3CH_2CH_2NH_2$ (*n*-propylamine) and $CH_3CH_2CH_2OH$ (*n*-propyl alcohol) in the same container of water and lowered the pH to 2 (by the addition of HCl), would anything happen to the structures of these compounds? If so, write the formulas for the species in solution at pH 2.

**15.26** A student looking in an old chemistry book found the following name and formula:

$$CH_3—\overset{\displaystyle CH_3}{\underset{}{CH}}—NH_2 \cdot HBr$$

Isopropylamine hydrobromide

What are the modern name and formula for this compound?

**15.27** The tertiary amine *N*,*N*-diethylaniline is insoluble in water. How could you dissolve it?

**15.28** How could you convert ethyl-*n*-propylamine, $CH_3CH_2—NH—CH_2CH_3$, to its salt?

**15.29** How could you make the following conversion?

$$CH_3-\underset{\underset{CH_3}{|}}{N}-CH_2CH_3 \longrightarrow$$

$$CH_3-\underset{\underset{CH_3}{|}}{\overset{\overset{CH_2CH_3}{|}}{\overset{+}{N}}}-CH_2CH_3 \quad Cl^-$$

**15.30** Write the structural formulas for the principal organic products (if no reaction say so):

(a) $CH_3CH_2-\underset{\underset{CH_2CH_3}{|}}{N}-CH_2CH_3 + CH_3I \longrightarrow$

(b) 

$+ CH_3CH_2Br \longrightarrow$

*(c) 

$+ CH_3Cl \longrightarrow$

**15.31** Tell which reagents and/or catalysts can be used to bring about the following conversions:

(a) $CH_3Br \longrightarrow CH_3-\underset{\underset{CH_3}{|}}{\overset{\overset{CH_3}{|}}{\overset{+}{N}}}-CH_3 \quad Br^-$

(b) 

$\longrightarrow$

**15.32** If you needed a sample of

$$CH_3CH_2-\underset{\underset{CH_3}{|}}{\overset{\overset{CH_3}{|}}{\overset{+}{N}}}-CH_3 \quad Cl^-$$

how could you make it in the laboratory?

## Structure and Nomenclature of Amides

**15.33** How does the structure of an amide differ from that of an amine?

**15.34** Tell whether each of the following is an amine, an amide, both, or neither:

(a) $CH_3-\underset{\underset{O}{\|}}{C}-NH_2$

(b) 

(c) $CH_3CH_2-NH_2$

(d) $CH_3-\overset{\overset{CH_3}{|}}{N}-CH_2CH_3$

(e) 

(f) 

(g) 

(h) $NH_2-\underset{\underset{O}{\|}}{C}-CH_2CH_2-\underset{\underset{O}{\|}}{C}-NH_2$

(i) 

(j) 

**15.35** Draw structural formulas for
(a) valeramide
(b) N-methylpropionamide
(c) N,N-diethylformamide

(d) *N*-phenylbenzamide

(e) *N*-methyl-*N*-phenyl-*meta*-hydroxybenza-
    mide

(f) *N*-(*para*-nitrophenyl)-*ortho*-chlorobenza-
    mide

(g) glutardiamide

**15.36** Name these amides:

(a) H—C—NH$_2$
         ‖
         O

(b) [benzene ring]—C—NH—CH$_2$CH$_3$
                    ‖
                    O

(c) CH$_3$—C—N—CH$_3$
         ‖    |
         O   CH$_3$

(d) [benzene ring]—NH—C—CH$_2$CH$_2$CH$_3$
                       ‖
                       O

(e) CH$_3$(CH$_2$)$_{14}$—C—NH—CH$_3$
                      ‖
                      O

(f)                O   CH$_2$CH$_3$
                   ‖   |
    [benzene ring]—C—N—CH$_2$CH$_2$CH$_3$
         |
         Br

## Acid–Base Properties of Organic Compounds

**15.37** Predict whether each of these compounds is
acidic, basic, or neutral:

(a) CH$_3$OH

(b) CH$_3$NH$_2$

(c) CH$_3$SH

(d) [benzene ring with Cl, Cl, OH]

(e) [benzene ring with Cl, Cl, CH$_2$OH]

(f) [benzene ring with Cl, Cl, CH$_2$NH$_2$]

(g) [benzene ring with Cl, Cl, CH$_2$—NH—C—CH$_3$ (‖O)]

(h) CH$_3$—C—H
         ‖
         O

(i) CH$_3$—C—CH$_3$
         ‖
         O

(j) CH$_3$—C—OH
         ‖
         O

(k) CH$_3$—C—OCH$_2$CH$_3$
         ‖
         O

(l) CH$_3$—C—NH$_2$
         ‖
         O

(m) CH$_3$—C—N—CH$_3$
         ‖   |
         O  CH$_3$

(n) [benzene ring with NH$_2$]

(o) [benzene ring with NH—CH$_3$]

(p) [benzene ring with N—CH$_3$, CH$_3$]

(q) [benzene ring with NH—C—H (‖O)]

(r) [benzene ring with SO$_2$OH]

(s)

$CH_3$

OH

$OCH_3$

(t)

$CH_3$

OH

$OCH_3$

## Preparation of Amides

**15.38** Is an amide formed when a carboxylic acid is added to an amine? If not, how can an amide be formed?

**15.39** Write the structural formulas for the principal products (if no reaction, say so):

(a)  $CH_3CH_2-\overset{\overset{\displaystyle O}{\|}}{C}-OH + NH_3 \longrightarrow$

(b)

$\overset{\overset{\displaystyle C-Cl}{|}}{\underset{\displaystyle O}{}}$  $+ CH_3NH_2 \longrightarrow$

(c)  $CH_3-\overset{\overset{\displaystyle O}{\|}}{C}-O-\overset{\overset{\displaystyle O}{\|}}{C}-CH_3 +$

$CH_3CH_2-NH-CH_2CH_3 \longrightarrow$

**\*15.40** Write the structural formulas of the principal organic products when putrescine (1,4-diaminobutane) reacts with (a) excess acetic acid   (b) excess acetyl chloride.

**15.41** Write the structural formulas for the principal organic products (if no reaction, say so):

(a)

$\overset{\displaystyle COOH}{}$  $+ CH_3-NH-CH_3$

(b)  $CH_3-\overset{\overset{\displaystyle O}{\|}}{C}-Cl +$

$CH_3CH_2-\overset{\overset{\displaystyle}{|}}{\underset{\displaystyle CH_2CH_3}{N}}-CH_3 \longrightarrow$

(c)  $CH_3-\overset{\overset{\displaystyle O}{\|}}{C}-O-\overset{\overset{\displaystyle O}{\|}}{C}-CH_3 +$

$CH_3NH_2 \longrightarrow$

(d)

## Hydrolysis of Amides

**15.42** Tell which reagents and/or catalysts can be used to bring about the following conversions:

(a)  $CH_3CH_2-\overset{\overset{\displaystyle}{\|}}{\underset{\displaystyle O}{C}}-NH_2 \longrightarrow$

$CH_3CH_2-\overset{\overset{\displaystyle}{\|}}{\underset{\displaystyle O}{C}}-O^-$

(b)  $CH_3CH_2-\overset{\overset{\displaystyle}{\|}}{\underset{\displaystyle O}{C}}-NH_2 \longrightarrow$

$CH_3CH_2-\overset{\overset{\displaystyle}{\|}}{\underset{\displaystyle O}{C}}-OH$

(c)

$\overset{\overset{\displaystyle C-NH-CH_3}{\|}}{\underset{\displaystyle O}{}} \longrightarrow$

$CH_3-\overset{+}{N}H_3$

(d)

$\overset{\overset{\displaystyle C-NH-CH_3}{\|}}{\underset{\displaystyle O}{}} \longrightarrow$

$CH_3-NH_2$

**15.43** For the reactions shown in Problem 15.42, draw the structural formula for the other product formed in each case.

## Alkaloids

**15.44** Define alkaloid.

**15.45** Which alkaloid
(a)  cures malaria
(b)  is in tobacco
(c)  is a rat poison
(d)  is converted to heroin
(e)  is in coffee
(f)  is medically used as an important painkiller
(g)  is an illegal drug that was taken by Sherlock Holmes
(h)  killed Socrates
(i)  is used to dilate the pupils of eyes

**15.46** Identify all the functional groups in (a) seconal (b) coniine (c) morphine (d) Methadone. If alcohol or amine groups are present, tell whether they are primary, secondary, or tertiary.

**\*15.47** Which has more rings, strychnine or reserpine?

## Boxes

**15.48** (Box 15A) What are the possible negative side effects of illegal usage of "crank"?

**15.49** (Box 15A) What is the basic structural difference between the natural hormone epinephrine and the synthetic pep pill methamphetamine?

**15.50** (Box 15B) Explain why the drug morphine is often administered in the form of its salt, morphine sulfate.

**15.51** (Box 15B) If you saw this label on a decongestant: phenylephrine·HCl, would you worry about being exposed to a strong acid such as HCl? Explain.

**15.52** (Box 15C) How does harmless nitrite ion get converted to carcinogenic nitrosamines in the body?

**15.53** (Box 15C) Some water wells have a high concentration of nitrite ions. Should these wells be considered health hazards and shut down? Explain.

**15.54** (Box 15D) What are the advantages and disadvantages of acetaminophen and ibuprofen as pain-relievers over aspirin?

**15.55** (Box 15D) What functional groups are present in (a) aspirin (b) acetaminophen (c) ibuprofen?

**\*15.56** (Box 15E) By analogy with the structure of nylon 66, draw the structure of nylon 46. If one of the starting compounds is adipoyl chloride, what is the name of the diamine used to make nylon 46?

**\*15.57** (Box 15E) As with polyesters, polyamides are also formed from acyl halides rather than from carboxylic acids. Draw the structure of the nylon that is formed from sebacyl chloride, $Cl-\overset{\overset{O}{\|}}{C}-(CH_2)_8-\overset{\overset{O}{\|}}{C}-Cl$, and hexamethylene diamine (1,6-diaminohexane).

**15.58** (Box 15F) Urea is soluble in water. Show how water can form hydrogen bonds with urea.

**15.59** (Box 15G) What is a synergistic effect? What is its importance when a patient takes barbiturates?

**15.60** (Box 15G) Barbiturates are derivatives of urea. Identify the portion of the structure in secobarbital that contains the urea.

**\*15.61** (Box 15G) Is there a urea portion in the structure of pentothal? Explain.

**15.62** (Box 15H) Does diazepam contain a urea portion, as the barbiturates do?

**15.63** (Box 15H) What structure is common to all the benzodiazepines?

**15.64** (Box 15I) Cocaine has two carboxylic ester groups in its structure. Identify both.

**15.65** (Box 15I) What is "crack"?

## Additional Problems

**15.66** Draw the structural formula of any particular quaternary ammonium salt.

**15.67** Show the principal products:

(a) $H-\overset{\overset{O}{\|}}{C}-NH-CH_3 + H_2O \xrightarrow{H^+}$

(b) $CH_3CH_2-\overset{\overset{O}{\|}}{C}-NH_2 + H_2O \xrightarrow{OH^-}$

(c) $CH_3(CH_2)_8-\overset{\overset{O}{\|}}{C}-\overset{\overset{|}{N}}{\underset{CH_3}{}}-CH_3 + H_2O \xrightarrow{OH^-}$

(d) ⬡$-CH_2-\overset{\overset{O}{\|}}{C}-NH-CH_2CH_3 +$ $H_2O \xrightarrow{H^+}$

**\*(e)** [cyclic structure] $+ H_2O \xrightarrow{OH^-}$

**\*15.68** Cetylpyridinium chloride has the structure

$CH_3-(CH_2)_{15}-{}^+N$⬡$Cl^-$.

(a) What makes this compound, which bears a long-chain alkyl group, water-soluble?

(b) What class of compounds does it belong to?

**15.69** When an ophthalmologist uses atropine to dilate the pupils for an eye examination, he or she warns the patient not to go into bright sunlight without sunglasses for a few hours because too much light will temporarily blind the patient. What conclusion can you draw from this regarding the effect of a topical application of atropine?

**15.70** Both *n*-decylamine and *n*-decyl alcohol are insoluble in water. If you were given a mixture of these two liquids, how could you separate them without distillation?

**15.71** In each case, tell which compound has the higher boiling point:
(a) (1) $CH_3CH_2CH_2CH_2NH_2$ vs.
    (2) $CH_3CH_2CH_2CH_2CH_3$
(b) (1) $CH_3CH_2CH_2CH_2NH_2$ vs.
    (2) $CH_3CH_2CH_2CH_2OH$
(c) (1) $CH_3CH_2CH_2CH_2NH_2$ vs.
    (2) $CH_3CH_2CH_2NH_2$
(d) (1) $CH_3CH_2CH_2CH_2NH_2$ vs.
    (2) $CH_3CH_2-\underset{\underset{CH_3}{|}}{N}-CH_3$

**15.72** Give IUPAC names for the following amphetamines:

(a)
Amphetamine

(b)
Methamphetamine

**15.73** There are four amines with the molecular formula $C_3H_9N$. Draw structural formulas for all of them and classify each as primary, secondary, or tertiary.

**15.74** Why is each of the following not considered to be an alkaloid?
(a) Demerol
(b) lidocaine
(c) heroin
(d) marijuana (tetrahydrocannabinol)

**15.75** Draw the structure of nicotinium chloride (nicotine hydrochloride).

**15.76** How could you convert butylammonium chloride, $CH_3CH_2CH_2CH_2\overset{+}{N}H_3$ $Cl^-$, to the free amine, $CH_3CH_2CH_2CH_2NH_2$?

**\*15.77** When a certain amide, of molecular formula $C_8H_9NO$, was hydrolyzed with an acid catalyst, two organic products were isolated, one of which was benzoic acid. Draw structural formulas (a) for the other organic product and (b) for the original amide.

# PART III

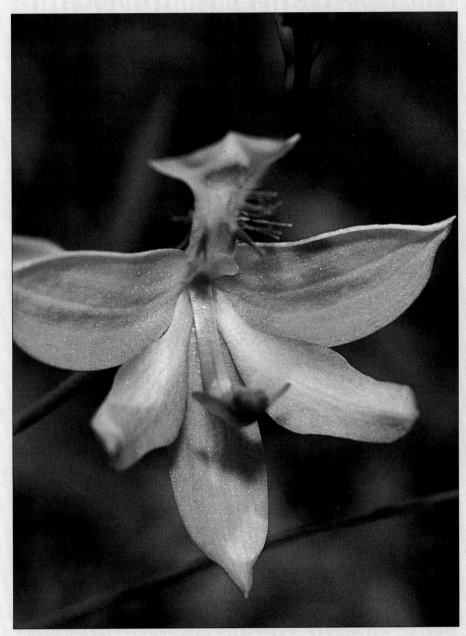

*Courtesy of USDA.*

# Biochemistry

# Jacqueline K. Barton

Dr. Jacqueline K. Barton is a native New Yorker and was educated in that city. She received her B.A. degree from Barnard College in 1974 and her Ph.D. from Columbia University in 1979. Following her Ph.D. work, Dr. Barton did further research at Yale University and Bell Laboratories and then joined the faculty of Hunter College. In 1983 she returned to Columbia University where she rose rapidly to the rank of full professor. In the fall of 1989 she assumed her present position as Professor of Chemistry at the California Institute of Technology.

Dr. Barton has done outstanding research in the field of biochemistry, particularly in the design of simple molecular probes to explore the variations in structure and conformation along the DNA helix (the subject of Chapter 23). In spite of having been a research scientist for a relatively short time, she has done important new work and has received many honors. In 1985 she received the Alan T. Waterman Award of the National Science Foundation as the outstanding young scientist in the United States. In 1987 she was the recipient of the American Chemical Soci-

ety's Eli Lilly Award in Biological Chemistry, and the following year she received the Society's Award in Pure Chemistry. That same year, 1988, she also received the Mayor of New York's Award of Honor in Science and Technology.

We met in Dr. Barton's office at Cal Tech. As we find with so many chemists, she is interested in art and has a painting by the Spanish artist Miro on her office wall as well as a print by the French artist Vasarely. As we were to see in the interview, this interest in form and color in art carries over into her research.

**A background in math—
but no chemistry**

We have been interested in how the scientists we interviewed came to their careers in chemistry. In Dr. Barton's case she said, "I never took chemistry in high school. Maybe one shouldn't publicize that, but it's the truth. However, I was always very interested in mathematics, so I took a lot of calculus when I was in high school. I also took a course in geometry, and that interest in geometry has carried over into my research, since the sort of science I do now is very much governed by structures and shapes.

"When I went to college I thought that, in addition to taking math, I should take some science courses. I walked into the freshman chemistry class, and there were about 150 people there. However, there was also a small honors class with about 10 students. Even though I hadn't had chemistry before I thought I would try it—and I loved it. What chemistry allowed me to do was to combine the abstract and the real. I was very excited by it."

But, Dr. Barton says, it was really the experience of the laboratory that got her interested in chemistry. Like many of us in chemistry, she was fascinated by color changes in reactions and the significance of these observations. However, she was also interested in trying to predict what would happen in a reaction and, if her prediction was not correct, to try to explain this and then to do more experiments that would solve the puzzle. As she said, "That's really what got me started in science."

In addition, Dr. Barton also had an inspirational teacher and role model, Bernice Segal. "She was an absolute inspiration to me. She gave a magnificient course, and she was a very tough lady who asked a lot of you—and you did it!"

### The platinum blues

Dr. Barton's Ph.D. thesis research was mostly on compounds known to chemists for years as the "platinum blues."

"Most platinum compounds are orange or red and yet there are these magnificient blue complexes. What are they? What are their structures, and why are they blue? We solved the structure of the molecule and found that it contains four platinum atoms in a line.

### Research in bioinorganic chemistry

After talking about how Dr. Barton came into chemistry, we discussed her current research in bioinorganic chemistry, the role of metals in biological systems. She has received numerous awards for this work, reflecting the importance that the chemistry community places on the field and her contributions to it.

"The interest of my group is to exploit inorganic chemistry as a tool to ask questions of biological interest and to explore biological molecules. A lot of the work in bioinorganic chemistry thus far has been the exploration of metal centers in biology. Why is blood red? Why does the iron [in heme] do what it does? That's just one example, but there are hundreds of others. Many enzymes and proteins within the body in fact contain metals, and the reason we've looked at blood and then the heme center within it has been because it's colored. An obvious tool that transition metal chemistry provides is color, and so things change color when reactions occur. That is one of the things that fascinated me in the first place.

"Another wonderful thing about transition metal chemistry is that it allows us to build molecules that have interesting shapes and structures depending upon the coordination geometry. In fact, you can create a wealth of different shapes, several of which are chiral, and that's something we take advantage of in particular.[1] What we want to do is make a variety of molecules of different shapes, target these molecules to sites on a DNA strand, and then ask questions such as 'Does DNA vary in its shape as a function of sequence?' If we think about how proteins bind to DNA, do they also take advantage of shape recognition in binding to one site to activate one gene or turn off another gene? When scientists first wondered about these and other such problems, they would write down a one-dimensional sequence of DNA and would think about it in one-dimensional terms. How does the protein recognize a particular DNA sequence? DNA is clearly not one dimensional. It has a three-dimensional structure, and different sequences of bases will generate different shapes and different forms. Therefore, we think we can build transition metal complexes of particular shapes, target them to particular sequences of bases in DNA, and then use these complexes to plot out the topology of DNA. We can then ask how nature takes advantage of this topology. We want to develop a true molecular understanding, a three-dimensional understanding, of the structure and the shapes of biologically important molecules such as DNA and RNA."

### A revolution in chemistry in the last ten years

One of the most important goals of these interviews has been to gain insight into the directions that science in general and chemistry in particular will move in this decade. Dr. Barton put her remarks in the context of recent developments in her field.

"I think our work may be an example of where chemistry is going in general. I think there has been a revolution in chemistry in the past 10 years. The revolution is at the interface between chemistry and biology where we can now ask chemical questions about biological molecules. First of all, we can make biological molecules that are pure. I can now go to a machine called a DNA synthesizer, and I can type in a sequence of DNA; from that sequence I can synthesize a pure material, with full knowledge of where all of the bonds are. Then I can run it through an HPLC and get it 100% pure.[2] Therefore, I can now talk about these biopolymers in chemical terms as molecules rather than as impure cellular extracts. I couldn't do that before."

---

[1] The chirality of molecules is discussed in Chapter 16.

[2] An HPLC is a "high pressure liquid chromatograph," an instrument capable of separating one type of molecule from another.

*Some double helical conformations of DNA (see Chapter 23). Photo by Jacqueline Barton.*

Not only do we have the ability to prepare biological molecules in pure form, but Dr. Barton stressed that we have the techniques to characterize them in ways that we chemists think about molecules. "The development of new techniques allows us to make a bridge between chemistry and biology and ask chemical questions with molecular detail. It's an exciting time to be doing chemistry, and that is why I see it as a new frontier area."

### New chemistry curricula

Dr. Barton stressed the point that it is chemists who are "making new materials and making and exploring biological systems. It's the chemist who looks at questions of molecular detail and asks about structure and its relationship to function." Since this involves so many areas of chemistry, she believes that "we are going to have to stop making divisions between inorganic, physical, analytical, and organic chemistry. We must all do a little bit of each." This is an attitude shared by many in chemical education today, and it means that we should perhaps rethink the curriculum in chemistry in particular and science in general.

No matter what the curricular structure, however, she believes what is important in the education of scientists is "to get across the excitement that now we can know what biologically important molecules look like. And, from knowing what they look like, we can manipulate them and change them a little. Then we ask how those changes affect the function, so we can relate the structure of the molecule and its macroscopic function."

"A protein molecule of average size is so small you could put more than a billion billion of them on the head of a pin. We now know we can manipulate molecules that are of those dimensions and can know exactly what they look like. I can't imagine that we can't get people interested in chemistry if we can get across the excitement that comes from the realization that we are looking at things so small and yet can do surgery on them."

### Chemistry is fun

There was one basic theme in all of the interviews that we did for this book, and Dr. Barton expressed it well. "The bottom line is that chemistry is fun, it's addictive, and, if one has a sense of curiosity, it can be tremendously entertaining and appealing. And it is not so difficult. It's difficult when one thinks about it as rote memorization, which *is* difficult and boring. But that isn't what chemistry is. Chemistry is trying to understand the world around us in some detail. For example, we are interested in knowing such things as what makes skin soft, what makes things different in color, why sugar is sweet, or why a particular pharmaceutical agent makes us feel better."

### Women in science

We were interested in Dr. Barton's perspective on the issue of women in science. Did she see special opportunities for women? Are there problems that women need to overcome or be aware of?

"Because I am a woman, and there are so few women currently in professional positions in chemistry, I'm asked those questions often. First of all, I am not an expert on the subject. What I like to think my best contribution to women in chemistry can be is to do the best science I can, and to be recognized for my science, not for being a woman in science. I think that it is generally important when women go into science that they should appreciate that there are no special opportunities; that is, you will be treated like any other person doing science. But just as there should be no special opportunities in that respect, happily—maybe this is naive of me—I think there are also no special detriments or obstacles that one need consider in this day and age. One shouldn't think that 'because I am a woman I can't do it.' That's patently false. In fact, everyone is extremely supportive of women who do science. However, I remember talking to Bernice Segal, my former teacher at Barnard College, and having her explain to me that when she was a graduate student she had to do things behind a curtain, because the women weren't supposed to be doing chemistry. Mildred Cohn, another one of my role models, took over 20 years to have her own independent position as a professor, as opposed to being a laboratory assistant working for someone else. The bottom line is that I don't have a story like that to tell. That's the good news. In my generation there are few such stories of blatant discrimination. Now the world is a much better place for a woman to do science."

Dr. Barton's enthusiasm for her work and for chemistry in general was obvious. It is evident that she will continue to do some of the most important work in science and that her infectious enthusiasm for chemistry will bring many more young people into our profession.

# CHAPTER 16

*A rice field in China. (Photograph by Beverly March.)*

# Carbohydrates

## 16.1 INTRODUCTION

**Carbohydrates** are either **polyhydroxy aldehydes or ketones, or compounds that yield polyhydroxy aldehydes or ketones on hydrolysis.** This definition will become clearer as we proceed in this chapter.

Carbohydrates are of major importance to both plants and animals. It has been estimated that more than half of all the organic carbon atoms in the world are in carbohydrate molecules. Simple carbohydrate molecules are synthesized chiefly by chlorophyll-containing plants as long as the sun is shining. In this process, called **photosynthesis,** plants combine carbon dioxide from the air with water from the soil to give simple carbohydrates (mainly glucose, $C_6H_{12}O_6$):

$$6CO_2 + 6H_2O \xrightarrow[\text{chlorophyll}]{\text{light}} C_6H_{12}O_6 + 6O_2$$

Plants have two main uses for the carbohydrates they make. They use them (1) as a means of storing energy (the energy comes, of course, from the sunlight) and (2) to provide supporting structures, such as the trunks of trees. Carbohydrates make up about three fourths of the dry weight of plants. Animals (including humans) get their carbohydrates by eating plants, but do not store much of what they consume. Less than 1 percent of the body weight of animals is made up of carbohydrates.

Because $H_2SO_4$ added to carbohydrates removes —OH and —H groups and leaves black carbon, early chemists wrongly thought these compounds might be made of carbon and water and hence called them carbohydrates (*hydro* means water).

It is the chlorophyll in leaves and grass that makes them green.

479

**Figure 16.1**
The roles of plants and animals in the oxygen–carbon dioxide cycle.

Carbohydrates + O₂ ← High-energy molecules

Plants take energy from the sun and use it to change low-energy molecules to high-energy molecules.

Animals get energy from reactions between high-energy molecules and use it for motion, breathing, and so on. They produce low-energy products.

$CO_2 + H_2O$ ← Low-energy molecules

We also use carbohydrates for clothing (cotton, rayon, linen), and we use wood (which is chiefly carbohydrates) for building, burning, and making paper.

Humans use carbohydrates for many purposes. The most important is, of course, food. Typically, carbohydrates constitute about 65 percent of our diet. Our bodies use these compounds mostly for energy (Chap. 21) but also as a source of carbon atoms for the synthesis of many other compounds. They are found in nucleic acids and in connective tissue and are an essential part of the energy cycle of living things (Fig. 16.1).

## 16.2 ENANTIOMERS AND CHIRALITY

In order to understand the structure of carbohydrates and other biological molecules, we must consider a new form of stereoisomerism.

Everything has a mirror image (except maybe a vampire). If we look at the world around us, we see that all objects can be divided into two classes: those that can be superimposed on their mirror image and those that cannot. **An object that cannot be superimposed on its mirror image** is called **chiral** (pronounced ky′ral). Chiral objects are nonsuperimposable on their mirror images.

Putting the two hands palm to palm is not superimposing them. For one thing, the fingernails are on opposite sides.

A good example of a chiral object is your left hand. If you hold it up to a mirror, you see in the mirror an image of your right hand (Fig. 16.2), and you cannot superimpose the two. Some objects that are not chiral are golf balls, spoons, baseball bats, and wine glasses, all of which are superimposable on their mirror images (Fig. 16.3).

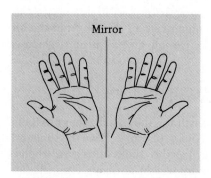

**Figure 16.2**
Left and right hands are nonsuperimposable mirror images and thus are chiral objects.

**Figure 16.3**
Some objects that are superimposable on their mirror image. Such objects are not chiral.

**Figure 16.4**
(a) A molecule of 2-bromobutane and its mirror image. The two cannot be superimposed and thus are a pair of enantiomers. (b) If we try to superimpose them by, say, putting the CH₃ over the CH₃ and the Br over the Br, we find that the H goes over the CH₃CH₂ and the CH₃CH₂ goes over the H. We can superimpose any two groups, but the other two groups will not be superimposed.

The same is true for molecules: Some molecules are chiral; others are not. What kinds of molecules are chiral? Figure 16.4 shows a molecule that contains a carbon atom connected to four different groups. (Remember that carbon points its four bonds to the corners of a tetrahedron; Sec. 3.6.) The molecule on the left in Figure 16.4 is the mirror image of the molecule on the right, just as a left hand is the mirror image of a right hand, but the two molecules cannot be superimposed, any more than we can superimpose a left hand on a right hand (try putting a left-hand glove on your right hand). **A carbon atom connected to four different groups** is called a **chiral carbon** or an **asymmetric carbon.**

Because the presence of a single chiral carbon atom always makes a molecule chiral, it is important that you be able to recognize one when you see it. It is really not difficult. If a carbon atom is connected to four different groups, *no matter how slight the differences,* it is a chiral carbon. If any two groups are identical, it is not a chiral carbon.

A chiral molecule is one that cannot be superimposed on its mirror image.

---

**EXAMPLE**

Which of the carbon atoms shown in color are chiral?

(a) CH₃—CH—CH₂—CH₃
　　　　|
　　　　OH

(b) CH₃—CH₂—CH—CH₂—CH₃
　　　　　　　|
　　　　　　　OH

(c) Br—CH₂—CH₂—CH₂—CH—CH₂—CH₂—CH₂—Cl
　　　　　　　　　　　|
　　　　　　　　　　　CH₃

(d) $CH_3-CH_2-\overset{\displaystyle}{\underset{\displaystyle O}{C}}-CH_3$

(f)

$$\begin{array}{c} H_2 \\ C \\ H_2C \qquad CH-Br \\ HC \qquad CH_2 \\ H_3C \qquad C \\ H_2 \end{array}$$

(e)

$$\begin{array}{c} O \\ \| \\ C \\ H_2C \qquad CH_2 \\ H_2C \qquad CH-CH_3 \\ C \\ H_2 \end{array}$$

**Answer**

(a) This is a chiral carbon. The four different groups are $-CH_3$, $-CH_2CH_3$, $-H$, and $-OH$.

(b) This carbon is not chiral. Two groups are identical: $-CH_2CH_3$.

(c) This is a chiral carbon. The four different groups are $-H$, $-CH_3$, $-CH_2CH_2CH_2Cl$, and $-CH_2CH_2CH_2Br$. Although the last two groups are similar, they are not identical. Even this small difference is enough to make the carbon chiral.

(d) This carbon is connected to only three groups. In order to be chiral, a carbon must be connected to *four* different groups. Thus, even though the three groups here are all different, they are not four, and this carbon is not chiral.

(e) The rules are no different for carbons in rings. This is a chiral carbon. $-H$ and $-CH_3$ are two of the groups. The third is $-CH_2-CO-$, which is different from the fourth, $-CH_2CH_2-$.

(f) Neither of these is a chiral carbon. Whether we proceed clockwise around the ring or counterclockwise, we do not encounter the slightest difference. The two groups are identical:

$$\begin{array}{c} H_2 \\ C \\ H_2C \qquad CH-Br \\ HC \qquad CH_2 \\ H_3C \qquad C \\ H_2 \end{array}$$

**PROBLEM 16.1** Which of the carbon atoms shown in color are chiral?

(a) $CH_3CH_2-\underset{\displaystyle Br}{CH}-CH_2CH_3$

(c)

$$\begin{array}{c} H_2 \\ C \\ H_2C \qquad CH-CH_3 \\ H_2C \qquad CH_2 \\ C \\ H_2 \end{array}$$

(b)

$$\underset{\displaystyle H_3C}{\bigcirc}-\underset{\displaystyle OH}{CH}-\bigcirc$$

(d) $CH_3-\underset{\displaystyle CH_3}{CH}-\underset{\displaystyle Br}{CH}-CH_3$

**Figure 16.5**
(a) A molecule of bromoethane and its mirror image. This molecule is superimposable on its mirror image. All we have to do is (b) rotate it so that the Br is placed over the Br (the $CH_3$ is already over the $CH_3$). The two Hs are over the two Hs, and the molecules are superimposed and therefore identical. Note that bromoethane does not contain a chiral carbon.

There are some chiral molecules that lack a chiral carbon, but these exceptions are not important enough for us to consider in this book. Also, some molecules that contain two or more chiral carbons are not chiral molecules, but these too are not very important. In Section 16.4 we will see what happens when molecules possess two or more chiral carbon atoms.

Now that we know how to recognize a chiral molecule, let us look once more at Figure 16.4. The two molecules of 2-bromobutane are not superimposable on each other. As we saw in Section 10.9, *superimposability is the same as identity.* If two objects are identical, we can superimpose one on the other (Fig. 16.5). Since the two molecules of 2-bromobutane are not superimposable, *they must be different.* They therefore meet our definition of stereoisomers (Sec. 10.9): two molecules with identical molecular and structural formulas but different three-dimensional shapes. Chiral molecules represent the most subtle kind of isomerism we have seen. Even the shapes of the molecules are identical in all respects except direction: One is "left-handed," the other "right-handed." **Isomers that are nonsuperimposable mirror images of each other** are called **enantiomers.**

## 16.3  PROPERTIES OF ENANTIOMERS. OPTICAL ACTIVITY

In all cases of isomerism that we have previously seen, both structural isomerism and stereoisomerism, the isomers differ in almost all properties, but this is not the case for enantiomers. **All physical and chemical properties of enantiomers are identical, except two.**

**Figure 16.6**
The waveforms of ordinary and polarized light.

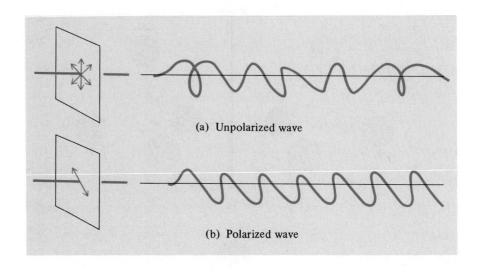

(a) Unpolarized wave

(b) Polarized wave

1. The two members of a pair of enantiomers rotate the plane of polarized light in opposite directions. **Polarized light** is **light that vibrates in only one plane,** in contrast to ordinary light, which vibrates in all planes (Fig. 16.6). When polarized light is passed through certain substances, the plane of its vibration is rotated through a certain angle, different for each substance (Fig. 16.7). An instrument that detects this rotation is called a **polarimeter.**

A substance that rotates the plane of polarized light is called **optically active.** Some substances rotate the plane to the right (clockwise). We call these **dextrorotatory** and use the symbol (+). Others rotate the plane to the left (counterclockwise). These are **levorotatory,** and their symbol is (−).

There is a basic difference between chiral and nonchiral substances. **All chiral substances rotate the plane of polarized light** (they are optically active). Substances that are not chiral do not rotate the plane of polarized light. As we saw in the previous section, chiral substances always exist as two enantiomers. Both enantiomers rotate the plane of polarized light, in equal amounts (Box 16A) **but in opposite directions.** In any pair of enantiomers, one always rotates the plane to the right and the other to the left. Only the direction is different; the amount of the rotation is the same.

These words are from the Latin *dextro,* right, and *levo,* left.

This principle was first stated by Louis Pasteur in 1848.

**Figure 16.7**
Operation of a polarimeter. Ordinary light, which vibrates in all planes (indicated by arrows), becomes polarized when passed through a polarizer. When the polarized light passes through a chiral sample, the plane of the light is rotated.

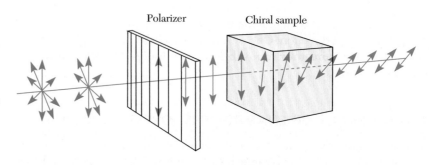

Polarizer   Chiral sample

When a chiral compound is placed in a polarimeter, it rotates the plane of polarized light either to the right (+) or to the left (−). But by how much? Not only is the amount of rotation different for different substances, but even for the same substance it depends on the number of molecules in the path of the beam—the more molecules, the more the plane is rotated. In order to make their measurements independent of this, chemists always report the **specific rotation, [α]**, which is

$$[\alpha] = \frac{\alpha}{c \cdot l}$$

where $\alpha$ is the actual rotation observed in the instrument, $c$ is the concentration (in grams per milliliter) if the compound is dissolved in a solvent or the density (in grams per milliliter) if the sample is a pure liquid, and $l$ is the length of the sample tube (in decimeters).

Specific rotation is a physical property, like density and melting point, and can be used to help identify an unknown optically active compound. Because it also depends on two other variables, the temperature and the wavelength of the polarized light, it is often reported like this (for example, for sucrose):

$$[\alpha]_D^{20} = +66.4°$$

In this case, the measurement was obtained at 20°C using yellow light from a sodium lamp, for which the letter D is used, following an old custom. This is the type of light used most often.

Of course, if $[\alpha]_D^{20}$ for sucrose is +66.4°, then we can be sure that $[\alpha]_D^{20}$ for its enantiomer must be −66.4°.

**2.** Enantiomers undergo chemical reactions, just as any other organic molecule would that has the same functional groups. When they react with another molecule *that is chiral*, the reaction rates are not the same for the two enantiomers. For example, (+)-1-phenyl-1-ethanol reacts with (+)-2-methylbutanoic acid to form a carboxylic ester:

Enantiomers react at the same rate with nonchiral molecules.

(+)-1-phenyl-1-ethanol        (+)-2-methylbutanoic acid

The reaction between (+)-1-phenyl-1-ethanol and (−)-2-methylbutanoic acid takes place at a different rate. This is a very important difference in property, with major biological consequences (Sec. 16.13).

In most cases, the only way to tell them apart is to pass polarized light through samples of the compounds and see if they rotate (+) or (−).

In all properties except these two, enantiomers are identical. They have identical melting points, boiling points, densities, rates of reactions with molecules that are not chiral, and all other properties.

Although each enantiomer of a pair rotates the plane of polarized light, a mixture of equal amounts of the two does not, because the rotations cancel each other. Such a mixture is called a **racemic mixture** or a **racemate.**

### Fischer Projections

Emil Fischer, who in 1902 became the second Nobel prize winner in chemistry, made many fundamental discoveries in the chemistry of carbohydrates, proteins, purines, and other areas of organic and biochemistry.

It is awkward to draw molecules in the three-dimensional shapes shown in Figures 16.4 and 16.5, but there is a way to represent these shapes correctly even if we are not very good artists (or don't want to take the time). This system was devised by the German organic chemist Emil Fischer (1852–1919; Fig. 16.8). In a **Fischer projection,** you hold the molecule in such a way that the two bonds coming out at you are horizontal and the two bonds going behind the paper are vertical, for example:

The —$CH_3$ and —OH are coming toward you, out of the page. The —CHO and —H are going away from you, into the page.

$$CH_3 \overset{\displaystyle CHO}{\underset{\displaystyle H}{\rule{0pt}{1em}\quad}} OH$$

(a)                                         (b)

When you see the formula shown in (b), the Fischer projection, you must realize that the molecule is oriented in space as shown in (a).

We can use Fischer projections to determine (1) whether two molecules are identical or different and (2) whether or not two formulas are mirror images. But there are two rules to follow in using these projections: (1) We must not take them out of the plane of the paper, and (2) we can rotate them 180° but not 90°.

**Figure 16.8**
Emil Fischer. (Courtesy of E. F. Smith Memorial Collection, Center for History of Chemistry, University of Pennsylvania.)

**EXAMPLE**

Are these mirror images? Are they identical (superimposable)?

(a)   HO—|—H     and     H—|—OH
with CH₃ top, CH₃ bottom

(b)   HO—|—H     and     H—|—HO
with CH₃ top and phenyl ring bottom

**Answer**

In both (a) and (b) the molecules are mirror images. If you held a mirror next to one you would see the other. In (a) they are superimposable. You cannot superimpose them by merely sliding one over onto the other, but you can if you turn one upside down (180°, which is allowed) and then slide it over. In (b) they are not superimposable. They would be if you could flip one over, but the rules say you must keep it in the plane of the paper.

**PROBLEM 16.2**   Are these mirror images? Are they identical?

(a)   $CH_3CH_2$—|—H     H—|—$CH_2CH_3$
with CH₃ top, Cl bottom

(b)   $CH_3CH_2$—|—$CH_2CH_3$     $CH_3CH_2$—|—$CH_2CH_3$
with H top, Cl bottom

## 16.4   COMPOUNDS WITH MORE THAN ONE CHIRAL CARBON

Let us use Fischer projections to see what happens when a molecule possesses two chiral carbon atoms. Such a molecule is 2,3-dihydroxybutanal. (Three-dimensional configurations of these molecules are shown in Figure 16.9.) One stereoisomer is **A.**

| CHO | CHO | CHO | CHO |
| H—OH | HO—H | H—OH | HO—H |
| H—OH | HO—H | HO—H | H—OH |
| CH₃ | CH₃ | CH₃ | CH₃ |
| A | B | C | D |
| A pair of enantiomers | | Another pair of enantiomers | |

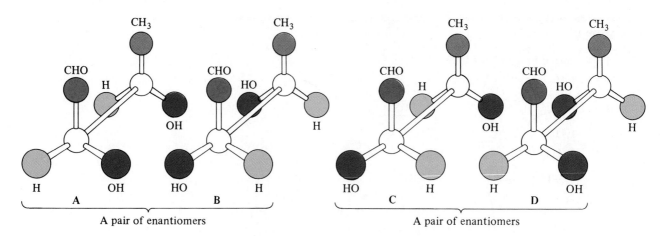

**Figure 16.9**
The three-dimensional configurations of the four stereoisomers of 2,3-dihydroxybutanal.

Make sure that **B** is indeed the mirror image of **A** and that they are not superimposable.

As can be seen from the formulas, **A** is not superimposable on its mirror image, **B.** Molecules **A** and **B** are therefore a pair of enantiomers and as such are identical in all properties except the two mentioned in the previous section.

But **C** is another stereoisomer of 2,3-dihydroxybutanal. If you inspect it carefully, you will see that it is not identical with (superimposable on) **A** or **B**, nor is it the mirror image of either of them. It is a third stereoisomer, different from **A** or **B**. Molecule **C** has its own mirror image, the compound **D.** Another inspection will show that **C** is not superimposable on its mirror image **D.** Molecules **C** and **D** are therefore a second pair of enantiomers. Their properties are also identical to each other except for the two properties mentioned earlier.

But are the properties of **A** and **B** identical to those of **C** and **D**? The answer is no. All physical and chemical properties of **A** and **B** are different from those of **C** and **D.** After all, **A** is not the mirror image of **C,** so there is no reason why their properties should be identical. The properties may be similar (they do have the same structural formula), but they are not identical.

Any molecule can have only one mirror image. Enantiomers always come in pairs; there are never more than two. **Stereoisomers that are not enantiomers** are called **diastereomers.** Molecule **C** is the enantiomer of **D** and a diastereomer of **A** and **B.**

What we saw for 2,3-dihydroxybutanal is also true for nearly all molecules that have two chiral carbon atoms: There are two pairs of enantiomers for a total of four stereoisomers. It does not matter whether the two chiral carbons are attached to each other or not.

In a similar manner, it can easily be shown (you may want to try it) that when three chiral carbons are present, there are eight stereoisomers (four pairs of enantiomers); four chiral carbons means 16 stereoisomers, and so forth. The general formula is

$$\text{number of stereoisomers} = 2^n,$$

The number $2^n$ is actually a maximum. In a few cases the actual number of stereoisomers is less.

where $n$ is the number of chiral carbon atoms.

---

**PROBLEM 16.3** How many stereoisomers are possible for

(a) $CH_3$—CH—CH—COOH
        |          |
        Br      $CH_3$

(b) $CH_3$—CH—CH—COOH
        |          |
        $CH_3$  Br

(c) $CH_2$—CH—CH—CH—CHO
        |        |        |        |
        OH    OH    OH    OH

---

## 16.5 GLUCOSE

Now that we have been introduced to chirality, we are ready to return to carbohydrates. Carbohydrates can be divided into classes based on the size of the molecules. **Monosaccharides,** the basic units of carbohydrates, are simple sugars having the general formula $C_nH_{2n}O_n$, where $n$ can vary from three to nine. **Disaccharides** consist of two monosaccharide units hooked together in a way that we shall discuss in Section 16.10. **Oligosaccharides** contain from three to ten units. **Polysaccharides** are also made of monosaccharides hooked together but may contain thousands of units (Fig. 16.10).

By far the most important monosaccharide is D-glucose. This sugar is not only the most important monosaccharide, but it is also the major constituent of disaccharides and polysaccharides.

D-Glucose is a white solid of molecular formula $C_6H_{12}O_6$. The structural formula of solid D-glucose is

$C_nH_{2n}O_n$ means that any monosaccharide molecule contains equal numbers of O and C atoms and twice as many H atoms.

D-Glucose is also called **dextrose, grape sugar,** and **blood sugar.**

A six-membered ring of this type is called a **pyranose ring.**

This kind of drawing is called a **Haworth formula.** Note the numbering of the carbon atoms. If you look at C-1, you will see that it is connected to two oxygen

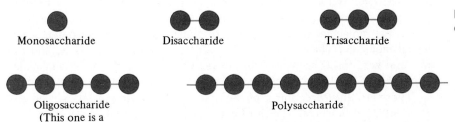

Monosaccharide

Disaccharide

Trisaccharide

Oligosaccharide
(This one is a
pentasaccharide)

Polysaccharide

**Figure 16.10**
Classification of carbohydrates.

## The Chair Form of Pyranose Rings

Although commonly used in carbohydrate chemistry, Haworth formulas do not really represent three-dimensional shapes of the molecules. The chair structure drawn here (see Box 10A), shows the three-dimensional pyranose ring of D-glucose. The solid lines projecting from the ring represent equatorial positions, and the dashed lines represent axial positions. Although the chair form is a more accurate picture of the molecule, Haworth forms are good enough for most purposes, and we shall use them in this book.

β-D-Glucose in the chair conformation

atoms: one in an —OH group, the other in an —OR group (the R begins with C-5). Such compounds are called **hemiacetals.** As we saw in Section 13.6, hemiacetals are not stable unless they are in a ring. Because of the ring, D-glucose is a perfectly stable molecule in the solid state. However, when it is dissolved in water, the instability of the hemiacetal shows itself, and the ring opens to turn C-1 into an aldehyde group and C-5 into an alcohol group:

$\qquad$ α-D-Glucose $\qquad$ Aldehyde form of D-glucose $\qquad$ β-D-Glucose

(16.1)

The hydrogen shown in color has traveled from the —OH oxygen to the —OR oxygen. But the aldehyde form of D-glucose is much less stable than the ring form. No sooner is it formed than it closes up again, but when it closes up there are two directions for the C-1 —OH to go: It can point down, which restores the original molecule, or it can point *up*, which gives a stereoisomer of the original molecule. The C-1 carbon is called an **anomeric carbon,** and the alpha and beta forms are called **anomers.** The α and β D-glucose molecules differ only at C-1. All other carbons have the same stereochemistry.

There are thus *three* forms of D-glucose. Two are ring forms; we call one alpha and the other beta. The third is the open-chain (aldehyde) form, which can also be drawn as a Fischer projection:

Anomers are two stereoisomers of a monosaccharide that differ only in the configuration at C-1 (aldoses) or C-2 (ketoses).

$$
\begin{array}{c}
\overset{1}{H} \\
| \\
\overset{1}{C}=O \\
| \\
H-\overset{2}{C}-OH \\
| \\
HO-\overset{3}{C}-H \\
| \\
H-\overset{4}{C}-OH \\
| \\
H-\overset{5}{C}-OH \\
| \\
\overset{6}{C}H_2OH
\end{array}
$$

Aldehyde form of D-glucose

By evaporating the water from D-glucose solutions at different temperatures, we can obtain either the alpha or the beta form in the solid state. As long as either is a solid, its formula does not change. The aldehyde form cannot be obtained in the solid state. When we dissolve either the alpha or beta solid in water, we get the equilibrium shown in Equation 16.1: a mixture of all three forms. Because the aldehyde form is the least stable, it is present in very small amounts, much less than 1 percent. The main reason we know it is there at all is that the alpha and beta forms cannot change into one another except by going through the open-chain form. All three forms are optically active; each rotates the plane of polarized light by a different amount. When either the alpha or the beta form is dissolved in water, the amount of rotation gradually changes because a single form is changing to a mixture of forms. Because of this, the process is called **mutarotation** (meaning "change of rotation").

Note that $\alpha$ and $\beta$ D-glucose are not enantiomers. To be enantiomers, two molecules must be mirror images at every part of the molecule, but these two have opposite configurations only at C-1. Their configurations are the same everywhere else. Therefore they are diastereomers, not enantiomers.

The alpha form has a specific rotation of +112.2°, the beta form +17.5°, and the equilibrium mixture +52.7°. This shows that at equilibrium there is more beta isomer than alpha.

## 16.6  MONOSACCHARIDES

### Aldohexoses

Let us consider the open-chain form of D-glucose:

$$
\begin{array}{c}
H \\
| \\
C=O \\
| \\
H-C-OH \\
| \\
HO-C-H \\
| \\
H-C-OH \\
| \\
H-C-OH \\
| \\
CH_2OH
\end{array}
$$

It has four chiral (asymmetric) carbon atoms (shown in color). By the formula given in Section 16.4, this means that there are $2^4 = 16$ isomers, all of which have the same structure, except that they differ in configuration at the four chiral carbons. These 16 compounds are called **aldohexoses:**

**aldo** because they have an aldehyde group

**hex** because they have a chain of six carbons

**ose** because this is an ending used for carbohydrates

Aldohexoses differ in whether the —OH groups point left or right in the Fischer projection. In the cyclic Haworth forms this means up or down.

The 16 aldohexoses constitute 8 pairs of enantiomers. For example, the enantiomer of D-glucose is

Two ways to show D-galactose:

— OH groups point left and right

— OH groups point up and down

The three most important aldohexoses are D-glucose, D-mannose, and D-galactose.

L-Glucose

By convention, the configuration of the highest-numbered chiral carbon of the chain determines whether a monosaccharide is put into the D or the L series. If, in the Fischer projection, the OH points to the right, the whole molecule is placed in the D series. If it points to the left, the molecule is in the L series. Almost all natural monosaccharides belong to the D series. The eight D aldohexoses shown in Figure 16.11 all have different names. Their enantiomers are given the same names except that the prefix is L instead of D.

## Other Aldoses

The aldohexoses have six-carbon chains. Similar compounds having chains of other lengths also exist. These are called aldotetroses, aldopentoses, aldoheptoses, and so on. The $2^n$ formula tells us there must be 8 aldopentoses, 32 aldoheptoses, and so forth. The D forms of some of these are shown in Figure 16.11. Most of these compounds are not found in nature, and only one is important enough to mention here, the aldopentose D-ribose:

$2^3 = 8$
$2^5 = 32$

D-Ribose

**Aldohexoses:** D-(+)-Allose, D-(+)-Altrose, D-(+)-Glucose, D-(+)-Mannose, D-(−)-Gulose, D-(−)-Idose, D-(+)-Galactose, D-(+)-Talose

**Aldopentoses:** D-(−)-Ribose, D-(−)-Arabinose, D-(+)-Xylose, D-(−)-Lyxose

**Aldotetroses:** D-(−)-Erythrose, D-(−)-Threose

**Aldotriose:** D-(+)-Glyceraldehyde

**Figure 16.11**
The family of D aldoses, shown in Fischer projections. D-Glyceraldehyde, the smallest monosaccharide with a chiral carbon, is the standard on which the whole series is based. A (+) symbol means the compound shown rotates the plane of polarized light to the right; (−) indicates rotation to the left.

As we shall see in Section 23.2, D-ribose in its cyclic form is a component of ribonucleic acid (RNA).

## Ketoses

Most monosaccharides are aldoses, but another type also exists, called **ketoses.** These compounds have a ketone group in the 2 position instead of an aldehyde group in the 1 position. Only one ketose is important; it has six carbons (so it is a **ketohexose**) and is called D-fructose:

$$
\begin{array}{c}
CH_2OH \\
| \\
C{=}O \\
| \\
HO{-}C{-}H \\
| \\
H{-}C{-}OH \\
| \\
H{-}C{-}OH \\
| \\
CH_2OH
\end{array}
$$

The highest-numbered chiral carbon of the chain

D-Fructose

Note that D-fructose has the same configuration as D-glucose at C-3, C-4, and C-5. It differs from D-glucose only at C-1 and C-2. D-Fructose is found in honey and in many sweet fruits. It is also known as *fruit sugar*. D-Fructose is also the main sugar in semen. The sperm use it as a source of energy to propel themselves.

## 16.7 PHYSICAL PROPERTIES OF MONOSACCHARIDES

### Sweetness

We have no mechanical way to measure sweetness. It is done by having a group of people taste solutions of varying sweetness.

Artificial sweeteners are discussed in Box 18B.

Most of the monosaccharides (and disaccharides) are sweet. For this reason they are called **sugars.** The degree of sweetness varies. As shown in Table 16.1, D-fructose is much sweeter than D-glucose. Ordinary table sugar, sucrose, is not a monosaccharide, but a disaccharide, the structure of which is given in Section 16.10.

### Solubility

At room temperature carbohydrates are solids. Because of the many —OH groups, as well as the oxygen of the aldehyde or ketone group, monosaccharides are extremely soluble in water. These groups form numerous hydrogen bonds with the solvating water molecules. In the body this solubility allows a fast transport of energy supply in the form of sugar to different parts of the body through the circulatory system. Very concentrated solutions of carbohydrates in water are thick, viscous liquids. We are all familiar with such solutions as honey, molasses, and maple syrup (Fig. 16.12).

### Optical Rotation

All three forms of D-glucose (alpha, beta, and open-chain) rotate the plane of polarized light to the right. For this reason it is often called D-(+)-glucose. Its enantiomer, L-glucose, must therefore rotate the plane to the left and is called

**TABLE 16.1  Comparative Sweetness of Some Sugars and Artificial Sweeteners**

| Sugar or Artificial Sweetener | Sweetness Relative to Sucrose | Type |
|---|---|---|
| Lactose | 0.16 | Disaccharide |
| Galactose | 0.32 | Monosaccharide |
| Maltose | 0.33 | Disaccharide |
| Glucose | 0.74 | Monosaccharide |
| Sucrose | 1.00 | Disaccharide (table sugar) |
| Invert sugar | 1.25 | Mixture of glucose and fructose |
| Fructose | 1.74 | Monosaccharide |
| Aspartame | 100–150 | Artificial sweetener |
| Acesulfame-K | 200 | Artificial sweetener |
| Saccharin | 450 | Artificial sweetener |

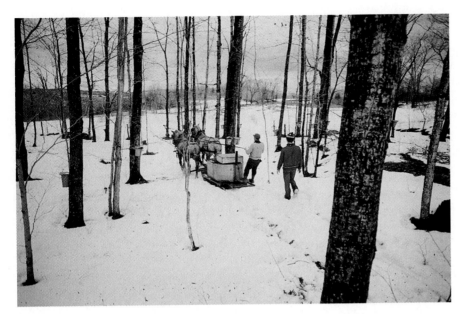

L-(−)-glucose. However, this does not mean that all monosaccharides in the D series rotate the plane to the right. The letters D and L refer to the configuration of the molecules, and **there is no simple relationship between configuration and rotation.** D-Glucose just happens to rotate the plane to the right. Some of the other D monosaccharides also rotate to the right, but others rotate to the left (see Fig. 16.11). The ketose D-fructose is one of the D monosaccharides that rotate the plane to the left and hence is named D-(−)-fructose. Because of their rotation, D-(+)-glucose is known as *dextrose* and D-(−)-fructose as *levulose*.

Of course, no matter which way any D monosaccharide rotates the plane of polarized light, its enantiomer always rotates it the opposite way.

## 16.8 CHEMICAL PROPERTIES OF MONOSACCHARIDES

### Cyclic Forms

Like D-glucose, all monosaccharides with at least five carbons exist predominantly in the cyclic forms rather than in the open-chain aldehyde or ketone forms. Most of these cyclic forms have six-membered rings, called **pyranose** rings, similar to those of alpha- and beta-glucose. However, some of the monosaccharides form five-membered rings, called **furanose** rings, rather than the six-membered kind. Two examples are D-fructose and D-ribose:

D-Fructose ($\alpha$ form)      D-Ribose ($\beta$ form)

Like D-glucose, all of the monosaccharides that have cyclic structures show mutarotation when dissolved in water; they form equilibrium mixtures of the $\alpha$, $\beta$, and open-chain forms.

## Oxidation

All aldoses have, in the open-chain form, an aldehyde group. Aldehydes are easily oxidized to carboxylic acids (Sec. 13.5), and so are aldoses. Because of this, aldoses are regarded as **reducing sugars,** meaning that they reduce oxidizing agents (this reaction is further considered in Section 16.9). Although ordinary ketones are not easily oxidized, ketoses are because they have an —OH group on the carbon next to the C=O group (Sec. 13.5). Therefore, ketoses are also reducing sugars, which means that all monosaccharides are reducing sugars. Because of this property, the most common tests for monosaccharides (and other reducing sugars) consist of treating them with oxidizing agents, such as Benedict's solution (Sec. 13.5).

## Glycoside Formation

We saw in Section 13.6 that aldehydes react with alcohols in the presence of $H^+$ to give acetals, with hemiacetals as intermediates. When D-glucose, a hemiacetal, is treated with methanol in the presence of HCl as a catalyst, it is converted to an acetal:

With glucose, the products are called **glucosides.**

$\alpha$-D-Glucose      Methyl $\alpha$-D-glucopyranoside      Methyl $\beta$-D-glucopyranoside

Glycosides

Only the C-1 —OH is changed because all the other —OH groups are ordinary alcohol groups. The C-1 —OH is part of a hemiacetal group. Other sugars also give this reaction, and the general name for all these products (including glucosides) is **glycosides.** Because glycosides are acetals rather than hemiacetals, they behave in some respects differently from monosaccharides. The main difference is that glycoside rings do not open when dissolved in water. (We say the ring is "locked.") This means that glycosides are *not* reducing sugars, because there is no aldehyde or keto group available, and do *not* exhibit mutarotation. It also means that $\alpha$ and $\beta$ glycosides do not interconvert when dissolved in water.

## 16.9 DERIVATIVES OF MONOSACCHARIDES

Monosaccharides in which one or more functional groups have been altered through chemical reactions are *monosaccharide derivatives.* The carbonyl group of a monosaccharide can be reduced to give the corresponding sugar alcohol. For example, reduction of D-glucose gives D-sorbitol:

Aldose reductases are the enzymes that specifically catalyze this reaction in the cells.

H
|
C=O
|
H—C—OH
|
HO—C—H            $\xrightarrow{\text{reduction}}$
|
H—C—OH
|
H—C—OH
|
CH₂OH

D-Glucose

CH₂OH
|
H—C—OH
|
HO—C—H
|
H—C—OH
|
H—C—OH
|
CH₂OH

D-Sorbitol

D-Sorbitol is also called D-glucitol. Analogously, the reduction product of D-galactose is D-galactitol, and so forth.

oxidation        oxidation by enzymes

COOH
|
H—C—OH
|
HO—C—H
|
H—C—OH
|
H—C—OH
|
CH₂OH

D-Gluconic acid, an aldonic acid

H
|
C=O
|
H—C—OH
|
HO—C—H
|
H—C—OH
|
H—C—OH
|
COOH

D-Glucuronic acid, a uronic acid

On the other hand, oxidation of the carbonyl group leads to the corresponding *aldonic acid*. Mild oxidizing agents and specific enzymes produce D-gluconic acid from D-glucose.

Another important group of oxidation products of monosaccharides are the *uronic acids*. Oxidation by enzymes of the primary alcohol group at C-6 yields these compounds, which are important building blocks of connective tissue acidic polysaccharides (Sec. 16.12).

Another group of acidic monosaccharides are phosphate esters of the sugars. Sugar phosphates are important intermediates in glycolysis (Sec. 21.2), for example:

Glucose 1-phosphate

Glucose 6-phosphate

Specific enzymes catalyze the esterification of the carbonyl group (C-1) or the primary alcohol group (C-6), to produce either glucose 1-phosphate or glucose 6-phosphate.

Amino sugars are produced by replacing the —OH group on C-2 of monosaccharides with an amino group. Amino sugars and their *N*-acetyl derivatives are important building blocks of the polysaccharides of connective tissue such as cartilage.

A typical *N*-acetyl derivative of an amino sugar.

$\beta$-D-*N*-Acetylglucosamine

## 16.10  DISACCHARIDES

We saw in Section 16.8 that monosaccharides can be converted to glycosides by treatment with an alcohol, ROH:

$$\xrightarrow[\text{H}^+]{\text{ROH}}$$

When the ROH is a second monosaccharide molecule, the product is a **disaccharide.**

### Maltose

The ether oxygen connects C-1 of one ring to C-4 of the other.

The disaccharide **maltose** contains two glucose rings connected in a $1 \rightarrow 4$ linkage:

Locked ring    Unlocked ring

Maltose, a disaccharide

The ring on the left is locked. C-1 is an acetal function (a glycoside, shown in blue), and this ring will not open when maltose is dissolved in water. Furthermore, the C-1 oxygen in maltose points downward; as we have seen, this is an $\alpha$ linkage, and we use the symbol $\alpha(1 \rightarrow 4)$. The second ring, however, is unlocked. It has a hemiacetal at C-1, shown in yellow, so this ring opens in water to give an aldehyde. Because it has one unlocked ring, maltose shows mutarotation and is a reducing sugar.

## Lactose

Another disaccharide abundant in food is **lactose.** It occurs in mammalian milk in approximately 5 percent concentration. It consists of a $\beta$-D-galactose ring linked to a D-glucose ring through a $\beta(1 \rightarrow 4)$ glycosidic linkage:

Lactose

Upon hydrolysis, lactose yields D-galactose and D-glucose.

## Sucrose

In maltose and lactose, the rings are linked $1 \rightarrow 4$ so that one ring is locked and the other is unlocked. In **sucrose** the linkage is from C-1 of D-glucose (an aldose) to C-2 of D-fructose (a ketose):

Sucrose

In this compound *both* rings are locked, so that sucrose is a nonreducing sugar and does not exhibit mutarotation. Sucrose, which is ordinary table sugar, is also called *cane sugar*, though it comes from sugar beets as well as from sugar cane and is also found in other plants. Sucrose is easily hydrolyzed by the enzyme invertase, producing a one-to-one mixture of D-glucose and D-fructose. This mixture, called *invert sugar*, is used as a food additive. It is sweeter than sucrose (Table 16.1).

The ABO blood group system, discovered by Karl Landsteiner (1868–1943) in 1900, is of primary importance in safe blood transfusion. Whether an individual belongs to the A, B, AB, or O group is genetically determined and depends on the particular oligosaccharide assembly on the surface of the red blood cells. These oligosaccharides act as *antigens*. The four different blood types have the following antigen arrangements on the surface of their red blood cells:

The A antigen is α-D-N-acetylgalactosamine:

The B antigen is α-D-galactose:

## 16.11  POLYSACCHARIDES

The unlocked ring of maltose is still a hemiacetal, so a third monosaccharide could be added, giving a **trisaccharide.** If the connection is 1 → 4, there would still be a hemiacetal, and so we could add a fourth and continue the process until thousands of units have been added. Nature does just this in forming **polysaccharides.** The two most important polysaccharides, both made entirely of D-glucose units, are starch and cellulose.

The H antigen, common to all four types, is L-fucose:

The blood carries antibodies against all foreign antigens. When a person receives a blood transfusion, the antibodies clump (aggregate) the foreign red blood cells (Fig. 16C). Thus, an individual having A-type blood has A antigens on the surface of the red blood cells and carries anti-B antibodies (against B antigens). A person of B-type blood carries B antigens on the surface of the red blood cells and has anti-A antibodies (against A antigens). Transfusion of A blood into a person with B blood would be fatal, and vice versa. O-type blood has neither A nor B antigens on the surface of the red blood cell but carries both anti-A and anti-B antibodies. The AB-type individual has both A and B antigens but no anti-A or anti-B antibodies. This gives rise to the following blood transfusion possibilities:

**Figure 16C**
Blood clumping resulting from mixing of blood types A and B. (Phototake.)

$$
\begin{array}{ccc}
 & \text{O} & \\
 & \downarrow & \\
 & \text{O} & \\
\nearrow & \downarrow & \nwarrow \\
\text{A} \longrightarrow \text{A} & & \text{B} \longleftarrow \text{B} \\
\searrow & \downarrow & \swarrow \\
 & \text{AB} & \\
 & \uparrow & \\
 & \text{AB} &
\end{array}
$$

Thus, people with O-type blood are universal donors, and those with AB-type blood are universal acceptors. People with A-type blood can accept only from A- or O-type donors. B-type individuals can accept only B- or O-type blood. AB-type persons can accept blood from all groups. O-type individuals can accept blood only from O-type donors.

## Starch

When the $1 \rightarrow 4$ condensation is repeated many times with $\alpha$-D-glucose molecules, a polyglucose named **amylose** is obtained. Amylose is one of the two components of **starch** and contains between 1000 and 2000 glucose units, all in the $\alpha(1 \rightarrow 4)$ glycosidic linkage (the same as in maltose). This linkage is flexible enough to allow the long-chain amylose molecules to curl up and form either random-coil or helical structures in solution (Fig. 16.13).

**Figure 16.13**
Helixes and random coils in amylose.

The second component of starch, called **amylopectin,** is also a polymer of $\alpha$-D-glucose. Amylopectin is not entirely a straight-chain molecule but has random branches. The branching point is the $\alpha(1 \rightarrow 6)$ glycosidic linkage:

The *m* and *n* refer to the number of glucose units in the linear portions of the chains.

Amylopectin

## Galactosemia

One child of every 18 000 is born with a genetic defect that makes the infant unable to utilize galactose. Galactose is part of milk sugar, lactose, and when the body cannot utilize it, it accumulates in the blood and in the urine (galactosuria). The accumulation is harmful because it leads to mental retardation, growth failure, cataract formation in the lens of the eye (Fig. 16D), and in some severe cases death due to liver damage. When galactose accumulation is due to a deficiency of the enzyme galactokinase, the disorder has only mild symptoms. When the enzyme galactose-1-phosphate uridinyltransferase is deficient, the disorder is called **galactosemia,** and its symptoms, as described above, are severe.

The deleterious effects of galactosemia can be avoided simply by giving the infant a milk formula in which sucrose is substituted for lactose. A galactose-free diet is critical only in infancy. With maturation the children develop another enzyme capable of metabolizing galactose. They are thus able to tolerate galactose as they mature.

**Figure 16D**
A lens removed from the eye of a patient who had a cataract. (a) Top view. (b) Side view.

There are, on the average, 20 to 25 glucose units in the straight-chain form, all connected by $\alpha(1 \rightarrow 4)$ linkages, between branching points (Fig. 16.14). Amylopectin contains as many as $10^5$ to $10^6$ glucose units in one gigantic macromolecule.

A **macromolecule** is a very large molecule.

An important test for the presence of starch is the addition of iodine, $I_2$, to give an intense, deep purple color (Fig. 16.15). In the presence of iodine,

Branching point

**Figure 16.14**
Schematic representation of an amylopectin molecule.

amylose and the straight-chain portion of amylopectin form helixes inside of which the iodine molecules assemble in long polyiodine chains. This assembly of iodine molecules gives the characteristic color.

### Glycogen

Amylose and amylopectin are storage polysaccharides in plants. **Glycogen,** also called **animal starch,** plays a similar role in animal tissue. Thus, when glucose is in abundance it is stored in the form of these polysaccharides in animal and plant tissues.

Glycogen is similar to amylopectin in that it has $\alpha(1 \rightarrow 6)$ branching points and $\alpha(1 \rightarrow 4)$ glycosidic linkages in the straight-chain portions. It is also a giant molecule, containing about $10^6$ glucose units. The structural difference between glycogen and amylopectin is in the degree of branching. Glycogen contains smaller numbers of glucose units in the straight-chain part between branching points (on the average, 10 to 12 units) and consequently has more branching points per molecule than amylopectin.

### Cellulose

**Cellulose** is a linear polymer containing only D-glucose units, all in the $1 \rightarrow 4$ linkage. Structurally, the difference between amylose and cellulose is that in cellulose all the $1 \rightarrow 4$ glycosidic linkages are $\beta$ instead of $\alpha$ (the C-1 oxygen points up instead of down in the Haworth drawing):

**Figure 16.15**
The starch in bread gives a positive iodine test. (Photograph by Beverly March.)

Cellulose

Cellulose is the most abundant molecule in living tissues. It makes up about 50 percent of the total organic carbon in the biosphere. Cotton is almost pure cellulose (95 percent), and wood is about 50 percent cellulose. Cellulose is a structural polysaccharide that provides strength and rigidity to plants. It may contain between 300 and 3000 glucose units in one macromolecule.

Cellulose molecules act very much like extended stiff rods. This enables them to align themselves side by side into well-organized, water-insoluble crystalline lattices (fibers) in which the alcohol groups form numerous intermolec-

## Polysaccharides as Fillers in Drug Tablets

Drugs that have the same generic name contain the same active ingredient (for example, aspirin). In spite of that, many physicians prefer one trade-name drug over another. In most cases there is some difference in the inactive ingredients, called **fillers.** Fillers are not completely inactive, but may affect the rate of drug delivery.

Polysaccharides have been used as fillers because of their solubility in water and their thickening and gelling properties. For example, starch is used in pills, tablets, and capsules for oral delivery, as a filler and as a disintegrating or binding agent. Disintegration is mostly due to the rapid uptake of water. As a result the tablet swells and falls apart, allowing the drug to be delivered in the stomach and the intestines.

ular hydrogen bonds with neighboring chains (Fig. 16.16). Thus, when a piece of cellulosic material is placed in water, there are not enough water molecules on the surface of the cellulose to pull one molecule away from the strongly hydrogen-bonded crystalline matrix. This is why cellulose is insoluble in water.

Cellulose is also a structural polysaccharide for humans since many of our houses, furniture, and other structures are made of wood. Large parts of the textile and plastic industries are based on cellulose (cotton) or cellulose derivatives (rayon) as well as on cellulose acetate and nitrocellulose (an explosive):

Cellulose acetate
(—OH groups
esterified at
random)

Nitrocellulose
(explosive grade)

**Figure 16.16**
Electron micrograph of cellulose fibers. (Biophoto Associates.)

## 16.12 ACIDIC POLYSACCHARIDES

Another group of polysaccharides, **acidic polysaccharides,** plays an important role in the structure and function of connective tissues. These tissues are the matrix between organs and cells that provides mechanical strength and also filters the flow of molecular information between cells. Connective tissues are usually made up of collagen, a structural protein, and a variety of acidic polysaccharides that interact with the collagen to form tight or loose networks.

There is no generalized connective tissue, but there are a large number of highly specialized forms, such as cartilage, bone, synovial fluid, skin, tendons, blood vessel walls, intervertebral disks, and cornea.

### Hyaluronic Acid

A typical and probably the simplest (structurally) acidic polysaccharide of connective tissue is **hyaluronic acid:**

**Figure 16.17**
The joints of this hand are swollen by rheumatoid arthritis. (Courtesy of Drs. P. A. Dieppe, P. A. Bacon, A. N. Bamji, I. Watt, and the Gower Medical Publishing Ltd., London, England.)

In rheumatoid arthritis inflammation of the synovial tissue results in swelling of the joints (Fig. 16.17).

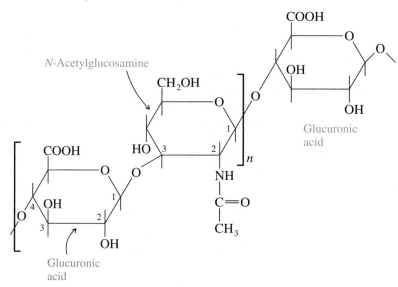

Hyaluronic acid

This giant molecule, with a molecular weight of between $10^5$ and $10^7$, may contain from 300 to 100 000 repeating units, depending on the organ in which it occurs. Hyaluronic acid is most abundant in embryonic tissues and in specialized connective tissues such as synovial fluid, where it acts as a lubricant, or the vitreous of the eye, where its function is to provide a clear, elastic gel that maintains the retina in its proper position (see Box 16F).

Hyaluronic acid is composed of D-glucuronic acid linked to N-acetylglucosamine by a $\beta(1 \rightarrow 3)$ linkage; the latter in turn is linked to the next glucuronic acid by a $\beta(1 \rightarrow 4)$ linkage.

### Heparin

**Heparin** (Fig. 16.18) is one of the most common acidic polysaccharides. It occurs in such tissues as lung, liver, skin, and intestinal mucosa. It has many biological functions, the best known of which is its anticoagulant activity; that is,

## Retina Detachment

The vitreous in a young person is a strong, clear, elastic gel located in the eyeball between the lens and the retina. In the vitreous, collagen fibers are entangled with random coil-like molecules of hyaluronic acid (Fig. 16F). As a person ages, the collagen and hyaluronic acid separate; the hyaluronic acid forms liquid pockets in the otherwise collagen-rich gel. When these pockets become large, the vitreous is not sufficiently elastic to maintain the retina flattened on the back of the eyeball. Under these circumstances the retina can become detached, and the result is blindness.

**Figure 16F**
The structure of vitreous.

A number of serious retinal operations have been performed, with success, in which the vitreous was removed and a pure, noninflammatory hyaluronic acid gel was used as a vitreous replacement.

it prevents blood clotting. Although the details of this function are still unknown, the general features are as follows.

For blood to clot, an active enzyme called thrombin is necessary. Heparin helps to inactivate this enzyme by forming a complex (Fig. 16.19) between itself, thrombin, and antithrombin (another blood constituent). Different heparin preparations have different anticoagulant activities. As is evident from its structure (Fig. 16.18), heparin is a heterogeneous polysaccharide in which the monosaccharide units are linked by $\alpha(1 \rightarrow 4)$ glycosidic bonds. The composi-

**Figure 16.18**
A tetrasaccharide repeating unit of heparin.

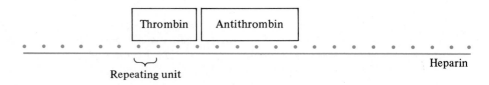

**Figure 16.19**
Schematic diagram of the complex between heparin, thrombin, and antithrombin.

tion of heparin is not constant; it varies slightly from repeating unit to repeating unit. The structure in Figure 16.18 shows a tetrasaccharide as a repeating unit. A heparin preparation with good anticoagulant activity has a minimum size of at least eight to ten such repeating units. The larger the molecule, the better its anticoagulant activity.

## 16.13 CHIRAL COMPOUNDS IN BIOLOGICAL ORGANISMS

Most organic compounds in the body contain one or more chiral carbons, which means that two or more stereoisomers are possible. The general rule is that, in most cases, **nature makes only one of the possible stereoisomers.** For example, the important steroid cholesterol (Sec. 17.9) contains eight chiral carbons. There are, therefore, $2^8 = 256$ stereoisomers of this compound. Of these 256, nature makes only one. No matter what the organism, the same isomer of cholesterol is found.

This is not an absolute rule, however. As we have seen, both D-glucose and D-galactose (diastereomers) are found in nature, and both the $(+)$ and the $(-)$ forms of lactic acid (enantiomers) are found in different organisms. There are even some cases where nature makes racemic mixtures (Sec. 16.3). But in most cases, where several isomers are possible, only one is made.

What is the reason for this? The answer is that biological molecules must often react with each other. We saw in Section 16.3 that enantiomers differ in the rates at which they react with other chiral molecules. Diastereomers differ even more. Thus it is likely that the other 255 isomers of cholesterol (including its enantiomer) would undergo cholesterol's biochemical reactions much more slowly than cholesterol itself.

It's a matter of *fit*. A chiral molecule fits another chiral molecule, where its enantiomer (or a diastereomer) does not fit. This behavior becomes extremely important when we consider that almost all biological reactions are catalyzed by enzymes. Enzymes, as we shall see in Chapter 19, are chiral molecules with chiral cavities. Figure 16.20 shows a typical situation. One enantiomer fits into the cavity in a way that allows the reaction to proceed (like a right hand slipping into a right-handed glove). As seen in Figure 16.20, the enantiomer (the left hand) does not fit. This slows the reaction so much that, for all practical purposes, it does not take place at all.

When we know all this, we are not surprised to learn that the biological properties of enantiomers can differ greatly. D-Glucose is sweet and nutritious and is an important component of our diets. The L isomer, on the other hand, is

CH₃—CH—COOH
|
OH
Lactic acid

The $(+)$ isomer of lactic acid is produced by muscle tissue in humans. The $(-)$ isomer is found in sour milk.

One tragic example of the different biological properties of enantiomers is the compound thalidomide:

Chiral carbon

This compound has a chiral carbon and therefore exists in two enantiomeric forms (+ and −). In 1957 the racemic mixture was introduced into several European countries (though not the United States) as a hypnotic: a sleeping pill and tranquilizer said to have no side effects. However, a major side effect soon came to light. Many pregnant women who took the drug gave birth to babies with deformities, especially of the limbs. For example, a large number of babies had mere stumps instead of arms, with 2, 3, or 4 fingers or none at all. Other birth defects were also present. In 1961 thalidomide was taken off the market and has not been sold since. Some years later it was discovered that only the (−) enantiomer has teratogenic properties while only the (+) enantiomer has the hypnotic action. (A *teratogen* is a substance that produces birth defects.) The testing of the separate enantiomers for teratogenic properties was done on rats, not humans.

**Box 16G**

# Thalidomide

tasteless, and the body cannot metabolize it. These substances have different tastes because taste arises from interactions between chiral molecules and chiral receptors on the taste buds of our tongues. Again, the enantiomers fit differently.

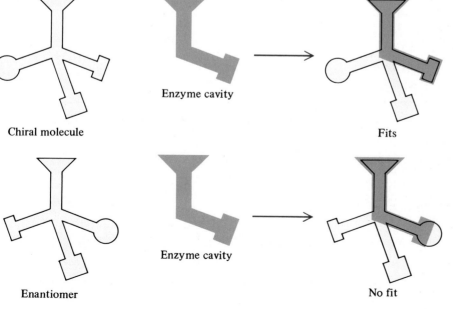

Chiral molecule

Enzyme cavity

Fits

Enantiomer

Enzyme cavity

No fit

**Figure 16.20**
Schematic diagram showing how one chiral molecule fits into an enzyme cavity but its enantiomer does not.

## SUMMARY

**Carbohydrates** are polyhydroxy aldehydes or ketones or compounds that, on hydrolysis, yield polyhydroxy aldehydes or ketones. Their chief source in nature is photosynthesis, a process carried out by plants.

Molecules that are not superimposable on their mirror images are called **chiral.** Most chiral molecules contain a **chiral carbon** (also called an **asymmetric carbon**), which is one connected to four different groups. Since a chiral molecule is not superimposable on its mirror image, the two are different molecules. They are **a pair of enantiomers,** or **optical isomers.** All chiral molecules rotate the plane of **polarized light** (light that vibrates in only one plane); molecules that are not chiral do not rotate the plane of polarized light.

Enantiomers have identical properties except for two: (1) they rotate the plane of polarized light in opposite directions, and (2) they react at different rates with other chiral compounds. This is very important in biological systems, where nature usually makes only one of a pair of enantiomers. A mixture of equal amounts of a $(+)$ and $(-)$ enantiomer is called a **racemic mixture** or a **racemate.** When a molecule has more than one chiral carbon, the total number of stereoisomers usually equals $2^n$, where $n$ is the number of chiral carbons. **Diastereomers,** which are stereoisomers that are not enantiomers, do not have identical properties.

**Monosaccharides,** the basic units of carbohydrates, are simple sugars of formula $C_nH_{2n}O_n$, where $n = 3$ to 8. The most important is D-glucose, which in the solid state has a six-membered (**pyranose**) ring

containing a hemiacetal group. In solution, the ring opens to give a mixture of one open-chain and two cyclic forms. Glucose is an **aldose.** All aldoses, in the open-chain form, are straight-chain compounds containing an aldehyde group at C-1 and an —OH group on all other carbons. The 16 **aldohexoses** differ in their configuration at C-2 to C-5. **Ketoses** are similar, but contain a keto group at C-2 instead of an aldehyde at C-1. Almost all monosaccharides are more stable in cyclic forms, either pyranose (six-membered) or **furanose** (five-membered) rings. Isomerism at the C-1 carbon produces $\alpha$ and $\beta$ cyclic **anomers.** Monosaccharides are placed in the D or L series depending on whether the —OH group on the highest-numbered chiral carbon of the chain points left or right in the Fischer projection.

Carbohydrates are soluble in water. Many mono- and disaccharides are sweet. All monosaccharides are **reducing sugars,** and almost all form **glycosides** when treated with alcohols and an acid catalyst. Derivatives of monosaccharides are formed by oxidation, reduction, and phosphorylation. When two or more monosaccharides are linked together by glycosidic bonds, we get **di-, tri-, oligo-,** and **polysaccharides.** Storage polysaccharides (starch and glycogen) are polymers of $\alpha$-D-glucose linked by $1 \rightarrow 4$ and $1 \rightarrow 6$ glycosidic bonds. Cellulose, a structural polysaccharide, has $\beta(1 \rightarrow 4)$ glycosidic linkages. Acidic polysaccharides are important components of connective tissues and play roles in many biochemical processes.

## KEY TERMS

Acidic polysaccharide (Sec. 16.12)
Aldohexose (Sec. 16.6)
Aldonic acid (Sec. 16.9)
Aldose (Sec. 16.6)
Anomers (Sec. 16.5)
Antibody (Box 16C)
Antigen (Box 16C)
Asymmetric carbon (Sec. 16.2)
Carbohydrate (Sec. 16.1)
Chiral carbon (Sec. 16.2)
Chirality (Sec. 16.2)

Dextrorotatory (Sec. 16.3)
Diastereomer (Sec. 16.4)
Disaccharide (Sec. 16.10)
Enantiomers (Sec. 16.2)
Fischer projection (Sec. 16.3)
Furanose ring (Sec. 16.8)
Galactosemia (Box 16D)
Glucoside (Sec. 16.8)
Glycoside (Sec. 16.8)
Haworth formula (Sec. 16.5)
Ketose (Sec. 16.6)

Levorotatory (Sec. 16.3)
Locked ring (Sec. 16.8)
Macromolecule (Sec. 16.11)
Monosaccharide (Sec. 16.5)
Mutarotation (Sec. 16.5)
Oligosaccharide (Sec. 16.5)
Optical activity (Sec. 16.3)
Photosynthesis (Sec. 16.1)
Polarimeter (Sec. 16.3)
Polarized light (Sec. 16.3)

Polysaccharide (Sec. 16.11)
Pyranose ring (Sec. 16.8)
Racemate (Sec. 16.3)
Racemic mixture (Sec. 16.3)
Reducing sugar (Sec. 16.8)
Specific rotation (Box 16A)
Sugar (Sec. 16.7)
Teratogen (Box 16G)
Unlocked ring (Sec. 16.10)
Uronic acid (Sec. 16.9)

## PROBLEMS

**16.4** (a) What percent of the dry weight of plants is carbohydrates? (b) How does that compare with humans?

**16.5** Which gas is produced (a) by plants during photosynthesis, (b) by animals during metabolism?

### Enantiomers and Chirality

**16.6** Give an example of an optically inactive compound containing one carbon atom.

**16.7** Identify all the chiral carbons:

(a) $CH_3CH_2CH_2CH-CH_3$
        |
        OH

(b) $CH_3CH_2CH_2CH_2OH$

(c) $CH_3CHCH_2CHCH_2CH_3$
        |        |
        OH      OH

(d) $CH_3CH_2$        Cl
              \      /
               C=C
              /      \
            H          $CH_3$

(e) $CH_3CH_2$        Cl
              \      /
               C=C
              /      \
            H          $CHCH_3$
                        |
                        Cl

(f) $CH_3CH_2CHCH_2CH_2Br$
              |
              OH

(g) Br

(h)

(i)

(j)

(k)

(l) $HO-CH_2CH_2-\overset{\overset{\displaystyle Br}{\overset{\displaystyle |}{\overset{\displaystyle CH_2}{\overset{\displaystyle |}{CH_2}}}}}{\underset{\underset{\displaystyle CH_3}{\underset{\displaystyle |}{CH_2}}}{\underset{\displaystyle |}{C}}}-CH_2CH_2-Cl$

(m) $CH_2-CH-C-CH-CH_2$
         |      |     ‖     |      |
         OH   OH   O   OH   OH

(n) $CH_2-CH-CH-CH-CH-C-H$
         |      |      |      |      |     ‖
         OH   OH   OH   OH   OH   O

## Properties of Enantiomers. Optical Activity

**16.8**   Here are some properties of D-isomenthol. Fill in the corresponding properties of its enantiomer, L-isomenthol.

| Property | D-Isomenthol | L-Isomenthol |
|---|---|---|
| Melting point | 82.5°C | |
| Boiling point | 218.5°C | |
| Solubility in water | 1.4 g/100 mL | |
| Density | 0.9040 g/mL | |
| Rotation of the plane of polar- ized light | +24.1° | |

**16.9**   You are given a sample of 2-butanol, a compound that contains a chiral carbon. Yet it fails to rotate the plane of polarized light. Explain how this is possible.

**16.10**   In which two properties does the (+) form of the amino acid arginine differ from the (−) form?

**\*16.11**   Pasteur noticed that a solution of a racemic mixture of tartaric acid in his laboratory was growing moldy. He filtered off the mold and found that the solution now rotated the plane of polarized light, though it had not done so before it became moldy. Explain how the mold caused the solution to become optically active.

## Compounds with More Than One Chiral Carbon

**16.12**   How many stereoisomers are possible for each of these compounds?

(a)   $CH_3-CH-CH-CH_3$
          $\quad\quad\;\; |\quad\;\; |$
          $\quad\quad\; OH\;\; Br$

(b)   $CH_3CH_2-CH-CH_2CH_3$
          $\quad\quad\quad\quad\;\; |$
          $\quad\quad\quad\quad\; Br$

(c)   $CH_2-CH-CH-CH_2$
          $\quad\; |\quad\;\; |\quad\;\; |\quad\;\; |$
          $\quad OH\;\; OH\;\; Br\;\; OH$

(d)

**16.13**   Define:

(a)   enantiomer         (c)   polarized light
(b)   diastereomer      (d)   racemic mixture

(e)   $CH_2-CH-CH-CH-C-H$
          $\quad |\quad\;\; |\quad\;\; |\quad\;\; |\quad\; ||$
          $\quad OH\;\; OH\;\; OH\;\; OH\;\; O$

(f)   $CH_2-CH-CH-C-CH_2$
          $\quad |\quad\;\; |\quad\;\; |\quad\; ||\quad\; |$
          $\quad OH\;\; OH\;\; OH\;\; O\;\; OH$

(g)

## Glucose

**16.14**   What is the difference between oligo- and polysaccharides?

**16.15**   In glucose which carbon is the anomeric carbon?

**16.16**   Define: (a) carbohydrate  (b) monosaccharide

**16.17**   What functional groups are present on each of these carbons of the glucose molecule?
(a) C-1   (b) C-2   (c) C-6

**16.18**   Are $\alpha$-D-glucose and $\beta$-D-glucose enantiomers? Explain.

**16.19**   Explain why the optical rotation of D-glucose changes when it is dissolved in water.

**16.20**   (a) The specific rotation of $\alpha$-D-glucose is +112.2°. What is the specific rotation of $\alpha$-L-glucose?   (b) When D-glucose is dissolved in water, the specific rotation changes to +52.7°. Does the specific rotation of L-glucose also change when it is dissolved in water? If so, what value does it change to?

## Monosaccharides

**16.21**   Tell whether each of these monosaccharides is D or L:

(a)

(b)

(c)

$$
\begin{array}{c}
\text{CHO} \\
| \\
\text{H}-\text{C}-\text{OH} \\
| \\
\text{H}-\text{C}-\text{OH} \\
| \\
\text{HO}-\text{C}-\text{H} \\
| \\
\text{CH}_2\text{OH}
\end{array}
$$

**16.22** The structure of D-altrose is given in Figure 16.11. Draw its Haworth formula in both the alpha and beta configurations.

**16.23** Which is the anomeric carbon in a ketopentose?

**16.24** In an aldoheptose, configurational change about which carbon produces D and L isomerism?

**\*16.25** Draw the structure of L-(−)-mannose both in the straight-chain and cyclic forms (use Fig. 16.11).

## Physical Properties of Monosaccharides

**16.26** Explain why all monosaccharides and disaccharides are soluble in water.

**16.27** The amino acid L-leucine rotates the plane of polarized light to the left. (a) In which direction does D-leucine rotate the light? (b) Can you predict in which direction L-valine (another amino acid) rotates the plane of the light?

## Chemical Properties of Monosaccharides

**16.28** Define: (a) aldose (b) ketose (c) pyranose (d) furanose

**\*16.29** The open-chain formula of D-arabinose is shown in Figure 16.11. Draw the structural formulas for the two furanose forms of this sugar.

**16.30** (a) Why is D-fructose given the designation D-(−)? (b) How is this expressed in its common name?

**16.31** What do we call sugars that give positive tests with Benedict's solution?

## Derivatives of Monosaccharides

**16.32** What is the product when an aldose reductase catalyzes the reduction of D-glucose?

**16.33** Draw the structural formula for each principal product:

(a)

$+ \text{CH}_3\text{CH}_2\text{OH} \xrightarrow{\text{H}^+}$

(b)

$$
\begin{array}{ccccc}
\text{CH}_2-\text{CH}-\text{CH}-\text{CH}-\text{C}-\text{H} \\
| \quad | \quad | \quad | \quad \| \\
\text{OH} \quad \text{OH} \quad \text{OH} \quad \text{OH} \quad \text{O}
\end{array}
$$

$+ [\text{Ag (NH}_3)_2]^+ + \text{OH}^- \longrightarrow$

(c)

$+ \text{CH}_3\text{OH} \xrightarrow{\text{H}^+}$

**\*16.34** Draw the structural formulas for (a) D-galactonic acid, (b) D-galacturonic acid

**16.35** In amino sugars the OH group on C-2 is replaced by an amino group. Write the structure of $\alpha$-D-$N$-acetylgalactosamine.

## Disaccharides

**16.36** (a) Draw the structure of a disaccharide made of D-riboses linked together by a $(1 \rightarrow 3)$ glycosidic linkage. (b) Is this a reducing sugar?

**\*16.37** A disaccharide often found in mushrooms and other fungi is called trehalose. It is formed by the condensation of two $\alpha$-D-glucose molecules through their anomeric carbons $(1 \rightarrow 1)$. Therefore, it is a nonreducing sugar. Draw the structure of trehalose.

**16.38**

(a) Does this compound have an acetal linkage? Find it. (b) Does it have a hemiacetal linkage? Find it. (c) Does it give a positive test with Benedict's solution? (d) How does it differ from maltose?

## Polysaccharides

**16.39** What is the difference between (a) cellulose and amylose, (b) amylopectin and glycogen?

**16.40** How does iodine interact with amylose?

**16.41** Draw the possible products of the reaction between $H_2SO_4$ and amylose:

## Acidic Polysaccharides

**\*16.42** Polygalacturonic acid (pectic acid) is part of the intercellular matrix in plants. It is used in the food industry as a gelling agent in preparing marmalades and jellies. The glycosidic linkage between the $\alpha$-D-galacturonic acid units is $\alpha(1 \rightarrow 4)$. Draw the structure of polygalacturonic acid.

**\*16.43** Identify the products when hyalobiuronic acid is completely hydrolyzed by 6 $M$ HCl:

**16.44** Hyaluronic acid acts as a lubricant in the synovial fluid of joints. In rheumatoid arthritis, inflammation breaks hyaluronic acid down to smaller molecules. What happens to the lubricating power of the synovial fluid?

**\*16.45** The anticlotting property of heparin is partly due to the negative charges it carries. (a) Identify the functional groups that provide the negative charges. (b) Which type of heparin is a better anticlotting agent, one with a high or a low degree of polymerization?

## Chiral Compounds in Biological Organisms

**16.46** Why is the study of chiral molecules important in biochemistry?

**16.47** Explain why sucrose is sweet but its enantiomer is not.

## Boxes

**16.48** (Box 16A) The specific rotation $[\alpha]$ of a particular compound is not a constant but depends on two factors. What are they?

**16.49** (Box 16A) A chemist finds that a sample with a concentration of 2.73 g/mL in a sample tube 2.00-dm long rotates the plane of polarized light by $-71.4°$. Calculate the specific rotation $[\alpha]$.

**16.50** (Box 16B) Are all the alcohol groups in $\alpha$-D-glucose in the equatorial position?

**16.51** (Box 16B) Draw the formula of (a) $\alpha$-L-glucose and (b) $\alpha$-L-mannose (Fig. 16.11) in the chair forms.

**16.52** (Box 16B) The chair form of the pyranose ring shows the three-dimensional shape of carbohydrate molecules. The axis of the molecule runs through its center vertically. Therefore, groups above or below the "plane" of the ring are called axial. Write the structure of the chair form of $\alpha$-D-galactose and indicate any axial $-$OH groups.

**16.53** (Box 16C) The H antigen that all blood types possess is L-fucose. (a) Is this a hexose? (b) Is this a pyranose or furanose ring? (c) What is unusual about this molecule?

**16.54** (Box 16C) (a) What is an antibody? (b) How does it act against foreign red blood cells?

**16.55** (Box 16C) An individual with AB-type blood is a universal acceptor. He or she can accept blood even from an individual with O-type blood. Why is this so?

**16.56** (Box 16C) Why can't a person with B-type blood donate to a person with A-type blood?

**16.57** (Box 16D) What is the difference between galactosemia and galactosuria?

**16.58** (Box 16D) Why can galactosemia be relieved by feeding a baby a formula containing sucrose?

**16.59** (Box 16D) Why does congenital galactosemia appear only in infants?

**16.60** (Box 16E) Why are polysaccharides such as starch or pectin good fillers?

**16.61** (Box 16E) How does a filler influence the effectiveness of a drug?

**16.62** (Box 16F) What is the most important biological function of the vitreous and its component, hyaluronic acid?

**16.63** (Box 16G) Thalidomide was a good sleeping pill. Why was it taken off the market?

**16.64** (Box 16G) Would you expect thalidomide to be a base? Explain.

## Additional Problems

**16.65** How many different aldononoses (9 carbons) can there be?

**16.66** Do glycosides exhibit mutarotation?

**16.67** What role does chlorophyll play in carbohydrate synthesis by plants?

**16.68** Which is the most important monosaccharide in the human body?

**16.69** How many chiral carbons are in (a) D-threose, (b) D-xylose? (See Fig. 16.11.)

*****16.70** Draw the structure of a disaccharide made of one unit of D-idose (locked pyranose) and one unit of D-altrose (unlocked pyranose) in a $1 \rightarrow 4$ linkage.

**16.71** Why are monosaccharides and disaccharides called sugars?

**16.72** What are the names of the compounds produced when mannose is (a) reduced at C-1, (b) oxidized at C-6?

*****16.73** The structure of chondroitin-4-sulfate, an important component of cartilage, is similar to that of hyaluronic acid. The repeating unit consists of a D-glucuronic acid unit connected by a $\beta(1 \rightarrow 3)$ linkage to an N-acetylgalactosamine unit, the C-4 of which has a sulfate group. The N-acetyl group is on C-2. These disaccharide units are connected by a $\beta(1 \rightarrow 4)$ linkage as in hyaluronic acid. Write the structure of chondroitin-4-sulfate.

**16.74** A monosaccharide is dissolved in water. Its optical rotation is $+17.5°$. An hour later the rotation of the same solution is $+15.1°$, and 48 hours later it is $-2.6°$. What molecular process is going on?

**16.75** Why is cellulose insoluble in water?

**16.76** Which of these compounds are chiral?

(a)

(b)

(c)

(d)

(e)

**16.77** Draw the structure of a disaccharide in which two D-glucose units are linked in a $\beta(1 \rightarrow 6)$ linkage.

**16.78** The shells of insects, crabs, and lobsters contain a structural polysaccharide called chitin, the structure of which is

In what ways is this polymer similar to cellulose, and in what ways is it different?

**16.79** Ribitol and ribose-1-phosphate are derivatives of $\beta$-D-ribose. Draw the structures of these compounds.

**16.80** (a) Which disaccharide contains only $\alpha$-D-glucose units? (b) Which polysaccharides?

# CHAPTER 17

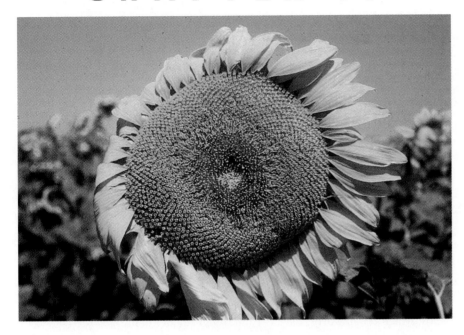

*Oil from sunflower seeds is polyunsaturated, owing to its high linoleic acid content. (Courtesy of the U.S. Department of Agriculture.)*

# Lipids

## 17.1   INTRODUCTION

**Lipids** are substances found in living organisms that are **insoluble in water but soluble in nonpolar solvents and solvents of low polarity.** This lack of solubility in water is an important property because our body chemistry is so firmly based on water. Most body constituents, including carbohydrates, are soluble in water. But the body also needs insoluble compounds for many purposes, including the separation of compartments containing aqueous solutions from each other, and that's where lipids come in.

Note that, unlike the case of carbohydrates, we define lipids in terms of a property and not in terms of structure.

The water-insolubility of lipids is due to the fact that the polar groups they contain are much smaller than their alkane-like (nonpolar) portions. These nonpolar portions provide the water-repellent, or *hydrophobic,* property.

"Hydrophobic" means water-hating.

An important use for lipids, especially in animals, is the storage of energy. As we saw in Section 16.1, plants store energy in the form of starch. Animals (including humans) find it more economical to use lipids (fats) instead. Although our bodies do store some carbohydrates in the form of glycogen for quick energy when we need it, energy stored in the form of fats is much more important. The reason is simply that the burning of fats produces more than twice as much energy (about 9 kcal/g) as the burning of an equal weight of carbohydrates (about 4 kcal/g).

For purposes of study, we can divide lipids into four groups: (1) fats and waxes, (2) complex lipids, (3) steroids, and (4) prostaglandins and leukotrienes.

$$R-\overset{\overset{O}{\|}}{C}-O-R'$$

Acid part    Alcohol part

## 17.2 THE STRUCTURE OF FATS

Fats are esters. We saw in Section 14.7 that esters are made up of an alcohol part and an acid part.

### The Alcohol Part

In fats the alcohol part is always glycerol:

$$
\begin{array}{l}
CH_2-OH \\
CH-OH \\
CH_2-OH
\end{array}
$$

Glycerol

### The Acid Part

In contrast to the alcohol part, the acid component of fats may be any number of acids, which do, however, have certain things in common:

**1.** They are practically all straight-chain carboxylic acids (virtually no branching).
**2.** They range in size from about 10 carbons to 20 carbons.
**3.** They have an even number of carbon atoms.
**4.** Apart from the —COOH group, they have no functional groups except that some do have double bonds.

The reason that only even-numbered acids are found in fats is that the body builds these acids entirely from acetic acid units and therefore puts the carbons in two at a time (Sec. 22.3). Table 17.1 shows the most important acids found in fats.

Since glycerol has three —OH groups, a single molecule of glycerol can be attached to three different acid molecules. Thus, a typical fat molecule might be

Remember (Sec. 14.2) that these are called fatty acids.

$$CH_2-O-\overset{\overset{O}{\|}}{C}-(CH_2)_{12}-CH_3$$  Myristic acid, 14 carbons

$$CH-O-\overset{\overset{O}{\|}}{C}-(CH_2)_7-CH=CH-(CH_2)_7-CH_3$$  Oleic acid, 18 carbons, 1 double bond

$$CH_2-O-\underset{\underset{O}{\|}}{C}-(CH_2)_{16}-CH_3$$  Stearic acid, 18 carbons

Such compounds are called **triglycerides** or **triacylglycerols:** All three —OH groups of glycerol are esterified. Triglycerides are the most common lipid materials, although **mono-** and **diglycerides** are not infrequent. In the latter two types, only one or two —OH groups of the glycerol are esterified by fatty acids.

Triglycerides are complex mixtures. Although some of the molecules have three identical fatty acids, in most cases two or three different acids are present.

**TABLE 17.1  The Most Important Acids in Fats**

| Acid | Number of Carbons | Number of Double Bonds | Formula | Schematic Structure | Melting Point (°C) |
|------|-------------------|------------------------|---------|---------------------|--------------------|
| Lauric | 12 | 0 | $CH_3(CH_2)_{10}COOH$ | | 44 |
| Myristic | 14 | 0 | $CH_3(CH_2)_{12}COOH$ | | 54 |
| Palmitic | 16 | 0 | $CH_3(CH_2)_{14}COOH$ | | 63 |
| Stearic | 18 | 0 | $CH_3(CH_2)_{16}COOH$ | | 70 |
| Oleic | 18 | 1 | $CH_3(CH_2)_7CH=CH(CH_2)_7COOH$ | | 4 |
| Linoleic | 18 | 2 | $CH_3(CH_2)_4CH=CHCH_2CH=CH(CH_2)_7COOH$ | | −5 |
| Linolenic | 18 | 3 | $CH_3CH_2CH=CHCH_2CH=CHCH_2CH=CH(CH_2)_7COOH$ | | −11 |
| Arachidonic | 20 | 4 | $CH_3(CH_2)_4CH=CHCH_2CH=CHCH_2CH=CHCH_2CH=CH(CH_2)_3COOH$ | | −49.5 |

The hydrophobic character of fats is caused by the long hydrocarbon chains. The ester linkages, though polar, are buried in a nonpolar environment, and this makes the fats insoluble in water.

The fatty acids can be divided into two groups: saturated and unsaturated. Saturated fatty acids have only single bonds in the hydrocarbon chain. Unsaturated fatty acids have at least one C=C double bond in the chains. All the unsaturated fatty acids listed in Table 17.1 are the cis isomers. This explains their physical properties, reflected in their melting points. Saturated fatty acids are solids at room temperature because the regular nature of their aliphatic chains allows the molecules to be packed in a close, parallel alignment:

The interactions (attractive forces) between neighboring chains are weak, but the regular packing allows these forces to operate over a large portion of the chain so that a considerable amount of energy is needed in order to melt them.

In contrast, unsaturated fatty acids are all liquids at room temperature because the cis double bonds interrupt the regular packing of the chains:

Thus much less energy is required to melt them. The greater the degree of unsaturation, the lower the melting point, because each double bond introduces more disorder into the packing of the molecules.

*The longer the aliphatic chain, the higher the melting point.*

## 17.3   PROPERTIES OF FATS

### Physical State

With some exceptions, fats that come from animals are generally solids at room temperature, and those from plants or fish are usually liquids. Liquid fats are often called **oils,** though they are esters of glycerol just like solid fats and should not be confused with petroleum, which is mostly alkanes.

What is the structural difference between solid fats and liquid oils? In most cases it is the degree of unsaturation. The physical properties of the fatty acids are carried over to the physical properties of the triglycerides. Solid animal fats contain mainly *saturated fatty acids,* and vegetable oils contain high amounts of *unsaturated fatty acids.* Table 17.2 shows the average fatty acid content of some common fats and oils. Note that even solid fats contain some unsaturated acids and that liquid fats contain some saturated acids. Some unsaturated fatty acids (linoleic and linolenic acids) are called *essential fatty acids* because the body cannot synthesize them from precursors; they must therefore be included in the diet.

**TABLE 17.2  Average Percentage of Fatty Acids and Iodine Number of Some Common Fats and Oils**

|  | Lauric | Myristic | Palmitic | Stearic | Oleic | Linoleic | Linolenic | Other | Iodine Number |
|---|---|---|---|---|---|---|---|---|---|
| **Animal Fats** | | | | | | | | | |
| Beef tallow | — | 6.3 | 27.4 | 14.1 | 49.6 | 2.5 | — | 0.1 | 50 |
| Butter | 2.5 | 11.1 | 29.0 | 9.2 | 26.7 | 3.6 | — | 17.9 | 36 |
| Human | — | 2.7 | 24.0 | 8.4 | 46.9 | 10.2 | — | 7.8 | 68 |
| Lard | — | 1.3 | 28.3 | 11.9 | 47.5 | 6.0 | — | 5.0 | 59 |
| **Vegetable Oils** | | | | | | | | | |
| Coconut | 45.4 | 18.0 | 10.5 | 2.3 | 7.5 | — | — | 16.3 | 10 |
| Corn | — | 1.4 | 10.2 | 3.0 | 49.6 | 34.3 | — | 1.5 | 123 |
| Cottonseed | — | 1.4 | 23.4 | 1.1 | 22.9 | 47.8 | — | 3.4 | 106 |
| Linseed | — | — | 6.3 | 2.5 | 19.0 | 24.1 | 47.4 | 0.7 | 179 |
| Olive | — | — | 6.9 | 2.3 | 84.4 | 4.6 | — | 1.8 | 81 |
| Palm | — | 1.4 | 40.1 | 5.5 | 42.7 | 10.3 | — | — | 54 |
| Peanut | — | — | 8.3 | 3.1 | 56.0 | 26.0 | — | 6.6 | 93 |
| Safflower | | ←——— | 6.8 | ———→ | 18.6 | 70.1 | 3.4 | 1.1 | 145 |
| Soybean | 0.2 | 0.1 | 9.8 | 2.4 | 28.9 | 52.3 | 3.6 | 2.7 | 130 |
| Sunflower | — | — | 5.6 | 2.2 | 25.1 | 66.2 | — | — | 126 |

Though most vegetable oils have high amounts of unsaturated fatty acids, there are exceptions. Note that coconut oil has only a small amount of unsaturated acids. This oil is a liquid not because it contains many double bonds but because it is rich in low-molecular-weight fatty acids (chiefly lauric).

Oils with an average of more than one double bond per fatty acid chain are called *polyunsaturated*. For some years there has been a controversy about whether a diet rich in unsaturated and polyunsaturated fats helps to prevent heart attacks. (See Box 17G.)

Pure fats and oils are colorless, odorless, and tasteless. This statement may seem surprising, since we all know the tastes and colors of such fats and oils as butter and olive oil. The tastes, odors, and colors are caused by substances dissolved in the fat or oil.

### Hydrogenation

In Section 11.6 we learned that double bonds can be reduced to single bonds by treatment with hydrogen ($H_2$) and a catalyst. It is therefore not difficult to convert unsaturated liquid oils to solids, for example,

It is useful to have a test for degree of unsaturation in fats and oils. One such test that has been used for many years is the iodine number test. Recall that halogens add to the carbons of a double bond (Sec. 11.6):

$$-\overset{|}{\underset{|}{C}}=\overset{|}{\underset{|}{C}}- + I_2 \longrightarrow -\overset{\overset{\displaystyle I}{|}}{\underset{|}{C}}-\overset{\overset{\displaystyle I}{|}}{\underset{|}{C}}-$$

The test consists of adding a halogen to a sample of fat or oil. The iodine number is defined as the number of grams of iodine (or iodine equivalent) that adds to 100 grams of the sample. Because iodine itself is not very reactive, the interhalogen compounds iodine monochloride, ICl, and iodine monobromide, IBr, are used instead. Iodine numbers for some common fats and oils are given in Table 17.2. As the table shows, liquid oils generally have higher iodine numbers than solid fats.

Margarine is also made by partial hydrogenation of vegetable oils. Because less hydrogen is used, margarine contains more unsaturation than hydrogenated shortenings.

This hydrogenation is carried out on a large scale to produce the solid shortening sold in stores under such brand names as Crisco, Spry, and Dexo. In making such products, manufacturers must be careful not to hydrogenate all the double bonds because a fat with no double bonds at all would be too solid. Partial, but not complete, hydrogenation results in a product with the right consistency for cooking.

**Saponification**

Glycerides, being esters, are subject to hydrolysis, which can be catalyzed by either acids or bases. As we saw in Section 14.10, base catalysis is more practical. An example of the saponification of a typical fat is

$$
\begin{array}{l}
H_2C-O-\overset{\overset{\displaystyle O}{\|}}{C}-(CH_2)_{12}CH_3 \\[2mm]
HC-O-\overset{\overset{\displaystyle O}{\|}}{C}-(CH_2)_{14}CH_3 \\[2mm]
H_2C-O-\overset{\overset{\displaystyle O}{\|}}{C}-(CH_2)_7CH=CHCH_2CH=CH(CH_2)_4CH_3
\end{array}
\quad + \ 3NaOH \xrightarrow{H_2O}
$$

$$
\begin{array}{ll}
H_2C-OH & CH_3(CH_2)_{12}-\overset{\overset{\displaystyle O}{\|}}{C}-O^-\ Na^+ \\[2mm]
HC-OH & + \ CH_3(CH_2)_{14}\overset{\overset{\displaystyle O}{\|}}{C}-O^-\ Na^+ \\[2mm]
H_2C-OH & CH_3(CH_2)_4CH=CHCH_2CH=CH(CH_2)_7\overset{\overset{\displaystyle O}{\|}}{C}-O^-\ Na^+
\end{array}
$$

The mixture of sodium salts of fatty acids produced by saponification of fats or oils is **soap.** Soap has been used for thousands of years, and saponification is one of the oldest known chemical reactions (Box 17C).

The double bonds in fats and oils are subject to oxidation by the air. When a fat or oil is allowed to stand out in the open, this reaction slowly turns some of the molecules into aldehydes and other compounds with bad tastes and odors. We say the fat or oil has become rancid and is no longer edible. Vegetable oils, which generally contain more double bonds, are more susceptible to this than solid fats, but even fats contain some double bonds, so rancidity can be a problem here, too.

Another cause of unpleasant taste is hydrolysis. The hydrolysis of triglycerides may produce short-chain fatty acids, which have bad odors. In order to prevent rancidity, fats and oils should be kept refrigerated (these reactions are slower at low temperatures) and in dark bottles (the oxidation is catalyzed by ultraviolet light). In addition to this, antioxidants are often added to fats and oils to prevent rancidity.

## 17.4   COMPLEX LIPIDS

The triglycerides discussed in the previous sections are significant components of fat storage cells. Other kinds of lipids, called **complex lipids,** are important in a different way. They constitute the main components of membranes (Sec. 17.5). Chemically, complex lipids can be divided into two groups: phospholipids and glycolipids.

**Phospholipids** contain an alcohol, fatty acids, and a phosphate group. There are two types: phosphoglycerides and sphingolipids. In **phosphoglycerides,** the alcohol is glycerol (see Sec. 17.6). In **sphingolipids,** the alcohol is sphingosine (see Sec. 17.7).

**Glycolipids** are complex lipids that contain carbohydrates (see Sec. 17.8). Figure 17.1 shows schematic structures for all of these.

**Figure 17.1**
Schematic diagram of simple and complex lipids.

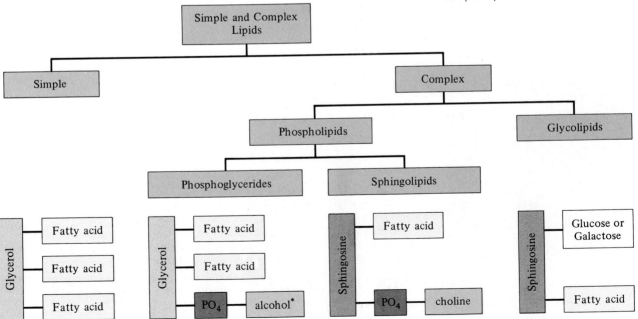

*The alcohol can be choline, serine, ethanolamine, or certain others.

**B o x   17C**

## Soaps and Detergents

Soaps clean because each soap molecule has a hydrophilic head and a hydrophobic tail. The —COO⁻ end of the molecule (the hydrophilic end), being ionic, is soluble in water. The other end (the hydrophobic end) is a long-chain alkane-like portion, which is insoluble in water but soluble in organic compounds. Dirt may contain water-soluble and water-insoluble portions. The water-soluble portion dissolves in water; no soap is needed for it. The soap is needed for the water-insoluble portion. The hydrophilic end of the soap molecule dissolves in the water; the hydrophobic end dissolves in the dirt. This causes an emulsifying action in which the soap molecules surround the dirt particles in an orderly fashion, the hydrophobic tail interacting with the hydrophobic dirt particle and the hydrophilic head providing the attraction for water molecules (Fig. 17C).

**Figure 17C**
The cleaning action of soap. (a) Without soap, water molecules cannot penetrate through molecules of hydrophobic dirt. (b and c) Soap molecules line up at the interface between dirt and water. (d) Soap molecules carry dirt particles away.

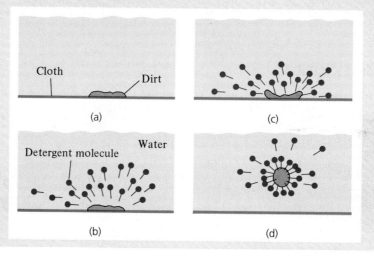

**B o x   17D**

## Waxes

**Waxes** are simple esters. They are solids because of their high molecular weights. As in fats, the acid portions of the esters consist of a mixture of fatty acids, but the alcohol portions are not glycerol but simple long-chain alcohols. For example, a major component of beeswax (Fig. 17D) is 1-triacontyl palmitate:

$$CH_3(CH_2)_{14}-\overset{\overset{\displaystyle O}{\|}}{C}-O-(CH_2)_{29}-CH_3$$

Palmitic acid portion          1-Triacontanol portion

Waxes generally have higher melting points than fats (60 to 100°C) and are harder. Animals and plants use them for protective coatings. The leaves of most plants are coated with wax, which helps to prevent microorganisms from attacking them and also allows them to conserve water. The feathers of birds and the fur of animals are also coated with wax. This is what allows ducks to swim.

Some important waxes are carnauba wax (from a Brazilian palm tree), lanolin (from lamb's wool), beeswax, and spermaceti (from whales). These are used to make cosmetics, polishes, candles, and ointments. Paraffin waxes are not esters. They are mixtures of high-molecular-weight alkanes. Neither is ear wax a simple ester. It is a gland secretion and contains a mixture of fats (triglycerides), phospholipids, and esters of cholesterol.

**Figure 17D**
Bees making beeswax. (Photograph by Charles D. Winters.)

The solid soaps and soap flakes with which we are all familiar are sodium salts of fatty acids. Liquid soaps are generally potassium salts of the same acids. Soap has the disadvantage of forming precipitates in *hard water*, which is water that contains relatively high amounts of $Ca^{2+}$ and $Mg^{2+}$ ions. Most community water supplies in the United States are hard water. Calcium and magnesium salts of fatty acids are insoluble in water. Therefore, when soaps are used in hard water, the $Ca^{2+}$ and $Mg^{2+}$ ions precipitate the fatty acid anions, with two unfortunate consequences: A certain amount of soap is wasted because the precipitated anions are not available for cleaning, and the precipitate itself is now a kind of dirt, which deposits on clothes, dishes, sinks, and bathtubs (it is the "ring" around the bathtub). Because of these difficulties, much of the soap used in the United States has been replaced in the last 40 years by **detergents** for both household and industrial cleaning. Two typical detergent molecules are

$$CH_3(CH_2)_{11}-O-SO_2O^-\ Na^+$$

Sodium dodecyl sulfate
(sodium lauryl sulfate)

$$CH_3(CH_2)_9-\underset{\underset{CH_3}{|}}{CH}-\hspace{-0.5em}\bigcirc\hspace{-0.5em}-SO_2O^-\ Na^+$$

Sodium *para*-(2-dodecyl) benzene sulfonate

It can be seen immediately that these molecules also have hydrophobic tails and hydrophilic heads and so can function in the same way as soap molecules. However, they do not precipitate with $Ca^{2+}$ and $Mg^{2+}$, and so work very well in hard water.

## 17.5 MEMBRANES

It is the complex lipids mentioned in Section 17.4 that form the **membranes** around body cells and around small structures inside the cells. Unsaturated fatty acids are important components of these lipids. The purpose of these membranes is to separate cells from the external environment and to provide selective transport for nutrients and waste products. That is, membranes allow the selective passage of substances into and out of cells.

These small structures inside the cell are called **organelles.**

These membranes are made of lipid bilayers (Fig. 17.2). In a **lipid bilayer** there are two rows (layers) of lipid molecules arranged tail to tail. The hydrophobic tails point toward each other because that enables them to get as far away from the water as possible. This leaves the hydrophilic heads projecting to the inner and outer surfaces of the membrane.

"Hydrophilic" means water-loving.

The unsaturated fatty acids prevent the tight packing of the hydrophobic chains in the lipid bilayer and thereby provide a liquid-like character to the membranes. This is of extreme importance because many products of the body's biochemical processes must cross the membrane, and the liquid nature of the lipid bilayer allows such transport.

Most lipid molecules in the bilayer contain at least one unsaturated fatty acid.

The lipid part of the membrane serves as a barrier against any movement of ions or polar compounds into and out of the cells. In the lipid bilayer, protein molecules are either suspended on the surface or partly or fully embedded in

This effect is similar to that which causes unsaturated fatty acids to have lower melting points than saturated fatty acids.

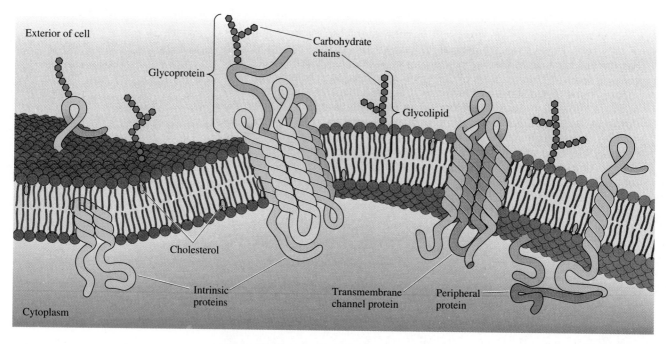

**Figure 17.2**
The fluid mosaic model of membranes. Note that proteins are embedded in the lipid matrix. (Adapted from J. R. Amend, B. P. Mundy, and M. T. Armold, *General, Organic, and Biochemistry.* Philadelphia: Saunders College Publishing, 1993.)

the bilayer. These proteins stick out either on the inside or on the outside of the membrane; others are thoroughly embedded, going through the bilayer and projecting from both sides. The model shown in Figure 17.2, called the **fluid mosaic model** of membranes, allows the passage of nonpolar compounds by diffusion, since these compounds are soluble in the lipid membranes. Polar compounds are transported either through specific channels through the protein regions or by another mechanism called active transport (see Fig. 24.3). For any transport process, the membrane must behave like a nonrigid liquid so that the proteins can move sideways within the membrane.

## 17.6   PHOSPHOGLYCERIDES

The structure of phosphoglycerides is very similar to that of simple fats. The alcohol is glycerol. Two of the three —OH groups are esterified by fatty acids. As with the simple fats, these fatty acids may be any long-chain carboxylic acids with or without double bonds. The third —OH group is esterified not by a fatty acid but by a phosphate group, which is also connected to another alcohol. If the other alcohol is *choline*, a quaternary ammonium compound,

$$HO—CH_2CH_2—\overset{+}{N}(CH_3)_3$$

Choline

*Phosphoglycerides are membrane components of cells throughout the body.*

the phosphoglycerides are called **phosphatidylcholines** (common name **lecithin**):

Glycerol portion

$$CH_2-O-\overset{\overset{\displaystyle O}{\|}}{C}-(CH_2)_{16}-CH_3$$

Stearic acid portion

$$CH-O-\overset{\overset{\displaystyle O}{\|}}{C}-(CH_2)_7-CH=CH-CH_2-CH=CH-(CH_2)_4-CH_3$$

$$CH_2-O-\overset{\overset{\displaystyle O}{\|}}{\underset{\underset{\displaystyle O-CH_2CH_2-\overset{+}{N}(CH_3)_3}{|}}{P}}-O^-$$

Phosphate portion

Choline portion

Linoleic acid portion

This typical lecithin molecule has stearic acid on one end and linoleic acid in the middle. Other lecithin molecules contain other fatty acids, but the one on the end is always saturated and the one in the middle unsaturated.

A typical lecithin

Note that lecithin has a negatively charged phosphate group and a positively charged quaternary nitrogen from the choline. These charged parts of the molecule provide a strongly hydrophilic head, while the rest of the molecule is hydrophobic. Thus, when a phospholipid such as lecithin is part of a lipid bilayer, the hydrophobic tail points toward the middle of the bilayer, and the hydrophilic heads line both the inner and outer surfaces of the membranes (Figs. 17.2 and 17.3).

**Figure 17.3**
Space-filling molecular models of complex lipids in a bilayer. (From L. Stryer, *Biochemistry,* 2nd ed. New York: W. H. Freeman, 1981. Copyright 1981 by W. H. Freeman and Co. All rights reserved.)

Lecithins are just one example of phosphoglycerides. Another is the **cephalins,** which are similar to the lecithins in every way except that, instead of choline, they contain other alcohols, such as ethanolamine or serine:

$$HOCH_2CH_2NH_2$$
Ethanolamine

$$HOCH_2CH-COOH$$
$$|$$
$$NH_2$$
Serine

R = glycerol + fatty acid portions

Phosphatidylethanolamine
(cephalin)

Phosphatidylserine
(cephalin)

## 17.7   SPHINGOLIPIDS

Johann Thudichum, who discovered sphingolipids in 1874, named these brain lipids after the monster of Greek mythology, the sphinx. Part woman and part winged lion, the sphinx devoured all who could not provide the correct answer to her riddles. Sphingolipids appeared to Thudichum as part of a dangerous riddle of the brain.

The coating of nerve axons **(myelin)** contains a different kind of complex lipid called **sphingolipids.** In sphingolipids the alcohol portion is sphingosine:

$$CH_3(CH_2)_{12}-CH=CH-CH-CH-CH_2$$
$$|\quad\quad|\quad\quad|$$
$$OH\quad NH_2\quad OH$$
Sphingosine

A long-chain fatty acid is connected to the $-NH_2$ group by an amide linkage, and the $-OH$ group at the end of the chain is esterified by phosphorylcholine:

Sphingomyelin (a sphingolipid)

Sphingomyelin (schematic diagram)

The combination of a fatty acid and sphingosine (shown in color) is often referred to as the **ceramide** part of the molecule, since many of these compounds are also found in cerebrosides (see Sec. 17.8). The ceramide part of complex lipids may contain different fatty acids; stearic acid occurs mainly in sphingomyelin.

The human brain and spinal cord can be divided into gray and white regions. Forty percent of the human brain is white matter. Microscopic examination reveals that the white matter is made up of nerve axons wrapped in a white lipid coating, the **myelin sheath,** which provides insulation and thereby allows the rapid conduction of electrical signals. The myelin sheath consists of 70 percent lipids and 30 percent proteins in the usual lipid bilayer structure.

A specialized cell, the **Schwann cell,** wraps itself around the peripheral nerve axons to form numerous concentric layers (Fig. 17E). In the brain, other cells do the wrapping in a similar manner.

**Box 17E**

## The Myelin Sheath and Multiple Sclerosis

**Figure 17E**
Myelination of a nerve axon out-side the brain by a Schwann cell. The myelin sheath is produced by the Schwann cell and is rolled around the nerve axon for insula-tion. (From "Myelin" by P. Morell and W. T. Norton. Copyright 1980 by Scientific American, Inc. All rights reserved.)

Multiple sclerosis affects 250 000 people in the United States. In this disease, a gradual degradation of the myelin sheath can be observed. The symptoms are muscle weariness, lack of coordination, and loss of vision. The symptoms may vanish for a time and return with greater severity. Autopsy of multiple sclerotic brains shows scar-like plaques of white matter, with bare axons not covered by myelin sheaths. The symptoms are produced because the demyelinated axons cannot conduct nerve impulses. A secondary effect of the demyelination is damage to the axon itself.

Similar demyelination occurs in the Guillain-Barré syndrome that follows certain viral infections. In 1976 fears of a "swine flu" epidemic prompted a vaccination program that precipitated a number of cases of Guillain-Barré syndrome. The main result is a paralysis that can cause death unless artificial breathing is supplied. The U.S. government assumed legal responsibility for the few bad vaccines and paid compensation to the victims and their families.

The phospholipids are not randomly distributed in membranes. In human red blood cells the sphingomyelin and phosphatidylcholine (lecithin) are on the outside of the membrane facing the blood plasma, while phosphatidylethanol-amine and phosphatidylserine (cephalins) are on the inside of the membrane. In viral membranes most of the sphingomyelin is on the inside of the membrane.

## 17.8 GLYCOLIPIDS

**Glycolipids** are complex lipids that contain carbohydrates. Among the glycolipids are the **cerebrosides,** which are ceramide mono- or oligosaccharides.

In cerebrosides the fatty acid of the ceramide part may contain either 18-carbon or 24-carbon chains, the latter found only in these complex lipids. A glucose or galactose carbohydrate unit forms a beta glycosidic bond with the ceramide portion of the molecule. The cerebrosides occur primarily in the brain (7 percent of the dry weight) and at nerve synapses.

β-D-Glucose

Ceramide

Glucocerebroside

Complex lipids are constantly being synthesized and decomposed in the body. Several genetic diseases are classified as lipid storage diseases. In these cases some of the enzymes needed to decompose the complex lipids are defective or altogether missing from the body. As a consequence, the complex lipids accumulate and cause enlarged liver and spleen, mental retardation, blindness, and, in certain cases, early death. Table 17F summarizes some of these diseases and indicates the missing enzyme and the accumulating complex lipid (Fig. 17F).

At present there is no treatment for these diseases. The best way to prevent them is by genetic counseling. Some of these diseases can be diagnosed during fetal development. For example, Tay–Sachs disease, which is carried by about 1 in every 30 Jewish Americans (versus 1 in 300 in the non-Jewish population), can be diagnosed from amniotic fluid obtained by amniocentesis.

However, there is now an experimental treatment available for patients with mild forms of certain lipid storage diseases such as Gaucher's disease. The human enzyme β-glucosidase can be extracted from placenta. It is encapsulated in a lipid bilayer and injected intravenously. The lipid bilayer capsule allows the drug to pass through the membranes of spleen cells and thus deliver the enzyme where it is needed. Unfortunately, because it is difficult to isolate the human enzyme from placenta, each injection costs thousands of dollars. If and when the human gene for the enzyme is isolated, the enzyme can be manufactured by the recombinant DNA technique (Sec. 23.14) and the cost drastically lowered.

**TABLE 17F  Lipid Storage Diseases**

| Name | Missing or Defective Accumulating Lipid | Enzyme |
|---|---|---|
| Gaucher's disease | Glucocerebroside | β-Glucosidase |
| Krabbe's leukodystrophy | Galactocerebroside | β-Galactosidase |
| Fabry's disease | Ceramide trihexoside | α-Galactosidase |
| Tay–Sachs disease | Ceramide oligosaccharide | Hexosaminidase A |
| Niemann–Pick disease | Sphingomyelin | Sphingomyelinase |

## 17.9   STEROIDS. CHOLESTEROL

The third major class of lipids is the **steroids,** which are compounds containing this ring system:

There are three cyclohexane rings (A, B, and C) connected in the same way as in phenanthrene (Sec. 11.12) and a fused cyclopentane ring (D). Steroids are thus completely different in structure from the lipids already discussed. Note that they are not necessarily esters, though some of them are.

### Cholesterol

The most abundant steroid in the human body, and the most important, is cholesterol:

Cholesterol

It serves as a membrane component, mostly in the plasma membranes of red blood cells and in the myelinated nerve cells. The second important function of cholesterol is to serve as a raw material for other steroids, such as the sex and adrenocorticoid hormones (Sec. 17.10) and bile salts (Sec. 17.11).

Cholesterol exists both in the free form and esterified with fatty acids. Gallstones contain free cholesterol (Fig. 17.4).

Because the correlation between high serum cholesterol levels and such diseases as atherosclerosis has received so much publicity, many people are afraid of cholesterol and regard it as some kind of poison. It should be apparent from this discussion that, far from being poisonous, cholesterol is necessary for human life. Without it, we would die. Fortunately, there is no chance of that, since, even if it were completely eliminated from the diet, our livers would make enough to satisfy our needs.

Cholesterol in the body is in a dynamic state. Most of the cholesterol ingested and that manufactured by the liver is used by the body to make other molecules, such as bile salts. The serum cholesterol level controls the amount of cholesterol synthesized by the liver.

Cholesterol, along with fat, is transported from the liver to the peripheral tissues by lipoproteins. There are two important kinds: **high density lipopro-**

When the cholesterol level goes above 150 mg/100 mL, cholesterol synthesis in the liver is reduced to half the normal rate of production.

**Figure 17.4**
Cholesterol crystals, taken from fluid in the elbow of a patient suffering from bursitis, as seen in a polarizing light microscopic photograph. (Courtesy of Drs. P. A. Dieppe, P. A. Bacon, A. N. Bamji, I. Watt, and the Gower Medical Publishing Co. Ltd., London, England.)

tein (HDL), which has a protein content of about 50 percent, and **low density lipoprotein (LDL),** in which the protein content is only about 25 percent. LDL contains about 45 percent cholesterol; HDL only 18 percent. High density lipoprotein transports cholesterol to the liver and also transfers cholesterol to LDL. Low density lipoprotein plays an important role in cholesterol metabolism. The core of LDL (Fig. 17.5) contains fats (triglycerides) and esters of cholesterol in which linoleic acid is an important constituent. On the surface of the LDL cluster are phospholipids, free cholesterol, and proteins. These molecules contain many polar groups so that water molecules in the blood plasma can solvate them by forming hydrogen bonds, making the LDL soluble in the blood plasma. The LDL carries cholesterol to the cells, where specific LDL-receptor molecules line the cell surface in certain concentrated areas called **coated pits.** One of the proteins on the surface of the LDL binds specifically to the LDL-receptor molecules in the coated pits. After such binding the LDL is taken inside the cell, where enzymes break it down, liberating the free choles-

**Lipoproteins** are spherically shaped clusters containing both lipid molecules and protein molecules.

Michael Brown and Joseph Goldstein of the University of Texas shared the Nobel Prize in Medicine for 1986 for the discovery of the LDL-receptor-mediated pathway.

- ▨ Unesterified cholesterol
- ▨ Phospholipid
- ▢ Cholesteryl ester
- ▨ Protein
- ▨ Hydrophobic tails

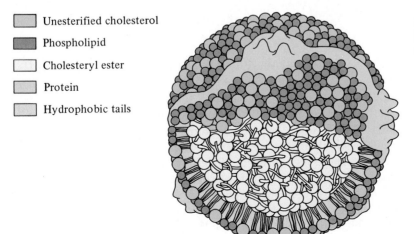

**Figure 17.5**
Low density lipoprotein shown schematically.

## Cholesterol and Heart Attacks

Like all lipids, cholesterol is insoluble in water, and if its level is elevated in the blood serum, plaque-like deposits may form on the inner surfaces of the arteries. This leads to a decrease in the diameter of the blood vessels, which may lead to a decrease in the flow of blood. *Atherosclerosis* (Fig. 17G) is the result, along with accompanying high blood pressure, which may lead to heart attack, stroke, or kidney dysfunction. Atherosclerosis enhances the possible complete blockage of some arteries by a clot at the point where the arteries are constricted by plaque. Furthermore, blockage may deprive cells of oxygen, and these may cease to function. The death of heart muscles due to lack of oxygen is called *myocardial infarction.*

The more general condition, *arteriosclerosis,* or hardening of the arteries with age, is also accompanied by increased levels of cholesterol in the blood serum. Whereas young adults have, on the average, 1.6 g of cholesterol per liter of blood, in people older than 55 this almost doubles to 2.5 g/L because the rate of metabolism slows with age. Diets low in cholesterol and saturated fatty acids usually reduce the serum cholesterol level, and a number of drugs are available that inhibit the synthesis of cholesterol in the liver. The commonly used drug, lovastatin, and related compounds act to inhibit one of the key enzymes in cholesterol synthesis, called HMG-CoA reductase. Thus, they block the synthesis of cholesterol inside the cells and stimulate the synthesis of LDL receptor proteins. In this way more LDL enters the cells, diminishing the amount of cholesterol that will be deposited on the inner walls of arteries. Although there is a good correlation between high serum cholesterol and various circulatory diseases, not everyone who suffers from hardening of the arteries has high serum cholesterol, nor do all patients with high serum cholesterol develop arteriosclerosis.

**Figure 17G**
Effect of atherosclerosis in arteries. (a) Minimal change. (b) Severe atherosclerosis. (Courtesy of Dr. William G. Marshall, Jr., Jewish Hospital of St. Louis, and *Annals of Thoracic Surgery.*)

(a)                                        (b)

terol from cholesterol esters. In this manner cholesterol can be used by the cell, for example, as a component of a membrane. This is the normal fate of LDL and the normal course of cholesterol transport.

In certain cases, however, there are not enough LDL receptors. In the disease called familial hypercholesterolemia, the cholesterol level in the plasma may be as high as 680 mg/100 mL as opposed to 175 mg/100 mL in normal subjects. These high levels of cholesterol can cause atherosclerosis and heart attacks (Box 17G). The high plasma cholesterol levels in these patients are caused because there are not enough functional LDL receptors, or, if there are enough, they are not concentrated in the coated pits. Thus high LDL content means high cholesterol content in the plasma, since LDL cannot get into the cells and be metabolized. Therefore a high LDL level together with a low HDL level is a symptom of faulty cholesterol transport and a warning for possible atherosclerosis.

At the present time, our knowledge of the role played by serum cholesterol in atherosclerosis is incomplete. The best we can say is that it probably makes good sense to reduce the amount of cholesterol and saturated fatty acids in the diet.

## 17.10   STEROID HORMONES

Cholesterol is the starting material for the synthesis of steroid hormones. In this process, the aliphatic side chain on the D ring is shortened by the removal of a six-carbon unit, and the secondary alcohol group on C-3 is oxidized to a ketone. The resulting molecule, *progesterone*, serves as the starting compound for both the sex hormones and the adrenocorticoid hormones (Fig. 17.6).

### Adrenocorticoid Hormones

The adrenocorticoid hormones are products of the adrenal glands. We divide them into two groups according to function: *Mineralocorticoids* regulate the concentrations of ions (mainly $Na^+$ and $K^+$), and *glucocorticoids* control carbohydrate metabolism.

*Aldosterone* is one of the most important mineralocorticoids. Increased secretion of aldosterone enhances the reabsorption of $Na^+$ and $Cl^-$ ions in the kidney tubules and increases the loss of $K^+$. Since $Na^+$ concentration controls water retention in the tissues, aldosterone also controls tissue swelling.

The term "adrenal" comes from *adjacent* to the *renal* (which refers to the kidney). The name "corticoid" indicates that the site of the secretion is the cortex (outer part) of the gland.

**Figure 17.6**
The biosynthesis of hormones from progesterone.

## Anabolic Steroids

Testosterone, the principal male hormone, is responsible for the buildup of muscles in men. Because of this, many athletes have taken this drug in an effort to increase muscular development. This is especially common among athletes in sports in which strength and muscle mass are important, including weight lifting, shot put, and hammer throw, but participants in other sports, such as running, swimming, and cycling, would also like larger and stronger muscles.

Although used by many athletes, testosterone has two disadvantages: (1) Besides its effect on muscles, it also affects secondary sexual characteristics, and too much of it can result in undesired side effects. (2) It is not very effective when taken orally, and must be injected for best results.

For these reasons, a large number of other anabolic steroids, all of them synthetic, have been developed. Some examples are the following:

Methandienone

Methenolone

Nandrolone decanoate

Of the three shown here, methandienone and methenolone can be taken orally, but nandrolone decanoate must be injected.

Anabolic steroids are also used by some women athletes. Since their bodies produce only small amounts of testosterone, women have much more to gain from anabolic steroids than men.

The use of anabolic steroids is forbidden in many sporting events, especially in international competition, largely for two reasons: (1) It gives some competitors an unfair advantage, and (2) these drugs can have many side effects, ranging from acne to liver tumors. Side effects can be especially disadvantageous for women, since they can include growth of facial hair, baldness, deepening of the voice, and menstrual irregularities. All athletes participating in the Olympic Games are required to pass a urine test for anabolic steroids. A number of winning athletes have had their victories taken away because they tested positive for steroid use. One tragic example is the world-class sprinter, the Canadian Ben Johnson, whose career was brought to an end at the 1988 Olympiad in Seoul, Korea. He had just won the 100-meter race in world record time when a urine analysis proved that he had taken steroid drugs. He was stripped of both his world record and his gold medal.

*Cortisol* is the major glucocorticoid. Its function is to increase the glucose and glycogen concentrations in the body. This is done at the expense of other nutrients. Fatty acids from fat storage cells and amino acids from body proteins are transported to the liver, which, under the influence of cortisol, manufactures glucose and glycogen from these sources.

Cortisol and its ketone derivative, *cortisone,* have remarkable anti-inflammatory effects. These or similar synthetic derivatives, such as prednisolone, are used to treat inflammatory diseases of many organs, rheumatoid arthritis, and bronchial asthma.

## Sex Hormones

The most important male sex hormone is *testosterone* (Fig. 17.6). This hormone, which promotes the normal growth of the male genital organs, is synthesized in the testes from cholesterol. During puberty, increased testosterone production leads to such secondary male sexual characteristics as deep voice and facial and body hair.

Female sex hormones, the most important of which is *estradiol* (Fig. 17.6), are synthesized from the corresponding male hormone (testosterone) by aromatization of the A ring. Estradiol, together with its precursor progesterone, regulates the cyclic changes occurring in the uterus and ovaries known as the **menstrual cycle.** As the cycle begins, the level of estradiol in the body rises, and this causes the lining of the uterus to thicken. Then another hormone, called luteinizing hormone, triggers ovulation. If the ovum is fertilized, increased progesterone levels will inhibit any further ovulation. Both estradiol and progesterone then promote further preparation of the uterine lining to receive the fertilized ovum. If no fertilization takes place, progesterone production stops altogether, and estradiol production decreases. This decreases the thickening of the uterine lining, and it is then sloughed off with accompanying bleeding. This is menstruation (Fig. 17.7).

Because progesterone is essential for the implantation of the fertilized ovum, blocking its action leads to termination of pregnancy. Progesterone interacts with a receptor (a protein molecule) in the nucleus of cells. The receptor changes its shape when progesterone binds to it (see Sec. 25.3). A new drug, widely used in France and China, called mifepristone, or RU486,

The conversion of the male hormone to the female hormone results from the loss of a —$CH_3$ and a —H:

acts as a competitor to progesterone. It binds to the same receptor site. Because the progesterone molecule is prevented from reaching the receptor molecule,

RU486 also binds to the receptors of glucocorticoid hormones. Its use as an antiglucocorticoid is also recommended to alleviate a disease known as Cushing syndrome, the overproduction of cortisone.

**(a) HORMONAL CYCLE**

Gonadotropins

Luteinizing hormone (LH)

Follicle-stimulating hormone (FSH)

Ovulation

Hormone concentration in the blood

**Surge of estrogen stimulates surge of LH, which stimulates ovulation**

Steroids

Estrogen

Progesterone

| Menstrual period | Proliferative phase (follicle growth) | Secretory phase (corpus luteum) | Menstrual period |

0    5    14    28

Day of menstrual cycle

**(b) OVARIAN PHASES**

Oocyte

Primary follicle   Growing follicle   Mature follicle   Ovulation: rupture of follicle   Development of corpus luteum

**(c) UTERINE PHASES**

Endometrium

Menstrual flow

Proliferative phase    Secretory phase    Menstrual phase

**Figure 17.7**

Events of the menstrual cycle. (a) Levels of sex hormones in the bloodstream during the phases of one menstrual cycle in which pregnancy does not occur. (b) Development of an ovarian follicle during the cycle. (c) Phases of development of the endometrium, the lining of the uterus. The endometrium thickens during the proliferative phase. In the secretory phase, which follows ovulation, the endometrium continues to thicken, and the glands secrete a glycogen-rich nutritive material in preparation to receive an embryo. When no embryo implants, the new outer layers of the endometrium disintegrate and the blood vessels rupture, producing the menstrual flow. (From K. Arms and P. Camp, *Biology: A Journey into Life*, 2nd ed. Philadelphia: Saunders College Publishing, 1991, p. 544.)

the uterus is not prepared for the implantation of the fertilized ovum, and the ovum is aborted. This is a "morning-after pill" (see also Box 17I). If taken orally within 72 hours after intercourse it prevents pregnancy. However, the drug fails in 20 percent of cases unless followed 36 to 48 hours later by an injection, by an intravaginal suppository, or by oral administration of a prostaglandin (see Sec. 17.12). This chemical form of abortion is politically controversial, and the Food and Drug Administration has only recently sanctioned clinical trials of this drug in the United States.

Estradiol and progesterone also regulate secondary female sex characteristics, such as the growth of breasts. Because of this, RU486, as an antiprogesterone, has also been reported to be effective against certain types of breast cancer.

Testosterone and estradiol are not exclusive to either males or females. A small amount of estradiol production occurs in males, and a small amount of testosterone production is normal in females. Only when the proportion of these two hormones (hormonal balance) is upset can one observe symptoms of abnormal sexual differentiation.

---

**Box 17I**

**Oral Contraception**

Because progesterone prevents ovulation during pregnancy, it occurred to investigators that progesterone-like compounds might be used for birth control (see pp. 280–283). Synthetic analogs of progesterone proved to be more effective than the natural compound. In "the Pill," a synthetic progesterone-like compound is supplied together with an estradiol-like compound (the latter prevents irregular menstrual flow). Triple-bond derivatives of testosterone, such as norethindrone, norethynodrel, and ethynodiol diacetate, are used most often in birth control pills:

Norethindrone

Norethynodrel

Ethynodiol diacetate

## 17.11 BILE SALTS

Bile salts are oxidation products of cholesterol. First the cholesterol is converted to the trihydroxy derivative, and the end of the aliphatic chain is oxidized to the carboxylic acid. The latter in turn forms an amide linkage with an amino acid, either glycine or taurine:

Glycine portion

Glycocholate

Taurine portion

Taurocholate

Bile salts are powerful detergents. One end of the molecule is strongly hydrophilic because of the negative charge, and the rest of the molecule is hydrophobic. Thus, bile salts can disperse dietary lipids in the small intestine into fine emulsions and thereby help digestion.

Since they are eliminated in the feces, bile salts also remove excess cholesterol in two ways: They themselves are breakdown products of cholesterol (thus cholesterol is eliminated via bile salts), and they solubilize deposited cholesterol in the form of bile salt–cholesterol particles.

## 17.12 PROSTAGLANDINS AND LEUKOTRIENES

This group of fatty acid–like substances was first discovered when it was demonstrated that seminal fluid caused a hysterectomized uterus to contract. The name implies that these substances are a product of the prostate gland, and in mature males the seminal gland does secrete 0.1 mg of prostaglandin per day. However, small amounts of prostaglandins are present throughout the body in both sexes.

**B o x  17J**

## Action of Anti-inflammatory Drugs

Anti-inflammatory steroids (such as cortisone, Sec. 17.10) exert their function by inhibiting the enzyme phospholipase $A_2$. Phospholipase $A_2$ is the enzyme that releases unsaturated fatty acids from complex lipids in the membranes. This is how arachidonic acid, one of the components of membranes, is made available to the cell. Since arachidonic acid is the precursor of both prostaglandins and leukotrienes, inhibiting its release stops the synthesis of these compounds and prevents inflammation.

Steroids such as cortisone have many undesirable side effects (duodenal ulcer and cataract formation, among others). Therefore, their use must be controlled. Nonsteroidal anti-inflammatory agents are available, among them aspirin, ibuprofen, and indomethacin.

Aspirin (Box 14D) inhibits the enzyme that causes ring closure in the synthesis of prostaglandin from arachidonic acid. Thus, aspirin reduces fever and inflammation by stopping the production of prostaglandins, but it does not stop the production of leukotrienes.

Asthmatic patients must be aware that large quantities of aspirin or indomethacin not only inhibit prostaglandin formation but allow the manufacture of more leukotrienes from the unused arachidonic acid. Thus, large doses of aspirin may precipitate a severe asthma attack in asthmatic patients.

Prostaglandins are synthesized in the body from arachidonic acid by a ring closure at C-8 and C-12:

Arachidonic acid

several steps

$PGF_{2\alpha}$

The prostaglandin E group (PGE) has a carbonyl group at C-9; the subscript indicates the number of double bonds. The prostaglandin F group (PGF) has two hydroxyl groups on the ring at positions C-9 and C-11.

Other prostaglandins (PGAs and PGBs) are derived from PGE.

$PGF_2$

The prostaglandins as a group have a wide variety of effects on body chemistry. They seem to act as mediators of hormones. For example, $PGE_2$ and $PGF_{2\alpha}$ induce labor and are used for therapeutic abortion in early pregnancy. $PGE_2$ lowers blood pressure, but $PGF_{2\alpha}$ causes hypertension (increase of blood pressure). $PGE_2$ in aerosol form is used to treat asthma; it opens up the bronchial tubes by relaxing the surrounding muscles. $PGE_1$ is used as a decongestant; it opens up nasal passages by constricting blood vessels.

Many prostaglandins cause inflammation and fever. The analgesic effect of aspirin results from the inhibition of prostaglandin synthesis (Box 17J). PGA and PGE inhibit gastric secretions.

Another group of substances that act as mediators of hormonal responses are the leukotrienes. Like prostaglandins, leukotrienes are derived from arachidonic acid by an oxidative mechanism. However, in this case there is no ring closure.

Most ulcer patients lack PGE. Ulcers are treated with PGA and PGE.

Arachidonic acid                    Leukotriene B4

Leukotrienes occur mainly in leukocytes but also in other tissues of the body. They produce long-lasting muscle contractions, especially in the lungs, and they cause asthma-like attacks. They are a hundred times more potent than histamines. Both prostaglandins and leukotrienes cause inflammation and fever. The inhibition of their production in the body is a major pharmacological concern.

## SUMMARY

**Lipids** are water-insoluble substances. They are divided into four groups: fats (glycerides), complex lipids, steroids, and prostaglandins and leukotrienes. **Fats** are made up of fatty acids and glycerol. In saturated fatty acids the hydrocarbon chains have only single bonds; unsaturated fatty acids have hydrocarbon chains with one or more double bonds, all in the cis configuration. Solid fats contain mostly saturated fatty acids, while **oils** contain substantial amounts of unsaturated fatty acids. The alkali salts of fatty acids are called **soaps.**

**Complex lipids** can be divided into two groups: phospholipids and glycolipids. **Phospholipids** are made of a central alcohol (glycerol or sphingosine), fatty acids, and a nitrogen-containing phosphate ester, such as phosphorylcholine. The **glycolipids** contain sphin-gosine and a fatty acid, which together are called the **ceramide** portion of the molecule, and a carbohydrate portion. Many phospholipids and glycolipids are important constituents of cell membranes.

**Membranes** are made of a **lipid bilayer** in which the hydrophobic parts of phospholipids (fatty acid residues) point to the middle of the bilayer and the hydrophilic parts point toward the inner and outer surfaces of the membrane.

The third major group of lipids is the **steroids.** The basic feature of the steroid structure is a fused four-ring nucleus. The most common steroid is **cholesterol,** which also serves as a raw material for other steroids, such as bile salts and sex and other hormones. Cholesterol is also an integral part of membranes, occupying the hydrophobic region of the lipid bilayer. Because of its low solubility in water, cholesterol deposits are implicated in the formation of gallstones and the plaque-like deposits of atherosclerosis. Cholesterol is transported in the blood plasma by two kinds of lipoprotein: **HDL** and **LDL.** The LDL plays an important role in cholesterol metabolism. It is soluble in blood plasma because of polar groups on its surface. An oxidation product of cholesterol is **progesterone,** which is a sex hormone and also gives rise to the synthesis of other sex hormones, such as testosterone and estradiol, as well as to the **adrenocorticoid hormones.** Among the latter,

the best known are cortisol and cortisone for their anti-inflammatory action. **Bile salts** are also oxidation products of cholesterol. They emulsify all kinds of lipids, including cholesterol, and are essential in the digestion of fats.

**Prostaglandins** and **leukotrienes** are derived from arachidonic acid. They have a wide variety of effects on body chemistry; among other things, they can lower or raise blood pressure, cause inflammation, and induce labor. They act generally as mediators of hormone action.

## KEY TERMS

Adrenocorticoid hormone (Sec. 17.10)
Anabolic steroid (Box 17H)
Bilayer (Sec. 17.5)
Bile salt (Sec. 17.11)
Cephalin (Sec. 17.6)
Ceramide (Sec. 17.7)
Cerebroside (Sec. 17.8)
Cholesterol (Sec. 17.9)
Coated pits (Sec. 17.9)
Complex lipid (Sec. 17.4)
Detergent (Box 17C)
Diglyceride (Sec. 17.2)
Fat (Sec. 17.2)
Fluid mosaic model (Sec. 17.5)
Glycolipid (Sec. 17.8)
HDL (Sec. 17.9)
Hydrophilic (Sec. 17.5)
Hydrophobic (Sec. 17.1)
Iodine number (Box 17A)
LDL (Sec. 17.9)
Lecithin (Sec. 17.6)
Leukotriene (Sec. 17.12)

Lipid (Sec. 17.1)
Lipid bilayer (Sec. 17.5)
Lipoprotein (Sec. 17.9)
Membrane (Sec. 17.5)
Monoglyceride (Sec. 17.2)
Myelin (Sec. 17.7)
Myelin sheath (Box 17E)
Oil (Sec. 17.3)
Organelle (Sec. 17.5)
Phosphoglyceride (Sec. 17.6)
Phospholipid (Sec. 17.4)
Polyunsaturated oils (Sec. 17.3)
Prostaglandin (Sec. 17.12)
Schwann cell (Box 17E)
Sex hormone (Sec. 17.10)
Soap (Sec. 17.3)
Sphingolipid (Sec. 17.7)
Steroid (Sec. 17.9)
Steroid hormone (Sec. 17.10)
Triacylglycerol (Sec. 17.2)
Triglyceride (Sec. 17.2)

## PROBLEMS

**17.1** How is the definition of lipid different from the definitions of other types of compounds we have considered, such as carbohydrate, ester, or alkene?

**17.2** What is the meaning of the term hydrophobic? Why is the hydrophobic nature of lipids important?

### The Structure of Fats

**17.3** Draw the structural formula of a fat molecule (triglyceride) made of myristic acid, oleic acid, palmitic acid, and glycerol.

**\*17.4** How many glycerides (mono-, di-, and tri-) can be formed from glycerol and myristic acid? Draw their structures.

**17.5**    Draw the six possible combinations of triglycerides you can make out of glycerol, palmitic acid, and oleic acid.

## Properties of Fats

**17.6**    Predict which acid in each pair has the higher melting point, and explain why: (a) palmitic acid or stearic acid  (b) arachidonic acid or arachidic acid (arachidic acid is the saturated 20-carbon acid)

**17.7**    For the six triglycerides in Problem 17.5, predict, on the basis of unsaturation, the order in which their melting points increase.

**\*17.8**    The insoluble zinc salt of a $C_{11}$ unsaturated fatty acid (undecylenic acid) is used in the treatment of athlete's foot. How would you prepare this powder from potassium undecylenate?

**17.9**    Which has the higher melting point: (a) a triglyceride containing only lauric acid and glycerol or  (b) a triglyceride containing only stearic acid and glycerol?

**17.10**    Explain why the melting points of the saturated fatty acids increase going from lauric to stearic acid (Table 17.1).

**17.11**    Predict the order of the melting points of triglycerides containing fatty acids, as follows:
(a)  palmitic, palmitic, stearic
(b)  oleic, stearic, palmitic
(c)  oleic, linoleic, oleic

**17.12**    Coconut oil is a liquid fat (oil), yet it is saturated. Explain why.

**\*17.13**    Rank, in order of increasing solubility in water (assuming that all are made with the same fatty acids) (a) triglycerides    (b) diglycerides (c) monoglycerides. Explain.

**17.14**    How many moles of $H_2$ are used up in the catalytic hydrogenation of one mole of a triglyceride containing glycerol, stearic acid, arachidonic acid, and linoleic acid?

**17.15**    Name the products of the saponification of this triglyceride:

$$CH_2-O-\underset{\underset{O}{\|}}{C}-(CH_2)_{14}CH_3$$

$$CH-O-\underset{\underset{O}{\|}}{C}-(CH_2)_{16}CH_3$$

$$CH_3CH_2-CH=CH$$

$$CH_2-O-\underset{\underset{O}{\|}}{C}-(CH_2)_7CH=CH-CH_2-CH=CH=CH-CH_2$$

**17.16**    Using the equation on page 522 as a guideline for stoichiometry, calculate the number of moles of NaOH it takes to saponify 5 moles of (a) triglycerides    (b) diglycerides    (c) monoglycerides.

## Membranes

**17.17**    (a) Where in the body are membranes found? (b) What functions do they serve?

**17.18**    How do the unsaturated fatty acids of the complex lipids contribute to the fluidity of a membrane?

**17.19**    Which type of lipid molecules are most likely to be present in membranes?

**17.20**    What is the function of protein molecules in membranes?

## Complex Lipids

**17.21**    Draw the structure of a cephalin that contains palmitic acid and oleic acid.

**17.22**    Complex lipids can act as emulsifying agents. The lecithin of egg yolk is used in making mayonnaise. Draw the structure of a lecithin. Identify the portion of the molecule that interacts with oil droplets and the portion that interacts with vinegar (acetic acid in water).

**\*17.23**    Among the phosphoglycerides containing palmitic acid and linolenic acid, which will have the greatest solubility in water: (a) phosphatidylcholine  (b) phosphatidylethanolamine  (c) phosphatidylserine? Explain.

**17.24**    List the names of all the groups of complex lipids that contain ceramides.

**17.25**    Are the various phospholipids randomly distributed in membranes? Give an example.

## Steroids

**\*17.26**    Cholesterol has a fused four-ring core and is a part of body membranes. The —OH group on C-3 is the polar head, and the rest of the molecule provides the hydrophobic tail that does not fit into the zig-zag packing of the hydrocarbon portion of the saturated fatty acids. Considering this structure, tell whether cholesterol contributes to the stiffening (rigidity) or to the fluidity of a membrane. Explain.

**17.27** (a) Is cholesterol necessary for human life?
(b) Why do many people restrict cholesterol intake in their diet?

**\*17.28** (a) Find all the chiral carbons in cholesterol.
(b) How many total isomers are possible?
(c) How many of these do you think are found in nature?

**17.29** Look at the structures of cholesterol and the hormones shown in Figure 17.6. Which ring of the steroid structure undergoes the most substitution?

**17.30** Cholesterol linoleate is found in the core of LDL. Draw its structural formula.

**17.31** What makes LDL soluble in blood plasma?

**17.32** How does LDL deliver its cholesterol to the cells?

## Steroid Hormones and Bile Salts

**17.33** What physiological functions are associated with cortisol?

**17.34** What is the role of progesterone in pregnancy?

**17.35** Describe the chemical difference between the male hormone testosterone and the female hormone estradiol.

**\*17.36** Considering that RU486 can bind to the receptors of progesterone as well as those of cortisone and cortisol, what can you say regarding the importance of the functional group on C-11 of the steroid ring in drug and receptor binding?

**17.37** (a) How does the structure of RU486 resemble that of progesterone? (b) How do the structures differ?

**17.38** Explain how the constant elimination of bile salts through the feces can reduce the danger of plaque formation in atherosclerosis.

## Prostaglandins and Leukotrienes

**17.39** What is the basic chemical difference (a) between arachidonic acid and prostaglandin $PGE_2$? (b) Between $PGE_2$ and $PGF_{2\alpha}$?

**17.40** Find and name all the functional groups in (a) glycocholate (b) cortisone (c) prostaglandin $PGE_2$ (d) leukotriene B4.

**17.41** Both leukotrienes and prostaglandins are oxidation products of arachidonic acid. Which requires a higher degree of oxidation?

## Boxes

**\*17.42** (Box 17A) A sample of 140 g of an unsaturated fatty acid consumed 1 mole of $I_2$ when titrated in ether. Calculate the molecular weight of the unsaturated fatty acids given in Table 17.1. Using these molecular weights, determine which of the unsaturated fatty acids was present in the sample.

**17.43** (Box 17B) What causes rancidity? How can it be prevented?

**17.44** (Box 17C) Calcium and magnesium salts of fatty acids are sparsely soluble in water, but sodium and potassium salts make good soap. Explain, with a stoichiometric equation, the chemical reaction that results in the "ring" around the tub when you take a bath in hard water.

**17.45** (Box 17C) Explain how soap acts as a cleaning agent.

**17.46** (Box 17C) Why have detergents replaced much of the soap used in the United States in recent years?

**17.47** (Box 17E) (a) What is the role of sphingomyelin in the conductance of nerve signals? (b) What happens to this process in multiple sclerosis?

**17.48** (Box 17F) In Gaucher's disease, one of the symptoms is abdominal pain due to an enlarged spleen. What causes this symptom?

**\*17.49** (Box 17F) Comparing the complex lipid structures listed for the lipid storage diseases (Table 17F) with the missing or defective enzymes, explain why in Fabry's disease the missing enzyme is $\alpha$-galactosidase and not $\beta$-galactosidase.

**17.50** (Box 17G) How does lovastatin prevent atherosclerosis?

**17.51** (Box 17H) (a) Why do many athletes take anabolic steroids? (b) Why is their use forbidden in many sporting events?

**17.52** (Box 17I) What is the role of progesterone and similar compounds in contraceptive pills?

**17.53** (Box 17J) How does cortisone prevent inflammation?

**17.54** (Box 17J) What is the role of phospholipase $A_2$ in inflammation?

**17.55** (Box 17J) How does indomethacin act in the body to reduce inflammation?

**17.56** (Box 17J) How does aspirin reduce inflammation and fever?

## Additional Problems

**17.57** How is cholesterol converted to cortisone?

**17.58** What are the constituents of sphingomyelin?

**17.59** (a) Classify aldosterone into as many chemical families as appropriate. (b) In what functional groups does aldosterone differ from cortisone?

**17.60** Fats and complex lipids cannot be absorbed through the intestinal walls. (a) What enzymes help to break down the lipids? (b) In what form are the lipids absorbed into the blood stream?

**17.61** (a) Where in the body is cholesterol synthesized? (b) In what form is cholesterol secreted from the body?

**17.62** What is the role of taurine in lipid digestion?

**17.63** Draw a schematic diagram of a lipid bilayer. Show how the bilayer prevents the passage by diffusion of a polar molecule such as glucose. Show why nonpolar molecules, such as $CH_3CH_2-O-CH_2CH_3$, can diffuse through the membrane.

**17.64** What is the cholesterol level in normal blood plasma?

**\*17.65** Suggest a reason why free cholesterol forms gallstones but the various esters of cholesterol do not.

**\*17.66** What is the major difference between aldosterone and all the other hormones listed in Figure 17.6?

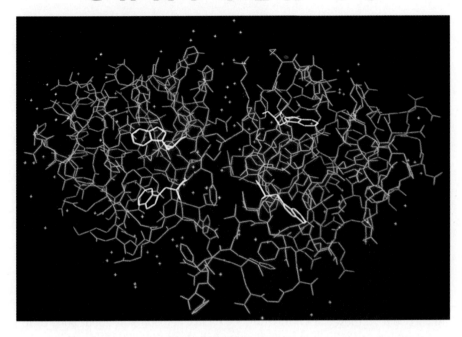

# CHAPTER 18

# Proteins

*The three-dimensional structure of a protein called γII-crystallin, that exists in the lens of the eye, as displayed by computer graphics. The four tryptophan residues are highlighted. (Courtesy of Brookhaven Data Bank.)*

## 18.1 INTRODUCTION

**Proteins** are by far the most important of all biological compounds. The very word "protein" is derived from the Greek *proteios,* meaning "of first importance," and the scientists who named these compounds more than 100 years ago chose an appropriate term. There are many types of proteins, and they perform many functions, some of which are as follows:

1. **Structure** We saw in Section 16.11 that the main structural material for plants is cellulose. For animals, it is structural proteins, which are the chief constituents of skin, bones, hair, and fingernails. Two important structural proteins are collagen and keratin.

2. **Catalysis** Virtually all the reactions that take place in living organisms are catalyzed by proteins called enzymes. Without enzymes, the reactions would take place so slowly as to be useless. We will discuss enzymes in Chapter 19.

3. **Movement** Every time we crook a finger, climb stairs, or blink an eye, we use our muscles. Muscle expansion and contraction are involved in every movement we make. Muscles are made up of protein molecules called myosin and actin.

4. **Transport** A large number of proteins fall into this category. Hemoglobin, a protein in the blood, carries oxygen from the lungs to the cells in which it is used and carbon dioxide from the cells to the lungs. Other proteins transport molecules across cell membranes.

The heart itself is a muscle, expanding and contracting about 70 to 80 times a minute.

**5.   Hormones**   Many hormones are proteins, among them insulin, oxytocin, and human growth hormone.

**6.   Protection**   When a protein from an outside source or other foreign substance (called an antigen) enters the body, the body makes its own proteins (called antibodies) to counteract the foreign protein. This is the major mechanism the body uses to fight disease. Blood clotting is another protective device carried out by a protein, this one called fibrinogen. Without blood clotting, we would bleed to death from any small wound.

**7.   Storage**   Some proteins are used to store materials, in the way that starch and glycogen store energy. Examples are casein in milk and ovalbumin in eggs, which store nutrients for newborn mammals and birds. Ferritin, a protein in the liver, stores iron.

**8.   Regulation**   Some proteins control the expression of genes, and thereby regulate the kind of proteins manufactured in a particular cell, and control when such manufacture takes place.

These are not the only functions of proteins, but they are among the most important. It is very easy to see that any individual needs a great many proteins to carry out all these varied functions. A typical cell contains about 9000 different proteins; an individual human being has about 100 000 different proteins.

We can divide proteins into two major types: **fibrous proteins,** which are insoluble in water and are used mainly for structural purposes, and **globular proteins,** which are more or less soluble in water and are used mainly for nonstructural purposes.

## 18.2   AMINO ACIDS

Although there are so many different proteins, they all have basically the same structure: They are linear chains of amino acids. As its name implies, an **amino acid** is an **organic compound containing an amino group and an acid group.** Organic chemists can synthesize many thousands of amino acids, but nature is much more restrictive and uses only 20 different amino acids to make up proteins. Furthermore, all but one of the 20 fit the formula

$$R-\underset{\underset{NH_2}{|}}{\overset{\overset{H}{|}}{C}}-COOH$$

and even the one that doesn't fit the formula (proline) comes pretty close. Proline would fit except that it has a bond between the R and the N. The 20 amino acids found in proteins are called **alpha** amino acids. They are shown in Table 18.1, which also shows the one- and three-letter abbreviations that chemists use for them.

The most important aspect of the R groups is their polarity, and on this basis we can classify amino acids into the four groups shown in Table 18.1: nonpolar, polar but neutral, acidic, and basic. This aspect of the R groups is very important in determining both the structure and the function of each protein molecule.

The antibodies are produced in the gamma globulin fraction of blood plasma.

Collagen, actin, and keratin are some fibrous proteins. Albumin, hemoglobin, and immunoglobulins are some globular proteins.

Proline

In these compounds, the —$NH_2$ is on the alpha carbon (the one next to the —COOH).

The one-letter abbreviations are more recent, but the three-letter abbreviations are still frequently used.

**TABLE 18.1**  **The 20 Amino Acids Commonly Found in Proteins (both the three-letter and one-letter abbreviations are shown)**

### Nonpolar R Groups

| | | | |
|---|---|---|---|
| Glycine | Gly | G | $H-CH-COOH$ with $NH_2$ |
| Alanine | Ala | A | $CH_3-CH-COOH$ with $NH_2$ |
| *Valine[a] | Val | V | $CH_3-CH-CH-COOH$ with $CH_3$ and $NH_2$ |
| *Leucine | Leu | L | $CH_3-CH-CH_2-CH-COOH$ with $CH_3$ and $NH_2$ |

| | | | |
|---|---|---|---|
| *Isoleucine | Ile | I | $CH_3-CH_2-CH-CH-COOH$ with $CH_3$ and $NH_2$ |
| Proline | Pro | P | (pyrrolidine ring with $-COOH$, N-H) |
| *Phenylalanine | Phe | F | (phenyl)$-CH_2-CH-COOH$ with $NH_2$ |
| *Methionine | Met | M | $CH_3-S-CH_2CH_2-CH-COOH$ with $NH_2$ |

### Polar but Neutral R Groups

| | | | |
|---|---|---|---|
| Serine | Ser | S | $HO-CH_2-CH-COOH$ with $NH_2$ |
| *Threonine | Thr | T | $CH_3-CH-CH-COOH$ with $OH$ and $NH_2$ |
| Cysteine | Cys | C | $HS-CH_2-CH-COOH$ with $NH_2$ |
| Tyrosine | Tyr | Y | $HO-$(phenyl)$-CH_2-CH-COOH$ with $NH_2$ |

| | | | |
|---|---|---|---|
| Asparagine | Asn | N | $NH_2-\overset{\parallel O}{C}-CH_2-CH-COOH$ with $NH_2$ |
| Glutamine | Gln | Q | $NH_2-\overset{\parallel O}{C}-CH_2CH_2-CH-COOH$ with $NH_2$ |
| *Tryptophan | Trp | W | (indole)$-CH_2-CH-COOH$ with $NH_2$ |

### Acidic R Groups

| | | | |
|---|---|---|---|
| Glutamic acid | Glu | E | $HO-\overset{\parallel O}{C}-CH_2CH_2-CH-COOH$ with $NH_2$ |
| Aspartic acid | Asp | D | $HO-\overset{\parallel O}{C}-CH_2-CH-COOH$ with $NH_2$ |

### Basic R Groups

| | | | |
|---|---|---|---|
| *Lysine | Lys | K | $NH_2-CH_2CH_2CH_2CH_2-CH-COOH$ with $NH_2$ |
| *Arginine | Arg | R | $NH_2-\overset{\parallel NH}{C}-NH-CH_2CH_2CH_2-CH-COOH$ with $NH_2$ |
| *Histidine | His | H | (imidazole)$-CH_2-CH-COOH$ with $NH_2$ |

[a] Those marked with an asterisk are essential amino acids (see Sec. 26.4).

When proteins are synthesized by the body, only L amino acids are built into the growing polypeptide chains. However, in proteins that remain in the body for decades, a number of changes take place. It has been observed that in the calcified proteins of human teeth as well as in the human lens proteins, more and more D-aspartic acid accumulates with age. In these two tissues, proteins, once they have been synthesized, are preserved for many years. In the centers of the lens in each of our eyes are the same proteins we are born with. This is in contrast, for example, to the hemoglobin molecules in the blood, which are resynthesized every 120 days. In preserved proteins we find that, as time goes by, racemization slowly takes place. That is, more and more of the D-aspartic acid accumulates. It has been calculated that every year 0.14 percent of the L-aspartic acid is converted to D-aspartic acid in a normal lens.

When we look at the general formula for the 20 amino acids

$$R-\underset{\underset{NH_2}{|}}{\overset{\overset{H}{|}}{C}}-COOH$$

we see at once that all of them (except glycine, in which R = H) have a chiral carbon, since R, H, COOH, and $NH_2$ are four different groups. This means, as we saw in Section 16.2, that each of the amino acids (except glycine) exists as two enantiomers. As is the case for most examples of this kind, nature makes only one of the two possible forms for each amino acid, and it is virtually always the L form. Except for glycine, which exists in only one form, all the amino acids in all the proteins in your body are the L form. D amino acids are extremely rare in nature; some are found, for example, in the cell walls of a few types of bacteria.

Two of the amino acids in Table 18.1 have a second chiral carbon. Can you find them?

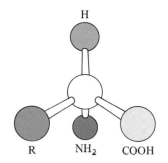

Configuration of L amino acids.

### 18.3 ZWITTERIONS

Up to now, we have shown the structural formula for amino acids as

$$R-\underset{\underset{NH_2}{|}}{\overset{\overset{H}{|}}{C}}-COOH$$

But in Section 14.7 we learned that carboxylic acids, RCOOH, cannot exist in the presence of a moderately weak base (such as $NH_3$). They donate a proton to become carboxylate ions, $RCOO^-$. Likewise, amines, $RNH_2$ (Sec. 15.4), cannot exist as such in the presence of a moderately weak acid (such as acetic acid). They gain a proton to become substituted ammonium ions, $RNH_3^+$.

An amino acid has —COOH and —$NH_2$ groups in the same molecule. Therefore, in water solution the —COOH donates a proton to the —$NH_2$, so that an amino acid actually has the structure

$$R-\underset{\underset{NH_3^+}{|}}{\overset{\overset{H}{|}}{C}}-COO^-$$

**Compounds that have a positive charge on one atom and a negative charge on another** are called **zwitterions.** Amino acids are zwitterions, not only in water solution but in the solid state as well. They are therefore ionic compounds, that is, internal salts. *Un-ionized RCH(NH$_2$)COOH molecules do not actually exist, in any form.*

The fact that amino acids are zwitterions explains their physical properties, which would otherwise be quite puzzling. All of them are solids with high melting points (for example, glycine melts at 262°C). This is just what we expect of ionic compounds (Table 10.1). The 20 amino acids are also fairly soluble in water, as ionic compounds generally are; if they had no charges, only the smaller ones would be expected to be soluble.

If we add an amino acid to water, it dissolves and then has the same zwitterionic structure that it has in the solid state. Let us see what happens if we change the pH of the solution, as we can easily do by adding a source of H$_3$O$^+$, such as HCl (to lower the pH), or a strong base such as NaOH (to raise the pH). Since H$_3$O$^+$ is a stronger acid than the typical carboxylic acid (Sec. 8.3), it donates a proton to the —COO$^-$ group,

$$R-\underset{\underset{NH_3^+}{|}}{\overset{\overset{H}{|}}{C}}-COO^- + H_3O^+ \longrightarrow R-\underset{\underset{NH_3^+}{|}}{\overset{\overset{H}{|}}{C}}-COOH + H_2O$$

turning the zwitterion into a positive ion. This happens to all amino acids if the pH is sufficiently lowered.

Addition of OH$^-$ to the zwitterion causes the —NH$_3^+$ to donate its proton,

$$R-\underset{\underset{NH_3^+}{|}}{\overset{\overset{H}{|}}{C}}-COO^- + OH^- \longrightarrow R-\underset{\underset{NH_2}{|}}{\overset{\overset{H}{|}}{C}}-COO^- + H_2O$$

turning the zwitterion into a negative ion. This happens to all amino acids if the pH is sufficiently raised.

Note that in both cases *the amino acid is still an ion,* so that it is still soluble in water. There is no pH at which an amino acid has no ionic character at all. If the amino acid is a positive ion at low pH and a negative ion at high pH, there must be some **pH at which all the molecules are in the zwitterionic form.** This pH is called the **isoelectric point.**

Every amino acid has a different isoelectric point, though most of them are not very far apart. Fifteen of the twenty have isoelectric points near 6 (for example, glycine 6.06, tryptophan 5.88, alanine 6.11, tyrosine 5.63). However, the three basic amino acids have higher isoelectric points (lysine's is 9.47), and the two acidic amino acids have lower values (aspartic acid's is 2.98).

*"Zwitterion" comes from the German zwitter, meaning hybrid.*

At or near the isoelectric point, amino acids exist in aqueous solution largely or entirely as zwitterions. As we have seen, they react with either a strong acid, by taking a proton (the —COO⁻ becomes —COOH), or a strong base, by giving a proton (the —NH₃⁺ becomes —NH₂). To summarize:

$$
\underset{\text{Acid solution}}{\overset{\displaystyle H}{R-\underset{\underset{\displaystyle NH_3^+}{|}}{\overset{|}{C}}-COOH}}
\underset{H_3O^+}{\overset{OH^-}{\rightleftharpoons}}
\underset{\substack{\text{Isoelectric}\\\text{point}}}{\overset{\displaystyle H}{R-\underset{\underset{\displaystyle NH_3^+}{|}}{\overset{|}{C}}-COO^-}}
\underset{H_3O^+}{\overset{OH^-}{\rightleftharpoons}}
\underset{\text{Basic solution}}{\overset{\displaystyle H}{R-\underset{\underset{\displaystyle NH_2}{|}}{\overset{|}{C}}-COO^-}}
$$

We learned in Section 8.3 that a compound that is both an acid and a base is called *amphoteric*. We also learned, in Section 8.10, that a solution that neutralizes both acid and base is a *buffer solution*. Amino acids are therefore amphoteric compounds, and aqueous solutions of them are buffers.

Proteins also have isoelectric points, and also act as buffers. This will be discussed in Section 18.6.

## 18.4 CYSTEINE: A SPECIAL AMINO ACID

We met these reactions in Section 12.8.

One of the 20 amino acids in Table 18.1 has a chemical property not shared by any of the others. This amino acid is cysteine. It can easily be dimerized by many mild oxidizing agents:

$$
2HS-CH_2-\underset{\underset{\displaystyle NH_2}{|}}{\overset{|}{CH}}-COOH
\underset{[H]}{\overset{[O]}{\rightleftharpoons}}
HOOC-\underset{\underset{\displaystyle NH_2}{|}}{\overset{|}{CH}}-CH_2-S-S-CH_2-\underset{\underset{\displaystyle NH_2}{|}}{\overset{|}{CH}}-COOH + H_2O
$$

Cysteine                  Cystine

A dimer is a molecule made up of two identical units.

The dimer of cysteine, which is called **cystine,** can in turn be fairly easily reduced to give two molecules of cysteine. As we shall see, the presence of cystine has important consequences for the chemical structure and shape of protein molecules it is part of. **The S—S bond** (shown in color) is called a **disulfide linkage.**

## 18.5 PEPTIDES AND PROTEINS

Each amino acid has a carboxylic acid group and an amino group. In Section 15.5 we saw that a carboxylic acid and an amine could be combined to form an amide:

$$
\underset{\text{Carboxylic acid}}{R-\underset{\underset{\displaystyle O}{\|}}{C}-OH} + \underset{\text{Amine}}{R'NH_2}
\xrightarrow[-H_2O]{}
\underset{\text{Amide}}{R-\underset{\underset{\displaystyle O}{\|}}{C}-NH-R'}
$$

In the same way, it is possible for the COOH group of one amino acid molecule, say glycine, to combine with the amino group of a second molecule, say alanine:

Many people restrict their sugar intake. Some are forced to do so by diseases such as diabetes, others by the desire to lose weight. Since most of us like to eat sweet foods, artificial sweeteners are added to many foods and drinks for those who must (or want to) restrict their sugar intake. Three noncaloric artificial sweeteners are now approved by the Food and Drug Administration. The oldest of these, saccharin, is 450 times sweeter than sucrose (Table 16.1), yet it has very little caloric content. Saccharin has been in use for about a hundred years. Unfortunately, some tests have shown that saccharin, when fed in massive quantities to rats, caused some of these rats to develop cancer. Although other tests have given negative results, the question about the safety of saccharin is still not settled, though it remains on the market.

Saccharin          Aspartame          Acesulfame-K

A newer artificial sweetener, aspartame, does not have the slight aftertaste that saccharin does. Aspartame is the methyl ester of a simple dipeptide, aspartyl-phenylalanine (Asp-Phe). Its sweetness was discovered in 1969, and after extensive biological testing it was approved by the U.S. Food and Drug Administration in 1981 for use in cold cereals, drink mixes, and gelatins and as tablets or powder to be used as a sugar substitute. Aspartame is 100 to 150 times sweeter than sucrose. It is made from natural amino acids, so that both the aspartic acid and the phenyl-alanine have the L configuration. The other possibilities have also been synthesized: the L-D, the D-L, and the D-D. They are all bitter rather than sweet. This is another example of the principle that, when a biological organism uses or makes a compound that has chiral carbons, in most cases only one stereoisomer is used or made. Aspartame is sold under such brand names as Equal and NutraSweet. The third artificial sweetener, acesulfame-K, is the most recent (approved in 1988). It is 200 times sweeter than sucrose and is used (under the name Sunette) in dry mixes. It is not metabolized in the body; that is, it passes through unchanged.

Glycine    Alanine

Glycylalanine (Gly-Ala), a dipeptide

The —C—NH— is called the **peptide linkage.**

This reaction takes place in the cells by a mechanism that we shall examine in Section 23.10. The product is an amide made up of two amino acids joined together, called a **dipeptide.**

It is important to realize that glycine and alanine could also be linked the other way:

$$\underset{\text{Alanine}}{NH_2-\underset{\underset{O}{\|}}{\overset{\overset{CH_3}{|}}{CH}-C}-OH} + \underset{\text{Glycine}}{NH_2-\underset{\underset{O}{\|}}{\overset{\overset{H}{|}}{CH}-C}-OH} \xrightarrow[-H_2O]{}$$

$$NH_2-\overset{\overset{CH_3}{|}}{CH}-\underset{\underset{O}{\|}}{C}-NH-\overset{\overset{H}{|}}{CH}-\underset{\underset{O}{\|}}{C}-OH$$

Alanylglycine (Ala-Gly), a dipeptide

In this case we get a *different* dipeptide. The two dipeptides are structural isomers, of course, but they are different compounds in all respects, with different properties.

Any two amino acids, the same or different, can be linked together to form dipeptides in a similar manner. Nor does it end there. Each dipeptide still contains a COOH and an amino group. We can therefore add a third amino acid to alanylglycine, say lysine:

$$NH_2-\overset{\overset{CH_3}{|}}{CH}-\underset{\underset{O}{\|}}{C}-NH-\overset{\overset{H}{|}}{CH}-\underset{\underset{O}{\|}}{C}-OH + NH_2-\overset{\overset{(CH_2)_4}{\overset{|}{\overset{NH_2}{|}}}}{CH}-\underset{\underset{O}{\|}}{C}-OH$$

Lysine

$$\Big\downarrow -H_2O$$

$$NH_2-\overset{\overset{CH_3}{|}}{CH}-\underset{\underset{O}{\|}}{C}-NH-\overset{\overset{H}{|}}{CH}-\underset{\underset{O}{\|}}{C}-NH-\overset{\overset{(CH_2)_4}{\overset{|}{\overset{NH_2}{|}}}}{CH}-\underset{\underset{O}{\|}}{C}-OH$$

Alanylglycyllysine (Ala-Gly-Lys), a tripeptide

The product is a **tripeptide.** Since it too contains a COOH and an NH₂ group, we can continue the process to get a tetrapeptide, a pentapeptide, and so on until we have a chain of hundreds or even thousands of amino acids. These **chains of amino acids** are the **proteins** that serve so many important functions in living organisms.

A word must be said about the terms used to describe these compounds. The shortest chains are often simply called **peptides,** longer ones are **polypeptides,** and still longer ones are **proteins,** but chemists differ about where to

**B o x   18C**

## Glutathione

A very important tripeptide, present in high concentrations in all tissues, is **gluta-thione.** It contains L-glutamic acid, L-cysteine, and glycine. Its structure is unusual because the glutamic acid is linked to the cysteine by its γ-carboxyl group rather than by the α-carboxyl group as is usual in most peptides and proteins:

$$NH_3^+—CH—CH_2—CH_2—\underset{\underset{O}{\|}}{C}—NH—CH—\underset{\underset{O}{\|}}{C}—NH—CH_2—COO^-$$

with $COO^-$ on Glu and $CH_2$—$SH$ on Cys.

Glu                    Cys        Gly

Glutathione functions in the cells as a general protective agent. Oxidizing agents that would damage the cells, such as peroxides, will oxidize glutathione instead (the cysteine portion; see also Sec. 18.4), thus protecting the proteins and nucleic acids. Many foreign chemicals also get attached to glutathione, and in this sense it acts as a detoxifying agent.

draw the line. Many chemists use the terms "polypeptide" and "protein" almost interchangeably. We shall consider a protein to be a polypeptide chain that contains a minimum of 30 to 50 amino acids.

The amino acids in a chain are often called **residues.** It is customary to use either the one-letter or the three-letter abbreviations shown in Table 18.1 to represent peptides and proteins. For example, the tripeptide shown above, alanylglycyllysine, is AGK or Ala-Gly-Lys. The amino acid residue with the free COOH group is called the **C-terminal** one (lysine in Ala-Gly-Lys), and the amino acid residue with the free amino group is the **N-terminal** one (alanine in Ala-Gly-Lys). It is the universal custom to write peptide and protein chains with the N-terminal residue on the left. No matter how long a protein chain gets—hundreds or thousands of units—it always has just two ends: one C-terminal and one N-terminal.

*The naming of peptides also begins with the N-terminal amino acid.*

## 18.6   SOME PROPERTIES OF PEPTIDES AND PROTEINS

The continuing pattern of

$$—\underset{\underset{H}{|}}{N}—CH—\underset{\underset{O}{\|}}{\overset{\overset{R}{|}}{C}}—\underset{\underset{H}{|}}{N}—CH—\underset{\underset{O}{\|}}{\overset{\overset{R}{|}}{C}}—$$

is called the **backbone** of the peptide or protein molecule; the R groups are called the **side chains.** The six atoms of the peptide linkage

$$—C—\underset{\underset{O}{\|}}{C}—\underset{\underset{H}{|}}{N}—C—$$

lie in the same plane, and two adjacent peptide bonds can rotate relative to one another about the C—N and C—C bonds.

*For an example of the planar structure, see Figure 18.6.*

The terminal —COOH and —NH$_2$ groups also ionize, but these are only 2 out of 50 or more residues.

The 20 different amino acid side chains supply variety and determine the physical and chemical properties of proteins. Among these properties, acid-base behavior is one of the most important. Like amino acids (Section 18.3), proteins behave as zwitterions. The side chains of glutamic and aspartic acids provide COOH groups, while lysine and arginine provide basic groups (histidine does, too, but this side chain is less basic than the other two). (See the structures of these amino acids in Table 18.1.)

The isoelectric point of a protein occurs at the pH at which the positive and negative charges are balanced. At their isoelectric points, proteins behave like zwitterions: The number of —COO$^-$ groups equals the number of —NH$_3^+$ groups. At any pH above the isoelectric point, the protein molecules have a net negative charge; at any pH below the isoelectric point, they have a net positive charge. Some proteins, such as hemoglobin, have an almost equal number of acidic and basic groups; the isoelectric point of hemoglobin is at pH 6.8. Others, like serum albumin, have more acidic groups than basic groups; the isoelectric point of this protein is 4.9. In each case, however, because proteins behave like zwitterions, they act as buffers, for example, in the blood (Fig. 18.1).

Carbonates and phosphates are also blood buffers (Sec. 8.10).

The water-solubility of large molecules such as proteins often depends on the repulsive forces between like charges on their surfaces. When protein molecules are at a pH at which they have a net positive or negative charge, the

**Figure 18.1**
Schematic diagram of a protein (a) at its isoelectric point and its buffering action when (b) H$^+$ or (c) OH$^-$ ions are added.

presence of these like charges causes the protein molecules to repel each other. These repulsive forces are smallest at the isoelectric point, when the net charges are close to zero. When there are no repulsive forces, the protein molecules tend to clump together to form aggregates of two or more molecules, reducing their solubility. Therefore, *proteins are least soluble in water at their isoelectric points and can be precipitated from their solutions.*

We pointed out in Section 18.1 that proteins have many functions. In order to understand these functions, we must look at four levels of organization in their structures. The *primary structure* describes the linear sequence of amino acids in the polypeptide chain. *Secondary structure* refers to certain repeating patterns, such as the $\alpha$-helix conformation or the pleated sheet (see Figs. 18.5 and 18.6). The *tertiary structure* describes the overall conformation of the polypeptide chain. A good analogy for all this is a coiled telephone cord (Fig. 18.2). The primary structure is the stretched-out cord. The secondary structure is the coil in the form of a helix. We can take the entire coil and twist it into various shapes. Any structure made by doing this is a tertiary structure. As we shall see, protein molecules twist and curl in a very similar manner.

*Quaternary structure* applies only to proteins with more than one chain and has to do with how the different chains are spatially related to each other.

(a)  (b)

(c)

**Figure 18.2**
The "structure" of a telephone cord: (a) primary, (b) secondary, (c) tertiary.

## 18.7  THE PRIMARY STRUCTURE OF PROTEINS

Very simply, the **primary structure** of a protein consists of the sequence of amino acids that make up the chain. Each of the very large number of peptide and protein molecules in biological organisms has a different sequence of amino acids—and it is that sequence that allows the protein to carry out its function, whatever it may be.

Is it possible that so many different proteins can arise from different sequences of only 20 amino acids? Let us look at a little arithmetic, starting with a dipeptide. How many different dipeptides can be made from 20 amino acids? There are 20 possibilities for the N-terminal amino acid, and for each of these 20 there are 20 possibilities for the C-terminal amino acid. This means that there are

$$20 \times 20 = 400$$

different dipeptides possible from the 20 amino acids. What about tripeptides? We can form a tripeptide by taking any of the 400 dipeptides and adding any of the 20 amino acids. Thus there are

$$20 \times 20 \times 20 = 8000$$

tripeptides, all different. It is easy to see that we can calculate the total number of possible peptides or proteins for a chain of $n$ amino acids simply by raising 20 to the $n$th power ($20^n$).

Taking a typical small protein to be one with 60 amino acid residues, the number of proteins that can be made from the 20 amino acids is $20^{60} = 10^{78}$. This is an enormous number, probably greater than the total number of atoms

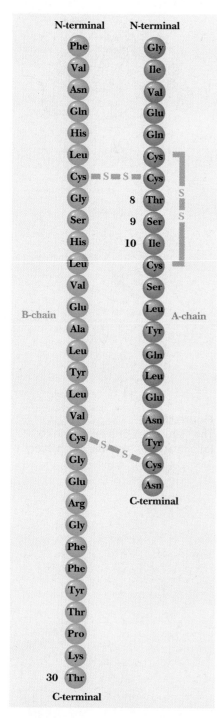

**Figure 18.3**
The primary structure (amino acid sequence) of human insulin. (From M. K. Campbell, *Biochemistry*. Philadelphia: Saunders College Publishing, 1991.)

in the universe. It is clear that only a tiny fraction of all possible protein molecules has ever been made by biological organisms.

Each peptide or protein in the body has its own sequence of amino acids. We mentioned that proteins also have secondary, tertiary, and in some cases also quaternary structures. We will deal with these in Sections 18.8 and 18.9, but here we can say that **the primary structure of a protein determines to a large extent the native** (most frequently occurring) **secondary, tertiary, and quaternary structures.** That is, it is the particular sequence of amino acids on the chain that enables the whole chain to fold and curl in such a way as to assume its final shape. As we shall see in Section 18.10, without its particular three-dimensional shape, a protein cannot function.

Just how important is the exact amino acid sequence? Can a protein perform the same function if its sequence is a little different? The answer to this question is that a change in amino acid sequence may or may not matter, depending on what kind of a change it is. As an example, the enzyme ribonuclease is a protein chain consisting of 124 amino acid residues. It is possible, by means of another enzyme called carboxypeptidase, to remove the C-terminal amino acid residue of ribonuclease, leaving the rest of the chain intact. When this is done, the 124th amino acid, valine, is removed, leaving a chain of 123 units. This modified chain has the same biological activity as the original ribonuclease. In this case, removal of one amino acid makes no difference.

Another example is the hormone *insulin*. Human insulin consists of two chains having a total of 51 amino acids, connected by disulfide linkages. The sequence of amino acids is shown in Figure 18.3. Insulin is necessary for proper utilization of carbohydrates (Sec. 21.2), and people with severe diabetes (Box 24G) must take insulin injections. The amount of human insulin available is far too small to meet the need, so bovine insulin (from cattle) or insulin from hogs or sheep is used instead. Insulin from these sources is similar to human insulin, but not identical. The differences are entirely in the 8, 9, and 10 positions of the A chain and the C-terminal position (30) of the B chain:

| | A Chain | | | B Chain |
|---|---|---|---|---|
| | 8 | 9 | 10 | 30 |
| Human | —Thr— | Ser— | Ile— | —Thr |
| Bovine | —Ala— | Ser— | Val— | —Ala |
| Hog | —Thr— | Ser— | Ile— | —Ala |
| Sheep | —Ala— | Gly— | Val— | —Ala |

The remainder of the molecule is the same in all four varieties of insulin. Despite the slight differences in structure, all these insulins can be used by humans and perform the same function as human insulin. However, none of the other three is quite as effective as human insulin.

Another factor showing the effect of substituting one amino acid for another is that sometimes patients become allergic to, say, bovine insulin and can switch to hog or sheep insulin without causing allergies.

In contrast to the previous examples, there are small changes in amino acid sequence that make a great deal of difference. First we can consider two pep-

## The Use of Human Insulin

Although human insulin, manufactured by recombinant DNA techniques (see Sec. 23.14), is on the market, many diabetic patients continue to use hog or sheep insulin because it is cheaper. Changing from animal to human insulin creates an occasional problem for diabetics. All diabetics experience an insulin reaction (hypoglycemia) when the insulin level in the blood is too high relative to blood sugar level. Hypoglycemia is preceded by symptoms of hunger, sweating, and poor coordination. These symptoms, called "hypoglycemic awareness," are a signal to the patient that hypoglycemia is coming and that it must be reversed, which the patient can do by eating sugar. Some diabetics who changed from animal to human insulin reported that the hypoglycemic awareness from recombinant DNA human insulin is not as strong as that from animal insulin. This can create some hazards. The effect is probably due to different rates of absorption in the body. The literature supplied with human insulin now incorporates a warning that hypoglycemic awareness may be altered.

tide hormones, oxytocin and vasopressin (Fig. 18.4). These nonapeptides have identical structures, including a disulfide bond, except for different amino acids in positions 3 and 8. Yet their biological functions are quite different. Vasopressin is an antidiuretic hormone. It increases the amount of water reabsorbed by the kidneys and raises blood pressure. Oxytocin has no effect on water in the kidneys and slightly *lowers* blood pressure. It affects contractions of the uterus in childbirth and the muscles in the breast that aid in the secretion of milk. Vasopressin also stimulates uterine contractions, but much less so than oxytocin.

Another instance where a minor change makes a major difference is in the blood protein hemoglobin. A change in only one amino acid in a chain of 146 is enough to cause a fatal disease—sickle cell anemia (Box 18E).

Both oxytocin and vasopressin are secreted by the pituitary gland.

Both hormones are used as drugs, vasopressin to combat loss of blood pressure after surgery and oxytocin to induce labor.

**Figure 18.4**
The structures of vasopressin and oxytocin. Differences are shown in color.

**B o x 18E**

## Sickle Cell Anemia

Normal adult human hemoglobin has two alpha chains and two beta chains (Fig. 18.15). Some people, however, have a slightly different kind of hemoglobin in their blood. This hemoglobin (called HbS) differs from the normal type only in the beta chains and only in one position on these two chains: The glutamic acid in the sixth position of normal Hb is replaced by a valine residue in HbS:

| | 4 | 5 | 6 | 7 | 8 | 9 |
|---|---|---|---|---|---|---|
| Normal Hb | —Thr— | Pro— | Glu— | Glu— | Lys— | Ala— |
| Sickle cell Hb | —Thr— | Pro— | Val— | Glu— | Lys— | Ala— |

This change affects only two positions in a molecule containing 574 amino acid residues. Yet it is enough to result in a very serious disease, *sickle cell anemia*.

Red blood cells carrying HbS behave normally when there is an ample oxygen supply. When the oxygen pressure decreases, the red blood cells become sickle-shaped (Fig. 18E). This occurs in the capillaries. As a result of this change in shape, the cells may clog the capillaries. The body's defenses destroy the clogging cells, and the loss of the blood cells causes anemia.

This change at only a single position of a chain consisting of 146 amino acids is severe enough to cause a high death rate. A child who inherits two genes programmed to produce sickle cell hemoglobin (a *homozygote*) has an 80 percent smaller chance of surviving to adulthood than a child with only one such gene (a *heterozygote*) or a child with two normal genes. In spite of the high mortality of homozygotes, the genetic trait survives. In central Africa, 40 percent of the population in malaria-ridden areas carry the sickle cell gene, and 4 percent are homozygotes.

It seems that the sickle cell genes help to acquire immunity against malaria in early childhood, so that in malaria-ridden areas the transmission of these genes is advantageous.

The first sequence of an important protein, insulin, was obtained by Frederick Sanger (1918– ) in England, for which he received the Nobel Prize in 1958.

Although in some cases slight changes in amino acid sequence make little or no difference in the functioning of peptides and proteins, it is clear that the sequence is highly important in most cases. The sequences of a large number of protein and peptide molecules have now been determined. The methods for doing it are complicated and will not be discussed in this book.

Linus Pauling (1901– ) won the 1954 Nobel Prize in Chemistry for determining these structures. He also discovered or contributed much to our understanding of certain fundamental concepts, including electronegativity (Sec. 3.7), hybrid orbitals (Sec. 10.4), and resonance (Sec. 11.9).

An *intra*molecular hydrogen bond goes from a hydrogen atom in a molecule to an O, N, or F atom in the same molecule.

## 18.8 THE SECONDARY STRUCTURE OF PROTEINS

Proteins can fold or align themselves in such a manner that certain patterns repeat themselves. These repeating patterns are referred to as **secondary structures.** The two most common secondary structures encountered in proteins are the $\alpha$-helix and the $\beta$-pleated sheet (Figs. 18.5 and 18.6) originally proposed by Linus Pauling (Fig. 18.7) and Robert Corey. In contrast, those protein conformations that do not exhibit a repeated pattern are called **random coils** (Fig. 18.8).

In the **$\alpha$-helix form,** a single protein chain twists in such a manner that its shape resembles a coiled spring, that is, a helix. The shape of the helix is maintained by numerous intramolecular hydrogen bonds that exist between the *backbone* —C≡O and H—N— groups. It can be seen from Figure 18.5 that there is a hydrogen bond between the —C≡O oxygen atom of each peptide

**Figure 18E**
Blood cells from a patient with sickle cell anemia. Both normal cells (round) and sickle cells (shriveled) are visible. (G. W. Willis, M.D./Biological Photo Service.)

In the black population of the United States today, the HbS gene frequency is about 10 percent, and 0.25 percent of the black population is homozygous. The sharp drop in the HbS gene frequency in the Afro-American population over a period of 200 to 300 years indicates the fast adaptation of human genetics to a new environment. People with only one HbS gene are said to carry the *sickle cell trait,* which is not generally harmful. However, when two people with sickle cell trait produce a child who inherits the HbS gene from both parents, that child is a homozygote and will have sickle cell anemia. Screening tests can determine whether prospective parents carry the sickle cell trait.

KEY: ● Carbon atom ● Oxygen atom ● Nitrogen atom ● Hydrogen atom ○ R group

**Figure 18.5**
(Left) The α-helix. (From C. A. Villee et al., *Biology,* 2nd ed. Philadelphia: Saunders College Publishing, 1989.)

**Figure 18.6**
(Right) The β-pleated sheet structure. (From C. A. Villee et al., *Biology,* 2nd ed. Philadelphia: Saunders College Publishing, 1989.)

The amount of cross-linking in collagen controls its mechanical strength. The collagen in the Achilles tendon is highly cross-linked, but more flexible tendons are cross-linked to a lesser extent. **Marfan's syndrome** is a human hereditary disease in which collagen is insufficiently cross-linked. Some patients with Marfan's syndrome have excessive bone growth and high stature, which encourages athletic activity. But the insufficiently cross-linked collagen in the blood vessels yields to stresses, especially during vigorous athletic activity, and leads to lethal rupture of the aorta.

**Figure 18.7**
Linus Pauling.

Keratin, a fibrous protein of hair, fingernails, horns, and wool, is one protein that does have a predominantly α-helix structure. Silk is made of fibroin, another fibrous protein, which exists mainly in the pleated-sheet form.

**Figure 18.8**
A random coil.

linkage and the —N—H hydrogen atom of another peptide linkage farther along the chain. These hydrogen bonds are in just the right position to cause the molecule (or a portion of it) to maintain a helical shape.

The other important orderly structure in proteins is called the **β-pleated sheet.** In this case, the orderly alignment of protein chains is maintained by **intermolecular** hydrogen bonds. The β-pleated sheet structure can occur between molecules when polypeptide chains run *parallel* (all the N-terminal ends on one side) or *antiparallel* (neighboring N-terminal ends on opposite sides). β-pleated sheets can also occur intramolecularly, when the polypeptide chain makes a U-turn, forming a hairpin structure, and the pleated sheet is antiparallel (Fig. 18.6).

**In all secondary structures the hydrogen bonding is between backbone —C═O and H—N— groups.** This is the distinction between secondary and tertiary structure. In the latter, as we shall see, the hydrogen bonding is between R groups on the side chains.

Few proteins have predominantly α-helix or pleated sheet structures. Most proteins, especially globular ones, have only certain portions of their molecules in these conformations. The rest of the molecule is random coil. Many globular proteins contain all three kinds of secondary structure in different parts of their molecules: α-helix, β-pleated sheet, and random coil. A schematic representation of such a structure is shown in Figure 18.9.

Another repeating pattern classified as a secondary structure is the **triple helix** of *collagen.* Collagen, which was mentioned in Section 16.12, is the structural protein of connective tissues (bone, cartilage, tendon, aorta, skin), where it provides strength and elasticity. It is the most abundant protein in the body,

**Figure 18.9**
Schematic structure of the enzyme carboxypeptidase. The β-pleated sheet portions are shown in blue, the green structures are the α-helix portions, and the orange strings are the random coil areas.

making up about 30 percent by weight of all the body's protein. The triple helix structure (Fig. 18.10) is unique to collagen. The triple helix units, called *tropocollagen,* constitute the soluble form of collagen; they are stabilized by hydrogen bonding between the backbones of the three chains. Collagen is made of many tropocollagen units.

## 18.9 THE TERTIARY AND QUATERNARY STRUCTURE OF PROTEINS

### Tertiary Structure

Tropocollagen is found only in fetal or young connective tissues. With aging, the triple helixes that organize themselves (Fig. 18.10) into fibrils cross-link and form insoluble collagen. This cross-linking of collagen is an example of the **tertiary structures** that stabilize the three-dimensional conformations of protein molecules. In collagen, the cross-linking consists of covalent bonds that link together two lysine residues on adjacent chains of the helix (Fig. 18.11).

A covalent bond that often stabilizes the tertiary structure of proteins is the **disulfide bridge.** In Section 18.4 we noted that the amino acid cysteine is easily converted to the dimer cystine. When a cysteine residue is in one chain and another cysteine residue is in another chain (or in another part of the same chain), formation of a disulfide bridge provides a covalent linkage that binds together the two chains or the two parts of the same chain:

$$\{\sim SH + HS\sim\} \xrightarrow{[O]} \{\sim S-S\sim\} + H_2O$$

Examples of both types are found in the structure of insulin (Fig. 18.3).

Besides covalent bonds, there are three other interactions that can stabilize tertiary structures: hydrogen bonding between side chains, salt bridges, and hydrophobic interactions (Fig. 18.12).

**1. Hydrogen Bonding** We saw in Section 18.8 that secondary structures are stabilized by hydrogen bonding between backbone $-C=O$ and $H-N-$ groups. **Tertiary structures are stabilized by hydrogen bonding between polar groups on side chains.**

**2. Salt Bridges** These occur only between two amino acids with ionized side chains, that is, between an acidic amino acid and a basic amino acid, each in its ionized form. The two are held together by simple ion-ion attraction.

**Figure 18.10**
The triple helix of collagen.

**Figure 18.11**
Covalent cross-linking in collagen. The double bond and the aldehyde group are the result of oxidative processes, which include removal of the amino groups of lysine residues.

**Figure 18.12**
Noncovalent interactions that stabilize the tertiary and quaternary structures of proteins: (a) hydrogen bonding, (b) salt bridge, (c) hydrophobic interaction.

(a)

(b)

(c)

**3. Hydrophobic Interactions**   In aqueous solution, globular proteins usually turn their *polar* groups *outward,* toward the aqueous solvent, and their *nonpolar* groups *inward,* away from the water molecules. The nonpolar groups prefer to interact with each other, excluding water from these regions. This is called **hydrophobic interaction.** Although this type of interaction is weaker than hydrogen bonding or salt bridges, it usually acts over large surface areas so that cooperatively the interactions are strong enough to stabilize a loop or some other tertiary structure formation.

The four types of interaction that stabilize the tertiary structures of proteins are shown in Figure 18.13.

Section 18.7 pointed out that the primary structure of a protein largely determines its secondary and tertiary structures. We can now see the reason for this. When the particular R groups are in the proper positions, all the hydrogen bonds, salt bridges, disulfide linkages, and hydrophobic interactions that stabilize the three-dimensional structure of that molecule are allowed to form.

The side chains of some protein molecules allow them to fold (form a tertiary structure) in only one possible way, but other proteins, especially those with long polypeptide chains, can fold in a number of possible ways. Certain proteins in living cells, called **chaperones,** help a newly synthesized polypeptide chain to assume the proper secondary and tertiary structures that are necessary for the functioning of that molecule, and prevent foldings that are not biologically active.

**Figure 18.13**
The four types of interaction that stabilize the tertiary structure of proteins, shown schematically.

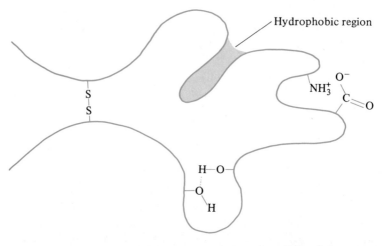

## The Structure of Collagen in Bones and Teeth

The triple helix units of collagen (Fig. 18.10) are further organized into a pattern called the **quarter-stagger arrangement** (Fig. 18G). In this arrangement, the units along a row are not spaced end to end. There is a gap between the end of one unit and the start of another. These gaps play an important role in bone formation. Collagen is one of the main constituents of bones and teeth, and the gaps in the quarter-stagger arrangement are essential for the deposition of inorganic crystals of calcium hydroxyapatite, $Ca_5(PO_4)_3OH$. The gaps serve as the nucleation sites for the growth of these crystals. The combination of hydroxyapatite crystals and collagen creates a hard material that still has some springiness, owing to the collagen present. *Dentine,* the main constituent of the internal part of a tooth, contains a higher percentage (about 75 percent) of inorganic crystals than does bone and is therefore harder. *Enamel,* the outer part of the tooth, has a still higher mineral content (about 95 percent) and is even harder.

Covalent bonds link adjacent rows

**Figure 18G**
The quarter-stagger arrangement of the triple helix units of collagen in a fibril.

### Quaternary Structure

The highest level of protein organization is **quaternary structure,** which applies to proteins with more than one chain. Quaternary structure determines how the different subunits of the protein fit into an organized whole. The hemoglobin molecule provides an important example. Hemoglobin in adult humans is made of four chains (called **globins**): two identical chains (called alpha) of 141 amino acid residues each and two other identical chains (beta) of 146 residues each. Figure 18.14 shows how the four chains fit together.

The hemoglobin molecule does not consist only of the four globin chains. Each globin chain surrounds a *heme* unit, the structure shown in Box 3C. Proteins that contain non-amino acid portions are called **conjugated proteins.** The non-amino acid portion of a conjugated protein is called a **prosthetic group.** In hemoglobin, the globins are the amino acid portions and the heme units are the prosthetic groups.

The quarter-stagger alignment of tropocollagen in collagen fibrils and the twisting of these fibrils into a five-strand or six-strand helix in the collagen fibers also represent quaternary structures. (See Box 18G.)

When a protein consists of more than one polypeptide chain, each is called a **subunit.**

The $O_2$- and $CO_2$-carrying functions of hemoglobin are discussed in Sections 25.3 and 25.4.

### 18.10 DENATURATION

Protein conformations are stabilized in their native states by secondary and tertiary structures and through the aggregation of subunits in quaternary structure. Any physical or chemical agent that destroys these stabilizing structures

**Box 18H**

## Some Applications of Protein Denaturation

Surgical instruments are sterilized by heat. The heat denatures the proteins of bacteria, killing them.

Alcohol coagulates proteins. A 70 percent alcohol solution is used to sterilize the skin before injections. At a concentration of 70 percent, ethanol penetrates bacteria and kills them by coagulating their proteins, whereas 95 percent alcohol denatures only surface proteins.

When milk is allowed to stand at room temperature, it turns sour because bacteria in the milk produce lactic acid. This acid is a denaturing agent and turns the casein (the principal milk protein) into a solid called a curd. This process is used in making cheese.

Heavy metal salts (for example, of $Hg^{2+}$, $Cu^{2+}$, or $Pb^{2+}$) are poisons because they denature proteins in the stomach. This very feature is taken advantage of in the antidote: raw egg whites and milk. The egg and milk proteins are denatured by the metal ions, forming insoluble precipitates in the stomach. These must be pumped out or removed by inducing vomiting *before* the digestive enzymes destroy the protein and release the heavy metal ions.

changes the conformation of the protein. We call this process **denaturation.** For example, heat cleaves hydrogen bonds, so boiling a protein solution destroys the $\alpha$-helical structure (compare Figs. 18.5 and 18.8). In collagen the triple helixes disappear upon boiling, and the molecules have a largely random-coil conformation in the denatured state, which is *gelatin*. In other proteins, especially globular proteins, heat causes the unfolding of the polypeptide chains and, because of subsequent intermolecular protein-protein interactions, precipitation or coagulation takes place. That is what happens when we boil an egg.

Similar conformational changes can be brought about by the addition of denaturing chemicals. Solutions such as 6 $M$ aqueous urea or guanidinium chloride break hydrogen bonds and cause the unfolding of globular proteins. Surface-active agents (detergents) change protein conformation by opening up the hydrophobic regions. Reducing agents, such as 2-mercaptoethanol ($HOCH_2CH_2SH$), can break the $-S-S-$ disulfide bridges, while acids,

**Figure 18.14**
Three-dimensional model of the hemoglobin molecule. The two alpha chains are in green, and the two beta chains are in blue. The red disks represent the four hemes. (From "The Hemoglobin Molecule" by M. F. Perutz. Copyright 1964 by Scientific American, Inc. All rights reserved.)

## Permanent Waves

The protein called keratin, which makes up human hair, has a high percentage of cystine. The disulfide bridges of cystine are primarily responsible for the shape of the hair. Hair is straight or curly because the keratin contains disulfide linkages that enable the molecules to hold their particular shapes. In the process of permanent waving, the hair is first treated with a reducing agent that cleaves some of the —S—S— bonds:

This allows the molecules to lose their rigid orientations and become more flexible. The hair is then set into the desired shape, using curlers or rollers, and an oxidizing agent is applied. The oxidizing agent reverses the above reaction, forming new disulfide links, which now hold the molecules together in the desired positions. The straightening of curly hair is done the same way.

Permanent waves are not really permanent. The hair keeps growing, and after some time the head contains a significant amount of new hair. This hair has not been treated, so it has the same disulfide linkages as the original hair.

bases, and salts affect the salt bridges as well as the hydrogen bonds. Ions of heavy metals ($Hg^{2+}$, $Pb^{2+}$) interact with the —SH groups.

Denaturation changes secondary, tertiary, and quaternary structure. It does not affect primary structure; that is, the sequence of amino acids that make up the chain. If these changes occur to a small extent, denaturation can be reversed. For example, in many cases when we remove a denatured protein from a urea solution and put it back into water it reassumes its secondary and tertiary structure. This is reversible denaturation. In living cells, some denaturation caused by heat can be reversed by chaperones (Sec. 18.9) called *heat-shock proteins*. These help a partially heat-denatured protein to regain its native secondary, tertiary, and quaternary structures. However, some denaturation is irreversible. We cannot unboil a hard-boiled egg.

## 18.11   GLYCOPROTEINS

Although many proteins such as serum albumin consist exclusively of amino acids, others also contain covalently linked carbohydrates and are therefore classified as **glycoproteins.** These include most of the plasma proteins (for example, fibrinogen), enzymes such as ribonuclease, hormones such as thyroglobulin, storage proteins such as casein and ovalbumins, and protective proteins such as immunoglobulins and interferon. The carbohydrate content of these proteins may vary from a few percent (immunoglobulins) up to 85 percent (blood group substances, Box 16C). Most of the proteins in membranes (lipid bilayers, Sec. 17.5) are glycoproteins.

## SUMMARY

**Proteins** are giant molecules made of amino acids linked together by **peptide bonds.** Proteins have many functions: structural (collagen), enzymatic (ribonuclease), carrier (hemoglobin), storage (casein), protective (immunoglobulin), and hormonal (insulin). **Amino acids** are organic compounds containing an amino (—NH$_2$) and a carboxylic acid (—COOH) group. The 20 amino acids found in proteins are classified by their side chains: nonpolar, polar but neutral, acidic, and basic. All amino acids in human tissues are L amino acids. Amino acids in the solid state, as well as in water, carry both positive and negative charges; they are called **zwitterions.** The pH at which the number of positive charges is the same as the number of negative charges is the **isoelectric point** of an amino acid or protein.

When the amino group of one amino acid condenses with the carboxyl group of another, an amide **(peptide)** linkage is formed, with the elimination of water. The two amino acids form a **dipeptide.** Three amino acids form a **tripeptide,** and so forth. Many amino acids form a **polypeptide** chain. Proteins are made of one or more polypeptide chains.

The linear sequence of amino acids is the **primary structure** of proteins. The repeating short-range conformations (**α-helix, β-pleated sheet, triple helix** of collagen, or **random coil**) are the **secondary structures.** The **tertiary structure** is the three-dimensional conformation of the protein molecule. Tertiary structures are maintained by covalent cross-links such as **disulfide bonds** and by **salt bridges, hydrogen bonds,** and **hydrophobic interactions.** The precise fit of polypeptide subunits into an aggregated whole is called the **quaternary structure.**

Secondary and tertiary structures stabilize the native conformation of proteins; physical and chemical agents, such as heat or urea, destroy these structures and **denature** the protein. Protein functions depend on native conformation; when a protein is denatured, it can no longer carry out its function. Some (but not all) denaturation is reversible; in some cases **chaperone** molecules reverse denaturation. Many proteins are classified as **glycoproteins** because they contain carbohydrate units.

## KEY TERMS

Alpha amino acid (Sec. 18.2)
Amino acid (Sec. 18.2)
Backbone (Sec. 18.6)
C-terminal amino acid (Sec. 18.5)
Chaperones (Sec. 18.9)
Conjugated protein (Sec. 18.9)
Cross-link (Sec. 18.9)
Denaturation (Sec. 18.10)
Disulfide linkage (Sec. 18.4)
Fibrous protein (Sec. 18.1)
Globular protein (Sec. 18.1)
Glycoprotein (Sec. 18.11)
Heat-shock protein (Sec. 18.10)
α-Helix (Sec. 18.8)
Hydrophobic interaction (Sec. 18.9)
Hypoglycemic awareness (Box 18D)
Intramolecular hydrogen bond (Sec. 18.8)
Isoelectric point (Sec. 18.3)
Marfan's syndrome (Box 18F)
N-terminal amino acid (Sec. 18.5)

Peptide (Sec. 18.5)
Peptide linkage (Sec. 18.5)
β-Pleated sheet (Sec. 18.8)
Polypeptide (Sec. 18.5)
Primary structure (Sec. 18.7)
Prosthetic group (Sec. 18.9)
Protein (Sec. 18.1)
Quarter-stagger arrangement (Box 18G)
Quaternary structure (Sec. 18.9)
Random coil (Sec. 18.8)
Residue (Sec. 18.5)
Salt bridge (Sec. 18.9)
Secondary structure (Sec. 18.8)
Sickle cell anemia (Box 18E)
Side chain (Sec. 18.6)
Subunit (Sec. 18.9)
Tertiary structure (Sec. 18.9)
Triple helix (Sec. 18.8)
Zwitterion (Sec. 18.3)

## PROBLEMS

**18.1**  What functions do fibrous proteins perform?

**18.2**  The human body has about 100 000 different proteins. Why do we need so many?

### Amino Acids

**18.3**  Name two acidic and two basic amino acids.

**18.4**  In what respect is proline different from the other amino acids that make up proteins?

**18.5**  Classify the following amino acids as nonpolar, polar but neutral, acidic, or basic:
(a)  arginine          (e)  tyrosine
(b)  leucine           (f)  phenylalanine
(c)  glutamic acid     (g)  glycine
(d)  asparagine

**18.6**  From Table 18.1 select the amino acid with the smallest and the largest side chain.

**18.7**  Which amino acids have aromatic side chains?

*__18.8__  Draw the structure of proline. Which class of heterocyclic compounds does this molecule belong to?

**18.9**  Why is it necessary to have proteins in our diets?

**18.10**  Which amino acid is also a thiol?

**18.11**  What are the similarities and differences in the structures of alanine and phenylalanine?

**18.12**  Which amino acids in Table 18.1 have more than one chiral carbon?

**18.13**  Draw the structures of D- and L-leucine.

*__18.14__  What special structural feature or property does each of the following amino acids have that makes it different from all the others:
(a)  glycine          (c)  tyrosine
(b)  cystine

### Zwitterions

**18.15**  Why are all amino acids solids at room temperature?

*__18.16__  Show how alanine, in solution at its isoelectric point, acts as a buffer (write equations to show why the pH does not change much if we add acid or base).

**18.17**  Draw the structure of lysine (a) above  (b) below and  (c) at its isoelectric point.

**18.18**  Explain why an amino acid cannot exist in an un-ionized form [$RCH(NH_2)COOH$] at any pH.

### Peptides and Proteins

**18.19**  Draw the structure of a tripeptide made of alanine, lysine, and tyrosine.

**18.20**  Draw the structure of the dipeptide serylleucine.

**18.21**  Write by symbols all the tripeptides that can be formed from alanine (Ala) and proline (Pro).

**18.22**  (a) Use the three-letter abbreviations to write the representation of the following tetrapeptide.  (b) Which amino acid is the C-terminal end and which the N-terminal end?

**18.23**  Show by chemical equations how serine and asparagine can be combined to give two different dipeptides.

**18.24**  (a) What is a protein backbone?  (b) What is an N-terminal end of a protein?

### Properties of Peptides and Proteins

**18.25**  (a) How many atoms of the peptide bond lie in the same plane?  (b) Which atoms are they?

**18.26**  (a) Draw the structural formula of the tripeptide Ala-Val-Phe.  (b) Draw the different ionic structures of this tripeptide at pH 2.0, 7.0, and 10.0.

**18.27**  Proteins are least soluble at their isoelectric points. What would happen to a protein precipitated at its isoelectric point if a few drops of dilute HCl were added?

**18.28**  How can a protein act as a buffer?

### Primary Structure of Proteins

**18.29**  How many different tripeptides can be made (a) using only leucine, threonine, and valine (b) using all 20 amino acids

*__18.30__  How many different tetrapeptides can be made (a) if the peptides contain one residue each of asparagine, proline, serine, and methionine (b) if all 20 amino acids can be used

**18.31** Based on your knowledge of the chemical properties of amino acid side chains, suggest a substitution for leucine in the primary structure of a protein that would probably not change the character of the protein very much.

**18.32** How many amino acid residues in the A chain of insulin are the same in insulin from humans, cattle (bovine), hogs, and sheep?

**18.33** Describe the functions of oxytocin and vasopressin.

## Secondary Structure of Proteins

**18.34** Is a random coil a (a) primary (b) secondary (c) tertiary or (d) quaternary structure?

**18.35** Decide whether these types of structures that exist in collagen are primary, secondary, tertiary, or quaternary: (a) tropocollagen (b) collagen fibril (c) collagen fiber (d) the proline–hydroxyproline–glycine repeating sequence.

**18.36** Look at the structure of carboxypeptidase in Figure 18.9. What kind of $\beta$-pleated sheets can you observe?

## Tertiary and Quaternary Structure of Proteins

**18.37** Polyglutamic acid (a polypeptide chain made only of glutamic acid residues) has an $\alpha$-helix conformation below pH 6.0 and a random-coil conformation above pH 6.0. What is the reason for this conformational change?

**18.38** Distinguish between inter- and intramolecular hydrogen bonding between backbone groups. Where in protein structures do you find one and where the other?

**18.39** Identify the different primary, secondary, and tertiary structures in the numbered boxes:

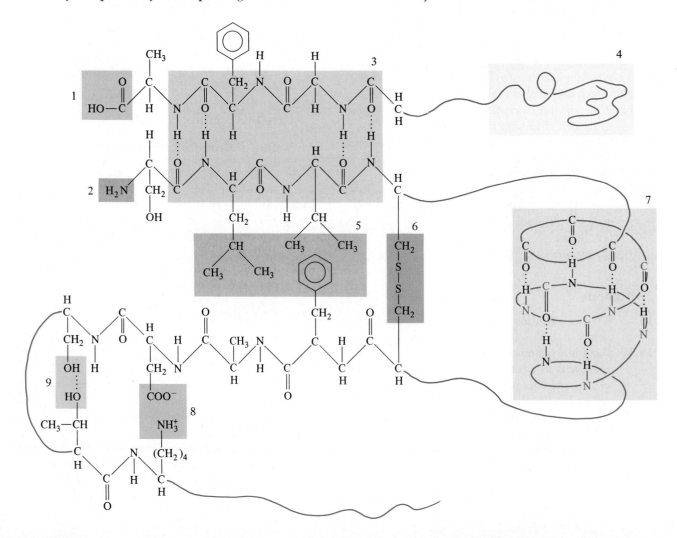

**18.40** Many of the plasma proteins (aqueous environment) are globular. On the basis of the solubilities of the side chains, where do you expect to find most of the leucine, isoleucine, valine, and alanine residues—inside or outside (on the surface of) the protein? Explain.

**18.41** Which amino acid side chains can form hydrophobic bonds?

**18.42** What kind of interaction operates when an aspartic acid residue is on one side chain and a lysine residue is on a neighboring one?

**18.43** What is a conjugated protein?

## Protein Denaturation

\*18.44 Which amino acid side chain is most frequently involved in denaturation by reduction?

**18.45** Does the primary structure of a protein always determine the secondary and tertiary structures?

**18.46** What happens to the tertiary structure of a globular protein if 2-mercaptoethanol is added to it?

## Boxes

**18.47** (Box 18A) Would you expect to find more D-aspartic acid in the teeth of (a) a newborn baby or (b) an adult?

**18.48** (Box 18B) (a) Describe the difference between the structure of aspartame and the methyl ester of phenylalanylaspartic acid. (b) Do you expect this second compound to be as sweet as aspartame?

**18.49** (Box 18C) How does glutathione protect the cell against oxidizing agents?

**18.50** (Box 18C) What is unusual about the peptide bond of glutathione?

**18.51** (Box 18D) Define hypoglycemic awareness.

**18.52** (Box 18E) In sickle cell anemia, Val is substituted for Glu in the sixth position of the beta chain of hemoglobin. Some individuals have hemoglobin in which Asp is substituted for Glu. Would you expect this substitution also to be detrimental to health? Explain.

**18.53** (Box 18E) Why has the mutation in the beta chain of hemoglobin survived in many generations of heterozygotes, even though it causes sickle cell anemia?

**18.54** (Box 18F) Marfan's syndrome involves changes in the conformation of collagen. Which structure is abnormal in this syndrome—primary, secondary, tertiary, or quaternary? Explain.

**18.55** (Box 18G) What property does collagen contribute to bone? Consider the differences in bone, dentine, and enamel.

**18.56** (Box 18H) Silver nitrate, $AgNO_3$, is sometimes put into the eyes of newborn infants as a disinfectant against gonorrhea. Silver is a heavy metal. Explain how this may work against the bacteria.

**18.57** (Box 18H) Why do nurses and physicians use 70 percent alcohol to wipe the skin before giving injections?

**18.58** (Box 18H) Silver nitrate is also used in cauterizing wounds. Explain the action of $Ag^+$ in this process.

**18.59** (Box 18I) What makes natural hair either straight or curly?

## Additional Problems

**18.60** Cysteine plays an important role in forming the tertiary structures of proteins. What is it?

**18.61** Considering the vast number of animal and plant species on earth (including those now extinct) and the large number of different protein molecules in each organism, have all possible protein molecules been used already by some species or other? Explain.

**18.62** Draw the structure of valine at pH 1 and at pH 12.

**18.63** Hydrogen bonds stabilize both $\alpha$-helix and $\beta$-pleated sheet structures. These bonds occur between the hydrogen of an N—H and the oxygen of a C=O. Can hydrogen bonding occur between two N—H groups?

**18.64** Aspartic acid has an isoelectric point of 2.98. Is the isoelectric point of asparagine higher, lower, or the same?

\*18.65 Collagen is a glycoprotein. Collagen from kidneys, which is involved in water filtration, has a high carbohydrate content, while collagen from tendons has a low carbohydrate content. What role do you think carbohydrate plays in collagen in the kidneys?

\*18.66 Denaturation is usually associated with transitions from helical structures to random coils. If an imaginary process were to transform the keratin in your hair from an $\alpha$-helix to a $\beta$-pleated sheet structure, would you call the process denaturation?

**18.67** Which amino acid does not rotate the plane of polarized light?

**18.68** How many different dipeptides can be made (a) using only alanine, tryptophan, glutamic acid, and arginine  (b) using all 20 amino acids

**18.69** Proline is often called an $\alpha$-helix terminator. That is, it is usually in the random-coil secondary structure following an $\alpha$-helix portion of a protein chain. Why does proline not fit easily into an $\alpha$-helix structure?

**18.70** Carbohydrates are integral parts of glycoproteins. Does the addition of carbohydrate to the protein core make the protein more or less soluble in water? Explain.

**18.71** Write the expected products of the acid hydrolysis of the following tripeptide:

$$NH_2-CH-\underset{\underset{\displaystyle CH_2}{|}}{C}=NH-CH-\overset{\overset{\displaystyle CH_3}{|}}{\underset{\underset{O}{\|}}{C}}-NH-CH-COOH$$

(tripeptide structure with side chains: $CH_2-S-CH_3$, $CH_2-CH(CH_3)-CH_3$, $CH_2-COOH$)

$$+ \; 2H_2O \xrightarrow{\;H^+\;}$$

**18.72** In diabetes, insulin is administered intravenously. Explain why this hormone protein cannot be taken orally.

**\*18.73** How many different decapeptides (peptides containing 10 amino acids each) can be made from the 20 amino acids?

# CHAPTER 19

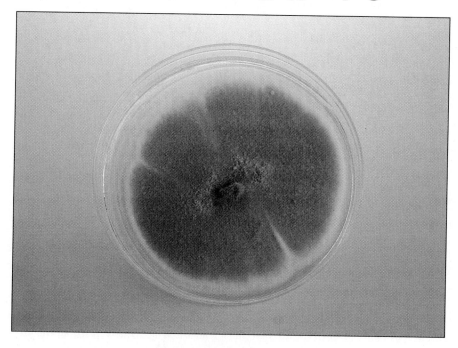

*Penicillin. (Photograph by Beverly March.)*

# Enzymes

## 19.1 INTRODUCTION

The cells in your body are chemical factories. Only a few of the thousands of compounds necessary for the operation of the human organism are obtained from the diet. Most of them are synthesized within the cells, which means that hundreds of chemical reactions are taking place in your cells every minute of your life.

Nearly all of these reactions are catalyzed by **enzymes,** which are **protein molecules that increase the rates of chemical reactions without themselves undergoing any change.** Without enzymes, life as we know it would not be possible.

Like all catalysts, enzymes do not change the position of equilibrium. That is, enzymes cannot make a reaction take place that would not take place without them. What they do is increase the rate; they cause reactions to take place faster. As catalysts, enzymes are remarkable in two respects: (1) They are extremely effective, increasing reaction rates by anywhere from $10^9$ to $10^{20}$ times, and (2) they are extremely specific.

As an example of their effectiveness, consider the oxidation of glucose. A lump of glucose or even a glucose solution exposed to oxygen under sterile conditions would show no appreciable change for months. In the human body the same glucose is oxidized within seconds.

Proteins are not the only biological catalysts. Ribonucleic acids, especially introns (Sec. 23.11), can also act as enzymes in certain cases. Such biocatalysts are called **ribozymes.** This was discovered by Sidney Altman (1939– ) of Yale University and Thomas R. Cech (1948– ) of the University of Colorado, who were awarded the 1989 Nobel Prize in Chemistry for this work.

$$\underset{\text{H}}{\overset{\text{H}}{>}}\text{N}-\text{CH}_2-\overset{\overset{\text{O}}{\|}}{\text{C}}-\text{NH}-\underset{\underset{\underset{\text{NH}_2}{|}}{\overset{|}{(\text{CH}_2)_4}}}{\overset{|}{\text{CH}}}-\overset{\overset{\text{O}}{\|}}{\text{C}}\downarrow\text{NH}-\underset{\overset{|}{\text{CH}_3}}{\text{CH}}-\overset{\overset{\text{O}}{\|}}{\text{C}}-\text{NH}-\underset{\underset{\text{OH}}{\overset{|}{\text{CH}_2}}}{\text{CH}}-\overset{\overset{\text{O}}{\|}}{\text{C}}-\text{NH}-\underset{\underset{\underset{\text{HN}=\text{C}-\text{NH}_2}{|}}{\overset{|}{\text{NH}}}}{\overset{|}{(\text{CH}_2)_3}}-\overset{\overset{\text{O}}{\|}}{\text{C}}\downarrow\text{NH}-\text{CH}_2-\overset{\overset{\text{O}}{\|}}{\text{C}}-\text{NH}-\underset{\underset{\text{COOH}}{\overset{|}{\text{CH}_2}}}{\text{CH}}-\overset{\overset{\text{O}}{\|}}{\text{C}}-\text{NH}-\text{CH}_2-\overset{\overset{\text{O}}{\|}}{\text{C}}\overset{}{<}_{\text{OH}}$$

Gly     Lys     Ala     Ser     Arg     Gly     Asp     Gly

**Figure 19.1**
A typical amino acid sequence. The enzyme trypsin catalyzes the hydrolysis of this chain only at the points marked with an arrow (the —COOH side of lysine and arginine).

Every organism has many enzymes—more than 3000 in a single cell. Presumably, each chemical reaction has one enzyme that catalyzes it. This means that enzymes are very specific, each of them speeding up only one particular reaction or class of reactions. For example, the enzyme urease catalyzes only the hydrolysis of urea and not that of other amides, even closely related ones. Another type of specificity can be seen with trypsin, an enzyme that cleaves the peptide linkages of protein molecules—but not every peptide linkage, only those on the carboxyl side of lysine and arginine residues (Fig. 19.1). The enzyme carboxypeptidase specifically cleaves only the last amino acid on a protein chain—the one at the C-terminal end. Lipases are less specific: They cleave any triglyceride, but they still don't affect carbohydrates or proteins.

The specificity of enzymes also extends to stereospecificity. The enzyme arginase converts the amino acid L-arginine (the naturally occurring form) to a compound called L-ornithine but has no effect on its mirror image, D-arginine.

Enzymes are distributed according to the body's need to catalyze specific reactions. A large number of protein-splitting enzymes are in the blood, ready to promote clotting. Digestive enzymes, which also split proteins, are located in the secretions of the stomach and pancreas. Even within the cells themselves, some enzymes are localized according to the need for specific reactions. The enzymes that help the oxidation of compounds that are part of the citric acid cycle (Sec. 20.4) are located in the mitochondria of the cells.

The enzyme that oxidizes D-glucose does not work on L-glucose.

Special organelles such as those called lysosomes contain an enzyme (lysozyme) that aids the dissolution of bacterial cell walls.

## 19.2 NAMING AND CLASSIFYING ENZYMES

Enzymes are commonly given names derived from the reaction they catalyze and/or the compound or type of compound they act on. For example, lactate dehydrogenase is the enzyme that speeds up the removal of hydrogen from lactate. Acid phosphatase helps to cleave phosphate ester bonds under acidic conditions. As can be seen from these examples, the names of most enzymes end in "-ase." Some enzymes, however, have older names, ones that were assigned before their actions were clearly understood. Among these are pepsin, trypsin, and chymotrypsin—all enzymes of the digestive tract.

Acetylcholine is a neurotransmitter (Sec. 24.1) that operates between the nerve endings and muscles. It attaches itself to a specific receptor in the muscle end plate. This transmits a signal to the muscle to relax. A specific enzyme, acetylcholinesterase, then catalyzes the hydrolysis of the acetylcholine,

$$H_2O + CH_3\overset{\overset{\displaystyle O}{\|}}{C}-O-CH_2CH_2\overset{\overset{\displaystyle CH_3}{|}}{\underset{\underset{\displaystyle CH_3}{|}}{N^+}}-CH_3 \xrightarrow{enzyme} CH_3\overset{\overset{\displaystyle O}{\|}}{C}-OH + \overset{\overset{\displaystyle OH}{|}}{CH_2CH_2}\overset{\overset{\displaystyle CH_3}{|}}{\underset{\underset{\displaystyle CH_3}{|}}{N^+}}-CH_3$$

Acetylcholine        Acetic acid     Choline

removing it from the receptor site. Succinylcholine,

$$\begin{array}{c} \overset{\overset{\displaystyle O}{\|}}{CH_2C}-O-CH_2CH_2-\overset{\overset{\displaystyle CH_3}{|}}{\underset{\underset{\displaystyle CH_3}{|}}{N^+}}-CH_3 \\ | \\ \\ CH_2C-O-CH_2CH_2-\overset{\overset{\displaystyle CH_3}{|}}{\underset{\underset{\displaystyle CH_3}{|}}{N^+}}-CH_3 \\ \underset{\displaystyle O}{\|} \end{array}$$

Succinylcholine

is sufficiently similar to acetylcholine so that it too attaches itself to the receptor of the muscle end plate and causes muscle relaxation. However, acetylcholinesterase can hydrolyze succinylcholine only very slowly, and thus the muscle stays relaxed for a long time.

This feature makes succinylcholine a good muscle relaxant during minor surgery, especially when a tube must be inserted into the bronchus (bronchoscopy). For example, after intravenous administration of 50 mg of succinylcholine, paralysis and respiratory arrest are observed within 30 seconds. While respiration is carried on artificially, the bronchoscopy can be performed within minutes.

Enzymes can be classified into six major groups according to the type of reaction they catalyze (see also Table 19.1):

**1.** *Oxidoreductases* catalyze oxidations and reductions.
**2.** *Transferases* catalyze the transfer of a group of atoms, such as $CH_3$, $CH_3CO$, or $NH_2$, from one molecule to another.
**3.** *Hydrolases* catalyze hydrolysis reactions.
**4.** *Lyases* catalyze the addition of a group to a double bond or the removal of a group to create a double bond.
**5.** *Isomerases* catalyze isomerization reactions.
**6.** *Ligases*, or *synthetases*, catalyze the joining of two molecules.

Besides their common names, enzymes are given classification numbers. For example, an enzyme called alcohol dehydrogenase is described as EC 1.1.1.1. The EC stands for Enzyme Commission. The first number designates

**TABLE 19.1 Classifications of Enzymes**

| Class | Typical Example | Reaction Catalyzed | Page Number in this Book |
|---|---|---|---|
| 1. Oxidoreductases | Lactate dehydrogenase | $CH_3-C(=O)-COOH \longrightarrow CH_3-CH(OH)-COOH$ (Pyruvate → L-(+)-Lactate) | 616 |
| 2. Transferases | Aspartate amino transferase or Aspartate transaminase | Aspartate + α-Ketoglutarate $\longrightarrow$ Oxaloacetate + Glutamate | 627 |
| 3. Hydrolases | Acetylcholinesterase | $CH_3-C(=O)-OCH_2CH_2N(CH_3)_3^+ + H_2O \longrightarrow CH_3COOH + HOCH_2CH_2N(CH_3)_3^+$ (Acetylcholine → Acetic acid + Choline) | 575 |
| 4. Lyases | Aconitase | cis-Aconitate $+ H_2O \longrightarrow$ Isocitrate | 600 |
| 5. Isomerases | Phosphohexose isomerase | Glucose-6-phosphate $\longrightarrow$ Fructose-6-phosphate | 616 |
| 6. Ligases | Tyrosine-tRNA synthase | $ATP + \text{L-tyrosine} + tRNA \longrightarrow \text{L-tyrosyltRNA} + AMP + PP_i$ | 667 |

the major group or class: 1 stands for oxidoreductases. The second number, also 1, designates the type of substance that donates the hydrogen, a primary or secondary alcohol. The third number designates the hydrogen acceptor: 1 in this example means the coenzyme $NAD^+$. The fourth number in the classification is an arbitrary serial number indicating the enzyme's place in its subclass.

## 19.3 COMMON TERMS IN ENZYME CHEMISTRY

Some enzymes, such as pepsin and trypsin, consist of polypeptide chains only. Other enzymes contain **nonprotein portions** called **cofactors.** The protein (polypeptide) portion of the enzyme is called an **apoenzyme.** The cofactors may be metallic ions, such as $Zn^{2+}$ or $Mg^{2+}$, or they may be organic compounds. Organic cofactors are called **coenzymes.** An important group of coenzymes are the B vitamins, which are essential to the action of many enzymes (see Sec. 20.3). Another important coenzyme is heme (Box 3C), which is part of a number of oxidoreductases as well as of hemoglobin. In any case, an apoenzyme cannot catalyze a reaction without its cofactor, nor can the cofactor function without the apoenzyme. When a metal ion is a cofactor, it can be bound directly to the protein or to the coenzyme, if the enzyme contains one.

The compound on which the enzyme works, the one whose reaction it speeds up, is called the **substrate.** The substrate usually adheres to the enzyme surface while it undergoes the reaction. There is a specific portion of the enzyme to which the substrate binds during the reaction. This part is called the **active site.** If the enzyme has coenzymes, they are located at the active site. Therefore the substrate is simultaneously surrounded by parts of the apoenzyme, coenzyme, and metal ion cofactor (if any), as shown in Figure 19.2.

Some enzymes have two kinds of cofactor: a coenzyme and a metallic ion.

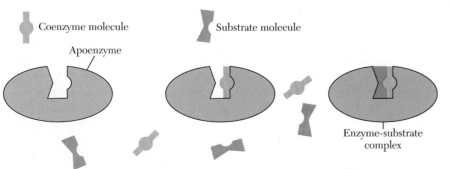

**Figure 19.2**
Schematic diagram of the active site of an enzyme and the participating components.

Some enzymes are manufactured by the body in an inactive form. In order to make them active, a small part of their polypeptide chain must be removed. These inactive forms of enzymes are called **proenzymes** or **zymogens.** Once the excess polypeptide chain is removed, the enzyme becomes active. For example, trypsin is manufactured as the inactive molecule trypsinogen (a zymogen). When a fragment containing six amino acid residues is removed from the N-terminal end, the molecule becomes a fully active trypsin molecule. Removal of the fragment not only shortens the chain but also changes the three-dimensional structure (the tertiary structure), allowing the molecule to achieve its active form.

The removal of the six-amino-acid fragment is of course also catalyzed by an enzyme.

Why does the body go to all this trouble? Why not just make the fully active trypsin to begin with? The reason is very simple. As we have seen, trypsin is a protease—it cleaves proteins (Fig. 19.1) and is therefore an important catalyst for the digestion of the proteins we eat. But it would not be good if it cleaved the proteins our own bodies are made of! Therefore, the body makes trypsin in an inactive form, and only after it has entered the digestive tract is it allowed to become active.

**Activation** is **any process that makes an inactive enzyme active.** This can be the simple addition of a cofactor to an apoenzyme or the cleavage of a polypeptide chain of a proenzyme. **Inhibition** is the opposite—**any process that makes an active enzyme less active or inactive** (Sec. 19.5). **Inhibitors** are compounds that accomplish this. Some inhibitors bind to the active site of the enzyme surface, thus preventing the binding of substrate. These are **competitive inhibitors.** Others, which bind to some other portion of the enzyme surface, may sufficiently alter the tertiary structure of the enzyme so that its catalytic effectiveness is slowed down. These are called **noncompetitive inhibitors.** Both competitive and noncompetitive inhibition are reversible, but there are some compounds that alter the structure of the enzyme *permanently* and thus make it *irreversibly* inactive.

**Figure 19.3**
The effect of enzyme concentration on the rate of reaction. Substrate concentration, temperature, and pH are constant.

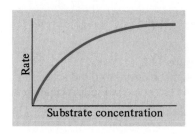

**Figure 19.4**
The effect of substrate concentration on the rate of an enzyme-catalyzed reaction. Enzyme concentration, temperature, and pH are constant.

## 19.4 FACTORS AFFECTING ENZYME ACTIVITY

**Enzyme activity** is a measure of how much reaction rates are increased. In this section we examine the effects of concentration, temperature, and pH on enzyme activity.

### Enzyme and Substrate Concentration

If we keep the concentration of substrate constant and increase the concentration of enzyme, the rate increases linearly (Fig. 19.3). That is, if the enzyme concentration is doubled, the rate also doubles; if the enzyme concentration is tripled, the rate also triples. This is the case in practically all enzyme reactions because the molar concentration of enzyme is almost always much lower than that of substrate (that is, there are almost always many more molecules of substrate present than molecules of enzyme).

On the other hand, if we keep the concentration of enzyme constant and increase the concentration of substrate, we get an entirely different type of curve, called a saturation curve (Fig. 19.4). In this case the rate does not increase continuously. Instead, a point is reached after which the rate stays the same even if we increase the substrate concentration further. This happens because, at the saturation point, substrate molecules are bound to all the available active sites of the enzymes. Since the active sites are where the reactions take place, once they are all occupied the reaction is going at its maximum rate. Increasing the substrate concentration can no longer increase the rate because the excess substrate cannot find any active sites to attach to.

### Temperature

Temperature affects enzyme activity because it changes the three-dimensional structure of the enzyme. In uncatalyzed reactions, the rate usually increases as the temperature increases. The effect of temperature on enzyme-catalyzed reactions is different. When we start at a low temperature (Fig. 19.5), an increase in temperature first causes an increase in rate. However, protein conformations are very sensitive to temperature changes. Once an optimum temperature is reached, any further increase in temperature causes changes in enzyme conformation. The substrate may then not fit properly onto the changed enzyme surface. Therefore the rate of reaction *decreases*.

After a *small* temperature increase above optimum, the (decreased) rate could still be increased again by lowering the temperature because, over a narrow temperature range, changes in conformation are reversible. However, at some higher temperature above optimum, we reach a point where the protein denatures (Sec. 18.10); the conformation is altered irreversibly, and the polypeptide chain cannot refold. At this point, the enzyme is completely inactivated. The inactivation of enzymes at low temperatures is used in the preservation of food by refrigeration.

### pH

Since the pH of its environment changes the conformation of a protein (Sec. 18.10), we expect effects similar to those observed when the temperature is changed. Each enzyme operates best at a certain pH (Fig. 19.6). Once again, within a narrow pH range, changes in enzyme activity are reversible. However, at extreme pH values (either acidic or basic), enzymes are denatured irreversibly, and enzyme activity cannot be restored by changing back to the optimal pH.

## 19.5  MECHANISM OF ENZYME ACTION

We have seen that the action of enzymes is highly specific. What kind of mechanism can account for such specificity? It was suggested by Arrhenius about 100 years ago that catalysts speed up reactions by combining with the substrate to form some kind of intermediate compound. In an enzyme-catalyzed reaction, the intermediate is the **enzyme-substrate complex.**

To account for the high specificity of most enzyme-catalyzed reactions, a number of models have been proposed. The simplest and most frequently

The effect of temperature on reaction rates is discussed in Section 7.4.

**Figure 19.5**
The effect of temperature on the rate of an enzyme-catalyzed reaction. Substrate and enzyme concentrations and pH are constant.

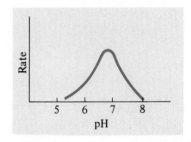

**Figure 19.6**
The effect of pH on the rate of an enzyme-catalyzed reaction. Substrate and enzyme concentrations and temperature are constant.

## Active Sites

The perception of the active site as either a rigid (lock-and-key model) or a partly flexible template (induced-fit model) is an oversimplification. Not only is the geometry of the active site important, but so are the specific interactions that take place between enzyme surface and substrate. To illustrate, we take a closer look at the active site of the enzyme pyruvate kinase (Fig. 19C). This enzyme catalyzes the transfer of the phosphate group from phosphoenol pyruvate (PEP) to ADP, an important step in glycolysis (Sec. 21.2):

Phosphoenol pyruvate       ADP                    Pyruvate            ATP

The active site of the enzyme binds both substrates, PEP and ADP. The enzyme has two cofactors, $K^+$ and $Mg^{2+}$. The $K^+$ binds the carboxyl group of the PEP, and the $Mg^{2+}$ anchors two phosphate groups, one from the PEP and one from the ADP. Other side chains of the apoenzyme bind the rest of the ADP into the active site. All these acids are in the form of their ions. There is also a hydrophobic area on the enzyme that binds the nonpolar $=CH_2$ unit.

**Figure 19C**
The active site and the substrates of pyruvate kinase.

Hydrophobic pocket

quoted is the **lock-and-key model** (Fig. 19.7). This model assumes that the enzyme is a rigid three-dimensional body. The surface that contains the active site has a restricted opening into which only one kind of substrate can fit, just as only the proper key can fit exactly into a lock.

According to the lock-and-key mechanism, an enzyme molecule has its own particular shape because that shape is necessary to maintain the active site in exactly the geometric alignment required for that particular reaction. An enzyme molecule is very large (typically 100 to 200 amino acid residues), but the active site is usually composed of only two or a few amino acid residues, which may well be located at different places in the chain. The other amino acids—those not part of the active site—are located in the sequence in which we find them because that is the sequence that causes the whole molecule to fold up in exactly the required way.

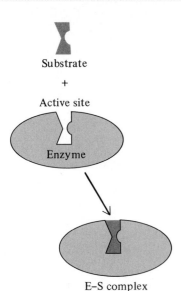

Substrate

+

Active site

Enzyme

E–S complex

**Figure 19.7**
The lock-and-key model of enzyme mechanism.

The vitamin folic acid is a coenzyme in a number of biosynthetic processes, such as the synthesis of amino acids and of nucleotides. Humans obtain folic acid from the diet or from microorganisms in the intestinal tract. These microorganisms can synthesize folic acid if *para*-aminobenzoic acid is available to them:

Folic acid

p-Aminobenzoic acid

Sulfapyridine

Sulfanilamide

Sulfathiazole

In the 1930s it was discovered that the compound sulfanilamide could kill many types of harmful bacteria and thus cure several diseases. Because sulfanilamide itself has an unacceptably high toxicity to humans, a number of derivatives of this compound, such as sulfapyridine and sulfathiazole, which act in a similar way, are used instead. These drugs work by "tricking" the bacteria, which normally use *para*-aminobenzoic acid as a raw material in the synthesis of folic acid. When the bacteria get a sulfa drug instead, they cannot tell the difference and use it to make a molecule that also has a folic acid type of structure but is not exactly the same. When they try to use this fake folic acid as a coenzyme, not only doesn't it work, but it is now a competitive inhibitor of the enzyme's action, so that many of the bacteria's amino acids and nucleotides cannot be made, and the bacteria die.

The lock-and-key model explains the action of many enzymes. But for other enzymes, there is evidence that this model is too restrictive. Enzyme molecules are in a dynamic state, not a static one. There are constant motions within them, so that the active site has some flexibility.

From x-ray diffraction we know that the size and shape of the active site cavity change when the substrate enters. The American biochemist Daniel Koshland introduced the **induced-fit model** (Fig. 19.8), in which he compared the changes occurring in the shape of the cavity upon substrate binding to the changes in the shape of a glove when a hand is inserted. That is, the enzyme modifies the shape of the active site to accommodate the substrate.

Both the lock-and-key and the induced-fit model explain the phenomenon of competitive inhibition (Sec. 19.3). The inhibitor molecule fits into the active

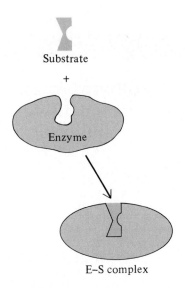

**Figure 19.8**
The induced-fit model of enzyme mechanism.

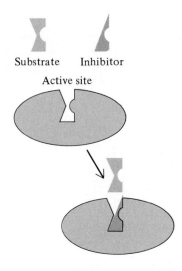

**Figure 19.9**
The mechanism of competitive inhibition. When a competitive inhibitor enters the active site, the substrate cannot get in.

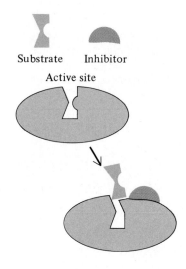

**Figure 19.10**
Mechanism of noncompetitive inhibition. The inhibitor attaches itself to a site other than the active site (allosterism) and thereby changes the conformation of the active site.

site cavity in the same way the substrate does (Fig. 19.9), preventing the substrate from entering. The result is that whatever reaction is supposed to take place on the substrate does not take place.

Many cases of noncompetitive inhibition can also be explained by either model. In this case, the inhibitor does not bind to the active site but to a different part of the enzyme. Nevertheless, the binding causes a change in the three-dimensional shape of the enzyme molecule, and this so alters the shape of the active site (the lock) that the substrate (the key) can no longer fit (Fig. 19.10).

If we compare enzyme activity in the presence and absence of an inhibitor, we can tell whether competitive or noncompetitive inhibition is taking place (Fig. 19.11). The maximum reaction rate is the same without an inhibitor and in

**Figure 19.11**
Enzyme kinetics in the presence and absence of inhibitors.

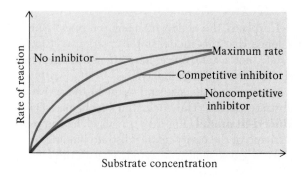

Many enzymes contain a number of —SH groups (cysteine residues) that are easily oxidized by heavy metal ions:

$$\begin{array}{ccc} | & & | \\ SH & & S \\ | & & | \\ & + Pb^{2+} \longrightarrow & Pb & + 2H^+ \\ | & & | \\ SH & & S \\ | & & | \end{array}$$

If these cross-linked cysteine residues are at or near active sites, the cross-linking can irreversibly inhibit enzyme activity. This is the basis of lead and mercury poisoning.

Mercury poisoning has become an acute problem in recent years because the mercury and mercury compounds that factories dumped into streams and lakes for decades have entered the food chain. Previously it was thought that mercury, as a heavy metal, would settle to the bottom of lakes and rivers and would be harmlessly buried there, covered by sand. However, certain microorganisms convert mercury metal to organic mercury compounds, mainly dimethylmercury, $(CH_3)_2Hg$. This compound enters the food chain when it is absorbed by algae. Later, it is concentrated in fish, and people who eat the contaminated fish can get mercury poisoning.

In Japan, where mercury was dumped in the Bay of Minamata, more than 100 people showed mercury poisoning from a diet high in seafood coming from the bay, and 44 people died. Once such contaminated food was removed from the markets, the number of mercury poisoning cases reported dropped drastically.

the presence of a competitive inhibitor. The only difference is that this maximum rate is achieved at a low substrate concentration with no inhibitor but at a high substrate concentration when an inhibitor is present. This is the true sign of competitive inhibition because here the substrate and the inhibitor are competing for the same active site. If the substrate concentration is sufficiently increased, the inhibitor will be displaced from the active site by Le Chatelier's principle.

If, on the other hand, the inhibitor is noncompetitive, it cannot be displaced by addition of excess substrate, since it is bound to a different site. In this case the enzyme cannot be restored to its maximum activity, and the maximum rate of the reaction is lower than it would be in the absence of the inhibitor.

## 19.6  ENZYME REGULATION AND ALLOSTERISM

Enzymes are often regulated by environmental conditions. The reaction product of one enzyme may control the activity of another, especially in a complex system in which enzymes work cooperatively. For example, in a system

$$A \xrightarrow{E_1} B \xrightarrow{E_2} C \xrightarrow{E_3} D$$

each step is catalyzed by a different enzyme. The last product in the chain, D, may inhibit the activity of enzyme $E_1$ (by competitive or noncompetitive inhi-

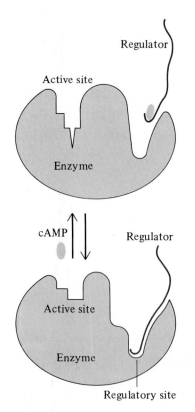

**Figure 19.12**
Interaction of the catalytic and regulatory subunits of protein kinase.

bition). When the concentration of D is low, the three reactions proceed rapidly, but as the concentration of D increases, the action of $E_1$ becomes inhibited and eventually stops. In this manner, the accumulation of D is a *message* telling enzyme $E_1$ to shut down because the cell has enough D for its present needs. Shutting down $E_1$ stops the whole process. Enzyme regulation by a process in which formation of a product inhibits an earlier reaction in the sequence is called **feedback control.**

Sometimes regulation takes place not by feedback control but by an event that occurs at a site other than the active site, but which eventually affects the active site. This type of interaction is called **allosterism,** and any enzyme regulated by this mechanism is called an **allosteric enzyme.** If a substance binds noncovalently and reversibly to a site *other than the active site,* it may affect the enzyme in either of two ways: It may inhibit enzyme action (**negative modulation**) or it may stimulate it (**positive modulation**).

The substance that binds to the allosteric enzyme is called a **regulator** or **modulator,** and the site it attaches to is called the **regulatory site.** In most cases, allosteric enzymes contain more than one polypeptide chain (subunits); the regulatory site is on one polypeptide chain and the active site on another.

Specific regulators can bind reversibly to the regulatory sites. For example, the enzyme protein kinase is an allosteric enzyme. In this case the enzyme has only one polypeptide chain, so it carries both the active site and the regulatory site at different parts of this chain (Fig. 19.12). The regulator is another protein molecule, one that binds reversibly to the regulatory site. As long as the regulator is bound to the regulatory site, the total enzyme-regulator complex (ER) is inactive. When the compound cyclic AMP (Sec. 24.4) approaches the inactive ER, it removes the regulator and makes the protein kinase active:

Cyclic AMP is a positive modulator. It stimulates enzyme action by removing the regulator that inhibits the enzyme. Thus **allosteric enzyme action is controlled by the modulator.**

In the body, protein kinase is in the inactive form most of the time. Whenever protein kinase activity is needed, the body sends cAMP to activate the protein kinase from its ER form.

Allosteric regulation may occur with proteins other than enzymes. Section 25.3 explains how the oxygen-carrying ability of hemoglobin, an allosteric protein, is affected by modulators.

Another form of regulation of enzyme activity occurs when the same enzyme appears in different forms in different tissues. Lactate dehydrogenase catalyzes the conversion of lactate to pyruvate, and vice versa (Fig. 21.3, step ⑪). The enzyme has four subunits. Two kinds of subunits exist, called H and M. The enzyme that dominates in the heart is an $H_4$ enzyme, meaning that all four subunits are of the H type, although some M-type subunits are also present. In the liver and skeletal muscles the M type dominates. Other types of tetramer (four units) combinations exist in different tissues: $H_3M$, $H_2M_2$, and $HM_3$. These different forms of the same enzyme are called **isozymes** or **isoenzymes.** The $H_4$ enzyme is allosterically inhibited by high levels of pyruvate, while the $M_4$ is not. In diagnosing the severity of heart attacks (Sec. 19.7), the release of $H_4$ isoenzyme is monitored in the serum.

## 19.7 ENZYMES IN MEDICAL DIAGNOSIS AND TREATMENT

Most enzymes are confined within the cells of the body. However, small amounts of enzymes can also be found in body fluids such as blood, urine, and cerebrospinal fluid. The level of enzyme activity in these fluids can easily be monitored. It has been found that abnormal activity (either high or low) of particular enzymes in various body fluids signals either the onset of certain diseases or their progression. Table 19.2 lists some enzymes used in medical diagnosis and their activities in normal body fluids.

A number of enzymes are assayed (measured) during myocardial infarction in order to diagnose the severity of the heart attack. Dead heart muscle cells spill their enzyme contents into the serum. Thus, the level of aspartate aminotransferase (AST) (formerly called glutamate-oxaloacetate transaminase, or GOT) in the serum rises rapidly after a heart attack. In addition to AST, lactate dehydrogenase (LD-P) and creatine kinase (CK) levels are also monitored. In infectious hepatitis, the alanine aminotransferase (ALT) (formerly called glutamate-pyruvate transaminase, or GPT) level in the serum can rise to ten times normal. There is also a concurrent increase in AST activity in the serum.

In some cases, the administration of an enzyme is part of therapy. After duodenal or stomach ulcer operations, patients are advised to take tablets containing digestive enzymes that are in short supply in the stomach after surgery. Such enzyme preparations contain lipases, either alone or combined with proteolytic enzymes, and are sold under such names as Pancreatin, Acro-lase, and Ku-zyme.

**TABLE 19.2  Enzyme Assays Useful in Medical Diagnosis**

| Enzyme | Normal Activity | Body Fluid | Disease Diagnosed |
|---|---|---|---|
| Alanine aminotransferase (ALT) | 3–17 U/L[a] | Serum | Hepatitis |
| Acid phosphatase | 2.5–12 U/L | Serum | Prostate cancer |
| Alkaline phosphatase (ALP) | 13–38 U/L | Serum | Liver or bone disease |
| Amylase | 19–80 U/L | Serum | Pancreatic disease or mumps |
| Aspartate aminotransferase (AST) | 7–19 U/L | Serum | Heart attack or hepatitis |
|  | 7–49 U/L | Cerebrospinal fluid |  |
| Lactate dehydrogenase (LD-P) | 100–350 WU/mL | Serum | Heart attack |
| Creatine kinase (CK) | 7–60 U/L | Serum |  |
| Phosphohexose isomerase (PHI) | 15–75 U/L | Serum |  |

[a] U/L = International units per liter; WU/mL = Wrobleski units per milliliter.

## Box 19F

## Penicillin: War of Enzyme Against Enzyme

Since World War II, the most widely used antibacterial agent has been penicillin, which is produced by a mold and prevents the growth of bacteria. Penicillin was discovered accidentally by Alexander Fleming (1881–1955) in 1929 and is now produced in large quantities. It inhibits an enzyme called transpeptidase. Bacteria need this enzyme to make their cell walls rigid and cross-linked. In the presence of penicillin, bacterial cell walls are not cross-linked, so that the contents of the bacterial cells cannot be maintained. The cytoplasm spills out, and the bacteria die.

Penicillin caused a revolution in medicine and paved the way for a host of antibiotics, all products of microorganisms directed against other microorganisms. In the last 40 years penicillin has been overused. In many countries it is sold without prescription and has also been used in animal feed. This happens in spite of the fact that some people are severely allergic to penicillin, so that administering the drug may cause shock and fatal coma.

New bacterial strains have appeared in the last 20 to 25 years that can resist penicillin. These new bacterial strains contain the enzyme penicillinase. Penicillin contains a *beta*-lactam ring. The penicillinase opens this ring and makes penicillin ineffective:

Penicillin G (R = $C_6H_5CH_2$)          Penicilloic acid

The rapid evolution of the new strains of bacteria threatened the effectiveness of this antibiotic, but research is always a step ahead in this continuous war. New synthetic penicillins have been developed with R groups that prevent the attack of penicillinase.

Even more effective is a drug called clavulinic acid. This compound has no antibacterial property; that is, it does not kill bacteria. However, it acts as an irreversible inhibitor of penicillinase. It can do so because, like penicillin, it has its own *beta*-lactam ring that irreversibly binds to the active site of penicillinase. Therefore, when given in combination with penicillin, it makes the penicillin effective against all strains of bacteria.

## SUMMARY

**Enzymes** are proteins that catalyze chemical reactions in the body. Most enzymes are very specific—they catalyze only one particular reaction. The compound whose reaction is catalyzed by an enzyme is called the **substrate.** Enzymes are classified into six major groups according to the type of reaction they catalyze. Most enzymes are named after the substrate and the type of reaction they catalyze, adding the ending **"-ase."**

Some enzymes are made of polypeptide chains only. Others have, besides the polypeptide chain (the **apoenzyme**), nonprotein **cofactors,** either organic compounds **(coenzymes)** or inorganic ions. Only a small part of the enzyme surface participates in the actual catalysis of chemical reactions. This part is called the **active site.** Cofactors, if any, are part of the active site. Some enzymes, called **proenzymes** or **zymogens,** must be activated by removing a small portion of the polypeptide chain.

Compounds that slow enzyme action are called **inhibitors. A competitive inhibitor** attaches itself to the active site. A **noncompetitive inhibitor** binds to other parts of the enzyme surface. The higher the enzyme and substrate concentrations, the higher the enzyme activity, except that, at sufficiently high substrate concentrations, a saturation point is reached. After this, increasing substrate concentration no longer increases the rate. Each enzyme has an optimum temperature and pH at which it has its greatest activity.

Two closely related mechanisms by which enzyme activity and specificity are explained are the **lock-and-key model** and the **induced-fit model.** Enzyme activity is regulated by feedback mechanisms and by **allosterism.** An allosteric interaction takes place at a position other than the active site but affects the active site, either positively or negatively. Enzyme activity is also regulated by **isozymes,** which are different forms of the same enzyme. Abnormal enzyme activity can be used to diagnose certain diseases.

## KEY TERMS

Activation (Sec. 19.3)
Active site (Sec. 19.3)
Allosterism (Sec. 19.6)
Apoenzyme (Sec. 19.3)
Coenzyme (Sec. 19.3)
Cofactor (Sec. 19.3)
Competitive inhibition (Sec. 19.3)
Enzyme (Sec. 19.1)
Enzyme activity (Sec. 19.4)
Enzyme-substrate complex (Sec. 19.5)
Feedback control (Sec. 19.6)
Induced-fit model (Sec. 19.5)

Inhibition (Sec. 19.3)
Isoenzyme (Sec. 19.6)
Isozyme (Sec. 19.6)
Lock-and-key model (Sec. 19.5)
Modulator (Sec. 19.6)
Noncompetitive inhibition (Sec. 19.3)
Proenzyme (Sec. 19.3)
Regulator (Sec. 19.6)
Ribozyme (Sec. 19.1)
Substrate (Sec. 19.3)
Zymogen (Sec. 19.3)

## PROBLEMS

### Enzymes

**19.1**  What is the difference between the terms catalyst and enzyme?

**19.2**  To what extent can enzymes increase the rate of reactions?

**19.3**  How many enzymes are in one single cell?

**19.4**  What does an enzyme do to the energy of activation of a reaction?

**19.5**  Why does the body need so many different enzymes?

**19.6**  Trypsin splits polypeptide chains at the carboxyl side of a lysine or arginine residue (Fig. 19.1).

Chymotrypsin splits polypeptide chains on the carboxyl side of an aromatic residue or any other nonpolar, bulky residue. Which enzyme is more specific?

### Naming and Classifying Enzymes

**\*19.7**  Lipases are less specific than trypsin. Explain why this is so.

**19.8**  Citrate synthetase is an enzyme of the citric acid cycle. Judging from its name, what is the function of this enzyme?

*19.9 On the basis of the classification given in Section 19.2, decide to which group each of the following enzymes belongs:

| Enzyme | Reaction |
| --- | --- |

(a) phosphoglyceromutase

$$^-OOC-CH-CH_2-OPO_3^{2-} \rightleftharpoons$$
$$\qquad\quad |$$
$$\qquad\quad OH$$

3-Phosphoglycerate

$$^-OOC-CH-CH_2OH$$
$$\qquad\quad |$$
$$\qquad\quad OPO_3^{2-}$$

2-Phosphoglycerate

(b) urease

$$\qquad\quad O$$
$$\qquad\quad \|$$
$$H_2N-C-NH_2 + H_2O \rightleftharpoons 2NH_3 + CO_2$$

Urea

(c) succinate dehydrogenase

$$^-OOC-CH_2-CH_2-COO^- + \; FAD \; \rightleftharpoons$$
$$\qquad\quad \text{Succinate} \qquad\qquad \text{Coenzyme}$$

$$^-OOC-CH=CH-COO^- + FADH_2$$
$$\qquad\qquad \text{Fumarate} \qquad\qquad \text{Reduced coenzyme}$$

(d) aspartase

$$^-OOC-CH=CH-COO^- + NH_4^+ \rightleftharpoons$$
$$\qquad\qquad \text{Fumarate}$$

$$^-OOC-CH_2-CH-COO^- + H^+$$
$$\qquad\qquad\quad |$$
$$\qquad\qquad\quad NH_2$$

L-Aspartate

19.10 What kind of reaction does each of the following enzymes catalyze?
(a) deaminases        (c) dehydrogenases
(b) hydrolases        (d) isomerases

19.11 What is the difference between a coenzyme and a cofactor?

19.12 What is the difference between a zymogen and proenzyme?

19.13 What is the difference between reversible and irreversible noncompetitive inhibition?

19.14 Which kind of inhibitor enters the active site: (a) a competitive inhibitor or (b) a noncompetitive inhibitor?

19.15 The enzyme trypsin is synthesized by the body in the form of a long chain (trypsinogen) from which a piece must be cut before the trypsin can be active. Why does the body not synthesize the trypsin directly?

## Factors Affecting Enzyme Activity

*19.16 At a very low concentration of a certain substrate we find that when the substrate concentration is doubled the rate of the enzyme-catalyzed reaction is also doubled. Would you expect to find the same at a very high substrate concentration?

19.17 If we wish to double the rate of an enzyme-catalyzed reaction, can we do this by increasing the temperature 10°C?

19.18 A bacterial enzyme has a temperature-dependent activity as follows:

(a) Is this enzyme more or less active at normal body temperature than when a person has a fever?  (b) What happens to the enzyme activity if the patient's temperature is lowered to 35°C?

19.19 The optimum temperature for the action of lactate dehydrogenase is 36°C. It is irreversibly inactivated at 85°C, but a yeast containing this enzyme can survive for months at −10°C. Explain how this can happen.

## Mechanism of Enzyme Action

19.20 Urease can catalyze the hydrolysis of urea,
$$H_2N-C-NH_2,$$ but not the hydrolysis of
$$\qquad\quad \|$$
$$\qquad\quad O$$

diethyl urea,

Explain why diethyl urea is not hydrolyzed.

**19.21** How does the lock-and-key model differ from the induced-fit model in explaining enzyme action?

**\*19.22** What is the maximum rate that can be achieved in competitive inhibition compared with noncompetitive inhibition?

**19.23** Enzymes are long protein chains, usually containing more than 100 amino acid residues. Yet the active site contains only a few amino acids. Explain why all the other amino acids of the chain are present and what would happen to the enzyme activity if significant changes were made in the structure.

## Enzyme Regulation and Allosterism

**19.24** The decomposition of glycogen to yield glucose is catalyzed by the enzyme phosphorylase. Epinephrine, a hormone, increases the activity of the phosphorylase by converting the inactive phosphorylase $b$ to the active $a$ form, without binding to the active site of phosphorylase. What kind of regulatory mechanism is at work?

**19.25** In the protein kinase system, how does the cAMP modulate the reaction?

**19.26** What is the difference between a regulator and a modulator?

**19.27** Can the product of a reaction in a sequence act as an inhibitor for another reaction of the sequence? Explain.

**19.28** Discuss three ways by which the body regulates enzyme action.

**19.29** What is an isozyme?

**19.30** Which LD-P isozyme is monitored in the case of a heart attack?

## Enzymes in Medical Diagnosis and Treatment

**\*19.31** The enzyme formerly known as GPT (glutamate-pyruvate transaminase) has a new name: ALT (alanine aminotransferase). Looking at the chemical equation on p. 630, which is catalyzed by this enzyme, what prompted this change of name?

**19.32** Chemists who have been exposed for years to organic vapors usually show higher-than-normal activity when given the alkaline phosphatase test. What organ in the body is affected by organic vapors?

**19.33** What enzyme preparation is given to patients after duodenal ulcer surgery?

**19.34** Chymotrypsin is secreted by the pancreas and passed into the intestine. The optimum pH for this enzyme is 7.8. If a patient's pancreas cannot manufacture chymotrypsin, would it be possible to supply it orally? What happens to the chymotrypsin activity during its passage through the gastrointestinal tract?

## Boxes

**19.35** (Box 19A) Acetylcholine causes muscles to contract. Succinylcholine, a close relative, is a muscle relaxant. Explain the different effects of these related compounds.

**19.36** (Box 19A) An operating team usually administers succinylcholine before bronchoscopy. What is achieved by this procedure?

**\*19.37** (Box 19B) Which enzyme resembles papain in its action: trypsin or carboxypeptidase? Explain.

**19.38** (Box 19C) What is the function of the hydrophobic pocket in the active site of pyruvate kinase?

**\*19.39** (Box 19C) Oxalate ion, $^-O-C-C-O^-$, acts

$$\underset{O}{\overset{\|}{}}\quad\underset{O}{\overset{\|}{}}$$

as a competitive inhibitor of pyruvate kinase. Identify the potential sites where oxalate ion can bind to the active site of the enzyme.

**19.40** (Box 19D) What part of folic acid is the key structure that sulfa drugs mimic?

**\*19.41** (Box 19D) Sulfa drugs kill certain bacteria by preventing the synthesis of folic acid, a vitamin. Why don't these drugs kill people?

**19.42** (Box 19E) Which functional group is affected in mercury poisoning?

**19.43** (Box 19E) Why doesn't mercury dumped into a lake settle harmlessly to the bottom?

**19.44** (Box 19F) (a) What enzyme does penicillin inhibit? (b) What is the function of this enzyme?

**19.45** (Box 19F) How does clavulinic acid accomplish the killing of penicillin-resistant bacteria?

## Additional Problems

**\*19.46** The activity of pepsin was measured at various pH values. The temperature and the concentrations of pepsin and substrate were held constant.

The following activities were obtained:

| pH | Activity |
|----|----------|
| 1 | 0.5 |
| 1.5 | 2.6 |
| 2 | 4.8 |
| 3 | 2.0 |
| 4 | 0.4 |
| 5 | 0.0 |

(a) Plot the pH dependence of pepsin activity. (b) What is the optimum pH? (c) Predict the activity of pepsin in the blood at pH 7.4.

*19.47 What would be the appropriate name for an enzyme that catalyzes each of the following reactions:

(a)  $CH_3CH_2OH \longrightarrow CH_3\overset{\displaystyle O}{\underset{\displaystyle \|}{C}}-H + H_2$

(b)  $CH_3\overset{\displaystyle O}{\underset{\displaystyle \|}{C}}-O-CH_2CH_3 + H_2O \longrightarrow$

$CH_3\overset{\displaystyle O}{\underset{\displaystyle \|}{C}}-OH + CH_3CH_2OH$

*19.48 What is the common characteristic of the amino acids of which the carboxyl groups of the peptide linkages can be hydrolyzed by trypsin?

19.49 How would you classify an enzyme that catalyzes the following reaction?

$CH_3-\overset{\displaystyle O}{\underset{\displaystyle \|}{C}}-COOH \longrightarrow CH_3-\overset{\displaystyle}{\underset{\displaystyle OH}{\overset{\displaystyle |}{CH}}}-COOH$

19.50 The enzyme chymotrypsin catalyzes the following type of reaction:

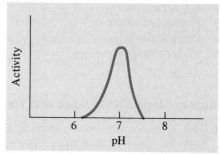

On the basis of the classification given in Section 19.2, which group of enzymes does chymotrypsin belong to?

*19.51 Food can be preserved by inactivation of enzymes that would cause spoilage, for example, by refrigeration. Give an example of food preservation in which the enzymes are inactivated by (a) heat and (b) lowering the pH.

19.52 Most enzymes operate best at one particular pH. Enzyme activity usually decreases at both higher and lower pH values. Give a possible explanation for this.

*19.53 Nerve gases operate by forming covalent bonds at the active site of cholinesterase. Is this an example of competitive inhibition? Can the nerve gas molecules be removed by simply adding more substrate (acetylcholine) to the enzyme?

19.54 Why is enzyme activity during myocardial infarction measured in the serum of patients rather than in the urine?

19.55 In which kind of inhibition is there an enzyme-substrate-inhibitor complex?

19.56 An enzyme has the following pH dependence:

Where do you think this enzyme works best?

*19.57 In an allosteric enzyme-regulator system, cAMP is a positive modulator. Predict what would happen to the enzyme-regulatory complex if a second enzyme were added that destroys cAMP.

# CHAPTER 20

*Hans Krebs.*

# Bioenergetics. How the Body Converts Food to Energy

## 20.1 INTRODUCTION

Living cells are in a dynamic state, which means that compounds are constantly being synthesized and then broken down into smaller fragments. Thousands of different reactions are taking place at the same time.

The same compounds may be synthesized in one part of a cell and broken down in a different part of the cell.

**The sum total of all the chemical reactions involved in maintaining the dynamic state of the cell** is called **metabolism.**

In general, we can divide metabolic reactions into two broad groups: (1) those in which molecules are broken down to provide the energy needed by the cell and (2) those that synthesize the compounds needed by the cell—both simple and complex.

591

> **The process of breaking down molecules to supply energy is catabolism.**
> **The process of building up molecules (synthesis) is anabolism.**

In spite of the large number of chemical reactions, there are only a few that dominate cell metabolism. In this chapter and the next, we focus our attention on the catabolic pathways that yield energy. **A biochemical pathway is a series of consecutive biochemical reactions.** We will see the actual reactions by means of which the chemical energy stored in our food is converted to the energy we use every minute of our lives—to think, to breathe, and to use our muscles to walk, write, eat, and everything else. In Chapter 22 we will look at some synthetic (anabolic) pathways.

The food we eat consists of many types of compounds, largely the ones we discussed: carbohydrates, lipids, and proteins. All of them can serve as fuel, and we derive our energy from them. To convert those compounds to energy, the body uses a different pathway for each type of compound. *However, all these diverse pathways converge to one* **common catabolic pathway,** which is illustrated in Figure 20.1. The diverse pathways are shown as different food streams. The small $C_2$ and $C_4$ molecules produced from the original large molecules in food drop into an imaginary collecting funnel that represents the common catabolic pathway. At the end of the funnel appears the energy carrier molecule adenosine triphosphate (ATP).

**The whole purpose of catabolic pathways is to convert the chemical energy in foods to molecules of ATP.** In this chapter we deal with the common catabolic pathway only. In Chapter 21 we will discuss the ways in which the different types of food (carbohydrates, lipids, and proteins) feed molecules into the common catabolic pathway.

## 20.2  MITOCHONDRIA

The enzymes that catalyze the common catabolic pathway in higher organisms are all located in special organelles within the cell. These organelles, called **mitochondria,** are made of two membranes (Fig. 20.2). Because the enzymes are located inside the mitochondria, the starting material of the reactions in the common pathway must get through the two membranes to enter the mitochondria, and products must leave the same way.

The inner membrane of a mitochondrion is quite resistant to the penetration of any ions and of most uncharged molecules. However, ions and molecules can still get through the membrane—they are transported across it by the numerous protein molecules embedded in it (Fig. 17.2, p. 526). The outer membrane, on the other hand, is quite permeable to small molecules and ions and does not need many different kinds of transporting membrane proteins.

The inner membrane is highly corrugated and folded, and the folds are called **cristae.** One can compare the organization of a mitochondrion to that of the galley of an ancient ship. The mitochondrion as a whole is the ship. The cristae are the benches to which the enzymes of the oxidative phosphorylation

*The common catabolic pathway consists of two sequences:*

1. the citric acid cycle (Sec. 20.4)
2. oxidative phosphorylation (Secs. 20.5, 20.6)

An **organelle** is a specialized structure within the cell.

Mitochondria is the plural form, and mitochondrion is the singular.

**Figure 20.1**

In this schematic diagram of the common catabolic pathway, an imaginary funnel represents what happens in the cell. The diverse catabolic pathways drop their products into the funnel of the common catabolic pathway, mostly in the form of $C_2$ fragments. The spinning wheel of the citric acid cycle breaks these molecules down further. The carbon atoms are released in the form of $CO_2$, and the hydrogen atoms and electrons are picked up by special compounds such as $NAD^+$ and FAD. Then the reduced NADH and $FADH_2$ cascade down into the stem of the funnel, where the electrons are transported inside the walls of the stem and the $H^+$ ions (represented by dots) are expelled to the outside. In their drive to get back, the $H^+$ ions form the energy carrier ATP. Once back inside, they combine with the oxygen that picked up the electrons and produce water.

**Figure 20.2**
A schematic drawing of a mitochondrion, cut to reveal the internal organization.

cycle are chained like ancient slaves, who provide the driving power. The space between the inner and outer membranes is like the space within the double hull of a ship.

The enzymes of the citric acid cycle are located in the matrix, which is the inner nonmembranous portion of a mitochondrion (Fig. 20.2). We shall soon see in detail how the specific sequence of the enzymes causes the chain of events in the common catabolic pathway. Beyond that, we must also discuss the ways in which nutrients and reaction products move into and out of the mitochondria.

## 20.3   THE PRINCIPAL COMPOUNDS OF THE COMMON CATABOLIC PATHWAY

There are two parts to the common catabolic pathway. The first is the **citric acid cycle** (also called the **Krebs cycle**), and the second is the **oxidative phosphorylation** pathway, also called the **electron transport chain** or the **respiratory chain.**

In order to understand what is actually happening in these reactions, we must first introduce the principal compounds participating in the common catabolic pathway. The most important of these are three rather complex compounds: **adenosine monophosphate (AMP), adenosine diphosphate (ADP),** and **adenosine triphosphate (ATP)** (Figs. 20.3 and 20.4). All three of

**Figure 20.3**
Adenosine monophosphate (AMP).

**Figure 20.4**
Hydrolysis of ATP produces ADP plus inorganic phosphate.

these molecules contain a heterocyclic amine portion called **adenine** and a sugar (furanose) portion called **ribose** (Sec. 16.8). Taken together, this portion of the molecules is called **adenosine.**

AMP, ADP, and ATP all contain adenosine connected to phosphate groups. The only difference between AMP, ADP, and ATP is the number of phosphate groups. As you can see from Figure 20.4, each phosphate is attached to the next by an anhydride linkage (Sec. 14.11). ATP contains three phosphates and two anhydride linkages. In all three molecules, the first phosphate is attached to the ribose by an inorganic ester linkage (Sec. 14.11).

A phosphoric anhydride linkage, P—O—P, contains more chemical energy than a phosphate-ester linkage, C—O—P. This means that when ATP and ADP are hydrolyzed to yield inorganic phosphate (Fig. 20.4), they release more energy per phosphate group than does AMP. Conversely, when inorganic phosphate bonds to AMP or ADP, greater amounts of energy are added to the chemical bond than when it bonds to adenosine. ADP and ATP contain *high-energy* phosphate bonds.

Of the three phosphate groups, the one closest to the adenosine is the least energetic; the other two possess a good deal more energy. Consequently, ATP releases the most energy and AMP the least when each gives up one phosphate group. This makes ATP a very useful compound for energy storage and release. The energy gained in the oxidation of food is stored in the form of ATP. However, this is only a short-term storage. ATP molecules in the cells normally do not last longer than about a minute. They are hydrolyzed to ADP and inorganic phosphate to yield energy for other processes, such as muscle contraction and nerve signal conduction. This means that ATP is constantly being formed and decomposed. Its turnover rate is very high. It has been estimated that during strenuous exercise the human body manufactures and degrades as much as 1 kg (more than 2 lb) of ATP every 2 minutes.

In summary, when the body takes in food, some of it goes to produce energy and some is used for building molecules and for other purposes. All of the energy that is extracted from the food is converted to ATP. This is the form in which the body stores its energy. In order to release this energy, the body hydrolyzes the ATP to ADP (sometimes to AMP). Exactly how these things happen is what we will be discussing in the rest of this chapter and in the next.

Adenine is a nucleotide base that frequently appears in biochemical compounds, especially nucleic acids.

The $PO_4^{3-}$ ion is generally called **inorganic phosphate.**

When one phosphate group is hydrolyzed from each, the following energy yields are obtained: ATP = 7.3 kcal/mol; ADP = 7.3 kcal/mol; AMP = 3.4 kcal/mol.

The body is able to extract only 40 to 60 percent of the total caloric content of food.

**Figure 20.5**
The structures of NAD$^+$ and FAD.

NAD$^+$
*N*icotinamide *a*denine *d*inucleotide

FAD
*F*lavin *a*denine *d*inucleotide

The + in NAD$^+$ refers to the positive charge on the nitrogen.

Nicotinamide and riboflavin are both members of the vitamin B group.

Two other actors in this drama are the coenzymes (Sec. 19.3) NAD$^+$ and FAD (Fig. 20.5), both of which contain an ADP core. In NAD$^+$ the operative part of the coenzyme is the nicotinamide part. In FAD the operative part is the flavin. In both molecules, the ADP is the *handle* by which the apoenzyme holds onto the coenzyme, and it is the other end of the molecule that carries out the actual chemical reaction. For example, when NAD$^+$ is reduced, it is the nicotinamide part of the molecule that gets reduced:

**B o x   20A**

**Nicotine and Nicotinic Acid**

In Section 15.9 we met the alkaloid nicotine, which is present in tobacco. Chemists in the nineteenth century found that, on strong oxidation, nicotine is converted to a carboxylic acid, which they called *nicotinic acid* because it was derived from nicotine.

Nicotine     strong oxidation →     Nicotinic acid     Nicotinamide

It was later discovered that nicotinic acid and its amide derivative, *nicotinamide,* are vitamins, members of the vitamin B family, essential to nutrition.

We have seen that nicotinic acid is part of the $NAD^+$ molecule. The body makes $NAD^+$ by using nicotinic acid or nicotinamide. As B vitamins, these compounds are present in vitamin pills and are also used as additives in such products as breakfast cereals. Because the marketers of these products felt that the public would be reluctant to buy anything containing "nicotinic acid" or "nicotinamide" (because of possible confusion with nicotine, which most people know to be harmful), they received permission from the U.S. Food and Drug Administration to call them "niacin" and "niacinamide," and it is these names that appear on the labels. Of course, chemists, biologists, and health workers know that nicotine and nicotinic acid are totally different compounds, with totally different biological effects.

$$+ H^+ + 2e^- \rightleftharpoons$$

$NAD^+$      NADH

The "R" stands for the rest of the $NAD^+$ molecule; "R'" stands for the rest of the FAD molecule.

The reduced form of $NAD^+$ is called NADH. The same reduction happens on two nitrogens of the flavin portion of FAD:

$$+ 2H^+ + 2e^- \rightleftharpoons$$

FAD      $FADH_2$

The reduced form of FAD is called $FADH_2$.

**Figure 20.6**
The structure of coenzyme A.

We view **NAD⁺** and **FAD coenzymes** as the **hydrogen ion** and **electron transporting molecules.**

The final principal compound in the common catabolic pathway is **coenzyme A** (CoA, Fig. 20.6), which is the **acetyl carrying group.** Coenzyme A also contains ADP, but here the next structural unit is pantothenic acid, another B vitamin. Just as ATP can be looked upon as an ADP molecule to which a —PO₃ is attached by a high-energy bond, so can **acetyl coenzyme A** be considered a CoA molecule linked to an acetyl group by another high-energy bond. The active part of coenzyme A is the mercaptoethylamine. The acetyl group of acetyl coenzyme A is attached to the mercapto (SH) group:

$$CoA-S-\underset{\underset{\displaystyle O}{\|}}{C}-CH_3$$

Acetyl coenzyme A

## 20.4   THE CITRIC ACID CYCLE

The process of catabolism begins when carbohydrates and lipids have been broken down into two-carbon pieces. The two-carbon fragments are the acetyl (CH₃CO—) portions of acetyl coenzyme A. The acetyl is now fragmented further in the **citric acid cycle** (named after the main component of the cycle), which is also called the **Krebs cycle** and, sometimes, the **tricarboxylic acid cycle.**

The details of the citric acid cycle are given in Figure 20.7. A good way to gain an insight is to use Figure 20.7 in connection with the simplified schematic diagram shown in Figure 20.8, which shows only the carbon balance.

We will now follow the two carbons of the acetate (C₂) through each step in the citric acid cycle. The circled numbers correspond to those in Figure 20.7.

Hans Krebs (1900–1981), Nobel laureate in 1953, established the relationships among the different components of the cycle.

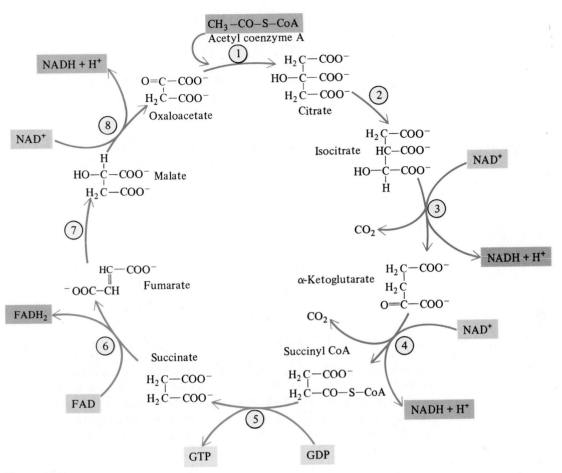

**Figure 20.7**
The citric acid (Krebs) cycle. The numbered steps are explained in detail in the text. The curved arrows are a shorthand way of showing the reactants and products. For example, in step ③ the NAD⁺ reacts with isocitrate to produce α-ketoglutarate, $CO_2$, NADH, and H⁺. The last two then leave the site of the reaction.

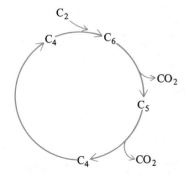

**Figure 20.8**
A simplified view of the citric acid cycle, showing only the carbon balance.

In step ⑧ we will see where the oxaloacetate comes from.

① Acetyl coenzyme A enters the cycle by combining with a $C_4$ compound called oxaloacetate:

$$
\underset{\text{Oxaloacetate}}{\begin{array}{c} COO^- \\ | \\ O{=}C \\ | \\ CH_2 \\ | \\ COO^- \end{array}}
\quad + \quad
\underset{\text{Acetyl CoA}}{\begin{array}{c} O \\ \parallel \\ CH_3{-}C{-}S{-}CoA \end{array}}
\quad \xrightarrow{\text{citrate synthetase}}
$$

$$
\underset{\text{Citryl CoA}}{\begin{array}{c} O \\ \parallel \\ C{-}S{-}CoA \\ | \\ CH_2 \\ | \\ HO{-}C{-}COO^- \\ | \\ CH_2 \\ | \\ COO^- \end{array}}
\quad \xrightarrow{H_2O} \quad
\underset{\text{Citrate}}{\begin{array}{c} COO^- \\ | \\ CH_2 \\ | \\ HO{-}C{-}COO^- \\ | \\ CH_2 \\ | \\ COO^- \end{array}}
\quad + \quad HS{-}CoA
$$

The first thing that happens is the addition of the $CH_3$ group of the acetyl CoA to the C=O bond of the oxaloacetate, catalyzed by the enzyme citrate synthetase. This is followed by hydrolysis to produce the $C_6$ compound citrate ion and CoA. Therefore, step ① is a building up rather than a degradation.

② The citrate ion is dehydrated to *cis*-aconitate:

$$
\underset{\text{Citrate}}{\begin{array}{c} COO^- \\ | \\ CH_2 \\ | \\ HO{-}C{-}COO^- \\ | \\ CH_2 \\ | \\ COO^- \end{array}}
\quad \xrightarrow{\text{aconitase}} \quad
\underset{\textit{cis}\text{-Aconitate}}{\begin{array}{c} COO^- \\ | \\ CH_2 \\ | \\ C{-}COO^- \\ \parallel \\ C{-}H \\ | \\ COO^- \end{array}}
\quad + \quad H_2O
$$

after which the *cis*-aconitate is hydrated again, but this time to isocitrate instead of citrate:

$$
\underset{\textit{cis}\text{-Aconitate}}{\begin{array}{c} COO^- \\ | \\ CH_2 \\ | \\ C{-}COO^- \\ \parallel \\ C{-}H \\ | \\ COO^- \end{array}}
\quad \xrightarrow[\text{aconitase}]{H_2O} \quad
\underset{\text{Isocitrate}}{\begin{array}{c} COO^- \\ | \\ CH_2 \\ | \\ CH{-}COO^- \\ | \\ HO{-}CH \\ | \\ COO^- \end{array}}
$$

In citric acid the alcohol is a tertiary alcohol. We learned in Section 12.3 that tertiary alcohols cannot be oxidized. The alcohol in the isocitrate is a secondary alcohol, which upon oxidation yields a ketone.

③ The isocitrate is now oxidized and decarboxylated at the same time:

$$
\begin{array}{c}
COO^- \\
| \\
CH_2 \\
| \\
CH-COO^- \\
| \\
HO-CH \\
| \\
COO^-
\end{array}
\; + \; \boxed{NAD^+}
\xrightarrow[\text{dehydrogenase}]{\text{isocitrate}}
\begin{array}{c}
COO^- \\
| \\
CH_2 \\
| \\
CH_2 \\
| \\
C=O \\
| \\
COO^-
\end{array}
\; + \; \boxed{NADH} \; + \; CO_2
$$

Isocitrate                    $\alpha$-Ketoglutarate

The oxidizing agent $NAD^+$ has removed two hydrogens. One of them has been used to reduce $NAD^+$ to NADH. The other has replaced the $COO^-$ that goes into making $CO_2$. Note that the $CO_2$ given off comes from the original oxaloacetate and not from the two carbons of the acetyl CoA. Both of these carbons are still present in the $\alpha$-ketoglutarate. Also note that we are now down to a $C_5$ compound, $\alpha$-ketoglutarate.

④ and ⑤ Next a complex enzyme system removes another $CO_2$, once again from the original oxaloacetate portion rather than from the acetyl CoA portion:

The $CO_2$ molecules given off in steps ③ and ④ are the ones we exhale.

$$
\begin{array}{c}
COO^- \\
| \\
CH_2 \\
| \\
CH_2 \\
| \\
C=O \\
| \\
COO^-
\end{array}
\; + \; \boxed{NAD^+} \; + \; \boxed{GDP} \; + \; P_i \; + \; H_2O
\xrightarrow[\text{system}]{\substack{\text{complex} \\ \text{enzyme}}}
\begin{array}{c}
COO^- \\
| \\
CH_2 \\
| \\
CH_2 \\
| \\
COO^-
\end{array}
\; + \; CO_2 \; + \; \boxed{NADH + H^+} \; + \; \boxed{GTP}
$$

$\alpha$-Ketoglutarate                              Succinate

The "$P_i$" is inorganic phosphate.

We are now down to a $C_4$ compound, succinate. This oxidative decarboxylation is more complex than the first. It occurs in many steps and requires a number of cofactors. For our purpose it is sufficient to know that, during this second oxidative decarboxylation, a high-energy compound called **guanosine triphosphate** (GTP) is also formed.

GTP is similar to ATP except that guanine replaces adenine. Otherwise, the linkages of the base to ribose and the phosphates are exactly the same as in ATP. The function of GTP is also similar to that of ATP, namely, to store energy in the form of high-energy phosphate bonds (chemical energy).

⑥ In this step, the succinate is oxidized by FAD, which removes two hydrogens to give fumarate (the double bond in this molecule is trans):

Guanine

$$
\begin{array}{c}
COO^- \\
| \\
CH_2 \\
| \\
CH_2 \\
| \\
COO^-
\end{array}
\; + \; \boxed{FAD}
\xrightarrow[\text{dehydrogenase}]{\text{succinate}}
\begin{array}{c}
\phantom{x}H \qquad COO^- \\
\diagdown \; / \\
C \\
\| \\
C \\
/ \; \diagdown \\
{}^-OOC \qquad H
\end{array}
\; + \; \boxed{FADH_2}
$$

Succinate                              Fumarate

This reaction cannot be carried out in the laboratory, but with the aid of an enzyme catalyst, the body does it easily.

⑦   The fumarate is now hydrated to give the malate ion:

Fumarate                                                    Malate

⑧   In the final step of the cycle, malate is oxidized by $NAD^+$ to give oxaloacetate (removal of $H_2$):

Malate                                        Oxaloacetate

Thus, the final product of the Krebs cycle is oxaloacetate, which is the compound that we started with in step ①.

What has happened in the entire process is that the original two acetate carbons of acetyl CoA were added to the $C_4$ oxaloacetate to produce a $C_6$ unit, which then lost two carbons in the form of $CO_2$, to produce, at the end of the process, the $C_4$ unit oxaloacetate. The *net* effect is the conversion of the two acetate carbons of acetyl CoA to two molecules of carbon dioxide.

How does this produce energy? We have already learned that one step in the process produces a high-energy molecule of GTP. But other steps contribute also. In several of the steps, the citric acid cycle converts $NAD^+$ to NADH and FAD to $FADH_2$. These reduced coenzymes carry the $H^+$ and electrons that eventually will provide the energy for the synthesis of ATP (discussed in detail in Secs. 20.5 and 20.6).

This stepwise degradation and oxidation of acetate in the citric acid cycle results in the most efficient extraction of energy. Rather than in one burst, the energy is released in small packets carried away step by step in the form of NADH and $FADH_2$.

But the cyclic nature of this acetate degradation has other advantages besides maximizing energy yield: (1) the citric acid cycle components also provide raw materials for amino acid synthesis as the need arises (Chap. 22), and (2) the many-component cycle provides an excellent method for regulating the speed of catabolic reactions. The regulation can occur at many different parts of the cycle, so that feedback information can be used at many points to speed up or slow down the process as necessary.

The following equation represents the overall reactions in the citric acid cycle.

For example, α-ketoglutaric acid is used to synthesize glutamic acid.

$$CH_3COOH + 2H_2O + \boxed{3NAD^+ + FAD} \longrightarrow$$

$$2CO_2 + \boxed{3NADH + FADH_2 + 3H^+} \quad (20.1)$$

The citric acid cycle is controlled by a feedback mechanism. When the essential products of the cycle, such as ATP and NADH + H$^+$, accumulate, they inhibit some of the enzymes in the cycle. Citrate synthase (step ①), isocitrate dehydrogenase (step ③), and $\alpha$-ketoglutarate dehydrogenase (step ④) are inhibited by ATP and/or by NADH + H$^+$. This slows down or shuts off the cycle. On the other hand, when the feed material, acetyl CoA, is in abundance, the cycle is speeded up. The enzyme isocitrate dehydrogenase (step ③) is stimulated by ADP and NAD$^+$.

## 20.5 ELECTRON AND H⁺ TRANSPORT (OXIDATIVE PHOSPHORYLATION)

The reduced coenzymes NADH and FADH$_2$ are end products of the citric acid cycle. They carry hydrogen ions and electrons and thus the potential to yield energy when these combine with oxygen to form water:

$$4H^+ + 4e^- + O_2 \rightleftharpoons 2H_2O$$

This simple exothermic reaction is carried out in many steps.

A number of enzymes are involved, all embedded in the inner membrane of the mitochondria. These enzymes are situated in a particular *sequence* in the membrane so that the product from one enzyme can be passed on to the next enzyme, in a kind of assembly line. The enzymes are arranged in order of increasing affinity for electrons, so electrons flow through the enzyme system. The sequence of electron-carrying enzymes in the mitochondrial membrane is

flavoprotein ⟶ FeS protein ⟶ Q (quinone) enzyme ⟶
    cytochrome b ⟶ cytochrome c$_1$ ⟶ cytochrome c ⟶
        cytochrome a ⟶ cytochrome a$_3$

The fact that these enzymes are so close together allows the electrons to pass from one to another.

In Section 20.4 we saw that NADH + H$^+$ is a product of the citric acid cycle. The **oxidative phosphorylation** process (Fig. 20.9) starts with the

The oxygen in this reaction is the oxygen we breathe.

The letters used to designate the cytochromes were given in order of their discovery.

**Figure 20.9**
Schematic diagram of oxidative phosphorylation, also called the electron transport chain.

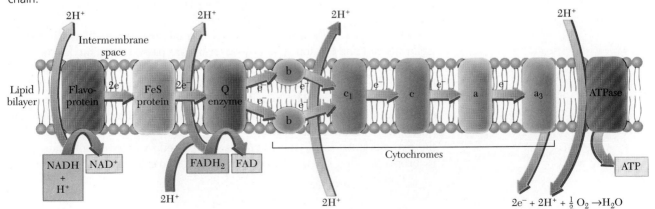

Mitochondrion

**Box 20B**

## 2,4-Dinitrophenol as an Uncoupling Agent

Nitrated aromatic compounds are highly explosive. Trinitrotoluene (TNT) is the best known. During World War I, many ammunition workers were exposed to 2,4-dinitrophenol (DNP), a compound used to prepare the explosive picric acid.

DNP        TNT        Picric acid

It was observed that these workers lost weight. As a consequence, DNP was used as a weight-reducing drug during the 1920s. Unfortunately, DNP eliminated not only the fat but sometimes also the patient, and its use as a diet pill was discontinued after 1929.

Today we know why DNP works as a weight-reducing drug. It is an effective *protonphore*, which is a compound that transports $H^+$ ions through a membrane passively, without the expenditure of energy. We have seen that $H^+$ ions accumulate in the intermembrane space of mitochondria and, under normal conditions, drive the synthesis of ATP while they are going back to the inside. This is Mitchell's chemiosmotic principle. When DNP is ingested, it transfers the $H^+$ back to the mitochondrion easily, and no ATP is manufactured. The energy of the electron separation is dissipated as heat and is not built in as chemical energy in ATP. The loss of this energy-storing compound makes the utilization of food much less efficient, resulting in weight loss.

Because of this property of DNP, it has been used in biochemical research on oxidative phosphorylation. It is an *uncoupling agent*—it uncouples the oxidative process from the phosphorylation and allows us to study the steps in the electron transport chain without the simultaneous phosphorylation. There exists in the medical literature a case history of a woman whose muscles contained mitochondria in which electron transport was not coupled to oxidative phosphorylation. This unfortunate woman could not utilize electron transport to generate ATP. As a consequence she was severely incapacitated and bedridden. When the uncoupling worsened, the heat generated by the uncontrolled oxidation (no ATP production) was so severe that the patient required continuous cooling.

NADH that arrives from the interior of the mitochondrion. This carries two electrons and one $H^+$ ion. The two electrons are passed on to the flavoprotein, which has a riboflavin coenzyme. The $H^+$ from the NADH and also another $H^+$ from the interior of the mitochondrion are expelled by the flavoprotein into the intermembrane space. The two electrons pass through the FeS protein and Q enzyme. At this point, two more $H^+$ ions are pumped out to the intermembrane space. The cytochromes in turn pick up only one electron at a time. Therefore, the two electrons from the Q enzyme are passed on to two molecules of cytochrome b. When the electrons are transferred to cytochrome $c_1$, another two

$H^+$ ions are pumped out from the matrix of the mitochondrion. The electrons finally end up, via cytochrome c $\rightarrow$ cytochrome a $\rightarrow$ cytochrome $a_3$, on the inside of the membrane, where they combine with oxygen, $O_2$, and $H^+$ ions to form water. During this process of transporting the two electrons from one molecule of NADH, a total of six $H^+$ ions are pumped out of the mitochondrion.

When the $FADH_2$ carries the electrons, it transfers them directly to the Q enzyme rather than to the flavoprotein. Otherwise, the electron flow follows the same route as with NADH. However, since the flavoprotein stage is skipped during the electron transport, only four $H^+$ ions are expelled into the intermembrane space for each $FADH_2$ molecule.

## 20.6  PHOSPHORYLATION AND THE CHEMIOSMOTIC PUMP

In 1961 Peter Mitchell (1920–1992), an English chemist, proposed the **chemiosmotic hypothesis:** The energy in the electron transfer chain creates a proton gradient. A **proton gradient** is a continuous variation in the $H^+$ concentration along a given region. In this case there is a higher concentration of $H^+$ in the intermembrane space than inside the mitochondrion. The protons pumped out of the mitochondrion provide a driving force. This driving force, which is the result of the spontaneous flow of ions from a region of high concentration to a region of low concentration, propels the protons back to the mitochondrion through a complex that is given the name of **proton translocating ATPase.** This compound is located on the inner membrane of the mitochondrion and is the active enzyme that catalyzes the conversion of ADP and inorganic phosphate to ATP (the reverse of the reaction shown in Fig. 20.4):

$$ADP + P_i \underset{\longleftarrow}{\overset{ATPase}{\longrightarrow}} ATP + H_2O$$

The proton translocating ATPase can catalyze the reaction in both directions. When protons that have accumulated on the outer surface of the mitochondrion stream inward, the enzyme manufactures ATP and stores the electrical energy in the form of chemical energy. On the other hand, the enzyme can also hydrolyze ATP and as a consequence pump out $H^+$ from the mitochondrion. Each pair of protons that is translocated gives rise to the formation of one ATP molecule.

The protons that enter a mitochondrion combine with the electrons transported through the **electron transport chain** and with oxygen to form water. The net result of the two processes (electron/$H^+$ transport and ATP formation) is that the oxygen we breathe in combines with four $H^+$ ions and four electrons to give two water molecules. The four $H^+$ ions and four electrons come from the NADH and $FADH_2$ molecules produced in the citric acid cycle. The functions of the oxygen, therefore, are (1) to oxidize NADH to $NAD^+$ and $FADH_2$ to FAD so that all these molecules can go back and participate in the citric acid cycle, and (2) to provide energy for the conversion of ADP to ATP.

Mitchell received the Nobel Prize in Chemistry in 1978.

Barbiturates produce sedative and sleep-inducing effects (Box 15G). However, in an unrelated way they also act as inhibitors of the succinate dehydrogenase system in the citric acid cycle. In high concentration (0.0007 $M$) they may inhibit the oxidation (dehydrogenation) of the succinate itself by as much as 50 percent. However, their main action consists of preventing the NADH liberated in this dehydrogenation from entering the electron transport chain. In small doses, barbiturates do not create a problem of lack of energy because $FADH_2$ and the other NADH molecules enter into the chain. In large doses, however, they can inhibit respiration enough to result in death.

Chlorpromazine, a common tranquilizer, also marketed as Thorazine, has a similar inhibiting effect.

The latter function is accomplished indirectly. The entrance of the $H^+$ ions into the mitochondrion drives the ATP formation, but the $H^+$ ions enter the mitochondrion because the $O_2$ depleted the $H^+$ ion concentration when water was formed. It is a rather complex process involving the transport of electrons along a whole series of enzyme molecules (which catalyze all these reactions), but without it the cell cannot utilize the $O_2$ molecules and eventually dies. The following equations represent the overall reactions in oxidative phosphorylation:

$$NADH + 3ADP + \tfrac{1}{2}O_2 + 3P_i + H^+ \longrightarrow NAD^+ + 3ATP + H_2O \quad (20.2)$$

$$FADH_2 + 2ADP + \tfrac{1}{2}O_2 + 2P_i \longrightarrow FAD + 2ATP + H_2O \quad (20.3)$$

## 20.7  THE ENERGY YIELD

The energy released during electron transport is now finally built into the ATP molecule. Therefore, it is instructive to look at the energy yield in the universal biochemical currency: the number of ATP molecules.

Each pair of protons entering a mitochondrion results in the production of one ATP molecule. For each NADH molecule, three pairs of protons are pumped into the intermembrane space in the electron transport process. Therefore, for each NADH molecule, we get three ATP molecules, as can be seen in Equation 20.2. For each $FADH_2$ molecule, we have seen that only four protons are pumped out of the mitochondrion. Therefore, only two ATP molecules are produced for each $FADH_2$, as seen in Equation 20.3.

Now we can produce the energy balance for the whole common catabolic pathway (citric acid cycle and oxidative phosphorylation combined). For each $C_2$ fragment entering the citric acid cycle, we obtain three NADH and one $FADH_2$ (Eq. 20.1) plus one GTP, which is equivalent in energy to one ATP. Thus, the total number of ATP molecules produced per $C_2$ fragment is

$$3NADH \times 3ATP/NADH = 9ATP$$

$$1FADH_2 \times 2ATP/FADH_2 = 2ATP$$

$$1GTP = \underline{1ATP}$$

$$12ATP$$

Each $C_2$ fragment that enters the cycle produces 12 ATP molecules and uses up two $O_2$ molecules. The total effect of the energy-production chain of reactions we have discussed in this chapter (the common catabolic pathway) is to oxidize one $C_2$ fragment with two molecules of $O_2$ to produce two molecules of $CO_2$ and 12 molecules of ATP:

$$C_2 + 2O_2 + 12ADP + 12P_i \longrightarrow 12ATP + 2CO_2$$

The important thing is not the waste product, $CO_2$, but the 12 ATP molecules, since these will now release their energy when they are converted to ADP.

## 20.8 CONVERSION OF CHEMICAL ENERGY TO OTHER FORMS OF ENERGY

As mentioned in Section 20.3, the storage of chemical energy in the form of ATP lasts only a short time. Usually within a minute the ATP is hydrolyzed and thus releases its chemical energy. How does the body utilize this chemical energy? To answer this question, let us look at the different forms in which energy is needed in the body.

### Conversion to Other Forms of Chemical Energy

The activity of many enzymes is controlled and regulated by phosphorylation. For example, the enzyme phosphorylase, which catalyzes the breakdown of glycogen (Box 21C), occurs in an inactive form, phosphorylase b. The enzyme has a single serine residue at its active site. When ATP transfers a phosphate group to this serine, the enzyme becomes active. Thus the chemical energy of ATP is used in the form of chemical energy to activate phosphorylase b so that glycogen can be utilized. We shall see several other examples of this in Chapters 21 and 22.

### Electrical Energy

The body maintains a high concentration of $K^+$ ions inside the cells despite the fact that outside the cells the $K^+$ concentration is low. The reverse is true for $Na^+$. In order that $K^+$ should not diffuse out of the cells and $Na^+$ should not penetrate in, there are special transport proteins in the cell membranes that constantly pump $K^+$ into the cells and $Na^+$ out. This pumping requires energy, which is supplied by the hydrolysis of ATP to ADP. With this pumping, the charges inside and outside the cell are unequal, and this generates an electric potential. Thus the chemical energy of ATP is transformed into electrical energy.

Chemical energy is converted to electrical energy in nerve transmission.

Thick filament (myosin)    Thin filament (actin)

(a)  Resting muscle

(b)  Contracted muscle

**Figure 20.10**
Schematic diagram of muscle contraction.

### Mechanical Energy

ATP is the immediate source of energy in muscle contraction. In essence, muscle contraction takes place when thick and thin filaments slide past each other (Fig. 20.10). The thick filament is *myosin,* an ATPase enzyme—that is, one that hydrolyzes ATP. The thin filament, *actin,* binds strongly to myosin in the contracted state. However, when ATP binds to myosin, the actin-myosin dissociates and the muscle relaxes. When myosin hydrolyzes ATP, it interacts with actin once more, and a new contraction occurs. Therefore, the hydrolysis of ATP drives the alternating association and dissociation of actin and myosin and consequently the contraction and relaxation of the muscle.

### Heat Energy

One molecule of ATP upon hydrolysis to ADP yields 7.3 kcal/mole. Some of this energy is released as heat and is used by the body to maintain body temperature. If we estimate that the specific heat of the body is about the same as that of water, a person weighing 60 kg would need to hydrolyze approximately 99 moles of ATP to raise the temperature of the body from room temperature, 25°C, to 37°C. Not all body heat is derived from ATP hydrolysis; some exothermic reactions in the body also contribute.

## SUMMARY

The sum total of all the chemical reactions involved in maintaining the dynamic state of cells is called **metabolism.** The breaking down of molecules is **catabolism;** the building up of molecules is **anabolism.** The **common metabolic pathway** uses a two-carbon $C_2$ fragment (acetyl) from different foods. Through the **citric acid cycle** and **oxidative phosphorylation (electron transport chain),** the $C_2$ fragment is oxidized. The products formed are water and carbon dioxide. The energy from oxidation is built into the high-chemical-energy-storing molecule ATP.

Both the citric acid cycle and oxidative phosphorylation take place in the **mitochondria.** The enzymes of the citric acid cycle are located in the matrix, while the enzymes of the oxidative phosphorylation chain are on the inner mitochondrial membrane. Some of them project into the intermembrane space.

The principal carriers in the common pathway are as follows: **ATP** is the phosphate carrier, **CoA** is the $C_2$ fragment carrier, and **NAD$^+$** and **FAD** carry the hydrogen ions and electrons. The unit common to all these carriers is **ADP.** This is the nonactive end of the carriers, which acts as a handle that fits into the active sites of the enzymes.

In the citric acid cycle the $C_2$ fragment first combines with a $C_4$ fragment (oxaloacetate) to yield a $C_6$ fragment (citrate). An oxidative decarboxylation yields a $C_5$ fragment. One $CO_2$ is released, and one $NADH + H^+$ is passed on to the electron transport chain. Another oxidative decarboxylation provides a $C_4$ fragment. Once again a $CO_2$ is released and another $NADH + H^+$ is passed on to the electron transport chain. Subsequently, two dehydrogenation (oxida-tion) steps yield one $FADH_2$ and one additional $NADH + H^+$, along with an analog of ATP called GTP. The cycle is controlled by a feedback mechanism.

The NADH enters the electron transport chain at the flavoprotein dehydrogenase stage. Two $H^+$ ions are expelled into the intermembrane space of the mitochondrion, and the electrons are passed along the enzymes of the electron transport chain. Each enzyme transfers electrons to the next in the series through redox reactions. As the electrons are transported, $H^+$ ions are expelled from the interior of the mitochondrion to the intermembrane space. For each NADH, six $H^+$ ions are expelled. Finally, the electrons inside the mitochondrion combine with oxygen and $H^+$ to form water. When the expelled $H^+$ ions stream back into the mitochondrion, they drive a complex enzyme called **proton translocating ATPase,** which makes one ATP molecule for each two $H^+$ ions that enter the mitochondrion. Therefore, for each $NADH + H^+$ coming from the citric acid cycle, three ATP molecules are formed. The $FADH_2$ enters the respiratory chain later. While the electrons of the $FADH_2$ are transported, only four $H^+$ ions are expelled into the intermembrane space. As a consequence, only two ATP molecules are formed for each $FADH_2$. The overall result is that for each $C_2$ fragment entering the citric acid cycle, 12 ATP molecules are produced.

The chemical energy is stored in ATP only for a short time. ATP is hydrolyzed, usually within a minute. This chemical energy is used to do chemical, mechanical, and electrical work in the body and to maintain body temperature.

## KEY TERMS

Acetyl coenzyme A (Sec. 20.3)
Anabolism (Sec. 20.1)
ATP (Sec. 20.3)
Biochemical pathway (Sec. 20.1)
Catabolism (Sec. 20.1)
Chemiosmotic hypothesis (Sec. 20.6)
Citric acid cycle (Sec. 20.4)
Coenzyme A (Sec. 20.3)
Common catabolic pathway (Sec. 20.1)
Cristae (Sec. 20.2)
Electron transport chain (Sec. 20.6)
Energy yield (Sec. 20.7)

FAD (Sec. 20.3)
Inorganic phosphate (Sec. 20.3)
Krebs cycle (Sec. 20.4)
Metabolism (Sec. 20.1)
Mitochondria (Sec. 20.2)
NAD$^+$ (Sec. 20.3)
Oxidative phosphorylation (Sec. 20.5)
Pathway (Sec. 20.1)
Proton gradient (Sec. 20.6)
Respiratory chain (Sec. 20.3)
Tricarboxylic acid cycle (Sec. 20.4)

## PROBLEMS

### Metabolism

**20.1**   Distinguish between catabolism and anabolism.

**20.2**   (a) How many sequences are there in the common catabolic pathway?   (b) Name them.

### Mitochondria

**20.3**   How do ions and molecules enter the mitochondria?

**20.4**   What is a crista?

**20.5**   (a) Where are the enzymes of the citric acid cycle located?   (b) Where are the enzymes of oxidative phosphorylation located?

### Principal Compounds of the Common Catabolic Pathway

**20.6**   Which is more energetic, the phosphate-ester linkage or the phosphoric anhydride linkage?

**20.7**   Draw the structure of ATP and label each part of the molecule.

**20.8**   What are the products of the reaction

$$AMP + H_2O \xrightarrow{H^+} ?$$

**20.9**   Which yields more energy, the hydrolysis (a) of ATP to ADP or   (b) of ADP to AMP?

**20.10**   ATP is often called the molecule for storing chemical energy. Does "storage" mean a long-term preservation of energy? Explain.

**\*20.11**   Does the amount of ATP produced by the metabolism of a given amount of food correspond to all the caloric energy that is in the food?

**\*20.12**   (a) What does the D in NAD and FAD stand for?   (b) What feature in the structure prompted this name?

**\*20.13**   When $NAD^+$ is reduced, two electrons enter the molecule, together with one $H^+$. Exactly where in the product will the two electrons be located?

**20.14**   Which atoms in the flavin portion of FAD are reduced to yield $FADH_2$?

**20.15**   In the common catabolic pathway, there are a number of important molecules that act as carriers (transfer agents). (a) What is the carrier of phosphate groups?   (b) Which are the coenzymes transferring hydrogen ions and electrons?   (c) What kind of groups does coenzyme A carry?

**20.16**   What is the common handle in all these carriers?

**20.17**   Name all the functional groups in the structure of CoA.

**20.18**   Name the vitamin B molecules that are a part of the structure of (a) $NAD^+$   (b) FAD   (c) coenzyme A

**\*20.19**   In both $NAD^+$ and FAD the vitamin B portion of the molecule is the active part. Is this also true for CoA?

**20.20**   What type of compound is formed when coenzyme A reacts with acetate?

**20.21**   The fats and carbohydrates metabolized by our bodies are eventually converted to a single compound. What is it?

### The Citric Acid Cycle

**20.22**   The first step in the citric acid cycle is abbreviated as

$$C_2 + C_4 = C_6$$

(a) What do these symbols stand for?   (b) What are the common names of the three compounds involved in this reaction?

**20.23**   What are the structural similarities and differences among succinate, fumarate, and malate?

**20.24**   Identify by number those steps of the citric acid cycle that are not redox reactions.

**20.25**   What kind of reaction occurs in the citric acid cycle when a $C_6$ compound is converted to a $C_5$ compound?

**20.26**   How many $CO_2$ molecules are produced in one citric acid cycle?

**20.27**   What is the role of succinate dehydrogenase in the citric acid cycle?

**20.28**   In the citric acid cycle, (a) how many times is a secondary alcohol oxidized?   (b) which molecules are decarboxylated?

**20.29**   Is ATP directly produced during any step of the citric acid cycle? Explain.

**20.30**   There are four dicarboxylic acid compounds, each containing four carbons, in the citric acid cycle. Which is (a) the least and   (b) the most oxidized?

**20.31**   Why is a many-step cyclic process more efficient in utilizing energy from food than a single-step combustion?

**20.32** Did the two $CO_2$ molecules given off in one citric acid cycle originate from the entering acetyl group?

**20.33** Name the enzyme involved in the oxidative decarboxylation of isocitrate. Does this name correspond to the nature of the reaction it catalyzes?

**20.34** Most of the reactions of the citric acid cycle can also be carried out in the laboratory, without the appropriate enzymes, albeit at a much slower rate. Which reaction in the citric acid cycle cannot be carried out without enzyme catalysis?

**\*20.35** Oxidation is defined as loss of electrons. When oxidative decarboxylation occurs, as in step ④, where do the electrons of the $\alpha$-ketoglutarate go?

## Oxidative Phosphorylation

**20.36** What is the main function of oxidative phosphorylation (the electron transport chain)?

**20.37** How many *pairs* of $H^+$ ions are expelled for each (a) NADH molecule (b) $FADH_2$ molecule?

**20.38** We breathe in $O_2$ and breathe out $CO_2$. Explain how the body uses the $O_2$ to convert the foods we eat to $CO_2$, $H_2O$, and energy.

**20.39** Where are the enzymes of the electron transport system (oxidative phosphorylation) located?

**20.40** At what points in oxidative phosphorylation are the $H^+$ ions and the electrons separated from each other?

**20.41** How many ATP molecules are generated for (a) each NADH + $H^+$, (b) each $FADH_2$ molecule, (c) each $C_2$ fragment that goes through the complete common catabolic pathway?

**20.42** Since $H^+$ is pumped out into the intermembrane space, what is the pH there compared with that in the matrix?

## The Chemiosmotic Pump

**20.43** What is the channel through which $H^+$ ions reenter the matrix of mitochondria?

**\*20.44** The proton gradient accumulated in the intermembrane area of a mitochondrion drives the ATP-manufacturing enzyme ATPase. Why do you think Mitchell called this the "chemiosmotic hypothesis"?

## The Energy Yield

**20.45** If each mole of ATP yields 7.3 kcal of energy upon hydrolysis, how many kcal of energy would you get from 1 g of $CH_3COO^-$ ($C_2$) entering the citric acid cycle?

**20.46** If a $C_{18}$ compound is metabolized and enters the citric acid cycle as nine $C_2$ fragments (acetyl groups), how many molecules of ATP are produced in the common metabolic pathway from the $C_{18}$ compound?

## Conversion of Chemical Energy to Other Forms

**20.47** (a) How do muscles contract? (b) Where does the energy in muscle contraction come from?

**20.48** Give an example of the conversion of the chemical energy of ATP to electrical energy.

**20.49** How is the enzyme phosphorylase activated?

## Boxes

**20.50** (Box 20A) (a) Why was nicotinic acid given its name? (b) Why is it often called niacin?

**\*20.51** (Box 20B) Oligomycin is an antibiotic that allows electron transport to continue but stops phosphorylation, in bacteria as well as in humans. Would you use this as an antibacterial drug for people? Explain.

**20.52** (Box 20B) 2,4-Dinitrophenol is an uncoupling agent. What does it uncouple, and how?

**20.53** (Box 20C) What enzyme system is inhibited by barbiturates?

## Additional Problems

**20.54** One of the end products of food metabolism is water. How many molecules of $H_2O$ are formed from the entry of each molecule of (a) NADH + $H^+$, (b) $FADH_2$? (Use Fig. 20.9.)

**\*20.55** In the structures of $NAD^+$ and FAD, pentose derivatives appear twice. What is the difference between the carbohydrate portions of these two coenzymes?

**\*20.56** Acetyl CoA is labeled with radioactive carbon as shown: $CH_3$*$CO$—$S$—$CoA$. This enters the citric acid cycle. If the cycle is allowed to progress only to the $\alpha$-ketoglutarate level, will the $CO_2$ expelled by the cell be radioactive?

*20.57 How many grams of $CH_3COOH$ molecules must be metabolized in the common metabolic pathway to yield 87.6 kcal of energy?

20.58 What happens to the end products of the common pathway?

20.59 In what part of the cell does the actual reduction of molecular oxygen to water take place?

20.60 (a) What is the difference between ATP and GTP? (b) Compared with ATP, would you expect GTP to carry more, less, or about the same amount of energy?

20.61 What is the chief energy-carrying molecule in the body?

*20.62 Two keto acids are important in the citric acid cycle. Identify the two keto acids and tell how they are manufactured.

*20.63 The action of the proton translocating ATPase enzyme is reversible. What happens to a cell when its $O_2$ supply is exhausted?

# C H A P T E R   2 1

*Micrograph of fat cells. (©CNRI/Phototake NYC.)*

# Specific Catabolic Pathways: Carbohydrate, Lipid, and Protein Metabolism

## 21.1 SPECIFIC PATHWAYS AND THEIR CONVERGENCE TO THE COMMON PATHWAY

The food we eat serves two main purposes: (1) It fulfills our energy needs, and (2) it provides the raw materials to build the compounds our bodies need. Before either of these processes can take place, the food—carbohydrates, fats, and proteins—must be broken down to small molecules that can be absorbed through the intestinal walls.

### Carbohydrates

Complex carbohydrates (di- and polysaccharides) are broken down by stomach acid and enzymes to produce monosaccharides (Sec. 26.7). Monosaccharides, the most important of which is glucose, may also come from the enzymatic breakdown of glycogen. As you may recall, this highly branched polymer stores carbohydrates in the liver and muscles until needed. Once monosaccharides are produced, they can be used either to build new oligo- and polysaccharides or to provide energy. The specific pathway by which energy is extracted from monosaccharides is called **glycolysis** (Secs. 21.2 and 21.3).

## Lipids

Ingested fats are broken down by lipases to glycerol and fatty acids or to monoglycerides, which are absorbed through the intestine (Sec. 26.8). In a similar way, complex lipids are also hydrolyzed to smaller units before absorption. As with carbohydrates, these smaller molecules (fatty acids, glycerol, and so on) can be used to build complex molecules needed in membranes, or they can be oxidized to provide energy, or they can be stored in fat storage depots (Fig. 21.1). The stored fats can then be broken down later to glycerol and fatty acids whenever they are needed as fuel.

Specialized cells that store fats are called **fat depots.**

The specific pathway by which energy is extracted from glycerol involves the same glycolysis pathway as that used for carbohydrates (Sec. 21.4). The specific pathway used by the cells to obtain energy from fatty acids is called **β-oxidation** (Sec. 21.5).

## Proteins

As you might expect from a knowledge of their structures, proteins are broken down by HCl in the stomach and by digestive enzymes in the stomach and intestines (pepsin, trypsin, chymotrypsin, and carboxypeptidases) to produce their constituent amino acids. The amino acids absorbed through the intestinal wall enter the **amino acid pool.** They serve as building blocks for proteins as needed and, to a smaller extent (especially during starvation), as a fuel for

The amino acid pool is the name used for the free amino acids found both inside and outside cells throughout the body.

Nucleus   Mitochondria

Fat globules

**Figure 21.1**
Storage of fat in a fat cell. As more and more fat droplets accumulate in the cytoplasm, they coalesce to form a very large globule of fat. Such a fat globule may occupy most of the cell, pushing the cytoplasm and the organelles to the periphery. (Modified from C. A. Villee, E. P. Solomon, and P. W. Davis, *Biology,* Philadelphia: Saunders College Publishing, 1985.)

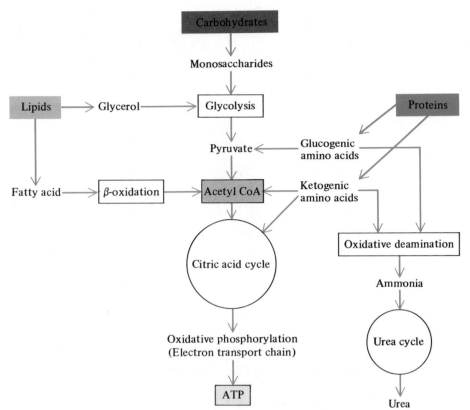

**Figure 21.2**
The convergence of the specific pathways of carbohydrate, fat, and protein catabolism into the common catabolic pathway, which is made up of the citric acid cycle and oxidative phosphorylation.

energy. In the latter case, the nitrogen of the amino acids is catabolized through **oxidative deamination** and the **urea cycle** and is expelled from the body as urea in the urine (Sec. 21.8). The carbon skeletons of the amino acids enter the common catabolic pathway (Chap. 20) either as $\alpha$-keto acids (pyruvic, oxaloacetic, and $\alpha$-ketoglutaric acids) or as acetyl coenzyme A (Sec. 21.9). In all cases, the **specific pathways of carbohydrate, fat, and protein degradation converge to the common catabolic pathway** (Fig. 21.2). This way, the body needs fewer enzymes to get energy from diverse food materials. Efficiency is thus achieved because a minimal number of chemical steps are required and also because the energy-producing factories of the body are localized in the mitochondria.

## 21.2 GLYCOLYSIS

**Glycolysis** is the **specific pathway by which the body gets energy from monosaccharides.** The detailed steps are shown in Figure 21.3, and the most important features are shown schematically in Figure 21.4.

In the first steps of glucose metabolism, energy is consumed rather than released. At the expense of two molecules of ATP (which are converted to ADP), glucose ($C_6$) is phosphorylated; first glucose 6-phosphate is formed ①, and then, after isomerization to fructose 6-phosphate ②, a second phosphate group is attached to yield fructose 1,6-diphosphate ③. We can consider these steps as the activation process.

616

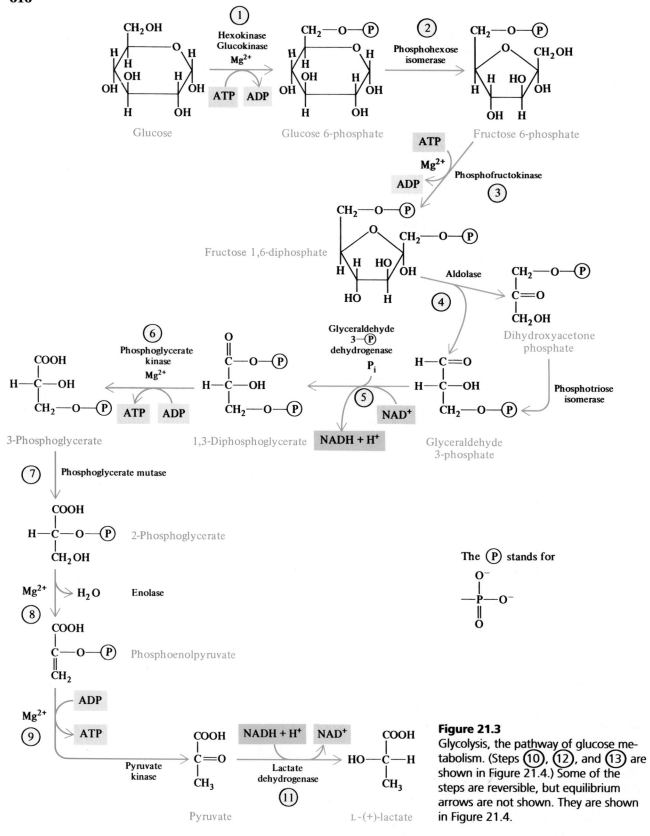

**Figure 21.3**
Glycolysis, the pathway of glucose metabolism. (Steps ⑩, ⑫, and ⑬ are shown in Figure 21.4.) Some of the steps are reversible, but equilibrium arrows are not shown. They are shown in Figure 21.4.

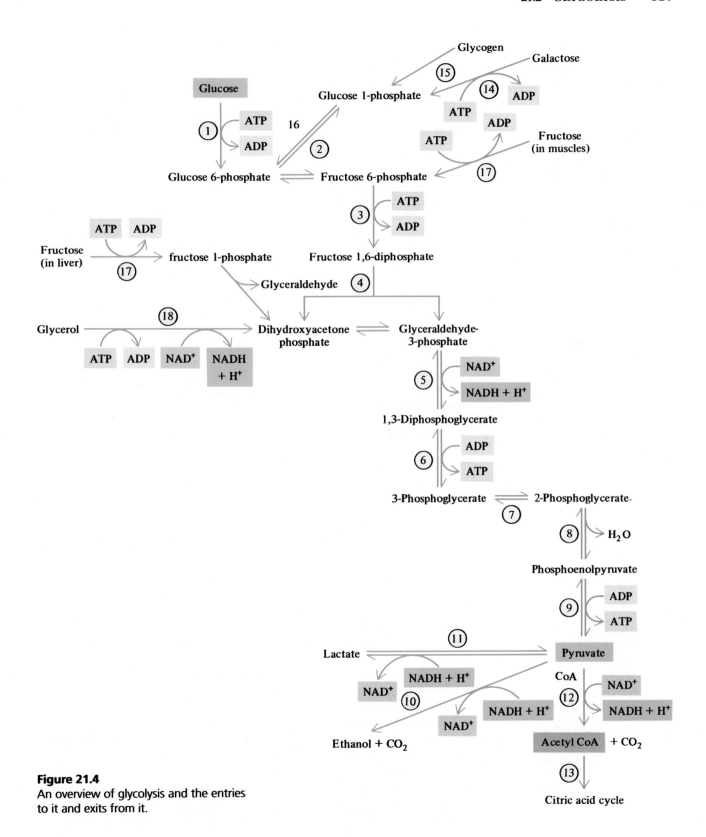

**Figure 21.4**
An overview of glycolysis and the entries
to it and exits from it.

We saw in Section 16.10 that table sugar, sucrose, is a disaccharide made up of one unit of D-glucose and one of D-fructose. When sucrose is digested, the body hydrolyzes it to glucose and fructose, and then each of these is converted to a phosphate. Some people have a genetic disease in which the enzyme fructose 1-phosphate aldolase is missing. This is the enzyme that catalyzes the breakdown of fructose 1-phosphate into two $C_3$ fragments, dihydroxyacetone phosphate and glyceraldehyde, which can enter normal glycolysis.

In the absence of this enzyme, the fructose 1-phosphate produced after ingestion of fructose or sucrose accumulates in the liver. The symptoms of the disease are sweating, absence of reflexes, and convulsions, all caused by *hypoglycemia*—low glucose concentration in the blood. In this case, however, the low blood glucose is produced not by excess insulin, as in most hypoglycemia cases, but by the accumulation of fructose 1-phosphate in the liver. This compound is an inhibitor of the fructose 1,6-diphosphate aldolase that catalyzes the breakdown of a $C_6$ to two $C_3$ fragments (Fig. 21.3). Thus, glucose is not catabolized.

Even more important is the fact that this same enzyme also catalyzes the synthesis of glucose from such sources as glycerol, lactic acid, and amino acids. Therefore, the inhibition of the enzyme lowers the glucose content in the blood while it raises the fructose content. Furthermore, the blood phosphate content is also lowered because the accumulating fructose 1-phosphate ties up the free phosphate. This accumulation in the liver also produces liver enlargement and jaundice. The disease can be treated by removing fructose and sucrose from the diet. All the symptoms of the disease are eliminated, and the patient can live an otherwise normal life.

In step ⑨, after hydrolysis of the phosphate, the resulting enol of pyruvic acid tautomerizes to the more stable keto form (Sec. 13.6).

The cytosol is the fluid inside the cell that surrounds the nucleus and organelles such as mitochondria.

In the second stage the $C_6$ compound fructose 1,6-diphosphate is broken ④ into two $C_3$ fragments. The two $C_3$ fragments, glyceraldehyde 3-phosphate and dihydroxyacetone phosphate, are in equilibrium (they can be converted to each other). Only the glyceraldehyde 3-phosphate is oxidized in glycolysis, but as this species is removed from the equilibrium mixture, Le Chatelier's principle ensures that the dihydroxyacetone phosphate is converted to glyceraldehyde 3-phosphate.

In the third stage, the glyceraldehyde 3-phosphate is oxidized to 1,3-diphosphoglycerate ⑤. The hydrogen of the aldehyde group is removed by the $NAD^+$ coenzyme. In the next step ⑥, the phosphate from the carboxyl group is transferred to ADP, yielding ATP and 3-phosphoglycerate. The latter, after isomerization ⑦ and dehydration ⑧, is converted to phosphoenolpyruvate, which loses its remaining phosphate ⑨ and yields pyruvate ion and another ATP molecule.

All these glycolysis reactions occur in the cytosol outside the mitochondria. They occur in the absence of $O_2$ and are therefore also called reactions of the **anaerobic pathway.** As indicated in Figure 21.4, the end product of glycolysis, pyruvate, does not accumulate in the body. In certain bacteria and yeast, pyruvate undergoes reductive decarboxylation ⑩ to produce ethyl alcohol. In some bacteria, and also in mammals in the absence of oxygen, pyruvate is reduced to lactate ⑪. But most importantly pyruvate goes through an oxidative decarboxylation in the presence of coenzyme A ⑫ to produce acetyl CoA:

Many athletes suffer muscle cramps when they undergo strenuous exercise. This is the result of a shift from normal glucose catabolism (glycolysis → citric acid cycle → oxidative phosphorylation) to that of lactate production (11) (Fig. 21.4). During exercise, oxygen is used up rapidly, and this slows down the rate of the common pathway. The demand for energy makes *anaerobic* glycolysis proceed at a high rate, but because the *aerobic* (oxygen-demanding) pathways are slowed down, not all the pyruvate produced in glycolysis can enter the citric acid cycle. The excess pyruvate ends up as lactate, which causes painful muscle contractions.

The same shift in catabolism also occurs in heart muscle when coronary thrombosis leads to cardiac arrest. The oxygen supply is cut off by the blockage of the artery to the heart muscles. The common pathway and its ATP production are shut off. Glycolysis proceeds at an accelerated rate, accumulating lactate. The heart muscle contracts, producing a cramp. Just as in skeletal muscle, massage of heart muscles can relieve the cramp and start the heart beating. Even if heartbeat is restored within 3 minutes (the amount of time the brain can survive without being damaged), acidosis develops as a result of the cardiac arrest. Therefore, at the same time that efforts are under way to start the heart beating by chemical, physical, or electrical means, an intravenous infusion of 8.4 percent bicarbonate solution is given to combat acidosis.

This reaction is catalyzed by a complex enzyme system, pyruvate dehydrogenase, that sits on the inner membrane of the mitochondrion. The reaction produces acetyl CoA, $CO_2$, and NADH + H$^+$. The acetyl CoA then enters the citric acid cycle (13) and goes through the common pathway.

In summary, after converting complex carbohydrates to glucose, the body gets energy from glucose by converting it to acetyl CoA (by way of pyruvate) and then using the acetyl CoA as a raw material for the common pathway.

## 21.3  THE ENERGY YIELD FROM GLUCOSE

In conjunction with Figure 21.4, we can sum up the energy derived from glucose catabolism in terms of ATP production. However, before we begin we must take into account that glycolysis takes place in the cytosol, while oxidative phosphorylation occurs in the mitochondria. Therefore, the NADH + H$^+$ produced in glycolysis must penetrate the mitochondrial membrane in order to be utilized in oxidative phosphorylation.

**TABLE 21.1 ATP Yield from Complete Glucose Metabolism**

| Step Number in Fig. 21.4 | Chemical Steps | Number of ATP Molecules Produced |
|---|---|---|
| ① ② ③ | Activation (glucose → 1,6-fructose diphosphate) | −2 |
| ⑤ | Oxidative phosphorylation 2(glyceraldehyde 3-phosphate → 1,3-diphosphoglycerate), producing 2 NADH + H⁺ in cytosol | 4 |
| ⑥ ⑨ | Dephosphorylation 2(1,3-diphosphoglycerate → pyruvate) | 4 |
| ⑫ | Oxidative decarboxylation 2(pyruvate → acetyl CoA), producing 2 NADH + H⁺ in mitochondrion | 6 |
| ⑬ | Oxidation of two $C_2$ fragments in citric acid and oxidative phosphorylation common pathway, producing 12 ATP for each $C_2$ fragment | $\underline{24}$ |
| | Total | 36 |

There are two routes available to get the NADH + H⁺ into the mitochondrion: (1) the **glycerol 3-phosphate transport** in muscle and nerve cells (Sec. 21.4) and (2) the **aspartate–malate shuttle** in the heart and liver. These two routes have different efficiencies. In the glycerol 3-phosphate transport, only two ATP molecules are produced for each NADH + H⁺. In the aspartate–malate shuttle, on the other hand, three ATP molecules are produced for each NADH + H⁺ produced in the cytosol, just as is the case in the mitochondrion (Sec. 20.7). Since most energy production takes place in skeletal muscle cells, when we construct the energy balance sheet we use two ATP molecules for each NADH + H⁺ produced in the cytosol.

With this knowledge we are ready to calculate the energy yield of glucose in terms of ATP molecules produced. This is shown in Table 21.1. In the first stage of glycolysis ① ② ③, two ATP molecules are used up, but this is more than compensated for by the production of 14 ATP molecules in ⑤, ⑥, ⑨, and ⑫ and in the conversion of pyruvate to acetyl CoA. The *net* yield of these steps is 12 ATP molecules. As we saw in Section 20.7, the oxidation of one acetyl CoA produces 12 ATP molecules, and one glucose molecule provides two acetyl CoA molecules. Therefore, the total net yield from metabolism of one glucose molecule in skeletal muscle,

$$C_6H_{12}O_6 + 6O_2 \longrightarrow 6CO_2 + 6H_2O$$

is 36 molecules of ATP.

This calculation applies to glucose metabolism in the skeletal muscle cells, which is what happens most frequently. The two NADH produced in the glycolysis yield only a total of four ATP because of the efficiency of the glycerol 3-phosphate transport. Thus, a total of 36 ATP molecules are produced from one glucose molecule. However, if the same glucose is metabolized in the heart or liver, the two NADH produced in the glycolysis are transported into the

Muscles attached to bones are called **skeletal muscles.**

Missing enzymes that metabolize glycogen give rise to a number of genetically inherited glycogen storage diseases. One is McArdle's disease. Patients with this disease have a limited capacity for exercise, suffering from painful muscle cramps. The genetic defect is the absence of phosphorylase in the muscles. Thus, glycogen cannot be utilized.

Another glycogen storage disease is Cori's disease. Patients suffering from this have no debranching enzyme in their liver and muscles. Thus, only the outer branches of glycogen are utilized and the rest accumulates, resulting in an enlarged liver. Hypoglycemia (Box 26D) and ketosis (high amounts of ketone bodies as a result of dominant fat metabolism) are other symptoms of Cori's disease.

Still another example of glycogen accumulation in the muscles is caused by a deficiency of the enzyme maltase (Fig. 21C).

**B o x 21C**

## Glycogen Storage Diseases

**Figure 21C**
Glycogen accumulation in muscles due to acid maltase enzyme deficiency. (Courtesy of Dr. Janice Anderson, Addenbrooke's Hospital, and MTP Press, Cambridge, England.)

mitochondrion by the aspartate–malate shuttle. Therefore, they yield a total of six ATP molecules, so that in this case, there are 38 ATP molecules produced for each glucose molecule.

Glucose is not the only monosaccharide that can be used as an energy source. Other hexoses, such as galactose ⑭ or fructose ⑰, enter the glycolysis pathway at the stages indicated in Figure 21.4. They also yield 36 molecules of ATP per hexose molecule. Furthermore, the glycogen stored in the liver and muscle cells and elsewhere can also be converted by enzymatic breakdown and phosphorylation to glucose 1-phosphate ⑮. This in turn isomerizes to glucose 6-phosphate, providing an entry to the glycolytic pathway.

## 21.4 GLYCEROL CATABOLISM

The glycerol hydrolyzed from neutral fats or complex lipids (Chap. 17) can also be a rich energy source. The first step in glycerol utilization is an activation step. The body uses one ATP molecule to form glycerol 1-phosphate:

Glycerol 1-phosphate is the same as glycerol 3-phosphate.

$$
\begin{array}{ccc}
CH_2OH & CH_2O{-}\textcircled{P} & CH_2O{-}\textcircled{P} \\
| & | & | \\
CHOH & \xrightarrow{\text{ATP} \quad \text{ADP}} \quad CHOH & \xrightarrow{\text{NAD}^+ \quad \text{NADH + H}^+} \quad C{=}O \\
| & | & | \\
CH_2OH & CH_2OH & CH_2OH \\
\text{Glycerol} & \text{Glycerol 1-phosphate} & \text{Dihydroxyacetone phosphate}
\end{array}
$$

The glycerol phosphate is oxidized by $NAD^+$ to dihydroxyacetone phosphate, yielding $NADH + H^+$ in the process. Dihydroxyacetone phosphate then enters the glycolysis pathway ⑱ and is isomerized to glyceraldehyde phosphate, as shown in Figure 21.4. A net yield of 20 ATP molecules is produced from each glycerol molecule, which is 6.7 ATP molecules per carbon atom.

## 21.5  β-OXIDATION OF FATTY ACIDS

As early as 1904, Franz Knoop in Germany proposed that the body utilizes fatty acids as an energy source by breaking them down into $C_2$ fragments. Prior to fragmentation, the β carbon is oxidized:

*The β carbon is the second carbon from the COOH group.*

$$-C-C-C-\overset{\beta}{C}-\overset{\alpha}{C}-COOH$$

Thus, the name **β-oxidation** has its origin in Knoop's prediction. It took about 50 years to establish the mechanism by which fatty acids are utilized as an energy source.

The overall process of fatty acid metabolism is shown in Figure 21.5. As is the case with the other foods we have seen, the first step is an activation step. This occurs in the cytosol, where the fat was previously hydrolyzed to glycerol and fatty acids. The activation ① converts ATP to AMP and inorganic phosphate. This is equivalent to the cleavage of two high-energy phosphate bonds. The chemical energy derived from the splitting of ATP is built into the compound acyl CoA, which is formed when the fatty acid combines with coenzyme A (Fig. 21.5). The fatty acid oxidation occurs inside the mitochondrion, so the acyl CoA must pass through the mitochondrial membrane.

*Transfer into the mitochondrion is accomplished by an enzyme system called carnitine acyltransferase.*

Once the fatty acid in the form of acyl CoA is inside the mitochondrion, the β-oxidation starts. In the first oxidation (dehydrogenation) ②, two hydrogens are removed, creating a double bond between the alpha and beta carbons of the acyl chain. The hydrogens and electrons are picked up by FAD.

In the next step ③, the double bond is hydrated. An enzyme specifically places the hydroxy group on C-3. The second oxidation (dehydrogenation) ④ requires $NAD^+$ as a coenzyme. The two hydrogens and electrons removed are transferred to the $NAD^+$ to form $NADH + H^+$. In the process, a secondary

---

▶ **B o x  21D**

## Diphtheria

In diphtheria, bacteria produce a toxin that causes fatty acid accumulation and infiltration of the myocardium (heart muscles). As a result, patients who have had diphtheria may have a weakened myocardium and be prone to heart failure. It is known that diphtheria toxin reduces the carnitine level in tissues. Carnitine is needed to transport fatty acids through the membrane to the inside of the mitochondria, where all the enzymes of β-oxidation are located. In the absence of carnitine, the fatty acids cannot enter and so cannot be metabolized. Thus, the diphtheria toxin's effect on carnitine causes the accumulation of fat in the heart tissues.

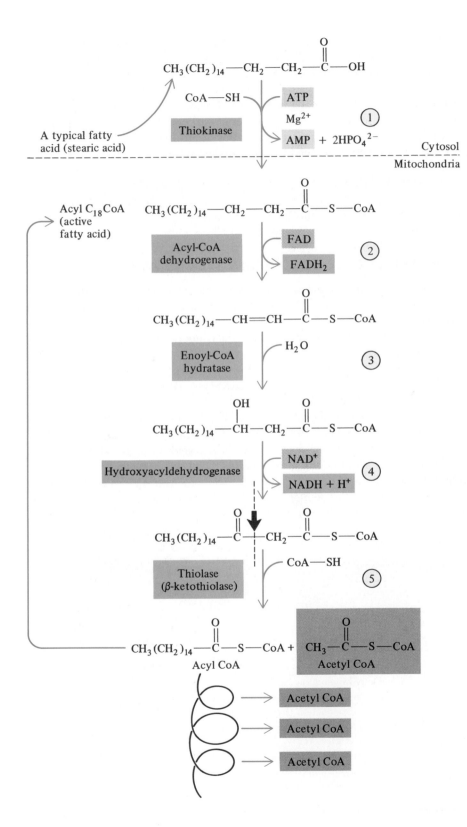

**Figure 21.5**
The β-oxidation spiral of fatty acids. Each loop in the spiral contains two dehydrogenations, one hydration, and one fragmentation. At the end of each loop, one acetyl CoA is released.

alcohol group is oxidized to a keto group at the beta carbon. In the final step ⑤, the enzyme thiolase cleaves the terminal $C_2$ fragment from the chain, and the rest of the molecule is attached to a new molecule of coenzyme A.

The cycle now starts again with the remaining acyl CoA, which is now two carbon atoms shorter. At each turn of the cycle, one acetyl CoA is produced. Most fatty acids contain an even number of carbon atoms. The cyclic spiral is continued until we reach the last four carbon atoms. When this fragment enters the cycle at the end, two acetyl CoA molecules are produced in the fragmentation step.

The $\beta$-oxidation of unsaturated fatty acids proceeds in the same way. There is an extra step involved, in which the cis double bond is isomerized to a trans bond, but otherwise the spiral is the same.

## 21.6 THE ENERGY YIELD FROM STEARIC ACID

In order to compare the energy yield from fatty acids with that of other foods, let us select a typical and quite abundant fatty acid: stearic acid, the $C_{18}$ saturated fatty acid.

We start with the initial step, in which energy is used up rather than produced. The reaction

$$ATP \longrightarrow AMP + 2P_i$$

breaks two high-energy phosphate bonds. This is equivalent to hydrolyzing two molecules of ATP to ADP. In each cycle of the spiral we obtain one $FADH_2$, one $NADH + H^+$, and one acetyl CoA. Stearic acid ($C_{18}$) goes through *seven cycles* in the spiral before it reaches the final $C_4$ stage. In the last (eighth) cycle one $FADH_2$, one $NADH + H^+$, and two acetyl CoA molecules are produced. Now we can add up the energy. Table 21.2 shows that for a $C_{18}$ compound we obtain a total of 146 ATP molecules.

**TABLE 21.2    ATP Yield from Complete Stearic Acid Metabolism**

| Step Number in Fig. 21.5 | Chemical Steps | Happens | Number of ATP Molecules Produced |
|---|---|---|---|
| ① | Activation (stearic acid → stearyl CoA) | Once | −2 |
| ② | Dehydrogenation (acyl CoA → transenoyl CoA), producing $FADH_2$ | 8 times | 16 |
| ④ | Dehydrogenation (hydroxyacyl CoA → keto acyl CoA), producing $NADH + H^+$ | 8 times | 24 |
|  | $C_2$ fragment (acetyl CoA → common catabolic pathway), producing 12 ATP for each $C_2$ fragment | 9 times | 108 |
|  |  | Total | 146 |

It is instructive to compare the energy yield from fats with that from carbohydrates, since both are important constituents of the diet. In Section 21.3 we saw that glucose, $C_6H_{12}O_6$, produces 36 ATP molecules, that is, 6 for each carbon atom. For stearic acid, there are 146 ATP molecules and 18 carbons, or $146/18 = 8.1$ ATP molecules per carbon atom. Since additional ATP molecules are produced from the glycerol portion of fats (Sec. 21.4), fats have a higher caloric value than carbohydrates.

## 21.7 KETONE BODIES

In spite of the high caloric value of fats, the body preferentially uses glucose as an energy supply. When an animal is well fed (plenty of sugar intake), fatty acid oxidation is inhibited, and fatty acids are stored in the form of neutral fat in fat depots. Only when the glucose supply dwindles, as in fasting or starvation, or when glucose cannot be utilized, as in the case of diabetes, is the $\beta$-oxidation pathway of fatty acid metabolism mobilized.

In some pathological conditions, it is possible for glucose not to be available at all.

Unfortunately, low glucose supply also slows down the citric acid cycle. This happens because oxaloacetate (a $C_4$ intermediate of the citric acid cycle, Fig. 20.7) is produced from the carboxylation of pyruvate. If there is no glucose, there will be no glycolysis, no pyruvate formation, and therefore no oxaloacetate production.

Thus, even though the fatty acids are oxidized, not all the resulting $C_2$ fragments (acetyl CoA) can enter the citric acid cycle because there is not enough oxaloacetate. Therefore, acetyl CoA builds up in the body, with the following consequences.

The liver is able to condense two acetyl CoA molecules to produce acetoacetyl CoA:

When the acetoacetyl CoA is hydrolyzed, it yields acetoacetate and its reduced form, $\beta$-hydroxybutyrate:

**Box 21E**

**Ketoacidosis in Diabetes**

In untreated diabetes the glucose concentration in the blood is high because the lack of insulin prevents utilization of glucose by the cells. Regular injections of insulin remedy this situation. However, in some stressful conditions ketoacidosis can still develop. A typical case was a diabetic patient admitted to the hospital in semicoma. He showed signs of dehydration, his skin was inelastic and wrinkled, his urine showed high concentrations of glucose and ketone bodies, and his blood contained excess glucose and had a pH of 7.0, a drop of 0.4 pH units from normal, which is an indication of severe acidosis. The urine also contained the bacteria *E. coli.* This indication of urinary tract infection explained why the normal doses of insulin were insufficient to prevent ketoacidosis.

The stress of infection can upset the normal control of diabetes by upsetting the balance between administered insulin and other hormones produced in the body. This happened during the infection mentioned above, and the body started to produce ketone bodies (Sec. 21.7) in large quantities. Both glucose and ketone bodies appear in the blood before they show up in the urine.

The acidic nature of ketone bodies (acetoacetic acid and $\beta$-hydroxybutyric acid) lowers the blood pH. A large drop in pH is prevented by the bicarbonate/carbonic acid buffer (Sec. 8.10), but even a drop of 0.3 to 0.5 pH units is sufficient to decrease the $Na^+$ concentration. The decrease of $Na^+$ in the interstitial tissues draws out $K^+$ ions from the cells. This in turn impairs brain function and leads to coma. During the secretion of ketone bodies and glucose in the urine, a lot of water is lost, and the body becomes dehydrated. This means the blood volume shrinks. Thus the blood pressure drops and the pulse rate increases to compensate. Smaller quantities of nutrients reach the brain cells, and this too can cause coma.

The patient mentioned above was infused with physiological saline solution to remedy the dehydration. Extra doses of insulin restored his glucose level to normal, and antibiotics cured the urinary infection.

**These two compounds, together with smaller amounts of acetone,** are collectively called **ketone bodies.** Under normal conditions, the liver sends these compounds into the blood stream to be carried to the tissues and utilized there via the common catabolic pathway. Normally the concentration of ketone bodies in the blood is low. But during starvation and in untreated diabetes mellitus, ketone bodies accumulate in the blood and can reach high concentrations. When this occurs, the excess is secreted in the urine. A check of urine for ketone bodies is used in the diagnosis of diabetes.

## 21.8 CATABOLISM OF THE NITROGEN OF AMINO ACIDS

The proteins of our foods are hydrolyzed to amino acids in digestion. These amino acids are primarily used to synthesize new proteins. However, unlike carbohydrates and fats, they cannot be stored, so excess amino acids are catabolized for energy production. What happens to the carbon skeleton of the amino acids is dealt with in the next section. Here we discuss the catabolic fate of the nitrogen.

In the tissues, amino ($-NH_2$) groups freely move from one amino acid to another. The enzymes that catalyze these reactions are the **transaminases.**

In essence there are three stages in nitrogen catabolism in the liver. The final product of the three stages is urea, which is excreted in the urine of mammals. The first stage is a **transamination.** Amino acids transfer their amino groups to $\alpha$-ketoglutaric acid:

Amino acid
(zwitterion form)        $\alpha$-Ketoglutarate            $\alpha$-Keto acid          Glutamate

The carbon skeleton of the amino acid remains behind as an $\alpha$-keto acid. The second stage of the nitrogen catabolism is the **oxidative deamination** of glutamate, which occurs in the mitochondrion:

The catabolism of the $\alpha$-keto acid is discussed in the next section.

Glutamate                                    $\alpha$-Ketoglutarate

The oxidative deamination yields $NH_4^+$ and regenerates $\alpha$-ketoglutarate, which can again participate in the first stage (transamination). The NADH + H$^+$ produced in the second stage undergoes oxidative phosphorylation and eventually produces three ATP molecules.

In the third stage the $NH_4^+$ is converted to urea through the **urea cycle** (Fig. 21.6). First the $NH_4^+$ is condensed with $CO_2$ in the mitochondrion to form an unstable compound, carbamoyl phosphate ①. This condensation occurs at the expense of two ATP molecules. In the next step ②, carbamoyl phosphate is condensed with ornithine, a basic amino acid similar in structure to lysine:

The body must get rid of $NH_4^+$ because both it and $NH_3$ are toxic.

Ornithine does not occur in proteins.

Ornithine                Carbamoyl              Citrulline
                         phosphate

The result is citrulline, which diffuses out of the mitochondrion into the cytosol.

**Figure 21.6**
The urea cycle.

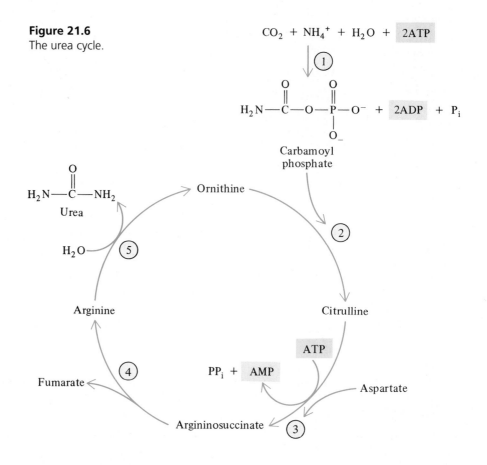

A second condensation reaction in the cytosol takes place between citrulline and aspartate, forming argininosuccinate ③:

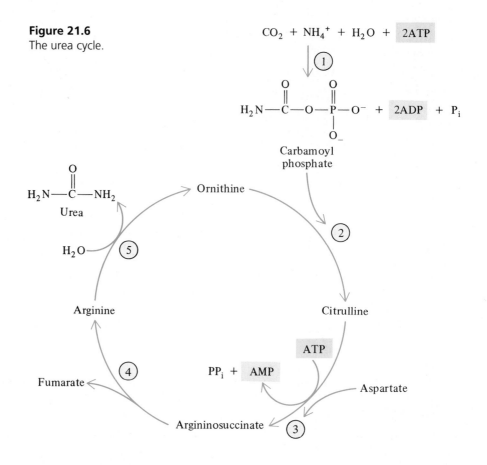

The energy for this reaction comes from the hydrolysis of ATP to AMP and pyrophosphate ($PP_i$).

In the fourth step, the argininosuccinate is split into arginine and fumarate ④:

The PP$_i$ stands for pyrophosphate:

Argininosuccinate     Arginine     Fumarate

In the final step ⑤, arginine is hydrolyzed to urea and ornithine:

Arginine     Urea     Ornithine

The urea is excreted in the urine. The ornithine re-enters the mitochondrion and thus completes the cycle. It is now ready to pick up another carbamoyl phosphate ②.

Hans Krebs, who elucidated the citric acid cycle, was also instrumental in establishing the urea cycle.

An important aspect of carbamoyl phosphate as an intermediate is that it can be used for synthesis of nucleotide bases (Chap. 23). Furthermore, the urea cycle is linked to the citric acid cycle because both involve fumarate.

## 21.9 CATABOLISM OF THE CARBON SKELETON OF AMINO ACIDS

Once the alpha amino group is removed from an amino acid by oxidative deamination (Sec. 21.8), the remaining carbon skeleton is used as an energy source (Fig. 21.7). We are not going to study the pathways involved except to point out the eventual fate of the skeleton. Not all the carbon skeletons of amino acids are used as fuel. Some of them may be degraded up to a certain point, and the resulting intermediate then used as a building block to construct another needed molecule. For example, if the carbon skeleton of an amino acid is catabolized to pyruvate, there are two possible choices for the body: (1) to use the pyruvate as an energy supply via the common pathway or (2) to use it as a building block to synthesize glucose (Sec. 22.2). **Those amino acids that yield a carbon skeleton that is degraded to pyruvate or another intermediate capable of conversion to glucose** are called **glucogenic**. One example is alanine. When alanine reacts with $\alpha$-ketoglutaric acid, the transamination produces pyruvate directly, as shown in Figure 21.7:

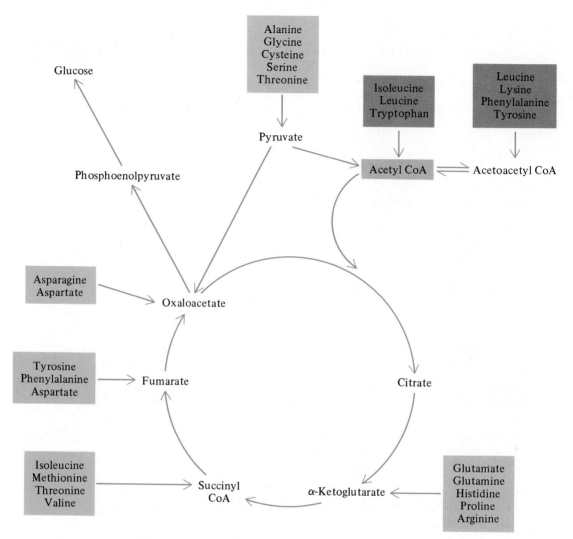

**Figure 21.7**
Catabolism of the carbon skeletons of amino acids. The glucogenic amino acids are in the blue boxes; the ketogenic ones in the brown boxes.

On the other hand, many amino acids are degraded to acetyl CoA and acetoacetic acid. These compounds cannot form glucose but are capable of

**B o x   21F**

## MSG and Chinese Restaurant Syndrome

Glutamic acid is one of the 20 amino acids found in proteins (Sec. 18.2). Its monosodium salt, **monosodium glutamate (MSG),** enhances the flavor of many foods without itself contributing any significant taste. It makes such foods as meats, vegetables, soups, and stews taste better. Because of this, MSG is added to many canned and frozen foods. In pure form it is sold in supermarkets under such brand names as Accent. The spice racks of many kitchens contain a bottle labeled MSG, next to the nutmeg, thyme, and basil. In particular, Chinese restaurants in this country use copious quantities of MSG.

Although most people can consume MSG with no problems, some are allergic to it and get symptoms that include dizziness, numbness spreading from the jaw to the back of the neck, hot flashes, and sweating. Collectively, these symptoms are known as Chinese restaurant syndrome. In the brain, an enzyme catalyzes the decarboxylation of glutamic acid to $\gamma$-aminobutyric acid:

$$
\begin{array}{c}
COO^- \\
| \\
(CH_2)_2 \\
| \\
H-C-COO^- \\
| \\
NH_3^+
\end{array}
\longrightarrow
\begin{array}{c}
COO^- \\
| \\
(CH_2)_2 \\
| \\
CH_2 \\
| \\
NH_3^+
\end{array}
+ CO_2
$$

Glutamic acid          $\gamma$-Aminobutyric acid

Both glutamic acid and $\gamma$-aminobutyric acid affect signal transmissions in the brain. Glutamic acid excites while $\gamma$-aminobutyric acid inhibits neurotransmission. Therefore excess consumption of glutamic acid in susceptible individuals may upset the balance between excitation and inhibition. (See also Section 24.4.)

yielding ketone bodies; they are called **ketogenic.** Leucine is an example of a ketogenic amino acid.

Both glucogenic and ketogenic amino acids, when used as an energy supply, enter the citric acid cycle at some point (Fig. 21.7) and are eventually oxidized to $CO_2$ and $H_2O$.

## 21.10 CATABOLISM OF HEME

Red blood cells are continuously being manufactured in the bone marrow. Their life span is relatively short—about four months. Aged red blood cells are destroyed in the phagocytic cells. When a red blood cell is destroyed, its hemoglobin is metabolized: The globin (Sec. 18.9) is hydrolyzed to amino acids, the heme is converted to *bilirubin,* and the iron is preserved in ferritin, an iron-carrying protein, and is reused. Bilirubin is the final product of heme destruction, which is shown in Figure 21.8. First the cyclic tetrapyrrole structure is broken at one of the CH bridges (marked in Fig. 21.8). *Biliverdin,* the green pigment that results from the ring opening and oxidation, is reduced to the orange pigment bilirubin. The bilirubin enters the liver via the blood and is then transferred to the gallbladder, where it is stored in the bile and finally excreted via the small intestine. The color of feces is provided by *urobilin,* an oxidation product of bilirubin.

Phagocytes are specialized blood cells that destroy foreign bodies.

**B o x  21G**

**Hereditary Defects in Amino Acid Catabolism. PKU**

Many hereditary diseases involve missing or malfunctioning enzymes that catalyze the breakdown of amino acids. The oldest-known of such diseases is cystinuria, which was described as early as 1810. In this disease, cystine shows up as flat hexagonal crystals in the urine. Stones form because of the low solubility of cystine in water. This leads to blockage in the kidneys or the ureters and requires surgery. One way to reduce the amount of cystine secreted is to remove as much methionine as possible from the diet. Beyond that, an increased fluid intake increases the volume of the urine, reducing the solubility problem. It has been found that penicillamine can also prevent cystinuria.

An even more important genetic defect is the absence of the enzyme phenylalanine hydroxylase, causing a disease called phenylketonuria (PKU). In normal catabolism, this enzyme helps degrade phenylalanine by converting it to tyrosine. If the enzyme is defective, phenylalanine is converted to phenylpyruvate (see the conversion of alanine to pyruvate in Section 21.9). Phenylpyruvate (a ketoacid) accumulates in the body and inhibits the conversion of pyruvate to acetyl CoA, thus depriving the cells of energy via the common pathway. This is most important in the brain, which gets its energy from the utilization of glucose. The result is mental retardation. This genetic defect can be detected early because phenylpyruvic acid appears in the urine. A federal regulation requires that all infants be tested for this disease. Once PKU is detected, mental retardation can be prevented by restricting the intake of phenylalanine in the diet. Patients with PKU should avoid the artificial sweetener aspartame (Box 18B) because it yields phenylalanine when hydrolyzed in the stomach.

**Figure 21.8**
The catabolism of heme.

In normal individuals there is a balance between the destruction of heme in the spleen and the removal of bilirubin from the blood by the liver. When this balance is upset, the bilirubin concentration in the blood increases. If this condition continues, the skin and whites of the eyes both become yellow, in a condition known as **jaundice.** Therefore jaundice can signal a malfunctioning of the liver, the spleen, or the gallbladder, where bilirubin is stored before excretion. Hemolytic jaundice is the acceleration of the destruction of red blood cells in the spleen to such a level that the liver cannot cope with the bilirubin production. When gallstones obstruct the bile ducts and the bilirubin cannot be excreted in the feces, the backup of bilirubin causes jaundice. In infectious hepatitis the liver is incapacitated, and bilirubin is not removed from the blood, causing jaundice.

## SUMMARY

The foods we eat are broken down into small molecules in the stomach and intestines before being absorbed. These small molecules serve two purposes: They can be the building blocks of new materials the body needs to synthesize (anabolism), or they can be used for energy supply (catabolism). Each group of compounds— carbohydrates, fats, and proteins—has its own catabolic pathway. All the different catabolic pathways converge to the common pathway.

The specific pathway of carbohydrate catabolism is **glycolysis.** In this process, hexose monosaccharides are activated by ATP and eventually converted to two $C_3$ fragments, dihydroxyacetone phosphate and glyceraldehyde phosphate. The glyceraldehyde phosphate is further oxidized and eventually ends up as pyruvate. All these reactions occur in the cytosol. Pyruvate is converted to acetyl CoA, which is further catabolized in the common pathway. When completely metabolized, a hexose yields the energy of 36 ATP molecules.

Fats are broken down to glycerol and fatty acids. Glycerol is catabolized in the glycolysis pathway and yields 20 ATP molecules.

Fatty acids are broken down into $C_2$ fragments in the **β-oxidation spiral.** At each turn of the spiral one acetyl CoA is released together with one $FADH_2$ and one $NADH + H^+$. These products go through the common pathway. Stearic acid, a $C_{18}$ compound, yields 146 molecules of ATP. In starvation and under certain pathological conditions, not all the acetyl CoA produced in the β-oxidation of fatty acids enters the common pathway. Some of it forms acetoacetate, β-hydroxybutyrate, and acetone, commonly called **ketone bodies.** Excess ketone bodies in the blood are secreted in the urine.

Proteins are broken down to amino acids. The nitrogen of the amino acids is first transferred to glutamate. This in turn is **oxidatively deaminated** to yield ammonia. Mammals get rid of the toxic ammonia by converting it to urea in the **urea cycle.** Urea is secreted in the urine. The carbon skeletons of amino acids are catabolized via the citric acid cycle. Some of these enter as pyruvate or other intermediates of the citric acid cycle; these are **glucogenic amino acids.** Others are incorporated into acetyl CoA or ketone bodies and are called **ketogenic amino acids.** Heme is catabolized to bilirubin, which is excreted in the feces.

## KEY TERMS

Amino acid pool (Sec. 21.1)
Anaerobic pathway (Sec. 21.2)
Aspartate–malate shuttle (Sec. 21.3)
Cytosol (Sec. 21.2)
Fat depot (Sec. 21.1)
Glucogenic amino acid (Sec. 21.9)
Glycerol 3-phosphate transport (Sec. 21.3)
Glycolysis (Sec. 21.2)

Ketoacidosis (Box 21E)
Ketogenic amino acid (Sec. 21.9)
Ketone bodies (Sec. 21.7)
β-Oxidation (Sec. 21.5)
Oxidative deamination (Sec. 21.8)
Transamination (Sec. 21.8)
Urea cycle (Sec. 21.8)

## PROBLEMS

### Specific Pathways

**21.1** What are the units into which large carbohydrate molecules are broken down in the digestive process?

**21.2** In what forms are fats absorbed through the intestinal wall?

**21.3** What is the main use of amino acids in the body?

### Glycolysis

**21.4** Although catabolism of a glucose molecule eventually produces a lot of energy, the first step uses up energy. Explain why this step is necessary.

**21.5** In one step of the glycolysis pathway, a $C_6$ chain is broken into two $C_3$ fragments, only one of which can be further degraded in the glycolysis pathway. What happens to the other $C_3$ fragment?

**21.6** Why is glycolysis called an anaerobic pathway?

**21.7** (a) Which steps in glycolysis of glucose need ATP? (b) Which steps in glycolysis yield ATP directly?

**21.8** What two $C_3$ fragments are obtained from splitting fructose 1,6-diphosphate?

**21.9** Find all the oxidation steps in the glycolysis pathway shown in Figure 21.3.

**21.10** The end product of glycolysis, pyruvate, cannot enter as such into the citric acid cycle. What is the name of the process that converts this $C_3$ compound to a $C_2$ compound?

**21.11** Pyruvate can be converted (in different organisms) to either lactate, ethanol, or acetyl CoA. Which product is formed under anaerobic conditions?

**\*21.12** Which of these steps yields energy and which consumes energy?
(a) pyruvate $\rightarrow$ lactate
(b) pyruvate $\rightarrow$ acetyl CoA + $CO_2$

**21.13** How many moles of lactate are produced from three moles of glucose?

**\*21.14** How many moles of net NADH + H$^+$ are produced from one mole of glucose going to (a) acetyl CoA (b) lactate

### Energy Yield from Glucose

**\*21.15** Why does the body need a shuttle, such as the aspartate–malate shuttle, to obtain ATP molecules from the NADH + H$^+$ of glycolysis?

**21.16** How many *net* ATP molecules are produced in the skeletal muscles for each glucose molecule
(a) in glycolysis alone (up to pyruvate)?
(b) in converting pyruvate to acetyl CoA?
(c) in the total oxidation of glucose to $CO_2$ and $H_2O$?

**\*21.17** (a) If fructose is metabolized in the liver, how many moles of net ATP are produced from each mole during glycolysis? (b) How many moles are produced if the same thing occurs in a muscle cell?

### Glycerol Catabolism

**21.18** Explain why dihydroxyacetone phosphate is called an oxidation product of glycerol-1-phosphate.

**\*21.19** Which yields more energy upon hydrolysis, ATP or glycerol 1-phosphate? Why?

### β-Oxidation of Fatty Acids

**21.20** Why is the breakdown of fatty acids called β-oxidation?

**21.21** (a) Which part of the cells contains the enzymes needed for the β-oxidation of fatty acids? (b) How does the activated fatty acid get there?

**21.22** Assume that stearic acid is metabolized through β-oxidation. What are the products of the oxidation after two turns of the spiral?

**21.23** What is the function of the enzyme enoyl-CoA hydratase?

**21.24** How many turns of the spiral are there in the β-oxidation of (a) lauric acid (b) palmitic acid

### Energy Yield from Fatty Acids

**21.25** Calculate the number of ATP molecules obtained in the β-oxidation of myristic acid, $CH_3(CH_2)_{12}COOH$.

**21.26** Two of the enzymes in the metabolism of fatty acids have *thio* in their names: thiolase and thiokinase. What processes do these enzymes catalyze?

**21.27** Assuming that both fats and carbohydrates are available, which does the body preferentially use as an energy source?

**21.28** If equal weights of fats and carbohydrates are eaten, which will give more calories? Explain.

## Ketone Bodies

**21.29** Why do starving people have ketone bodies in their urine?

**21.30** Do ketone bodies have nutritional value?

## Catabolism of Amino Acids

**21.31** What kind of reaction is this and what is its function in the body?

$$
\begin{array}{c}
\text{H}_3\text{C} \\
\diagdown \\
\quad\quad \text{CH}\!-\!\underset{\underset{\text{NH}_3^+}{|}}{\overset{\overset{\text{H}}{|}}{\text{C}}}\!-\!\text{COO}^- + \underset{\underset{\underset{\text{COO}^-}{|}}{\overset{\text{CH}_2}{|}}}{\underset{|}{\overset{\overset{\overset{\text{COO}^-}{|}}{\text{C}=\text{O}}}{\text{CH}_2}}} \longrightarrow \\
\text{H}_3\text{C}
\end{array}
$$

$$
\begin{array}{c}
\text{H}_3\text{C} \\
\diagdown \\
\quad\quad \text{CH}\!-\!\underset{\underset{\text{O}}{\|}}{\text{C}}\!-\!\text{COO}^- + \underset{\underset{\underset{\text{COO}^-}{|}}{\overset{\text{CH}_2}{|}}}{\underset{|}{\overset{\overset{\text{H}\!-\!\text{C}\!-\!\text{NH}_3^+}{|}}{\text{CH}_2}}} \\
\text{H}_3\text{C}
\end{array}
$$

**21.32** Write an equation for the oxidative deamination of aspartate.

**21.33** Write a balanced equation for the transamination reaction of tyrosine by $\alpha$-ketoglutarate.

**21.34** Ammonia, $NH_3$, and ammonium ion, $NH_4^+$, are both soluble in water and could easily be excreted in the urine. Why does the body convert them to urea rather than excreting them directly?

**21.35** (a) What is the toxic product of the oxidative deamination of glutamate? (b) How does the body get rid of it?

**\*21.36** (a) How many moles of ATP are used up in the conversion of one mole of ammonia to urea? (b) How many moles of high-energy bonds are hydrolyzed in the same conversion?

**21.37** The metabolism of the carbon skeleton of tyrosine yields pyruvate. Why is tyrosine a glucogenic amino acid?

**\*21.38** If leucine is the only carbon source in the diet of an experimental animal, what would you expect to find in the urine of such an animal?

## Catabolism of Heme

**\*21.39** A high bilirubin content in the blood may indicate liver disease. Explain why.

**21.40** When hemoglobin is fully metabolized, what happens to the iron in it?

**21.41** Heme, biliverdin, and bilirubin are colored compounds. Is the color due to the iron in their structure?

## Boxes

**21.42** (Box 21A) In cases of fructose intolerance, what causes the accumulation of fructose 1-phosphate in the liver?

**21.43** (Box 21A) Why does hypoglycemia appear as a symptom in fructose intolerance?

**21.44** (Box 21B) How does coronary thrombosis precipitate cardiac arrest?

**21.45** (Box 21C) (a) What are the symptoms of Cori's disease? (b) What causes it? (c) Answer the same questions for McArdle's disease.

**21.46** (Box 21D) How does diphtheria toxin cause the accumulation of fat in the heart muscles?

**21.47** (Box 21E) What system counteracts the acidic effect of ketone bodies in the blood?

**21.48** (Box 21F) What causes the accumulation of $\gamma$-aminobutyric acid in the Chinese restaurant syndrome?

**21.49** (Box 21F) Draw the structure of monosodium glutamate.

**21.50** (Box 21G) How can you prevent mental retardation in infants once PKU has been diagnosed?

**\*21.51** (Box 21G) Draw structural formulas for each reaction component, and complete the equation.

$$\text{phenylalanine} \longrightarrow \text{phenylpyruvate} + \;?$$

**21.52** (Box 21H) How does infectious hepatitis cause jaundice?

## Additional Problems

**21.53** Write the products of the transamination reaction between alanine and oxaloacetate:

$$\text{CH}_3\!-\!\underset{\underset{\text{NH}_3^+}{|}}{\text{CH}}\!-\!\text{COO}^- + \underset{\underset{\underset{\text{COO}^-}{|}}{\overset{\displaystyle\text{CH}_2}{|}}}{\overset{\displaystyle\text{COO}^-}{\overset{|}{\text{C}}}}\!\!=\!\!\text{O} \longrightarrow$$

**\*21.54** What is the net energy yield in moles of ATP produced when yeast converts one mole of glucose to ethanol?

**\*21.55** Can the intake of alanine, glycine, and serine relieve hypoglycemia caused by starvation? Explain.

**\*21.56** Phosphoenolpyruvate (PEP) has a high-energy phosphate bond that has more energy than the anhydride bonds in ATP. What step in glycolysis suggests that this is so?

**\*21.57** Suppose that a fatty acid labeled with radioactive carbon-14 is fed to an experimental animal. Where would you look for the radioactivity?

**21.58** Ornithine is a basic amino acid similar to lysine. (a) What is the structural difference between the two? (b) Is ornithine a constituent of proteins?

**21.59** How is sucrose metabolized through glycolysis?

**21.60** If you received a laboratory report showing the presence of a high concentration of ketone bodies in the urine of a patient, what disease would you suspect?

**21.61** Are ketone bodies produced in normal metabolism?

**21.62** What compound is common to both the urea and citric acid cycles?

**\*21.63** (a) At which step of the glycolysis pathway does NAD$^+$ participate (Figs. 21.3 and 21.4)? (b) At which step does NADH + H$^+$ participate? (c) As a result of the overall pathway, is there a net increase of NAD$^+$, of NADH + H$^+$, or of neither?

# C H A P T E R   2 2

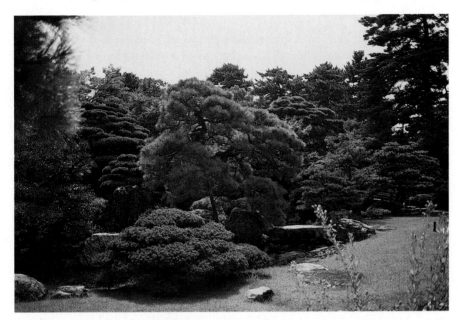

*Photosynthesis is the process responsible for the growth of all green plants. (Photograph by Beverly March.)*

# Biosynthetic Pathways

## 22.1  INTRODUCTION

In the human body, and in most other living tissues, the pathways by which a compound is synthesized (anabolism) are usually different from the pathways by which it is degraded (catabolism). There are several reasons why it is biologically advantageous for anabolic and catabolic pathways to be different. We will give two of them:

Anabolic pathways are also called **biosynthetic pathways.**

**1.  Flexibility**  If the normal biosynthetic pathway is blocked, the body can often use the reverse of the degradation pathway instead (remember that most steps in degradation are reversible), thus providing another way to make the necessary compounds.

**2.  Overcoming the effect of Le Chatelier's principle**  This can be illustrated by the cleavage of a glucose unit from a glycogen molecule, an equilibrium process:

$$(\text{Glucose})_n + \text{P}_i \underset{\phantom{phosphorylase}}{\overset{\text{phosphorylase}}{\rightleftharpoons}} (\text{glucose})_{n-1} + \text{glucose 1-phosphate} \qquad (22.1)$$

Glycogen                          Glycogen
                                     (one unit smaller)

Phosphorylase catalyzes not only glycogen degradation (the forward reaction) but also glycogen synthesis (the reverse reaction). However, in the body there is a large excess of inorganic phosphate, $\text{P}_i$. This would drive the reaction, on the

The structure of UDP glucose is shown in Section 22.2.

basis of Le Chatelier's principle, to the right, which represents glycogen degradation. In order to provide a method for the *synthesis* of glycogen even in the presence of excess inorganic phosphate, a different pathway is needed in which $P_i$ is not a reactant. Thus the body uses the following synthetic pathway:

$$(\text{Glucose})_{n-1} + \text{UDP glucose} \longrightarrow \text{UDP} + (\text{glucose})_n \qquad (22.2)$$

Glycogen    Glycogen
(one unit larger)

Not only are the synthetic pathways different from the catabolic pathways, but the energy requirements are also different. We shall not go into the energy balances of the biosynthetic processes as we did for catabolism. However, it must be kept in mind that, while energy (in the form of ATP) is *obtained* in the degradative processes, biosynthetic processes *consume* energy.

## 22.2  BIOSYNTHESIS OF CARBOHYDRATES

We discuss the biosynthesis of carbohydrates under three headings: (a) conversion of atmospheric $CO_2$ to glucose in plants, (b) synthesis of glucose in animals and humans, and (c) conversion of glucose to other carbohydrate molecules in animals and humans.

### Conversion of Atmospheric $CO_2$

The most important biosynthesis of carbohydrates takes place in plants. This is **photosynthesis.** In this process the energy of the sun is built into chemical bonds; the overall reaction is

$$n\text{H}_2\text{O} + n\text{CO}_2 \xrightarrow{\substack{\text{energy in} \\ \text{the form} \\ \text{of sunlight}}} (\text{CH}_2\text{O})_n + n\text{O}_2 \qquad (22.3)$$

where $(\text{CH}_2\text{O})_n$ is a general formula for carbohydrates. This is a very complicated process and takes place only in plants, not in animals (Box 22A). We shall not discuss it further here except to note that the carbohydrates of plants— starch, cellulose, and other mono- and polysaccharides—serve as the basic carbohydrate supply of all animals, including humans.

### Synthesis of Glucose

Gluconeogenesis means synthesis of "new" glucose.

We saw in Chapter 21 that when the body needs energy, carbohydrates are broken down via glycolysis. When energy is not needed, glucose can be synthesized from the intermediates of the glycolytic and citric acid pathways. This process is called **gluconeogenesis.** As shown in Figure 22.1, a large number of intermediates—pyruvate, lactate, oxaloacetate, malate, and several amino acids (the glucogenic amino acids we met in Sec. 21.9)—can serve as starting compounds. Gluconeogenesis proceeds in reverse order from glycolysis, and many

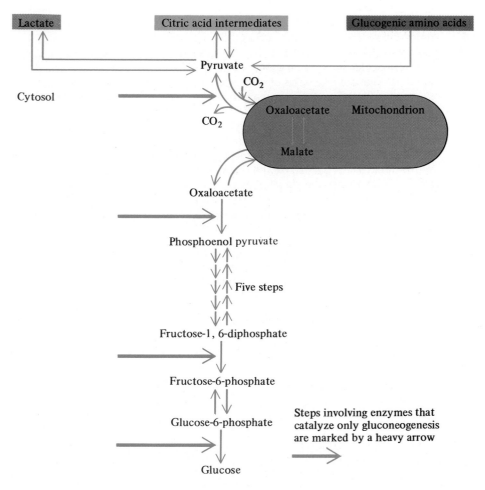

**Figure 22.1**
Gluconeogenesis. All reactions take place in the cytosol, except for those shown in the mitochondria.

of the enzymes of glycolysis also catalyze gluconeogenesis. However, at four points there are unique enzymes (marked in Fig. 22.1) that catalyze only gluconeogenesis and not the breakdown reactions. These four enzymes make gluconeogenesis a pathway that is distinct from glycolysis.

ATP is used up in gluconeogenesis and produced in glycolysis.

## Conversion of Glucose to Other Carbohydrates

The third important biosynthetic pathway for carbohydrates is the conversion of glucose to other hexoses and to hexose derivatives and the synthesis of di-, oligo-, and polysaccharides. The common step in all of these is the activation of glucose by uridine diphosphate (UDP) to form UDP glucose:

## Photosynthesis

Photosynthesis requires sunlight, water, $CO_2$, and pigments found in plants, mainly chlorophyll. The overall reaction shown in Equation 22.3 in the text actually occurs in two distinct steps. First, the light interacts with the pigments that are located in highly membranous organelles of plants, called **chloroplasts.** Chloroplasts resemble mitochondria (Sec. 20.2) in many respects: They contain a whole chain of oxidation–reduction enzymes similar to the cytochrome and iron–sulfur complexes of mitochondrial membranes, and they also contain a proton translocating ATPase. In a manner similar to mitochondria, in chloroplasts too the proton gradient accumulated in the intermembrane region drives the synthesis of ATP (see the chemiosmotic pump, Sec. 20.6). The chlorophylls themselves, buried in a complex protein that traverses the chloroplast membranes, are molecules similar to the heme we have already encountered in hemoglobin (Box 3C). In contrast to heme, the chlorophylls contain $Mg^{2+}$ instead of $Fe^{2+}$:

Chlorophyll a

The first reaction that takes place in photosynthesis is called the light reaction because chlorophyll captures the energy of sunlight and with its aid strips the electrons and protons from water to form oxygen, ATP, and NADPH + H$^+$ (See Fig. 22.3):

$$H_2O + ADP + P_i + NADP^+ + \text{sunlight} \longrightarrow O_2 + ATP + NADPH + H^+$$

The second reaction, called the dark reaction because it does not need light, in essence converts $CO_2$ to carbohydrates:

$$CO_2 + ATP + NADPH + H^+ \longrightarrow \text{carbohydrates} + ADP + P_i + NADP^+$$

In this step the energy, now in the form of ATP, is used to help NADPH + H$^+$ reduce carbon dioxide to carbohydrates. Thus, the protons and electrons stripped in the light reaction are added to the carbon dioxide in the dark reaction. This reduction takes place in a multistep cyclic process called the **Calvin cycle,** named after its discoverer, Melvin Calvin, who was awarded the 1961 Nobel Prize in Chemistry for this work. In this cycle the $CO_2$ is first attached to a $C_5$ fragment, making a $C_6$ fragment, which breaks down to two $C_3$ fragments (triose phosphates) that are eventually, through a complex series of steps, converted to a $C_6$ compound (glucose).

UDP glucose

UDP is similar to ADP except that the base is uracil instead of adenine. UTP, an analog of ATP, contains two high-energy phosphate bonds. For example, when the body has excess glucose and wants to store it as glycogen (this process is called **glycogenesis**), the glucose is first converted to glucose 1-phosphate, but then a special enzyme catalyzes the reaction:

The biosynthesis of many other di- and polysaccharides and their derivatives also uses the common activation step: forming the appropriate UDP compound.

## 22.3 BIOSYNTHESIS OF FATTY ACIDS

The body can synthesize all the fatty acids it needs except for linoleic and linolenic acids (essential fatty acids, Sec. 17.3). The source of carbon in this synthesis is acetyl CoA. Since acetyl CoA is also a degradation product of the $\beta$-oxidation spiral of fatty acids (Sec. 21.5), we might expect that the synthesis is the reverse of the degradation. This is not the case. For one thing, fatty acid synthesis occurs in the cytosol, while degradation takes place in the mitochondria. Fatty acid synthesis is catalyzed by a multienzyme system.

However, there is one aspect of fatty acid synthesis that is the same as in fatty acid degradation: Both processes involve acetyl CoA, and therefore both proceed in units of two carbons. Fatty acids are built up two carbons at a time, just as they are broken down two carbons at a time (Sec. 21.5).

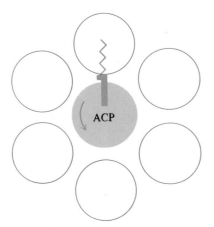

**Figure 22.2**
The biosynthesis of fatty acids. The ACP (central dark sphere) has a long side chain (━■) that carries the growing fatty acid (⌇⌇). The ACP rotates counterclockwise, and its side chain sweeps over a multienzyme system (empty spheres). As each cycle is completed, a $C_2$ fragment is added to the growing fatty acid chain.

Most of the time, fatty acids are synthesized when excess food is available. That is, when we eat more carbohydrates than we need for energy, our bodies turn the excess acetyl CoA (produced by catabolism of carbohydrates—Sec. 21.2) into fatty acids and then to fats. The fats are stored in the fat depots, which are specialized fat-carrying cells (see Fig. 21.1).

The key to fatty acid synthesis is an **a**cyl **c**arrier **p**rotein called **ACP.** This can be looked upon as a merry-go-round—a rotating protein molecule to which the growing chain of fatty acids is attached. As the growing chain rotates with the ACP, it sweeps over the multienzyme complex, and at each enzyme one reaction of the chain is catalyzed (Fig. 22.2).

At the beginning of this cycle, the ACP picks up an acetyl group from acetyl CoA and delivers it to the first enzyme, called synthase:

$$\underset{\text{O}}{\overset{\text{O}}{CH_3\overset{\parallel}{C}-S-CoA}} + HS-ACP \longrightarrow HS-CoA + CH_3\overset{\overset{\text{O}}{\parallel}}{C}-S-ACP$$

$$CH_3\overset{\overset{\text{O}}{\parallel}}{C}-S-ACP + synthase-SH \longrightarrow$$

$$CH_3\overset{\overset{\text{O}}{\parallel}}{C}-S-synthase + HS-ACP$$

The $C_2$ fragment on the synthase is condensed with a $C_3$ fragment attached to the ACP in a process in which $CO_2$ is given off:

$$synthase-S-\overset{\overset{\text{O}}{\parallel}}{C}-CH_3 + ACP-S-\overset{\overset{\text{O}}{\parallel}}{C}-CH_2-COOH \rightleftharpoons$$
Malonyl ACP

$$ACP-S-\overset{\overset{\text{O}}{\parallel}}{C}-CH_2-\overset{\overset{\text{O}}{\parallel}}{C}-CH_3 + CO_2 + synthase-SH$$
Acetoacetyl ACP

The result is a $C_4$ fragment, which is reduced twice and dehydrated before it becomes a fully saturated $C_4$ group. This is the end of one cycle of the merry-go-round.

In the next cycle, the $C_4$ fragment is transferred to the synthase, and another malonyl ACP ($C_3$ fragment) is added. When $CO_2$ splits out, a $C_6$ fragment is obtained. The merry-go-round continues to turn. At each turn, another $C_2$ fragment is added to the growing chain. Chains up to $C_{16}$ (palmitic acid) can be obtained in this process. If the body needs longer fatty acids — for example, stearic ($C_{18}$) — another $C_2$ fragment is added to palmitic acid by a different enzyme system.

Unsaturated fatty acids are obtained from saturated fatty acids by an oxidation step in which hydrogen is removed and combined with $O_2$ to form water:

$$R-\overset{\overset{\displaystyle H}{|}}{\underset{\underset{\displaystyle H}{|}}{C}}-\overset{\overset{\displaystyle H}{|}}{\underset{\underset{\displaystyle H}{|}}{C}}-(CH_2)_n COOH + O_2 + \boxed{NADPH} + H^+ \xrightarrow{\text{enzyme}}$$

$$\boxed{NADP^+} + 2H_2O + R-\overset{\overset{\displaystyle H}{|}}{C}=\overset{\overset{\displaystyle H}{|}}{C}-(CH_2)_n-COOH$$

The structure of $NADP^+$ is the same as that of $NAD^+$, except that an additional phosphate group is attached to one of the ribose units (Fig. 22.3).

## 22.4 BIOSYNTHESIS OF MEMBRANE LIPIDS

The different membrane lipids (Secs. 17.6 to 17.8) are assembled from their constituents. We just saw how fatty acids are synthesized in the body. These fatty acids are activated by CoA, forming acyl CoA. The compound glycerol 3-phosphate, which is obtained from the reduction of dihydroxyacetone (a $C_3$ fragment of glycolysis, Fig. 21.4), is the second building block of phosphoglycerides. This compound combines with two acyl CoA molecules, which can be the same or different:

Glycerol 3-phosphate     Acyl CoA     Phosphatidate

To complete the molecule, an activated serine or an activated choline is added to the $-OPO_3^{2-}$ group (see the structures in Sec. 17.6; a model of phosphatidylcholine is shown in Fig. 22.4).

**Figure 22.3**
The structure of $NADP^+$.

We encountered glycerol 3-phosphate as a vehicle for transporting electrons in and out of mitochondria (Sec. 21.3).

**Figure 22.4**
A model of phosphatidylcholine, commonly called lecithin.

Sphingolipids (Sec. 17.7) are similarly built up from smaller molecules. The alcohol sphingosine is made by oxidation of the product formed when serine is condensed with an acyl CoA in which the acyl group is palmitic acid:

$$CH_3(CH_2)_{14}\overset{\overset{O}{\|}}{C}-S-CoA + \underset{\underset{COO^-}{|}}{\overset{\overset{CH_2OH}{|}}{HC}-NH_3{}^+} \xrightarrow{H^+}$$

Palmitoyl CoA  Serine

$$CO_2 + CoA-SH + \underset{\underset{CH_3}{|}}{\overset{\overset{CH_2OH}{|}}{\underset{(CH_2)_{14}}{\overset{HC-NH_3{}^+}{|}}}}HC-OH \xrightarrow{[O]} \underset{\underset{CH_3}{|}}{\overset{\overset{CH_2OH}{|}}{\underset{(CH_2)_{12}}{\overset{\overset{HC-NH_3{}^+}{|}}{\underset{CH}{\overset{HC-OH}{\underset{\|}{CH}}}}}}} + H_2O$$

Sphingosine

---

Hyaline membrane disease, which affects the breathing of infants and has caused numerous sudden infant deaths, is not completely understood at this time. Analysis has shown that there is a great difference in the phospholipid composition of the lung between normal and affected individuals. The phospholipid of the lung is mainly phosphatidylcholine, which in normal lungs contains large quantities of unsaturated fatty acids. In contrast, the phosphatidylcholine in the lungs of patients with hyaline membrane disease is largely saturated. Furthermore, these patients have less phosphatidylcholine in their lung tissue than normal individuals. This indicates a faulty biosynthetic apparatus, but the exact cause has not yet been determined.

**Figure 22.5**
Biosynthesis of cholesterol. The circled carbon atoms come from the —CH$_3$ group, and the others come from the —COOH group of the acetate.

Another fatty acid is added to the amine group by an acyl CoA to form *N*-acyl-sphingosine (ceramide). An activated phosphocholine is added to this alcohol to make sphingomyelin.

The glycolipids are made in a similar fashion. Ceramide is assembled as above, and the carbohydrate is added one unit at a time in the form of activated monosaccharides (UDP glucose and so on).

Cholesterol, the molecule that controls the fluidity of membranes and is a precursor of all steroid hormones, is also synthesized by the human body. It is assembled in the liver from C$_2$ fragments in the form of acetyl CoA. All the carbon atoms of cholesterol come from the carbons of acetyl CoA molecules (Fig. 22.5).

Each constituent is added in its activated form (that is, combined with CoA).

## 22.5 BIOSYNTHESIS OF AMINO ACIDS

The human body needs 20 different amino acids to make its protein chains—all 20 are found in a normal diet. Some of the amino acids can be synthesized from other compounds: These are the nonessential amino acids. Others cannot be synthesized by the human body and must be supplied in the diet. These are the **essential amino acids** (see Chap. 26). Most nonessential amino acids are syn-

Table 18.1 shows the essential and nonessential amino acids.

---

**Box 22C**

## Amino Acid Transport and Blue Diaper Syndrome

Amino acids, both essential and nonessential, are usually more concentrated inside the cells than in the surroundings. Because of this, special transport mechanisms are available to carry the amino acids into the cells. Sometimes these transport mechanisms are faulty, and the result is a lack of an amino acid inside the cells despite the fact that the particular amino acid is being provided by the diet.

An interesting example of this is blue diaper syndrome, in which infants excrete blue urine. Indigo blue, a dye, is an oxidation product of the amino acid tryptophan. But how does this oxidation product get into the system to be excreted in the urine? The patient's tryptophan transport mechanism is faulty. Although enough tryptophan is being supplied by the diet, most of it is not absorbed through the intestine. It accumulates there and is oxidized by bacteria in the gut. The oxidation product is moved into the cells, but because the cells cannot use it, it is excreted in the urine. Much of the blue diaper syndrome can be eliminated with antibiotics. This treatment kills many of the gut's bacteria, and the nonabsorbed tryptophan is then excreted in the feces because the bacteria cannot now oxidize it to indigo blue.

thesized from some intermediate of either glycolysis (Sec. 21.2) or the citric acid cycle (Sec. 20.4). Glutamic acid plays a central role in the synthesis of five nonessential amino acids. Glutamic acid itself is synthesized from $\alpha$-ketoglutaric acid, one of the intermediates in the citric acid cycle:

$$\boxed{NADH} + H^+ + NH_4^+ + \begin{array}{c} COO^- \\ | \\ C=O \\ | \\ CH_2 \\ | \\ CH_2 \\ | \\ COO^- \end{array} \rightleftharpoons \begin{array}{c} COO^- \\ | \\ HC-NH_3^+ \\ | \\ CH_2 \\ | \\ CH_2 \\ | \\ COO^- \end{array} + \boxed{NAD^+} + H_2O$$

$\alpha$-Ketoglutaric acid     Glutamic acid

The forward reaction is the synthesis, and the reverse reaction is the oxidative deamination (degradation) reaction we encountered in the catabolism of amino acids (Sec. 21.8). This is one case in which the synthetic and degradative pathways are exactly the reverse of each other.

Glutamic acid can serve as an intermediate in the synthesis of alanine, serine, aspartic acid, asparagine, and glutamine. For example, the transamination reaction we saw in Section 21.8 leads to alanine formation:

$$\begin{array}{c} CH_3 \\ | \\ C=O \\ | \\ COO^- \end{array} + \begin{array}{c} COO^- \\ | \\ HC-NH_3^+ \\ | \\ CH_2 \\ | \\ CH_2 \\ | \\ COO^- \end{array} \rightleftharpoons \begin{array}{c} CH_3 \\ | \\ HC-NH_3^+ \\ | \\ COO^- \end{array} + \begin{array}{c} COO^- \\ | \\ C=O \\ | \\ CH_2 \\ | \\ CH_2 \\ | \\ COO^- \end{array}$$

Pyruvate     Glutamate     Alanine     $\alpha$-Ketoglutarate

Besides being the building blocks of proteins, amino acids also serve as intermediates for a large number of biological molecules. We have already seen that serine is needed in the synthesis of membrane lipids (Sec. 22.4). Certain amino acids are also intermediates in the synthesis of heme and of the purines and pyrimidines that are the raw materials for DNA and RNA (Chap. 23).

## SUMMARY

For most biochemical compounds, the biosynthetic pathways are different from the degradation pathways. Carbohydrates are synthesized in plants from $CO_2$ and $H_2O$, using sunlight as an energy source (**photosynthesis**). Glucose can be synthesized by animals from the intermediates of glycolysis, from those of the citric acid cycle, and from glucogenic amino acids. This process is called **gluconeogenesis.** When glucose or other monosaccharides are built into di-, oligo-, and polysaccharides, each monosaccharide unit in its activated form is added to a growing chain.

Fatty acid biosynthesis is accomplished by a multienzyme system. The key to this process is the **acyl carrier protein, ACP,** which acts as a merry-go-round transport system; it carries the growing fatty acid chain over a number of enzymes, each of which catalyzes a specific reaction. With each complete turn of the merry-go-round, a $C_2$ fragment is added to the growing fatty acid chain. The source of the $C_2$ fragment is malonyl ACP, a $C_3$ compound attached to the ACP. This becomes $C_2$ by loss of $CO_2$. Membrane lipids are synthesized in the body by assembling the constituent parts. The fatty acids are activated by conversion to acyl CoA.

Many nonessential amino acids are synthesized in the body from the intermediates of glycolysis or of the citric acid cycle. In half of these, glutamic acid is the donor of the amino group in transamination. Amino acids serve as building blocks for proteins.

## KEY TERMS

Acyl carrier protein (Sec. 22.3)
Biosynthetic pathway (Sec. 22.1)
Chloroplasts (Box 22A)
Essential amino acid (Sec. 22.5)

Gluconeogenesis (Sec. 22.2)
Glycogenesis (Sec. 22.2)
Photosynthesis (Sec. 22.2)

## PROBLEMS

### Biosynthesis

**22.1** Define anabolism.

**22.2** What compound provides the energy for biosynthesis?

**22.3** Glycogen can be synthesized in the body by the same enzymes that degrade it. Why is this process utilized in glycogen synthesis only to a small extent, while most glycogen biosynthesis occurs via a different synthetic pathway?

### Carbohydrate Biosynthesis

**22.4** Photosynthesis is the reverse of what process?

**22.5** In photosynthesis, what is the source of (a) carbon  (b) hydrogen  (c) energy

**22.6** What pathway is the reverse of gluconeogenesis?

**22.7** Name a compound that can serve as a raw material for gluconeogenesis and is
(a) from the glycolytic pathway
(b) from the citric acid cycle
(c) an amino acid

**22.8** How does the structure of UDP differ from that of ADP?

**22.9** Which steps in gluconeogenesis are not reversible?

**22.10** Are the enzymes that combine two $C_3$ compounds to a $C_6$ compound in gluconeogenesis the same as or different from those that cut the $C_6$ compound into two $C_3$ compounds in glycolysis?

**22.11** Which part of gluconeogenesis occurs in the mitochondria?

**22.12** Glycogen is written as $(glucose)_n$. (a) What does the $n$ stand for?  (b) What is the value of $n$?

**22.13** What phosphate-containing products appear in glycogenesis?

### Fatty Acid Biosynthesis

**22.14** What is the source of carbon in fatty acid synthesis?

**22.15** (a) Where in the body does fatty acid synthesis occur?  (b) Does fatty acid degradation occur in the same location?

**22.16** When you eat more food than your body needs, excess acetyl CoA is synthesized by the pathways outlined in Chapter 21. What does the body do with this excess acetyl CoA?

**22.17** Is ACP itself an enzyme?

**\*22.18** (a) What is the name of the first enzyme in fatty acid synthesis?
(b) What does it do?

**22.19** From what compound is the $CO_2$ released in fatty acid synthesis?

**22.20** What is the functional group of the synthase to which the $C_2$ fragment is attached?

**\*22.21** In the synthesis of unsaturated fatty acids, NADPH + H$^+$ is converted to NADP$^+$. Yet this is an oxidation step and not a reduction step. Explain.

**22.22** Which of these fatty acids can be synthesized by the multienzyme fatty acid synthesis complex alone?
(a) oleic
(d) arachidonic
(b) stearic
(e) lauric
(c) myristic

**22.23** What is the difference between the structures of $NAD^+$ and $NADP^+$?

**\*22.24** Write equations to show how the body converts activated capric acid,

$$ACP—S—\overset{\displaystyle O}{\underset{\|}{C}}—(CH_2)_8CH_3,$$

to activated lauric acid,

$$ACP—S—\overset{\displaystyle O}{\underset{\|}{C}}—(CH_2)_{10}CH_3$$

**\*22.25** Linoleic and linolenic acids cannot be synthesized in the human body. Does this mean that the human body cannot make an unsaturated fatty acid from a saturated one?

## Biosynthesis of Membrane Lipids

**22.26** When the body synthesizes this membrane lipid, from what ingredients does it assemble it?

$$H_2C—O—\overset{O}{\underset{\|}{C}}—(CH_2)_{14}CH_3$$
$$HC—O—\overset{O}{\underset{\|}{C}}—(CH_2)_{10}CH_3$$
$$H_2C—O—\overset{O}{\underset{\underset{O^-}{\|}}{P}}—O—CH_2—\overset{H}{\underset{\underset{NH_2}{|}}{C}}—COOH$$

**\*22.27** Name the activated constituents necessary to form the glycolipid glucoceramide.

**22.28** Would the cell membranes of a person on a diet containing no cholesterol stiffen?

## Biosynthesis of Amino Acids

**\*22.29** What compound reacts with glutamate in a transamination process to yield serine?

**22.30** What reaction is the reverse of the synthesis of glutamate from $\alpha$-ketoglutarate, ammonia, and $NADH + H^+$?

**\*22.31** Which amino acid will be synthesized by this process?

$$NADH + H^+ + NH_4^+ + COO^- \longrightarrow$$
$$\overset{\displaystyle |}{\underset{\displaystyle |}{C{=}O}}$$
$$\overset{\displaystyle |}{\underset{\displaystyle |}{CH_2}}$$
$$COO^-$$

**\*22.32** Draw the structure of the compound needed to synthesize asparagine from glutamate by transamination.

**22.33** Which amino acids are synthesized from glutamic acid?

## Boxes

**22.34** (Box 22A) What is the major difference in structure between the chlorophylls and heme?

**22.35** (Box 22A) What happens in the light reaction of photosynthesis?

**22.36** (Box 22C) (a) What compound produces the colored urine of blue diaper syndrome? (b) Where does this compound come from?

## Additional Problems

**22.37** Can the complex enzyme system participating in every fatty acid synthesis manufacture fatty acids of any length?

**22.38** When glutamate transaminates phenylpyruvate, what amino acid is produced?

$$\overset{\displaystyle COO^-}{\underset{\displaystyle COO^-}{\overset{\displaystyle |}{\underset{\displaystyle |}{\overset{\displaystyle HC—NH_3^+}{\overset{\displaystyle |}{\underset{\displaystyle |}{\overset{CH_2}{CH_2}}}}}}} \ + \ \overset{\displaystyle CH_2}{\underset{\displaystyle COO^-}{\overset{\displaystyle |}{\underset{\displaystyle |}{C{=}O}}}} \longrightarrow$$

**22.39** (a) What feature is common to the structures of ACP and CoA? (b) How do these compounds differ?

**22.40** Would a person on a completely cholesterol-free diet totally lack the cholesterol necessary to synthesize steroid hormones?

**\*22.41** Write an equation to show the formation of a ceramide from oleyl CoA and sphingosine.

**\*22.42** Each activation step in the synthesis of complex lipids occurs at the expense of one ATP molecule. How many ATP molecules are used up in the synthesis of one molecule of lecithin?

**\*22.43** What $C_3$ fragment carried by ACP is used in fatty acid synthesis?

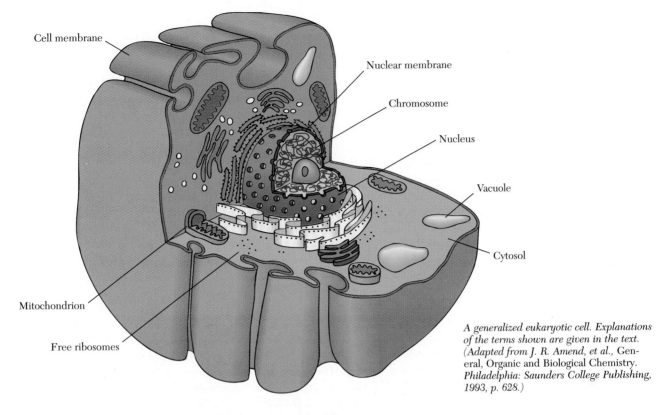

# CHAPTER 23

Cell membrane

Nuclear membrane

Chromosome

Nucleus

Vacuole

Cytosol

Mitochondrion

Free ribosomes

*A generalized eukaryotic cell. Explanations of the terms shown are given in the text. (Adapted from J. R. Amend, et al., General, Organic and Biological Chemistry. Philadelphia: Saunders College Publishing, 1993, p. 628.)*

# Nucleic Acids and Protein Synthesis

## 23.1 INTRODUCTION

Each cell of our bodies contains thousands of different protein molecules. We recall from Chapter 18 that all these molecules are made up of the same 20 amino acids, but in different sequences. A lobster has a different set of protein molecules than a mushroom or a zebra. Though the lobster and the zebra have some proteins in common, each has many protein molecules not found in the other. Within a species also, different individuals may have some differences in their proteins, though the differences are much less than those between species. This shows up graphically in cases where people have such conditions as hemophilia, albinism, or color-blindness, because they lack certain proteins that normal people have.

Once scientists understood this, the next question was, how do the cells know which proteins to synthesize out of the extremely large number of possible amino acid sequences? The answer to this question is that an individual gets

*There are two parents if reproduction is sexual; one parent if it is asexual.*

the information from its parents; this is called *heredity*. We all know that a pig gives birth to a pig and a mouse to a mouse. A pig does not give birth to a mouse. When an egg cell joins a sperm cell, a new cell called a *zygote* is produced. A pig zygote looks very much like a mouse zygote. Yet the pig zygote will produce the enzymes necessary to cause the zygote to grow into a pig and not a mouse.

It was easy to determine that the information is obtained from the parent or parents, but what form does this information take? During the last 40 years, revolutionary developments have led to an answer to this question—the transmission of heredity on the molecular level.

From about the end of the nineteenth century, biologists suspected that the transmission of hereditary information from one generation to another takes place in the *nucleus* of the cell. More precisely, it was felt that structures within the nucleus, called **chromosomes,** have something to do with heredity. Each species has a different number of chromosomes in the nucleus. The information that determines external characteristics (red hair, blue eyes) and internal characteristics (blood group, hereditary diseases) was thought to reside in **genes** located inside the chromosomes.

*Some simple organisms, such as bacteria, do not have a nucleus. Their chromosomes are condensed in a central region.*

Studies on the nuclei of fruit flies in the early 1930s revealed that the genes that carry the different traits lie in sequences along the chromosomes. Chemical analysis of nuclei showed that they are largely made up of special basic proteins called *histones* and a type of compound called **nucleic acids.** By 1940 it became clear through the work of Oswald Avery (1877–1955) that of all the material in the nucleus, only a nucleic acid called deoxyribonucleic acid (DNA) carries the hereditary information. That is, the genes are located in the DNA. Other work in the 1940s by George Beadle (1903–1989) and Edward Tatum (1909–1975) demonstrated that each gene controls the manufacture of one enzyme, and through this the external and internal characteristics are expressed. This came to be known as the **one-gene-one-enzyme hypothesis.** Thus, the expression of the gene (DNA) in terms of an enzyme (protein) led to the study of protein synthesis and its control. **The information that tells the cell which proteins to manufacture is carried in the molecules of DNA.**

In the following sections we provide some of the highlights of this rapidly developing field.

## 23.2  COMPONENTS OF NUCLEIC ACIDS

There are two kinds of nucleic acids found in cells—each has its own role in the transmission of hereditary information. The two types are **ribonucleic acid (RNA)** and **deoxyribonucleic acid (DNA).** As we just saw, DNA is present in the chromosomes of the nucleus. RNA is not found in the chromosomes. It is located elsewhere in the nucleus and even outside the nucleus, in the cytoplasm. As we will see in Section 23.5, there are three types of RNA, all with similar structures.

Both DNA and RNA are polymers. Just as proteins consist of chains of amino acids and polysaccharides of chains of monosaccharides, nucleic acids are also chains. The building blocks (monomers) of nucleic acid chains are **nucleotides.** Nucleotides themselves, however, are composed of three simpler units: a *base,* a *sugar,* and *phosphate.* We will look at each of these in turn.

## Bases

The bases found in DNA and RNA are chiefly those shown in Figure 23.1. All of them are basic because they are heterocyclic amines (Sec. 15.4). Two of these bases, adenine and guanine, are purines (they have the same ring system as the purines shown in Sec. 15.9), and the other three—cytosine, thymine, and uracil—are pyrimidines (Sec. 11.13). The two purines (A and G) and one of the pyrimidines (C) are found in both DNA and RNA, but uracil (U) is found *only* in RNA, while thymine (T) is found *only* in DNA. Note that thymine differs from uracil only in the methyl group in the 5 position. Thus, both DNA and RNA contain four bases: two pyrimidines and two purines. For DNA the bases are A, G, C, and T; for RNA the bases are A, G, C, and U.

The three pyrimidines and guanine are in their keto rather than their enol forms (see Sec. 13.6).

The initial letter of each base is used as an abbreviation for that base.

**Figure 23.1**
The five principal bases of DNA and RNA. Note how the rings are numbered. The hydrogens shown in blue are lost when the bases are connected to the sugars.

Purines

Adenine (A)          Guanine (G)

Pyrimidines

Cytosine (C)     Thymine (T)      Uracil (U)
                 (DNA only)       (RNA only)

## Sugars

The sugar component of RNA is D-ribose (Sec. 16.6). In DNA it is D-deoxyribose (hence the name deoxyribonucleic acid):

The only difference between these molecules is that ribose has an OH group in the 2 position not found in deoxyribose.

D-Ribose

D-Deoxyribose

The combination of sugar and base is known as a **nucleoside.** The purine bases are linked to the C-1 of the sugar through one of the nitrogens of the five-membered ring (the N connected to an H in color in Fig. 23.1):

Adenine

Ribose

$+ H_2O$

Adenosine

The nucleoside made of guanine and ribose is called guanosine. The names of the other nucleosides are given in Table 23.1.

The pyrimidine bases are linked to the C-1 of the sugar through their N-1 nitrogen:

Uridine

A major difference between cancer cells and most normal cells is that the cancer cells divide much more rapidly. Rapidly dividing cells require a constant new supply of DNA. One component of DNA is the nucleoside deoxythymidine, which is synthesized in the cell by the methylation of uridine. If fluorouracil is administered to a cancer patient as part of chemotherapy, the body converts it to fluorouridine, a compound that irreversibly inhibits the enzyme that manufactures thymidine from uridine, greatly decreasing DNA synthesis.

Fluorouracil

Since this affects the fast-dividing cancer cells more than the healthy cells, the growth of the tumor and the spread of the cancer are arrested. However, chemotherapy with fluorouracil or other anticancer drugs weakens the body because it also interferes with DNA synthesis in normal cells. Therefore, chemotherapy is used intermittently to give the body time to recover from the side effects of the drug. During the period after chemotherapy, special precautions must be taken so that bacterial infections do not debilitate the already weakened body.

**TABLE 23.1    The Names of the Eight Nucleosides and Eight Nucleotides in DNA and RNA**

| Base | Nucleoside | Nucleotide |
|------|-----------|-----------|
| | | **DNA** |
| Adenine (A) | Deoxyadenosine | Deoxyadenosine 5′-monophosphate (dAMP)[a] |
| Guanine (G) | Deoxyguanosine | Deoxyguanosine 5′-monophosphate (dGMP) |
| Thymine (T) | Deoxythymidine | Deoxythymidine 5′-monophosphate (dTMP) |
| Cytosine (C) | Deoxycytidine | Deoxycytidine 5′-monophosphate (dCMP) |
| | | **RNA** |
| Adenine (A) | Adenosine | Adenosine 5′-monophosphate (AMP) |
| Guanine (G) | Guanosine | Guanosine 5′-monophosphate (GMP) |
| Uracil (U) | Uridine | Uridine 5′-monophosphate (UMP) |
| Cytosine (C) | Cytidine | Cytidine 5′-monophosphate (CMP) |

[a] The d indicates that the sugar is *deoxyribose*.

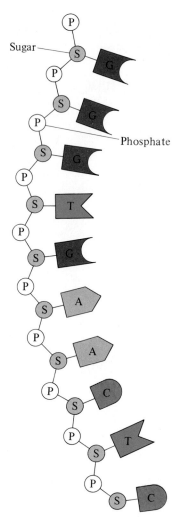

**Figure 23.2**
Schematic diagram of a nucleic acid molecule (DNA or RNA). The four bases of each nucleic acid are arranged in various specific sequences.

Primed numbers are used for the ribose and deoxyribose portions of nucleosides, nucleotides, and nucleic acids. Unprimed numbers are used for the bases.

### Phosphate

The third component of nucleic acids is phosphoric acid. When this group is linked to the —$CH_2OH$ group of a nucleoside, the result is a compound known as a **nucleotide.** For example, adenosine combines with phosphate to form the nucleotide adenosine monophosphate, AMP:

AMP

The names of the other nucleotides are given in Table 23.1. We are already familiar with some of these nucleotides. We have seen how they are part of the structure of key coenzymes, cofactors, and activators (Secs. 20.3 and 22.2). In Section 23.3 we will see how DNA and RNA are chains of nucleotides. In summary:

| | |
|---|---|
| A nucleoside | = base + sugar |
| A nucleotide | = base + sugar + phosphoric acid |
| A nucleic acid | = a chain of nucleotides |

## 23.3 STRUCTURE OF DNA AND RNA

In Chapter 18 we saw that proteins have primary, secondary, and higher structures. Nucleic acids, which are also chains of monomers, also have primary and secondary structures.

### Primary Structure

Nucleic acids are chains of nucleotides, as shown schematically in Figure 23.2. As the figure shows, the structure can be divided into two parts: (1) the backbone of the molecule and (2) the bases that are the side-chain groups. The backbone in DNA consists of alternating deoxyribose and phosphate groups. Each phosphate group is linked to the 3′ carbon of one deoxyribose unit and simultaneously to the 5′ carbon of the next deoxyribose unit (Fig. 23.3). Similarly, each sugar unit is linked to one phosphate at the 3′ position and to another at the 5′ position. The primary structure of RNA is the same except that each sugar is ribose (so there is an —OH group in the 2′ position) rather than deoxyribose.

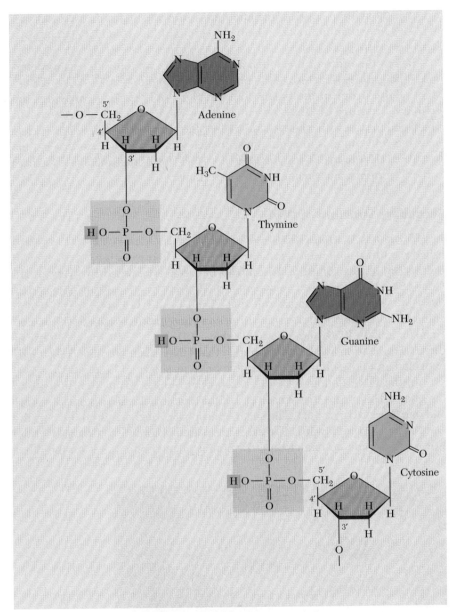

**Figure 23.3**
Primary structure of the DNA back-bone. The hydrogens shown in blue cause the acidity of nucleic acids. In the body, at neutral pH, the H's are replaced by $Na^+$ and $K^+$.

Thus, the backbone of the DNA and RNA chains has two ends: a 3' —OH end and a 5' —OH end. These two ends have roles similar to those of the C-terminal and N-terminal residues in proteins. This backbone provides the structural stability of the DNA and RNA molecules.

The bases that are linked, one to each sugar unit, are the side chains and carry all the information necessary for protein synthesis. Analysis of the base composition of DNA molecules from many different species was done by Erwin Chargaff (1905–    ), who showed that in DNA taken from many different species, the quantity of adenine (in moles) is always approximately equal to that of thymine, and the quantity of guanine is always approximately equal to that of cytosine, though the adenine/guanine ratio varies widely from species to species. This important information helped to establish the secondary structure of DNA, as we shall see below.

Just as the order of the amino acid residues of protein side chains determines the primary structure of the protein (for example, —Ala—Gly—Glu—Met—), **the order of the bases** (for example, —ATTGAC—) **provides the primary structure of DNA.** As with proteins, we need a convention to tell us which end to start with when we are writing the sequence of bases. For nucleic acids the convention is to begin the sequence with the nucleotide that has the free 5′—OH terminal. Thus, the sequence AGT means that adenine is the base at the 5′ terminal and thymine at the 3′ terminal.

The sequence TGA is not the same as AGT, but its opposite.

### Secondary Structure of DNA

In 1953 James Watson (1928–   ) and Francis Crick (1916–   ) established the three-dimensional structure of DNA. Their work is a cornerstone in the history of biochemistry. The model of DNA established by Watson and Crick (Fig. 23.4) was based on two important pieces of information obtained by other workers: (1) the Chargaff rule that (A and T) and (G and C) are present in equimolar quantities and (2) x-ray diffraction photographs obtained by Rosalind Franklin (1920–1958) and Maurice Wilkins (1916–   ). By the clever use of these facts, Watson and Crick concluded that DNA is composed of *two* strands entwined around each other in a double helix, as shown in Figure 23.5.

In the DNA double helix the two polynucleotide chains run in opposite directions. This means that at each end of the double helix there is one 5′—OH and one 3′—OH terminal. The sugar–phosphate backbone is on the outside, and the bases point inward. The bases are paired according to Chargaff's rule: For each adenine on one chain a thymine is aligned opposite it on the other

Watson, Crick, and Wilkins were awarded the 1962 Nobel Prize in Medicine for their discovery. Franklin had died in 1958.

Recall from Section 18.8 that a helix has a shape like a coiled spring or a spiral staircase. It was Pauling's discovery that human hair protein is helical that led Watson and Crick to look for helixes in DNA.

**Figure 23.4**
Watson and Crick with their model of the DNA molecule. (From M. M. Jones, et al., *Chemistry, Man, and Society,* 4th ed. Philadelphia: Saunders College Publishing, 1983, p. 256.)

**Figure 23.5**
Three-dimensional structure of the DNA double helix.

chain; each guanine on one chain has a cytosine aligned with it on the other chain. **The bases so paired form hydrogen bonds with each other, thereby stabilizing the double helix (Fig. 23.6).** They are called **complementary base pairs.**

The important thing here, as Watson and Crick realized, is that *only* adenine could fit with thymine and *only* guanine with cytosine. Let us consider the other possibilities. Can two purines (AA, GG, or AG) fit opposite each other? Figure 23.7 shows that they would overlap. How about two pyrimidines (TT, CC, or CT)? As shown in Figure 23.7, they would be too far apart. *There must be a pyrimidine opposite a purine.* But could A fit opposite C or G opposite T? Figure 23.8 shows that the hydrogen bonding would be much weaker.

The entire action of DNA—and of the heredity mechanism—depends on the fact that, **wherever there is an adenine on one strand of the helix, there must be a thymine on the other strand because that is the only base that fits and forms strong hydrogen bonds, and similarly for G and C.** The entire heredity mechanism rests on these slender hydrogen bonds (Fig. 23.6), as we shall see in Section 23.4.

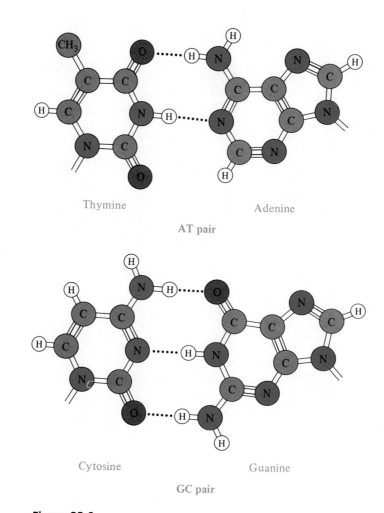

Thymine                                        Adenine

**AT pair**

Cytosine                                       Guanine

**GC pair**

**Figure 23.6**
A and T pair up by forming two hydrogen bonds; G and C, by forming three hydrogen bonds.

**Figure 23.7**
The bases of DNA cannot stack properly in the double helix if a purine is opposite a purine or a pyrimidine is opposite a pyrimidine.

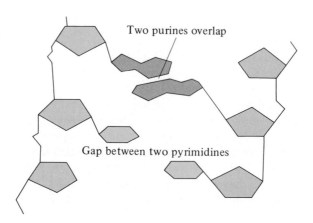

Two purines overlap

Gap between two pyrimidines

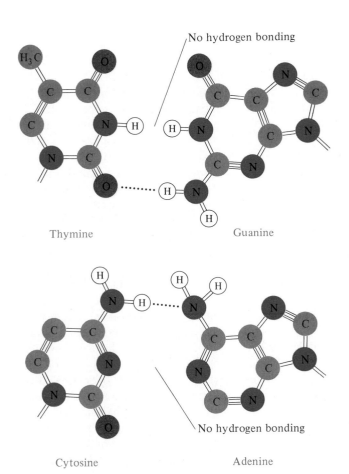

No hydrogen bonding

Thymine                    Guanine

Cytosine                   Adenine

No hydrogen bonding

**Figure 23.8**
Only one hydrogen bond is possible for GT or CA. These combinations are not
found in DNA. Compare this with Figure 23.6.

The beauty of establishing the three-dimensional structure of the DNA
molecule was that the knowledge of this structure immediately led to the explanation for the transmission of heredity: how the genes transmit traits from one
generation to another. Before we look at the mechanism of DNA replication (in
the next section), let us summarize the three differences in structure between
DNA and RNA:

1.   DNA has the four bases A, G, C, and T. RNA has three of these—A, G, and
C—but the fourth base is uracil, not thymine.
2.   In DNA the sugar is deoxyribose. In RNA it is ribose.
3.   DNA is almost always double-stranded, with the helical structure shown in
Figure 23.5. There are several kinds of RNA (as we shall see in Sec. 23.5); all of
them are single-stranded, though base-pairing can occur within a chain (see, for
example, Fig. 23.10). When it does, adenine pairs with uracil, since thymine is
not present.

## 23.4   DNA REPLICATION

The DNA in the chromosomes carries out two functions: (1) It reproduces itself, and (2) it supplies the information necessary to make all the proteins in the body, including enzymes. The second function is covered in Sections 23.6 to 23.10. Here we are concerned with the first, **replication.**

Each gene is a section of a DNA molecule that contains a specific sequence of the four bases A, G, T, and C, typically containing about 1000 to 2000 nucleotides. The base sequence of the gene carries the information necessary to produce one protein molecule. If the sequence is changed (for example, if one A is replaced by a G, or if an extra T is inserted), a different protein is produced instead, which might mean that the individual would have brown eyes instead of blue or perhaps would not have some vital metabolic protein such as insulin.

But consider the task that must be accomplished by the organism. When an individual is conceived, the egg and sperm cells unite to form the zygote. This cell, which is very tiny in most mammals, contains a small amount of DNA, but this DNA contains *all* the genetic information the individual will ever have. A fully grown large mammal, such as a human being or a horse, may contain more than a trillion cells. Each cell (except the egg and sperm cells) contains the same amount of DNA as the original single cell. Furthermore, cells are constantly dying and being replaced. Thus, there must be a mechanism by which DNA molecules can be copied (just as we can copy a letter on a photocopying machine) over and over again, millions of times, *without error.* In Section 23.13 we shall see that such errors sometimes do happen and can have serious consequences, but here we want to examine this remarkable mechanism that takes place every day in billions of organisms, from microbes to whales, and has been taking place for billions of years—with only a tiny percentage of errors.

The DNA double helix contains thousands of bases. One DNA strand may carry many inheritable genes, each of which is a stretch of DNA a few hundred or thousand bases long. Genetic information is transmitted from one cell to the next when cell division occurs. The two new cells carry all the information that the original cell possessed. Where originally there was one set of DNA molecules, there will now be two sets of DNA molecules, one set in each new cell.

The replication of DNA molecules starts with the unwinding of the double helix. This can occur at either end or in the middle. Special molecules called **unwinding proteins** attach themselves to one DNA strand (Fig. 23.9) and

The pigments responsible for the blue or brown color of eyes are synthesized with the help of specific enzymes. If one of these enzymes is lacking, the eye color may be different.

**Figure 23.9**
The replication of DNA. The two strands of the DNA double helix are shown separating. (Modified from P. W. Davis and E. P. Solomon, *The World of Biology.* Philadelphia: Saunders College Publishing, 1986.)

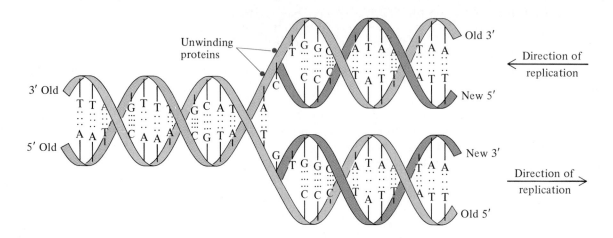

An interesting detail of DNA replication is that the two daughter strands are synthesized in different ways. One of the syntheses is continuous in the 5′ → 3′ direction, and the enzymes involved are capable of linking millions of ester linkages continuously in this direction. However, since the strands in the double helix run in opposite directions, only one strand of the double helix runs 5′ → 3′; the other runs 3′ → 5′. Along this second strand there is no possibility of continuous synthesis. What happens is that along the 3′ → 5′ strand, the enzymes can synthesize only short fragments because the only way they can work is from 5′ to 3′. These short fragments consist of about 1000 nucleotides each and are called **Okazaki fragments,** after their discoverer. These short fragments are near the replication fork of the open loop or bubble (Fig. 23B). They are eventually joined together by the enzyme DNA ligase. The two newly formed strands run in opposite directions. The new strand synthesized continuously is called the **leading strand;** the one assembled from Okazaki fragments is called the **lagging strand.**

**B o x   23B**

## Okazaki Fragments in DNA Replication

**Figure 23B**
How the two strands of DNA are replicated.

cause the separation of the double helix. All four kinds of free DNA nucleotide molecules are present in the vicinity. These nucleotides constantly move into the area and try to fit themselves into new chains. The key to the process is that, as we saw in Section 23.3, **only thymine can fit opposite adenine and only cytosine opposite guanine.** Wherever a cytosine, for example, is present on one of the strands of an unwound portion of the helix, all four nucleotides may approach, but three of them are turned away because they do not fit. Only the nucleotide of guanine fits.

While the bases of the newly arrived nucleotides are being hydrogen-bonded to their partners, enzymes called polymerases join the nucleotide backbones. At the end of the process, there are two double-stranded DNA molecules, each exactly the same as the original one because only T fits opposite A and only G against C. The process is called **semiconservative** because only two of the four strands are new; the other two were present in the original molecule.

If the unwinding begins in the middle, the synthesis of new DNA molecules on the old templates continues in both directions until the whole molecule is duplicated. This is the more common pathway. The unwinding can also start at one end and proceed in one direction until the whole double helix is unwound.

The base sequence of each newly synthesized DNA chain is **complementary** to the chain already there.

### 23.5   RNA

We previously noted that there are three types of RNA. These are

In DNA–RNA interactions, the complementary bases are

| DNA | RNA |
|---|---|
| A | U |
| G | C |
| C | G |
| T | A |

1.   **Messenger RNA (mRNA)** carries the genetic information from the DNA in the nucleus directly to the cytoplasm, where the protein is synthesized. It consists of a chain of nucleotides whose sequence is exactly complementary to that of one of the strands of the DNA. This type of RNA is not very stable. It is synthesized as needed and then degraded. Thus its concentration at any time is rather low.

2.   **Transfer RNA (tRNA)**   These molecules are relatively small, containing from 73 to 93 nucleotides per chain. There is at least one different tRNA molecule for each of the 20 amino acids from which the body makes its proteins. The tRNA molecules are L-shaped, but they can be represented as a cloverleaf in two dimensions. A typical one is shown in Figure 23.10. Transfer RNA

**Figure 23.10**

Structure of tRNA. (a) Two-dimensional simplified cloverleaf structure. (b) Three-dimensional structure. (From *Biochemistry,* 2nd ed., by Lubert Stryer. Copyright 1981 by W. H. Freeman and Co. All rights reserved. Part (b) also courtesy of Dr. Sung-Hou Kim.)

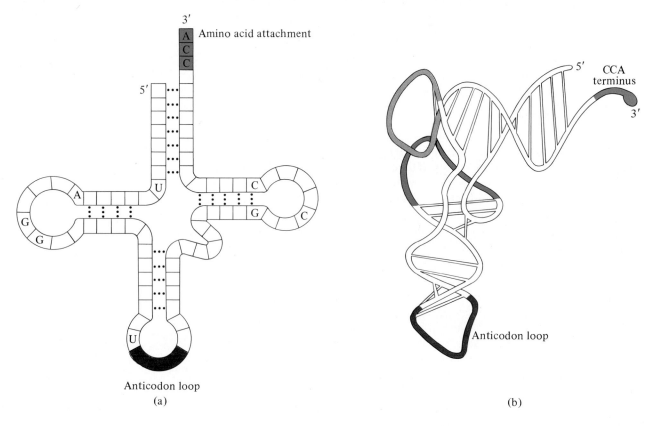

molecules contain not only cytosine, guanine, adenine, and uracil but also several other modified nucleotides.

**3.  Ribosomal RNA (rRNA)** is found in the **ribosomes,** which are small spherical bodies located in the cells but outside the nuclei. They consist of about 35 percent protein and 65 percent of a type of RNA called ribosomal RNA. These are large molecules with molecular weights up to 1 million. As we shall see in Section 23.10, protein synthesis takes place on the ribosomes.

## 23.6  TRANSMISSION OF INFORMATION

We have seen that the DNA molecule is a storehouse of information. We can compare it to a loose-leaf cookbook, each page of which contains one recipe. The pages are the genes. In order to prepare a meal, we use a number of recipes. Similarly, to provide a certain inheritable trait, a number of genes, segments of DNA, are needed.

However, the recipe itself is not the meal. The information in the recipe must be expressed in the proper combination of food ingredients. Similarly, the information stored in DNA must be expressed in the proper combination of amino acids representing a particular protein. The way this works is now so well established that it is called the **central dogma** of molecular biology. The dogma states that the **information contained in DNA molecules is transferred to RNA molecules, and then from the RNA molecules the information is expressed in the structure of proteins.** Two steps are involved.

**1.  Transcription**  Since the information (that is, the DNA) is in the nucleus of the cell and the amino acids are assembled outside the nucleus, the information must first be carried out of the nucleus. This is analogous to copying the recipe from the cookbook. All the necessary information is copied, though in a slightly different format, as if we were converting the printed page into handwriting. On the molecular level this is accomplished by transcribing the information from the DNA molecule onto a molecule of messenger RNA, so named because it carries the message from the nucleus to the site of protein synthesis. The transcribed information on the mRNA molecule is then carried out of the nucleus.

**2.  Translation**  The mRNA serves as a template on which the amino acids are assembled in the proper sequence. In order for this to happen, the information that is written in the language of nucleotides must be translated into the language of amino acids. The translation is done by the second type of RNA, transfer RNA. There is an exact word-to-word translation. Each amino acid in the protein language has a corresponding word in the RNA language. Each word in the RNA language is a sequence of three bases. This correspondence between three bases and one amino acid is called the **genetic code** (we will discuss the code in Sec. 23.9).

A summary of the process is

$$\text{DNA} \xrightarrow{\text{replication}} \text{DNA} \xrightarrow{\text{transcription}} \text{mRNA} \xrightarrow{\text{translation}} \text{protein}$$

**Box 23C**

## The Action of Antibiotics

An antibiotic is a drug manufactured by microorganisms (mold) that cures diseases either by killing bacteria or by stopping their reproduction (Box 19F). One way to do this is to interfere with the transcription process. For example, actinomycin D, an antibiotic produced by *Streptomyces* microorganisms, is a potent inhibitor of protein synthesis. Actinomycin works at the template level. It forms a strong complex with DNA and therefore prevents the transcription of a gene into RNA.

A group of bacterial antibiotics called rifamycins inhibits protein synthesis in a different manner. They do not interact with the DNA template, but inhibit the enzyme RNA polymerase that makes mRNA. Furthermore, rifamycins inhibit only bacterial enzymes and have little effect on the RNA polymerases of higher organisms. Thus rifamycin kills bacteria without interfering with human protein synthesis. A toxin of the poisonous mushroom (toadstool) *Amanita phalloides* has exactly the opposite effect. This toxin inhibits RNA polymerase of higher organisms but does not affect bacterial or plant RNA polymerase.

Still other antibiotics act by inhibiting the translation process. Among these are streptomycin and the related antibiotics neomycin and kanamycin. These molecules bind to the smaller ribosomal unit and thereby do not allow protein synthesis to begin.

It has been estimated that antibiotics have added almost ten years to the average life expectancy in the United States.

## 23.7   TRANSCRIPTION

The copying of the information (the recipe from the cookbook) is done with the help of an enzyme called RNA polymerase, which catalyzes the synthesis of mRNA. First the DNA double helix begins to unwind at a point near the gene that is to be transcribed. Only one strand of the DNA molecule is transcribed. Ribonucleotides assemble along the unwound DNA strand in the complementary sequence. Opposite each C on the DNA there is a G on the growing mRNA, and the other complementary bases follow the patterns $G \rightarrow C$, $A \rightarrow U$, and $T \rightarrow A$.

On the DNA strand, there is always a sequence of bases that the RNA polymerase recognizes as an **initiation signal,** saying, in essence, "Start here." At the end of the gene, there is a **termination sequence** that tells the enzyme, "Stop the synthesis." Between these two signals, the enzyme zips up the complementary bases by combining each ribose to the next phosphate. The enzyme synthesizes the mRNA molecule from the 5′ to the 3′ end (the zipper can move only in one direction). But because the complementary chains (RNA and DNA) run in opposite directions, the enzyme must move along the DNA template in the 3′ → 5′ direction of the DNA (Fig. 23.11). Once the mRNA molecule has been synthesized, it moves away from the DNA template, which then rewinds to the original double-helix form.

Transfer RNA and ribosomal RNA are also synthesized in this manner, on DNA templates.

These are not the deoxyribonucleotides used in DNA replication.

Note again that RNA contains no thymine but has uracil instead.

After RNA molecules are synthesized, they move out of the nucleus and into the cytoplasm.

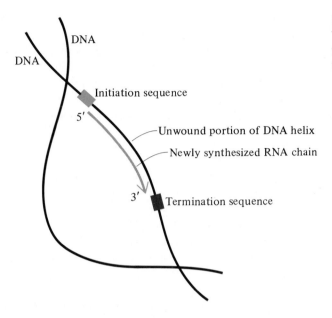

**Figure 23.11**
Transcription of a gene. The information in one DNA strand is transcribed to a strand of RNA.

## 23.8 THE ROLE OF RNA IN TRANSLATION

Translation is the process by which the genetic information preserved in the DNA and transcribed into the mRNA is converted to the language of proteins, that is, the amino acid sequence. All three types of RNA participate in the process.

The synthesis of proteins takes place on the ribosomes (Sec. 23.5). These spheres dissociate into two parts—a larger and a smaller body. Each of these bodies contains rRNA and some polypeptide chains that act as enzymes, speeding up the synthesis. In higher organisms, including humans, the larger ribosomal fragment is called the 60S ribosome and the smaller one the 40S ribosome. As the mRNA is being made on the DNA template (Sec. 23.7), the 5′ end of the mRNA, coming off this assembly, is first attached to the smaller ribosomal body and later joined by the larger body. Together they form a unit on which the mRNA is stretched out. Once the mRNA is attached to the ribosome in this way, the 20 amino acids are brought to the site, each carried by its own particular tRNA molecule.

The most important segments of the tRNA molecule are (1) the site to which enzymes attach the amino acids and (2) the recognition site. Figure 23.10 shows that the 3′ terminal of the tRNA molecule carries the amino acid. As we have said, each tRNA is specific for one amino acid only. How does the body make sure that, say, alanine attaches only to the one tRNA molecule that is specific for alanine? The answer is that each cell carries 20 specific enzymes for this purpose. Each of these enzymes recognizes only one amino acid and only one tRNA. The enzyme attaches the activated amino acid to the 3′ terminal of the tRNA.

The second important segment of the tRNA molecule carries the *recognition site,* which is a sequence of three bases called an **anticodon** (Fig. 23.10). This triplet of bases can align itself in a complementary fashion to another triplet on mRNA. The triplets of bases on the mRNA are called **codons.**

The S, or Svedberg unit, is a measure of the size of these bodies.

The 20 amino acids are always available in the cytoplasm, near the site of protein synthesis.

The recognition site is at the opposite end of the molecule in the three-dimensional structure of tRNA.

**Figure 23.12**
Marshall Nirenberg. (Courtesy of National Institutes of Health.)

## 23.9 THE GENETIC CODE

By 1961 it was apparent that the order of bases in a DNA molecule corresponds to the order of amino acids in a particular protein. But the code was unknown. Obviously, it could not be a one-to-one code. There are only four bases, so if, say, A coded for glycine, G for alanine, C for valine, and T for serine, there would be 16 amino acids that could not be coded.

In 1961 Marshall Nirenberg (1927–    ; Fig. 23.12) and his coworkers attempted to break the code in a very ingenious way. They made a synthetic molecule of mRNA consisting of uracil bases only. They put this into a cellular system that synthesized proteins, and then supplied the system with all 20 amino acids. The only polypeptide produced was a chain consisting solely of the amino acid phenylalanine. This showed that the code for phenylalanine must be UUU or some other multiple of U.

A series of similar experiments by Nirenberg and other workers followed, and by 1967 the entire genetic code had been broken. **Each amino acid is coded for by a sequence of three bases,** called a **codon.** The complete code is shown in Table 23.2.

The first important aspect of **the genetic code** is that it is **universal.** In every organism, from a bacterium to an elephant to a human, the same sequence of three bases codes for the same amino acid. The universality of the genetic code implies that all living matter on earth arose from the same primordial organisms. This is perhaps the strongest evidence for Darwin's theory of evolution.

There are 20 amino acids in proteins, but there are 64 possible combinations of four bases into triplets. All 64 codons (triplets) have been deciphered. Three of them, UAA, UAG, and UGA, are stop signs. They terminate protein synthesis. The remaining 61 codons all code for amino acids. Since there are only 20 amino acids, there must be more than one codon for each amino acid.

**TABLE 23.2   The Genetic Code**

| | | | | | | | |
|---|---|---|---|---|---|---|---|
| UUU | Phe | UCU | Ser | UAU | Tyr | UGU | Cys |
| UUC | Phe | UCC | Ser | UAC | Tyr | UGC | Cys |
| UUA | Leu | UCA | Ser | UAA | END[a] | UGA | END[a] |
| UUG | Leu | UCG | Ser | UAG | END[a] | UGG | Trp |
| CUU | Leu | CCU | Pro | CAU | His | CGU | Arg |
| CUC | Leu | CCC | Pro | CAC | His | CGC | Arg |
| CUA | Leu | CCA | Pro | CAA | Gln | CGA | Arg |
| CUG | Leu | CCG | Pro | CAG | Gln | CGG | Arg |
| AUU | Ile | ACU | Thr | AAU | Asn | AGU | Ser |
| AUC | Ile | ACC | Thr | AAC | Asn | AGC | Ser |
| AUA | Ile | ACA | Thr | AAA | Lys | AGA | Arg |
| AUG[b] | Met | ACG | Thr | AAG | Lys | AGG | Arg |
| GUU | Val | GCU | Ala | GAU | Asp | GGU | Gly |
| GUC | Val | GCC | Ala | GAC | Asp | GGC | Gly |
| GUA | Val | GCA | Ala | GAA | Glu | GGA | Gly |
| GUG | Val | GCG | Ala | GAG | Glu | GGG | Gly |

[a] END refers to signals indicating chain endings.
[b] This codon also signals the beginning of the chain.

Indeed, some amino acids are coded for by as many as six codons. Alanine, for example, is coded for by four: GCU, GCC, GCA, and GCG.

As there are three stop signs in the code, there is also an initiation sign. The initiation sign is AUG, which is also the codon for the amino acid methionine. This means that, in all protein synthesis, the first amino acid is always methionine. Methionine can also be put into the middle of the chain because there are two kinds of tRNA for it.

Although all protein synthesis starts with methionine, most proteins in the body do not have a methionine residue at the beginning of the chain. In most cases the initial methionine is removed by an enzyme before the polypeptide chain is completed. The code on the mRNA is always read in the $5' \rightarrow 3'$ direction, and the first amino acid to be linked to the initial methionine is the N-terminal end of the translated polypeptide chain.

Because of this multiple coding, the genetic code is called a **multiple** code or a **degenerate** code.

## 23.10 TRANSLATION AND PROTEIN SYNTHESIS

So far we have met the molecules that participate in protein synthesis (Sec. 23.8) and the dictionary of the translation, the genetic code. Now let us look at the actual mechanism by which the polypeptide chain is assembled.

There are four major stages in protein synthesis: activation, initiation, elongation, and termination.

All protein synthesis takes place outside the nucleus, in the cytosol.

### Activation

Each amino acid is first activated by reacting with a molecule of ATP:

In the activation step energy is used up: Two high-energy phosphate bonds are broken for each amino acid added to the chain: $ATP \rightarrow AMP + PP_i$ and $PP_i \rightarrow 2P_i$.

ATP                          Amino acid

Amino acid-AMP               Pyrophosphate

The activated amino acid is then attached to its own particular tRNA molecule with the aid of an enzyme (a synthetase) that is specific for that particular amino acid and that particular tRNA molecule:

Amino acid-AMP    +    tRNA    →    Amino acid-tRNA    +    AMP

**Figure 23.13**
The initiation of protein synthesis.
(a) The mRNA attaches to the 40S ribosomal body. (b) The first tRNA anticodon binds to the initiation mRNA codon. (c) The 60S ribosomal body joins the unit.

(a)                    (b)                    (c)

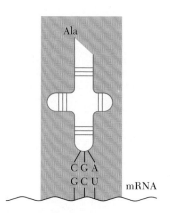

**Figure 23.14**
Alanine tRNA aligning on the ribosome with its complementary codon.

This enzyme, which is part of the 60S ribosome unit, is not a protein but a ribozyme.

## Initiation

This stage consists of three steps: (a) The mRNA molecule, which carries the information necessary to synthesize one protein molecule, attaches itself to the 40S ribosome (the smaller body; see Sec. 23.8). This is shown in Figure 23.13(a). (b) The anticodon of the first tRNA molecule (which is always a methionine tRNA) binds to the codon of the mRNA that represents the initiation signal. (c) The 60S ribosome (the larger portion) now combines with the 40S body, as shown in Figure 23.13(c). The 60S body carries two binding sites. The one shown on the left in Figure 23.13(c) is called the **P site** because that is where the growing peptide chain will bind. The one right next to it is called the **A site** because that is where the incoming tRNA will bring the next amino acid. When the 60S ribosome attaches itself to the 40S one, it does so in such a way that the P site is right where the methionine tRNA already is.

## Elongation

At this point the A site is vacant, and each of the 20 tRNA molecules can come in and try to fit itself in. *But only one of the 20 carries exactly the right anticodon that corresponds to the next codon on the mRNA.* (In Figures 23.14 and 23.15 we have made this an alanine tRNA.) The binding of this tRNA to the A site takes place with the aid of proteins called **elongation factors.** Energy for the process is obtained by hydrolyzing a guanosine triphosphate (GTT) to GDP and inorganic phosphate. Once at the A site, the new amino acid (Ala) is linked to the Met in a peptide bond by the enzyme **transferase.** The empty tRNA remains on the P site.

In the next phase of elongation, the whole ribosome moves one codon along the mRNA. Simultaneously with this move, the dipeptide is **translocated** from the A site to the P site, as shown in Figure 23.15(d), while the empty tRNA dissociates and goes back to the tRNA pool to pick up another amino acid. After the translocation, the A site is associated with the next codon on the mRNA, which is UUG in Figure 23.15(d). Once again, each tRNA can try to fit itself in, but only the one whose anticodon is AAC can align itself with UUG. This one, the tRNA that carries Leu, now comes in. The transferase now establishes a new peptide linkage between Leu and Ala, moving the dipeptide from the P site to the A site and forming a tripeptide. These elongation steps are repeated until the last amino acid is attached.

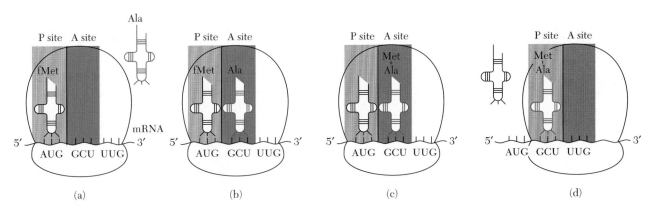

**Figure 23.15**
Phases of elongation in protein synthesis. (a) After initiation, the tRNA of the second amino acid (alanine in this case) approaches the ribosome. (b) The tRNA binds to the A site. (c) The enzyme transferase connects methionine to alanine, forming a peptide bond. (d) The peptide tRNA is moved (translocated) from the A to the P site while the ribosome moves to the right and simultaneously releases the empty tRNA. In this diagram, the alanine tRNA is shown in blue.

### Termination

After the last translocation, the next codon reads "STOP" (UAA, UGA, or UAG). No more amino acids can be added. Releasing factors then cleave the polypeptide chain from the last tRNA in a mechanism not yet fully understood. The tRNA itself is released from the P site, and the whole mRNA is released from the ribosome.

Their work done, the two parts of the ribosome separate.

## 23.11  GENES, EXONS, AND INTRONS

A **gene** is a stretch of DNA that carries one particular message; for example, "make a globin molecule." In bacteria this message is continuous. The series of codons (triplets of bases) spell out which amino acids must be assembled to make the globin molecule, and in what sequence. This series of codons lies between an initiation signal and a termination signal.

In higher organisms the message is not continuous. Stretches of DNA that spell out the amino acid sequence to be assembled are interrupted by long stretches that seemingly do not code for anything. The coding sequences are called **exons** and the noncoding sequences, **introns.** For example, the globin gene has three exons broken up by two introns. Since DNA contains exons and introns, the mRNA transcribed from it also contains exons and introns. The introns are cut out by enzymes (and the exons spliced together) before the mRNA is actually used to synthesize a protein. In other words, the introns function as spacers. Introns can also function as enzymes catalyzing the splicing of exons into "mature mRNA" (that is, with no intervening sequences). They can also catalyze the splicing of tRNAs from larger precursor molecules.

One DNA molecule may have between 1 million and 100 million bases. Therefore, there are many genes in one DNA molecule. If a DNA molecule were fully stretched out, its length would be perhaps 1 cm. However, the DNA

Philip A. Sharp (1944–    ) and Richard J. Roberts (1943–    ) were awarded the 1993 Nobel Prize in Medicine for the discovery of the noncoding nature of introns.

**Box 23D**

# Viruses

**Figure 23D.1**
The protein coat of a virus, in a computer graphics picture. (Courtesy of Drs. D. M. Salunke and D. L. D. Caspar, Brandeis University, and R. L. Garcea, Dana Farber Cancer Institute, Boston, Massachusetts.)

Nucleic acids are essential for life as we know it. No living thing can exist without them because they carry the information necessary to make protein molecules. The smallest form of life, the viruses, consist only of a molecule of nucleic acid surrounded by a ''coat'' of protein molecules (Fig. 23D.1). In some viruses the nucleic acid is DNA; in others it is RNA. No virus has both. The shapes and sizes of viruses vary greatly. Some of them are shown in Figure 23D.2.

Because their structures are so simple, viruses are unable to reproduce themselves in the absence of other organisms. They carry DNA or RNA but do not have the nucleotides, enzymes, amino acids, and other molecules necessary to replicate their nucleic acid (Sec. 23.4) or to synthesize proteins (Sec. 23.10). Instead, viruses invade the cells of other organisms and cause those cells (the hosts) to do these tasks for them. Typically, the protein coat of a virus remains outside the host cell, attached to the cell wall, while the DNA or RNA is pushed inside. Once the viral nucleic acid is inside the cell, the cell stops replicating its own DNA and making its own proteins and now replicates the viral nucleic acid and synthesizes the viral protein, according to the instructions on the viral nucleic acid. One cell can make many copies of the virus.

In many cases, the cell bursts when a large number of new viruses have been synthesized, sending the new viruses out into the intercellular material, where they can infect other cells. This kind of process causes the host organisms to get sick, perhaps to die. Among the many human diseases caused by viruses are measles, hepatitis, mumps, influenza, the common cold, rabies, and smallpox. There is no cure for most viral diseases. Antibiotics (Box 23C), which can kill bacteria, have no effect on viruses. So far, the best defense against these diseases has been immunization (Box 24H), which under the proper circumstances can work spectacularly well. Smallpox, once one of the most dreaded diseases, has been totally eradicated from this planet by many years of vaccination, and comprehensive programs of vaccination against such diseases as polio and measles have greatly reduced the incidence of these diseases.

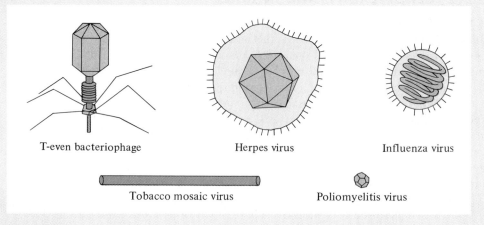

**Figure 23D.2**
The shape of various viruses. (From *The Structure of Viruses* by R. W. Horne. Copyright 1963 by Scientific American, Inc. All rights reserved.)

Lately, a number of antiviral agents have been developed that completely stop the reproduction of viral nucleic acids (DNA or RNA) inside infected cells without preventing the DNA of normal cells from replicating. One such drug is called vidarabine, or Ara-A, and is sold under the trade name Vira-A. Antiviral agents often act like the anticancer drugs (Box 23A), in that they have structures similar to one of the nucleotides necessary for the synthesis of nucleic acids. Vidarabine is the same as adenosine, except that the sugar is arabinose instead of ribose. Vidarabine is used to fight a life-threatening viral illness, herpes encephalitis. It is also effective in neonatal herpes infection and chicken pox. But as with many other anticancer and antiviral drugs, vidarabine is toxic, causing nausea and diarrhea. In some cases it has caused chromosomal damage.

Many other antiviral drugs are now in clinical trials. It is hoped that ways will be found to reduce their toxicity and improve their specificity so that they will act against one particular virus only.

molecules in the nuclei are not stretched out. They are coiled around basic protein molecules called **histones.** The acidic DNA and the basic histones attract each other by electrostatic (ionic) forces. The DNA and histones combine to form units called **nucleosomes.** A nucleosome is a core of eight histone molecules around which the DNA double helix is wrapped (Fig. 23.16).

Aaron Klug (1926–     ) won the 1982 Nobel Prize in Chemistry in part for elucidating the structure of nucleosomes.

**Figure 23.16**
Schematic diagram of a nucleosome. The band-like DNA double helix winds around cores consisting of eight histones. (Modified from P. W. Davis and E. P. Solomon, *The World of Biology.* Philadelphia: Saunders College Publishing, 1986.)

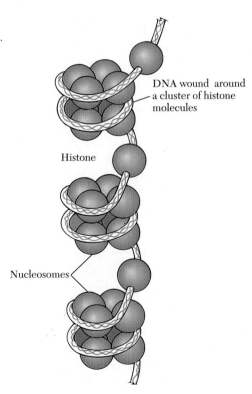

DNA wound around a cluster of histone molecules

Histone

Nucleosomes

## 23.12 GENE REGULATION

Every embryo that is formed by sexual reproduction inherits its genes from the parent sperm and egg cells. But the genes in its chromosomal DNA are not active all the time. They are switched on and off during development and growth of the organism. Soon after formation of the embryo the cells begin to differentiate. Some cells become neurons, some cells become muscle cells, some liver cells, and so on. Each cell is a specialized unit that uses only some of the many genes it carries in its DNA. This means that each cell must switch some of its genes on and off—either permanently or temporarily. How this is done is the subject of **gene regulation.** We know less about gene regulation in eukaryotes than in the simpler prokaryotes. Even with our limited knowledge, however, we can state that organisms do not have a single, unique way of controlling genes. Many gene regulations occur at the transcriptional level (DNA → RNA). Some operate at the translational level (mRNA → protein). Among these ways are the following:

The turning on, or activation, of a gene is called **gene expression.**

**Prokaryotes** are simple unicellular organisms with no true nucleus and no organelles surrounded by membranes (example: bacteria). **Eukaryotes** are organisms with cells that have visibly evident nuclei, each surrounded by a membrane, and also have organelles.

1.  **Operon, a case of negative regulation**   In the common species of bacteria *Escherichia coli,* the genes for a number of enzymes are organized in a unit. This unit contains DNA sequences that code for proteins and also sequences that play a part in gene regulation. Sequences that code for proteins are called **structural genes.** Preceding the structural genes is a DNA sequence called the **control sites,** made up of two parts called **operator site** and **promoter site.** Preceding the control sites is a DNA sequence called the **regulatory gene.** The structural genes, the control sites, and the regulatory gene together form a unit called the **operon.**

Since RNA synthesis proceeds in one direction (5′ → 3′) (see Fig. 23.11), the gene (DNA) to be transcribed runs from 3′ → 5′. Thus the control sites are in front of, or upstream of, the 3′ end of the structural gene.

One of the operons of *E. coli,* called the *lac* operon, contains three structural genes that code for three enzymes, called β-galactosidase, lactose permease, and thiogalactosidase transacetylase (Fig. 23.17). There is an allosteric protein, called a repressor protein, that prevents the transcription of the structural genes. This binds to the operator site of the control sites. The RNA polymerase that actually synthesizes the mRNA must be bound to the promoter site in order to transcribe the gene. When the repressor occupies the operator site, the RNA polymerase cannot bind to the promoter site. Thus, the gene will be inactive. This is negative regulation. However, when the proper inducer is present—in this case, lactose—it will form a complex with the repressor. As a consequence, the operator site will be empty, and the RNA polymerase can bind to the promoter site to perform the transcription.

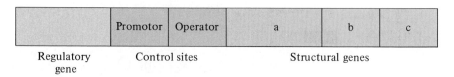

| | Promotor | Operator | a | b | c |
|---|---|---|---|---|---|
| Regulatory gene | Control sites | | Structural genes | | |

**Figure 23.17**
A schematic view of the *lac* operon in *E. coli.* The a, b, and c correspond, respectively, to the genes that code for the proteins β-galactosidase, lactose permease, and thiogalactosidase transacetylase.

**2. The epigenetic process in eukaryotes** During development of the embryo, some genes are switched on while others are switched off in a predestined program. This allows the embryonic cells to differentiate. The process that controls the developmental program is called the **epigenetic process.** In multicellular organisms, besides the switching on and off of the genes, there must be a process that insures that the daughter cells will also inherit the capability for switching genes on and off. Many of the cytosines in the DNA of higher organisms (eukaryotes) are methylated on C-5. Of special importance is the sequence CG, which in mammals is 70 percent methylated. Since base-pairing (see Fig. 23.6) requires that a C be opposite a G when DNA is duplicated, opposite every CG sequence on the parent DNA will be a GC sequence on the daughter DNA. There are enzymes that recognize the methylated CG on the parent strand and then methylate the GC on the daughter strand. In this way the particular methylation pattern is transmitted to the daughter cells.

Chemical evidence indicates that methylation may interfere with the binding of certain regulatory proteins to the DNA surface. These regulatory proteins are needed to activate the RNA polymerase. Thus, genes that have a high degree of methylation are inactive and prevent the transcription of the structural gene. Demethylation, on the other hand, is associated with the activation of the gene.

**3. Human genes** Humans have about a thousand times as much DNA as bacteria such as *E. coli*. However, humans have only about 50 times as many genes as the bacteria have. Only 2 to 4 percent of human DNA codes for proteins. Other parts of the DNA code for RNAs, play a part in regulatory systems, and act as special structural DNAs during cell division. However, the function of the greatest part of the DNA (about 3 billion base pairs) is at present unknown.

Many of the coding sequences in eukaryotes are repetitious. Many genes that code for rRNA are repeated more than 100 times in clusters such as

ABC————ABC————ABC————ABC————

The clusters are separated by base sequences that are not transcribed and seem to act simply as spacers. Histone genes are also clustered and repeated 300 to 1000 times. These repeated sequences enhance the rapid synthesis of rRNA and histones. In contrast, many genes of enzymes and structural proteins have only a few copies.

In eukaryotes, including humans, there are specific proteins, called **transcription factors,** that regulate gene expression by positive enhancement. One such protein, called transcription factor **IIIA,** or **TFIIIA,** has many finger-shaped areas (at least 9 or 10) that project and bind to specific groups present in the double-stranded DNA that is to be transcribed. These finger shapes are created by zinc ions, which form coordinate covalent bonds with amino acid side chains of the protein (Fig. 23.18). Therefore they are called **metal-binding fingers.** The specific binding of the fingers to the DNA allows the transcription to go much faster. Besides metal-binding fingers, at least two other prominent transcription factors exist, called helix-turn-helix and leucine zipper. Transcription factors can also be repressors, reducing the rate of transcription. Some can serve as both activators and repressors.

**Figure 23.18**
Schematic view of a metal-binding finger (a region of a protein called a transcription factor). The zinc ion forms coordinate-covalent bonds to two cysteines and two histidines.

These spacers are not introns. Introns are transcribed but are spliced out when mature mRNA is formed.

**Box 23E**

# Mutations, Biochemical Evolution, and Molecular Disease

We can trace the genetic relationship of different species through the variability of their amino acid sequences in different proteins. For example, the blood of all mammals contains hemoglobin, but the amino acid sequences of the hemoglobins are not identical. In Table 23E we see that the first ten amino acids in the β-globin of humans and gorillas are exactly the same. As a matter of fact, there is only one amino acid difference, at position 104, between us and apes. The β-globin of the pig differs from ours at ten positions, of which two are in the N-terminal decapeptide. That of the horse differs from ours in 26 positions, of which four are in this decapeptide.

β-Globin seems to have gone through many mutations during the evolutionary process, since only 26 of the 146 sites are invariant, that is, exactly the same in all species studied so far.

The relationship between different species can also be established by similarities in protein or mRNA primary structures. The proteins of closely related species, such as humans and apes, have very similar primary structures, presumably because these two species diverged on the evolutionary tree only recently. On the other hand, species far removed from each other diverged long ago and have undergone more mutations, which show up in differences in primary structures of their proteins.

Not only the *number* of amino acid substitutions is significant in the evolutionary process caused by mutation, but the *kind* of substitution is even more important. If the substitution is by an amino acid with physicochemical properties similar to those of the amino acid in the ancestor protein, the mutation is most probably viable. For example, whereas in human and gorilla β-globin position 4 is occupied by threonine, in the pig and horse it is occupied by serine. Both amino acids provide an —OH-carrying side chain.

**TABLE 23E   Amino Acid Sequence of the N-Terminal Decapeptides of β-Globin in Different Species**

| Species | Position | | | | | | | | | |
|---------|---|---|---|---|---|---|---|---|---|----|
| | 1 | 2 | 3 | 4 | 5 | 6 | 7 | 8 | 9 | 10 |
| Human | Val | His | Leu | Thr | Pro | Glu | Glu | Lys | Ser | Ala |
| Gorilla | Val | His | Leu | Thr | Pro | Glu | Glu | Lys | Ser | Ala |
| Pig | Val | His | Leu | Ser | Ala | Glu | Glu | Lys | Ser | Ala |
| Horse | Val | Glu | Leu | Ser | Gly | Glu | Glu | Lys | Ala | Ala |

## 23.13   MUTATIONS, MUTAGENS, AND GENETIC DISEASES

In Section 23.4 we saw that the base-pairing mechanism provides an almost perfect way to copy a DNA molecule during replication. The key word here is "almost." No machine, not even the copying mechanism of DNA replication, is totally without error. It has been estimated that, on average, there is one error for every $10^{10}$ bases (that is, one in ten billion). An error in the copying of a sequence of bases is called a **mutation.** Mutations can occur during replication. Base errors can also occur during transcription in protein synthesis (a nonheritable error). These errors may have widely varying consequences. For example,

# Mutagens and Carcinogens

There are many chemicals (both synthetic and natural) that cause cancer when introduced into the body. These are called carcinogens (Box 11F). One of the main tasks of the U.S. Food and Drug Administration and the Environmental Protection Agency is to identify these chemicals and eliminate them from our food, drugs, and environment. To prove that a chemical is a carcinogen requires long and costly laboratory testing with animals. Because we know that most carcinogens are also mutagens (that is, they induce mutations in bacteria), we now have available a much faster and cheaper screening test (the Ames test). In this test, one takes a bacterial strain, for example, *Salmonella,* that needs the amino acid histidine in its diet. When the bacteria are grown on a medium that does not contain histidine, very few survive. If a mutagen is added to the medium, however, some of the *Salmonella* may undergo mutations that can live without a supply of histidine. These will multiply and show up as a heavy growth of bacterial colonies. With such a simple test, many chemicals can be tested for mutagenic activity. Those that are found to be mutagens then can be further tested in animals to see if they are also carcinogens.

the codon for valine in mRNA can be GUA, GUG, GUC, or GUU. In DNA these correspond to CAT, CAC, CAG, and CAA, respectively. Let us assume that the original codon in the DNA is CAT. If during replication a mistake is made and the CAT was spelled as CAG in the copy, there will be no harmful mutation because when a protein is synthesized the CAG will be transcribed onto the mRNA as GUC, which also codes for valine. Therefore, although a mutation occurred, the same protein is manufactured.

On the other hand, assume that the original sequence in the DNA is CTT, which transcribes onto mRNA as GAA and which codes for glutamic acid. If during replication a mutation occurs and CTT becomes ATT, the new cells will probably die. The reason is that ATT transcribes to UAA, which does not code for any amino acid but is a stop signal. Thus, instead of continuing to build a protein chain with glutamic acid, the synthesis stops altogether. An important protein is not manufactured, and the organism may die. In this way, *very* harmful mutations are not carried over from one generation to the next.

Ionizing radiation (x-rays, ultraviolet light, gamma rays) can cause mutations. Furthermore, a large number of chemicals can induce mutation by reacting with DNA. Such chemicals are called **mutagens.** Many changes caused by radiation and chemicals do not become mutations because the cell has its own repair mechanism, which can prevent mutations by cutting out damaged areas and resynthesizing them. In spite of this defense mechanism, certain errors in copying that result in mutations do slip by.

Not all mutations are harmful. Certain ones are beneficial because they enhance the survival rate of the species. For example, mutation is used to develop new strains of plants that can withstand pests.

If a mutation is harmful, it results in an inborn genetic disease. This may be carried as a recessive gene from generation to generation with no individual demonstrating the symptoms of the disease. Only when both parents carry recessive genes does an offspring have a 25 percent chance of inheriting the disease. If the defective gene is dominant, on the other hand, every carrier will develop symptoms.

Sickle cell anemia (Box 18E) is caused by a single amino acid change: valine for glutamic acid. In this case CTT is mutated to CAT.

This is discussed in Section 9.5.

Examples of mutagens are benzene, carbon tetrachloride, and vinyl chloride.

A **recessive gene** is one that expresses its message only if both parents transmit it to the offspring.

Specific diseases that result from defective or missing genes are discussed in Boxes 17F, 18E, 18F, 21A, 21C, and 21G.

An **oncogene** is a gene that in some way or other participates in the development of cancer. Cancer cells differ from normal cells in a variety of structural and metabolic ways; however, the most important difference is their uncontrolled proliferation. Most ordinary cells are quiescent. When they lose their controls they give rise to tumors, benign or malignant. The uncontrolled proliferation allows these cells to spread, invade other tissues and colonize them, a process called **metastasis.** Malignant tumors can be caused by a variety of agents: chemical carcinogens such as benzo(e)pyrene (Box 11F), ionizing radiation such as x-rays, and viruses. However, for a normal cell to become a cancer cell certain transformations must occur. Many of these transformations take place on the level of the genes. Certain genes in normal cells have counterparts in viruses, especially in retroviruses (which carry their genes in the form of RNA rather than DNA). This means that a normal cell in a human being and in a virus both have genes that code for similar proteins. Usually these genes code for proteins that control cell growth. Such a protein is epidermal growth factor (EGF). When a retrovirus gets into the cell it takes control of the cell's own EGF gene. At that point the EGF gene becomes an oncogene, a cancer-causing gene, because it produces large amounts of the EGF protein, which in turn allows the epidermal cells to grow in an uncontrolled manner.

Because the normal EGF gene in the normal cell was turned into an oncogene, we call it a **proto-oncogene.** Such conversions of a proto-oncogene to an oncogene can occur not just by viral invasion but also by a mutation of the gene. For example, a form of cancer called human bladder carcinoma is caused by a single mutation of a gene; a change from guanine to thymine. The resulting protein has a valine in place of a glycine. This change is sufficient to transform the cell to a cancerous state, because the protein for which the gene is coded is a protein that amplifies signals (Sec. 24.4). Thus, because of the mutation that caused the proto-oncogene to become an oncogene, the cell stays "on"—its metabolic process is not shut off, and it is now a cancer cell that proliferates without control. There are a number of other mechanisms by which a proto-oncogene is transformed to an oncogene, but in each case the gene product has something to do with growth, hormonal action, or some other kind of cell regulation. About 50 viral and cellular oncogenes have been identified.

## 23.14   RECOMBINANT DNA

There are no cures for the inborn genetic diseases that we discussed in the preceding section. The best we can do is detect the carriers and, through genetic counseling of prospective parents, try not to perpetuate the defective genes. However, a technique called the **recombinant DNA technique** gives some hope for the future. At this time recombinant DNA techniques are used mostly in bacteria, but it is possible that they may someday be extended to human cells as well.

As used today, the recombinant DNA technique begins with certain circular DNA molecules found in the cells of the bacteria *E. coli.* These molecules, called **plasmids** (Fig. 23.19), consist of double-stranded DNA arranged in a ring. Certain highly specific enzymes called **restriction endonucleases** cleave DNA molecules at specific locations (a different location for each enzyme). For example, one of these enzymes may split a double-stranded DNA as follows:

**Figure 23.19**
Plasmid from a bacterium used in the recombinant DNA technique. (Thomas Broker/Cold Spring Harbor Laboratory.)

The enzyme is so programmed that whenever it finds this specific sequence of bases in a DNA molecule, it cleaves it as shown. Since a plasmid is circular, cleaving it this way produces a double-stranded chain with two ends (Fig. 23.20). These are called "sticky ends" because each has on one strand several free bases that are ready to pair up with a complementary section if they can find one.

**Figure 23.20**
The recombinant DNA technique can be used to turn a bacterium into an insulin "factory." (From P. B. Berlow, et al., *Introduction to the Chemistry of Life.* Philadelphia: Saunders College Publishing, 1982.)

# DNA Fingerprinting

9  8  7  6  5  4  3  2  1

**Figure 23H**
DNA fingerprint. (Courtesy of Dr. Lawrence Kobilinsky.)

Every person has a genetic makeup consisting of about 3 billion pairs of nucleotides, distributed over 46 chromosomes. The base sequence in the nucleus of every one of our billions of cells is identical. However, except for people who have an identical twin, the base sequence in the total DNA of one person is different from that of every other person. This makes it possible to identify suspects in criminal cases from a bit of skin or a trace of blood left at the scene of the crime and to prove the identity of a child's father in paternity cases. The nuclei of these cells are extracted, and restriction enzymes are used to cut the DNA molecules at specific points. The resulting DNA fragments are put on a gel and undergo a process called electrophoresis. In this process the DNA fragments move with different velocities; the smaller fragments move faster and the larger fragments slower. After a sufficient amount of time the fragments separate, and when they are made visible in the form of an autoradiogram, one can see bands in a lane. This is called a **DNA fingerprint.**

When the DNA fingerprint made from a sample taken from a suspect matches that from a sample obtained at the scene of the crime, the police have a positive identification. Figure 23H shows DNA fingerprints derived by using one particular restriction enzyme. A total of nine lanes can be seen. Three (numbers 1, 5, and 9) are control lanes. They contain the DNA fingerprint of a virus, using one particular restriction enzyme. Three other lanes (2, 3, and 4) were used in a paternity suit: These contain the DNA fingerprints of the mother, the child, and the alleged father. This is a positive identification. The child's DNA fingerprint contains six bands. The alleged father's DNA fingerprint also contains six bands, of which three match those of the child. The mother's DNA fingerprint has five bands, all of which match those of the child. In such cases one cannot expect a perfect match even if the man is the actual father because the child has inherited only half of its genes from the father. Thus, of six bands only three are expected to match. In the case just described the paternity suit was won on the basis of the DNA fingerprint matching.

In the left area of the radiogram are three more lanes (6, 7, and 8). These DNA fingerprints were used in an attempt to identify a rapist. In lanes 7 and 8 are the DNA fingerprints of semen obtained from the rape victim. In lane 6 is the DNA fingerprint of the suspect. The DNA fingerprints of the semen do not match that of the suspect. This is a negative identification and excluded the suspect from the case. When positive identification occurs, the probability that a positive match is due to chance is 1 in 100 billion.

However, a certain caution has developed lately in court cases involving DNA fingerprinting. Scientific committees and expert witnesses now demand that in addition to matching or not matching autoradiograms, some internal controls should appear on the gel. This is demanded in order to prove that the two lanes to be compared were run and processed under identical conditions. Furthermore, identifications in courts do not rely solely on DNA fingerprinting. Additional evidence is provided by analysis of blood group substances (see Box 16C), either directly from blood or from secretions such as saliva or cervical mucus samples, which also contain them.

The next step is to give them one. We do this by adding a gene from some other species. The gene is a strip of double-stranded DNA that has the necessary base sequence. For example, we can put in the human gene that manufactures insulin, which we can get in two ways: (1) It can be made in a laboratory by chemical synthesis; that is, chemists can combine the nucleotides in the proper sequence to make the gene, or (2) we can cut a human chromosome with the same restriction enzyme. Since it is the same enzyme, it cuts the human gene so as to leave the same sticky ends:

$$\begin{array}{l} H-GAATTC-H \\ H-CTTAAG-H \end{array} \longrightarrow \begin{array}{l} H-G \\ H-CTTAA \end{array} + \begin{array}{l} AATTC-H \\ G-H \end{array}$$

We use "H" to indicate a human gene.

The human gene must be cut at two places, so that a piece of DNA that carries two sticky ends is freed. To splice the human gene into the plasmid, the two are mixed in the presence of DNA ligase, and the sticky ends come together:

$$\begin{array}{l} H-G \\ H-CTTAA \end{array} + \begin{array}{l} AATTC-B \\ G-B \end{array} \longrightarrow \begin{array}{l} H-GAATTC-B \\ H-CTTAAG-B \end{array}$$

We use "B" to indicate a bacterial plasmid.

This reaction takes place at both ends of the human gene, and the plasmid is a circle once again (Fig. 23.20).

The modified plasmid is then put back into a bacterial cell, where it replicates naturally every time the cell divides. Bacteria multiply quickly, and soon we have a large number of bacteria, all containing the modified plasmid. All of these cells now manufacture human insulin by transcription and translation. We can thus use bacteria as a factory to manufacture specific proteins. This new industry has tremendous potential for lowering the price of drugs that are now manufactured by isolation from human or animal tissues (for example, human interferon, a molecule that fights infection). Not only bacteria but also plant cells can be used (Fig. 23.21). Ultimately, if recombinant DNA techniques can be applied to humans and not just to bacteria, it is possible that genetic diseases might someday be cured by this powerful technique. An infant or fetus missing a gene might be given that gene. Once in the cells, the gene would reproduce itself for an entire lifetime (and be passed down to offspring).

At the time of writing (early 1993), such gene therapy has been sanctioned by the National Institutes of Health for some patients suffering from cystic fibrosis, a hereditary disease. Seventy percent of cystic fibrosis patients have one identified faulty gene. The plan is to spray a solution of virus into which the

Millions of copies of selected DNA fragments can be made within a few hours with high precision by a technique called **polymerase chain reaction,** discovered by Kary B. Mullis (1945– ), who shared the 1993 Nobel Prize in Chemistry for this discovery.

Human insulin is now marketed by the Lilly Corporation in two forms, called Humulin R and Humulin N (see Box 18D). The R form has a faster onset action. Both of them are manufactured by the recombinant DNA technique.

(a)

(b)

**Figure 23.21**
(a) Injection of an aqueous DNA solution into the nucleus of a protoplast. (Courtesy of Dr. Anne Crossway, Calgene, Inc., Davis, California, and *Biotechniques,* Eaton Publishing, Natick, Massachusetts.) (b) A luminescent tobacco plant. The gene of the enzyme luciferase (from a firefly) has been incorporated into the genetic material of the tobacco plant. (Courtesy of Dr. Marlene DeLuca, University of California at San Diego.)

healthy gene has been cloned into the nose or lungs of the patient. The virus, which is otherwise harmless, enters the cells of the patients, carrying with it the healthy gene. The hope is that the product of the healthy gene will compete with the product of the faulty gene, alleviating the symptoms of the disease or curing it altogether.

## SUMMARY

**Nucleic acids** are composed of sugars, phosphates, and organic bases. There are two kinds: **ribonucleic acid (RNA)** and **deoxyribonucleic acid (DNA).** In DNA the sugar is deoxyribose; in RNA it is ribose. In DNA the bases are adenine (A), guanine (G), cytosine (C), and thymine (T). In RNA they are A, G, C, and uracil (U). Nucleic acids are giant molecules with backbones made of alternating units of sugar and phosphate. The bases are side chains linked to the sugar units.

DNA is made of two strands that form a double helix. The sugar–phosphate backbone runs on the outside of the double helix, and the bases point inward. There is **complementary pairing** of the bases in the double helix. Each A on one strand is hydrogen-bonded to a T on the other, and each G is hydrogen-bonded to a C. No other pairs fit. The DNA molecule carries, in the sequence of its bases, all the information necessary to maintain life. When cell division occurs and this information is passed from parent cell to daughter cells, the sequence of the parent DNA is copied.

A **gene** is a segment of a DNA molecule that carries the sequence of bases that directs the synthesis of one particular protein. The information stored in the DNA is transcribed onto RNA and then expressed in the synthesis of a protein molecule. This is done in two steps: **transcription** and **translation.** There are three kinds of RNA: **messenger RNA (mRNA), transfer RNA (tRNA),** and **ribosomal RNA (rRNA).** In transcription, the information is copied from DNA onto mRNA by complementary base-pairing (A → U, T → A, G → C, C → G). There are also start and stop signals. The mRNA is strung out along the ribosomes. Transfer RNA carries the individual amino acids. Each tRNA goes to a specific site on the mRNA. A sequence of three bases (a triplet) on mRNA constitutes a **codon.** It spells out the particular amino acid that the tRNA brings to this site. Each tRNA has a recognition site, the **anticodon,** that pairs up with the codon. When two tRNA molecules are aligned at adjacent sites, the amino acids that they carry are linked by an enzyme. The pro-

cess continues until the whole protein is synthesized. The genetic code is multiple. In most cases there is more than one codon for each amino acid. DNA in higher organisms contains sequences, called **introns,** that do not code for proteins. The sequences that do code for proteins are called **exons.**

There are a number of mechanisms for **gene regulation,** the switching on and off of the action of genes. In *E. coli,* portions of DNA called **operons** contain not only structural genes but also control sites and a regulatory gene. A certain protein inactivates the control sites by binding to one of them. In one of the operons of *E. coli* (the *lac* operon) this is reversed by lactose, which removes the repressor, freeing the control sites. Embryo development in eukaryotes is controlled by an epigenetic process, which involves methylation of the cytosine in DNA. Finally, **transcription factors** regulate gene expression by positive enhancement. Most of human DNA (96 to 98 percent) does not code for proteins. Many DNA sequences in eukaryotes are repetitious.

A change in the sequence of bases is called a **mutation.** Mutations can be caused by an internal mistake or can be induced by chemicals or radiation. A change in just one base can cause a mutation. This may be harmful or beneficial or may cause no change whatsoever in the amino acid sequence. If a mutation is very harmful, the organism may die. Chemicals that cause mutations are called **mutagens.**

With the discovery of **restriction enzymes** that can cut DNA molecules at specific points, scientists have found ways to splice DNA segments together. In this manner a human gene—for example, the one that codes for insulin—can be spliced into a bacterial **plasmid.** Then the bacteria, when multiplied, transmit this new information to the daughter cells. Therefore, the ensuing generations of bacteria are able to manufacture human insulin. This powerful method is called the **recombinant DNA technique.**

## KEY TERMS

A site (Sec. 23.10)
Activation (Sec. 23.10)
Anticodon (Sec. 23.8)
Base (Sec. 23.2)
Central dogma (Sec. 23.6)
Chromosome (Sec. 23.1)
Codon (Sec. 23.9)
Complementary sequence (Sec. 23.4)
Control sites (Sec. 23.12)
Degenerate code (Sec. 23.9)
Deoxyribonucleic acid (Sec. 23.2)
DNA (Sec. 23.2)
DNA fingerprinting (Box 23H)
Double helix (Sec. 23.3)
Elongation (Sec. 23.10)
Elongation factor (Sec. 23.10)
Epigenetic process (Sec. 23.12)
Eukaryote (Sec. 23.12)
Exon (Sec. 23.11)
Gene (Sec. 23.1, 23.11)
Gene expression (Sec. 23.12)
Gene regulation (Sec. 23.12)
Genetic code (Sec. 23.9)
Histone (Sec. 23.1, 23.11)
Initiation (Sec. 23.10)
Initiation signal (Sec. 23.7)
Intron (Sec. 23.11)
Lagging strand (Box 23B)
Leading strand (Box 23B)
Messenger RNA (mRNA) (Sec. 23.5)
Metal-binding fingers (Sec. 23.12)
Metastasis (Box 23G)
Multiple code (Sec. 23.9)

Mutagen (Sec. 23.13)
Mutation (Sec. 23.13)
Nucleic acid (Sec. 23.1)
Nucleoside (Sec. 23.2)
Nucleosome (Sec. 23.11)
Nucleotide (Sec. 23.2)
Okazaki fragment (Box 23B)
Oncogene (Box 23G)
Operator site (Sec. 23.12)
Operon (Sec. 23.12)
P site (Sec. 23.10)
Plasmid (Sec. 23.14)
Polymerase chain reaction (Sec. 23.14)
Prokaryote (Sec. 23.12)
Promoter site (Sec. 23.12)
Proto-oncogene (Box 23G)
Recognition site (Sec. 23.8)
Recombinant DNA (Sec. 23.14)
Regulatory gene (Sec. 23.12)
Replication (Sec. 23.4)
Restriction endonuclease (Sec. 23.14)
Ribonucleic acid (Sec. 23.2)
Ribosomal RNA (rRNA) (Sec. 23.5)
Ribosome (Sec. 23.5)
RNA (Sec. 23.2)
Structural genes (Sec. 23.12)
Termination (Sec. 23.10)
Termination sequence (Sec. 23.7)
Transcription (Sec. 23.6)
Transfer RNA (tRNA) (Sec. 23.5)
Translation (Sec. 23.6)
Translocation (Sec. 23.10)
Unwinding protein (Sec. 23.4)

## PROBLEMS

### Nucleic Acids and Heredity

**23.1**  Explain, in terms of DNA, why a pig cannot give birth to a mouse.

**23.2**  State the one-gene-one-enzyme hypothesis.

**23.3**  What structures of the cell, visible in a microscope, contain hereditary information?

**23.4**  Name one hereditary disease.

### Components of Nucleic Acids

**23.5**  Name the two types of nucleic acids present in cells.

**23.6**  What are the components of
(a) a nucleotide
(b) a nucleoside

**23.7**  Draw the structures of (a) guanine  (b) adenine

**23.8** What is the difference in structure between thymine and uracil?

**23.9** Which DNA and RNA bases contain a carbonyl group?

**23.10** Which DNA and RNA bases are primary amines?

**23.11** What is the difference in structure between ribose and deoxyribose?

**23.12** What is the difference between a nucleoside and a nucleotide?

**23.13** Draw the structures of (a) cytidine   (b) deoxycytidine

## Structure of DNA and RNA

**23.14** In DNA and RNA, which carbon atom of the sugar is linked directly to the base?

**23.15** Draw the structures of (a) UMP   (b) dAMP

**23.16** In DNA and RNA, which carbon atoms of the sugar are linked to the phosphate groups?

**23.17** The sequence of a short DNA segment is ATGGCAATAC. (a) What name do we give to the two ends (terminals) of a DNA molecule? (b) In this segment, which end is which?

**\*23.18** Chargaff showed that in samples of DNA from many different species, the molar quantity of A was always approximately equal to the molar quantity of T, and the same for C and G. How did this information help to establish the structure of DNA?

**23.19** What makes nucleic acids acidic?

**23.20** What interactions stabilize the three-dimensional structure of DNA?

**23.21** Which nucleic acid is single-stranded?

## DNA Replication

**23.22** A DNA molecule normally replicates itself millions of times, with almost no errors. What single fact about the structure is most responsible for this?

**23.23** What functional groups on the bases form hydrogen bonds in the DNA double helix?

**23.24** Draw the structures of adenine and thymine, and show with a diagram the two hydrogen bonds that stabilize A—T pairing in DNA.

**23.25** Draw the structures of cytosine and guanine, and show with a diagram the three hydrogen bonds that stabilize C—G pairing in nucleic acids.

**23.26** How many bases are there in a DNA double helix?

**23.27** Where does the unwinding of DNA occur?

**23.28** What is the function of the unwinding proteins?

**23.29** What do we call the enzymes that join nucleotides into a DNA strand?

## RNA, Transmission, Transcription, and Translation

**23.30** How many nucleotides are there in a tRNA chain?

**23.31** Which has the longest chains: tRNA, mRNA, or rRNA?

**23.32** What is the central dogma of molecular biology?

**23.33** Which kind of RNA has a sequence complementary to that of DNA?

**23.34** Which kind of RNA is associated with proteins?

**23.35** (a) What is the shape of a tRNA molecule in a two-dimensional diagram?   (b) In a three-dimensional diagram?

**23.36** Which end of the DNA contains the initiation signal?

**23.37** In what part of the cell does transcription occur?

**23.38** (a) Which ribosome portion has specific A and P sites?   (b) What happens at each site during translation?

**23.39** What are the two most important sites on tRNA molecules?

**23.40** Which kind of RNA has codons and which kind has anticodons?

## The Genetic Code

**23.41** (a) If a codon is GCU, what is the anticodon? (b) What amino acid does this codon code for?

**\*23.42** If a segment of DNA is 981 units long, how many amino acids appear in the protein this DNA segment codes for? (Assume that the entire segment is used to code for the protein and that there is no methionine at the N-terminal end of the protein.)

**23.43** Which codons are stop signals?

## Translation and Protein Synthesis

**23.44** (a) What is the main role of the 40S ribosome? (b) of the 60S ribosome?

**23.45** What is the function of elongation proteins?

## Genes, Exons, and Introns

**23.46** What is the nature of the interaction between histones and DNA in nucleosomes?

**23.47** Define (a) intron  (b) exon

**23.48** Does mRNA also have introns and exons? Explain.

**\*23.49** (a) What happens to the introns of DNA?  (b) of RNA?

## Gene Regulation

**23.50** When gene regulation occurs at the transcriptional level, what molecules are involved?

**23.51** What is a structural gene?

**23.52** Name the two kinds of control sites.

**23.53** What is an operon?

**23.54** What is the function of the methylation of cytosines in the DNA of higher organisms?

**23.55** (a) What percentage of human DNA codes for proteins?  (b) What is the function of the rest of the DNA?

**23.56** What is the function of the metal-binding fingers on transcription factor IIIA?

## Mutations

**23.57** Using Table 23.2, give an example of a mutation that (a) does not change anything in a protein molecule  (b) might cause fatal changes in a protein

## Recombinant DNA

**23.58** How do restriction endonucleases operate?

**23.59** What are sticky ends?

## Boxes

**23.60** (Box 23A) Draw the structure of the fluorouridine nucleoside that inhibits DNA synthesis.

**23.61** (Box 23A) Give an example of how anticancer drugs work in chemotherapy.

**23.62** (Box 23B) Define: (a) replication  (b) daughter DNA  (c) unwinding protein  (d) Okazaki fragment

**23.63** (Box 23B) In which direction is the DNA molecule synthesized continuously?

**23.64** (Box 23B) Why does the body synthesize the leading strand continuously and the lagging strand discontinuously (that is, in Okazaki fragments)?

**23.65** (Box 23C) Explain how actinomycin D inhibits protein synthesis.

**23.66** (Box 23D) Why are viruses considered to be parasites?

**\*23.67** (Box 23D) What is a viral "coat"? Where do the ingredients—amino acids, enzymes, and so forth—necessary to synthesize the "coat" come from?

**23.68** (Box 23F) If 1-amino-2-cyanoethane were found to be mutagenic by the Ames test, how could you prove that it is not carcinogenic?

**23.69** (Box 23G) What is a proto-oncogene?

**23.70** (Box 23G) What kind of mutation transforms the EGF proto-oncogene to an oncogene in human bladder carcinoma?

**23.71** (Box 23H) After having been cut by restriction enzymes, how are the DNA fragments separated from each other?

**23.72** (Box 23H) How is DNA fingerprinting used in paternity suits?

## Additional Problems

**23.73** Why do we call the genetic code degenerate?

**23.74** How does the cell make sure that a specific amino acid (say valine) attaches itself only to the one tRNA molecule that is specific for valine?

**23.75** In the tRNA structure there are stretches where complimentary base pairing is necessary and other areas where it is absent. Describe two functionally critical areas (a) where base pairing is mandatory and  (b) where it is absent.

**23.76** Is there any way to prevent a hereditary disease? Explain.

**23.77** Some structural genes are repetitious; that is, they have more than one copy. The genes of which type of proteins have the most copies?

**23.78** Draw the structures of (a) uracil  (b) uridine

**\*23.79** Which nuclear superstructure is stabilized by an acid-base interaction?

**23.80** How does an inducer such as lactose enhance the production of the enzyme $\beta$-galactosidase?

**23.81** Why is it important that a DNA molecule be able to replicate itself millions of times without error?

**23.82** (a) What is a plasmid?  (b) How does it differ from a gene?

# C H A P T E R   2 4

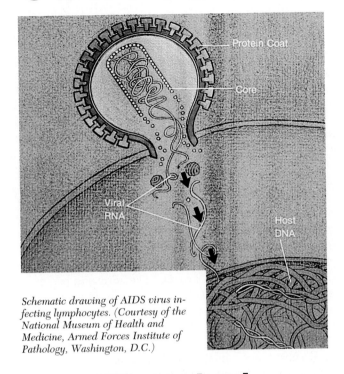

*Schematic drawing of AIDS virus infecting lymphocytes. (Courtesy of the National Museum of Health and Medicine, Armed Forces Institute of Pathology, Washington, D.C.)*

# Chemical Communication: Neurotransmitters, Hormones, and Immunoglobulins

## 24.1  INTRODUCTION

Each cell in the body is an isolated entity enclosed in its own membrane. The thousands of reactions in each cell would be uncoordinated unless cells could communicate with each other. Such communication allows the activity of a cell in one part of the body to be coordinated with the activity of cells in a different part of the body. The body uses two principal types of molecules for these communications: (1) relatively small molecules called **chemical signals** or **mediators** and (2) protein molecules on the surface of cell membranes called **receptors**.

Nerve cells are present throughout the body and, together with the brain, constitute the nervous system.

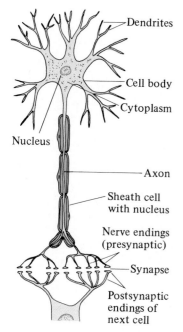

**Figure 24.1**
Neuron and synapse.

When your house is on fire and the fire is threatening your life, external signals—light, smoke, and heat—register alarm at specific receptors in your eyes, nose, and skin. From there the signals are transmitted by specific compounds to nerve cells, or **neurons.** In the neurons the signals travel as electric impulses along the axons (Fig. 24.1). When they reach the end of the neuron, the signals are transmitted to adjacent neurons by specific compounds called **neurotransmitters.** Communication between the eyes and the brain, for example, is by neural transmission.

Once the danger signals are processed in the brain, other neurons carry messages to the muscles (Fig. 24.2) and to the endocrine glands. The message to the muscles is to run away or to take some other action in response to the fire (save the baby or run to the fire extinguisher, for example). In order to do one of these things, the muscles must be activated. Again, neurotransmitters carry the messages from the neurons to the muscle cells and to endocrine glands. The endocrine glands are stimulated, and a different chemical signal, called a **hormone,** is secreted into your blood stream. "The adrenalin begins to flow." The danger signal carried by adrenalin (Box 15A) makes quick energy available so that the muscles can contract and relax rapidly, allowing your body to take quick action to avoid the danger.

If during the process you get cut and foreign bodies enter your blood stream, special cells in your blood, the **lymphocytes,** recognize the foreign bodies and manufacture still other communication compounds, called **immunoglobulins,** to fight and eliminate the foreign bodies.

Without these chemical communicators, the whole organism—you— would not survive because there is a constant need for coordinated efforts to face a complex outside world.

In the following sections we will investigate the chemistry and the mode of communication achieved by the three groups of communication chemicals:

**Figure 24.2**
Contact of neuron (dark shape) with muscle cells (light color). (Courtesy of Dr. Janice R. Anderson, Addenbrooke's Hospital and MTP Press, Cambridge, England.)

neurotransmitters, hormones, and immunoglobulins. (The classification is made on the basis of how each works rather than on chemical structure.)

## 24.2   NEUTROTRANSMITTERS

**Neurotransmitters** are compounds that communicate between two nerve cells or between a nerve cell and another cell (such as a muscle cell). If we look at a nerve cell (Fig. 24.1), we see that it consists of a main cell body from which projects a long, fiber-like part called an **axon.** Coming off the other side of the main body are hair-like structures called **dendrites.**

Neurons do not typically touch each other. Between the axon end of one neuron and the cell body or dendrite end of the next, there is a space filled with an aqueous fluid. This fluid-filled space is called a **synapse.** If the chemical signal travels, say, from top to bottom, we call the nerve ends on the top **presynaptic** and those on the bottom **postsynaptic.**

The neurotransmitters are stored at the presynaptic site in **vesicles,** which are small, membrane-enclosed packages. Events begin when a message is transmitted from one neuron to the next by neurotransmitters. They travel across the synapse and are adsorbed onto specific receptor sites. There are, broadly speaking, three classes of neurotransmitters: *cholinergic, adrenergic,* and *peptidergic neurotransmitters.* This classification is based on the chemical nature of the most important neurotransmitters in each group.

The neurotransmitters fit into the receptor sites in a manner reminiscent of the lock-and-key model mentioned in Section 19.5.

## 24.3   ACETYLCHOLINE

The main **cholinergic neurotransmitter** is acetylcholine:

$$CH_3-\underset{\underset{O}{\parallel}}{C}-O-CH_2CH_2-\underset{\underset{CH_3}{\mid}}{\overset{\overset{CH_3}{\mid}}{N}}{}^+-CH_3$$

Acetylcholine

When an electric nerve impulse moves along the neuron and reaches the vesicles, it causes acetylcholine molecules to be released into the synapse. These molecules travel across the short synapse to the next neuron, where they are adsorbed onto specific receptor sites on the postsynaptic membrane. The presence of the acetylcholine molecules at the postsynaptic receptor site then triggers chemical reactions in which ions can freely cross membranes. Because it involves ions, which carry electric charges, this process is translated into an electric signal that now travels along this neuron until it reaches the other end, where the process is repeated.

By this means, the message moves from neuron to neuron until finally it gets transmitted, again by acetylcholine molecules, to the muscles or endocrine glands that are the ultimate target of the message.

Meanwhile, what happens to the postsynaptic receptor sites of the neurons after the message gets transmitted? If the acetylcholine molecules remained there, no signal would be transmitted even though a message was being re-

B o x 24A

## Alzheimer's Disease and Acetylcholine Transferase

Alzheimer's disease is the name given to the symptoms of senile behavior (second childhood) that afflict about 1.5 million older people in the United States. In patients with this disease, the nerve fibers in the cerebral cortex are entangled, and some of the nerve endings degenerate and form plaques. The brain becomes smaller, and part of the cortex atrophies (Fig. 24A).

It has been found that patients with Alzheimer's disease have significantly diminished acetylcholine transferase activity in their brains. This enzyme synthesizes acetylcholine by transferring the acetyl group from acetyl CoA to choline:

$$CH_3\overset{\overset{O}{\|}}{C}-S-CoA + HO-CH_2CH_2-\overset{\overset{CH_3}{|}}{\underset{\underset{CH_3}{|}}{N^{\pm}}}-CH_3 \longrightarrow CH_3\overset{\overset{O}{\|}}{C}-O-CH_2CH_2-\overset{\overset{CH_3}{|}}{\underset{\underset{CH_3}{|}}{N^{\pm}}}-CH_3 + CoA-SH$$

The lack of acetylcholine in the neurons makes people with Alzheimer's disease forgetful, especially about recent events. As the disease advances, they become confused and in severe cases lose their ability to speak; they then need total care.

There is no cure for this disease. However, certain drugs, which act as acetylcholinesterase inhibitors, have been shown to improve memory and other cognitive functions in some people with the disease. One such drug, Tacrine, is already on the market. The alkaloid huperzine A, an active ingredient of Chinese herb tea, used for centuries to improve memory, is also a potent inhibitor of acetylcholinesterase. These drugs act by inhibiting the enzyme that decomposes the neurotransmitter, leaving more neurotransmitter molecules available for brain function.

(a)

(b)

**Figure 24A**
(a) Normal brain. (b) Brain of a person with Alzheimer's disease. (Courtesy of Anatomical Collections, National Museum of Health and Medicine, Armed Forces Institute of Pathology, Washington, D.C.)

ceived from the previous neuron. Therefore, the acetylcholine must be removed so that the neuron is reactivated. The acetylcholine is removed rapidly from the receptor site by the enzyme acetylcholinesterase, which hydrolyzes it.

$$H_2O + CH_3\overset{\overset{O}{\|}}{C}-O-CH_2CH_2\overset{\overset{CH_3}{|}}{\underset{\underset{CH_3}{|}}{N^{\pm}}}-CH_3 \xrightarrow[\text{esterase}]{\text{acetylcholin-}} CH_3\overset{\overset{O}{\|}}{C}-OH + HO-CH_2CH_2\overset{\overset{CH_3}{|}}{\underset{\underset{CH_3}{|}}{N^{\pm}}}-CH_3$$

Acetylcholine                                            Acetic acid          Choline

**B o x   24B**

## Nerve Gases and Antidotes

Most nerve gases in the military arsenal exert their lethal effect by binding to acetylcholinesterase. Under normal conditions, this enzyme decomposes the synaptic neurotransmitter acetylcholine within a few milliseconds after it is released at the nerve endings. Nerve gases such as Sarin (agent GB, also called Tabun), Soman (agent GD), and agent VX are organic phosphonates related to such pesticides as parathion (the latter being much less lethal, of course):

$$CH_3-CH(CH_3)-O-P(=O)(CH_3)-F$$

Sarin
(agent GB)

$$CH_3-C(CH_3)(CH_3)-CH(CH_3)-O-P(=O)(CH_3)-F$$

Soman
(agent GD)

$$(CH_3)_2CH-N(CH(CH_3)_2)-CH_2CH_2-S-P(=O)(CH_3)-O-CH_2CH_3$$

Agent VX

If any of these phosphonates bind to acetylcholinesterase, the enzyme is completely inactivated and the transmission of nerve signals stops. The result is a cascade of symptoms: sweating, bronchial constriction due to mucus buildup, dimming of vision, vomiting, choking, convulsions, paralysis, and respiratory failure. Direct inhalation of as little as 0.5 mg can cause death within a few minutes. If the dosage is less or if the nerve gas is absorbed through the skin, the lethal effect may take several hours. In warfare, protective clothing and gas masks are effective countermeasures. Also, first aid kits containing antidotes that can be injected are available. These antidotes contain the alkaloid atropine.

The nations of the world, with the exception of a few Arab countries, recognizing the dangers of proliferating these and other chemicals used for warfare, signed a treaty in 1993. The treaty obliges all the signatory governments to destroy their stockpiles of chemical weapons, and to dismantle the plants that manufacture them within 10 years.

The removal of acetylcholine from the receptor site opens the site once again and gets it ready for the next message. This rapid removal enables the nerves to transmit more than 100 signals per second.

The action of the acetylcholinesterase enzyme is obviously essential to the whole process. When this enzyme is inhibited, the removal of acetylcholine is incomplete, and nerve transmission ceases. For example, the plant extract *curare* inhibits acetylcholinesterase and in large doses can cause death by paralysis. This is how the poisoned arrows of the Amazon Indians work. In small doses, curare is used as a muscle relaxant. A less dangerous muscle relaxant is decamethionium:

$$H_3C-\overset{+}{N}(CH_3)(CH_3)-CH_2(CH_2)_8CH_2-\overset{+}{N}(CH_3)(CH_3)-CH_3$$

Decamethionium

This molecule resembles the choline end of acetylcholine and therefore acts as a competitive inhibitor of acetylcholinesterase. Succinylcholine, another competitive inhibitor (Box 19A), is also an excellent muscle relaxant.

In Japan the fugu, a puffer fish, is regarded as a great culinary delicacy. But fugu carries a deadly poison, tetrodotoxin, so great care must be taken in the preparation of this dish. Special chefs are licensed to prepare and serve fugu, usually only in the season when the level of poison production is low. In spite of all these precautions, fugu occasionally takes its toll.

The fugu poison is very specific—it blocks $Na^+$ ion channels in nerve membranes that are necessary for nerve transmission. If these channels are irreversibly blocked, $Na^+$ ions cannot diffuse into the neuron when acetylcholine is adsorbed at cholinergic receptors. The consequence is that nerve signals are blocked. Fugu poisoning results in respiratory paralysis and death even if the intake is only in microgram quantities, making this one of the most deadly poisons known.

## 24.4 MONOAMINES AND AMINO ACIDS: ADRENERGIC NEUROTRANSMITTERS

The second class of neurotransmitters, the adrenergic neurotransmitters, includes such monoamines as epinephrine (Box 15A), serotonin, and dopamine. These monoamines transmit nerve signals by a mechanism whose beginning is similar to the action of acetylcholine. But once the monoamine neurotransmitter (for example, norepinephrine) is adsorbed onto the receptor site, it activates a secondary messenger inside the cell, called **cyclic AMP (cAMP).** The manufacture of cAMP activates processes that result in the transmission of an electrical signal. The cAMP is manufactured by adenylate cyclase from ATP:

Norepinephrine

ATP

cAMP

Pyrophosphate

The action of cAMP is depicted in Figure 24.3. The activation of adenylate cyclase accomplishes two important goals: (1) It converts an event occurring at the outer surface of the target cell (adsorption onto receptor site) to a change inside the target cell (release of cAMP). Thus the primary messenger (neurotransmitter or hormone) does not have to cross the membrane. (2) It *amplifies* the signal. One molecule adsorbed on the receptor triggers the adenylate cyclase to make many cAMP molecules. Thus the signal is amplified many thousands of times.

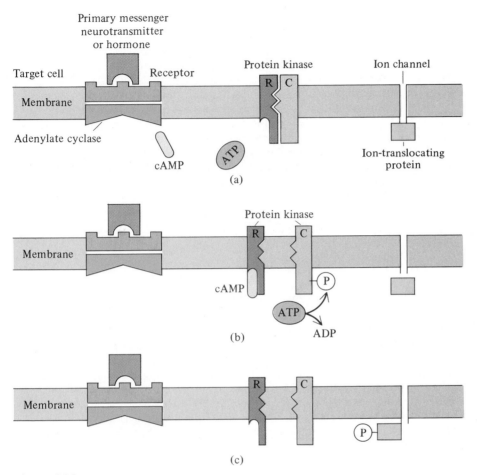

**Figure 24.3**

The sequence of events in the postsynaptic membrane when norepinephrine is adsorbed onto the receptor site. (a) A molecule of cAMP is formed when, upon adsorption of monoamine neurotransmitter or hormone, adenylate cyclase cleaves ATP (not shown) into cAMP and pyrophosphate. (b) Cyclic AMP activates protein kinase by dissociating the regulatory (R) unit from the catalytic unit (C). A second molecule of ATP, shown in pink in (a), has phosphorylated the catalytic unit and has been converted to ADP. (c) The catalytic unit phosphorylates the ion-translocating protein that blocked the channel for ion flow. The phosphorylated ion-translocating protein changes its shape and position and opens the ion gates.

How does this signal amplification stop? When the neurotransmitter or hormone dissociates from the receptor, the adenylate cyclase stops the manufacture of cAMP. The cAMP already produced is destroyed by the enzyme phosphodiesterase.

The amplification through the secondary messenger (cAMP) is a relatively slow process. It may take from 0.1 second to a few minutes. Therefore, in cases where the transmission of signals must be fast, in milli- or microseconds, a neurotransmitter such as acetylcholine acts on membrane permeability directly, without the mediation of a secondary messenger.

The inactivation of the adrenergic neurotransmitters is somewhat different from that of the cholinergic transmitters. While acetylcholine is decomposed by

**Box 24D**

## Parkinson's Disease —Depletion of Dopamine

Parkinson's disease is characterized by spastic motion of the eyelids as well as rhythmic tremors of the hands and other parts of the body, often when the patient is at rest. The posture of the patient changes to a forward, bent-over position; walking becomes slow, with shuffling footsteps. This is a degenerative nerve disease. The neurons affected contain, under normal conditions, mostly dopamine as a neurotransmitter. People with Parkinson's disease have depleted amounts of dopamine in their brains. However, the dopamine receptors are not affected. Thus, when L-dopa is administered, many of these patients are able to synthesize dopamine and resume normal nerve transmission. In these patients, L-dopa reverses the symptoms of Parkinson's disease, although the respite is only temporary. In other patients the L-dopa regimen provides little benefit.

Dopamine cannot be administered directly because it cannot penetrate the blood-brain barrier and therefore does not reach the tissue where its action is needed. L-Dopa, on the other hand, is transported through the arterial wall and is converted to dopamine in the brain:

L-Dopa
(dihydroxyphenylalanine)

Dopamine

Recently, a new drug treatment has shown promise as a possible route for slowing neuron degeneration. Deprenyl, a monoamine oxidase (MAO) inhibitor, given together with L-dopa, reduced the symptoms of Parkinson's disease and even increased the life span of patients. It appears that Deprenyl not only increases the level of dopamine by preventing its oxidation by MAOs, but also reduces the degeneration of neurons.

acetylcholinesterase, most of the adrenergic neurotransmitters are inactivated in a different way. The body inactivates monoamines by oxidizing them to aldehydes. Enzymes that catalyze these reactions are very common in the body. They are called *monoamine oxidases* (MAOs). For example, there is an MAO that converts both epinephrine and norepinephrine to the corresponding aldehyde:

Epinephrine
(salt form)

Norepinephrine
(salt form)

There are a number of compounds that cause headaches because they dilate the arteries in the head. This dilation occurs as a response to signals generated either at certain receptor sites in the wall of the arteries or in neurons in the brain stem. In any case, after the compounds bind to these receptors, muscle relaxation occurs in the arterial wall, which causes the dilation. Some of these compounds are nitroglycerine; nitrites in smoked and cured meats; monosodium glutamate, a seasoning used especially in Chinese food (Box 21F); tyramine, in cheeses; and phenylethylamine in red wines (frequently the cause of hangover headaches).

Many drugs that are used as antidepressants or antihypertensive agents are inhibitors of monoamine oxidases. These inhibitors prevent MAOs from converting monoamines to aldehydes, thus increasing the concentration of the active adrenergic neurotransmitters.

Shortly after adsorption onto the postsynaptic membrane, the neurotransmitter comes off the receptor site and is reabsorbed through the presynaptic membrane and stored again in the vesicles.

The neurotransmitter histamine is present in mammalian brains:

Histidine          Histamine

Histamine cannot readily pass the blood-brain barrier (Sec. 25.1) and must be synthesized in the brain neurons by a one-step decarboxylation of the amino acid histidine.

The action of histamine as a neurotransmitter is very similar to that of other monoamines. There are two kinds of receptors for histamine. One receptor, $H_1$, can be blocked by classic antihistamines such as dimenhydrinate (Dramamine) or diphenhydramine (Benadryl). The other receptor, $H_2$, can be blocked by ranitidine and cimetidine. $H_1$ receptors are found in the respiratory tract. They affect the vascular, muscular, and secretory changes associated with hay fever and asthma. Therefore, the classic antihistamines relieve these symptoms. The $H_2$ receptors are mainly in the stomach and affect the secretion of HCl. Cimetidine, an $H_2$ blocker, reduces acid secretion and thus is an effective drug for ulcer patients.

Sir James W. Black of England received the 1988 Nobel Prize in Medicine for the invention of cimetidine and such other drugs as propranolol (Table 24.1).

Amino acids distributed throughout the neurons also act as neurotransmitters. Some of them, such as glutamic acid, aspartic acid, and cysteine, act as excitatory neurotransmitters similar to acetylcholine and norepinephrine. Others, such as glycine, $\beta$-alanine, taurine, and, mainly, $\gamma$-aminobutyric acid (GABA)

$$H_2NCH_2CH_2SO_2OH \qquad H_2NCH_2CH_2COOH \qquad H_2NCH_2CH_2CH_2COOH$$

Taurine                    $\beta$-Alanine                 $\gamma$-Aminobutyric acid
                                                           (GABA)

reduce neurotransmission. Note that some of these neurotransmitter amino acids are not found in proteins.

## Nitric Oxide and Long-Term Memory

It has been known for some time that a primitive part of the brain, the *hippocampus,* has something to do with long-term memory. In the 1950s a physician found that a patient suffering from frequent epileptic seizures was cured by removal of the hippocampus. However, the patient also permanently lost his ability to remember new things. Faces or facts learned before the operation could be recalled with ease, but not those encountered after the operation. Since then the hippocampus has been studied not so much as a suspected storehouse of memories but for its "long-term potentiation" ability, or LTP. LTP is turned on by a brief experience and persists for a long time.

The molecular mechanism proposed for LTP is a process in which the primary messenger (the neurotransmitter glutamic acid) that shuttles between two nerve cells across the synapse also elicits the production of a secondary messenger. This secondary messenger goes back to the first nerve cell and prompts it to release more glutamic acid. Such a process would strengthen the connections between the two nerve cells, thereby inducing LTP.

The most likely candidate for this secondary messenger is nitric oxide, NO. This simple molecule is synthesized in the cells when arginine is converted to citrulline. (These two compounds appear in the urea cycle, Sec. 21.8.) Nitric oxide is a relatively nonpolar molecule, and once it has been produced in the second nerve cell, it quickly diffuses across the lipid bilayer membrane. Within its short half-life, 4 to 6 seconds, it can reach the neighboring first nerve cell and deliver its message. Four seconds are sufficient to induce LTP. It has been found that NO can act as an intercellular messenger in other parts of the body as well. It appears in the intercellular communication between blood vessels and the smooth muscles surrounding them. Thus it controls the constriction and relaxation of blood vessels.

Furthermore, it appears that nitric oxide is used by the immune system (Sec. 24.8) to fight infections by viruses. Studies have shown that NO is a potent agent against certain viruses, including the herpes simplex type 1 virus, which causes cold sores. When the viral infection begins the immune system manufactures the protein interferon, which in turn activates a gene that produces the enzyme nitric oxide synthase.

## 24.5 PEPTIDERGIC NEUROTRANSMITTERS

In the last few years, scientists have isolated a number of brain peptides that have affinity for certain receptors and therefore act as if they were neurotransmitters in the classic sense. Some 25 or 30 such peptides are now known, and not all of them behave as true neurotransmitters.

The first brain peptides isolated were the enkephalins. These pentapeptides are present in certain nerve cell terminals. They bind to specific pain receptors and seem to control pain perception. Since they bind to the receptor site that also binds the pain-killing alkaloid morphine (Sec. 15.9), it is assumed that it is the N-terminal end of the pentapeptide that fits the receptor (Fig. 24.4).

Table 24.1 lists some typical examples of the three main classes of neurotransmitters and drugs that affect their action.

**Figure 24.4**
Similarities between the structure of morphine and that of the brain's own pain regulators, the enkephalins.

Morphine          Methionine enkephalin

**TABLE 24.1  Drugs That Affect Nerve Transmission**

| Neurotransmitter | Drugs That Affect Receptor Sites | | Drugs That Affect Available Concentration of Neurotransmitters or Their Removal from Receptors | |
|---|---|---|---|---|
| | *Agonists (Activate Receptor Sites)* | *Antagonists (Block Receptor Sites)* | *Increase Concentration* | *Decrease Concentration* |
| Acetylcholine (cholinergic) | Nicotine Pilocarpine Carbachol | Curare Succinylcholine Atropine Propantheline (Pro-Banthine) | Malathion Nerve gases | *Clostridium botulinum* toxin |
| Norepinephrine (adrenergic) | Phenylephrine (Neo-Synephrine) Epinephrine (adrenalin) | Methyldopa (Aldomet) Propranolol (Inderal, Lopresor) | Amphetamines Iproniazide Antidepressants (Tofranil, Elavil) | Reserpine |
| Dopamine (adrenergic) | | Clozapine (Clorazil) | Deprenyl | |
| Serotonin (adrenergic) | | | Antidepressant Fluoxetin (Prozac) | |
| Histamine (adrenergic) | 2-Methylhistamine Betazole Pentagastin | Terfenadine (Seldane) Promethazine (Phenergan) Diphenylhydramine (Benadryl) Ranitidine (Zantac) Cimetidine (Tagamet) | Histidine | Hydrazino histidine |
| Enkephalin (peptidergic) | Opiate | Morphine Heroin Demerol | | Naloxone (Narcan) |

**TABLE 24.2   The Principal Hormones and Their Action**

| Gland | Hormone | Action | Structures Shown On Page |
|---|---|---|---|
| Parathyroid | Parathyroid hormone | Increases blood calcium | |
| | | Excretion of phosphate by kidney | |
| Thyroid | Thyroxine (T$_4$) | Growth, maturation, and metabolic rate | 383 |
| | Triiodothyronine (T$_3$) | Metamorphosis | |
| Pancreatic islets | | | |
| Beta cells | Insulin | Hypoglycemic factor | 558 |
| | | Regulation of carbohydrates, fats, and proteins | |
| Alpha cells | Glucagon | Liver glycogenolysis | |
| Adrenal medulla | Epinephrine | Liver and muscle glycogenolysis | 446 |
| | Norepinephrine | | |
| Adrenal cortex | Cortisol | Carbohydrate metabolism | 535 |
| | Aldosterone | Mineral metabolism | 535 |
| | Adrenal androgens | Androgenic activity (esp. females) | |
| Kidney | Renin | Hydrolysis of blood precursor protein to yield angiotensin | |
| Anterior pituitary | Luteinizing hormone | Causes ovulation | |
| | Interstitial cell-stimulating hormone | Formation of testosterone and progesterone in interstitial cells | |
| | Prolactin | Growth of mammary gland | |
| | Mammotropin | Lactation | |
| | | Corpus luteum function | |
| Posterior pituitary | Vasopressin | Contraction of blood vessels | 559 |
| | | Kidney reabsorption of water | |
| | Oxytocin | Stimulates uterine contraction and milk ejection | 559 |
| Ovaries | Estradiol | Estrous cycle | 535 |
| | Progesterone | Female sex characteristics | 535 |
| Testes | Testosterone | Male sex characteristics | 535 |
| | Androgens | Spermatogenesis | |

## 24.6   HORMONES

Glands that secrete hormones are called **endocrine glands.**

**Hormones** are diverse compounds secreted by specific tissues (glands), released into the blood stream, and then adsorbed onto specific receptor sites, usually relatively far from their source. This is the classic definition of a hormone. Some of the principal hormones are tabulated in Table 24.2. Figure 24.5 shows the location of the major hormone-secreting glands, and Figure 24.6 shows the target organs of hormones secreted by the pituitary gland.

This classic definition is too rigid, however, for what modern science knows about hormones. For example, in Section 24.4 we saw that epinephrine and norepinephrine are neurotransmitters. But these compounds also fit the classic definition of a hormone. This shows that the distinction between hormones and neurotransmitters is physiological (hormones are secreted into the blood stream) and not chemical. Whether a certain compound is considered to be a neurotransmitter or a hormone depends on whether it acts over a short distance across a synapse ($2 \times 10^{-6}$ cm), in which case it is a neurotransmitter, or over a long distance (20 cm) from secretory gland through the blood stream to target cell, in which case it is a hormone.

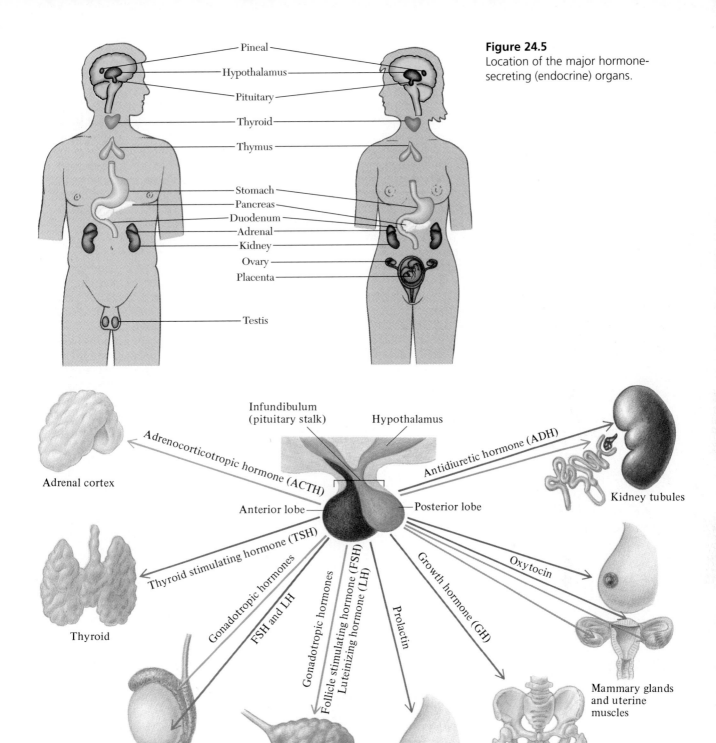

**Figure 24.5**
Location of the major hormone-secreting (endocrine) organs.

Pineal
Hypothalamus
Pituitary
Thyroid
Thymus
Stomach
Pancreas
Duodenum
Adrenal
Kidney
Ovary
Placenta
Testis

Infundibulum (pituitary stalk)
Hypothalamus
Adrenocorticotropic hormone (ACTH)
Antidiuretic hormone (ADH)
Adrenal cortex
Kidney tubules
Anterior lobe
Posterior lobe
Thyroid stimulating hormone (TSH)
Oxytocin
Gonadotropic hormones
FSH and LH
Gonadotropic hormones
Follicle stimulating hormone (FSH)
Luteinizing hormone (LH)
Prolactin
Growth hormone (GH)
Thyroid
Testis
Ovary
Mammary gland
Bones
Mammary glands and uterine muscles

**Figure 24.6**
The pituitary gland is suspended from the hypothalamus by a stalk of neural tissue. The hormones secreted by the anterior and posterior lobes of the pituitary gland and the target tissues they act upon are shown. (Modified from P. W. Davis and E. P. Solomon, *The World of Biology.* Philadelphia: Saunders College Publishing, 1986.)

**Box 24G**

## Diabetes

The disease *diabetes mellitus* affects about 10 million people in the United States. In a normal person the pancreas, a large gland behind the stomach, secretes the hormone insulin, as well as other hormones. Diabetes usually results from low insulin secretion. Insulin is necessary for glucose molecules to penetrate such cells as brain, muscle, and fat cells, where they can be used. Insulin accomplishes this task by being adsorbed onto the receptors in the target cells. This adsorption triggers the manufacture of cyclic GMP (not cAMP), and this secondary messenger increases the transport of glucose molecules into the target cells.

In diabetic patients the glucose level rises to 600 mg/100 mL of blood or higher (normal is 80 to 100 mg/100 mL). There are two kinds of diabetes. In insulin-dependent diabetes, patients do not manufacture enough of this hormone in the pancreas. This disease develops early, before the age of 20, and must be treated with daily injections of insulin. Even with daily injection of insulin the blood sugar level fluctuates, and the fluctuation may cause other disorders, such as cataracts, blindness, kidney disease, heart attack, and nervous disorders.

In non-insulin-dependent diabetes the patient has enough insulin in the blood but cannot utilize it properly because there is an insufficient number of receptors in the target cells. These patients usually develop the disease after age 40 and are likely to be obese. Overweight people usually have a lower-than-normal number of insulin receptors in their adipose (fat) cells.

There are several oral drugs that help this second type of diabetic patient. These are mostly sulfonyl urea compounds, such as

Tolbutamide (Orinase)

Chlorpropamide (Diabinese)

These drugs seem to control the symptoms of diabetes, but it is not known exactly how.

Chemically, hormones can be classified into three groups: (1) *small molecules* derived from amino acids, for example, epinephrine and thyroxine; (2) *peptides* and *proteins*, for example, insulin, glucagon, and vasopressin; (3) *steroids* derived from cholesterol, such as testosterone, cortisol, and aldosterone (Sec. 17.10).

Hormones can also be classified according to how they work. Some of them—epinephrine, for example—activate enzymes. Others *affect the synthesis of enzymes and proteins* by working on the transcription of genes (Sec. 23.7); steroid hormones work in this manner. Finally, some hormones affect the *permeability of membranes;* insulin and glucagon belong to this class.

Still another way of classifying hormones is according to their potential to act directly or through a secondary messenger. The steroid hormones can penetrate the cell membrane and also pass through the membrane of the nucleus. For example, estradiol stimulates uterine growth.

Other hormones act through secondary messengers, as mentioned in Section 24.4. For example, epinephrine, glucagon, luteinizing hormone, norepinephrine, and vasopressin use cAMP as a secondary messenger.

There are several diseases for which vaccines are available (polio, measles, small-pox, and others). A vaccine may be made up of dead or weakened viruses or bacteria. Small doses of the vaccine eventually confer immunity. For example, the Salk polio vaccine is a polio virus that has been made harmless by treatment with formaldehyde; it is given by intramuscular injection. In contrast, the Sabin polio vaccine is a live virus that must be taken orally so that the gastric juice can attenuate it, rendering it harmless by denaturing it before it penetrates the blood stream.

Vaccines change lymphocytes into plasma cells that produce large quantities of antibodies to fight any invading antigens. This is, however, only the immediate, short-term response. Some lymphocytes become *memory cells* rather than *plasma cells*. These memory cells do not secrete the antibody; they store it to serve as a detecting device for future invasion of the same foreign cells. In this way, long-term immunity is conferred. If a second invasion occurs, these memory cells divide directly into antibody-secreting plasma cells as well as into more memory cells. This time the response is faster, because it does not have to go through the process of activation and differentiation into plasma cells, which usually takes two weeks.

Smallpox, which was once one of humanity's worst scourges, has been totally wiped out, and smallpox vaccination is no longer required.

**Box 24H**

# Immunization

## 24.7 IMMUNOGLOBULINS

The third group of molecules that act as chemical communicators are the **immunoglobulins.** Foreign substances that invade the body are called **antigens.** Molecules that counteract these substances are called **antibodies.** Immunoglobulins are antibodies.

This chemical communication is the slowest of the three types. While neurotransmitters act within a millisecond and hormones within seconds, minutes, or hours, immunoglobulins respond to an antigen over a longer span of time: weeks and months.

Immunoglobulins are glycoproteins, that is, carbohydrate-carrying protein molecules. Not only do the different classes of immunoglobulins vary in molecular weight and carbohydrate content, but there is considerable variation in their concentration in the blood. The IgG and IgM antibodies are the most important antibodies in the blood. They interact with antigens and trigger the swallowing up (phagocytosis) of these cells by such specialized cells as the microphages and the leukocytes (white blood cells).

The IgA molecules are found mostly in secretions: tears, milk, and mucus. Therefore these immunoglobulins attack the invading material before it gets

The classes of immunoglobulins are shown in Table 24.3.

**TABLE 24.3 Immunoglobulin Classes**

| Class | MW | Carbohydrate Content (%) | Concentration in Serum (mg/100 mL) |
|---|---|---|---|
| IgA | 200 000–700 000 | 7–12 | 90–420 |
| IgD | 160 000 | <1 | 1–40 |
| IgE | 190 000 | 10–12 | 0.01–0.1 |
| IgG | 150 000 | 2–3 | 600–1800 |
| IgM | 950 000 | 10–12 | 50–190 |

**Figure 24.7**
Antigen, antibody, and antigen-antibody complex. Note that the antibody molecule is composed of two light chains and two heavy chains, which are joined together by disulfide bonds. The constant and variable regions of the chains are indicated. (Modified from P. W. Davis and E. P. Solomon, *The World of Biology*. Philadelphia: Saunders College Publishing, 1986.)

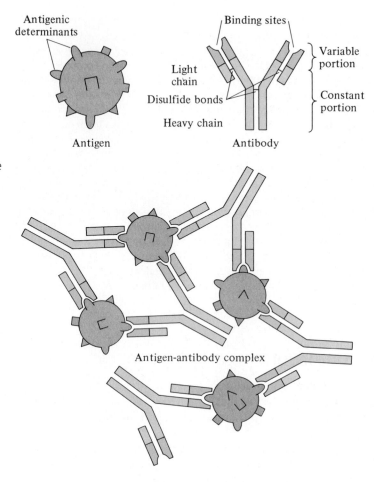

Antigen-antibody complex

In immunoglobulin IgG, the light chains have a molecular weight of 25 000 and the heavy chains of 50 000.

into the blood stream. IgE immunoglobulins play a part in allergic reactions: asthma and hay fever.

Each immunoglobulin molecule is made of four polypeptide chains. There are two identical light chains and two identical heavy chains in each molecule. The four polypeptide chains are arranged symmetrically, forming a "Y" shape (Fig. 24.7). Three disulfide bridges link the four chains into a unit. Both light

**B o x   24l**

**AIDS**

In recent years a deadly virus, commonly called the AIDS virus (for Acquired Immune Deficiency Syndrome), or HIV (for Human Immunodeficiency Virus), has spread alarmingly all over the world. The AIDS virus invades the human immune system, and especially enters the T (lymphocyte) cells and kills them. The human body has two kinds of responses to an invading foreign substance (an antigen). Cells known as B (lymphocyte) cells manufacture immunoglobulins as receptors on their surfaces. They can secrete these immunoglobulins, which bind directly to the antigen and thus act at a distance. In contrast, the T cells come into action when they touch another cell. They kill the invading foreign cell at short range or by cell-to-cell contact.

The AIDS virus, by invading the T cells and killing them, decreases the population of these cells and thus allows other opportunistic invaders, such as the protozoan *Pneumocystis carinii,* to proliferate, causing pneumonia and eventual death.

The progressive paralysis of the immune system makes the AIDS virus a lethal threat. Those who are ill with AIDS or have been exposed to the virus carry antibodies against the virus in their blood. One can detect these antibodies through patented tests, and all blood banks are now testing donated blood for AIDS infection. In the United States currently two to three out of 10 000 blood samples are found to carry antibodies against AIDS. With such screening, infections through blood transfusions are minimized.

At the time of writing, there is no known cure for AIDS. At least two approaches have been tried. One involves administering a drug called interleukin-2, which is supposed to stimulate the immune system to produce more T cells. This would increase the number of targeted cells, thereby enabling the immune system to function, although with diminished capacity.

In the other approach, an antiviral agent, azidothymidine (AZT), is used as an anti-AIDS drug. Clinical trials have shown that AIDS patients with *Pneumocystis carinii* pneumonia show some improvement after taking this drug. However, AZT is not considered a cure.

AZT is toxic, and most patients cannot tolerate therapeutic doses. After 26 weeks of therapy, bone marrow suppression occurs. Even more serious is that AZT provides only transient improvements—after two years the virus develops resistance to the drug. Two new drugs, dideoxyinosine (DDI) and dideoxycytidine (DDC), have been approved by the Food and Drug Administration for treatment of HIV infections.

**Figure 24I**
Scanning electron micrograph of the AIDS virus (HIV). (Courtesy of the National Library of Medicine).

Dideoxyinosine (DDI)

Dideoxycytidine (DDC)

DDI and DDC have been found to be beneficial in patients who cannot tolerate AZT or whose HIV becomes resistant to AZT. AZT is still the most frequently used anti-HIV drug. Like AZT, neither DDI nor DDC provides a cure. They only slow the development of the disease.

A vaccine would be the ideal remedy. But because of the deadly nature of the virus, one cannot use attenuated or even "killed" virus for immunization as was done with polio vaccine. Even if only one virus out of a million survived, it could kill the vaccinated person.

People with advanced AIDS often suffer from a blood infection called mycobacterium avium complex (MAC). The mycobacteria, related to the tuberculosis bacillus, are opportunistic invaders of the immunochemically weakened system, causing fever, severe anemia, and liver dysfunction, among other symptoms. Recently, the Food and Drug Administration approved a new drug, Rifabutin, that effectively blocks or delays the MAC blood infection.

and heavy chains have constant and variable regions. The constant regions have the same amino acid sequences in different antibodies, and the variable regions have different amino acid sequences in each antibody.

Studies have shown that many other molecules that function as communication signals between cells also have the same basic structural design as the immunoglobulins. They belong to what is known as the **immunoglobulin superfamily.** Glycoprotein molecules called cell-adhesion molecules, from nerve or liver cells, operate between cells and cause loose cells to adhere to each other to form layers and then tissues such as epithelium. They also play a significant part in the development of embryos.

## 24.8   HOW ANTIBODIES WORK

Any macromolecule, whether protein, carbohydrate, or nucleic acid, that is foreign to the body will cause an immune response. If the foreign material is a bacterium or virus, then macromolecules on the surfaces of these microorganisms initiate the immune response (Box 16C).

### Box 24J

### Using Antibodies to Locate Specific Cell Components

The immune response of the body is very specific. If a protein isolated from one animal, such as a rat, is injected into the blood stream of another animal (say, a goat), the protein is perceived as a foreign body, and the goat will produce specific antibodies against the foreign protein. Such antibodies, when isolated from the animal, can be made to fluoresce by covalently attaching to them a fluorescent dye. (Fluorescence is a phenomenon by which a substance is made to give off light of certain wavelengths when activated by light of shorter wavelengths.)

This phenomenon can be used to locate specific components within a cell. An example is shown in Figure 24J, where the protein, isolated from a rat, is called tubulin. In this case, we want to locate the position of tubulin in a rat kidney cell. A microscopic slide is prepared and stained with an antibody (goat anti-rat-tubulin) to which a fluorescent dye has been attached. The stain interacts with the tubulin that is present in the microtubules and, because of the specificity of the immune response, does not react with any other compound present. The display of brilliant fluorescent colors shows the location and the distribution of this specific protein within the cell.

**Figure 24J**
Fluorescent staining of the antibody that interacts with tubulin. (Courtesy of Dr. M. Kirscher and Dr. T. Mitchinson.)

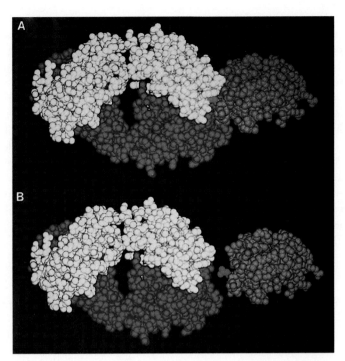

**Figure 24.8**
(a) An antigen-antibody complex. The antigen (shown in green) is lysozyme. The heavy chain of the antibody is shown in blue; the light chain in yellow. The most important amino acid residue (glutamine in the 121 position) on the antigen is the one that fits into the antibody groove (shown in red). (b) The antigen-antibody complex has been pulled apart. Note how they fit each other. (Courtesy of Dr. R. J. Poljak, Pasteur Institute, Paris, France.)

The variable regions of the antibody recognize the foreign substance (the antigen) and bind to it (Fig. 24.8). Since each antibody contains two variable regions, it can bind two antigens, and this results in a large aggregate that precipitates from the serum (Fig. 24.7). This is the **precipitin reaction.**

The binding of the antigen to the variable region of the antibody is not by covalent bonds but by much weaker forces such as London forces, dipole-dipole interactions, and hydrogen bonds. It is similar to the binding of neurotransmitters and hormones to a receptor site. The antigen has to fit into the antibody surface. Since there are a large number of different antigens against which the human body must fight, there are more than 10 000 different antibodies in our systems.

These forces are discussed in Section 5.7.

## SUMMARY

There are three kinds of molecules whose function is to communicate between different parts of the body. The neurotransmitters transmit **chemical signals** across a short distance—the **synapse** between two **neurons** or between a neuron and a muscle or endocrine gland cell. This communication occurs in milliseconds. The hormones transmit their signals more slowly and over a longer distance, from the source of their secretion (**endocrine gland**) through the blood stream into target cells. Immunoglobulins are antibodies that interact with large molecules that are foreign to the body (antigens).

Chemical communicators are called **mediators.** They interact with specific molecules called **receptors.**

There are three kinds of **neurotransmitters: cholinergic, adrenergic,** and **peptidergic.** Acetylcholine belongs to the first class, epinephrine (adrenalin) and norepinephrine to the second class, and enkephalins to the third class. Nerve transmission starts with the neurotransmitters packaged in **vesicles** in the **presynaptic end** of neurons. When these neurotransmitters are released, they cross the membrane and the synapse and are adsorbed onto receptor sites on the

**postsynaptic membranes.** This adsorption triggers an electrical response. Some neurotransmitters act directly, while others act through a secondary messenger, **cyclic AMP.** After the electrical signal is triggered, the neurotransmitter molecules must be removed from the postsynaptic end. In the case of acetylcholine, this is done by an enzyme called acetylcholinesterase; in the case of monoamines, by enzymes (MAOs) that oxidize them to aldehydes.

**Hormones** can be classified into three groups: small molecules, peptides and proteins, and steroids. The first two classes bind to receptors on the target cell membrane and use secondary messengers to exert their influence. The third class penetrates the cell membrane, and their receptors are in the cytoplasm. To-gether with their receptors, they penetrate the cell nucleus. Hormones can act in three ways: (1) They activate enzymes, (2) they affect the gene transcription of an enzyme or protein, and (3) they change membrane permeability.

**Immunoglobulins (antibodies)** are protein molecules containing small amounts of carbohydrates. They are made of two heavy chains and two light chains. All four are linked together by disulfide bridges. Immunoglobulins contain variable regions in which the amino acid composition of each antibody is different. These regions interact with **antigens.** Immunoglobulins interact with antigens to form insoluble large aggregates called **precipitin.**

## KEY TERMS

Adrenergic neurotransmitter (Sec. 24.4)
Antibody (Sec. 24.7)
Antigen (Sec. 24.7)
Axon (Sec. 24.2)
Chemical signal (Sec. 24.1)
Cholinergic neurotransmitter (Sec. 24.3)
Dendrite (Sec. 24.2)
Endocrine gland (Sec. 24.6)
Hormone (Sec. 24.6)
Immunoglobulin (Sec. 24.7)
Immunoglobulin superfamily (Sec. 24.7)

Mediator (Sec. 24.1)
Neuron (Sec. 24.1)
Neurotransmitter (Sec. 24.2)
Peptidergic neurotransmitter (Sec. 24.5)
Postsynaptic (Sec. 24.2)
Precipitin reaction (Sec. 24.8)
Presynaptic (Sec. 24.2)
Receptor (Sec. 24.1)
Synapse (Sec. 24.2)
Vesicle (Sec. 24.2)

## PROBLEMS

### Chemical Communication

**24.1**   What are the names given to the two principal types of molecules used in cell-to-cell communication?

**24.2**   What kind of signal travels along the axon of a neuron?

### Neurotransmitters

**24.3**   What is the difference between the presynaptic and postsynaptic ends of a neuron?

**24.4**   Define: (a) synapse  (b) receptor  (c) presynaptic  (d) postsynaptic  (e) mediator  (f) vesicle

### Cholinergic Neurotransmitters

**24.5**   List all functional groups in the products of the reaction catalyzed by acetylcholinesterase.

**24.6**   How do curare-poisoned arrows work?

**24.7**   (a) After a nerve signal is transmitted, how are the neurons reactivated (in the case of cholinergic neurotransmitters)?   (b) How long does this take?

**\*24.8**   Which end of the acetylcholine molecule fits into the receptor site?

**24.9**   Explain how succinylcholine acts as a muscle relaxant (see Box 19A).

**24.10**   Which carbons of cAMP contain the phosphate linkages?

## Adrenergic Neurotransmitters

**24.11** (a) Find two monoamine neurotransmitters in Table 24.1. (b) Explain how they act. (c) What medication controls the particular diseases caused by the lack of monoamine neurotransmitters?

**\*24.12** How much energy is used up in manufacturing one mole of cAMP?

**\*24.13** How is the catalytic unit of protein kinase activated in adrenergic neurotransmission?

**24.14** What is the advantage of neurotransmission through secondary messengers?

**24.15** The formation of cyclic AMP is described in Section 24.4. Show by analogy how cyclic GMP is formed from GTP.

**24.16** What happens to the cAMP in the cell once the neurotransmitter is removed from the receptor?

**24.17** Explain how adrenergic neurotransmission is affected by (a) amphetamines (b) reserpine (see Table 24.1)

**\*24.18** Which step in the events depicted in Figure 24.3 provides an electrical signal?

**24.19** What kind of product is epinephrine oxidized to by MAO?

**24.20** How is histamine removed from the receptor site.

**24.21** Cyclic AMP affects the permeability of membranes for ion flow. (a) What blocks the ion channel? (b) How is this blockage removed? (c) What is the direct role of cAMP in this process?

**24.22** Dramamine and cimetidine are both antihistamines. Would you expect Dramamine to cure ulcers and cimetidine to relieve the symptoms of asthma? Explain.

**24.23** How do alpha and beta alanine differ in structure?

**24.24** Is there any relation between the structure of bile salts (Sec. 17.11) and that of one of the amino acid neurotransmitters?

**24.25** What is unique in the structure of GABA that distinguishes it from all the amino acids that are present in proteins?

## Peptidergic Neurotransmitters

**24.26** What is the chemical nature of enkephalins?

**24.27** What is the mode of action of Demerol as a pain killer? (See Table 24.1.)

## Hormones

**\*24.28** The structures of vasopressin and oxytocin are given in Figure 18.4. They have completely different functions and different receptors. At which amino acid position, 3 or 8, is the difference greater? Explain.

**24.29** What gland controls lactation?

**24.30** To which of the three chemical groups do these hormones belong? (a) norepinephrine (b) thyroxine (c) oxytocin (d) progesterone

## Immunoglobulins, Antibodies, and Antigens

**24.31** Distinguish among the roles of IgA, IgE, and IgG immunoglobulins.

**24.32** (a) Which immunoglobulin has the highest carbohydrate content and the lowest concentration in the serum? (b) What is its main function?

**\*24.33** Box 16C states that the antigen in the red blood cells of a person with B-type blood is a galactose unit. Show schematically how the antibody of a person with A-type blood would aggregate the red blood cells of a B-type person if such a transfusion were made by mistake.

**24.34** In the immunoglobulin structure there is a region called the "hinge region" that joins the stem of the **Y** to the arms. The hinge region can be cleaved by a specific enzyme to yield one Fc fragment (the stem of the **Y**) and two Fab fragments (the two arms). Which of these two kinds of fragments can interact with an antigen? Explain.

**24.35** How are the light and heavy chains of an antibody held together?

**24.36** What do we mean by the term immunoglobulin superfamily?

**24.37** What is the function of cell adhesion molecules?

**24.38** What kind of interaction takes place between an antigen and an antibody?

## Boxes

**24.39** (Box 24A) Alzheimer's disease causes loss of memory. What kind of drugs may provide some relief for, if not cure, this disease? How do they act?

**24.40** (Box 24A) How is the acetyl group activated before the formation of acetylcholine?

**24.41** (Box 24B) (a) What is the effect of nerve gases? (b) What molecular substance do they affect?

**24.42** (Box 24C) How does tetrodotoxin paralyze the body?

**24.43** (Box 24D) Why would a dopamine pill be ineffective in treating Parkinson's disease?

**24.44** (Box 24D) How does the drug Deprenyl help Parkinson's disease patients?

**24.45** (Box 24E) How can a common food such as cheese cause headaches?

**24.46** (Box 24F) What are the most likely primary and secondary messengers in long-term potentiation ability?

**24.47** (Box 24F) How is nitric oxide synthesized in the cells?

**24.48** (Box 24G) What is the difference between insulin-dependent and non-insulin-dependent diabetes?

**24.49** (Box 24G) What is the common chemical feature of oral antidiabetic drugs?

**24.50** (Box 24H) What is the difference between memory cells and plasma cells?

**24.51** (Box 24I) What are the target cells for the AIDS virus?

**\*24.52** (Box 24I) What class of substances does DDI belong to?

**24.53** (Box 24I) What is the similarity in the structures of DDI and DDC?

**Additional Problems**

**24.54** What feature of the immunoglobulins enables them to interact with thousands of different antigens?

**24.55** Endorphin, a potent pain-killer, is a peptide containing 22 amino acids, among them the same five N-terminal amino acids found in the enkephalins. Does this explain its pain-killing action?

**\*24.56** By analogy to the action of MAO on epinephrine, write the structural formula of the product of the corresponding oxidation of dopamine.

**24.57** (a) In terms of their action, what do the hormone vasopressin and the neurotransmitter dopamine have in common? (b) What is the difference in their mode of action?

**24.58** Which immunoglobulins form the first line of defense against invading bacteria?

**24.59** What is the function of the ion-translocating protein in adrenergic neurotransmission?

**\*24.60** Considering its chemical nature, how does aldosterone (Sec. 17.10) affect mineral metabolism (Table 24.2)?

**\*24.61** The variable regions of immunoglobulins bind the antigens. How many polypeptide chains carry variable regions in one immunoglobulin molecule?

**\*24.62** Decamethionium acts as a muscle relaxant. If an overdose of decamethionium occurs, can paralysis be prevented by administering large doses of acetylcholine? Explain.

# CHAPTER 25

*Blood cells from a patient with sickle cell anemia. Both normal (round) and sickle (shriveled) cells are visible. (G. W. Willis, M.D./Biological Photo Service.)*

# Body Fluids

## 25.1 INTRODUCTION

Single-cell organisms receive their nutrients directly from the environment and discard waste products directly into it. In multicellular organisms, the situation is not so simple. There, too, each cell needs nutrients and produces wastes, but most of the cells are not directly in contact with the environment.

Body fluids serve as a medium for carrying nutrients to and waste products from the cells and also for carrying the chemical communicators (Chapter 24) that coordinate activities among cells.

All the body fluids that are not inside the cells are collectively known as **extracellular fluids.** These fluids make up about one quarter of a person's body weight. The most abundant is **interstitial fluid,** which directly surrounds most cells and fills the spaces between them. It makes up about 17 percent of body weight. Another body fluid is **blood plasma,** which flows in the arteries and veins. This makes up about 5 percent of body weight. Other body fluids that occur in lesser amounts are urine, lymph, cerebrospinal fluid, aqueous humor, and synovial fluid. All body fluids are aqueous solutions. Water is the only solvent in the body.

The blood plasma circulates in the body and is in contact with the other body fluids through the semipermeable membranes of the blood vessels. Therefore, blood can exchange chemical compounds with other body fluids (Fig. 25.1) and, through them, with the cells and organs of the body.

There is, however, only a limited exchange between blood and cerebrospinal fluid, on the one hand, and blood and the interstitial fluid of the brain, on the other. The limited exchange between blood and interstitial fluid is referred

The fluid inside the cells is called **intracellular** fluid.

**Figure 25.1**

Exchange of compounds among three body fluids: blood, interstitial fluid, and lymph. (After J. R. Holum, *Fundamentals of General, Organic and Biological Chemistry*. New York: John Wiley & Sons, 1978, p. 569.)

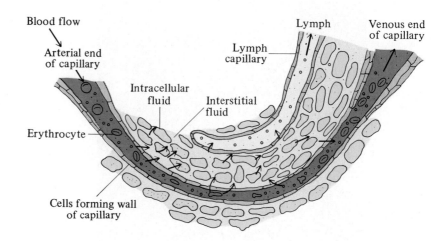

to as the **blood-brain barrier.** This is permeable to water, oxygen, carbon dioxide, glucose (Fig. 25.2), alcohols, and most anesthetics, but only slightly permeable to electrolytes such as $Na^+$, $K^+$, and $Cl^-$ ions. Many higher-molecular-weight compounds are also excluded.

The blood-brain barrier protects the cerebral tissue from detrimental substances in the blood and allows it to maintain low $K^+$ concentration, which is needed to generate the high electrical potential essential for neurotransmission.

Body fluids have special importance for the health care professions. Samples of body fluid can be taken with relative ease. The chemical analysis of blood plasma, blood serum, urine, and occasionally cerebrospinal fluid is of major importance in diagnosing disease.

**Figure 25.2**

Positron emission tomography brain scans showing that glucose can cross the blood-brain barrier. The second scan in each case shows the result of visual stimulation. Both the local cerebral blood flow (48 percent increase) and the glucose metabolism (56 percent increase) go up during visual stimulation. (Courtesy of Dr. Peter T. Fox, Washington University School of Medicine.)

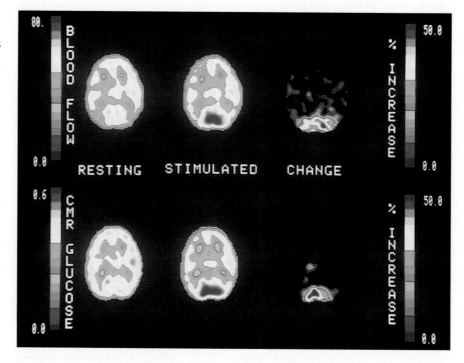

Many drugs have undesirable side effects. For example, many antihistamines, such as Dramamine and Benadryl (see p. 693 and Table 24.1), cause drowsiness. These antihistamines are supposed to act on the peripheral $H_1$ histamine receptors to relieve seasickness, hay fever, or asthma. But because they penetrate the blood-brain barrier, they also act as antagonists to the $H_1$ receptors in the brain, causing sleepiness. A drug that also acts on the peripheral $H_1$ receptors, terfenadine (sold under the trade name Seldane), cannot penetrate the blood-brain barrier. This antihistamine alleviates seasickness and asthma in the same way as the old antihistamines, but it does not cause drowsiness as a side effect.

**B o x 25A**

## Using the Blood-Brain Barrier to Eliminate Undesirable Side Effects of Drugs

Terfenadine

## 25.2 FUNCTIONS AND COMPOSITION OF BLOOD

It has been known for centuries that "life's blood" is essential to human life. The blood has many functions, including the following:

1. It carries $O_2$ from the lungs to the tissues.
2. It carries $CO_2$ from the tissues to the lungs.
3. It carries nutrients from the digestive system to the tissues.
4. It carries waste products from the tissues to the excretory organs.
5. With its buffer systems, it maintains the pH of the body (with the help of the kidneys).
6. It maintains a constant body temperature.
7. It carries hormones from the endocrine glands to wherever they are needed.
8. It fights infection.

The rest of this chapter describes how the blood carries out some of these functions.

Figure 25.3 shows some of these functions.

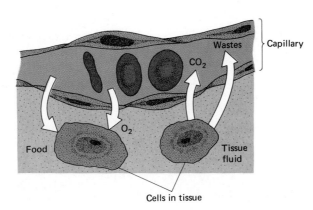

**Figure 25.3**
Nutrients, oxygen, and other materials diffuse out of the blood and through the tissue fluid that bathes the cells. Carbon dioxide and other waste products diffuse out of the cells and enter the blood through the capillary wall. (From C. A. Villee, E. P. Solomon, and P. W. Davis, *Biology*. Philadelphia: Saunders College Publishing, 1985.)

**Figure 25.4**
Some cellular elements of blood. The dimensions are shown in microns ($\mu$m; also called micrometers). (From Villee, Solomon, and Davis, *Biology*. Philadelphia: Saunders College Publishing, 1985.)

7 $\mu$m

ERYTHROCYTES

1 to 2 $\mu$m

THROMBOCYTES
(PLATELETS)

LEUKOCYTES

Granulocytes                                    Agranular leukocytes

10 to 14 $\mu$m

Neutrophil

15 to 20 $\mu$m

Monocyte

8 to 10 $\mu$m

Lymphocyte

A 150-pound (68-kg) man has about 6 L of whole blood, 50 to 60 percent of which is plasma.

Whole blood is a complicated mixture. It contains several types of cells (Fig. 25.4) and a liquid, noncellular portion called plasma, in which many substances are dissolved (Table 25.1). There are three main types of **cellular elements** of blood: erythrocytes, leukocytes, and platelets.

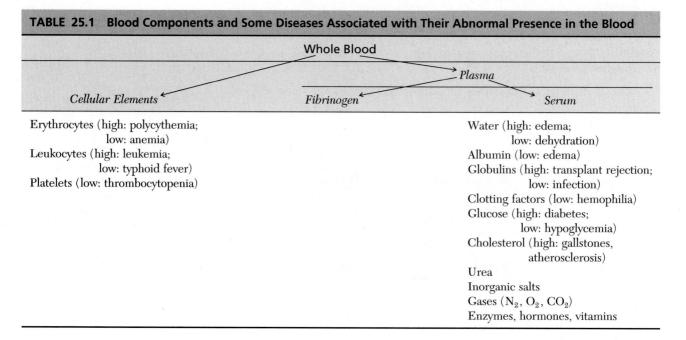

**TABLE 25.1   Blood Components and Some Diseases Associated with Their Abnormal Presence in the Blood**

Whole Blood

*Plasma*

*Cellular Elements*                    *Fibrinogen*                    *Serum*

| Cellular Elements | Serum |
|---|---|
| Erythrocytes (high: polycythemia; low: anemia) | Water (high: edema; low: dehydration) |
| Leukocytes (high: leukemia; low: typhoid fever) | Albumin (low: edema) |
| Platelets (low: thrombocytopenia) | Globulins (high: transplant rejection; low: infection) |
| | Clotting factors (low: hemophilia) |
| | Glucose (high: diabetes; low: hypoglycemia) |
| | Cholesterol (high: gallstones, atherosclerosis) |
| | Urea |
| | Inorganic salts |
| | Gases ($N_2$, $O_2$, $CO_2$) |
| | Enzymes, hormones, vitamins |

### Erythrocytes

The most prevalent cells of blood are the red blood cells, which are called **erythrocytes.** There are about 5 million red blood cells in every cubic millimeter of blood, or roughly 100 million in every drop. Erythrocytes are very specialized cells. They have no nuclei and hence no DNA. Their main function is to carry oxygen to the cells and carbon dioxide away from the cells. Erythrocytes are formed in the bone marrow. They stay in the blood stream for about 120 days. Old erythrocytes are removed by the liver and spleen and destroyed. The constant formation and destruction of red blood cells maintain a steady number of erythrocytes.

In an adult male, there are approximately 30 trillion erythrocytes.

### Leukocytes

**Leukocytes** (white blood cells) are relatively minor cellular components (minor in numbers, not in function). For each 1000 red blood cells, there are only one or two white blood cells. Most of the different leukocytes destroy invading bacteria or other foreign substances by devouring them (**phagocytosis**). Like erythrocytes, leukocytes are manufactured in the bone marrow. Specialized white blood cells formed in the lymph nodes and in the spleen are called **lymphocytes.** They synthesize immunoglobulins (antibodies, Sec. 24.7) and store them.

### Platelets

When a blood vessel is cut or injured, the bleeding is controlled by a third type of cellular element: the **platelets,** or **thrombocytes.** They are formed in the bone marrow and spleen and are more numerous than leukocytes but less numerous than erythrocytes.

There are about 300 000 platelets in each cubic millimeter of blood, or one for every 10 or 20 red blood cells.

### Plasma

If all the cellular elements from whole blood are removed by centrifugation, the resulting liquid is the **plasma.** The cellular elements—mainly red blood cells, which settle at the bottom of the centrifuge tube—occupy between 40 and 50 percent of the blood volume. The measured volume percentage of cellular elements, known as the **hematocrit value,** is used for diagnosis.

Blood plasma is 92 percent water. The dissolved solids in the plasma are mainly proteins (7 percent). The remaining 1 percent contains glucose, lipids, enzymes, vitamins, hormones, and such waste products as urea and $CO_2$. Of the plasma proteins, 55 percent is albumin, 38.5 percent is globulin, and 6.5 percent is fibrinogen. If plasma is allowed to stand, it forms a clot, a gel-like substance. We can squeeze out a clear liquid from the clot. This fluid is the **serum.** It contains all the components of the plasma except the fibrinogen. This protein is involved in the complicated process of clot formation (Box 25B).

A dried clot becomes a scab.

As for the other plasma proteins, most of the globulins take part in the immune reactions (Sec. 24.8), and the albumin provides proper osmotic pressure. If the albumin concentration drops (from malnutrition or from kidney disease, for example), the water from the blood oozes into the interstitial fluid and creates the swelling of tissues called **edema.**

# Blood Clotting

When body tissues are damaged, the blood flow must be stopped or else enough of it will pour out to cause death. The mechanism used by the body to stem leaks in the blood vessels is **clotting.** It is a complicated process involving many factors. Here we mention only a few important steps.

When a blood vessel is injured, the first line of defense is the platelets, which are constantly circulating in the blood. These rush to the site of injury, adhere to the collagen molecules in the capillary wall that are exposed by the cut, and form a gel-like plug. This plug is porous, however, and in order to seal the site a firmer gel (a **clot**) is needed. The clot is a three-dimensional network of fibrin molecules that also contains the platelets (Fig. 25B). The fibrin network is formed from the blood fibrinogen by the enzyme thrombin. Together with the embedded platelets this constitutes the blood clot.

**Figure 25B**
Components of a blood clot. (From ''The Functioning of Blood Platelets'' by M. B. Zucker. Copyright 1980 by Scientific American, Inc. All rights reserved.)

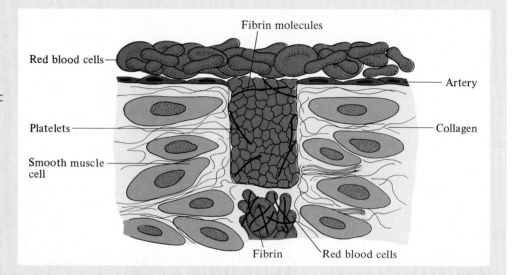

The question arises: Why doesn't the blood clot in the blood vessels under normal conditions (with no injury or disease)? The reason is that the enzyme that starts clot formation, thrombin, exists in the blood only in its inactive form, called prothrombin. Prothrombin itself is manufactured in the liver, and vitamin K is needed for its production. Even when prothrombin is in sufficient supply, a number of proteins are needed to change it to thrombin. These proteins are given the collective name *thromboplastin.* Any thromboplastic substance can activate prothrombin in the presence of $Ca^{2+}$ ions. Thromboplastic substances exist in the platelets, in the plasma, and in the injured tissue itself.

Clotting is nature's way of protecting us from loss of blood. However, we don't want the blood to clot during blood transfusions because this would stop the flow. To prevent this, we add sodium citrate. Since sodium citrate interacts with $Ca^{2+}$ ions and removes them from the solution, the thromboplastic substances cannot activate prothrombin in the absence of $Ca^{2+}$ ions, and no clot forms. After surgery, anticoagulant drugs are occasionally administered to prevent clot formation. A clot is not dangerous if it stays near the injury, because once the body repairs the tissue, the clot is digested and removed. However, a clot formed in one part of the body may break loose and travel to other parts, where it may lodge in an artery. This is **thrombosis.** If the clot then blocks oxygen and nutrient supply to the heart and brain, it can result in paralysis and death. The most common anticoagulants are bishydroxycoumarin (Dicumarol) and heparin. Heparin enhances the inhibition of thrombin by antithrombin (Fig. 16.19), and bishydroxycoumarin blocks the transport of vitamin K to the liver, preventing prothrombin formation.

## 25.3   BLOOD AS A CARRIER OF OXYGEN

One of the most important functions of blood is to carry oxygen from the lungs to the tissues. This is done by hemoglobin molecules located inside the erythrocytes. As we saw in Section 18.9, hemoglobin is made up of two alpha and two beta protein chains, each attached to a molecule of heme (Fig. 18.14).

The active sites are the hemes, and at the center of each heme is an iron(II) ion. The heme with its central $Fe^{2+}$ ion forms a plane. Because each hemoglobin molecule has four hemes, it can hold a total of four $O_2$ molecules. However, the ability of the hemoglobin molecule to hold $O_2$ depends on how much oxygen is in the environment. To see this, look at Figure 25.5, which shows how the oxygen-carrying ability of hemoglobin depends on oxygen pressure. When oxygen enters the lungs, the pressure is high (100 mm Hg). At this pressure, all the $Fe^{2+}$ ions of the hemes bind oxygen molecules; they are fully saturated. By the time the blood reaches the muscles through the capillaries, the oxygen pressure in the muscle is only 20 mm Hg. At this pressure only 30 percent of the binding sites carry oxygen. Thus, 70 percent of the oxygen carried will be released to the tissues. The S shape of the binding (dissociation) curve (Fig. 25.5) implies that

$$HbO_2 \rightleftharpoons Hb + O_2$$

is not a simple equilibrium reaction. Each heme has a cooperative effect on the other hemes. This cooperative action allows hemoglobin to deliver twice as much oxygen to the tissues as it would if each heme acted independently. The reason for this is as follows: When a hemoglobin molecule is carrying no oxygen, the four globin units coil into a certain shape (Fig. 18.14). When the first oxygen molecule attaches to one of the heme subunits, it changes the shape not only of that subunit but also of a second one, making it easier for the second one to bind an oxygen. When the second heme binds, the shape changes once again, and the oxygen-binding ability of the two remaining subunits is increased still more.

The $O_2$ picked up by the hemoglobin binds to the $Fe^{2+}$.

This is an allosteric effect (Sec. 19.6) and explains the S-shaped curve shown in Figure 25.5.

**Figure 25.5**
An oxygen dissociation curve. Saturation (%) means the percentage of $Fe^{2+}$ ions that carry $O_2$ molecules.

Lowering the pH decreases the oxygen-binding capacity of hemoglobin.

This is how the body delivers more oxygen to those tissues that need it.

The oxygen-delivering capacity of hemoglobin is also affected by its environment. A slight change in the pH of the environment changes the oxygen-binding capacity. This is called the **Bohr effect.** An increase in $CO_2$ pressure also decreases the oxygen-binding capacity of hemoglobin. When a muscle contracts, both $H^+$ ions and $CO_2$ are produced. The $H^+$ ions, of course, lower the pH of the muscle. Thus, at the same pressure (20 mm Hg), more oxygen is released for an active muscle than for a muscle at rest. Similarly, an active contracting muscle also produces $CO_2$, which then accumulates and further enhances the release of oxygen.

## 25.4 TRANSPORT OF CARBON DIOXIDE IN THE BLOOD

At the end of the previous section we saw that the waste products of tissue cells, $H^+$ ions and $CO_2$, facilitate the release of oxygen from hemoglobin. This enables the cells to receive the oxygen they need. What happens to the $CO_2$ and $H^+$? They bind to the hemoglobin. There is an equilibrium in which, upon the release of oxygen, **carbaminohemoglobin** is formed:

$$HbO_2 + H^+ + CO_2 \rightleftharpoons Hb\big\langle\substack{H^+ \\ CO_2} + O_2$$

Carbaminohemoglobin

The $CO_2$ is bound to the terminal $-NH_2$ groups of the four polypeptide chains so that each hemoglobin can carry a maximum of four $CO_2$ molecules, one for each chain. How much $CO_2$ each hemoglobin actually carries depends on the pressure of $CO_2$. The higher the $CO_2$ pressure, the more carbaminohemoglobin is formed.

The remaining 5 percent of the $CO_2$ is transported as $CO_2$ gas dissolved in the plasma.

But only 25 percent of the total $CO_2$ produced by the cells is transported to the lungs in the form of carbaminohemoglobin. Another 70 percent is converted in the red blood cells to carbonic acid by the enzyme carbonic anhydrase:

$$CO_2 + H_2O \underset{\text{anhydrase}}{\overset{\text{carbonic}}{\rightleftharpoons}} H_2CO_3$$

The reaction proceeds to the left because loss of $CO_2$ from the lungs causes this equilibrium to shift to the left (Le Chatelier's principle).

A large part of this $H_2CO_3$ is carried as such to the lungs, where it is converted back to $CO_2$ by carbonic anhydrase and released.

## 25.5 BLOOD BUFFERS

The average pH of human blood is 7.4. Any change larger than 0.1 pH unit in either direction is pathological, and if the pH goes below 6.8 or above 7.8, death can result. When the pH drops below normal, the condition is called **acidosis;** when it rises above 7.45, it is called **alkalosis** (Box 8E). There are three buffer systems that maintain an average blood pH of 7.4. The most important is the $HCO_3^-/H_2CO_3$ buffer (Sec. 8.10). The other two are the $HPO_4^{2-}/H_2PO_4^-$ buffer (Sec. 8.10) and the protein molecules dissolved in the plasma (Sec. 18.6).

## 25.6  BLOOD CLEANSING—A KIDNEY FUNCTION

We have seen that one of the functions of the blood is to carry waste products away from the cells. $CO_2$, one of the principal waste products of respiration, is carried by the blood to the lungs and exhaled. The other waste products are filtered out by the kidneys and eliminated in the urine. The kidney is a superfiltration machine. In the event of kidney failure, filtration can be done by hemodialysis (Fig. 25.6). About 100 L of blood passes through a normal human kidney daily. Of this, only about 1.5 L is excreted as urine. Obviously, we do not have here just a simple filtration system in which small molecules are lost and large ones are retained. The kidneys also reabsorb from the urine those small molecules that are not waste products.

The biological units inside the kidneys that perform these functions are called **nephrons,** and each kidney contains about a million of them. A nephron is made up of a filtration head called a **Bowman's capsule** connected to a tiny, U-shaped tube called a **tubule.** The part of the tubule close to the Bowman's capsule is the **proximal tubule,** the U-shaped twist is called **Henle's loop,** and the part of the tubule farthest from the capsule is the **distal tubule** (Fig. 25.7).

Blood vessels penetrate the kidney throughout. The arteries branch into capillaries, and one tiny capillary enters each Bowman's capsule. Inside the capsule the capillary first branches into even smaller vessels, called **glomeruli,** and then leaves the capsule. The blood enters the glomeruli with every heartbeat, and the pressure forces the water, ions, and all small molecules (urea, sugars, salts, amino acids) through the walls of the glomeruli and the Bowman's capsule. These molecules and ions enter the proximal tubule. Blood cells and large molecules (proteins) are retained in the capillary and exit with the blood.

Hemodialysis is discussed in Box 6G.

The balance between filtration and reabsorption is controlled by a number of hormones.

Arteries are blood vessels that carry oxygenated blood away from the heart.

The singular is glomerulus.

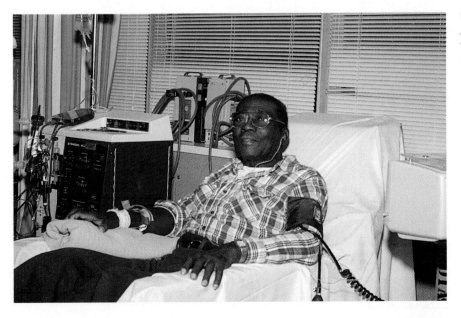

**Figure 25.6**
A patient undergoing hemodialysis for kidney disease. (Photograph by Beverly March; courtesy of Long Island Jewish Hospital.)

## Composition of Body Fluids

**Figure 25C**
The composition of body fluids.

In Section 25.1 it was pointed out that there are three main body fluids: intracellular fluid, interstitial fluid, and blood plasma. Figure 25C shows the average concentrations of the substances dissolved in these three fluids. Blood plasma and interstitial fluid have similar compositions (the main difference is that the blood has a higher protein concentration), but intracellular fluid is quite different. On the cation side, the intracellular fluid has a high concentration of $K^+$ and a low concentration of $Na^+$; in the other two fluids, these concentrations are reversed. Intracellular fluid also has a higher $Mg^{2+}$ concentration. On the anion side, there is virtually no $Cl^-$ inside cells, while this is the most abundant anion outside the cells. On the other hand, the concentrations of phosphate (mostly $HPO_4^{2-}$) and, to a lesser extent, of sulfate are much greater inside the cells. As might be expected, the protein concentration is much higher inside the cells.

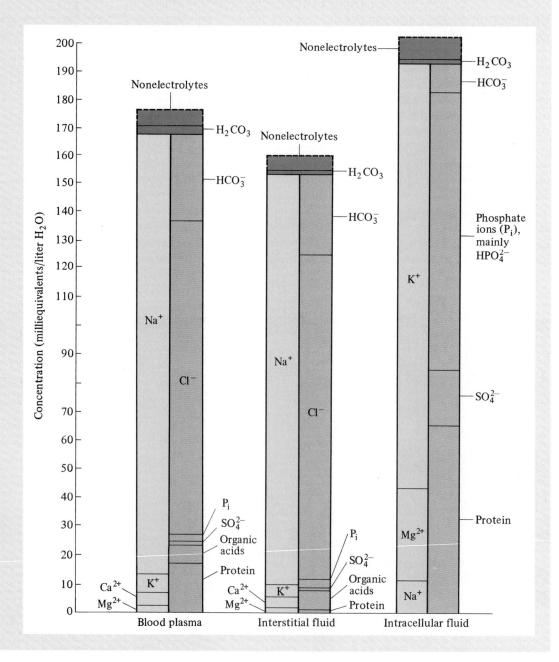

As shown in Figure 25.7, the tubules and Henle's loop are surrounded by blood vessels; these blood vessels reabsorb vital nutrients. Eighty percent of the water is reabsorbed in the proximal tubule. Almost all the glucose and amino acids are reabsorbed here. When excess sugar is present in the blood (diabetes, Box 24G), some of it passes into the urine, and the measurement of glucose concentration in the urine is used in diagnosing diabetes.

By the time the glomerular filtrate reaches Henle's loop, the solids and most of the water have been reabsorbed, and only wastes (such as urea, creatinine, uric acid, ammonia, and some salts) pass into the collecting tubules that bring the urine to the ureter, from which it passes to the bladder, as shown in Figure 25.7(a).

## Urine

Normal urine contains about 4 percent dissolved waste products. The rest is water. The daily amount of urine varies greatly but averages about 1.5 L per day. The pH varies from 5.5 to 7.5. The main solute is urea, the end product of

**Figure 25.7**
Excretion through the kidneys. (a) The human urinary system. (b) A kidney nephron and its components, with the surrounding circulatory system. (After T. M. Graham, *Biology: The Essential Principles.* Philadelphia: Saunders College Publishing, 1982, pp. 607 and 609.)

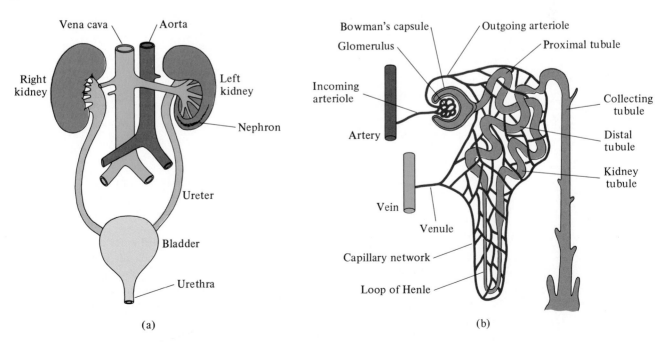

(a)                                                                 (b)

The same sex hormones that provide a healthy reproductive life in the young can and do cause problems in the aging body. For male fertility the prostate gland, among others, provides the semen. The prostate gland is situated at the exit of the bladder surrounding the ureter. Its growth is promoted by the conversion of the male sex hormone, testosterone, to 5-alpha-dihydrotestosterone by the enzyme 5-alpha-reductase. The majority of men over the age of 50 have enlarged prostate glands, a condition known as benign prostatic hyperplasia. This condition affects more than 90 percent of men over the age of 70. The enlarged prostate gland constricts the ureter and causes a restricted flow of urine.

The size of the prostate gland can be reduced by the oral intake of a synthetic steroid drug called finasteride (marketed as Proscar), which is a competitive inhibitor of 5-alpha-reductase. Thus, less conversion of testosterone takes place, reducing the growth of the prostate gland.

Estradiol and progesterone control sexual characteristics in the female body (Sec. 17.10). With the onset of the menopause, smaller amounts of these and other sex hormones are secreted. This not only stops fertility but occasionally creates medical problems. The most serious is *osteoporosis,* which is the loss of bone tissue through absorption. This results in brittle and occasionally deformed bone structure. Estrogens (a general name for female sex hormones other than progesterone) in the proper amount help to absorb calcium from the diet and thus prevent osteoporosis. The lack of enough female sex hormones can also result in atrophic vaginitis and atrophic urethritis. A large number of postmenopausal women take pills that contain a synthetic progesterone analog and natural estrogens. However, continuous medication with these pills, for example, Premarin (mixed estrogens) and Provera (a progesterone analog), is not without danger. A slight but statistically significant increase in cancer of the uterus has been reported in postmenopausal women who use such medication. A physician's advice is needed in weighing the benefits against the risks.

protein metabolism (Sec. 21.8). Other nitrogenous waste products, such as creatine, creatinine, ammonia, and hippuric acid,

Creatine

Creatinine

Hippuric acid

are also present, though in much smaller amounts. In addition, normal urine contains inorganic ions such as $Na^+$, $Ca^{2+}$, $Mg^{2+}$, $Cl^-$, $PO_4^{3-}$, $SO_4^{2-}$, and $HCO_3^-$.

Under certain pathological conditions, other substances can appear in the urine. It is because of this that urine analysis is such an important part of diagnostic medicine. We have already noted that the presence of glucose and ketone bodies is an indication of diabetes. Among other abnormal constituents may be proteins, which can indicate such kidney diseases as nephritis.

## 25.7 BUFFER PRODUCTION—ANOTHER KIDNEY FUNCTION

Among the waste products sent into the blood by the tissues are $H^+$ ions. These are neutralized by the $HCO_3^-$ ions that are a part of the blood's buffer system:

$$H^+ + HCO_3^- \rightleftharpoons H_2CO_3$$

When the blood reaches the lungs, the $H_2CO_3$ is decomposed by carbonic anhydrase, and the $CO_2$ is exhaled. If the body had no mechanism for replacing $HCO_3^-$, we would lose most of the bicarbonate buffer from the blood, and the blood pH would eventually drop (acidosis). The lost $HCO_3^-$ ions are continuously replaced by the kidneys—this is another principal kidney function. The replacement takes place in the distal tubules. The cells lining the walls of the distal tubules reabsorb the $CO_2$ that was lost in the glomeruli. With the aid of carbonic anhydrase, carbonic acid forms quickly and then dissociates to $HCO_3^-$ and $H^+$ ions:

$$CO_2 + H_2O \rightleftharpoons H_2CO_3 \rightleftharpoons HCO_3^- + H^+$$

The $H^+$ ions move from the cells into the urine in the tubule, where they are partially neutralized by a phosphate buffer. To compensate for the lost positive ions, $Na^+$ ions from the tubule enter the cells. When this happens, $Na^+$ and $HCO_3^-$ ions move from the cells into the capillary. Thus, the $H^+$ ions picked up at the tissues and temporarily neutralized in the blood by $HCO_3^-$ are finally pumped out into the urine. At the same time, the $HCO_3^-$ ions lost in the lungs are regained by the blood in the distal tubules.

## 25.8 WATER AND SALT BALANCE IN BLOOD AND KIDNEYS

We mentioned in Section 25.6 that the balance in the kidneys between filtration and reabsorption is under hormonal control. The reabsorption of water is promoted by vasopressin, a small peptide hormone manufactured in the hypophysis (Sec. 18.7, Fig. 18.4). In the absence of this hormone, only the proximal tubules—not the distal tubules or the collecting tubules—reabsorb water. As a consequence, without vasopressin too much water passes into the urine. In the presence of vasopressin, water is also reabsorbed in these parts of the nephrons; thus, vasopressin causes the blood to retain more water and produces a more concentrated urine. The production of urine is called **diuresis.** Any agent that reduces the volume of urine is called an **antidiuretic.**

Patients with diabetes insipidus suffer from this condition.

Usually, the vasopressin level in the body is sufficient to maintain the proper amount of water in tissues under various levels of water intake. However, when severe dehydration occurs (as a result of diarrhea, excessive sweating, or insufficient water intake), another hormone also helps to maintain proper fluid level. This hormone is aldosterone (Sec. 17.10); it controls the $Na^+$ ion concentration in the blood. In the presence of aldosterone, the reabsorption of $Na^+$ ions is increased. When more $Na^+$ ions enter the blood, more $Cl^-$ ions follow (to maintain electroneutrality), as well as more water to solvate these ions. Thus, increased aldosterone production enables the body to retain more water. When the $Na^+$ ion and water levels in the blood return to normal, aldosterone production stops.

For this reason vasopressin is also called ADH, for *antidiuretic hormone.*

## Hypertension and Its Control

Nearly 60 million people in the United States have high blood pressure **(hypertension)**. Epidemiologic studies have shown that even mild hypertension (diastolic pressure between 90 and 104 mm Hg) brings about the risk of cardiovascular disease. This can lead to heart attack, stroke, or kidney failure. Hypertension can be managed effectively with diet and drugs. As noted in the text, blood pressure is under a complex control system, so hypertension must be managed on more than one level. Recommended dietary practices are low $Na^+$ ion intake and abstention from caffeine and alcohol.

The most common drugs for lowering blood pressure are the **diuretics.** There are a number of synthetic organic compounds that increase urine excretion. By doing so, they decrease the blood volume as well as the $Na^+$ ion concentration and thus lower blood pressure. Other drugs affect the nerve control of blood pressure. Propranolol blocks the adrenoreceptor sites of adrenergic neurotransmitters (Sec. 24.4). By doing so, it reduces the flow of nerve signals and thereby reduces the blood output of the heart; it affects the responses of baroreceptors and chemoreceptors in the central nervous system and the adrenergic receptors in the smooth muscles surrounding the blood vessels. Propranolol and other blockers not only lower blood pressure but also reduce the risk of myocardial infarction. Similar lowering of blood pressure is achieved by drugs that block calcium channels (one example is the drug Cardizem). For proper muscle contraction, calcium ions must pass through muscle cell membranes. This is facilitated by calcium channels. If these channels are blocked by the drug, the heart muscles will contract less frequently, pumping less blood and reducing blood pressure.

Other drugs that reduce hypertension are the vasodilators that relax the vascular smooth muscles. However, many of these have side effects causing fast heartbeat (tachycardia) and also promote $Na^+$ and water retention. Some new drugs on the market act on one specific site rather than on many sites as the blockers do. This obviously reduces the possible side effects. A recent addition is an enzyme inhibitor that prevents the production of angiotensin. Captopril and similar specific enzyme inhibitors lower blood pressure only if the sole cause of hypertension is angiotensin production. Prostaglandins PGE and PGA (Section 17.12) lower the blood pressure and, taken orally, have prolonged duration of action. Since hypertension is such a common problem, and it is manageable by drugs, there is great activity in drug research to come up with even more effective and safer antihypertension drugs.

## 25.9 BLOOD PRESSURE

Blood pressure is produced by the pumping action of the heart. The blood pressure at the arterial capillary end is about 32 mm Hg. This pressure is higher than the osmotic pressure of the blood (18 mm Hg). The osmotic pressure of the blood is caused by the fact that more solutes are dissolved in the blood than in the interstitial fluid. At the capillary end, therefore, nutrient solutes flow from the capillary into the interstitial fluid and from there into the cell (Fig. 25.1). On the other hand, in the venous capillary the blood pressure is about 12 mm Hg. This is less than the osmotic pressure, so solutes (waste products) from the interstitial fluid flow into the capillaries.

The blood pressure is maintained by the total volume of blood, by the pumping of the heart, and by the muscles that surround the blood vessels and provide the proper resistance to blood flow. Blood pressure is controlled by several very complex systems, some of them acting within seconds and some that take days to respond after a change in blood pressure occurs. For example, if a patient hemorrhages, three different nervous control systems begin to function within seconds. The first is the **baroreceptors** in the neck, which detect the consequent drop in pressure and send appropriate signals to the heart to pump harder and to the muscles surrounding the blood vessels to contract and thus restore pressure. Chemical receptors on the cells that detect less $O_2$ delivery or $CO_2$ removal also send nerve signals. Finally, the central nervous system also reacts to oxygen deficiency with a feedback mechanism.

Hormonal controls act somewhat more slowly and may take minutes or even days. The kidneys secrete an enzyme called **renin,** which acts on an inactive blood protein called **angiotensinogen,** converting it to **angiotensin,** a potent vasoconstrictor. The action of this peptide increases blood pressure. Aldosterone (Sec. 25.8) also increases blood pressure by increasing $Na^+$ ion and water reabsorption in the kidneys.

Finally, there is a long-term renal blood control for volume and pressure. When blood pressure falls, the kidneys retain more water and salt, thus increasing blood volume and pressure.

## SUMMARY

The most important body fluid is **blood plasma,** which is whole blood from which cellular elements have been removed. Other important body fluids are **urine** and the **interstitial fluids** directly surrounding the cells of the tissues. The **cellular elements** of the blood are the **red blood cells (erythrocytes),** which carry $O_2$ to the tissues and $CO_2$ away from the tissues; the **white blood cells (leukocytes),** which fight infection; and the **platelets,** which control bleeding. The plasma contains **fibrinogen,** which is necessary for blood clot formation. When fibrinogen is removed from the plasma, what remains is the **serum.** The serum contains albumins, globulins, nutrients, waste products, inorganic salts, enzymes, hormones, and vitamins dissolved in water.

The red blood cells carry the $O_2$. The Fe(II) in the heme portion of hemoglobin binds the oxygen, which is then released at the tissues by a combination of factors: low $O_2$ pressure in the tissue cells, high $H^+$ ion concentration (**Bohr effect**), and high $CO_2$ concentration. Of the $CO_2$ in the venous blood, 25 percent binds to the terminal $-NH_2$ of the polypeptide chains of hemoglobin to form carbaminohemoglobin, 70 percent is carried in the plasma as $H_2CO_3$, and 5 percent is carried as dissolved gas.

Waste products are removed from the blood in the **kidneys.** The filtration units—**nephrons**—contain an entering blood vessel that branches out into fine vessels called **glomeruli.** Water and all the small molecules in the blood diffuse out of the glomeruli, enter the **Bowman's capsules** of the nephrons, and from there go into the **tubules.** The blood cells and large molecules remain in the blood vessels. The water, inorganic salts, and organic nutrients are then reabsorbed into the blood vessels. The waste products plus water form the urine, which goes from the tubule into the ureter and finally into the bladder.

The kidneys not only filter out waste products but also produce $HCO_3^-$ ions for buffering, replacing those lost through the lungs. The water balance and salt balance of the blood and urine are under hormonal control. Vasopressin helps to retain water, and aldosterone increases the blood $Na^+$ ion concentration by reabsorption. Blood pressure is controlled by a large number of factors.

## KEY TERMS

Antidiuretic (Sec. 25.8)
Baroreceptor (Sec. 25.9)
Blood-brain barrier (Sec. 25.1)
Blood plasma (Sec. 25.2)
Bohr effect (Sec. 25.3)
Cellular elements (Sec. 25.2)
Diuresis (Sec. 25.8)
Diuretic (Box 25E)
Edema (Sec. 25.2)
Erythrocyte (Sec. 25.2)
Extracellular fluid (Sec. 25.1)
Glomeruli (Sec. 25.6)
Hematocrit value (Sec. 25.2)

Hypertension (Box 25E)
Interstitial fluid (Sec. 25.1)
Intracellular fluid (Sec. 25.1)
Leukocyte (Sec. 25.2)
Lymphocyte (Sec. 25.2)
Nephron (Sec. 25.6)
Phagocytosis (Sec. 25.2)
Plasma (Sec. 25.2)
Platelet (Sec. 25.2)
Serum (Sec. 25.2)
Thrombocyte (Sec. 25.2)
Thrombosis (Box 25B)

## PROBLEMS

### Body Fluids

**25.1**   What is the common solvent in all body fluids?

**25.2**   What is the most abundant body fluid?

**25.3**   (a) Is the blood plasma in contact with the other body fluids?   (b) If so, in what manner?

### Functions and Composition of Blood

**25.4**   Name the three principal types of blood cells.

**25.5**   What is the function of erythrocytes?

**25.6**   Which blood cells are the largest?

**25.7**   Define:
(a) interstitial fluid      (c) serum
(b) plasma                  (d) cellular elements

**25.8**   Where are the following manufactured?
(a) erythrocytes      (c) lymphocytes
(b) leukocytes        (d) platelets

**25.9**   What is the nonprotein part of hemoglobin?

**25.10**   State the function of
(a) albumin          (c) fibrinogen
(b) globulin

**25.11**   Which blood plasma component protects against infections?

**25.12**   How is serum prepared from blood plasma?

### Blood as a Carrier of Oxygen

**25.13**   How many molecules of $O_2$ are bound to a hemoglobin molecule at full saturation?

**\*25.14**   A chemist breathing air contaminated in an industrial accident analyzes it and finds that the partial pressure is only 38 mm Hg. How many $O_2$ molecules, on the average, will be transported by each hemoglobin molecule? (Consult Figure 25.5.)

**25.15**   How is the $O_2$ bound in the hemoglobin molecule?

**\*25.16**   From Figure 25.5, predict the $O_2$ pressure at which the hemoglobin molecule binds 4, 3, 2, 1, and zero $O_2$ molecules to its heme.

**25.17**   Explain how the absorption of $O_2$ on one binding site in the hemoglobin increases the absorption capacity at other sites.

**25.18**   (a) What happens to the oxygen-carrying capacity of hemoglobin when the pH is lowered? (b) What is the name of this effect?

**\*25.19**   If blood fully saturated with oxygen arrives to two different muscles, one active and the other rested, which muscle will receive more oxygen and why?

### Transport of $CO_2$ in the Blood

**25.20**   (a) Where does $CO_2$ bind to hemoglobin? (b) What is the name of the complex formed?

**25.21**   What are the three ways in which $CO_2$ is transported in the blood, and what is the percentage of each way?

**\*25.22**   If the plasma carries 70 percent of its carbon dioxide in the form of $H_2CO_3$ and 5 percent in the form of dissolved $CO_2$ gas, predict the equilibrium constant of the reaction catalyzed by carbonic anhydrase:

$$H_2O + CO_2 \rightleftharpoons H_2CO_3$$

(Assume that the $H_2O$ concentration is included in the equilibrium constant.)

## Blood Buffers

**25.23** After reviewing Box 8E, explain why emphysema may lower the pH of the blood, causing acidosis.

**25.24** Which is the most important blood buffer?

## Kidney Functions and Urine

**25.25** What nitrogenous waste products are eliminated in the urine?

**25.26** Besides nitrogeneous waste products, what other components are normal constituents of urine?

**25.27** (a) Write the structure of urea. (b) What functional groups does urea have in common with creatine?

**25.28** What blood components stay in the glomeruli and do not pass through the Bowman's capsule?

## Water and Salt Balance in the Blood and Kidneys

**25.29** In which part of the kidneys does vasopressin change the permeability of membranes?

**25.30** How does aldosterone production counteract the excessive sweating that usually accompanies a high fever?

## Blood Pressure

**25.31** What would happen to the glucose in the arterial capillary if the blood pressure at the end of the capillary were 12 mm Hg rather than the normal 32 mm Hg?

**\*25.32** Some drugs used as antihypertensive agents, for example, Captopril (Box 25E), inhibit the enzyme that activates angiotensinogen. What is the mode of action of such drugs in lowering blood pressure?

**25.33** Which nervous control system is the fastest to respond to changes in blood pressure?

## Boxes

**25.34** (Box 25A) What side effect of antihistamines is avoided by use of the drug terfenadine?

**25.35** (Box 25B) What is thrombosis?

**25.36** (Box 25B) What is the first line of defense when a blood vessel is cut?

**25.37** (Box 25B) What are the functions of (a) thrombin, (b) vitamin K, and (c) thromboplastin in clot formation?

**25.38** (Box 25B) If a blood vessel is damaged, a porous plug is initially formed. What is the difference between the porous plug and a clot?

**25.39** (Box 25C) (a) Which body fluid(s) is (are) much richer in $K^+$ ions than in $Na^+$ ions? (b) Which is (are) much richer in $Na^+$ ions than in $K^+$ ions?

**25.40** (Box 25C) (a) In which body fluid(s) is (are) the negative ions mostly $Cl^-$? (b) In which is $Cl^-$ essentially absent?

**25.41** (Box 25D) What are the symptoms of enlarged prostate gland (benign prostatic hyperplasia)?

**25.42** (Box 25D) What causes osteoporosis in postmenopausal women?

**25.43** (Box 25E) How do diuretics act in lowering blood pressure?

**25.44** (Box 25E) How does propranolol normalize hypertension?

**25.45** (Box 25E) People with high blood pressure are advised to have a low-sodium diet. How does such a diet lower blood pressure?

## Additional Problems

**25.46** What happens to the oxygen-carrying ability of hemoglobin when acidosis occurs in the blood?

**25.47** Describe the action of renin.

**\*25.48** The oxygen dissociation curve of fetal hemoglobin is lower than that of adult hemoglobin. How does the fetus obtain its oxygen supply from the mother's blood when the two circulations meet?

**25.49** Which nitrogenous waste product of the body can be classified as an amino acid?

**25.50** Which hormone controls water retention in the body?

*25.51 When kidneys fail the body swells, a condition called edema. Why and where does the water accumulate?

25.52 Albumin makes up 55 percent of the protein content of plasma. Is the albumin concentration of the serum higher, lower, or the same as that of the plasma? Explain.

25.53 What is meant by "hematocrit value"?

25.54 Does the heme of hemoglobin participate in $CO_2$ transport as well as in $O_2$ transport?

*Vitamins sold for use as dietary supplements. (Photograph by Beverly March.)*

# Nutrition and Digestion

## 26.1 INTRODUCTION

In Chapters 20 and 21 we saw what happens to the food that we eat in its final stages—after the proteins, lipids, and carbohydrates have been broken down into their components. In this chapter we discuss the earlier stages—nutrition and diet—and then the digestive processes that break down the large molecules to the small ones that undergo metabolism. We have seen that the purpose of food is to provide energy and new molecules to replace those that the body uses. An additional purpose is to supply compounds needed for the period during which a child becomes an adult.

## 26.2 NUTRITION

**The components of food and drink that provide growth, replacement, and energy** are called **nutrients.** Not all components of food are nutrients. Some components of food and drinks, such as those that provide flavor, color, or aroma, enhance our pleasure in the food but are not themselves nutrients.

Nutritionists classify nutrients into six groups:

1. carbohydrates
2. lipids
3. proteins
4. vitamins
5. minerals
6. water

A healthy body needs the proper intake of all nutrients. However, nutrient requirements vary from one person to another. For example, it requires more energy to maintain the body temperature of an adult than that of a child. For this reason, nutritional requirements are usually given per kilogram of body weight. Furthermore, the energy requirements of a physically active body are greater than those of people in sedentary occupations. Therefore, when average values are given, as in recommended daily allowances (RDA), one should be aware of the wide range that these average values represent.

The public interest in nutrition and diet changes with time and geography. Sixty or seventy years ago the main nutritional interest of most Americans was getting enough food to eat and avoiding diseases caused by vitamin deficiency, such as scurvy or beriberi. This is still the main concern of the large majority of the world's population. Today, in affluent societies such as ours, the nutritional message is no longer "eat more" but rather "eat less and discriminate more in your selection of food." For example, a sizable percentage of the American population avoids foods containing substantial amounts of cholesterol (Box 17G) and saturated fatty acids to reduce the risk of heart attacks.

Along with discriminatory curtailment diets came many different faddish diets. Diet faddism is an exaggerated belief in the effects of nutrition upon health and disease. This is not new; it has been prevalent for many years (Fig. 26.1). A recommended food is rarely as good and a condemned food is never as bad as faddists claim. Scientific studies prove, for instance, that food grown with chemical fertilizers is just as healthy as food grown with organic (natural) fertilizers.

Discriminatory curtailment diets are those that avoid certain food ingredients that are considered harmful to the health of an individual; for example, low-sodium diets for people with high blood pressure.

**Figure 26.1**

A late nineteenth-century advertisement for a cure-all tonic. (Courtesy of the National Library of Medicine.)

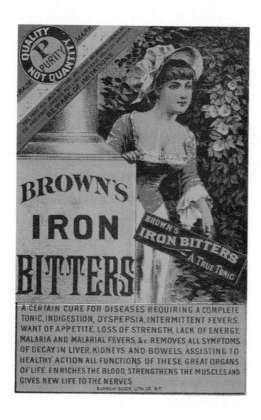

| TABLE 26.1 Nutritional Values in Three Commercial Food Products, per Ounce of Product | | | |
|---|---|---|---|
| | Breakfast Cereal | Canned Green Beans | Solid White Tuna |
| Calories | 110 | 5 | 77 |
| Protein | 2 g | 0.25 g | 7 g |
| Carbohydrates | 26 g | 1 g | 0 |
| Fat | 0 | 0 | 5 g |
| Potassium | — | 28 mg | — |
| Sodium | 195 mg | 120 mg | — |
| Iron | 10% | 1% | 1% |
| Calcium | 2% | 0.5% | 0 |
| Magnesium | 2% | — | 2% |
| Phosphorus | 2% | 0.5% | 6% |
| Zinc | 2% | — | — |
| Copper | 2% | — | — |
| Vitamin C | 25% | 1% | 0 |
| Vitamin A | 25% | 0.5% | 0 |
| Vitamin D | 10% | — | — |
| Thiamine | 25% | 0.5% | 0 |
| Riboflavin | 25% | 1% | 1% |
| Niacin | 25% | 0.5% | 19% |
| Vitamin $B_6$ | 25% | — | 6% |
| Vitamin $B_{12}$ | — | — | 9% |
| Folic acid | 25% | — | — |

% indicates the percentage of Recommended Daily Allowance (RDA).
— indicates no listing on the label.

Each food contains a large variety of nutrients. For example, a typical breakfast cereal lists as its ingredients: milled corn, sugar, salt, malt flavoring, vitamins A, B, C, and D, plus flavorings and preservatives. Consumer laws require that most packaged food be labeled to show the nutritional values of the food. Table 26.1 gives the nutritional values per 1-oz (28.4-g) serving of three different commercial foods, as listed on the labels. Thus we can see that nutritional requirements can be satisfied by a great number of combinations of different foods.

## 26.3 CALORIES

The largest part of our food supply goes to provide energy for our bodies. As we saw in Chapters 20 and 21, the energy comes from the oxidation of carbohydrates, fats, and proteins. The energy derived from food is usually measured in calories. One nutritional calorie (Cal) equals 1000 cal or 1 kcal. Thus, when we say that the average daily nutritional requirement for a young adult male is 2900 Cal, we mean the same amount of energy needed to raise the temperature of 2900 kg of water by 1°C or 29 kg (64 lb) of water by 100°C. A young adult female needs 2100 Cal per day. These are peak requirements. Children and older people, on the average, require less energy. Keep in mind that these

One calorie is the amount of heat necessary to raise the temperature of 1 g of water by 1°C (Sec. 1.8).

> **Box 26A**
>
> ## Iatrogenic Malnutrition

Many kinds of treatment by physicians can provoke nutritional deficiencies. **Iatrogenic malnutrition** is malnutrition that is inadvertently induced by therapy. Surgery, especially the removal of part of the stomach and/or duodenum in the treatment of ulcers, induces a host of deficiencies. The most curious is the depletion of vitamin $B_{12}$ that develops long after surgery, sometimes as much as six years later. Similarly, gastric surgery may induce bone diseases through decreased calcium absorption, increased calcium excretion in feces, and low levels of vitamin D. Iron deficiency and the resulting anemia are also associated with gastric surgery.

Various drugs can also induce deficiencies. Estrogen-containing oral contraceptives have been reported to reduce vitamin $B_{12}$ and pyridoxine levels in the serum. Thiazide diuretics, frequently prescribed for patients with hypertension, cause loss of sodium, potassium, magnesium, zinc, and chloride ions.

energy requirements are for active people. For bodies completely at rest, the corresponding energy requirement for young adult males is 1800 Cal/day, and that for females is 1300 Cal/day. The requirement for a resting body is called the **basal caloric requirement.**

An imbalance between the caloric requirement of the body and the caloric intake creates health problems. Chronic caloric starvation exists in many parts of the world, where people simply do not have enough food to eat because of prolonged drought, devastation by war, natural disasters, or simply overpopulation. Famine particularly affects infants and children. Chronic starvation, called **marasmus,** increases infant mortality up to 50 percent. It results in arrested growth, muscle wasting, anemia, and general weakness. Even if starvation is later alleviated, it leaves permanent damage, insufficient body growth, and lowered resistance to disease.

At the other end of the caloric spectrum is excessive caloric intake. This results in obesity, or the accumulation of body fat. Obesity increases the risk of hypertension, cardiovascular disease, and diabetes. Reducing diets aim at decreased caloric intake without sacrificing any essential nutrients. A combination of exercise and lower caloric intake can eliminate obesity, but usually these diets must achieve their goal over an extended period. Crash diets give the illusion of quick weight loss, but most of this is due to loss of water, which can be regained very quickly. To reduce obesity we must lose body fat and not water; this takes a lot of effort, because fats contain so much energy. A pound of body fat is equivalent to 3500 Cal. Thus, to lose 10 pounds it is necessary either to consume 35 000 fewer Calories, which can be achieved if one reduces caloric intake by 350 Cal every day for 100 days (or by 700 Cal daily for 50 days), or to use up, through exercise, the same numbers.

## 26.4  CARBOHYDRATES, FATS, AND PROTEINS IN THE DIET

Carbohydrates are the major source of energy in the diet. They also furnish important compounds for the synthesis of cell components (Chapter 22). The main dietary carbohydrates are the polysaccharide starch, the disaccharides lactose and sucrose, and the monosaccharides glucose and fructose. Nutritionists recommend a minimum carbohydrate intake equivalent to 500 Cal per day.

**Box 26B**

## Parenteral Nutrition

In clinical practice it occasionally happens that a patient cannot be fed through the gastrointestinal tract. Under these conditions intravenous feeding or **parenteral nutrition** is required. In the past, patients could be maintained on parenteral nutrition only for short periods. Calories were provided in the form of glucose, and amino acids were supplied in the form of protein hydrolysates. Over long periods, however, the use of parenteral nutrition resulted in deterioration and progressive emaciation of the patient. More recently, the solutions used in parenteral nutrition are being manufactured to contain both essential and nonessential amino acids in fixed proportions. Calories now do not come only from glucose. Glucose supplies only about 20 percent of the caloric needs. Fat emulsions provide the rest of the caloric requirements and the essential fatty acids. Finally, the full vitamin and mineral requirements are also supplied in the parenteral solution.

In addition to these digestible carbohydrates, cellulose in plant materials provides dietary fiber. This is a nondigestible material that provides dietary bulk to prevent constipation. Artificial sweeteners (Box 18B) can be used to reduce mono- and disaccharide intake.

Fats are the most concentrated source of energy. About 98 percent of dietary fats are triglycerides; the remaining 2 percent consists of complex lipids and cholesterol. Only two fatty acids are essential in higher animals, including humans: linolenic and linoleic acids (Sec. 17.3). These fatty acids are needed in the diet because our bodies cannot synthesize them. Nutritionists occasionally list arachidonic acid as an essential fatty acid. However, our bodies can synthesize arachidonic acid from linoleic acid.

Although the proteins in our diet can be used for energy (Sec. 21.9), their main use is to furnish amino acids from which the body synthesizes its own proteins (Sec. 23.10). The human body is incapable of synthesizing ten of the amino acids needed to make proteins. These ten (the **essential amino acids**) must be obtained from our food; they are shown in Table 18.1. The body breaks down food proteins into their amino acid constituents and then puts the amino acids together again to make body proteins. For proper nutrition, the foods we eat should contain about 20 percent protein.

A dietary protein that contains all the essential amino acids is called a **complete protein.** Casein, the protein of milk, is a complete protein, as are most other animal proteins—those found in meat, fish, and eggs. People who eat adequate quantities of meat, fish, eggs, and dairy products get all the amino acids they need to keep healthy. About 50 g per day of complete proteins constitutes an adequate quantity.

An important animal protein that is *not* complete is gelatin, which is made by denaturing collagen (Sec. 18.10). Gelatin lacks tryptophan and is low in several other amino acids, including isoleucine and methionine. Many people on quick reducing diets take "liquid protein." This is nothing but denatured and partially hydrolyzed collagen (gelatin). Therefore, if this is the only protein source in the diet, some essential amino acids will be lacking.

Most plant proteins are incomplete. For example, corn protein lacks lysine and tryptophan; rice protein lacks lysine and threonine; wheat protein lacks lysine; and even soy protein, one of the best plant proteins, is very low in

Some foods rich in dietary fiber are celery, carrots, and lettuce.

Arachidonic acid is needed for the synthesis of prostaglandins and leukotrienes (Sec. 17.12).

One might suppose that, since vitamins are essential nutrients, the more you take the better. Therefore, some people consider that the RDAs listed for each vitamin and mineral in Table 26.2 are only a minimum, and larger quantities will be even more beneficial. No scientific proof exists for such claims. Larger doses are usually harmless, because the body excretes the excess. But not in all cases.

High dosages can do unexpected harm in certain instances, while the RDA can prevent harm not usually associated with the vitamin. The case of folic acid illustrates this point. Table 26.2 shows an RDA of 400 $\mu$g for this vitamin, as the amount necessary to prevent anemia. Lately it has been shown that this amount of folic acid is especially important during pregnancy. It can prevent birth defects that result in *spina bifida,* the birth of a child with an open spine, and *anencephaly,* the birth of a child without most of the brain. Since vitamins are not stored in the body, it is important that the RDA be taken every day. *But not more.* An intake of 1000 $\mu$g or more of folic acid daily is harmful because it can hide the symptoms of certain anemias without actually preventing these diseases. The Food and Drug Administration now requires that food labeled as enriched must carry certain amounts of folic acid. The amount for a slice of enriched bread is 40 to 70 $\mu$g, and that for enriched cereals is 100 $\mu$g per serving.

Similarly, high dosages of niacin and of vitamin A have been shown to cause health problems.

methionine. Adequate amino acid nutrition is possible with a vegetarian diet, but only if a range of different vegetables is eaten.

In many underdeveloped countries, protein deficiency diseases are widespread because the people get their protein mostly from plants. Among these is a disease called kwashiorkor, the symptoms of which are a swollen stomach, skin discoloration, and retarded growth.

## 26.5 VITAMINS, MINERALS, AND WATER

Vitamins and minerals are essential for good nutrition. Animals maintained on diets that contain sufficient carbohydrates, fats, and proteins and that are provided with an ample water supply cannot survive on these alone. They also need the essential organic components called **vitamins** and inorganic ions called **minerals.** Deficiencies in vitamins and minerals lead to many nutritionally controllable diseases (Fig. 26.2). Vitamins and minerals, together with their sources, recommended daily allowances, and the diseases their deficiency may cause, are listed in Table 26.2 on pages 732–735.

Water makes up 60 percent of our body weight. Most of the compounds in our body are dissolved in water, which also serves as a transporting medium to carry nutrients and waste materials. These functions are discussed in Chapter 25. We must maintain a proper balance between water intake and water excretion via urine, feces, sweat, and exhalation of breath. A normal diet requires about 1200 to 1500 mL of water per day. This is in addition to the water content of our foods.

The name "vitamin" comes from a mistaken generalization. Casimir Funk (1884–1967) discovered that certain diseases such as beriberi, scurvy, and pellagra are caused by lack of certain nutrients. He found that the "antiberiberi factor" compound is an amine, and mistakenly thought that all such nutrients are amines. Hence he coined the name "Vitamine."

(a)

**Figure 26.2**
(a) Symptoms of rickets, a vitamin D deficiency in children. The nonmineralization of the bones of the radius and the ulna results in prominence of the wrist. (Courtesy of Drs. P. G. Bullogh and V. J. Vigorita and the Gower Medical Publishing Co., New York.) Histology of (b) normal and (c) osteomalacic bone. The latter shows the accumulation of osteoids (red stain) due to vitamin D deficiency. (Courtesy of Drs. P. A. Dieppe, P. A. Bacon, A. N. Bamji, and I. Watt and Gower Medical Publishing Ltd., London, England.)

(b)

(c)

## 26.6 DIGESTION

In order for food to be used in our bodies, it must be absorbed through the intestinal walls into the blood stream or lymph system. Some nutrients, such as vitamins, minerals, glucose, and amino acids, can be absorbed directly. Others, such as starch, fats, and proteins, must first be broken down into smaller components before they can be absorbed. **This breakdown process** is called **digestion.**

## 26.7 DIGESTION OF CARBOHYDRATES

Before the body can absorb carbohydrates, it must break down di-, oligo-, and polysaccharides into monosaccharides because only monosaccharides can pass into the blood stream.

The monosaccharide units are connected to each other by glycosidic (acetal) linkages. As we saw in Section 13.6, the cleavage of acetals by water is called

*(text continues on page 736)*

**TABLE 26.2  Vitamins and Minerals: Sources, Functions, Deficiency Diseases, and Daily Requirements**

*Fat-soluble Vitamins*

| Name | Structure | Best Food Source | Function | Deficiency Symptoms and Diseases | Recommended Daily Allowance[a] |
|---|---|---|---|---|---|
| A | *(structure)* | Liver, butter, egg yolk, carrots, spinach, sweet potatoes | Vision; to heal eye and skin injuries | Night blindness; blindness; keratinization of epithelium and cornea | 750 $\mu$g (1500 $\mu$g)[b] |
| D | *(structure)* | Salmon, sardines, cod liver oil, cheese, eggs, milk | Promotes calcium and phosphate absorption and mobilization | Rickets (in children): pliable bones; osteomalacia (in adults); fragile bones | 10 $\mu$g; exposure to sunlight |
| E | *(structure)* | Vegetable oils, nuts, potato chips, spinach | Antioxidant | In cases of malabsorption such as in cystic fibrosis: anemia. In premature infants: anemia | 4–9 mg |
| K | *(structure)* | Spinach, potatoes, cauliflower, beef liver | Blood clotting | Uncontrolled bleeding (mostly in newborn infants) | 70–140 $\mu$g |

*Water-soluble Vitamins*

| Name | Structure | Best Food Source | Function | Deficiency Symptoms and Diseases | Recommended Daily Allowance[a] |
|---|---|---|---|---|---|
| B$_1$ (thiamine) | *(structure)* | Beans, soybeans, cereals, ham, liver | Coenzyme in oxidative decarboxylation and in pentose phosphate shunt | Beriberi. In alcoholics: heart failure; pulmonary congestion | 1 mg |
| B$_2$ (riboflavin) | *(structure)* | Kidney, liver, yeast, almonds, mushrooms, beans | Coenzyme of oxidative processes | Invasion of cornea by capillaries; cheilosis; dermatitis | 1.4 mg |

| Vitamin | Structure | Sources | Function | Deficiency | U.S. RDA[a] |
|---|---|---|---|---|---|
| Nicotinic acid (niacin) | | Chickpeas, lentils, prunes, peaches, avocados, figs, fish, meat, mushrooms, peanuts, bread, rice, beans, berries | Coenzyme of oxidative processes | Pellagra | 18 mg |
| B₆ (pyridoxal) | | Meat, fish, nuts, oats, wheat germ, potato chips | Coenzyme in transamination; heme synthesis | Convulsions; chronic anemia; peripheral neuropathy | 2 mg |
| Folic acid | | Liver, kidney, eggs, spinach, beets, orange juice, avocados, cantaloupe | Coenzyme in methylation and in DNA synthesis | Anemia | 400 μg |
| B₁₂ | | Oysters, salmon, liver, kidney | Part of methyl-removing enzyme in folate metabolism | Patchy demyelination; degradation of nerves, spinal cord, and brain | 3 μg |

[a] The U.S. RDAs are set by the Food and Nutrition Board of the National Research Council. The numbers given here are based on the latest recommendations (1985). The RDA varies with age, sex, and level of activity; the numbers given are average values for both sexes between the ages of 18 and 54.
[b] Toxic if doses above the level shown in parentheses are taken.

733

**TABLE 26.2** Vitamins and Minerals: Sources, Functions, Deficiency Diseases, and Daily Requirements (*Continued*)

| Name | Structure | Best Food Source | Function | Deficiency Symptoms and Diseases | Recommended Daily Allowance[a] |
|---|---|---|---|---|---|
| Pantothenic acid | $CH_3$ $\quad$ O $\quad\quad\quad\quad\quad$ ‖ $HOCH_2C$—CHCNHCH$_2$CH$_2$COOH $\quad\mid\quad\quad\quad\mid$ $\quad CH_3\quad\quad OH$ | Peanuts, buckwheat, soybeans, broccoli, lima beans, liver, kidney, brain, heart | Part of CoA; fat and carbohydrate metabolism | Gastrointestinal disturbances; depression | 7 mg |
| Biotin | (structure: biotin ring with $(CH_2)_4COOH$) | Yeast, liver, kidney, nuts, egg yolk | Synthesis of fatty acids | Dermatitis; nausea; depression | 300 $\mu$g |
| C (ascorbic acid) | (structure: ascorbic acid with CH—OH, CH$_2$OH, OH groups) | Citrus fruit, berries, broccoli, cabbage, peppers, tomatoes | Hydroxylation of collagen; wound healing; bone formation; antioxidant | Scurvy; capillary fragility | 60 mg |
| *Minerals* | | | | | |
| Potassium | | Apricots, bananas, dates, figs, nuts, raisins, beans, chickpeas, cress, lentils | Provides membrane potential | Muscle weakness | 2000–5600 mg |
| Sodium | | Meat, cheese, cold cuts, smoked fish, table salt | Osmotic pressure | None | 1100–3300 mg |
| Calcium | | Milk, cheese, sardines, caviar | Bone formation; hormonal function; blood coagulation; muscle contraction | Muscle cramps; osteoporosis; fragile bones | 800 mg |
| Chloride | | Meat, cheese, cold cuts, smoked fish, table salt | Osmotic pressure | None | 1700–5100 mg |

| Mineral | Sources | Function | Deficiency | RDA[a] |
|---|---|---|---|---|
| Phosphorus | Lentils, nuts, oats, grain flours, cocoa, egg yolk, cheese, meat (brain, sweetbreads) | Balancing calcium in diet | Excess causes structural weakness in bones | 800–1500 mg |
| Magnesium | Cheese, cocoa, chocolate, nuts, soybeans, beans | Cofactor in enzymes | Hypocalcemia | 400 mg |
| Iron | Raisins, beans, chickpeas, parsley, smoked fish, liver, kidney, spleen, heart, clams, oysters | Oxidative phosphorylation; hemoglobin | Anemia | 12 mg; in menstruating women, 17–23 mg |
| Zinc | Yeast, soybeans, nuts, corn, cheese, meat, poultry | Cofactor in enzymes, insulin | Retarded growth; enlarged liver | 15 mg |
| Copper | Oysters, sardines, lamb, liver | Oxidative enzymes cofactor | Loss of hair pigmentation; anemia | 2–3 mg |
| Manganese | Nuts, fruits, vegetables, whole-grain cereals | Bone formation | Low serum cholesterol levels; retarded growth of hair and nails | 2.5–5.0 mg |
| Chromium | Meat, beer, whole wheat and rye flours | Glucose metabolism | Glucose not available to cells | 0.05–0.2 mg |
| Molybdenum | Liver, kidney, spinach, beans, peas | Protein synthesis | Retarded growth | 0.15–0.5 mg |
| Cobalt | Meat, dairy products | Component of vitamin $B_{12}$ | Pernicious anemia | 0.05–1.8 mg (20–30 mg)[b] |
| Selenium | Meat, seafood | Fat metabolism | Muscular disorders | 0.05–0.2 mg (2.4–3.0 mg)[b] |
| Iodine | Seafood, vegetables, meat | Thyroid glands | Goiter | 150 $\mu$g (1000 $\mu$g)[b] |
| Fluorine | Fluoridated water; fluoridated toothpaste | Enamel formation | Tooth decay | 1.5–4.0 mg (8–20 mg)[b] |

[a] The U.S. RDAs are set by the Food and Nutrition Board of the National Research Council. The numbers given here are based on the latest recommendations (1985). The RDA varies with age, sex, and level of activity; the numbers given are average values for both sexes between the ages of 18 and 54.

[b] Toxic if doses above the level shown in parentheses are taken.

**Figure 26.3**
The action of different enzymes on glycogen and starch.

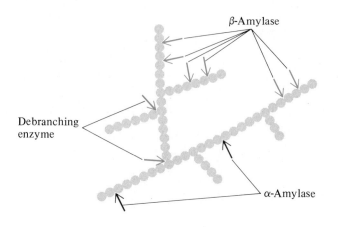

The three storage polysaccharides are amylose, amylopectin, and glycogen.

*hydrolysis.* In the body this hydrolysis is catalyzed by acids and by enzymes. When the metabolic need arises, storage polysaccharides are hydrolyzed to yield glucose and maltose.

The hydrolysis is aided by a number of enzymes: $\alpha$-amylase attacks all three storage polysaccharides at random, hydrolyzing the $\alpha(1 \rightarrow 4)$ glycosidic linkages, and $\beta$-amylase also hydrolyzes the $\alpha(1 \rightarrow 4)$ glycosidic linkages but in an orderly fashion, cutting disaccharidic maltose units one by one from the nonreducing (locked) end of a chain. A third enzyme, called the **debranching enzyme,** attacks and hydrolyzes the $\alpha(1 \rightarrow 6)$ glycosidic linkages (Fig. 26.3). In acid-catalyzed hydrolysis, storage polysaccharides are attacked at random points, although acid catalysis is slower than enzyme hydrolysis at body temperature.

The digestion (hydrolysis) of starch and glycogen in our food supply starts in the mouth, where $\alpha$-amylase is one of the main components of saliva. The hydrochloric acid in the stomach and the other hydrolytic enzymes in the intestinal tract decompose starch and glycogen to produce mono- and disaccharides (D-glucose and maltose).

The D-glucose produced by hydrolysis of the di-, oligo-, and polysaccharides enters the blood stream and is carried to the cells to be utilized (Sec. 21.2). For this reason, D-glucose is often called blood sugar (Box 26D). In healthy people little or none of this sugar ends up in the urine except for short periods of time. In the condition known as **diabetes mellitus,** however, glucose is not completely metabolized and does appear in the urine (Box 24G). Because of this, it is necessary to test the urine of diabetic patients for the presence of glucose.

Cellulose can be hydrolyzed to glucose by strong acids. Human beings and other mammals do not have enzymes that are capable of hydrolyzing the $\beta(1 \rightarrow 4)$ linkages, so cellulose goes through the intestinal tract largely undigested. However, many bacteria and fungi possess cellulase, an enzyme that hydrolyzes cellulose to glucose. Ruminant mammals, such as the cow, can digest grass because microorganisms in their rumen produce the necessary cellulose. Termites eat wood, and in this case too it is not the termites that have the enzymes but microorganisms inside them.

The glucose content of normal blood varies between 65 and 105 mg/100 mL. Usually this is measured after 8 to 12 hours of fasting, and therefore this range is known as the normal fasting level. **Hypoglycemia** develops when the glucose concentration falls below this level. Since the brain cells normally use glucose as their only energy source, hypoglycemic patients experience dizziness and fainting spells. In more severe cases convulsion and shock may occur.

In **hyperglycemia,** when the blood glucose level reaches 140 to 160 mg/100 mL, the kidneys begin to filter glucose out of the blood stream and it appears in the urine.

## 26.8  DIGESTION OF LIPIDS

The lipids in the food we eat must be broken down into smaller components before they can be absorbed into the blood or lymph system through the intestinal walls. The enzymes that promote this breakdown are located in the small intestine and are called **lipases.** However, since lipids are insoluble in the aqueous environment of the gastrointestinal tract, they must be dispersed into fine colloidal particles before the enzymes can act on them.

The bile salts (Sec. 17.11) perform this important function. Bile salts are manufactured in the liver from cholesterol and stored in the gallbladder. From there they are secreted through the bile ducts into the intestine. The emulsion produced by the bile salts is acted upon by the lipases. These break fats down into glycerol and fatty acids and complex lipids into fatty acids, alcohols (glycerol, choline, ethanolamine, sphingosine), and carbohydrates. All these breakdown products are absorbed through the intestinal walls.

## 26.9  DIGESTION OF PROTEINS

The digestion of dietary proteins begins with cooking, which denatures proteins. (Denatured proteins are broken down more easily by the hydrochloric acid in the stomach and by digestive enzymes than are native proteins.) Stomach acid contains about 0.5 percent HCl. The HCl both denatures the proteins and cleaves the peptide linkages randomly. Pepsin, the proteolytic enzyme of stomach juice, breaks peptide bonds on the $C=O$ side of three amino acids only: tryptophan, phenylalanine, and tyrosine (Fig. 26.4).

**Figure 26.4**
Different enzymes hydrolyze peptide chains in different but specific ways.

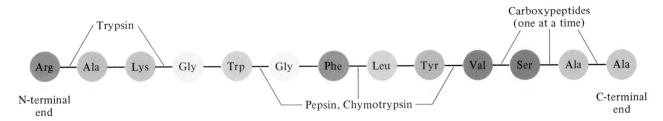

Most protein digestion occurs in the small intestine. There, the enzyme chymotrypsin breaks internal peptide bonds at the same positions as does pepsin, while another enzyme, trypsin, breaks them only on the $C=O$ side of arginine and lysine. Other enzymes, such as carboxypeptidase, cut amino acids one by one from the C-terminal end of the protein. The amino acids and small peptides are then absorbed through the intestinal walls.

## SUMMARY

**Nutrients** are components of foods that provide growth, replacement, and energy. Nutrients are classified into six groups: carbohydrates, lipids, proteins, vitamins, minerals, and water. Each food contains a variety of nutrients. The largest part of our food intake is used to provide energy for our bodies. A typical young adult needs 2900 Cal (male) or 2100 Cal (female) as an average daily caloric intake. **Basal caloric requirements** are the energy need when the body is completely at rest. These are less than the normal requirements. Imbalance between need and caloric intake may create health problems; chronic starvation increases infant mortality, while obesity leads to hypertension, cardiovascular diseases, and diabetes.

Carbohydrates are the major source of energy in our diet. Fats are the most concentrated source of energy. Essential fatty and amino acids are needed as building blocks because our bodies cannot synthesize them. Vitamins and minerals are essential constituents of diets that are needed in small quantities. Water makes up 60 percent of body weight.

The digestion of carbohydrates, fats, and proteins is aided by the stomach acid, HCl, and by enzymes that reside in the mouth, stomach, and intestines.

## KEY TERMS

Basal caloric requirement (Sec. 26.3)
Complete protein (Sec. 26.4)
Debranching enzyme (Sec. 26.7)
Hyperglycemia (Box 26D)
Hypoglycemia (Box 26D)
Iatrogenic malnutrition (Box 26A)
Kwashiorkor (Sec. 26.4)

Lipase (Sec. 26.8)
Marasmus (Sec. 26.3)
Mineral (Sec. 26.5)
Nutrient (Sec. 26.2)
Parenteral nutrition (Box 26B)
RDA (Sec. 26.2)
Vitamin (Sec. 26.5)

## PROBLEMS

### Nutrition

**26.1**  What are nutrients?

**26.2**  (a) Is raspberry flavoring a nutrient?   (b) What is the chemical nature of this flavoring? (See Box 14B.)

**\*26.3**  If sodium benzoate, a food preservative, is excreted as such, and if calcium propionate, another food preservative (see Box 14A), is hydrolyzed in the body and metabolized to $CO_2$ and $H_2O$, would you consider either of these preservatives as nutrients? If so, why?

**26.4**  Is corn grown solely with organic fertilizers more nutritious than that grown with chemical fertilizers?

**26.5**  If solid white tuna were the only source of vitamin $B_{12}$ in your diet, how many 3.5-ounce cans must you consume to fill your daily requirement (Table 26.1)?

**26.6**  (a) Which of the three commercial foods listed in Table 26.1 provides the most vitamins per ounce?   (b) Which provides the most protein?

## Calories

**26.7**  Define basal caloric requirement.

**26.8**  A young adult female needs a caloric intake of 2100 Cal/day. Her basal caloric requirement is only 1300 Cal/day. Why is the extra 800 Cal needed?

**26.9**  Are diuretics (compounds that cause loss of water) effective weight-reducing drugs? Explain.

**26.10**  Assume that you want to lose 20 pounds of body fat in 60 days. Your present dietary intake is 3000 Cal/day. What should your caloric intake be, in Cal/day, in order to achieve this goal, assuming no change in exercise habits?

## Carbohydrates, Fats, and Proteins in the Diet

**26.11**  What kind of carbohydrate provides dietary fiber?

**26.12**  Which nutrient provides energy in its most concentrated form?

**26.13**  Would it be advisable to exclude all carbohydrates from a reducing diet?

**26.14**  How many (a) essential fatty acids and  (b) essential amino acids do humans need in their diets?

**26.15**  What is the precursor of arachidonic acid in the body?

**26.16**  Suggest a way to cure kwashiorkor.

## Vitamins, Minerals, and Water

**26.17**  In a prison camp during a war, the prisoners are fed plenty of rice and water but nothing else. What would the result of such a diet be in the long run?

**26.18**  (a) How many mL of water per day does a normal diet require?  (b) How many calories does this contribute?

**26.19**  Which fat-soluble vitamin acts in our body as an antioxidant?

**26.20**  What are the symptoms of vitamin A deficiency?

**26.21**  What is the function of vitamin K?

**26.22**  (a) Which vitamin contains cobalt?  (b) What is the function of this vitamin?

**26.23**  Vitamin C is recommended in megadoses by some people for all kinds of symptoms from colds to cancer. What disease has been scientifically proven to be prevented when sufficient daily doses of vitamin C are in the diet?

**26.24**  What are the best dietary sources of calcium, phosphorus, and cobalt?

**26.25**  Which vitamin contains a sulfur atom?

**26.26**  What are the symptoms of vitamin $B_{12}$ deficiency?

## Digestion

**26.27**  What chemical processes take place during digestion?

**26.28**  Explain why human beings can digest starch but not cellulose.

**\*26.29**  Do lipases degrade (a) cholesterol  (b) fatty acids

**26.30**  Does the debranching enzyme help in digesting amylose?

**26.31**  Does HCl in the stomach hydrolyze both the $(1 \rightarrow 4)$ and $(1 \rightarrow 6)$ glycosidic linkages?

**26.32**  What is the difference between protein digestion by trypsin and by HCl?

## Boxes

**26.33**  (Box 26A) What is iatrogenic malnutrition?

**26.34**  (Box 26B) Why cannot glucose supply all the caloric needs in parenteral nutrition over a long period?

**26.35**  (Box 26C) The RDA for niacin is 18 mg. Would it be still better to take a 100-mg tablet of this vitamin every day?

**26.36**  (Box 26D) A patient complains of dizziness. Blood analysis shows that he has 30 mg of glucose in 100 mL of blood. What would you recommend to alleviate the dizziness?

**26.37**  (Box 26D) What happens to the blood glucose in patients with severe diabetes when, because of the lack of insulin, the cells cannot utilize glucose?

## Additional Problems

**26.38**  Which vitamin is part of coenzyme A (CoA)?

**26.39**  Name three essential amino acids.

**\*26.40**  As an employee of a company that markets walnuts, you are asked to provide information for an ad that would stress the nutritional value of walnuts. What information would you provide?

**26.41**  (a) Name the essential fatty acids.  (b) Are they saturated or unsaturated?

*26.42 What kind of supplemental enzyme would you recommend for a patient after a peptic ulcer operation?

*26.43 In diabetes, insulin is administered intravenously. Explain why this hormone protein cannot be taken orally.

26.44 Why is it necessary to have proteins in our diets?

*26.45 Which vitamin is part of the coenzyme $NAD^+$?

26.46 Is it possible to get a sufficient supply of nutritionally adequate proteins by eating only vegetables? Explain.

26.47 In a trial a woman was accused of poisoning her husband by adding arsenic to his meals. Her attorney stated that this was done to promote her husband's health, arsenic being an essential nutrient. Would you accept this argument?

# Exponential Notation

In this text we have often dealt with numbers that are very large or very small. Many years ago, scientists devised an easy way to handle such numbers. For example, an ordinary copper penny (before 1982, when pennies in the United States were still made of copper) contains approximately

29 500 000 000 000 000 000 000 atoms of copper

This is a very large number, but chemists and other scientists must often deal with very large and also with very small numbers. For example, a single copper atom weighs

0.0000000000000000000000023 pound

The **exponential notation** system is based on powers of 10 (Table A.1). For example, if we multiply $10 \times 10 \times 10 = 1000$, we express this as $10^3$. The 3 in this expression is called the **exponent** or the **power,** and it indicates how many times we multiplied 10 by itself and how many zeros follow the 1.

There are also negative powers of 10. For example, $10^{-3}$ means 1 divided by $10^3$:

$$10^{-3} = \frac{1}{10^3} = \frac{1}{1000} = 0.001$$

Numbers are frequently expressed like this: $6.4 \times 10^3$. In a number of this type, 6.4 is the **coefficient** and 3 is the exponent, or power of 10. This number means exactly what it says:

$$6.4 \times 10^3 = 6.4 \times 1000 = 6400$$

Similarly, we can have coefficients with negative exponents:

$$2.7 \times 10^{-5} = 2.7 \times \frac{1}{10^5} = 2.7 \times 0.00001 = 0.000027$$

For numbers greater than 10 in exponential notation, we proceed as follows: *Move the decimal point to the left,* to just after the first digit. The (positive) exponent is equal to the number of places we moved the decimal point.

Exponential notation is also called scientific notation.

For example, $10^6$ means a one followed by six zeros, or 1 000 000, and $10^2$ means 100.

| TABLE A.1 | Examples of Exponential Notation |
|---|---|
| $10\ 000 = 10^4$ | |
| $1000 = 10^3$ | |
| $100 = 10^2$ | |
| $10 = 10^1$ | |
| $1 = 10^0$ | |
| $0.1 = 10^{-1}$ | |
| $0.01 = 10^{-2}$ | |
| $0.001 = 10^{-3}$ | |

**EXAMPLE**

$$37\ 500 = 3.75 \times 10^4 \quad \text{4 because we went four places to the left}$$

Four places to the left          Coefficient

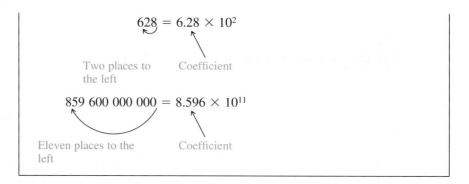

We don't really have to place the decimal point after the first digit, but by doing so we get a coefficient between 1 and 10, and that is the custom.

Using exponential notation, we can say that there are $2.95 \times 10^{22}$ copper atoms in a copper penny. For large numbers the exponent is always *positive*. Note that we do not usually write out the zeros at the end of the number.

For small numbers (less than 1) we move the decimal point *to the right*, to just after the first nonzero digit, and use a *negative exponent*.

---

**EXAMPLE**

$$0.00346 = 3.46 \times 10^{-3}$$

Three places to
the right

$$0.000004213 = 4.213 \times 10^{-6}$$

Six places to
the right

---

In exponential notation a copper atom weighs $2.3 \times 10^{-25}$ pounds.

To convert exponential notation into fully written-out numbers, we do the same thing backward.

---

**EXAMPLE**

Write out in full: (a) $8.16 \times 10^{7}$   (b) $3.44 \times 10^{-4}$

**Answer**

(a) $8.16 \times 10^{7} = 81\ 600\ 000$

Seven places to the right
(add enough zeros)

(b) $3.44 \times 10^{-4} = 0.000344$

Four places to the left

---

When scientists add, subtract, multiply, and divide, they are always careful to express their answers with the proper number of digits, called significant figures. This method is described in Appendix B.

## Adding and Subtracting Numbers in Exponential Notation

We are allowed to add or subtract numbers expressed in exponential notation *only if they have the same exponent*. All we do is add or subtract the coefficients and leave the exponent as it is.

---

**EXAMPLE**

Add $3.6 \times 10^{-3}$ and $9.1 \times 10^{-3}$.

**Answer**

$$
\begin{array}{r}
3.6 \times 10^{-3} \\
+ \quad 9.1 \times 10^{-3} \\
\hline
12.7 \times 10^{-3}
\end{array}
$$

The answer could also be written in other, equally valid ways:

$$12.7 \times 10^{-3} = 0.0127 = 1.27 \times 10^{-2}$$

---

When it is necessary to add or subtract two numbers that have different exponents, we first must change them so that the exponents are the same.

A calculator with exponential notation changes the exponent automatically.

---

**EXAMPLE**

Add $1.95 \times 10^{-2}$ and $2.8 \times 10^{-3}$.

**Answer**
In order to add these two numbers, we make both exponents $-2$. Thus, $2.8 \times 10^{-3} = 0.28 \times 10^{-2}$. Now we can add:

$$
\begin{array}{r}
1.95 \times 10^{-2} \\
+ \quad 0.28 \times 10^{-2} \\
\hline
2.23 \times 10^{-2}
\end{array}
$$

---

## Multiplying and Dividing Numbers in Exponential Notation

To multiply numbers in exponential notation, we first multiply the coefficients in the usual way and then algebraically *add* the exponents.

---

**EXAMPLE**

Multiply $7.40 \times 10^5$ by $3.12 \times 10^9$.

**Answer**

$7.40 \times 3.12 = 23.1$

Add exponents:

$10^5 \times 10^9 = 10^{5+9} = 10^{14}$

Answer: $23.1 \times 10^{14} = 2.31 \times 10^{15}$

---

**EXAMPLE**

Multiply $4.6 \times 10^{-7}$ by $9.2 \times 10^4$.

**Answer**

$4.6 \times 9.2 = 42$

Add exponents:

$10^{-7} \times 10^4 = 10^{-7+4} = 10^{-3}$

Answer: $42 \times 10^{-3} = 4.2 \times 10^{-2}$

---

To divide numbers expressed in exponential notation, the process is reversed. We first divide the coefficients and then algebraically *subtract* the exponents.

---

**EXAMPLE**

Divide: $\dfrac{6.4 \times 10^8}{2.57 \times 10^{10}}$

**Answer**

$6.4 \div 2.57 = 2.5$

Subtract exponents:

$10^8 \div 10^{10} = 10^{8-10} = 10^{-2}$

Answer: $2.5 \times 10^{-2}$

---

**EXAMPLE**

Divide: $\dfrac{1.62 \times 10^{-4}}{7.94 \times 10^7}$

**Answer**

$1.62 \div 7.94 = 0.204$

Subtract exponents:

$10^{-4} \div 10^7 = 10^{-4-7} = 10^{-11}$

Answer: $0.204 \times 10^{-11} = 2.04 \times 10^{-12}$

---

For problems relating to exponential notation, see Chapter 1, Problems 1.52 to 1.59.

# A P P E N D I X   B

# Significant Figures

If you measure the volume of a liquid in a graduated cylinder (Fig. 1.6), you might find that it is 36 mL, to the nearest milliliter, but you cannot tell if it is 36.2, or 35.6, or 36.0 mL because this measuring instrument does not give the last digit with any certainty. A buret (Fig. 1.6) gives more digits, and if you use one you should be able to say, for instance, that the volume is 36.3 mL and not 36.4 mL. But even with a buret, you could not say whether the volume is 36.32 or 36.33 mL. For that, you would need an instrument that gives still more digits. This example should show you that *no measured number can ever be known exactly.* No matter how good the measuring instrument, there is always a limit to the number of digits it can measure with certainty.

Scientists have found it useful to establish a method for telling to what degree of certainty any measured number is known. The method is very simple; it consists merely of writing down all the digits that are certain and not writing down any that are not certain. We define the number of **significant figures** in a measured number as **the number of digits that are known with certainty.**

What do we mean by this? Assume that you are weighing a small object on a laboratory balance that can weigh to the nearest 0.1 g, and you find that the object weighs 16 g. Because the balance weighs to the nearest 0.1 g, you can be sure that the object does not weigh 16.1 g or 15.9 g. In this case, you would write the weight as 16.0 g. To a scientist, there is a difference between 16 g and 16.0 g. Writing 16 g says that you don't know the digit after the 6. Writing 16.0 g says that you do know it: It is 0. However, you don't know the digit after that.

There are several rules governing the use of significant figures in reporting measured numbers.

### Determining the Number of Significant Figures

All digits written down are significant except two types:

1.  Zeros that come before the first nonzero digit are *not* significant.

---

**EXAMPLE**

How many significant figures are there in each of these numbers?

| Number | Number of Significant Figures |
|---|---|
| 23.742 | 5 |
| 332 | 3 |
| 0.023 | 2 |
| 0.230 | 3 |
| 0.000023 | 2 |
| 3.004 | 4 |
| 0.050008 | 5 |

---

**2.** Zeros that come after the last nonzero digit are significant if the number is a decimal, but if it is a whole number (no decimal point) they may or may not be significant. We cannot tell without knowing something about the number. This is the ambiguous case.

---

**EXAMPLE**

How many significant figures are there in each of these numbers?

| Number | Number of Significant Figures |
|---|---|
| 32.0400 | 6 |
| 0.0002300 | 4 |
| 1.02000 | 6 |
| 32500 | 3, 4, or 5 |

---

The ambiguous case is the only flaw in the system. If you know that a certain small business made a profit of $36,000 last year, you can be sure that the 3 and 6 are significant, but what about the rest? It might have been $36,126, or $35,786.53, or maybe even exactly $36,000. We just don't know because it is customary to round such numbers off. On the other hand, if the profit were reported as $36,000.00, then all seven digits are significant.

In science, we often get around the ambiguous case by using exponential notation (Appendix A). Suppose a measurement comes out to be 2500 g. If we made the measurement, we of course know whether the two zeros are significant, but we need to tell others. If these digits are *not* significant, we write our number as $2.5 \times 10^3$. If one zero is significant, we write $2.50 \times 10^3$. If both zeros are significant, we write $2.500 \times 10^3$. Since we now have a decimal point, all the digits shown are significant.

## Multiplying and Dividing

This means that most of those beautiful digits on your calculator display are usually meaningless (insignificant).

The rule in multiplication and division is that the final answer should have the *same* number of significant figures as there are in the number with the *fewest* significant figures.

---

**EXAMPLE**

Do the following multiplications and divisions:

(a) $3.6 \times 4.27$

(b) $0.004 \times 217.38$

(c) $\dfrac{42.1}{3.695}$

(d) $\dfrac{0.30652 \times 138}{2.1}$

**Answer**

(a)  15 (3.6 has two significant figures)
(b)  0.9 (0.004 has one significant figure)
(c)  11.4 (42.1 has three significant figures)
(d)  $2.0 \times 10^1$ (2.1 has two significant figures)

## Adding and Subtracting

In addition and subtraction, the rule is completely different. The number of significant figures in each number doesn't matter. The answer is given to the same number of *decimal places* as the term with the fewest decimal places.

---

**EXAMPLE**

Add or subtract:

(a)  320.0|84      (b)  61|.4532
    80.4|7          13|.7
  200.2|3          22|      (c)     14.26|
   20.0|          0|.003      $-$   1.05|041
  620.8|          97|          13.21|

**Answer**

In each case, we add or subtract in the normal way but then round off so that the only digits that appear in the answer are those in the columns in which every digit is significant.

---

## Rounding Off

When we have too many significant figures in our answer, it is necessary to round off. In this book we have used the rule that *if the first digit dropped* is 5, 6, 7, 8, or 9, we raise *the last digit kept* to the next higher number; otherwise we do not.

---

**EXAMPLE**

In each case, drop the last two digits:

        33.679      2.4715      1.1145
        0.00130      3.529

**Answer**

      33.679      =      33.7

      2.4715      =      2.47

      1.1145      =      1.11

      0.001309      =      0.0013

      3.52      =      4

---

## Counted or Defined Numbers

All the preceding rules apply to *measured* numbers and **not** to any numbers that are *counted* or *defined*. Counted and defined numbers are known exactly. For example, a triangle has 3 sides, not 3.1 or 2.9. Here, we treat the number 3 as if it has an infinite number of zeros following the decimal point.

---

### EXAMPLE

Multiply 53.692 (a measured number) $\times$ 6 (a counted number).

### Answer

$$322.152$$

Because 6 is a counted number, we know it exactly, and 53.692 is the number with the fewest significant figures. All we really are doing is adding 53.692 six times.

---

For problems relating to significant figures, see Chapter 1, Problems 1.60 to 1.66.

# A N S W E R S *

---

## CHAPTER 1 Matter, Energy, and Measurement

1.1  (a) 147°F  (b) 8.3°C
1.2  59.9 kg
1.3  4200 m
1.4  743 mph
1.5  78.5 g
1.6  2.43 g/mL
1.7  1.016 g/mL
1.8  $48 \times 10^3$ cal = 48 kcal
1.9  540 cal
1.10  0.0430 cal/g·deg
1.11  It deals with the structure and properties of matter and the transformations from one form of matter to another.
1.13  A fact is a statement based on experience. A hypothesis is a statement offered to explain the facts.
1.15  Scientists will accept a theory if it agrees with all known facts. But if new facts arise that conflict with the theory, the theory is wrong.
1.17  (a) 1000 m  (b) 0.001 m
1.19  (a) 75 kg  (b) 50 mg  (c) 40 g  (d) 100 cm
(e) 100 mm  (f) 230 mL  (g) 15 mL
1.21  (a) 160°C, 433 K  (b) 100°C, 373 K  (c) −18°C, 255 K  (d) −157°C, 116 K
1.23  (a) 10.3 L  (b) 86 cm  (c) 1.18 fl oz  (d) 93.9 lb
(e) 2.2 oz (avdp)  (f) 11.5 gal  (g) 735 g
(h) 1.8 km  (i) 23.1 miles  (j) 31.80 L
1.25  A student weighing 127 lb weighs 57.6 kg. A student 5 feet 5 inches tall (65.0 in) is 1.65 m tall.
1.27  $2.99 \times 10^{10}$ cm/s
1.29  bottom: manganese; top: sodium acetate; middle: calcium chloride
1.31  1.023 g/cc
1.33  melting point, density, specific heat
1.35  water

1.37  They are equally good buys. Butter A costs $5.00 for 1135 g, or 0.44 cents/g; butter B costs $4.00 for 908 g, or 0.44 cents/g.
1.39  $8.3 \times 10^3$ cal
1.41  0.34 cal/g·deg
1.43  They did three tests. See Box 1A.
1.45  20.6 mg/kg of body weight
1.47  13.6 g
1.49  The body shivers. Further temperature lowering results in unconsciousness and then death.
1.51  Methyl alcohol, because its higher specific heat allows it to retain the heat longer.
1.53  (a) 403 000  (b) 3200  (c) 0.0000713
(d) 0.000000000555
1.55  (a) $2.00 \times 10^{18}$  (b) $1.37 \times 10^5$  (c) $2.36 \times 10^{-10}$
(d) $8.8 \times 10^1$  (e) $1.56 \times 10^{-8}$
1.57  (a) $2.80 \times 10^5$  (b) $5.03 \times 10^3$  (c) $1.572 \times 10^{-7}$
1.59  $3.25 \times 10^{-5}$
1.61  (a) 3  (b) 2  (c) 1  (d) 4  (e) 5
1.63  (a) 25 000  (b) 4.1  (c) 15.5
1.65  (a) 10963.1  (b) 244  (c) 172.34
1.67  The largest is 41 g. The smallest is $4.1 \times 10^{-8}$ kg.
1.69  0.732
1.71  (a) mass  (b) volume  (c) temperature  (d) volume  (e) heat  (f) velocity  (g) density  (h) heat
1.73  (a) $6.18 \times 10^4$ in  (b) 0.347 qt  (c) 6.58 kg
(d) 0.119 lb
1.75  Malpighi
1.77  $1.24 \times 10^{10}$ erg
1.79  5.9 lb/gal
1.81  (a) 1.57 g/mL  (b) 1.25 g/mL
1.83  6.07 hr or 6 hr 4 min
1.85  No. mL is a unit of volume, g a unit of mass.

---

## CHAPTER 2 Atoms

2.1  (a) 109 amu  (b) 230 amu
2.2  (a) silver  (b) protactinium
2.3  They were not based on evidence.
2.5  Chemical changes: (a) (e) (f) (g)
Physical changes: (b) (c) (d)

2.7  The mixture is heated; the liquid with the lower boiling point evaporates first and is condensed. The other liquid remains in the flask.
2.9  the law of conservation of mass
2.11  (a) +1  (b) −1  (c) no charge

* Answers to in-text and odd-numbered end-of-chapter problems.

2.13   (a) nucleus  (b) outside nucleus  (c) nucleus

2.15   (a) zirconium  (b) magnesium  (c) bromine
    (d) radon  (e) uranium

2.17   the same, since the number of protons would not change

2.19   All contain 86 protons. They have 124, 132, and 136 neutrons, respectively.

2.21   (a) 17p, 18n  (b) 10p, 12n  (c) 46p, 58n  (d) 3p, 4n  (e) 52p, 76n  (f) 92p, 146n

2.23   barium-136, barium-138, barium-139

2.25   (a) $Na^+$  (b) Kr  (c) $F^-$  (d) $Te^{2-}$  (e) Zn  (f) $Fe^{3+}$  (g) $Ca^{2+}$  (h) $H^+$

2.27   63.62 amu

2.29   56.5% $^{107}Ag$ and 43.5% $^{109}Ag$

2.31   (a) (c) (d) (f)

2.33   metals: (a) (e) (j); nonmetals: (b) (c) (d) (i); metalloids: (f) (g) (h) (k)

2.35   (a) 2  (b) 8  (c) 18

2.37   (a) $1s^22s^1$  (b) $1s^2$  (c) $1s^22s^2$  (d) $1s^22s^22p^2$
    (e) $1s^22s^22p^63s^23p^3$  (f) $1s^22s^22p^6$
    (g) $1s^22s^22p^63s^23p^64s^23d^5$  (h) $1s^22s^22p^63s^23p^5$

2.39   (a) Mg: $1s^22s^22p^63s^2$; $Mg^{2+}$: $1s^22s^22p^6$
    (b) Na: $1s^22s^22p^63s^1$; $Na^+$: same as $Mg^{2+}$
    (c) Al: $1s^22s^22p^63s^23p^1$; $Al^{3+}$: same as $Mg^{2+}$

2.41   (a) $1s^22s^22p^63s^23p^63d^{10}4s^24p^65s^2$
    (b) $1s^22s^22p^63s^23p^63d^{10}4s^24p^64d^{10}5s^25p^66s^1$

2.43   (c) $1s^22s^22p^63s^23p^63d^{10}4s^24p^64d^{10}5s^25p^5$

2.43   6s: 2; 6p: 6; 6d: 10; 6f: 14; 32 altogether

2.45   The properties are similar because all of them have the same outer-shell configurations. They are not identical because each has a different number of filled inner shells.

2.47   C, H, O, N

2.49   Calcium is an essential element in human bones and teeth. Since strontium behaves very much like calcium, strontium-90 gets into our bones and teeth and gives off radioactivity for many years directly into our bodies.

2.51   copper and tin

2.53   hydrogen and carbon

2.55   (a) 18  (b) fifth

2.57   (a) 17p, 17e, 18n  (b) 42p, 42e, 56n  (c) 20p, 18e, 24n  (d) 53p, 54e, 74n  (e) 64p, 64e, 94n  (f) 83p, 80e, 129n

2.59   (a) $1s^22s^22p^63s^23p^4$  (b) $1s^22s^22p^63s^23p^6$

2.61   (a) 135  (b) 141  (c) 148  (d) 150

2.63   47.9 amu

2.65   $5.3 \times 10^{24}$ atoms

2.67   (a) 2  (b) 6  (c) 4  (d) 7  (e) 1  (f) 6  (g) 4  (h) 2  (i) 5  (j) 8

2.69   76.5% rubidium-85 and 23.5% rubidium-87

2.71   (d)

---

## CHAPTER 3   Chemical Bonds

3.1   (a) KCl  (b) $CaF_2$  (c) $Fe_2O_3$

3.2

3.3

3.4

3.5   (a) F  (b) Br

3.6   (a) nonpolar  (b) polar covalent  (c) ionic

3.7   magnesium oxide, barium iodide, sodium bromide

3.8   SrS, NaF

3.9   copper(I) bromide or cuprous bromide; mercury(II) bromide or mercuric bromide; iron(III) oxide or ferric oxide

3.10   potassium hydrogen phosphate, aluminum sulfate, iron(II) sulfite or ferrous sulfite

3.11   nitrogen dioxide, phosphorus tribromide, sulfur dichloride

3.13   The patient gets calcium ions, $Ca^{2+}$.

3.15   (a) $Li^+$  (b) $Cl^-$  (c) $Ca^{2+}$  (d) $Na^+$  (e) $Al^{3+}$  (f) $Ba^{2+}$

3.17   $Li^-$ has two electrons in its outer shell, so it does not have a complete octet.

3.19   $C^{4+}$, $C^{4-}$, $Si^{4+}$, and $Si^{4-}$ all have charges that are too concentrated for small ions.

3.21   No. Solid KF consists of potassium ions, $K^+$, and fluoride ions, $F^-$, held together in a three-dimensional crystal similar to that shown in Figure 3.2.

3.23   The 2 means that in the crystal there are two $Na^+$ ions for every $SO_4^{2-}$ ion.

3.25   (a) NaBr  (b) $K_2O$  (c) $AlCl_3$  (d) $BaCl_2$  (e) CaO

3.27   (a) $Ca^{2+}$ $Cl^-$  (b) $Mg^{2+}$ $O^{2-}$  (c) $Al^{3+}$ $S^{2-}$  (d) $Rb^+$ $O^{2-}$

3.29   (a) 3  (b) 1  (c) 4  (d) 1  (e) 2

3.31

(e) $H-\overset{\overset{\displaystyle H}{|}}{\underset{\underset{\displaystyle O}{\|}}{C}}-H$   (f) $Cl-\overset{\overset{\displaystyle Cl}{|}}{\underset{\underset{\displaystyle Cl}{|}}{C}}-\overset{\overset{\displaystyle Cl}{|}}{\underset{\underset{\displaystyle Cl}{|}}{C}}-\overset{\overset{\displaystyle Cl}{|}}{\underset{\underset{\displaystyle Cl}{|}}{C}}-Cl$

3.33 (a) $Br-\overset{}{\underset{\underset{\displaystyle H}{|}}{C}}=\overset{}{\underset{\underset{\displaystyle H}{|}}{C}}-Br$   (b) $CH_3-\overset{}{\underset{\underset{\displaystyle O}{\|}}{C}}-H$

(c) $H-\overset{}{\underset{\underset{\displaystyle H}{|}}{C}}=NH$

3.35 An argon atom has a complete octet, so (a) it does not lose or gain any electrons, and (b) it has no need to share any.

3.37 (a) ·C̈·  (b) :S̈·  (c) ·N̈·  (d) Na·  (e) :Ö·
(f) ·P̈·  (g) :B̈r·

3.39 (a) :B̈r—B̈r:  (b) H—S̈—H (with H below)  (c) H—N̈=N̈—H
(d) [C≡N̈]⁻  (e) ammonium  (f) [BF₄]⁻

3.41 These are completely different species, whose Lewis structures are

bromine atom :B̈r·
bromine molecule :B̈r—B̈r:
bromide ion [:B̈r:]⁻

3.43 $H-\overset{\overset{\displaystyle H}{|}}{\underset{\underset{\displaystyle H}{|}}{N}}: + H^+ \longrightarrow \left[H-\overset{\overset{\displaystyle H}{|}}{\underset{\underset{\displaystyle H}{|}}{N}}-H\right]^+$

3.45 (a) tetrahedral (b) pyramidal (c) planar triangular (d) linear (e) angular (f) planar triangular (g) each carbon is planar triangular (h) angular (i) pyramidal (j) planar triangular (k) angular (l) pyramidal

3.47 Because each molecule is a dipole, the oppositely charged ends attract each other.

3.49 $CS_2$ has no dipole moment because the individual charges cancel:
$\overset{\delta-}{S}=\overset{\delta+}{C}=\overset{\delta-}{S}$

3.51 (a), (c), and (d) have dipole moments.
3.53 (a) yes, covalent (b) no (c) yes, ionic (d) yes, covalent (e) no (f) no (g) no (h) no
3.55 (a) perchlorate ion structure
(b) $Ca(ClO_4)_2$
3.57 (a) sodium fluoride (b) magnesium sulfide (c) aluminum oxide (d) barium chloride (e) calcium bisulfite (f) potassium iodide (g) strontium phosphate (h) iron(II) hydroxide or ferrous hydroxide (i) sodium dihydrogen phosphate (j) lead acetate (k) barium hydride (l) ammonium hydrogen phosphate
3.59 (a) sulfur dioxide (b) dihydrogen sulfide (c) iodine fluoride (d) iodine heptafluoride (e) selenium trioxide (f) nitrogen oxide (g) dinitrogen trioxide
3.61 hemoglobin
3.63 $Na^+, Cl^-$
3.65 It is needed for the thyroid gland.
3.67 It is bound to two nitrogen atoms by covalent bonds and to the other two nitrogen atoms by coordinate covalent bonds.
3.69 tetrahedral
3.71 (a) This ion has one more electron than is needed for a complete octet. (b) The charge density is too high. (c) This ion does not have a complete octet.
3.73 coordinate covalent
3.75 (a) The carbon on the right has only six electrons in its outer shell. (b) There are too many electrons. The compound $C_2H_4$ has a total of 12 valence electrons, but this formula shows 14.
3.77 $BF_3 + PH_3 \longrightarrow F_3B-PH_3$
3.79 No. Ionic bonds point in every direction in space (they are spherically symmetrical).
3.81 $[Fe(CN)_6]^{3-}$ structure
3.83 (·) + (·) ⟶ (··)

## CHAPTER 4   Chemical Reactions

4.1   (a) 180.0 amu  (b) 601.9 amu
4.2   5.545 moles
4.3   222 g
4.4   15 moles of C atoms, 30 moles of H atoms, 15 moles of O atoms
4.5   $4.9 \times 10^{-4}$ moles of $Cu^+$ ions
4.6   $7.86 \times 10^{24}$ molecules of water
4.7   $As_2O_5 + 3H_2O \longrightarrow 2H_3AsO_4$
4.8   $2C_6H_{14} + 19O_2 \longrightarrow 14H_2O + 12CO_2$
4.9   8.3 moles of $CH_4$
4.10  $1.25 \times 10^3$ g of $C_2H_2Br_4$
4.11  17.1 g of $H_2$
4.12  80.7%
4.13  $Cu^{2+} + S^{2-} \longrightarrow CuS$
4.15  (a) 74.6 amu  (b) 164.0 amu  (c) 89.8 amu
       (d) 342.0 amu  (e) 210.2 amu  (f) 77.0 amu
       (g) 342.3 amu  (h) 96.0 amu  (i) 354.5 amu
4.17  (a) 81.4 g $NO_2$  (b) 50 g 2-propanol
       (c) $1.33 \times 10^3$ g $WCl_5$  (d) 62.6 g galactose
       (e) 8.6 g vitamin C
4.19  (a) 3.71 moles $Ag^+$ ions  (b) 2.72 moles $Na^+$ ions
       (c) 0.74 mole $SO_4^{2-}$ ions
4.21  (a) $2.0 \times 10^{-21}$ g  (b) $3.0 \times 10^{-20}$ g
       (c) $1.0 \times 10^{-17}$ g  (d) $1.0 \times 10^{-4}$ g
4.23  (a) $1.7 \times 10^{24}$ molecules of TNT
       (b) $1.67 \times 10^{21}$ molecules of water
4.25  $1.1 \times 10^{-19}$ g
4.27  (a) $HI + NaOH \longrightarrow NaI + H_2O$
       (b) $Ba(NO_3)_2 + H_2S \longrightarrow BaS + 2HNO_3$
       (c) $CH_4 + 2O_2 \longrightarrow CO_2 + 2H_2O$
       (d) $2C_4H_{10} + 13O_2 \longrightarrow 8CO_2 + 10H_2O$
       (e) $2Fe + 3CO_2 \longrightarrow Fe_2O_3 + 3CO$
       (f) $2Al + 6HBr \longrightarrow 2AlBr_3 + 3H_2$
       (g) $P_4 + 5O_2 \longrightarrow P_4O_{10}$
4.29  (a) 2/3 mole $N_2$  (b) 2/3 mole $N_2O_3$  (c) 12 moles $O_2$
4.31  $1.44 \times 10^3$ g $Cl_2$
4.33  879 g $O_2$
4.35  4660 g $CO_2$
4.37  111 g of aspirin
4.39  (a) $H_2$  (b) 14.3 g $N_2$  (c) 18.8 g $NH_3$

4.41  94.0%
4.43  decomposition
4.45  (a) $2K + 2H_2O \longrightarrow 2KOH + H_2$
       (b) $Sr + 2AgBr \longrightarrow SrBr_2 + 2Ag$
       (c) $Al + Fe(NO_3)_3 \longrightarrow Al(NO_3)_3 + Fe$
4.47  (a) $K^+$, $Cl^-$  (b) $Ba^{2+} + SO_4^{2-} \longrightarrow BaSO_4$
4.49  $Pb^{2+} + 2Cl^- \longrightarrow PbCl_2$
4.51  $2H^+ + SO_3^{2-} \longrightarrow SO_2(g) + H_2O$
4.53  Soluble: (a) (b) (d) (e)
       Insoluble: (c) (f)
4.55  (a) $Ca_3(PO_4)_2$  (c) $BaCO_3$  (d) $Fe(OH)_2$
       (f) $Sb_2S_3$  (g) $PbSO_4$
       No precipitate will form in (b) and (e).
4.57  No; one gains electrons and the other loses electrons. Electrons are not destroyed, but transferred.
4.59  (a) $C_7H_{12}$ gets oxidized; $O_2$ gets reduced
       (b) $O_2$ is the oxidizing agent; $C_7H_{12}$ the reducing agent.
4.61  exothermic: (b) (c) (e); endothermic: (a) (d)
4.63  160 000 cal
4.65  (1) The higher the percentage yield, the smaller the percentage of starting materials that are wasted.
       (2) Any of the starting materials that are not converted to the desired product are either still there or are converted to an undesired product. In either case they must be removed, and this usually costs money.
4.67  The $CH_4$ is oxidized to $CO_2$. The $H_2O_2$ is reduced to $H_2O$.
4.69  Breakfast (a) supplies 360 Cal; breakfast (b) 330 Cal. (b) supplies the lower number.
4.71  0.00271 mole aspirin; $1.63 \times 10^{21}$ molecules aspirin
4.73  More than 90 percent of the energy needed to operate our cars, trucks, airplanes, ships, and farm and factory machinery and to light, heat, and cool our buildings comes from the combustion of oil, coal, and natural gas.
4.75  a mole of platinum
4.77  9.961 g
4.79  (a) $AlCl_3$ is in excess  (b) 82.5%
4.81  (a) 5.68 moles $CO_2$  (b) 7.00 moles $CO_2$

## CHAPTER 5   Gases, Liquids, and Solids

5.1   0.4 atm
5.2   16.4 atm
5.3   13 atm
5.4   7.8 atm
5.5   0.422 mole
5.6   MW = 82

5.7   0.107 atm
5.8   (a) yes  (b) no
5.9   Strong forces in the nucleus, electrostatic attractions between nucleus and electrons, covalent and ionic bonds within molecules, and intermolecular forces between molecules.

5.11

5.13 The barometric pressure drops because the air column above us weighs less than before.

5.15 1.7 L

5.17 1.87 L

5.19 475°C

5.21 836 ml

5.23 61.0 mL

5.25 1.87 atm

5.27 1.17 atm

5.29 22 L

5.31 8040 mm Hg

5.33 93.2 mL

5.35 12 L

5.37 MW = 52

5.39 Extensive: $n$, $V$; Intensive: $P$, $T$

5.41 324 mm Hg

5.43 (c)

5.45 London dispersion forces provide attraction that at low temperatures overcomes the kinetic energy of helium gas.

5.47 When liquids are boiled the molecules separate from each other. Water molecules are attracted to each other by hydrogen bonds, which are absent in $H_2S$. Therefore, more energy (and a higher boiling point) is needed to separate $H_2O$ molecules than for $H_2S$ molecules.

5.49 See Sec. 5.9.

5.51 (a) 83°C  (b) 63°C  (c) 44°C

5.53 Molecules of water escape the surface and go into the vapor phase, but many fewer come back. Eventually all of them are in the vapor.

5.55 water

5.57 (a)

5.59 In diamond each carbon atom is bonded to four other carbon atoms, forming a large network crystal ($10^{22}$ to $10^{23}$ atoms); a molecule of buckminsterfullerene contains only 60 atoms.

5.61

5.63 solid carbon dioxide

5.65 The systolic pressure is 140 mm Hg; the diastolic pressure is 110 mm Hg. The first tapping sound is at 140 mm Hg; the last is at 110 mm Hg.

5.67 CO is bound to hemoglobin and does not allow it to carry oxygen. Oxygen under high pressure in the hyperbaric chamber is dissolved in the plasma and is carried to the tissues without the aid of hemoglobin.

5.69 The volume of water increases when it freezes. This expansion will crack the bottle.

5.71 graphite

5.73 MW = 91.9

5.75 112 mL

5.77 40°C

5.79 0.633 g/L

5.81 MW = 28

5.83 pentane

---

## CHAPTER 6  Solutions and Colloids

6.1 To 11 g KBr add enough water to make 250 mL of solution.

6.2 1.9% w/v

6.3 To 158.2 g KCl (1.06 mole × 2) add enough water to make 2.0 L of solution.

6.4 0.055 $M$

6.5 185 mL

6.6 15 g

6.7 To 4.00 mL of stock solution add enough water to make 200.0 mL of solution.

6.8 To 0.13 mL of the concentrated solution add enough water to make 20.0 mL solution.

6.9 − 12.5°C

6.10 − 18.6°C

6.11 0.80 osmol

6.13 Glucose is the solute and water the solvent.

6.15 (a) solid in solid  (b) solid in liquid  (c) gas in gas
(d) gas (carbon dioxide) and liquid (ethyl alcohol) in liquid (water)

6.17 True solutions. They mix in all proportions.

6.19 unsaturated

6.21 (b), (a), (c), (d)

6.23 By cooling it without stirring.

6.25 The warm water contains less oxygen than cold water. The fish die because of lack of oxygen.

6.27 (a) no  (b) yes. Only the solubilities of gases are affected by pressure, not those of solids or liquids.

6.29 (a) To 110 mL of ethyl alcohol add enough water to make a total volume of 410 mL.
(b) To 5.9 mL of ethyl acetate add enough water to make a total volume of 330 mL.
(c) To 308 mL of benzene add enough chloroform to make a total volume of 3.85 L.

6.31 (a) 2.11% w/v  (b) 5.9% w/v  (c) 0.723% w/v

6.33 In each case add enough water to the given weight of solute to make the total volume specified: (a) 19.5 g of $NH_4Br$—175 mL of solution  (b) 167 g NaI—1.35 L of solution  (c) 2.4 g $C_2H_5OH$—330 mL of solution

6.35 68 mL

6.37 0.405 $M$

6.39 0.568 $M$

6.41 To 31 mL of the concentrated solution add enough water to make 750 mL of solution.

6.43 To 2.08 mL of the stock solution add enough water to make 250 mL of solution.

6.45 In each case add enough water to the given volume of concentrated solution to make the total volume specified: (a) 0.12 L—550 mL  (b) 0.36 L—1.75 L  (c) 1.8 mL—385 mL

6.47 No, the concentration is 0.01 ppb.

6.49 See Sec. 6.6.

6.51 (b)

6.53 soluble: (b), (c), (d)

6.55 homogeneous: (a); heterogeneous: (d); colloidal: (b), (c), (e), (f)

6.57 The surface charges repel each other and do not allow the particles to aggregate.

6.59 (a) −3.72°C  (b) −5.58°C  (c) −5.58°C
(d) −7.44°C

6.61 344 g methanol

6.63 5.4 moles

6.65 (a) B  (b) B  (c) A  (d) B  (e) A  (f) no rise. In each case, the side with the higher osmolarity rises.

6.67 (a) 15.6  (b) 1.32  (c) 5.6  (d) 0.036 osmol

6.69 (b)

6.71 $HNO_2$, $HNO_3$

6.73 846 g

6.75 $CaCO_3$

6.77 Because of perspiration they lose potassium ions, which are needed for proper muscle action.

6.79 (a) casein  (b) egg yolk

6.81 Osmolarities: 0.6% NaCl = 0.21; 0.04% KCl = 0.011; 0.2% $NaHCO_3$ = 0.048; 0.72% glucose = 0.040. Total osmolarity = 0.31.

6.83 Methanol is more efficient. It has a lower molecular weight; hence there are more moles in the same mass.

6.85 4 ppm

6.87 To 25.0 mL of the stock solution add enough water to make the volume of the solution 250 mL.

6.89 (a) 26.3% w/v  (b) 25.0% w/w

6.91 At least one of the solids must be melted. The other solid is then dissolved in the newly formed liquid; the mixture is stirred and allowed to solidify.

6.93 $CO_2 + H_2O \longrightarrow H_2CO_3$; $SO_2 + H_2O \longrightarrow H_2SO_3$

6.95 1.25 osmol; 0.625 $M$

## CHAPTER 7    Reaction Rates and Equilibrium

7.1 $K = \dfrac{[H_2SO_3]}{[SO_2][H_2O]}$

7.2 $K = \dfrac{[N_2][H_2]^3}{[NH_3]^2}$

7.3 $K = 0.602$

7.5 $6.14 \times 10^{-3}$ moles/L per min

7.7 (a) the amount of energy colliding molecules must possess for a reaction to take place  (b) a collision that leads to a reaction

7.9 Two are broken and two are formed.

7.11 The activation energy is low because the breaking of covalent bonds is not required.

7.13 The spoiling is caused by chemical reactions initiated by bacteria. These reactions are much faster at room temperature than at refrigerator temperature.

7.15 50°C

7.17 (1) increase the concentrations  (2) increase the temperature  (3) add a catalyst

7.19 It provides an alternate pathway, of lower activation energy.

7.21 digesting a piece of candy; explosion of TNT; rusting of iron; addition of Na or K metal to water

7.23 (a) $K = \dfrac{[COCl_2]}{[CO][Cl_2]}$  (c) $K = \dfrac{[H_2O]^2[SO_2]^2}{[H_2S]^2[O_2]^3}$
(b) $K = \dfrac{[NO_2]^2}{[N_2O_4]}$  (d) $K = \dfrac{[CO_2]^4[H_2O]^6}{[C_2H_6]^2[O_2]^7}$

7.25 $K = 1.5$

7.27 $K = 9.9 \times 10^{-2}$

7.29 products: (a), (b), (c); reactants: (d), (e)

7.31 (a) No; the speed of reaction depends on the energy of activation, not on the exothermic character.
(b) Yes; endothermic reactions always have high energies of activation (Fig. 7.5) and so must be very slow.

7.33 (a) left  (b) right  (c) right  (d) right  (e) no shift

7.35 It shifts to the left.

7.37 (a) no change  (b) no change  (c) $K$ smaller

7.39 because it makes combustible materials catch fire faster, especially if a spark is present

7.41 The higher temperature causes chemical reactions in the body to go faster. If the reactions go too fast, it is dangerous to the health of the patient. At low temperature, the reactions are too slow.

7.43 The silver chloride is converted to silver and chlorine (the equation is given in Box 7D).

7.45 The heat from the match provides the necessary energy of activation. Once it starts, the combustion is exothermic; the heat liberated provides the energy of activation for the other molecules.

7.47 A, because the orientation does not matter, but rod-like molecules need special orientations in order for effective collisions to occur.

7.49 No; see Fig. 7.5.

7.51 $0.8\ M$

7.53 $0.030$ mole/L per min

7.55 the nature of the reactants, the concentration, the presence of a catalyst

7.57 $0.4\ M$

7.59 $2.0 \times 10^{-3}$ mole of $N_2O_4$ per L per sec

---

# CHAPTER 8   Acids and Bases

8.1 (b)

8.2 $1 \times 10^{-2}\ M$

8.3 5

8.4 $1 \times 10^{-4}\ M$

8.5 2

8.6 8.1

8.7 $4 \times 10^{-8}\ M$

8.8 (a) 9.14  (b) 10.25

8.9 Use $CH_3CH(OH)COO^-/CH_3CH(OH)COOH$ in a ratio of 2 to 1.

8.10 4.27

8.11 (a) 60.0  (b) 85.7  (c) 32.7

8.12 (a) 1.79 equiv  (b) 0.0203 equiv

8.13 $0.093\ N$

8.14 $0.0988\ N$

8.15 (a) $HF + H_2O \rightleftharpoons H_3O^+ + F^-$
(b) $HBr + H_2O \rightleftharpoons H_3O^+ + Br^-(100\%)$
(c) $H_2SO_3 + H_2O \rightleftharpoons H_3O^+ + HSO_3^-$
(d) $H_2SO_4 + H_2O \rightleftharpoons H_3O^+ + HSO_4^-(100\%)$
(e) $HCO_3^- + H_2O \rightleftharpoons H_3O^+ + CO_3^{2-}$

8.17 (a) An acid is a substance that produces $H_3O^+$ ions in aqueous solution. A base is a substance that produces $OH^-$ ions in aqueous solution.
(b) An acid is a proton donor. A base is a proton acceptor.

8.19 (a) $H_2O$  (b) $NH_3$  (c) $NH_4^+$  (d) $HCO_3^-$
(e) $H_2CO_3$  (f) $HPO_4^{2-}$  (g) $H_3O^+$  (h) $HBr$
(i) $CH_3—NH_3^+$  (j) $HCOOH$

8.21 No. The Brønsted-Lowry definitions are simply an extension of the Arrhenius definitions.

8.23 (a) $HF$  (b) $CH_3COOH$  (c) $HCOOH$

8.25 (a) $CO_3^{2-} + 2H_3O^+ \rightleftharpoons H_2CO_3 + 2H_2O$ or
$Na_2CO_3 + 2HCl \rightleftharpoons H_2CO_3 + 2NaCl$

(b) $Mg + 2H_3O^+ \rightleftharpoons Mg^{2+} + H_2 + 2H_2O$ or
$Mg + 2HCl \rightleftharpoons MgCl_2 + H_2$
(c) $OH^- + H_3O^+ \rightleftharpoons 2H_2O$ or
$NaOH + HCl \rightleftharpoons H_2O + NaCl$
(d) $NH_3 + H_3O^+ \rightleftharpoons NH_4^+ + H_2O$ or
$NH_3 + HCl \rightleftharpoons NH_4^+ + Cl^-$
(e) $CH_3NH_2 + H_3O^+ \rightleftharpoons CH_3NH_3^+ + H_2O$ or
$CH_3NH_2 + HCl \rightleftharpoons CH_3NH_3^+ + Cl^-$

8.27 (a) $10^{-6}\ M$  (b) $10^{-10}\ M$  (c) $10^{-7}\ M$  (d) $10^{-14}\ M$

8.29 (a) 3  (b) 11  (c) 1  (d) 10  (e) 7
acidic: (a) (c); basic: (b) (d); neutral: (e)

8.31 (a) $1 \times 10^{-11}$  (b) $1 \times 10^{-3}$  (c) $1 \times 10^{-7}$
(d) $1 \times 10^{-1}$  (e) $4 \times 10^{-3}$  (f) $8 \times 10^{-14}$
(g) $3 \times 10^{-7}$
acidic: (b) (d) (e) (g); basic: (a) (f); neutral: (c)

8.33 strongest: lactic acid; weakest: barbituric acid

8.35 $Na^+$ does not react with water, but $CO_3^{2-}$ does:

$$CO_3^{2-} + H_2O \rightleftharpoons HCO_3^- + OH^-$$

The $OH^-$ ions make the solution basic.

8.37 Buffer action is the result of Le Chatelier's principle. Adding acid or base disturbs an equilibrium, which shifts in the direction that minimizes the effect of the addition.

8.39 (a) $H_3O^+ + HPO_4^{2-} \rightleftharpoons H_2PO_4^- + H_2O$
(b) $OH^- + H_2PO_4^- \rightleftharpoons HPO_4^{2-} + H_2O$

8.41 The ability of the buffer to neutralize added amounts of $H_3O^+$ or $OH^-$. The greater the ability, the greater the capacity.

8.43 carbonate and phosphate; see Sec. 8.10

8.45 (a) 127.9  (b) 85.7  (c) 100.5  (d) 32.0  (e) 17.0
(f) 41.0  (g) 102.5  (h) 47.3

8.47 $0.348\ N$

8.49    5.60 $N$

8.51    (a) 3.0 $N$  (b) 9.0 $N$  (c) 0.34 $N$  (d) 5.4 $N$
(e) 6.15 $N$

8.53    13.0 mL

8.55    0.00330 equiv

8.57    To 2.0 g of $NaHCO_3$ add enough water to make a total volume of 1 L.

8.59    nitric acid

8.61    immediate washing with a steady stream of cold water

8.63    $NaHCO_3$ and an acid

8.65    See Box 8E.

8.67    (a) 3.6  (b) 8.6  (c) 4.0

8.69    Solving the Henderson-Hasselbalch equation,

$$4.55 = 4.75 + \log(CH_3COO^-/CH_3COOH)$$

we obtain a $CH_3COOH : CH_3COO^-$ ratio of 1.6 : 1. We therefore make up a solution that contains 1.6

times as many moles of $CH_3COOH$ as of $CH_3COO^-$ ions.

8.71    The concentrations, $1 \times 10^{-7}\ M$ for $H_3O^+$ and $OH^-$, are too low.

8.73    The HCl solution is more acidic because its $[H_3O^+]$ is higher.

8.75    EW = 172

8.77    strongest: uric acid; weakest: caproic acid

8.79    No. HF, being a stronger acid, reacts with water to a greater extent, producing a higher $[H_3O^+]$, so the pH of its solution is lower.

8.81    0.00625 $N$

8.83    Yes. The ethoxide ion will not react with $NH_3$, but it will react with water, since the latter is a stronger acid than ethyl alcohol (Table 8.2):

$$C_2H_5O^- + H_2O \rightleftharpoons C_2H_5OH + OH^-$$

# CHAPTER 9    Nuclear Chemistry

9.1    xenon-139

9.2    0.31 g

9.3    1.5 mL

9.5    Both are positively charged, but an alpha particle consists of two protons and two neutrons, while a proton is a single particle.

9.7    $5.2 \times 10^9\ s^{-1}$

9.9    X-rays are the same as gamma rays except that they are lower in frequency, energy, and penetrating power.

9.11    (a) $^{19}_{9}F$  (b) $^{32}_{15}P$  (c) $^{87}_{37}Rb$

9.13    none

9.15    (a) $^{159}_{64}Gd$  (b) $^{141}_{57}La$  (c) $^{242}_{96}Cm$

9.17    (a) $^{206}_{81}Tl$  (b) $^{234}_{92}U$  (c) $^{170}_{70}Yb$

9.19    $^{29}_{15}P \longrightarrow {}^{0}_{+1}e + {}^{29}_{14}Si$

9.21    $^{17}_{8}O$; $^{14}_{7}N + {}^{4}_{2}He \longrightarrow {}^{17}_{8}O + {}^{1}_{1}H$

9.23    $^{238}_{92}U \longrightarrow {}^{234}_{90}Th + {}^{4}_{2}He$; $^{234}_{90}Th \longrightarrow {}^{234}_{91}Pa + {}^{0}_{-1}e$

9.25    (a) $^{28}_{14}Si$  (b) $^{143}_{54}Xe$  (c) $^{1}_{0}n$  (d) $^{210}_{84}Po$  (e) $^{13}_{7}N$

9.27    140 years

9.29    No; the conversion $Ra \longrightarrow Ra^{2+}$ involves electrons from the electron cloud. Radioactivity is a nuclear phenomenon.

9.31    19.5 hr

9.33    We cannot see or feel it, nor can our senses detect it in any other way.

9.35    (a)

9.37    (a), (e) the amount of radiation absorbed by tissues
(b), (g) the effective dose absorbed by humans
(c) the effective dose delivered
(d), (f) intensity of radiation

9.39    gamma rays

9.41    $1 \times 10^{-5}$ sieverts

9.43    $^{99m}_{43}Tc \longrightarrow {}^{99}_{43}Tc + \gamma$

9.45    The half-life is too short. It would not last long enough to reach a hospital.

9.47    A small dose is used for diagnosis, so no radioactive material is left in the body. For therapy a large dose is used, so it can stay in the thyroid gland and kill the cancer cells.

9.49    $^{1}_{1}H$

9.51    neutrons

9.53    See Sec. 9.8.

9.55    two spots (the neutrons and gamma rays are uncharged)

9.57    one-tenth of a half-life

9.59    about 6%

9.61    They would lick the tips of their brushes. These tips contained radium, which emitted alpha particles that caused cancer.

9.63    See Box 9E.

9.65    The reaction between a deuterium and a tritium nucleus produces a helium nucleus and a neutron. The total mass of the helium nucleus and neutron is *less than* the total mass of the deuterium and tritium nuclei. The difference in mass is converted to energy.

9.67    (a) 82%  (b) 11%  (c) 0.1%

9.69    from the small amount of mass converted to energy during the breakup of heavy atomic nuclei

9.71    fluorine-19 and neon-20

9.73    (a) because a radioactive sample decays, and the decay products are always mixed with the starting isotope  (b) because a neutron in the nucleus is converted to a proton

9.75    two neutrons

9.77    Oxygen-16 is stable because it has an equal number of protons and neutrons. The others are unstable because the numbers are unequal; in this case the less equal, the faster the isotope decays.

9.79    zirconium-90

9.81    x-rays

## CHAPTER 10   Organic Chemistry. Alkanes

**10.1** (a) The carbon uses one $sp^3$ orbital to overlap with a $p$ orbital of the fluorine. Each of the other three $sp^3$ orbitals of the carbon overlaps with an $s$ orbital of a hydrogen.
(b) Two $sp^3$ orbitals, one from each carbon, overlap to form the C—C sigma ($\sigma$) bond. Each of the remaining $sp^3$ orbitals of the left carbon overlaps with one $s$ orbital of a hydrogen. One of the three remaining $sp^3$ orbitals of the right carbon overlaps with a $p$ orbital of a fluorine; each of the other two overlaps with an $s$ orbital of a hydrogen.

**10.2** (a) butane (b) 2-methylbutane (c) 3-ethylheptane

**10.3** (a) 2,3-dimethylpentane (b) 2,2,4-trimethylpentane

**10.4** 4-ethyl-3,4-dimethylheptane

**10.5** (a) 2,3-dichloropentane (b) 3-bromo-2-chloro-2,3-diiodohexane

**10.6** (a) 2-chloropropane; isopropyl chloride
(b) 1-iodo-2,2-dimethylpropane; neopentyl iodide

**10.7** (a) chlorocyclopentane (b) 1,1-diethylcyclobutane
(c) 1,2,4-trimethylcyclohexane

**10.9** No, they are nonelectrolytes.

**10.11** organic

**10.13** no difference

**10.15** (1): (b), (c); (2): (a), (e), (f), (g), (i); (3): (d), (h)

**10.17** (1) $CH_3CH_2CH_2$—OH   (2) $CH_3$—CH—$CH_3$
                                                            |
                                                            OH

(3) $CH_3$—O—$CH_2$—$CH_3$

**10.19** C 4, N 3, O 2, S 2, H 1, F 1, Br 1, Cl 1, I 1

**10.21** (a) $CH_3$—CH=CH—CH—$CH_3$
                                             |
                                             $CH_3$

(b)
                                                            $CH_3$
                                                            |
$NH_2$—$CH_2$—$CH_2$—O—$CH_2$—CH—Br

(c) Cl—CH=CH—C≡C—N—$CH_3$
                                          |
                                          $CH_3$

**10.23** (a) spherical (b) dumbbell shape (c) a mixture of the spherical and dumbbell shapes

**10.25** (a) Each of the 4 $sp^3$ orbitals of the carbon overlaps with a $p$ orbital of one bromine.
(b) Each carbon uses 4 $sp^3$ orbitals. One $sp^3$ orbital of one carbon overlaps with one $sp^3$ orbital of the other carbon. One $sp^3$ orbital of one carbon overlaps with the $p$ orbital of the bromine. Each of the other 5 $sp^3$ orbitals overlaps with the $s$ orbital of a different hydrogen.

**10.27** (a) $CH_3$—CH—$CH_2CH_3$   (b) $CH_3$—C—$CH_3$
                                |                                             |
                                $CH_3$                                    $CH_3$
                                                                               |
                                                                               $CH_3$

(c) $CH_3$ ◁

**10.29** (1) $CH_3CH_2CH_2CH_2CH_2CH_3$   hexane
(2) $CH_3$—CH—$CH_2CH_2CH_3$   2-methylpentane
             |
             $CH_3$

(3) $CH_3CH_2$—CH—$CH_2CH_3$   3-methylpentane
                      |
                      $CH_3$

(4) $CH_3$—C—$CH_2CH_3$   2,2-dimethylbutane
             |
             $CH_3$   (with $CH_3$ above)

(5) $CH_3$—CH—CH—$CH_3$   2,3-dimethylbutane
             |       |
             $CH_3$ $CH_3$

**10.31** (a) heptane (b) 2-methylpentane (c) 2,4-dimethylhexane (d) 2,5-dimethylheptane (e) 2,3,3-trimethylpentane (f) 3,5-dimethylheptane (g) 4,6-diethyl-2,7-dimethyl-6-propylnonane

**10.33** (a) 2-bromopropane (b) 1,4-diiodo-2-methylpentane (c) 1,1,1,3-tetrabromopropane (d) 2-chloro-4-iodo-3-methylpentane (e) 1,2-dichloro-4-bromopentane (f) 1,3,3-triiodo-2-methylbutane

**10.35** (a) methylpropane (b) 2-bromopropane (c) 2-fluorobutane (d) 1-chloro-2,2-dimethylpropane (e) 2-iodo-2-methylpropane

**10.37** (a) chlorocyclohexane (b) 1-iodo-2-methylcyclobutane (c) 1-bromo-3-chloro-2-ethyl-3-methylcyclopentane (d) 1,1,3,4-tetrachlorocyclohexane

**10.39** (a) ⬠   (b) (cyclopropane with $CH_3$, Br, $CH_3$ substituents)

(c) $CH_3CH_2$ (cyclohexane with $H_3C$, $H_3C$, $CH_3$ substituents)

**10.41** (a) *cis*-1-bromo-3-methylcyclopentane (b) *trans*-1-chloro-4-ethylcyclohexane (c) *cis*-1,2-dimethylcyclopropane

**10.43** They burn but are otherwise inert.

**10.45** (a) halide (b) alcohol (c) alkyne (d) amine (e) aldehyde (f) alkane (g) carboxylic acid (h) ketone (i) carboxylic ester

**10.47** alcohol, ether, aldehyde, ketone, carboxylic acid, carboxylic ester, amide

**10.49** (a) carboxylic acid group, carboxylic ester group
(b) halide, ether, carboxylic acid group
(c) (d) carboxylic acid group, amine

**10.51** Gasoline dissolves body oils and extracts them from the skin, leaving the skin "chapped." Petroleum jelly does not dissolve body oils and so they stay in the skin, but it protects the skin from moisture.

**10.53** Petroleum is a mixture of compounds with a wide range of boiling points. Each use requires a much narrower boiling point range.

**10.55** Most of the $CO_2$ in the atmosphere is converted to $H_2CO_3$ in rain and in oceans, and then to solid carbonates.

**10.57** alcohol, ether, ketone, carboxylic ester, amide, C=C double bond.

**10.59** one at a tertiary, two at secondary carbons

**10.61** (a) Pentane has five carbons. There cannot be a group only in the 4 position, since a lower number would be obtained counting from the other end.
(b) There cannot be a methyl group at the end of a parent chain, since that makes the parent chain longer.
(c) The prefix di requires two numbers. Only one is given.
(d) If two numbers are given, the prefix di must be used.
(e) If there is only one group on a ring, it gets no number (the number 1 is understood).
(f) There cannot be an ethyl group in the 2 position of an alkane, since that only makes the parent chain longer.
(g) same answer as (f)
(h) There cannot be a two-membered ring.

**10.63** (a) Two $sp^3$ orbitals, one from each carbon, overlap to form the C—C $\sigma$ bond. Each of the three remaining orbitals of the carbon on the right overlaps with a $p$ orbital of the fluorine, as does one of the three remaining $sp^3$ orbitals of the carbon on the left. Each of the last two remaining $sp^3$ orbitals of the left carbon overlaps with the $s$ orbital of a hydrogen.

(b) The two C—C $\sigma$ bonds are formed by overlap of one $sp^3$ orbital from each of two carbons. The central carbon has two $sp^3$ orbitals left: Each of these overlaps with an $s$ orbital of a hydrogen. Each of the end carbons has three $sp^3$ orbitals left. In both cases, one of these overlaps with a $p$ orbital of a chlorine; each of the other two, with the $s$ orbital of a hydrogen.

**10.65** (a) $CH_3$—CH—CH—$CH_2CH_2CH_3$
  with Cl   Cl below the two CH's

(b)

(c) (structure: cyclobutane with C($CH_3$)($CH_3$))

(d) (cycloheptane ring with substituents: CH—$CH_3$ / $CH_3$, CH—$CH_2CH_3$ / $CH_3$, $CH_2$—C($CH_3$)($CH_3$)—$CH_3$)

**10.67** the conversion of ammonium chloride and silver cyanate to urea

**10.69** $CH_3$—NH—$CH_3$    $CH_3CH_2NH_2$

**10.71** (a) alcohol, ketone (b) carboxylic acid, carboxylic ester (c) aldehyde, alcohol

**10.73** They look the same. Neither has any color. You could probably tell them apart by measuring a physical property, such as boiling point.

---

# CHAPTER 11   Alkenes, Alkynes, and Aromatic Compounds

**11.1** 2,5-dichloro-2-hexene

**11.2** 3-ethyl-2-hexene

**11.3** (a) 1,4-dimethylcyclopentene (b) 5-chlorocycloheptene

**11.4** (a) yes (b) no (c) no

**11.5** $CH_3CH_2CH_2CH_2CH_3$

**11.6** (cyclopentane ring with Cl, Cl and $CH_3$, H) + (cyclopentane ring with Cl, H and $CH_3$, Cl)

**11.7** $CH_3CH_2$—C(Br)($CH_3$)—$CH_3$ with Br above and $CH_3$ below

**11.8** $CH_3$—CH—$CH_3$ with OH below

**11.9** —$CH_2$—CH—$CH_2$—CH—$CH_2$—CH— with CN, CN, CN below

**11.10** (a) 2-hexyne (b) 4-chloro-3-methyl-1-pentyne

11.11  (a)  $CH_3-C=C-CH_2CH_2CH_3$
              $\quad\quad\ \ |\ \ |$
              $\quad\quad\ Cl\ \ Cl$

$$CH_3-\underset{\underset{Cl}{|}}{\overset{\overset{Cl}{|}}{C}}-\underset{\underset{Cl}{|}}{\overset{\overset{Cl}{|}}{C}}-CH_2CH_2CH_3$$

(b)  $H_2C=C-CH_2CH_2CH_3$
          $\quad\quad\ \ |$
          $\quad\quad\ Br$

$$CH_3-\underset{\underset{Br}{|}}{\overset{\overset{Br}{|}}{C}}-CH_2CH_2CH_3$$

11.12  (a)  *meta*-bromochlorobenzene  (b)  *para*-bromotoluene  (c)  *ortho*-nitrophenol

11.13  (a)  2,4-dichloroaniline  (b)  5-bromo-2-ethylbenzoic acid

11.15  (a)  propene  (b)  2,3-dibromo-2-butene
        (c)  4,4,4-trichloro-1-butene
        (d)  2-methyl-1-butene
        (e)  2-ethyl-4-methyl-1-pentene
        (f)  3-methylcyclopentene
        (g)  4-chlorocyclohexene

11.17  (a)  1,3-butadiene  (b)  1,4-heptadiene
        (c)  4-chloro-2,3-hexadiene
        (d)  3-methyl-1,4-cyclohexadiene

11.19  90°

11.21  (a)  because the position of the double bond determines the numbering of carbons (1-propene and not 3-propene)

(b)  $CH_2=CH-CH-Cl$  (c)  no
                    $\quad\quad\quad\ |$
                    $\quad\quad\quad Cl$

11.23  reactions in which something is added to a double or triple bond (for examples, see Sec. 11.6)

11.25  (a)  $CH_3CH_2CH_2CH_3$

(b)  $CH_3-\underset{\underset{CH_3}{|}}{\overset{\overset{Br}{|}}{C}}-CH_2CH_3$

(c)

(d)  $CH_3-\underset{\underset{OH}{|}}{CH}-\underset{\underset{CH_3}{|}}{CH}-CH_3$

(e)  $CH_3-\underset{\underset{CH_3}{|}}{\overset{\overset{Cl}{|}}{C}}-CH_2CH_2CH_3$

(f)

11.27  (a)  $CH_3-\underset{\underset{Cl}{|}}{CH}-\underset{\underset{Cl}{|}}{CH}-CH_3$

(b), (c)  no reaction

(d)

11.29  (a)  $Br_2$  (b)  HBr  (c)  $H_2$, Pt  (d)  $H_2O$, $H_2SO_4$
        (e)  HI

11.31  Add a drop or two of $Br_2$ dissolved in $CCl_4$. If the brown color of $Br_2$ disappears, the compound is cyclohexene. If it does not disappear, the compound is cyclohexane.

11.33  (a)  $CH_3CH=CHCl$  (b)  $CH_2=CH-COOCH_3$

11.35  (a)  2-pentyne  (b)  3,4,4-tribromo-1-butyne
        (c)  1,3,5-trichloro-1-pentyne

11.37  because the two triple-bonded carbons and the two atoms connected to them form a straight line

11.39  one that contains an aromatic loop (sextet) of electrons

11.41  Benzene can cause liver damage and cancer.

11.43  If they underwent addition, the very stable aromatic sextet would be destroyed.

11.45  (a)  iodobenzene  (b)  phenol  (c)  *meta*-dibromobenzene  (d)  *meta*-ethylphenol  (e)  *para*-bromonitrobenzene  (f)  *meta*-*tert*-butylbenzoic acid
        (g)  *ortho*-isopropylaniline

11.47  (a)   (b)

(c)

11.49

11.51

11.53  (a)  a compound that contains a ring in which at least one atom is not carbon  (b)  a heterocyclic compound with an aromatic loop of electrons

**11.55**

**11.57** β-carotene: 40 carbons; vitamin A: 20 carbons

**11.59** the all-*trans* isomer   (b) the all-*trans* isomer

**11.61** (1) $CH_3-\overset{\underset{\textstyle CH_3}{|}}{C}=CH_2 + H^+ \longrightarrow CH_3-\overset{\underset{\textstyle CH_3}{|}}{\overset{+}{C}}-CH_3$

(2) $CH_3-\overset{\underset{\textstyle CH_3}{|}}{\overset{+}{C}}-CH_3 + \overset{..}{O}H_2 \longrightarrow CH_3-\overset{\underset{\textstyle CH_3}{|}}{\overset{\overset{\textstyle OH_2^+}{|}}{C}}-CH_3$

(3) $CH_3-\overset{\underset{\textstyle CH_3}{|}}{\overset{\overset{\textstyle OH_2^+}{|}}{C}}-CH_3 \longrightarrow CH_3-\overset{\underset{\textstyle CH_3}{|}}{\overset{\overset{\textstyle OH}{|}}{C}}-CH_3 + H^+$

**11.63** Raw rubber is too soft and tacky.

**11.65**

$-CH_2-\overset{\overset{\textstyle Cl}{|}}{C}=CH-CH_2-CH_2-\overset{\overset{\textstyle Cl}{|}}{C}=CH-CH_2-$

**11.67** aromatic hydrocarbons with at least four rings and at least one angular junction

**11.69** (a) $CH_2{=}CH_2$   ethylene, ethene

(b) $HC{\equiv}CH$   acetylene, ethyne

(c)

benzene

**11.71**

**11.73** An unsaturated compound has at least one double or triple bond. A saturated compound has none.

**11.75** $HC{\equiv}CH$

**11.77** Aromatic rings do not have double bonds, nor are they unsaturated. They have a closed loop of electrons.

**11.79**

---

# CHAPTER 12   Alcohols, Phenols, Ethers, and Halides

**12.1** (a) secondary  (b) primary  (c) tertiary

**12.2** (a) 2,3-dimethyl-3-pentanol  (b) 5-ethyl-3-octanol

**12.3** $CH_3CH_2-CH{=}CH-CH_3$

**12.4**

**12.5** $CH_3-\overset{\overset{\textstyle }{|}}{\underset{\underset{\textstyle CH_3}{|}}{CH}}-\overset{\underset{\underset{\textstyle O}{\|}}{}}{C}-H$     $CH_3-\underset{\underset{\textstyle CH_3}{|}}{CH}-\overset{\underset{\underset{\textstyle O}{\|}}{}}{C}-OH$

**12.6** $CH_3-\underset{\underset{\textstyle CH_3}{|}}{CH}-\overset{\underset{\underset{\textstyle O}{\|}}{}}{C}-CH_2CH_3$

**12.7** (a) di-*sec*-butyl ether   (b) ethyl phenyl ether

**12.9** (a) 1-butanol; *n*-butyl alcohol

(b) 2-butanol; *sec*-butyl alcohol

(c) 2-methyl-2-propanol; *tert*-butyl alcohol

(d) phenylmethanol; benzyl alcohol

(e) 2-chloro-3,3-difluoro-1-butanol

(f) 2,4-dimethyl-3-hexanol

(g) 5-bromo-4-methyl-2-pentanol

(h) 4-chloro-2-propyl-1-butanol

(i) 3-methyl-3-pentanol

(j) 3-bromo-2,6-dimethylcyclohexanol

(k) 1,3-dimethylcyclopentanol

(l) 1,2,3-propanetriol; glycerol

**12.11** See Sec. 12.2.

**12.13** (a) $CH_3-\underset{\underset{\textstyle OH}{|}}{CH}-CH_3$

(b) $CH_3-\underset{\underset{\textstyle CH_3}{|}}{CH}-CH_2OH$

(c)   (d)

(e) $\underset{\underset{\textstyle OH}{|}}{CH_2}-\underset{\underset{\textstyle OH}{|}}{CH}-\underset{\underset{\textstyle OH}{|}}{CH_2}$   (f) $CH_3OH$

**12.15**  $CH_3CH_2CH_2OH + H_2SO_4 \longrightarrow$
$CH_3-CH=CH_2$ or
$CH_3CH_2CH_2-O-CH_2CH_2CH_3$

**12.17**  (a) $CH_3-O-CH_3$   (b)

(c) $CH_3CH_2-\underset{\underset{O}{\|}}{C}-CH_2CH_3$

(d) $CH_3CH_2CH_2-\underset{\underset{O}{\|}}{C}-OH$   (e) NR

(f)

**12.19**  (a) methanol  (b) isopropyl alcohol  (c) glycerol
(d) ethylene glycol  (e) glycerol  (f) ethanol
(g) methanol

**12.21**  (a)

$+ H_2O$

(b)

$+ H_2O$

(c)

$+ H_3O^+$

**12.23**  (a) ethyl methyl ether  (b) diethyl ether  (c) diisopropyl ether  (d) isopropyl methyl ether  (e) *n*-
pentyl phenyl ether  (f) tetrahydrofuran

**12.25**  They are inert.

**12.27**  (a) (1)(2)  (b) (1)(5)  (c) (4)(5)  (d) (2)(5)
(e) (3)(5)  (f) (1)(2)  (g) (1)(4)  (h) (1)(2)(5)

**12.29**  (1) $CH_3CH_2CH_2CH_2-OH$

(2) $CH_3-\underset{\underset{OH}{|}}{CH}-CH_2CH_3$

(3) $CH_3-\underset{\underset{CH_3}{|}}{CH}-CH_2-OH$   (4) $CH_3-\underset{\underset{CH_3}{|}}{\overset{\overset{CH_3}{|}}{C}}-OH$

(5) $CH_3-O-CH_2CH_3CH_3$

(6) $CH_3-O-\underset{\underset{CH_3}{|}}{CH}-CH_3$

(7) $CH_3CH_2-O-CH_2CH_3$

**12.31**  (c) (a) (b) (d)

**12.33**  They are soluble. Two and three —OH groups, respectively, form extensive hydrogen bonds with water.

**12.35**  (a) ethanethiol  (b) 2-propanethiol

**12.37**  (a) $2\ CH_3CH_2-SH \xrightarrow{[O]}$
$CH_3CH_2-S-S-CH_2CH_3$

(b) $CH_3CH_2CH_2-S-S-CH_2CH_2CH_3 \xrightarrow{[H]}$
$2\ CH_3CH_2CH_2-SH$

**12.39**  (a) chloroform  (b) carbon tetrachloride
(c) methyl chloride, ethyl chloride  (d) triiodomethane (iodoform)  (e) dichlorodifluoromethane

**12.41**  (a) perchloroethylene  (b) 1,1,1,2-tetrafluoroethane

**12.43**  (a) 34 proof  (b) 82 proof  (c) 22 proof

**12.45**  Congeners are distillation products other than ethanol, or extracts from the containers (wooden casks,
for example).

**12.47**  It causes nausea, sweating, and other unpleasant
symptoms because acetaldehyde is built up. See Box
12B.

**12.49**  It blinds and kills.

**12.51**  No. One is —O—NO$_2$; the other —O—NO.

**12.53**  2-chloro-1,1,2-trifluoroethyl difluoromethyl ether

**12.55**  the action of certain skin bacteria on natural body
secretions

**12.57**

**12.59**  A biodegradable substance can be decomposed by
natural processes in the environment.

**12.61**  $C_6H_{12}O_6 \xrightarrow{zymase} 2CH_3CH_2OH + 2CO_2$

**12.63**  (a)

(b)

(c)

(d)

**12.65**  A Cl group increases the acidity of a phenol; a CH$_3$
group decreases it.

**12.67**  Because they have more —OH groups, ethylene glycol and (even more) glycerol molecules are held
more closely together by hydrogen bonds, making it
harder for the molecules to slide past each other.

**12.69** (a) $CH_3CH_2$—$S^-$ $Na^+$

(b) Mostly ⬠ , but some

⬠—O—⬠ can be produced,

depending on the temperature

(c) NR

**12.71** Molecules of 1-butanol form hydrogen bonds with each other. Molecules of methyl propyl ether cannot do this.

**12.73** Acidic: phenols, thiols. Neutral: alcohols, ethers, halides, thioethers, disulfides

**12.75** with care; do not drop it

---

# CHAPTER 13    Aldehydes and Ketones

**13.1** (a) 2-chlorobutanal  (b) 4-chloro-2-butanone

**13.2** (a) $CH_3CH_2CH_2$—$\overset{\displaystyle O}{\underset{\|}{C}}$—OH

(b) NR

**13.3** $CH_3$—CH—$CH_2$—CH—$OCH_3$
$\qquad\quad$ | $\qquad\qquad$ |
$\qquad\quad CH_3 \qquad\quad$ OH

Hemiacetal

$CH_3$—CH—$CH_2$—CH—$OCH_3$
$\qquad\quad$ | $\qquad\qquad$ |
$\qquad\quad CH_3 \qquad\quad OCH_3$

Acetal

**13.4** (a) hemiacetal  (b) acetal  (c) hemiacetal  (d) ketal

**13.5**

(a) ⬡CHO + $CH_3OH$

(b) ⬠CHO + $CH_3OH$ + $CH_3CH_2OH$

(c) $CH_3$—$\overset{\displaystyle O}{\underset{\|}{C}}$—H + $CH_3CH_2CH_2OH$

(d) (cyclohexanone with CH₃) + $CH_3CH_2OH$

**13.6** (a) ⬡$CH_2OH$

(b) $CH_3CH_2$—CH—$CH_2CH_3$
$\qquad\qquad\quad$ |
$\qquad\qquad\quad$ OH

**13.7** (a) (c) (d) (e)

**13.9** aldehyde: (c) ketone: (a) (e)  both: (d) neither: (b) (f)

**13.11** (a) 3-bromobutanal  (b) 2-hydroxy-3-hexanone  (c) 2-methylcyclopentanone  (d) 4-pentenal

**13.13** (a) $CH_3CH_2CH_2CHO$  (b) H—$\overset{\displaystyle O}{\underset{\|}{C}}$—H

(c) $CH_3$—CH—$CH_2$—CH—CHO
$\qquad\quad$ | $\qquad\qquad$ |
$\qquad\quad CH_3 \qquad\quad CH_3$

(d) $CH_3CH_2CHO$

**13.15** (a) 2  (b) 2  (c) 1  (d) 2  (e) 1

**13.17** Acetone is more polar than methyl ethyl ether.

**13.19** $CH_3$—$\overset{\displaystyle O\cdots H-O}{\underset{\|}{C}}$—H$\quad\quad$ $CH_3$—$\overset{\displaystyle O\cdots H-O}{\underset{\|}{C}}$—$CH_3$

**13.21** Benedict's or Tollens' reagents will give positive tests with the aldehyde, but not with the ketone.

**13.23** In Benedict's test $Cu^{2+}$ is reduced to red $Cu_2O$; in Tollens' test $Ag(NH_3)_2^+$ is reduced to metallic silver that forms a mirror inside the test tube.

**13.25** $\overset{\delta+ \quad \delta-}{-C=O}$

**13.27** (a) $CH_3$—CH—$CH_3$  (b) $CH_3CH_2CH_2OH$
$\qquad\qquad$ |
$\qquad\qquad$ OH

(c) ⬠—OH

(d) ⬡$\overset{\displaystyle C-CH_3}{\underset{\|}{\,}}\overset{}{O}$ + $CH_3OH$  (e) NR

**13.29** acetals: (a) (d) hemiacetals: (b) (e) neither: (c) (f)

**13.31** hemiacetals: (a) (b) (d)  none of these: (c)

**13.33** (a) $CH_3CH_2-\overset{\overset{\displaystyle O}{\|}}{C}-H + CH_3CH_2OH$

(b) $CH_3CH_2-\overset{\overset{\displaystyle O}{\|}}{C}-CH_3 + CH_3OH$

(c)

$+ CH_3OH$

(d) $HO-CH_2-\underset{\underset{\displaystyle OH}{|}}{CH}-CH_2CH_2-\overset{\overset{\displaystyle O}{\|}}{C}-H$

$+ CH_3CH_2OH$

(e) $CH_3CH_2-\underset{\underset{\displaystyle O}{\|}}{C}-\underset{\underset{\displaystyle CH_3}{|}}{CH}-CH_2CH_2-OH$

(f) $HO-CH_2CH_2-\underset{\underset{\displaystyle CH_3}{|}}{CH}-CH_2-\overset{\overset{\displaystyle O}{\|}}{C}-H$

**13.35** because an aldehyde carbon is always a primary carbon and a ketone carbon is always a secondary carbon

**13.37** (a) $CH_3-CH=CH-OH$

(b) two enol forms: $CH_3-CH=\underset{\underset{\displaystyle OH}{|}}{C}-CH_3$

and $CH_3CH_2-\underset{\underset{\displaystyle OH}{|}}{C}=CH_2$

(c) $CH_3-CH=\underset{\underset{\displaystyle OH}{|}}{C}-$ $-CH_3$

(d)

**13.39** to make polymers

**13.41** It indicates diabetes.

**13.43** Do not open the bottle inside a room; only in the open air or in a fume hood. Even in the open air, great care is needed.

**13.45** sleeping pill

**13.47** $\underset{\underset{\displaystyle OH}{|}}{CH_2}-\underset{\underset{\displaystyle OH}{|}}{CH}-\underset{\underset{\displaystyle OH}{|}}{CH_2}$ glycerol

**13.49** a three-dimensional polymer network that is formed at high temperatures and becomes rigid once cooled so that it cannot be melted again

**13.51** muscone, $\beta$-ionone

**13.53** (a) benzaldehyde (b) butanedione (c) cinnamaldehyde

**13.55** A molecule of propanol forms hydrogen bonds with other molecules of propanol. Molecules of propanal cannot do this.

**13.57** for example:

(a)

(b)

(c) $CH_3-\overset{\overset{\displaystyle OH}{|}}{\underset{\underset{\displaystyle OCH_3}{|}}{C}}-CH_3$

**13.59** (a) $CH_3-\overset{\overset{\displaystyle O}{\|}}{C}-CH_2CH_2CH_3 + CH_3OH$

(b) NR

(c) $CH_3-\underset{\underset{\displaystyle OCH_2CH_3}{|}}{CH}-OCH_2CH_3$

(d) $CH_3CH_2CH_2CH_2-\overset{\overset{\displaystyle O}{\|}}{C}-CH_3$

**13.61** Some of the butanal had become oxidized on standing, to give butanoic acid.

**13.63** Acetone, $CH_3-CO-CH_3$, does not have an H connected to an O, N, or F.

**13.65** A 37 percent solution of formaldehyde in water. It is used as a disinfectant and to embalm cadavers.

**13.67** (a)

(b)

(c)

and

## CHAPTER 14 Carboxylic Acids and Esters

**14.1** (a) heptanoic acid (b) 2-ethyl-3-hydroxy-5-methyl-hexanoic acid

**14.2** $CH_3CH_2CH_2CH_2—\overset{\displaystyle O}{\underset{\displaystyle \|}{C}}—O^- + H_2O$

**14.3** (a) potassium laurate (b) sodium salicylate

**14.4** (a) ethyl valerate (b) isopropyl benzoate

**14.5** (a) $H—\overset{\displaystyle O}{\underset{\displaystyle \|}{C}}—OCH_3$

(b)

(c)

mono

di

**14.6** (a) $CH_3CH_2CH_2—\overset{\displaystyle O}{\underset{\displaystyle \|}{C}}—O—$

(b)

The alcohol or phenol portions are circled.

**14.7** (a) $CH_3CH_2CH_2—\overset{\displaystyle O}{\underset{\displaystyle \|}{C}}—O—CH_2—$

(b) $CH_3CH_2—\overset{\displaystyle O}{\underset{\displaystyle \|}{C}}—OCH_3$

**14.8** $CH_3COOH + CH_3CH_2OH$

**14.9** $CH_3—\underset{\displaystyle \underset{\textstyle CH_3}{|}}{CH}—COO^- + CH_3CH_2CH_2OH$

**14.11** (a) $CH_3CH_2CH_2—\underset{\displaystyle \underset{\textstyle OH}{|}}{CH}—COOH$

(b) $CH_3—\underset{\displaystyle \underset{\textstyle Br}{|}}{CH}—\underset{\displaystyle \underset{\textstyle CH_3}{|}}{CH}—COOH$

(c)

(d) $CH_3CH_2CH_2—CH{=}\underset{\displaystyle \underset{\textstyle CH_3}{|}}{C}—CH_2COOH$

**14.13** (a) $HCOOH$ (b) $CH_3(CH_2)_{16}COOH$

(c) $HOOCCH_2CH_2COOH$ (d) $HOOCCH_2COOH$

(e) (f) $CH_3(CH_2)_4COOH$

**14.15** It was first isolated from rancid butter (Latin: *butyrum*).

**14.17** A propanoic acid molecule forms two hydrogen bonds with another propanoic acid molecule (Sec. 14.3). Two 1-butanol molecules form only one hydrogen bond.

**14.19**

**14.21** Bad odors: thiols, carboxylic acids
Pleasant odors: ketones

**14.23** (a) $CH_3—\underset{\displaystyle \underset{\textstyle OH}{|}}{CH}—COOH$

(b) $HOOCCH_2—\underset{\displaystyle \underset{\textstyle COOH}{\overset{\textstyle OH}{|}}}{C}—CH_2COOH$

**14.25** to cauterize gums and canker sores

**14.27** $CH_3CH_2OH + O_2 \longrightarrow CH_3COOH + H_2O$

**14.29** (a) $HCOO^- \, K^+$

(b) $CH_3CH_2CH_2COO^- \, Li^+$

(c) $HOOCCOO^- \, Na^+$

(d) $Ca^{2+}$

(e) $[CH_3(CH_2)_4COO^-]_2 \ Mg^{2+}$

(f) $\left( \begin{array}{c} {}^-OOC \phantom{xx} H \\ \phantom{xx}C=C \\ H \phantom{xx} COO^- \end{array} \right) 2K^+$

**14.31** (a) $CH_3COO^- + H_2O$

(b) $HCOO^- \ NH_4^+$

(c) $^-OOC-CH_2-COO^- + 2H_2O$

(d) $CH_3-\underset{\underset{OH}{|}}{CH}-COO^- + H_2CO_3$

(e) $CH_3COOH + Cl^-$

(f) ⟨benzene ring⟩—COOH $+ \ HSO_4^-$

**14.33** (a) $CH_3CH_2CH_2-\underset{\underset{O}{\|}}{C}-OH$

(b) It is converted to its salt:
$CH_3CH_2CH_2-\underset{\underset{O}{\|}}{C}-O^-$

(c) same as (a)

**14.35** a compound in which the H of an O—H acid is replaced by an R

**14.37** (a) ethyl formate  (b) methyl propionate
(c) ethyl acetate  (d) isopropyl caprylate
(e) diethyl malonate  (f) phenyl benzoate
(g) *tert*-butyl palmitate  (h) methyl heptanoate

**14.39** (b) (d) (e) (a) (c)

**14.41** (a) $CH_3-\underset{\underset{O}{\|}}{C}-OCH_2CH_3$

(b) ⟨benzene ring⟩—$\underset{\underset{O}{\|}}{C}-OCH_3$

(c) $CH_3-\underset{\underset{CH_3}{|}}{CH}-CH_2-\underset{\underset{O}{\|}}{C}-OCH_2CH_2CH_3$

(d) ⟨benzene ring with $CH_3$⟩—$\underset{\underset{O}{\|}}{C}-O-CH_2-$⟨benzene ring⟩

(e) $CH_3CH_2-O-\underset{\underset{O}{\|}}{C}-(CH_2)_3-\underset{\underset{O}{\|}}{C}-OCH_2CH_3$

(f) $Br-CH_2CH_2-\underset{\underset{O}{\|}}{C}-O-$⟨benzene ring⟩

**14.43** (a) ⟨benzene ring⟩—$\underset{\underset{O}{\|}}{C}-OCH_2CH_3$ $+ \ H_2O$

(b) $CH_3(CH_2)_6-\underset{\underset{O}{\|}}{C}-OCH_3 + HCl$

(c) $CH_3-\underset{\underset{O}{\|}}{C}-O-\underset{\underset{CH_3}{|}}{CH}-CH_3 + CH_3COOH$

(d) ⟨cyclopentane⟩—$\underset{\underset{O}{\|}}{C}-O-$⟨benzene ring with Br⟩ $+ \ HCl$

(e) NR

(f) $CH_3-\underset{\underset{O}{\|}}{C}-O-CH_2CH_2CH_2-O-\underset{\underset{O}{\|}}{C}-CH_3$
$+ \ 2CH_3COOH$

**14.45** the base-catalyzed hydrolysis of carboxylic esters

**14.47** (a) $HO-CH_2CH_2CH_2CH_2COOH$

(b) $HO-CH_2CH_2-\underset{\underset{CH_3}{|}}{CH}-COO^-$

**14.49** $HO-CH_2-\underset{\underset{O}{\|}}{C}-CH_2-OH + H_3PO_4$

$\longrightarrow HO-CH_2-\underset{\underset{O}{\|}}{C}-CH_2-O-\underset{\underset{OH}{|}}{\overset{\overset{O}{\|}}{P}}-OH + H_2O$

**14.51** (a) $HO-\underset{\underset{OH}{|}}{\overset{\overset{O}{\|}}{P}}-O-\underset{\underset{OH}{|}}{\overset{\overset{O}{\|}}{P}}-OH$

(b) $HO-\underset{\underset{OH}{|}}{\overset{\overset{O}{\|}}{P}}-O-\underset{\underset{OH}{|}}{\overset{\overset{O}{\|}}{P}}-O-\underset{\underset{OH}{|}}{\overset{\overset{O}{\|}}{P}}-OH$

**14.53** $HO-\underset{\underset{OH}{|}}{\overset{\overset{O}{\|}}{P}}-O-\underset{\underset{OH}{|}}{\overset{\overset{O}{\|}}{P}}-O-\underset{\underset{OH}{|}}{\overset{\overset{O}{\|}}{P}}-OCH_2CH_3$

$^-O-\underset{\underset{O^-}{|}}{\overset{\overset{O}{\|}}{P}}-O-\underset{\underset{O^-}{|}}{\overset{\overset{O}{\|}}{P}}-O-\underset{\underset{O^-}{|}}{\overset{\overset{O}{\|}}{P}}-OCH_2CH_3$

**14.55**

$$\left( CH_3CH_2 - \overset{\displaystyle O}{\underset{\displaystyle \|}{C}} - O^- \right)_2 Ca^{2+}$$

**14.57** Salicylic acid has a sour taste and irritates the mouth and stomach lining. Aspirin does not have these effects.

**14.59** to reduce or prevent the occurrence of strokes and heart attacks

**14.61**

**14.63** (a) yes

(b) $CH_3 - \overset{\displaystyle CH}{\underset{\displaystyle \diagdown O \diagup}{}} - C=O$   $\overset{\displaystyle CH_2}{\underset{\displaystyle \diagdown O \diagup}{}} - C=O$

(c) The polymers would not dissolve in water; the lactones would.

**14.65** thiols, phenols, carboxylic acids, sulfonic acids

**14.67**

$$CH_3O - \overset{\displaystyle O}{\underset{\displaystyle \underset{\displaystyle O}{\|}}{\overset{\displaystyle \|}{S}}} - OCH_3$$

**14.69** (a) $CH_3(CH_2)_6 - \overset{O}{\underset{\|}{C}} - O - \overset{O}{\underset{\|}{C}} - (CH_2)_6CH_3 + \big\langle\text{—}\big\rangle - OH \longrightarrow \big\langle\text{—}\big\rangle - O - \overset{O}{\underset{\|}{C}} - (CH_2)_6CH_3 + CH_3(CH_2)_6COOH$

(b)

**14.71** for example:

$$HO - \overset{\displaystyle O}{\underset{\displaystyle \underset{\displaystyle OH}{|}}{\overset{\displaystyle \|}{P}}} - O - \overset{\displaystyle O}{\underset{\displaystyle \underset{\displaystyle OH}{|}}{\overset{\displaystyle \|}{P}}} - OCH_3$$

**14.73** only (a)

---

# CHAPTER 15   Amines and Amides

**15.1** (a) tertiary (b) secondary (c) primary

**15.2** (a) ethylamine (b) ethylmethylamine (c) diethyl-methylamine

**15.3** *para*-nitro-*N*-ethylaniline

**15.4** 1,4-diaminocyclohexane

**15.5** (a) $CH_3CH_2CH_2NH_3^+$  $HCOO^-$

(b)

15.6

$$
\left[ \begin{array}{c} CH_3 \\ | \\ CH_3CH_2CH_2-N-CH_2CH_3 \\ | \\ CH_2CH_3 \end{array} \right]^+ Cl^-
$$

15.7 butyramide

15.8 (a) *N,N*-diphenylvaleramide
(b) *para*-bromo-*N*-ethylbenzamide

15.9 (a) $CH_3CH_2-\overset{\displaystyle O}{\underset{\|}{C}}-NH_2$

(b) $CH_3CH_2-\overset{\displaystyle O}{\underset{\|}{C}}-\overset{\displaystyle CH_3}{\underset{|}{N}}-\overset{\displaystyle }{\underset{|}{CH}}-CH_3$ with $CH_3$

15.10 acid: $CH_3CH_2CH_2COOH + CH_3NH_3^+$
base: $CH_3CH_2CH_2COO^- + CH_3NH_2$

15.11 ammonia

15.13 one

15.15 (a) ethylamine (b) *tert*-butylamine
(c) methyl-*n*-propylamine (d) isopropyl-*n*-propyl-amine (e) triphenylamine (f) aniline (g) *para*-bromoaniline (h) *N*-ethylaniline (i) *para*-nitro-*N,N*-dimethylaniline (j) aminocyclohexane (or cyclohexylamine) (k) 2,6-dimethylcyclohexylamine (or 1-amino-2,6-dimethylcyclohexane)

15.17 (a) (1) (2) (b) (2) (c) (3) (d) (1) (2) (e) (1) (2)
(f) (3) (g) (2) (h) (2) (i) (3) (j) (2)

15.19 (d) (b) (a) (c)

15.21 Trimethylamine is the only one of the three that has no N—H bond, and so cannot form a hydrogen bond with another molecule of itself.

15.23 (a) ethylammonium bromide (b) ethylmethyl-ammonium acetate (c) *N*-ethylanilinium chloride (d) tetramethylammonium hydroxide

15.25 Nothing would happen to the alcohol. The amine would be converted to $CH_3CH_2CH_2NH_3^+ Cl^-$.

15.27 It will dissolve in a solution of a dilute strong acid such as HCl.

15.29 Add chloroethane $CH_3CH_2Cl$.

15.31 (a) $CH_3-\overset{\displaystyle }{\underset{\underset{\displaystyle CH_3}{|}}{N}}-CH_3$ (b) $CH_3Cl$

15.33 In an amide the nitrogen is connected directly to a C=O group. In an amine the three nitrogen bonds are connected to any combination of hydrogens, alkyl, or aryl groups (at least one alkyl or aryl), but not to a C=O group.

15.35 (a) $CH_3(CH_2)_3-\overset{\displaystyle O}{\underset{\|}{C}}-NH_2$

(b) $CH_3CH_2-\overset{\displaystyle O}{\underset{\|}{C}}-NH-CH_3$

15.6 (c) $H-\overset{\displaystyle }{\underset{\underset{\displaystyle O}{\|}}{C}}-\overset{\displaystyle }{\underset{\underset{\displaystyle CH_2CH_3}{|}}{N}}-CH_2CH_3$

(d) (benzene ring)$-\overset{\displaystyle }{\underset{\underset{\displaystyle O}{\|}}{C}}-NH-$(benzene ring)

(e) (benzene ring with OH)$-\overset{\displaystyle }{\underset{\underset{\displaystyle O}{\|}}{C}}-\overset{\displaystyle CH_3}{\underset{|}{N}}-$(benzene ring)

(f) (benzene ring with Cl)$-\overset{\displaystyle O}{\underset{\|}{C}}-NH-$(benzene ring)$-NO_2$

(g) $NH_2-\overset{\displaystyle O}{\underset{\|}{C}}-(CH_2)_3-\overset{\displaystyle O}{\underset{\|}{C}}-NH_2$

15.37 acidic: (c) (d) (j) (r) (t)
basic: (b) (f) (n) (o) (p)
neutral: (a) (e) (g) (h) (i) (k) (l) (m) (q) (s)

15.39 (a) $CH_3CH_2-COO^- NH_4^+$

(b) (benzene ring)$-\overset{\displaystyle }{\underset{\underset{\displaystyle O}{\|}}{C}}-NH-CH_3 + HCl$

(c) $CH_3-\overset{\displaystyle }{\underset{\underset{\displaystyle O}{\|}}{C}}-\overset{\displaystyle }{\underset{\underset{\displaystyle CH_2CH_3}{|}}{N}}-CH_2CH_3 + CH_3COOH$

15.41 (a) (benzene ring)$-COO^-$  $CH_3-\overset{+}{N}H_2-CH_3$

(b) NR

(c) $CH_3-\overset{\displaystyle }{\underset{\underset{\displaystyle O}{\|}}{C}}-NH-CH_3 + CH_3COOH$

(d) (cyclopentane)$-COO^-$  (benzene ring)$-\overset{+}{N}H_2-CH_3$

15.43 (a) $NH_3$ (b) $NH_4^+$ (c) (benzene ring)$-COOH$

(d) (benzene ring)$-COO^-$

15.45 (a) quinine (b) nicotine (c) strychnine (d) morphine (e) caffeine (f) morphine (g) cocaine (h) coniine (i) atropine

15.47  Strychnine has seven rings; reserpine has six.

15.49  Epinephrine has three —OH groups not present in methamphetamine.

15.51  No. The HCl designation indicates that the phenylephrine was converted to its ammonium salt by the addition of HCl to make it water-soluble.

15.53  It is probably prudent to shut them down, if an alternative supply of water is available, but we are not yet certain that nitrites are a health hazard.

15.55  (a) carboxylic acid, carboxylic ester  (b) amide, phenol  (c) carboxylic acid

15.57  $-NH-\underset{\underset{O}{\|}}{C}-(CH_2)_8-\underset{\underset{O}{\|}}{C}-NH-(CH_2)_6-NH-\underset{\underset{O}{\|}}{C}-$

15.59  The enhancement of the effect of one drug by another. Barbiturates can be lethal when taken with alcohol.

15.61  There is no urea portion, but there is a thiourea portion: $-NH-\underset{\underset{S}{\|}}{C}-NH-$

Thiourea is urea in which an S replaces the O.

15.63  a seven-membered ring containing two nitrogens

15.65  a cheap, smokable form of cocaine

15.67  (a)  $HCOOH + CH_3NH_3^+$

(b)  $CH_3CH_2COO^- + NH_3$

(c)  $CH_3(CH_2)_8COO^- + CH_3-NH-CH_3$

(d)  $CH_2COOH$

$+ CH_3CH_2NH_3^+$

(e)  $NH_2CH_2CH_2CH_2CH_2COO^-$

15.69  The effect (enlarged pupils) lasts for hours.

15.71  (a) (1)  (b) (2)  (c) (1)  (d) (1)

15.73  primary: $CH_3CH_2CH_2NH_2$,

$CH_3-\underset{\underset{CH_3}{|}}{CH}-NH_2$

secondary: $CH_3CH_2-NH-CH_3$

tertiary: $CH_3-\underset{\underset{CH_3}{|}}{N}-CH_3$

15.77  (a)  $CH_3NH_3^+$  (b)  $\underset{\underset{O}{\|}}{C}-NH-CH_3$

---

## CHAPTER 16   Carbohydrates

16.1  (b); the right carbon in (d)

16.2  (a)  They are mirror images. They are not identical.
(b)  They are mirror images. They are identical.

16.3  (a) 4  (b) 2  (c) 8

16.5  (a) $O_2$  (b) $CO_2$

16.7  The chiral carbons are indicated. (b), (d), (g), and (i) have no chiral carbons.

(a)  $CH_3CH_2CH_2\underset{\underset{OH}{|}}{CH}-CH_3$

(c)  $CH_3\underset{\underset{OH}{|}}{CH}CH_2\underset{\underset{OH}{|}}{CH}CH_2CH_3$

(e)

(f)  $CH_3CH_2\underset{\underset{OH}{|}}{CH}CH_2CH_2Br$  (h)

(j)   (k)

(l)  $HO-CH_2CH_2-\underset{\underset{\underset{CH_3}{|}}{\overset{\overset{\overset{Br}{|}}{CH_2}}{\overset{|}{CH_2}}}{\overset{|}{C}}-CH_2CH_2-Cl$

(m)  $\underset{\underset{OH}{|}}{CH_2}-\underset{\underset{OH}{|}}{CH}-\underset{\underset{O}{\|}}{C}-\underset{\underset{OH}{|}}{CH}-\underset{\underset{OH}{|}}{CH_2}$

(n) CH₂—CH—CH—CH—CH—C—H
     |     |     |     |     |   ‖
   OH  OH  OH  OH  OH  O

16.9   It is a racemic mixture.

16.11  The mold destroyed one enantiomer and did not touch the other. The one enantiomer that remained was optically active.

16.13  (a) isomers that are nonsuperimposable mirror images of each other
(b) stereoisomers that are not enantiomers
(c) light that vibrates only in one plane
(d) a mixture of equal amounts of two enantiomers

16.15  C-1

16.17  (a) hemiacetal (b) secondary alcohol (c) primary alcohol

16.19  See Sec. 16.5.

16.21  (a) L (b) D (c) L

16.23  C-2

16.25

CHO
|
H—C—OH
|
H—C—OH
|
HO—C—H
|
HO—C—H
|
CH₂OH

(structures: two pyranose ring forms shown "and")

16.27  (a) to the right (b) No, it must be measured.

16.29 (two furanose ring structures with HOCH₂, O, HO, OH)

16.31  reducing sugars

16.33  (a) (two furanose ring structures with HOCH₂, O, OCH₂CH₃, OH OH) ... and ...

(b) CH₂—CH—CH—CH—C—O⁻
     |     |     |     |   ‖
   OH  OH  OH  OH  O

(c) (two pyranose ring structures with CH₂OH, HO, O, OCH₃, OH) ... and ...

16.35 (pyranose ring structure with CH₂OH, HO, O, OH, OH, NH, C=O, CH₃)

16.37 (disaccharide structure with CH₂OH, O, HO, OH, OH, OH, HOCH₂, OH, O)

16.39  (a) Amylose has α and cellulose has β(1 → 4) linkages. (b) The degree of branching and the molecular weights are different. Glycogen is more heavily branched and has a greater molecular weight than amylopectin.

16.41 (polysaccharide structure with CH₂OSO₃H, O, OH, HO, OSO₃H units; repeating [CH₂OSO₃H, O, OH, OH, O]ₙ; CH₂OH, O, OH, OSO₃H, OSO₃H)

16.43  (1) glucuronic acid  (2) D-glucosamine  (3) acetic acid

16.45  (a) COO⁻, OSO₃⁻, NHSO₃⁻  (b) high degree

16.47  Sweetness depends on the proper fitting of molecules into receptors on the taste buds. Sucrose fits the receptors. Its enantiomer does not.

16.49  −13.1°

16.51  (a)

(b)

16.53  (a) yes  (b) pyranose  (c) it has no OH group on C-6.

16.55  because this person's blood has no alpha or beta antibodies

16.57  Galactosuria is a symptom: the accumulation of galactose in the urine. Galactosemia is a deficiency of the enzyme galactose 1-phosphate uridinyltransferase.

16.59  With maturation children develop another enzyme that metabolizes galactose.

16.61  By influencing the rate of delivery. If the filler disintegrates rapidly in the stomach, it can deliver the drug rapidly.

16.63  It was found to cause birth defects.

16.65  $2^7 = 128$

16.67  It is a catalyst.

16.69  (a) 2  (b) 3

16.71  because most of them are sweet

16.73

16.75  See Sec. 16.11.

16.77

16.79

---

## CHAPTER 17  Lipids

17.1  Lipids are defined in terms of a physical property, not in terms of structure.

17.3  There are three. One of them is

CH₂—O—C—(CH₂)₁₂CH₃
            ‖
            O

CH—O—C—(CH₂)₇—CH=CH—(CH₂)₇CH₃
          ‖
          O

CH₂—O—C—(CH₂)₁₄CH₃
            ‖
            O

This one has oleic acid in the middle. The other two have myristic and palmitic acids in the middle.

17.5

CH₂—O—C—palm
          ‖
          O

CH—O—C—palm
          ‖
          O

CH₂—O—C—palm
            ‖
            O

1

CH₂—O—C—palm
          ‖
          O

CH—O—C—palm
          ‖
          O

CH₂—O—C—oleic
            ‖
            O

2

$$CH_2-O-\overset{\overset{\displaystyle O}{\parallel}}{C}-palm$$
$$CH-O-\overset{\overset{\displaystyle O}{\parallel}}{C}-oleic$$
$$CH_2-O-\overset{\overset{\displaystyle O}{\parallel}}{C}-palm$$

3

$$CH_2-O-\overset{\overset{\displaystyle O}{\parallel}}{C}-oleic$$
$$CH-O-\overset{\overset{\displaystyle O}{\parallel}}{C}-oleic$$
$$CH_2-O-\overset{\overset{\displaystyle O}{\parallel}}{C}-palm$$

4

$$CH_2-O-\overset{\overset{\displaystyle O}{\parallel}}{C}-oleic$$
$$CH-O-\overset{\overset{\displaystyle O}{\parallel}}{C}-palm$$
$$CH_2-O-\overset{\overset{\displaystyle O}{\parallel}}{C}-oleic$$

5

$$CH_2-O-\overset{\overset{\displaystyle O}{\parallel}}{C}-oleic$$
$$CH-O-\overset{\overset{\displaystyle O}{\parallel}}{C}-oleic$$
$$CH_2-O-\overset{\overset{\displaystyle O}{\parallel}}{C}-oleic$$

6

17.7   lowest: 6; then 4 and 5; then 2 and 3; highest: 1

17.9   (b) because its molecular weight is higher

17.11  lowest (c); then (b); highest (a)

17.13  the more —OH groups, the more hydrogen bonds; lowest (a); then (b); highest (c)

17.15  glycerol, sodium palmitate, sodium stearate, and sodium linolenate.

17.17  (a) They are found around cells and around small structures inside cells.  (b) They separate cells from the external environment and allow selective passage of nutrients and waste products into and out of cells.

17.19  complex lipids and cholesterol

17.21
$$CH_2-O-\overset{\overset{\displaystyle O}{\parallel}}{C}-(CH_2)_{14}CH_3$$
$$CH-O-\overset{\overset{\displaystyle O}{\parallel}}{C}-(CH_2)_7CH=CH(CH_2)_7CH_3$$
$$CH_2-O-\overset{\overset{\displaystyle O}{\parallel}}{P}-O-CH_2-\underset{\underset{\displaystyle COO^-}{|}}{CH}-NH_3^+$$
$$\underset{\displaystyle O^-}{|}$$

17.23  (c) because it has three charges, while the others have only two.

17.25  No. For example, in red blood cells lecithins are on the outside, facing the plasma; cephalins are on the inside.

17.27  (a) yes  (b) because high serum cholesterol has been correlated with such diseases as atherosclerosis

17.29  the D ring

17.31  It contains polar phospholipids, cholesterol, and proteins.

17.33  It increases the glucose and glycogen concentrations in the body. It is also an anti-inflammatory agent.

17.35  Loss of a methyl group at the junction of rings A and B, together with a hydrogen at position 1, produces a double bond. This causes tautomerization of the keto group at C-3, resulting in an aromatic ring A, with a phenolic group at the C-3 position.

17.37  (a) They both have a steroid ring structure.
(b) RU486 has a *para*-aminophenyl group on ring C and a triple-bond group on ring D.

17.39  (a) $PGE_2$ has a ring, a ketone group, and two OH groups  (b) The C=O group on $PGE_2$ is an OH group on $PGF_{2\alpha}$.

17.41  prostaglandins

17.43  (a) oxidation of the double bond to aldehydes and other compounds  (b) exclude oxygen and sunlight; keep refrigerated

17.45  The hydrophobic tail of the soap penetrates the fat globules, and the hydrophilic head stays on the surface to interact with the water and solubilize the grease particles.

17.47  (a) Sphingomyelin acts as an insulator.
(b) The insulator is degraded, impairing nerve conduction.

17.49  Degradation of the complex lipid must start with the monosaccharide at the end of the chain. In Fabry's disease this is $\alpha$-D-galactose.

17.51  (a) They enhance muscular development.
(b) They give some competitors unfair advantage, and they have harmful side effects.

17.53  It inhibits the enzyme phospholipase $A_2$.

17.55  It inhibits prostaglandin formation by preventing ring closure.

17.57  See Sec. 17.10.

17.59  (a) aldehyde, ketone, alcohol, C=C double bond
(b) Cortisone has all these groups except the aldehyde.

17.61  (a) in the liver  (b) as bile salts

17.63  (See Fig. 17.2.) Polar molecules cannot penetrate the bilayer. They are insoluble in lipids. Nonpolar molecules can interact with the interior of the bilayer (like dissolves like).

17.65  The various esters of cholesterol have different fatty acids (saturated, unsaturated, etc.). This prevents them from packing into a crystal lattice. Cholesterol, being uniform, does form a crystal lattice.

## CHAPTER 18  Proteins

**18.1**  They perform structural functions.

**18.3**  acidic: glutamic acid, aspartic acid; basic: lysine, arginine, histidine

**18.5**  nonpolar: (b) (f) (g)  polar but neutral: (d) (e)  acidic: (c)  basic: (a)

**18.7**  phenylalanine, tyrosine, tryptophan, and histidine

**18.9**  They supply most of the amino acids we need in our bodies.

**18.11**  Their structures are the same, except that a hydrogen of alanine is replaced by a phenyl group in phenylalanine.

**18.13**

$$CH_3{-}CH{-}CH_2{-}\overset{\overset{\displaystyle H}{|}}{C}{-}COOH$$
$$\underset{CH_3}{|}\qquad\underset{NH_2}{|}$$

D-Leucine

$$HOOC{-}\overset{\overset{\displaystyle H}{|}}{C}{-}CH_2{-}CH{-}CH_3$$
$$\underset{NH_2}{|}\qquad\underset{CH_3}{|}$$

L-Leucine

**18.15**  Ionic compounds are solids at room temperature. Amino acids, being zwitterions, are ionic compounds.

**18.17**  (a) $NH_2{-}(CH_2)_4{-}\underset{\underset{NH_2}{|}}{CH}{-}COO^-$

(b) $\overset{+}{N}H_3{-}(CH_2)_4{-}\underset{\underset{NH_3^+}{|}}{CH}{-}COOH$

(c) $NH_2{-}(CH_2)_4{-}\underset{\underset{NH_3^+}{|}}{CH}{-}COO^-$

**18.21**  Ala-Ala-Ala   Ala-Ala-Pro   Ala-Pro-Ala   Pro-Ala-Ala
Ala-Pro-Pro   Pro-Ala-Pro   Pro-Pro-Ala   Pro-Pro-Pro

**18.23**  $NH_2{-}CH{-}COOH \quad NH_2{-}CH{-}COOH \longrightarrow$
$$\underset{\underset{\underset{\underset{NH_2}{|}}{C=O}}{|}}{CH_2}\qquad\qquad\underset{\underset{OH}{|}}{CH_2}$$
$$+$$

$$NH_2{-}CH{-}\overset{\overset{\displaystyle O}{\|}}{C}{-}NH{-}CH{-}COOH$$
$$\underset{\underset{OH}{|}}{CH_2}\quad\qquad\underset{\underset{\underset{NH_2}{|}}{C=O}}{CH_2}$$

or

$$NH_2{-}CH{-}\overset{\overset{\displaystyle O}{\|}}{C}{-}NH{-}CH{-}COOH$$
$$\underset{\underset{\underset{NH_2}{|}}{C=O}}{CH_2}\qquad\underset{\underset{OH}{|}}{CH_2}$$

**18.25**  (a) six  (b) $C{-}\underset{\underset{O}{\|}}{C}{-}\underset{\underset{H}{|}}{N}{-}C$

**18.27**  It would acquire a net positive charge and become more water-soluble.

**18.29**  (a) $3^3 = 27$  (b) $20^3 = 8000$

**18.31**  valine or isoleucine

**18.33**  See Sec. 18.7.

**18.35**  (a) secondary  (b) tertiary and quaternary  (c) quaternary  (d) primary

**18.37**  Above pH 6.0 the COOH groups are converted to COO⁻ groups. The negative charges repel each other, disrupting the compact $\alpha$-helix and converting it to a random coil.

**18.39**  (1) C-terminal end  (2) N-terminal end
(3) pleated sheet  (4) random coil
(5) hydrophobic interaction
(6) disulfide bridge  (7) $\alpha$-helix
(8) salt bridge  (9) hydrogen bonds

**18.19**  There are six. One of them is:

$$NH_2{-}CH{-}\overset{\overset{\displaystyle O}{\|}}{C}{-}NH{-}CH{-}\overset{\overset{\displaystyle O}{\|}}{C}{-}NH{-}CH{-}COOH$$
$$\underset{CH_2}{|}\qquad\qquad\underset{CH_3}{|}\qquad\qquad\underset{CH_2CH_2CH_2CH_2NH_2}{|}$$

(benzene ring with OH)

18.41   valine, phenylalanine, alanine, leucine, isoleucine

18.43   a protein that contains a non–amino acid portion called a prosthetic group

18.45   No. Sometimes a helper molecule, called a chaperone, is needed to attain the biologically active species.

18.47   (b)

18.49   By becoming oxidized itself it uses up the oxidizing agents.

18.51   See Box 18D.

18.53   because this mutation provided resistance to malaria in heterozygotes

18.55   The gaps in the quarter-stagger arrangement provide the nucleation sites for the hydroxyapatite crystals. The more inorganic the crystal, the harder the tissue: enamel is hardest, bone is softest, and dentine is intermediate.

18.57   to destroy bacteria on the surface of the skin.

18.59   the particular shape of the keratin molecules of the hair that are stabilized by disulfide linkages.

18.61   No; that number is $20^{60} = 10^{78}$, even if we limit it to 60 amino acids. This number is larger than the total number of atoms in the universe.

18.63   yes

18.65   It permits water molecules to surround the collagen, thus allowing them to pass through the kidneys.

18.67   glycine

18.69   The side chain of proline is part of a ring. This restricts its flexibility.

18.71

$$\overset{+}{N}H_3-CH-COOH \atop CH_2CH_2-S-CH_3$$

$$+$$

$$\overset{+}{N}H_3-CH-COOH \atop \underset{\underset{CH_3}{\overset{|}{CH}-CH_3}}{\overset{|}{CH_2}}$$

$$+$$

$$\overset{+}{N}H_3-CH-COOH \atop CH_2-COOH$$

18.73   $20^{10} = 1.0 \times 10^{13}$

---

## CHAPTER 19   Enzymes

19.1   A catalyst is a substance that speeds up a reaction without being used up itself. An enzyme is a catalyst that is a protein or a nucleic acid.

19.3   more than 3000

19.5   because enzymes are highly specific, and there are a great many different reactions that must be catalyzed

19.7   Trypsin catalyzes the hydrolysis of peptides only at specific points (for example, at a lysine), while lipases catalyze the hydrolysis of all kinds of triglycerides.

19.9   (a) isomerase  (b) hydrolase  (c) oxidoreductase  (d) lyase

19.11   A cofactor is a nonprotein portion of an enzyme. A coenzyme is an organic cofactor.

19.13   See Sec. 19.3.

19.15   It would degrade the proteins of our bodies.

19.17   No. An increase of 10°C may decrease enzyme activity or even inactivate the enzyme.

19.19   At 85°C the enzyme is irreversibly denatured. At −10°C the structure of the enzyme is not changed; reversible inhibition occurs because not enough energy is available for the reaction.

19.21   See Sec. 19.5.

19.23   The other amino acids are necessary to hold the enzyme into exactly that 3-dimensional shape into which the substrate can fit. If significant changes were made in the structure, the enzyme activity would decrease or disappear.

19.25   See Sec. 19.6.

19.27   Yes. This is feedback control.

19.29   an enzyme that is active in more than one form

19.31   The enzyme is now named for the compound (alanine) that donates the amino group. The former name refers to the products.

19.33   digestive enzymes such as Pancreatin or Acro-lase

19.35   Succinylcholine is a competitive inhibitor of acetylcholinesterase and prevents acetylcholine from triggering muscle contractions.

19.37   Trypsin. Both trypsin and papain convert proteins to peptides. Carboxypeptidase yields amino acids by cleaving peptide bonds starting at the C-terminal end.

19.39   It binds to the $K^+$ and to the $Mg^{2+}$ cofactors.

19.41   Humans do not synthesize their folic acid; they get it in their diets.

19.43   Certain microorganisms convert it to organic mercury compounds.

19.45   It inhibits penicillinase, which bacteria synthesize to deactivate penicillin.

19.47   (a) ethanol dehydrogenase  (b) esterase

19.49   it is an oxidoreductase

19.51   (a) pasteurized milk, canned tomatoes  (b) yoghurt, pickles

19.53   No; the nerve gas binds irreversibly by a covalent bond, so it can't be removed simply by adding substrate.

19.55   noncompetitive inhibition

19.57   Destruction of the cAMP allows the first enzyme to bind more tightly to the regulator, decreasing its overall activity.

## CHAPTER 20 Bioenergetics. How the Body Converts Food to Energy

20.1 Catabolism is the breaking down of molecules to supply energy. Anabolism is the building up of molecules.

20.3 See Sec. 20.2.

20.5 (a) in the matrix  (b) They are imbedded in the inner membranes of the mitochondria.

20.7 See Fig. 20.4.

20.9 Neither; they yield the same energy.

20.11 No. The body is able to extract only from 40 to 60 percent of the caloric value in the form of ATP.

20.13 The two electrons are in the bond connecting the newly arrived hydrogen to the 4 position of the pyridine ring.

20.15 (a) adenosine  (b) $NAD^+$ and FAD  (c) acetyl groups

20.17 thiol, amide, hydroxy, phosphoric anhydride, phosphoric ester, ether, amino

20.19 No. The pantothenic acid portion is not the active part.

20.21 acetyl coenzyme A

20.23 All are salts of dicarboxylic acids with four carbons. Succinate and malate are saturated; fumarate is unsaturated; malate has an OH group.

20.25 oxidative decarboxylation

20.27 It removes two hydrogens from succinate to produce fumarate.

20.29 No, but GTP is produced in step ⑤.

20.31 It allows the energy to be released in small packets.

20.33 Isocitrate dehydrogenase. Yes, a dehydrogenation is an oxidation.

20.35 to $NAD^+$, which becomes $NADH + H^+$

20.37 (a) 3  (b) 2

20.39 in the inner mitochondrial membrane

20.41 (a) 3  (b) 2  (c) 12

20.43 the proton translocating ATPase

20.45 1.5 kcal

20.47 (a) by sliding the thick filaments (myosin) and the thin filaments (actin) past each other  (b) from the hydrolysis of ATP

20.49 ATP transfers a phosphate group to the serine residue at the active site of the enzyme.

20.51 No. It would harm humans because they would not synthesize enough ATP molecules.

20.53 succinate dehydrogenase

20.55 $NAD^+$ has two ribose units, both in the furanose form. FAD has one ribose unit in the ring form; the other is in the reduced open-chain ribitol form.

20.57 60 g or 1 mole of $CH_3COOH$

20.59 in the matrix of the mitochondrion

20.61 ATP

20.63 See Sec. 20.6.

## CHAPTER 21 Specific Catabolic Pathways: Carbohydrates, Lipid, and Protein Metabolism

21.1 monosaccharides

21.3 They serve as building blocks for the synthesis of proteins.

21.5 The two $C_3$ fragments are in equilibrium, and as the glyceraldehyde phosphate is used up, the equilibrium shifts and converts the other $C_3$ fragment (dihydroxyacetone phosphate) to glyceraldehyde phosphate.

21.7 (a) steps ① and ③  (b) steps ⑥ and ⑨

21.9 step ⑤

21.11 lactate

21.13 6 moles

21.15 ATP is produced in the mitochondria, but glycolysis and its production of $NADH + H^+$ occur in the cytosol. In order to be used for energy production, $NADH + H^+$ must be carried into the mitochondria by a shuttle mechanism.

21.17 (a) 3  (b) 2

21.19 ATP, because hydrolysis of an anhydride (P—O—P) linkage yields more energy than hydrolysis of an ester (C—O—P) linkage

21.21 (a) the mitochondria  (b) they are transported by carnitine acyltransferase

21.23 It catalyzes the addition of water to the double bond of the fatty acid chain in the $\beta$-oxidation process.

21.25 112

21.27 carbohydrates

21.29 See Sec. 21.7.

21.31 It is a transamination, the first step in amino acid catabolism.

21.33

21.35 (a) $NH_4^+$ (b) by converting it to urea in the urea cycle

21.37 because the body can synthesize glucose from pyruvate

21.39 The breakdown product of heme, bilirubin, is normally removed from the blood stream by the liver, later to be excreted via the gallbladder and intestines. High bilirubin levels in the blood indicate that the liver is malfunctioning because it is not sufficiently removing the bilirubin.

21.41 No; biliverdin and bilirubin do not contain iron.

21.43 See Box 21A.

21.45 (a) enlarged liver, hypoglycemia, and ketosis (b) missing debranching enzyme (c) symptoms: muscle cramps and limited capacity to exercise; missing enzyme: phosphorylase

21.47 the bicarbonate/carbonic acid buffer

21.49 HOOC—$CH_2CH_2$—CH—$COO^-$ $Na^+$ or
$\qquad\qquad\qquad\quad$ |
$\qquad\qquad\qquad\quad NH_3^+$

$Na^+$ $^-OOC$—$CH_2CH_2$—CH—COOH
$\qquad\qquad\qquad\qquad\quad$ |
$\qquad\qquad\qquad\qquad\quad NH_3^+$

21.51

Phenylalanine $\qquad$ $\alpha$-Ketoglutarate $\qquad$ Phenylpyruvate $\qquad$ Glutamate

21.53

21.55 Yes; they are glucogenic and can be converted to glucose.

21.57 in the carbon dioxide exhaled by the animal

21.59 It is first hydrolyzed to glucose and fructose. Metabolism of these compounds is shown in Figure 21.4.

21.61 Yes; in small quantities.

21.63 (a) in steps ⑤ and ⑫
(b) in steps ⑪ and ⑩
(c) If enough $O_2$ is present in muscles, a net increase of $NADH + H^+$ occurs. If not enough $O_2$ is present (as in yeast), there is no net increase of either $NAD^+$ or $NADH + H^+$.

# CHAPTER 22 Biosynthetic Pathways

22.1 the process of building up molecules

22.3 because the presence of a large inorganic phosphate pool would shift the reaction to the degradation process so that no substantial amount of glycogen would be synthesized

22.5 (a) carbon dioxide (b) water (c) sunlight

22.7 (a) pyruvate (b) oxaloacetate (c) alanine

22.9 conversions of (1) pyruvate to oxaloacetate (2) oxaloacetate to phosphoenol pyruvate (3) fructose 1,6-diphosphate to fructose 6-phosphate (4) glucose 6-phosphate to glucose

22.11 the conversion of oxaloacetate to malate

22.13 UDP, pyrophosphate

22.15 (a) the cytosol (b) no

22.17 No. It is a carrier protein.

22.19 malonyl ACP

22.21 It is an oxidation step because hydrogen is removed from the substrate. The oxidizing agent is $O_2$. NADPH is also oxidized during this step.

22.23 $NADP^+$ has an extra phosphate.

22.25 No; the body makes other unsaturated fatty acids such as oleic and arachidonic.

22.27 palmitoyl CoA, serine, acyl CoA, UDP-glucose

22.29 $CH_2$—OH
$\quad$ |
$\quad$ C=O
$\quad$ |
$\quad COO^-$

22.31 aspartic acid

22.33 alanine, serine, aspartic acid, asparagine, and glutamine

22.35 The chlorophyll captures the energy of light and with its aid strips the electrons and protons from water.

22.37 No; it can add $C_2$ fragments to a growing chain up to $C_{18}$. After that another enzyme system takes over.

22.39 (a) An SH group (b) ACP is a protein; CoA is not.

**22.41**

$$\begin{array}{ccc}
CH_2-OH & & CH_2-OH \\
| & & | \\
CH-NH_3^+ & & CH-NH-\overset{\displaystyle O}{\underset{\displaystyle \|}{C}}-R \\
| & & | \\
CH-OH & & CH-OH \quad O \\
| & +\ R-\overset{O}{\underset{\|}{C}}-S-CoA\ \longrightarrow & | \\
CH & & CH \\
\| & & \| \\
CH & & CH \\
| & & | \\
(CH_2)_{12} & & (CH_2)_{12} \\
| & & | \\
CH_3 & & CH_3
\end{array}$$

$$R = CH_3(CH_2)_7CH=CH(CH_2)_7-$$

**22.43**  malonyl ACP

---

## CHAPTER 23  Nucleic Acids and Protein Synthesis

**23.1** because it can transmit only its own hereditary information (DNA molecules) to the offspring

**23.3** chromosomes

**23.5** deoxyribonucleic acid and ribonucleic acid

**23.7** See Fig. 23.1.

**23.9** guanine, thymine, uracil, cytosine

**23.11** Ribose has an —OH in the 2 position; deoxyribose has an H there.

**23.13** (a)

(b) the same, but without the —OH in the 2 position.

**23.15** (a)

(b)

**23.17** (a) the 3′ —OH end and the 5′ —OH end
(b) A is the 5′ —OH end. C is the 3′ —OH end.

**23.19** the phosphoric acid groups in the backbone

**23.21** RNA

**23.23** amino and carbonyl groups

**23.25** See Fig. 23.6.

**23.27** at either end of the chain or in the middle

**23.29** polymerases

**23.31** rRNA

**23.33** mRNA

**23.35** (a) cloverleaf  (b) L-shaped

**23.37** in the nucleus

**23.39** the recognition site (anticodon) and the 3′ terminal to which the specific amino acid is attached

**23.41** (a) CGA  (b) alanine

**23.43** UAA, UAG, UGA

**23.45** They enable the tRNA to bind to the A site of the rRNA.

**23.47** An exon is a portion of a DNA molecule that codes for proteins; an intron is a portion that does not.

23.49 (a) They are transcribed onto mRNA. (b) They are cut out by enzymes before the mRNA is used for protein synthesis.

23.51 a gene that codes for a certain protein

23.53 a functional unit containing DNA sequences that are structural genes, regulatory genes and control sites

23.55 (a) 2 to 4 percent (b) Some codes for RNA, and some are regulatory sequences, but the function of most of it is unknown.

23.57 (a) UAU and UAC. Both code for tyrosine. (b) If UAU is changed to UAA, then instead of tyrosine, the chain will be terminated. This might well be fatal.

23.59 the ends of a strand of DNA that has been cut by endonucleases to leave several free bases that can pair up with a complementary section

23.61 The anticancer drug fluorouracil interferes with the methylating enzyme that manufactures thymidine, thereby inhibiting new DNA synthesis.

23.63 in the $5' \rightarrow 3'$ direction

23.65 It forms a strong complex with DNA, preventing its transcription into RNA.

23.67 It is the protein envelope of the virus. The ingredients come from the host cell.

23.69 a gene that can be turned into an oncogene

23.71 by a technique called gel electrophoresis

23.73 because for most amino acids there is more than one codon

23.75 (a) The 5' end and the section just before the 3' end must be base paired. (b) There must be no base pairing in the anticodon loop.

23.77 histones

23.79 the nucleosomes

23.81 because every time an error is introduced it may cause a mutation that can be harmful or even detrimental to the propagation of the species

## CHAPTER 24 Chemical Communication: Neurotransmitters, Hormones, and Immunoglobulins

24.1 (1) chemical signals or mediators (2) receptors

24.3 The presynaptic end stores vesicles containing neurotransmitters; the postsynaptic end contains receptors.

24.5 acetic acid: carboxylic acid; choline: primary alcohol, quaternary ammonium ion

24.7 (a) The acetylcholine is removed from the receptor site by an enzyme. (b) one-hundredth of a second

24.9 See Box 19A.

24.11 (a) norepinephrine and histamine (b) They activate a secondary messenger, cAMP, inside the cell. (c) amphetamines and histidine

24.13 It is phosphorylated by an ATP molecule.

24.15 the same, except that guanine replaces adenine

24.17 (a) Amphetamines increase and (b) reserpine decreases the concentration of the adrenergic neurotransmitter.

24.19 the corresponding aldehyde

24.21 (a) the ion-translocating protein (b) It gets phosphorylated and changes its shape. (c) It activates the protein kinase that does the phosphorylation of the ion-translocating protein.

24.23 In alpha-alanine the $NH_2$ group is on the 2 position of propionic acid; in beta-alanine it is on the 3 position.

24.25 The amino group in GABA is in the gamma position; proteins contain only alpha amino acids.

24.27 It blocks the pain receptors.

24.29 the anterior pituitary

24.31 See Sec. 24.7.

24.33 See Fig. 24.7.

24.35 by disulfide bridges

24.37 They allow cells to adhere to each other to form layers and eventually tissues.

24.39 Acetylcholinesterase inhibitors, such as Tacrine, inhibit the enzyme that decomposes the neurotransmitter.

24.41 (a) They stop nerve transmission. (b) They bind to acetylcholinesterase.

24.43 Dopamine cannot penetrate the blood-brain barrier.

24.45 Cheese contains tyramine, which dilates the arteries in the head, causing headaches.

24.47 in the conversion of arginine to citrulline

24.49

24.51 T (lymphocyte) cells

24.53 Both are dideoxynucleosides.

24.55 Yes. It fits into the same receptor site as enkephalins.

24.57 (a) Both deliver a message to a receptor. (b) Neurotransmitters act rapidly over short distances. Hormones act more slowly, and over longer distances. They are carried by the blood from the source to the target cell.

24.59 It opens the ion gates.

24.61 four

## CHAPTER 25    Body Fluids

25.1    water
25.3    (a) yes  (b) through semipermeable membranes
25.5    They carry oxygen to the cells and carbon dioxide away from the cells.
25.7    See Secs. 25.1 and 25.2.
25.9    heme
25.11   globulins
25.13   four
25.15   by coordinate covalent bonds to the Fe(II) ion
25.17   Each heme has a cooperative effect on the other hemes.
25.19   the active muscle, because it has a somewhat lower pH
25.21   25% as carbaminohemoglobin, 70% as carbonic acid, and 5% as $CO_2$ gas dissolved in plasma
25.23   In emphysema the lungs fill with water, and so cannot accept $CO_2$ from the blood. The accumulation of $CO_2$ in the blood forms carbonic acid, which lowers the pH.
25.25   urea, creatine, creatinine, hippuric acid, and ammonia
25.27   (a) $H_2N\!-\!\overset{\displaystyle \|}{\underset{\displaystyle O}{C}}\!-\!NH_2$

        (b) amino and carbonyl groups

25.29   the distal tubule and the collecting tubule
25.31   No glucose reaches the tissues or the cells. Because 12 mm Hg is lower than the osmotic pressure, nutrients flow from the interstitial fluid into the blood.
25.33   baroreceptors
25.35   blood clots lodged in arteries
25.37   See Box 25B.
25.39   (a) intracellular fluid  (b) blood plasma, interstitial fluid
25.41   restricted flow of urine
25.43   by lowering the volume of blood
25.45   Sodium ions are surrounded by water molecules (hydration). The lower the $[Na^+]$ in the body, the less water is needed, decreasing the amount of water and the blood pressure.
25.47   It converts angiotensinogen to angiotensin, which increases blood pressure.
25.49   creatine
25.51   When kidneys fail, no filtration of blood occurs, and urine excretion diminishes. Water accumulates in the interstitial fluid; hence the swelling.
25.53   the volume percentage of cellular elements in the blood

## CHAPTER 26    Nutrition and Digestion

26.1    the components of food and drink that provide growth, replacement, and energy
26.3    only the calcium propionate: When this is metabolized it yields energy.
26.5    3.2 cans
26.7    the average daily requirement of calories for a resting body
26.9    No. Weight loss due to loss of water is quickly regained.
26.11   cellulose
26.13   No. A minimum of 500 Cal/day is essential.
26.15   linoleic acid
26.17   Dietary deficiency diseases, since rice cannot supply the essential amino acids lysine or threonine, or essential fatty acids.
26.19   vitamin E
26.21   It assists blood clotting.
26.23   scurvy

26.25   biotin, thiamine
26.27   hydrolysis of acetal, ester, and amide linkages
26.29   neither
26.31   yes, randomly
26.33   malnutrition inadvertently induced by therapy
26.35   No. High dosages of niacin have been shown to cause health problems.
26.37   It will be filtered out by the kidneys and will appear in the urine.
26.39   See Table 18.1.
26.41   (a) linoleic acid, linolenic acid
        (b) unsaturated
26.43   The body would treat it as a food protein and digest it (hydrolyze it) before it could get into the blood serum.
26.45   nicotinic acid (niacin)
26.47   No. Arsenic is not an essential nutrient.

# G L O S S A R Y – I N D E X

**A**

Absolute alcohol, 370

Absolute temperature scale, 17, 137

**Absolute zero** The lowest possible temperature; the temperature at which molecules stop moving, 17

Absorbable staples, 432

Abundance of elements, 58

Accumulation of $CO_2$ in the atmosphere, 273

Acesulfame–K, 553

**Acetal** A compound that contains the linkage $-CH-O-C$, $O-C$ 397–403

cyclic, 400–3, 495–6

Acetaldehyde, 366, 370, 393, 397, 405

Acetaminophen, 452

Acetic acid, 220–1, 223, 366, 415–8, 575

properties, 220–1

reaction with ethanol, 210–1

Acetic anhydride, 429

Acetoacetate, 625

Acetoacetic acid, 625

Acetoacetyl ACP, 642

Acetoacetyl CoA, 625

Acetone, 367, 393–4, 405–6, 626

Acetylcholine, 575, 687–90, 695

Acetylcholinesterase, 575, 688–90

Acetylcholine transferase, 688

Acetyl coenzyme A, 598–600, 619, 622–5, 630, 645–51

in glycolysis, 615, 617–21

in $\beta$-oxidation, 622–5

Acetylene, 80, 341–3

$\alpha$-D-N-Acetylgalactosamine, 500

N-Acetylglucosamine, 498, 506

Acetylsalicyclic acid (*See* Aspirin)

Achilles tendon, 562

**Acid** (Arrhenius definition) A substance that produces $H_3O^+$ ions in aqueous solution; (Brønsted-Lowry definition) a proton donor, 219–46

burns, 227

properties, 226–7

reactions, 227–30

strength, 220–3

(*See also* Amino acid, Carboxylic acid, Nucleic acid)

Acid dissociation constants, 225–6

Acidic polysaccharides, 506–8

Acidic salts, 237

**Acidic solution** An aqueous solution with a pH of less than 7, 223–5, 237

Acidity, of carboxylic acids, 419–22

Acidity constant, 226

**Acidosis** A condition in which the pH of the blood is lower than 7.35, 237, 371, 626, 714

Acid rain, 165, 177, 273

Acne, 421

Aconitase, 576, 600

*cis*-Aconitate, 600

*cis*-Aconitic acid, 365

ACP, 642

Acquired Immune Deficiency Syndrome (AIDS), 341, 700–1

Acrylic plastics, 340

Actin, 547, 608

Actinomycin D, 664

**Activated complex** The reacting molecules at the transition state, 199

**Activation, of enzymes** Any process that makes an inactive enzyme active, 578

**Activation energy** The minimum energy necessary for a reaction to take place, 197–204

Active metals, 277–8

**Active site** The specific portion of an enzyme to which the substrate bonds when the reaction catalyzed by the enzyme takes place, 577–84

Actual yield, 107

**Acyclic compounds** Organic compounds without rings, 305

Acyl carrier protein (ACP), 642

**Acyl chloride** A compound that contains the group $-C-Cl$, $O$ 415, 429, 454, 456

**Acyl group** A group $R-C-$ $O$ derived from carboxylic acids, 415

Addition polymers, 388–401

**Addition reaction** A reaction in which a compound adds to a double or triple bond

of alkenes, 333–41

of alkynes, 343

Adenine, 594–5, 651–3

Adenosine, 594–5, 652–3

Adenosine diphosphate (ADP), 580, 594–8, 605–6

Adenosine monophosphate (AMP), 594–5, 654

Adenosine triphosphate (ATP), 435, 580, 594–5, 605–8

activation of fatty acid catabolism, 622–5

of glycerol catabolism, 621–2

of glycolysis, 615–21

of protein synthesis, 667

in biosynthesis, 638–9

conversion to cyclic AMP, 690

to other forms of energy, 607–8

as an end product of catabolic pathway, 592, 595

production of, 605–7, 620–2, 624–5

turnover rate, 595

Adenylate cyclase, 690

Adipoyl chloride, 455

ADP (*See* Adenosine diphosphate)

Adrenal gland, 535

Adrenalin (*See* Epinephrine)

Adrenergic neurotransmitters, 690–3, 695, 720

Adrenocorticoid hormones, 535, 537

Advil, 452

Agents GB, GD, and VX, 689

AIDS, 341, 685, 700–1

Air pollution, 117, 177
Alanine, 549, 553, 646
β-Alanine, 693
Alanine aminotransferase, 585
Albinism, 649
**Albumin**  A protein that helps to maintain the osmotic pressure of the blood, 548, 710–11
Alchemists, 366
**Alcohol**  A compound in which the —OH group is connected to an aliphatic carbon atom, 312–5, 361–75
  chemical properties of, 364–8, 384
  conversion to carboxylic esters, 425–31
  formation of acetals and ketals, 397–403
  nomenclature of, 362–4
  oxidation of, 366–8, 395, 419
  physical properties of, 378, 380
  preparation of, 336, 403, 432–4
  (See also Ethanol)
Alcoholic beverages, 170, 369
Alcoholism, 370
**Aldehyde**  A compound that contains the —C—H group, 391–407
   ‖
   O
  chemical properties of, 395–405, 408
  nomenclature of, 392–4
  oxidation of, 395–7
  physical properties of, 394–5
  preparation of, 366–7
Aldohexoses, 491–2
Aldonic acid, 497
**Aldose**  A monosaccharide containing an aldehyde group in the open-chain form, 491–3
Aldosereductase, 496
Aldosterone, 535, 721
*Alice in Wonderland*, 133
**Aliphatic hydrocarbon**  A hydrocarbon that contains no benzene rings, 344
**Alkali**  A basic substance, 219
**Alkali metal**  A metal in Group IA of the periodic table, 46
**Alkaloid**  A basic compound obtained from a plant. Virtually all are amines and many have physiological action, 459–64
**Alkalosis**  A condition in which the

pH of the blood is higher than 7.45, 237, 714
**Alkane**  A hydrocarbon with only single bonds, 293–315
**Alkene**  A hydrocarbon containing one or more double bonds, 323–41
  addition reactions of, 333–41
  chemical properties of, 333–41
  formation of, 365
  geometry of, 328–32
  nomenclature of, 323–7
  physical properties of, 332
  polymerization of, 338–41
  stereoisomerism of, 328–32
  summary of reactions, 353
Alkyl diphosphates, 435
**Alkyl group**  A group derived from an alkane by loss of one hydrogen, 299, 303, 314
**Alkyl halide**  A compound in which one or more hydrogens of an alkane has been replaced by halogen atoms, 302, 381–3
  nomenclature, 298–304
  preparation of, 333–5
Alkyl triphosphates, 435
**Alkyne**  A hydrocarbon containing one or more triple bonds, 341–3
**Allosterism**  Regulation of an enzyme or other protein by an event that occurs at a site other than the active site, but which affects the active site, 583–4, 713
Alloys, 164
Almonds, 391, 407
**Alpha amino acid**  An amino acid in which the amino group is on the second carbon of the chain, 548
5-alpha-dihydrotestosterone, 718
Alpha emission, 254, 256
Alpha helix, 560–1, 566
**Alpha particles**  Helium nuclei; they consist of two protons and two neutrons, 252, 254, 264
  damage from, 262–4
ALT, 585
Altman, Sidney, 573
Aluminum acetate, 421
Alzheimer's disease, 688
*Amanita phalloides*, 664
Ames test, 675

Amethystine quartz, 155
**Amide**  A compound that contains the linkage —C—N—, 451–9
   ‖  |
   O
  hydrogen bonding of, 453
  hydrolysis of, 456, 458–9
  neutrality of, 453–4
  nomenclature of, 451–3
  physical properties of, 453
  preparation of, 454–6
  (*See also* Peptide)
**Amine**  A compound of the form $RNH_2$, RNHR′, or RNR′R″, 443–51, 459–65
  basicity of, 229, 448–9
  chemical properties of, 229–30, 448–51
  classification of, 443
  hydrogen bonding of, 446–7
  nomenclature of, 443–6
  physical properties of, 446–7
  preparation of, 456–9
  reaction with acids, 229–30, 448–9
Amine salts, 448–51
**Amino acid**  An organic compound containing an amino group and an acid group, 548–51
  biosynthesis of, 581, 645–6
  catabolism of, 626–31
  essential, 549, 645
  as neurotransmitters, 693
  in peptides and proteins, 552–60
  in protein synthesis, 667–9
Amino acid pool, 614
Amino acid sequence, 557–60, 580, 663
Amino acid transport, 645
p-Aminobenzoic acid, 581
γ-Aminobutyric acid, 631, 693
Amino group, 443
Amino sugars, 498
Ammonia
  as a base, 220–1, 229
  diffusion of, 143
  excretion from body, 627
  in eyes, 227
  Lewis structure of, 74
  manufacture of, 104, 213
  properties, 220–6
  reaction with acids, 229–30, 448
  shape of molecule, 81
Ammonium carbamate, 108
Ammonium chloride, 143

Ammonium cyanate, 287

Ammonium hydroxide, 220

Ammonium ion, 87

**Amorphous solid** One whose atoms are randomly arranged, 155

AMP (*See* Adenosine monophosphate)

Amphetamines (Pep pills), 446

**Amphoteric substance** One that can act either as an acid or a base, 224, 552

Amu, 41

Amylases, 585, 736

Amyl nitrite, 373

**Amylopectin** A branched-chain polymer of glucose, found in starch, 501–4

**Amylose** A linear polymer of glucose, found in starch, 501–4

Anabolic steroids, 536

**Anabolism** The process of building up molecules, 592, 637

Anacin, 429

**Anaerobic pathway** A metabolic process that takes place in the absence of oxygen, 618

Anemia, 67, 728, 730

sickle cell, 560, 675

Anencephaly, 730

Anesthetics, 376

chloroform, 382

ethyl chloride, 151, 382

flammability of, 376

methyl chloride, 382

Angina pectoris, 373

Angiotensin, 720–1

Angiotensinogen, 720–1

Angular molecules, 79–80

Anydrides,

of carboxylic acids, 415, 429, 456

of phosphoric acid, 434–6, 594–5

**Anhydrous** Having no water, 178

Aniline, 347, 445

**Anion** A negatively charged ion, 66

Anode, 179

Anomeric carbon, 490

**Anomers** Two stereoisomers of a monosaccharide that differ only in the configuration at C-1 (aldoses) or C-2 (ketoses), 490–1

Antabuse, 370

Antacids, 228

Anthracene, 350

Antibiotics, 586, 664

**Antibody** Protein molecule synthesized by the body to counteract foreign substances, 500–1, 699–703

Anticancer drugs, 653

**Anticodon** A triplet of bases on a tRNA molecule that serves as a recognition site, 665, 668

Antidepressant drugs, 693, 695

Antidiabetic drugs, 698

Antidiuretic hormones, 559, 720–1

Antidiuretics, 720

**Antigen** A foreign substance that stimulates the production of antibodies, 500–1, 548, 699–700, 703

Antihistamines, 693, 709

Antihypertensive drugs, 720

Anti-inflammatory drugs, 429

Antilogarithms, 233

Antiseptics, 117, 374

Antistatic tiles, 200

Antithrombin, 507, 508, 612

Ants, 416

**Apoenzyme** The protein portion of an enzyme that also has one or more cofactors, 577

Apples, 415

**Aqueous solution** A solution in which the solvent is water, 110

Ara-A, 671

Arachidonic acid, 519, 541–2, 729

Arginase, 574

Arginine, 549, 556, 574, 628–9

Argininosuccinate, 628–9

Aristotle, 9

**Aromatic hydrocarbon** One that contains a closed loop of six electrons, 344–6

fused, 350

nomenclature of, 346–9

reactions of, 349–50

Aromatic sextet, 345–6

Aromatic substitutions, 344, 349

Arrhenius, Svante, 219

Arrhenius acids and bases, 219–22

Arteries, hardening of, 534

replacement of, 430

Arteriosclerosis, 534

Arthritis, 429

Artificial isotopes, 255, 269

Artificial kidney machine, 189, 715

Artificial respiration, 136

Artificial sweeteners, 494, 553

**Aryl halide** A compound in which one or more halogen atoms is connected to an aromatic ring, 381–3

A site, 668–9

Aspartame, 553, 632

Aspartate, 628

Aspartate amino transferase, 576, 585

Aspartate–malate shuttle, 620

Aspartic acid, 549–50, 556, 693

racemization of, 550

Aspartylphenylalanine, 553

Asphalt, 309

Aspirin, 101, 195, 415, 429, 452, 542

to prevent heart attacks and strokes, 429

AST, 585

Asthma, 542, 693, 700, 709

**Asymmetric carbon** A carbon atom connected to four different atoms or groups, 481

Atherosclerosis, 534–5

Athlete's foot, 421

Ativan, 459

**Atmosphere** A unit of pressure, 133

Atmospheric gases, 131

**Atmospheric pressure** The pressure exerted by the air. The standard unit is 760 mm Hg, 133

**Atom** The smallest particle of an element that shows the chemical behavior of that element, 39–42

early theories of, 33

mass of, 59

size of, 59

Atomic bomb, 259, 270–1

Atomic energy, 270–1

**Atomic mass unit (amu)** A unit of mass used for atoms and molecules. 1 amu = $1.67 \times 10^{-24}$ g, 41

Atomic nucleus, 42, 253

**Atomic number** The number of protons in an atom, 42

**Atomic weight** Weighted average of the masses of naturally occurring isotopes of an element, 44

ATP (*See* Adenosine triphosphate)

Atropine, 461, 463, 689

Attractive forces, 130, 145–9

Autoclaves, 136–7

Avery, Oswald, 650

Avogadro, Amedeo, 98

**Avogadro's law** Equal volumes of gases at the same temperature and pressure contain equal numbers of molecules, 140

**Avogadro's number** The number of particles (atoms, ions, molecules, etc.) equal to 1 mole: $6.02 \times 10^{23}$ particles, 98

Axial position, 306, 490

**Axon** A fiberlike part of a nerve cell, 529, 686–7

Azidothymidine (AZT), 701

**B**

Backbone, of nucleotides, 654–5
 of peptides and proteins, 555

**Background radiation** Radiation, including radioactivity and cosmic rays, that comes from the natural environment, 265

Baekeland, Leo, 406

Baeyer, Adolph von, 457

Bakelite, 406

Baking powder, 229

Balance, 15–6

Balancing equations, 102–4

Balloons, 129, 135

Barbital, 457

Barbiturates, 457–7
 overdose of, 606

Barbituric acid, 457

**Barometer** A device that measures atmospheric pressure, 132

Baroreceptors, 721

Barton, Jacqueline, 474–7

**Basal caloric requirement** The caloric requirement for a resting body, 728

**Base** (Arrhenius definition) A substance that produces $OH^-$ ions in aqueous solution: (Brønsted-Lowry definition) a proton acceptor, 219–46
 burns, 227
 properties, 226–7
 strength, 220–2

**Base** (of nucleic acids) A heterocyclic purine or pyrimidine that is a component of DNA, RNA, and nucleotides, 651–4

Base pairing, 656–60

Base triplets, 665

Basic salts, 236–7

**Basic solution** An aqueous solution with a pH of more than 7, 231, 233, 234–5

Basic types of reactions, 109–10

Battery, 116, 118, 179

Bayer Company, 429

B cells, 700–1

Beadle, George, 650

**Becquerel** A unit of radiation intensity; 1 Bq = 1 disintegration per second, 261–2

Becquerel, Henri, 251

Beer, 165, 369

Beeswax, 524

Belladonna, 461, 463

Benadryl, 693, 709

Bends, 169

**Benedict's test** A test for aldehydes and reducing sugars that uses $Cu(OH)_2$ and sodium citrate, 396–7, 496

Benzaldehyde, 407

Benzedrine, 446

Benzene, 344–9
 structure of, 344–6
 toxicity of, 344

Benzodiazepines, 459

Benzoic acid, 347

Benzopyrenes, 351

Benzyl group, 348

Beriberi, 730

Beta emission, 255–6

**Beta oxidation** The pathway by which fatty acids are metabolized, 614, 622–6

**Beta particle** Electron emitted from a radioactive nucleus, 252, 254, 255–6
 damage from, 262

Beta pleated sheet, 562

Biacetyl, 407

Bicarbonate buffer, 242, 714, 719

Bicarbonate ion, 111, 223–4, 226, 229
 balance in the body, 719
 reaction with carboxylic acids, 419–20

Bierce, Ambrose, 463

Big bang theory, 269

**Bile salts** Steroid molecules that act as detergents, dispensing lipid molecules so that enzymes can act on them, 540, 737

Bilirubin, 631–3

Biliverdin, 631–2

**Binary compound** One that contains just two elements, 88–9

Biochemical evolution, 674

**Biodegradable** Decomposable by natural processes, 381

Bioenergetics, 591

Biological effects of alkanes, 308

Biosynthesis, 288, 637–47

Birth-control pills, 280–2, 539

Bishydroxycoumarin, 712

Black, James W., 693

Bladder, 432

Bleaching, 118

Bleeding, 8

Blood, 707–721
 as a carrier of carbon dioxide, 714
  of oxygen, 713–4
 circulation, 9–11, 144
 clumping, 501
 composition of, 709–12
 functions of, 709
 osmolarity of, 187
 pH of, 234, 237–8, 714
 precipitation of, 184
 salt balance in, 719
 sickle cells, 560–1
 transfusion, 500–1

Blood-brain barrier, 693, 708–9

Blood buffers, 242, 556, 714, 719

Blood cleansing, 715–7, 719

Blood clotting, 507, 548, 712

Blood groups, 500–1, 567

**Blood plasma** The noncellular portion of blood, 707–11
 composition of, 710–11, 716

Blood pressure, 134, 534, 720–1
 measurement of, 134

**Blood serum** A fluid that contains all the components of blood plasma except fibrinogen, 173, 534, 710–11

Blood sugar, 489, 736

Blue diaper syndrome, 645

Boat conformation, 306

Body fluids, 707–21
 composition of, 716

Body temperature, 16, 202

Bohr, Neils, 49

Bohr effect, 714

**Boiling point** The temperature at which the vapor pressure of a

liquid equals the pressure of the atmosphere in contact with its surface, 152–3
of elements, 46
Boiling point elevation, 185
Bombykol, 329
Bond, chemical (*See* Chemical bonds)
Bone marrow, 631, 711
Boric acid, 221
Boron, 270
Boron trifluoride, 77, 80
Bottled gas, 294, 308
Bowman's capsule, 715, 717
Boyle, Robert, 136
**Boyle's law** At constant temperature, the volume of any sample of gas decreases as its pressure increases, and vice versa, 135–6
and breathing, 136
Brain, 528, 686, 688, 694
Brain tumors, 266
**Branched alkane** An alkane that contains carbon atoms attached to the longest linear chain, 303–4, 309
Branching point in polysaccharides, 503
Breathing (*See* Respiration)
Bromine
addition to multiple bonds, 334, 343
oxidation of alcohols, 367
reaction with benzene, 344, 349
test for unsaturation, 334, 343
2-Bromobutane, 481
Bronchoscopy, 575
Brønsted, Johannes, 223
**Brønsted-Lowry acid-base theory** A theory in which an acid is defined as a proton donor and a base a proton acceptor, 223–5
Bronze, alloy, 48
Bronze Age, 48
Brown, Michael, 533
Brown, Robert, 183
**Brownian motion** The random motion of colloidal particles, 183
Buckminsterfullerene, 155–6
**Buffer** A solution whose pH does not change very much when $H_3O^+$ or $OH^-$ ions are added, 238–42
amino acids as, 552
in blood, 242, 556, 714, 719

Buffer capacity, 241
Bufferin, 429
Burets, 14, 243, 245
Burning fossil fuels, 311
Butane, 295
2-Butene, stereoisomerism of, 325, 330
*t*-Butylamine, 444
*sec*-Butyl group, 303
*tert*-Butyl group, 303
Butyric acid, 416–7

**C**

Cadaverine, 447
Caffeine, 461
Cake rising, 229
Calcium, 36, 46
Calcium carbide, 342
Calcium carbonate, 177, 310
Calcium channel(s), 720
Calcium channel blocker, 720
Calcium chloride, 184
Calcium EDTA, 10
Calcium hydroxyapatite, 565
Calcium ions, 67, 712
Calcium propionate, 421
Calcium sulfate, 177
dihydrate, 177
Caloric value of foods, 121
**Calorie** The amount of heat necessary to raise the temperature of 1 g of water by 1°C, 25
in food, 121, 727–8
Calvin, Melvin, 640
Calvin cycle, 640
Cancer
caused by compounds, 351, 461, 675
caused by radioactivity, 265, 272
caused by ultraviolet light, 675
treatment of, 268, 653
Cane sugar (*See* Sucrose)
*Cannabis sativa*, 379
Capillaries, blood, 11, 560, 708, 715, 717
Captopril, 720
**Carbaminohemoglobin** Hemoglobin to which carbon dioxide is attached, 714
Carbamoyl phosphate, 627–9
**Carbocation** An ion with a positive charge on a carbon atom, 336
**Carbohydrate** A polyhydroxy aldehyde or ketone, or a compound that yields polyhydroxy alde-

hydes or ketones on hydrolysis, 479–80, 489–508
biosynthesis of, 638–41
configuration of, 495
in the diet, 728–9
digestion of, 613–4, 731, 736
energy yield from, 620–1, 625
metabolism of, 613–22
Carbolic acid, 374
Carbon
cage-like C-60, C-72, C-80 structures, 155
crystal forms, 156
Carbon atoms
bonding in organic compounds, 285–6, 290–1
Carbon-14 dating, 258–9
Carbonate buffer, 237, 242
Carbonates, formation of, 310
reaction with acids, 229, 419
Carbon dioxide, 85, 229, 310
in acidosis, 237
in acid rain, 165
in alcoholic beverages, 369
in blood, 709–10, 714
and breathing, 144–5
product of metabolism, 599–602, 607, 642
reaction with $H_2$, 204–5
shape of the molecule, 78–9
solid, 158
solubility in water, 165
sublimation of, 158
uses of, 131
Carbonic acid, 229, 310
in blood, 242, 714
Carbonic anhydrase, 714, 719
Carbonium ions, 336
Carbon monoxide, 310
as a combustion product, 117, 310
as a poison, 78, 139, 310
pollution, 377
reaction with water, 204–6
Carbon skeleton catabolism, of amino acids, 629–31
Carbon tetrachloride, 21, 81, 382
**Carbonyl group** The $-\overset{\displaystyle \|}{\underset{\displaystyle O}{C}}-$ group, 391, 397, 415, 451
**Carboxylate ion** The ion $R-COO^-$ that remains when a carboxylic acid donates its proton, 420–2

Carboxylate ion *(Continued)*
preparation of, 395–7, 419, 433, 456, 458

**Carboxyl group**  The group —COOH, 415

**Carboxylic acid**  A compound that contains the group —C—OH,
$$\overset{\parallel}{\underset{O}{}}$$
415–22
acidity of, 419–20
conversion to carboxylic esters, 425–31, 437
metabolism of, 622–5
nomenclature of, 416–7
physical properties of, 417–8
preparation of, 366–7, 395–7, 419, 425–31, 437, 456, 458–9
salts of, 420–2

**Carboxylic ester**  A compound that contains the linkage —C—O—C, 422–34
$$\overset{\parallel}{\underset{O}{}}$$
as flavoring agents, 425
hydrolysis of, 431–4, 437
nomenclature of, 422–4
physical properties of, 424–5
preparation of, 425–31

Carboxypeptidase, 558, 562, 574, 614, 738

**Carcinogen**  A compound that causes cancer, 351, 406, 450, 461
Carcinoma, 351
Carnitine, 622
Carnitine acyl transferase, 622
β-Carotene, 325
Casein, 183, 548, 567, 729

**Catabolism**  The process of breaking down molecules to supply energy, 591–633
common pathway, 591–608

**Catalyst**  A substance that increases the rate of a reaction without being used up, 203–4
effect on equilibrium, 214
*(See also* Enzyme)
Catalytic hydrogenation, 333, 343, 403
of fats, 521
Cataract formation, 139, 503, 541
Cathode, 179

**Cation**  A positively charged ion, 66
CAT scans, 266
Cauterization, 419

Cave painting, 258
Cech, T. R., 573
Cell adhesion molecules, 702
Cell membranes *(See* Membranes)
Cellophane membrane, 185, 189
Cellular elements, 710

**Cellulose**  A polymer of D-glucose in which all linkages are $\beta[1 \rightarrow 4]$, 504–5, 729, 736
Cellulose acetate, 505
Celsius scale, 16–7
Cement, manufacture of, 110
Centigrade scale, 16–7

**Central dogma of molecular biology**  Information contained in DNA molecules is transferred to RNA molecules, and from there it is expressed in the structure of proteins, 663
Cephalins, 528–9
Ceramide, 528, 531, 645
Cerebrosides, 530–1
Cerebrospinal fluid, 585, 707–8
CFCs, 382
Chain reaction, 270
Chair conformation, 306, 490

**Chaperones**  Protein molecules that help polypeptide chains to attain the proper secondary and tertiary structures, 564, 567
Chargaff, Erwin, 655–6
Charges, electrical, 183
Charles, Jacques, 135

**Charles's law**  At constant pressure, the volume of any sample of gas increases as its temperature increases, and vice versa, 135–7
Cheese, 421, 566
Chest cavity, 136

**Chemical bonds**  The attractive forces that hold atoms and ions together, 65–90

**Chemical change**  A process by which substances are converted to other substances, 38–9
Chemical communication, 685–703

**Chemical energy**  The energy stored within chemical substances, 24
change to other forms in body, 607–8

**Chemical equation**  An expression that shows all the reactants and products of a chemical reaction, 101
balancing, 102–4
Chemical equilibrium *(See* Equilibrium)

**Chemical kinetics**  The study of reaction rates, 195–204

**Chemical properties**  The chemical reactions undergone by a substance, 39, 308, 311–2, 333–8

**Chemical reaction**  A chemical change, 38
Chemical signals, 685

**Chemical species**  Any type of particle: atom, ion, or molecule, 66
Chemiosmotic hypothesis, 605–7

**Chemistry**  The science that deals with the structure and properties of matter and the transformations from one form of matter to another, 9
Chemotherapy, 653
Chernobyl, nuclear accident, 259, 272–3
Chinese restaurant syndrome, 631

**Chiral**  An object that cannot be superimposed on its mirror image, 480

**Chiral carbon**  A carbon atom connected to four different groups, 480–1
in amino acids, 550
in carbohydrates, 489–94
more than one in a molecule, 487–9
Chiral compounds, 480–9
in biological organisms, 508–9
Chitin, 515
Chloral, 398
Chloral hydrate, 398
Chlordane, 381
Chlordiazepoxide hydrochloride, 459
Chloride ions, 67, 196
balance in the body, 716
Chlorinated hydrocarbon insecticides, 381
Chlorine, 117, 312, 334, 349
Chloroacetophenone, 406
Chloroethane, 382
Chloroform, 22, 81, 382
Chloromethane, 196, 382
Chlorophyll, 479
Chlorophyll *a*, 640
Chloroplast, 640
Chlorpromazine, 606
Chlorpropamide, 698
Cholesterol, 540, 552–5
biosynthesis, 645

and heart attacks, 534, 726

Cholesterol esters, 533–4

Choline, 526, 575, 643, 688

Cholinergic neurotransmitters, 687–9

Chromatography, 166

**Chromosome** A structure in the nucleus of cells that carries the genes, 650

Chronic osteomyelitis, 139

Chymotrypsin, 614, 738

Cigarette smoke, 351, 461

Cimetidine, 693

Cinchona tree, 460, 464

Cinnamaldehyde, 407

Cinnamon, 391, 407

Cirrhosis, 370

**Cis isomer** A stereoisomer in which two similar or identical groups are on the same side
  of alkenes, 329–32
  of cycloalkanes, 307
  of fatty acids, 520
  in vision, 330

Citrate, 599–600

Citrate synthetase, 600, 603

Citric acid, 365, 415, 418

**Citric acid cycle** A series of biochemical reactions in which acetyl CoA is converted to $CO_2$ and energy-rich molecules, 594, 598–603
  energy yield of, 606–7
  regulation of, 603

Citrulline, 627–8

CK, 585

Clavulinic acid, 586

Clear solutions, 166

Clorazepate, 459

Clozapine, 695

Clotting, of blood, 507, 548, 712

Coagulation, 566

Coal, 288

Coated pits, 533

Cobalt-60, 265, 268

Cocaine, 462

Coconut oil, 521

Codeine, 463

**Codon** A triplet of bases in mRNA that codes for a particular amino acid, 665–9

Coefficients
  in balancing equations, 102
  in exponential numbers, A.1

**Coenzyme** An organic molecule that is necessary for the action of an enzyme, 577

Coenzyme A, 598

**Cofactor** A substance that is necessary for the action of an enzyme. It may be a metallic ion or an organic molecule (coenzyme), 577

Cold compresses, 127

Collagen, 507, 547–8, 562–5, 712, 729
  abnormalities, 562
  fibers, fibrils, 565

**Colligative property** A property that depends only on the number of particles dissolved in a given amount of solvent, and not on their nature, 184

Collision theory, 196–8, 202–3

**Colloid** A mixture in which particles of sizes ranging from about 1 to 1000 nm are dispersed in a medium, 164, 181–4

Color blindness, 649

**Combination reaction** A reaction in which two or more compounds come together to form a larger compound, 109–10

Combined gas law, 137

**Combustion** A reaction with atmospheric oxygen that gives off light and heat; burning, 117, 121, 308, 311, 337

**Common catabolic pathway** The series of biochemical reactions that convert the acetyl group of acetyl CoA to $CO_2$ and energy, 591–608

**Competitive inhibitor** A compound that binds to the active site of an enzyme, preventing the substrate from entering, 578, 581–3

**Complementary base pairs** Bases that fit opposite each other on the two strands of DNA, or on one strand of DNA and one of mRNA, 657–62

**Complete protein** A protein that contains all the essential amino acids, 729

Complete shell, 54

**Complex lipid** A lipid made of an alcohol, fatty acids, and either

phosphates or carbohydrates, 523, 526–31

**Compound** A pure substance made of two or more elements in a fixed proportion by weight, 36

Computer-assisted tomography, 266

**Concentrated solution** One that contains a large amount of solute per unit volume of solvent, 168

**Concentration** The amount of solute dissolved in a given quantity of solvent or solution, 168
  effect, on reaction rates, 200–1
  on equilibrium, 205, 209–11
  units, 168–75

**Condensation** Conversion of a gas to a liquid, 149–53

Condensation polymers, 338, 430, 455

Condensed structural formulas, 294

Conductance, 179

**Conformation** The three-dimensional shape of a molecule, 298

**Congener** A component of alcoholic beverages other than ethanol and water, 369

Coniine, 460

**Conjugate acid** The acid formed when a base accepts a proton, 223–5

**Conjugate base** The base that remains when an acid gives up a proton, 223–5

**Conjugated protein** One that contains non-amino acid portions, 565

Connective tissues, 506

Conservation of energy, 24

Conservation of mass, 40

Control rods, 270

Control sites, 672

**Conversion factor** A fraction that is a ratio of two units, 18

**Coordinate covalent bond** A covalent bond in which one atom supplies both electrons, 77–8

Coordination compounds, 78

Copolymers, 339, 432

Copper Age, 48

Copper ions, 115–6, 396–7, 566

Copper sulfate, 178, 396–7

Corey, Robert, 560

Cori's disease, 621

Cormack, Allan M., 266
Cornea, 227
Coronary thrombosis, 619
Cortisol, 535, 537
Cortisone, 535, 537, 541
Cosmic rays, 252, 264
Cotton, 504
**Covalent bond**   A bond in which two atoms share a pair of electrons, 72–6
  coordinate, 77–8
  nonpolar, 86
  polar, 86
Covalent compounds
  solubility of, 181
"Crack," 462
"Crank," 446
Crash diets, 728
Cream of tartar, 229
Creatine, 718
Creatine kinase, 585
Creatinine, 718
**Crenation**   The shriveling of red blood cells, 188
Crick, Francis, 656–7
**Cristae**   Folds of the inner membranes of mitochondria, 592, 594
Cross linking, 339, 562–3
Crude oil, 309
**Crystalline solid**   One whose atoms, ions, or molecules are arranged in a regular order, 154–6
Crystallins, 547
Crystallization, 154, 158
Cubic centimeter, 14
Curare, 689
**Curie**   A unit of radiation intensity; 1 Ci = $3.7 \times 10^{10}$ disintegrations per second, 261
Curie, Marie, 252, 261
Curie, Pierre, 252
Cushing syndrome, 537
Cyclic AMP, 584, 690–1
C-terminal, 555
Cyanide poisoning, 139
**Cycloalkane**   An alkane that includes a ring of carbon atoms, 304–7
Cycloalkenes, 327
Cyclohexane ring, 306
Cyclopropane, 103, 304
Cysteine, 549, 552, 563, 693
Cystine, 549, 563, 567, 632
Cystinuria, 632

Cytidine, 653
Cytochromes, 603–5
Cytosine, 651
Cytosol, 586, 618–20

**D**

Dacron, 340, 430
Dakin's solution, 117
Dalton, unit of atomic weight, 44
Dalton, John, 39
**Dalton's atomic theory**   The theory that matter is made up of atoms, and that all atoms of an element are identical and different from those of all other elements, 39–41
**Dalton's law of partial pressures**   The total pressure of a mixture of gases is the sum of the individual partial pressures, 143
  and breathing, 144–5
Dark reaction, 640
Darwin's theory, 666
DDT, 381
Debranching enzymes, 621, 736
Decamethonium, 689
Decay curve, 257
**Decomposition reaction**   A reaction in which a compound splits into two or more pieces, 109–10
DDI, 701
Degenerate code, 667
Dehydration, of alcohols, 365–6
Demerol, 463
Demethylation of genes, 673
Democritus, 33
**Denaturation**   Any process that changes the native conformation of a protein by destroying secondary and tertiary structures, 405, 565–7
Denatured alcohol, 370
**Dendrites**   Hairlike structures that are parts of nerve cells, 686–7
**Density**   Mass per unit volume, 21–3
  of alkanes, 308
  effect of temperature on, 22
  of ice and water, 147
Dentine, 565
Deodorants, 377
Deoxyribonucleic acid (See DNA)
Deoxyribose, 652
Deprenyl, 692, 695

Detergents, 221, 436, 524, 540
Deuterium, 255, 269–70
**Dextrorotatory**   The rotation of the plane of polarized light to the right, 484
Dextrose, 489
Diabetes insipidus, 719
**Diabetes mellitus**   A disease in which the body lacks sufficient insulin, allowing glucose to accumulate, 237, 406, 559, 626, 698, 717, 736
Diabinese, 698
Diagnosis with radioactive isotopes, 265, 267
**Dialysis**   The passage of solvent and small solute, but not large solute, molecules through a semipermeable membrane, 188–9
Diamines, 447
Diamond, 154, 156, 157
Diaper rash, 421
Diaphragm, 136
Diarrhea, 189
**Diastereomers**   Stereoisomers that are not enantiomers, 488, 508
Diastolic blood pressure, 134
Diatomic elements, 40
Diazepam, 459
Dibenzanthracene, 351
Dicarboxylic acids, 416, 430
*para*-Dichlorobenzene, 382
Dichlorodifluoromethane, 382
Dichloromethane, 382
Dicumerol, 712
Dideoxycytidine (DDC), 701
Dideoxyinosine (DDI), 701
Diene, 339
Diesel fuel, 309
Dietary fiber, 729
Diethyl ether, 167, 366, 375–6
  as an anesthetic, 376
Diethyl malonate, 457
**Diffusion**   The process of mixing of gases, 143
**Digestion**   The process in which the body breaks down large molecules into smaller ones that can then be metabolized, 613–5, 725, 731, 736–8
  of carbohydrates, 613–4, 731, 736
  of lipids, 614, 737
  of proteins, 614–5, 737–8

Diglycerides, 518

Dihydrogen phosphate ion, 226, 240–2, 434

Dihydroxyacetone phosphate, 616–8, 622

**Dilute solution** One that contains a small amount of solute per unit volume of solvent, 168

Dilution calculations, 173–5

Dimenhydrinate, 693

**Dimer** A species made up of two identical units, 418, 552

1,2-Dimethylcyclopentane, 305–7

Dimethyl ether, 289–90, 375

Dimethylmercury, 583

Dimethylpropane, 152

2,4-Dinitrophenol, 604

Dioxane, 375

Dioxin, 175

**Dipeptide** A compound made of two amino acids joined in an amide linkage, 554

Diphenhydramine, 693

1,3-Diphosphoglycerate, 616–8

Diphtheria, 622

Dipole-dipole attractions, 145–6

**Dipole moment** Extent of charge separation, 84

Dipoles, 82–5

**Diprotic acid** An acid that can give up two protons, 224

**Disaccharide** A carbohydrate made of two monosaccharide units, connected by an acetal linkage, 494, 498–500

Discriminatory curtailment diets, 726

Disinfectants, 117

Dispersion forces, 148, 152

Distal tubule, 715, 717

Distillation, 38, 166, 309, 369

Distribution of speeds, 202

**Disulfide bridge** A linkage in which two organic groups, most often cysteine molecules, are connected: C—S—S—C, 552, 563, 566

Disulfides, 380

Disulfuram, 370

Diuresis, 719

Diuretics, 720

Djerassi, Carl, 280–3

**DNA (Deoxyribonucleic acid)** A polymer made of nucleotide units. It is what constitutes genes, and is responsible for heredity, and for carrying the information necessary for protein synthesis, 475–6, 649–80

base pairing in, 657–61

components of, 650–4

double helix of, 656–61

fingerprinting, 678

in genes, 669, 671, 674

genetic diseases and, 674–6

primary structure of, 654–6

recombinant, 676–80

replication of, 660–1

Okazaki fragments in, 661

secondary structure of, 656–9

transcription, 663–4

in viruses, 670–1

DNA ligase, 661

DNA replication, 660–1

L-Dopa, 692

Dopamine, 692, 695

Dosimetry, of radiation, 261–5

**Double bond** A bond in which two atoms share four electrons, (*See also* Alkene) 74, 291, 323–4

Double bonds, C=C; C=O; C=N, 327–8

**Double displacement reaction** A reaction in which two compounds exchange partners, 110

**Double helix** The arrangement in which two strands of DNA entwine about each other in a helical configuration, 656–61

**Downhill reaction** Exothermic reaction, 198

Dramamine, 693, 709

Dronabinol, 379

Drug dosage, 15

Drug fillers, 505

Drugs, solubility of, 449

Drug testing, 267

Dry ice, 131, 157

Ductility, of metals, 47–8

Dust, 182

Dynamic equilibrium (*See* Equilibrium)

Dynamite, 373

**E**

Ear wax, 524

Earth, 7

**Edema** A swelling of the tissues caused by water oozing from the blood into the interstitial fluid, 67, 711

EDTA, 10

**Effective collision** A collision that results in a reaction, 196–8, 202

Effective dose delivered, 262

Effects of radiation, 261–5

Egg yolk, 183

Einstein, Albert, 270

Electric charges, 41

Electric conductance, 179

Electric current, 116, 179

Electrical neutrality, principle of, 71

**Electrolyte** A substance that conducts an electric current when dissolved in water or in the molten state, 179–80

in body fluids, 716, 719

strong and weak, 180, 222

Electromagnetic radiation, 252–3

Electromagnetic spectrum, 252–3

**Electron** A tiny negatively charged particle that is a constituent of atoms, 41

mass of, 41, 59

**Electron configuration** The arrangement of electrons in orbitals of atoms, 54–6

**Electronegativity** The attraction an atom has for the electrons in a covalent bond connecting it to another atom, 82–5, 560

Electronic buildup, 52–4

Electronic structure of atoms, 49–56

Electron transport chain, 594, 603–6

schematic diagram, 603

**Element** A pure substance that cannot be broken down into simpler substances by chemical reactions, 34

abundance of, 58

necessary for human life, 36

trace, 36

Elimination reactions, 365

Elongation factors, 668

Embryo, 672

**Emulsion** A stable colloidal dispersion, 183

**Emulsifying agent** A material that can cause two insoluble substances to form an emulsion, 183

Enamel, of teeth, 565

**Enantiomers**   Isomers that are non-superimposable mirror images of each other, 480–7
  properties of, 483–7, 509
**Endergonic reaction**   A reaction that requires energy, 119–20
**Endocrine gland**   A gland that secretes hormones into the blood, 686, 696–7
**Endothermic reaction**   A reaction that requires heat, 119–20
  effect on equilibrium, 211–2
  energy diagram, 198–204
**End point**   The point in a titration when an equivalent amount of acid or base has been added, as signaled by a sharp pH change, 245
**Energy**   The capacity to do work, 23–4
  from mass, 270
  of radiation, 261–2
Energy diagrams, 198–204
**Energy level**   A region where an electron in an atom is likely to be found, 49–51
Energy requirements, in food, 726–8
Energy yield,
  of ATP, ADP, AMP, 595
  of common catabolic pathway, 606–7
  of fatty acids, 624–5
  of glucose, 620–1
  of glycerol, 622
Enflurane, 376
Enkephalins, 694–5
Enol form, 404, 651
Environmental problems, 273
**Enzyme**   A protein molecule that catalyzes reactions in the body, 547, 573–86, 650
  active site, 577
  activity of, effect, of pH on, 579
    of concentration on, 578–9
    of temperature on, 579
  classification of, 574–7
  in medical diagnosis, 585
  regulation, 583–5
  specificity, 573–4
**Enzyme-substrate complex**   The weak association formed between an enzyme and its substrate while the enzyme catalyzes the reaction, 579–85

Epidermal growth factor, EGF, 676
Epigenetic processes, 673
Epinephrine (adrenalin), 446, 690, 692, 695
Equal, 553
**Equation, balanced**   A representation of a chemical reaction in which the number of atoms of each element is the same on both sides, 102–4
Equatorial position, 306, 490
**Equilibrium**   A situation in which two opposing processes take place at the same rate, resulting in no change in the system, 204–14
  effect of catalyst on, 214
  effect of concentration on, 206, 209–11
  effect of temperature on, 211–2
  liquid-vapor, 150
  and reaction rates, 209
  of water, 230–1
**Equilibrium constant**   A constant that relates the concentrations of the species in an equilibrium, 206–9
  of acids, 225–6
Equilibrium expressions, 206–9
**Equivalent**   The equivalent weight expressed in grams, 244
Equivalent dose of radiation, 263
**Equivalent weight**   The formula weight divided by the number of $H^+$ or $OH^-$ ions produced, 243–4
**Erythrocytes**   Red blood cells, 708–11
**Essential amino acid**   One that the body cannot synthesize in the required amounts, and so must be obtained in the diet, 549, 645, 729
Essential fatty acids, 520, 729
Essential oils, 324
**Ester**   The compound that results when the H of an —OH acid is replaced by an R group, 422–36
**Esterification**   A reaction in which an ester is formed, usually from an acid and an alcohol, 425–31, 434–6
Estradiol, 535, 537–9

Estrogens, 718
Ethane, 204, 293–4
Ethanol, 314, 315, 368–70
  in alcoholic beverages, 368–9
  in cold compresses, 27
  as gasoline additive, 377
  oxidation of, 221, 366, 419
  production of, 336, 368–9
  properties of, 290
  reaction with acetic acid, 209–10
  solubility of, 166
  specific heat of, 26–7
  as a sterilizer, 566
  structure of, 292
Ethanolamine, 528
Ethene   (*See* Ethylene)
**Ether**   A compound that contains the linkage —C—O—C—, 375–80
  formation of, 366
  nomenclature of, 375
  physical properties of, 378–80
Ethrane, 376
Ethyl acetate, 210–1
Ethyl alcohol (*See* Ethanol)
Ethyl chloride, as an anesthetic, 151, 382
Ethylene, 80, 204, 324, 329, 332
Ethylene glycol, 184, 364, 371, 379, 430
Ethylene oxide, 375
Ethyl ether (*See* Diethyl ether)
Ethyl group, 299
Ethynodiol diacetate, 539
Eukaryotes, 672, 673
**Evaporation**   The conversion of a liquid to a gas, 149–53, 158
Evolution, 666, 674
Excess of reactant, 107
Excited states, 50–1, 260
**Exergonic reaction**   A reaction that gives off energy, 119–20
Exchange reactions, 110
**Exons**   Portions of DNA molecules that code for proteins, 669
**Exothermic reaction**   A reaction in which heat is given off, 119–20, 198
  effect on equilibrium, 211–2
  energy diagram, 198
Expanding universe, 269
Exponential notation, A.1–5
**Extensive property**   A property that

depends on the quantity of a substance, 23

**Extracellular fluids**    Body fluids outside of cells, 707

### F

Fabricius, 10

Fabry's disease, 531

**Fact**    A statement based on direct experience, 9

Factor-label method, 18–9

FAD and $FADH_2$, 596–7
  in citric acid cycle, 599, 603, 606
  in fatty acid catabolism, 622–5
  in oxidative phosphorylation, 605–6
  structures of, 596–7

Fahrenheit scale, 16–7

Fallout (*See* Radioactive fallout)

Faraday, Michael, 344

Fat cell, 613–4

Fat depots, 614

**Fats**    Esters composed of glycerol and long-chain carboxylic acid molecules, 518–23, 614
  in the diet, 728–9
  energy yield from, 624–5

**Fatty acid**    An unbranched carboxylic acid having between 3 and 20 carbons; their esters are found in fats, 416, 518–23
  biosynthesis of, 641–3
  catabolism of, 622–6

**Feedback control**    Enzyme regulation by a process in which formation of a product inhibits an earlier reaction in the sequence, 584

Fermentation, 369

Fermi, Enrico, 270

Ferritin, 548, 631

Fertilizer, 104, 213, 456

FeS protein, 603–5

Fever, 201, 429

Fiber, in the diet, 729

Fiberscope, 25

Fibrin, 712

Fibrinogen, 548, 567, 710–2

Fibroin, 562

**Fibrous protein**    A protein insoluble in water and used mainly for structural purposes, 548

Fillers, drug, 505

Filtration, 165

Finasteride (Proscar), 718

Fire extinguishers, 229

Firefly, 679

Fire hazards, 200

Fischer, Emil, 486

Fischer projection, 486–7

Fission (*See* Nuclear fission)

Fixation of nitrogen, 213

Flaking of stone buildings and monuments, 177

Flavin adenine dinucleotide (*See* FAD)

Flavin adenine dinucleotide, reduced form (*See* $FADH_2$)

Flavoprotein, 603–5

Flavoring agent, 425

Fleming, Alexander, 586

Fluidity, 149

Fluid mosaic model, of membrane, 525–6

Fluorescence, 702

Fluorine, 72
  bonding of, 72
  fluoridation, 68
  Lewis structure of, 74

Fluorite, 5, 154

Fluorouracil, 653

Fluoxetin, 695

Folic acid, 581, 730

Food, caloric values of, 121

Formaldehyde, 371, 405–6, 699
  Lewis structure of, 75
  odor of, 395

**Formalin**    An aqueous solution of formaldehyde, 405

Formica, 406, 416

Formic acid, 371, 416

**Formula weight**    The sum of the atomic weights of all the atoms in the molecular formula, 97

Fortified wines, 369

Franklin, Rosalind, 656

Free rotation, 298
  lack of, 305, 329

Freezing biological tissue, 147

Freezing point, 156–7
  depression of, 184–5

Freons, 382

**Frequency**    The number of crests of a wave that pass a given point in one second, 252–3

Frostbite, 147

D-Fructose, 493–4, 495, 499, 617, 621, 728
  in semen, 494

Fructose 1,6-diphosphate, 616–8

Fructose intolerance, 618

Fructose 1-phosphate, 617–8, 621

Fructose 6-phosphate, 616–8

L-Fucose, 501

Fuel oil, 309

Fugu poison, 690

Fumarase, 336, 601–2

Fumarate, 599, 601–2, 629

Fumaric acid, 337

**Functional group**    A combination of atoms that is not part of the alkane-like portion of a molecule, 312–5, 362

Funk, Casimir, 730

Furanose, 495

Fused aromatic rings, 350

Fusion (*See* Nuclear fusion)

### G

GABA ($\gamma$-Aminobutyric acid), 693

Galactitol, 497

Galactocerebroside, 531

Galactokinase, 503

Galactose, 492, 497, 503, 617, 621

Galactosemia, 503

Galactose-1-phosphate uridinyl-transferase, 503

Galactosuria, 503

Galen, 9

Gallstones, 532

Gamma carboxyl group, 555

Gamma crystallin, 547

Gamma emission, 526–7

Gamma globulin, 548

**Gamma ray**    Electromagnetic radiation emitted by radioactive nuclei, 254, 256–7, 259–61
  damage from, 262–3

Garnet, 154

**Gas**    The state of matter in which substances have no definite shapes, and expand to fill their containers, 20, 131–49
  mixtures of, 142–3, 164
  solubility in liquids, 168–9

Gas constant, R, 141

Gas gangrene, 139

Gas laws, 135–45

Gasoline, 308–9

Gastric fluid, 221, 228

Gaucher's disease, 531

Gay-Lussac, Joseph, 135

**Gay-Lussac's law**   At constant volume, as the temperature increases the pressure of any sample of gases increases, and vice versa, 135–7

Geiger-Müller counter, 260

Gelatin, 566, 729

**Gene**   A segment of a DNA molecule that codes for one protein, 650, 660, 663, 669, 671

Gene expression, 672

Gene regulation, 672–3

**Genetic code**   The code by which a triplet of 3 bases on mRNA corresponds to one specific amino acid in protein synthesis, 663, 666–7

Genetic diseases, 674–5

Geometrical isomers, 307

Glacial acetic acid, 221

Glass, 155

**Globin**   One of the protein chains that make up the hemoglobin molecule, 565, 674, 710, 713

**Globular protein**   A globular-shaped protein that is more or less soluble in water, 548

Glomeruli, 189, 715

Glucitol, 497

Glucocerebrosides, 530–1

Glucocorticoids, 535, 537

**Glucogenic amino acids**   Amino acids that yield a carbon skeleton that is degraded to pyruvate or another intermediate capable of conversion to glucose, 629–31

**Gluconeogenesis**   The process by which glucose is synthesized in the body, 638–9

D-Gluconic acid, 497

D-Glucose, 489–98, 508, 625–6, 728, 736–7

activation of, 616–8

biosynthesis of, 479, 618, 638–40

in the diet, 728

and diabetes, 698

isotonic solutions of, 188

metabolism of, 573, 615–9

open-chain and cyclic forms, 489–91

L-Glucose, 492, 494, 508, 574

Glucose 1-phosphate, 497, 617, 621, 641

Glucose 6-phosphate, 497, 616–22

Glucosides, 496

D-Glucuronic acid, 497, 506

Glutamate, 549, 627, 631

Glutamate oxaloacetate transaminase (GOT), 585

Glutamate pyruvate transaminase (GPT), 585

Glutamic acid, 549, 555–6, 560, 602, 646, 675, 693

Glutathione, 555

Glyceraldehyde, 493

Glyceraldehyde 3-phosphate, 616–8, 620, 622

Glycerol, 364, 372

catabolism, 615–7, 621

in fats, 518, 521–2

to make nitroglycerine, 373

viscosity, 378

Glycerol 1-phosphate, 434, 621

Glycerol 3-phosphate, 621, 643

Glycerol 3-phosphate transport, 620

Glycine, 167, 540, 549–50, 555, 693

Glycocholate, 540

Glycogen, 504, 621

degradation of, 613, 617, 621, 637–8

synthesis of, 638–41

**Glycogenesis**   The conversion of glucose to glycogen in the living organism, 641

Glycogen storage diseases, 621

Glycolic acid, 432

Glycolipids, 530–1, 645

**Glycolysis**   Specific pathway by which the body gets energy from monosaccharides, 580, 615–22

Glycoproteins, 567, 699, 702

**Glycoside**   An acetal or ketal made by the reaction between a monosaccharide and an alcohol, 496

Goiter, 383

Goldstein, Joseph, 533

GOT, 585

Gout, 178

GPT, 585

Graduated cylinder, 14

Graham, Thomas, 143

Graham's law, 143

Grain alcohol (*See* Ethanol)

**Gram**   The basic unit of mass in the metric system, 14

Grape sugar, 489

Graphite, 154–6

Gravity, 15

**Gray**   A unit of radiation, 262

Grease, 309

Greenhouse effect, 273, 311

Ground state, 49–50

of nucleus, 260

Group, in nomenclature, 299, 303

Growth hormone, 548

GTP, 601, 668

Guanidinium chloride, 566

Guanine, 601, 651

Guanosine, 653

Guanosine triphosphate (GTP), 601, 668

Guillain-Barré syndrome, 529

Guldberg, Cato, 206

Gypsum, 177


**H**

Haber, Fritz, 213

Haber process, 104, 213

Haddam Neck reactor, 270

Hahn, Otto, 270

**Half-life**   The time required for one-half of any sample of radioactive material to decay, 257–9

*Halobates,* 153

**Halogen**   An element in Group VII of the periodic table, 44–5

Halogenation, of aromatic rings, 349

Halothane, 376

**Hard water**   Natural water containing relatively high concentrations of $Ca^{2+}$ and $Mg^{2+}$ ions, 525

Harvey, William, 10–1

Hashish, 379

**Haworth formula**   A formula that shows three-dimensional structures of carbohydrates, 489

Hay fever, 693, 700, 709

HDL (high density lipoprotein), 532–5

Headache, 693

Heart, action of, 10–1, 134, 429

Heart attack, 429, 534, 585, 619, 720, 728

**Heat**   A form of energy that usually accompanies chemical reactions, 25

contrast with temperature, 25

Heating curves, 157–8

Heat denaturation

prevention by chaperones, 567

**Heat of combustion** The heat given off in a combustion reaction, 121

**Heat of fusion** The heat necessary to melt 1 g of solid, 157

**Heat of reaction** The heat lost or gained in a chemical reaction, 119–21

**Heat of vaporization** The heat required to evaporate 1 g of a liquid, 158

Heat shock proteins, 564

Heavy metal poisoning, 10, 566, 583

Helium, 260, 269

Helix, 502

Hematocrit value, 711

Heme, 78, 565–6, 713
  catabolism of, 631–3

**Hemiacetal** A compound that contains the linkage,
  —CH—O—C, 397–403
      |
      OH
  cyclic, 399–403, 490, 496

**Hemiketal** A compound that contains the linkage
          C
          |
  C—C—O—C, 399–403
          |
          OH

Hemlock, 460

**Hemodialysis** An artificial process that substitutes for the action of kidneys, 189, 715

**Hemoglobin** The oxygen-carrying protein of the blood, 10, 67, 78, 310, 547–8, 560–1, 565–6, 631, 674
  carbon dioxide carrying ability, 714
  isoelectric point, 556
  normal, 560
  oxygen carrying ability, 78, 711, 713
  poisoning by CO, 78, 139
  sickle cell, 560–1
  three-dimensional structure of, 565–6

**Hemolysis** The swelling and bursting of red blood cells, 187

Hemophilia, 649

Hemorrhoids, 189

Hemp, 379

Henbane, 461

Henderson-Hasselbalch equation, 240–1

Henle's loop, 717

**Henry's law** The solubility of a gas is directly proportional to the pressure, 168

Heparin, 506–8, 712

Hepatitis, 585, 633, 670

1-Heptanol, 380

Herbs, 443

Heredity, 650

Heroin, 463

Herpes, 670

Hess's law, 121

**Heterocyclic compound** One that contains a ring in which at least one atom is not carbon, 350, 352

**Heterogeneous catalyst** A catalyst that is in a separate phase from the reactants, 204

**Heterogeneous mixture** A mixture that is not uniform throughout, 163

Heterozygote, 560

Hexachlorophene, 377

Hexamethylenediamine, 455

Hexoses, synthesis of, 639–41

4-*n*-Hexylresorcinol, 374

High blood pressure, 720–1

High density lipoprotein (HDL), 532–5

High density polyethylene, 341

High dosage of vitamins, 730

High-energy linkages, 435–6, 595, 601

Hippocampus, 694

Hippocrates, 429

Hippuric acid, 718

Histamine, 542, 693, 695

Histidine, 549, 556, 673, 693

Histone, 650, 671–3

Historical landmarks, 48

HMG-CoA reductase, 534

Hoffmann, Roald, 2–5

**Homogeneous catalyst** A catalyst in the same phase as the reactants, 204

**Homogeneous mixture** A mixture in which the molecules or ions are thoroughly mixed; one that is uniform throughout, 163–8

Homozygote, 560

**Hormone** A compound secreted by specific glands, released into the blood stream and then absorbed onto specific receptor sites, 548, 686, 696–8
  peptide hormones, 698

steroid hormones, 535–9

Hospitals
  fire hazards in, 200

Hounsfield, Godfrey N., 266

Housefly, 329

Human genes, 673

Human insulin, 558

Humulins, 679

Hund's rule, 51, 53

Hyaline membrane disease, 644

Hyalobiuronic acid, 514

Hyaluronic acid, 506–7

Hybrid orbital, 291–3

Hybrid, resonance, 346

Hydrates, 177, 398

Hydration, 176–9
  of alkenes, 335–6

**Hydrocarbon** A compound that contains only carbon and hydrogen
  aliphatic, 293–312
  aromatic, 344–50

Hydrochloric acid, 221, 227–9
  diffusion of, 143
  solubility in water, 180

Hydrogen
  addition to multiple bonds, 333, 365, 403, 521
  fusion of, 269–70
  isotopes of, 255
  reaction, with carbon dioxide, 204–6
    with ethylene, 204
    with iodine, 208
    with nitrogen, 199, 213
    with oxygen, 195, 198–9, 211–2
  uses of, 131

Hydrogenation
  of aldehydes and ketones, 403
  of alkenes, 333
  of alkynes, 343
  of fats and oils, 521

Hydrogen atom, 49–51

Hydrogen bomb, 269

**Hydrogen bond** A weak attractive force between a hydrogen and a nearby oxygen, nitrogen, or fluorine atom, 146
  of alcohols and phenols, 378–80
  of aldehydes and ketones, 395
  of amides, 453
  of amines, 446
  between backbones of proteins, 560, 562
  between DNA chains, 656–61

Hydrogen bond (*Continued*)
  between side-chains of proteins, 563, 566
  of carboxylic acids, 418
  of carboxylic esters, 424
  in cellulose, 504–5
  and solubility, 181
  in water and ice, 147
Hydrogen chloride (*See* Hydrochloric acid)
Hydrogen halides
  addition to multiple bonds, 334–5, 343
  mechanism of, 336–7
Hydrogen iodide, 208
Hydrogen ion transport, 605–6
Hydrogen peroxide, 117
  Lewis structure, 75
Hydrogen phosphate ion, 239
Hydrolases, 575–6
**Hydrolysis**   A reaction with water
  of acetals, 399
  of amides, 456, 458–9
  of carboxylic esters, 432–4
  of fats, 522–3, 737
  of phosphoric anhydrides, 435–6
  of salts, 236–7
**Hydronium ion**   $H_3O^+$, 180, 219–26, 228–31
  and pH, 231–7
**Hydrophilic**   Water-loving, 524–5
**Hydrophobic**   Water-hating, 380, 517, 524–5
Hydrophobic interactions, 564, 566, 580
**Hydroxide ion**   $OH^-$, 219–22, 230–1
  and pH, 232
  reaction with acids, 228
    with carboxylic acids, 419–20
*p*-Hydroxyacetanilide, 452
$\beta$-Hydroxybutyrate, 625–6
$\beta$-Hydroxybutyric acid, 625–6
Hydroxy group, 312
$\alpha$-Hydroxy ketone, 397
**Hygroscopic**   The property of taking water from the air, 178
Hyperbaric medicine, 139, 168
Hypercholesterolemia, 534
**Hyperglycemia**   A condition in which blood sugar level is too high, 737
Hyperosmotic solution, 187
Hypertension, 542, 720, 728
**Hyperthermia**   Sustained unusually high body temperature, 26

**Hypertonic solution**   One that has a higher osmolarity than human blood, 188
**Hyperventilation**   Rapid or heavy breathing, 237
Hypnotic action, 509
Hypodermic syringe, 14
**Hypoglycemia**   A condition in which the body output of insulin is elevated, and blood sugar is depleted, 180, 559, 618, 621, 737
Hypoglycemic awareness, 559
Hypo-osmotic solution, 187
Hypothalamus, 697
**Hypothermia**   Sustained unusually low body temperature, 26
**Hypothesis**   A statement offered to explain facts, 9
Hypothyroidism, 383
**Hypotonic solution**   One that has a lower osmolarity than red blood cells, 187–8
**Hypoventilation**   Difficulty in breathing, 237
**Hypoxia**   Insufficiency of oxygen in the tissues, 139
Hysterectomy, 432

**I**

Iatrogenic malnutrition, 728
Ibuprofen, 452, 541
"Ice," 446
Ice,
  heat of fusion, 157
  melting of, 157–8, 184
  structure of, 147
**Ideal gas**   A gas that perfectly obeys all the gas laws, 132
Ideal gas law, 141
Immune response, 699, 702–3
Immunization, 670, 699
**Immunoglobulins**   Glycoproteins used by the body to counteract foreign substances, 548, 567, 686, 699–703, 711
Immunoglobulin superfamily, 702
**Indicator**   A substance that changes color at a certain pH, 234–5, 243
Indigo blue, 645
Indomethacin, 541
Indoor radon problem, 263
**Induced-fit model**   A mechanism of enzyme action in which the

shape of the enzyme cavity changes to accommodate the substrate, 581–3
Industrial diamonds, 157
Industrial production, 108
Infrared light, emission of, 311
**Inhibition, of enzymes**   Any process that makes an active enzyme less active or inactive, 578, 579–83
Initiation factors, 668
Initiation, of protein synthesis, 664, 668
Inner shell, 53
**Inner transition elements**   Elements 58 to 71 and 90 to 103, 47, 57
**Inorganic compound**   One that does not contain the element carbon, 285–6
  naming of, 88–9
Inorganic esters, 436
Insect control, 281–2
Insecticides, 381
Insulin, 548, 558, 626, 679, 698
  primary structure of, 558
Intensity of radiation, 259–61
**Intensive property**   A property that does not depend on the quantity of a substance, 23
Interferon, 567
Interleukin-2, 701
**Intermediate**   A species formed in a reaction and used up immediately, 336–7
**Intermolecular forces**   The forces between one molecule and another, 130, 145–9
  effect on boiling point, 153
International System of Units (SI), 12
International Union of Pure and Applied Chemistry (*See* IUPAC)
**Interstitial fluid**   The fluid that surrounds the cells and fills the spaces between them, 707, 716
Intestines, 432
Intracellular fluid, 707
Intravenous feeding, 729
**Intron**   A segment of a DNA molecule that does not code for anything, 573, 669, 673
**Invert sugar**   A mixture of D-fructose and D-glucose, made by hydrolyzing sucrose, 499
Iodine, 117, 208, 334, 383
Iodine-131, 259, 265, 267, 272

Iodine number, 522
Iodized salt, 383
Iodoform, 382
Iodomethane, 196
**Ion** A particle with an unequal number of protons and electrons, 43, 65–72
biologically important, 67, 716
Ion channel, 691
**Ionic bond** The attraction between positive and negative ions, 69–72
Ionic compounds, 69–72, 155–6
in medicine, 70
predicting formulas of, 71–2
solubility of, 113–4, 176
Ionic dissociation, 110
Ionic equations, 110–4
Ionization constant, 226
**Ionization energy** The energy necessary to remove an electron from an atom in the gaseous state, 56
**Ionizing radiation** Radiation of sufficient energy to ionize neutral matter by knocking electrons out of the electron clouds, 259, 675
Ionone, 407
**Ion product of water** The concentration of $H_3O^+$ times the concentration of $OH^-$: $[H_3O^+][OH^-] = 1 \times 10^{14}$, 230
Ion translocating protein, 691
Iron Age, 48
Iron(II) ion, 67
Isoamyl nitrite, 373
Isobutane, 296
Isobutyl group, 303
Isocitrate, 600–1
Isocitrate dehydrogenase, 601
**Isoelectric point** The pH at which all the amino acid or protein molecules are in the zwitterionic form, 552, 556
Isoenzyme, 585
Isoleucine, 729
Isomerases, 575–6
**Isomers** Compounds that have the same molecular formulas but different structural formulas, 289–91, 296–8, 325
Iso-osmotic solution, 187
Isoprene, 324, 339

Isoprene rule, 324
Isopropyl alcohol, 371
Isopropyl group, 303
**Isotonic solution** One that has the same osmolarity as red blood cells, 187–8
Isotope generators, 269
**Isotopes** Atoms with the same number of protons but different numbers of neutrons, 43–4
artificial, 255, 259, 269
naturally occurring, 255
radioactive, 255–9
Isozyme, 585
IUPAC (International Union of Pure and Applied Chemistry), 298
IUPAC nomenclature system, 298
for alcohols, 362–4
for aldehydes and ketones, 392–4
for alkanes, 298–305
for alkenes, 323–7
for alkyl halides, 302
for alkynes, 341
for carboxylic acids, 416–7
for carboxylic acid salts, 421–2
for carboxylic esters, 422–4
for cycloalkanes, 304–5

**J**

Japanese beetles, 329
Jaundice, 618, 633
Joule, 26

**K**

$K_a$ values, 226
of organic compounds, 420
$K_w$, 230
Kanamycin, 664
Kekulé, Friedrich, 287, 345
Kekulé formulas, 346
**Kelvin scale** The temperature scale in which absolute zero is set at 0 K = −273°C, 17, 137–8
Keratin, 547–8, 562, 567
Kerosene, 309
**Ketal** A compound that contains the linkage
C
|
C—C—O—C,
|
O—C
399–403
**Ketoacidosis** Acidosis resulting from

lack of utilization of glucose in diabetes, 626
Keto-enol tautomerism, 404–5
Keto form, 404, 651
**Ketogenic amino acids** Amino acids that are degraded to ketone bodies, 630–1
$\alpha$-Ketoglutarate, 601, 627, 630, 646
$\alpha$-Ketoglutaric acid, 602, 646
**Ketone** A compound that contains the linkage C—C—C,
||
O
391–408
chemical properties of, 397–405, 408
nomenclature of, 392–4
oxidation of, 396–7
physical properties of, 394–5
preparation of, 367, 395
**Ketone bodies** Acetoacetic acid, $\beta$-hydroxybutyric acid (and their salts) and acetone, produced by the body during diabetes or starvation, 237, 625–6
**Ketose** A monosaccharide that contains a ketone group, 493–4
Ketosis, 621
Kidneys, 715–21
action of, 189, 715–21
buffer production, 719
cleansing function of, 715–8
diseases, 711, 715, 721
**Kinetic energy** The energy of motion, 23, 130
**Kinetic-molecular theory** The theory that explains the behavior of matter on the basis of the movement of molecules, 131
**Kinetics** The study of reaction rates, 195–205
Klug, Aaron, 671
**Knocking** Premature explosion of the gasoline-air mixture in an automobile cylinder, 309
Knockout drops, 398
Knoop, F., 622
Known compound, 287
Koshland, Daniel, 581
Krabbe's leucodystrophy, 531
Krebs, Hans, 591, 598, 629
Krebs cycle (*See* Citric acid cycle)
**Kwashiorkor** A disease caused by protein deficiency, 730

## L

Lachrymators, 406
*Lac* operon, 672
*beta*-Lactam ring, 586
Lactate
  accumulation of, 619
  in muscles, 404
  production of, 616–8, 619
Lactate dehydrogenase (LDH), 576, 584–5
Lactic acid, 237, 418, 432, 508, 566
Lactomer, 432
Lactones, 423
Lactose, 499, 728
Lagging strand, 661
Lake Nyos, 310
Landsteiner, Karl, 500
Lascaux, France, 258
Laser beam, 182
Laser surgery, 25
Latex, 339
Laughing gas, 376
Lavoisier, Antoine, 33
**Law of conservation of energy**   Energy can neither be created nor destroyed, 24
**Law of conservation of mass**   Matter can neither be created nor destroyed, 40
**Law of constant composition**   A compound is always made up of elements in the same proportion by weight, 40
Laxatives, 221
Lead-206, 255, 258
Lead-208, 59
Leading strand, 661
Lead poisoning, 10, 566, 583
Lead shielding, for radiation, 261
Lead storage battery, 119
Lead tetraethyl, 309
Le Chatelier, Henri, 209
**Le Chatelier's principle**   If an external stress is placed on a system in equilibrium, the system reacts in such a way as to partially relieve the stress, 209–14, 365, 637
  and sunglasses, 211
Lecithins, 527, 644
Leeches, 8
LDL (low density lipoprotein), 533–5
LDL receptor, 533–5

Lemons, 415
Length, 12–3
Lens, of the eye, 507, 547
Leucine, 549, 631
Leukemia, 264, 710
**Leukocyte**   A white blood cell, 542, 710–1
Leukotrienes, 540–2
**Levorotatory**   The rotation of the plane of polarized light to the left, 484
Levulose, 495
Lewis, Gilbert N., 72
Lewis structures, 74–7
  of benzene, 345
Librium, 459
Lidocaine, 462
Ligases, 575–6
Light reaction, 640
Light, speed of, 252, 270
Like dissolves like rule, 166
Limestone, 110
Limonene, 358
Lincoln Cathedral, 177
Linear molecules, 79–80
Linoleic acid, 519, 521, 641, 729
Linolenic acid, 519, 521, 641, 729
**Lipase**   An enzyme that digests lipids, 434, 574, 585, 737
**Lipid bilayer**   Structure found in membranes, consisting of two rows of lipid molecules arranged tail to tail, 525–7
**Lipids**   Substances found in living organisms that are insoluble in water but soluble in nonpolar solvents, (*See also* Fats) 517–42
  biosynthesis of, 641–5
  digestion of, 614, 737
  metabolism of, 621–6, 737
**Lipid storage disease**   A genetic disease in which an enzyme needed to decompose a complex lipid is missing, resulting in accumulation of that lipid, 530–1
Lipoprotein, 533–5
**Liquid**   The state of matter with a definite volume but no definite shape, 20, 130, 149–53
Liquid protein, 729
Lister, Joseph, 374
**Liter**   The basic unit of volume in the metric system, 13–4

Lithium, 68
Lithium carbonate, 68
Litmus paper, 233
**Lock and key model**   A theory of enzyme action in which the substrate fits exactly into an opening in the enzyme, 580–3
Locked rings, 496, 498–9
Logarithms, 231–3
London, Fritz, 148
**London dispersion forces**   Very weak intermolecular forces between nonpolar molecules, 146, 148–9, 703
Long term potentiation (LTP), 694
Lorazepam, 459
Low density lipoprotein (LDL), 533–5
Low density polyethylene, 340
Lowry, Thomas, 223
Lovastatin, 534
Lubricating oil, 309
Luciferase, 679
Lung, 136
Lung cancer, 263
Luteinizing hormone, 538, 696–7
Lyases, 675–6
Lye, 221
Lymph, 708
Lymphocytes, 686, 699, 711
Lysine, 549, 551, 554, 555–6, 574, 729, 738
Lysozyme, 574, 703

## M

Mace, 406
**Macromolecule**   A very large molecule, 503
Mad hatter, 133
Magnesium hydroxide, 221
Magnesium ions, 66
  in the body, 67, 716
  in hypertonic solutions, 189
Magnet, 254
Magnetic resonance imaging (mri), 266
Main chain, 297
Malaria, 8, 381, 460, 464, 560
Malate, 602
Malate dehydrogenase, 602
Malic acid, 337, 415
Malleability, of metals, 47–8
Malnutrition, 728
Malonyl ACP, 642
Malpighi, 11

Maltase, 621

Maltose, 498

Mannose, 492–3

**Manometer** A device for measuring the pressure of a gas in a closed container, 133–4

MAOs (*See* monoamine oxidases)

**Marasmus** Chronic starvation, 728

Marble monuments, 177

Marfan's syndrome, 562

Margarine, 407, 522

Marijuana, 379

Markovnikov, Vladimir, 335

**Markovnikov's rule** In addition of a reagent HX to a double bond or triple bond, the H goes to the carbon that already has more hydrogens, 335–6, 343

Marsh gas, 294

**Mass** The quantity of matter in an object. The basic unit is the gram, 14, 22

energy from, 270

**Mass number** The total number of protons and neutrons in the nucleus of an atom, 42

Matrix, 594

**Matter** Anything that has mass and takes up space, 9

organization of, 130

Mayonnaise, 181, 183

McArdle's disease, 621

Measles, 670, 699

Measurements, 11–7

Meat tenderizer, 577

**Mechanism, reaction** The steps by which a reaction takes place, 336–7

Mediators, 685

Meitner, Lisa, 270

Melamine-formaldehyde, 406

Melmac, 406

**Melting point** The temperature at which a solid becomes a liquid, 130, 156–7

of elements, 46

Membranes, 463, 525–7

Memory cells, 699

Mendeleev, Dmitri, 45, 54

Menopause, 718

Menstrual cycle, 537–8

Mental retardation, 383, 632

Menthol, 324, 371

Mercaptan (*See* Thiol)

2-Mercaptoethanol, 566

Mercaptoethylamine, 598

Mercury

in barometers and manometers, 133

as a poison, 133, 566, 583

**Messenger RNA (mRNA)** A form of RNA that carries genetic information from DNA to the ribosome, and acts as a template for protein synthesis, 662–5

Metabolic acidosis, 237

Metabolic alkalosis, 237

**Metabolism** The sum total of all chemical reactions involved in maintaining the dynamic state of the cell, 591–633

Meta compounds, 347

**Metal** An element with a shiny surface whose atoms tend to give up electrons, 47

reaction with acids, 227–8

Metal-binding fingers, 673

Metals, historical, 48

**Metalloid** An element that shares properties of metals and nonmetals, 47

Metastatic state, 268, 676

**Meter** The basic unit of length in the metric system, 12

Meteorite, 48

Methadone, 463

Methamphetamine, 446

Methandienone, 536

Methane, 81, 291, 294–5, 308

Methanethiol, 380

Methanol, 313, 370–1

Metathesis reactions, 110

Methedrine, 446

Methenolone, 536

Methionine, 549, 632, 668, 729

Methionine enkephalin, 694–5

Methyl alcohol (*See* Methanol)

Methylamine, 229, 444

Methylated cytosine, 673

2-Methyl-1,3-butadiene, 324, 339

Methyl *tert*-butyl ether, 377

Methyl chloride, 196, 382

Methylene chloride, 382

Methyl group, 299

Methyl orange, 235

Methylpentane, 296–300

Methyl salicylate, 374, 425

**Metric system** A system of measurements in which units are related to each other by powers of ten. The system is used by scientists throughout the world, 11–3

Mickey Finn, 398

Microtubules, 314

Microwaves, 253

Mifepristone, or RU486, 537, 539

Milk of magnesia, 221

**Milliequivalent** One-thousandth of an equivalent, 244

**Milliliter** One-thousandth of a liter; equal to 1 cubic centimeter, 14

**Millimeter of mercury** A unit of pressure equal to 1/760 of an atmosphere, 133

Minamata, Bay of, 583

Mineralocorticoids, 535

Mineral oil, 309

Minerals, in the diet, 730

list of, 734–5

Mirror image, 480–91

Miscible, 166

Mitchell, Peter, 605

**Mitochondrion** An organelle, located in cells, in which the common catabolic pathway takes place, 574, 592–4, 603–6, 618–20, 623, 627

**Mixture** A system with more than one component, 37–8, 163

Models, molecular, 290, 295, 298

**Modulator** A substance that regulates the action of an enzyme, 584

**Molarity** Moles of solute per liter of solution, 170–3

relation to normality, 245

**Mole** The formula weight of a substance, expressed in grams, or $6.02 \times 10^{23}$ of anything, 98–101

value of (ideal gas), 141

Molecular collisions, 196–8, 200–2

**Molecular compound** A compound that does not contain ions, 89

**Molecular formula** A formula that shows how many of each atom there are in a molecule, 74, 89–90, 97, 289

Molecular orbital, 73

Molecular solids, 155

**Molecular weight** For molecular compounds, the sum of the atomic weights of all the atoms in a molecule, 97
  calculation of, 142
  effect on boiling point, 152, 307
**Molecule** A tightly bound group of atoms that acts as a unit, 39
  motion of, 129–31, 196–8
  shape of, 77–81
Monatomic elements, 40
Monoamine neurotransmitters, 690–3, 695
Monoamine oxidases (MAO), 692–3
Monoamine oxidase inhibitors, 692–3
Monocalcium phosphate, 229
Monoglycerides, 518
**Monomer** The basic unit that makes up a polymer, 338
**Monoprotic acid** An acid that gives up only one proton, 224
**Monosaccharide** A compound of the general formula $C_nH_{2n}O_n$ where $n$ can vary from 3 to 9, 489–96
  derivatives of, 496–8
Monosodium glutamate, 631, 693
Montgolfier brothers, 129
Moon, 16
Morning after pill, 537, 539
Morphine, 463, 695
Motrin, 452
mRNA (*See* Messenger RNA)
MSG, 631, 693
MTBE, 377
Mullis, Kary B., 679
Multiple code, 666
Multiple sclerosis, 529
Muscle action, 542, 547, 608, 686
Muscle cramps, 180
Muscle relaxants, 575, 689
Muscone, 407
Muscovite, 155
Mushrooms, 664
**Mutagen** A compound that induces mutation by reacting with DNA, 674–5
**Mutarotation** Change in rotation of the plane of polarized light, 491, 496
**Mutation** An error in the copying of the base sequence of DNA resulting in a change in heredity, 264, 674–5

Mycobacterium avium complex (MAC), 701
Myelin sheath, 529
Mylar, 260, 430
Myocardial infarction, 534, 585
Myocardium, 622
Myosin, 547, 608

**N**

$NAD^+$ and NADH, 404, 596–7
  in amino acid biosynthesis, 646
  in amino acid catabolism, 627
  in citric acid cycle, 598–603
  in fatty acid catabolism, 622–5
  in glycerol catabolism, 621–2
  in glycolysis, 616–20
  in oxidative phosphorylation, 603–6
  structure of, 596–7
NADPH and $NADP^+$, 646
Naming (*See* Nomenclature)
Nandrolone decanoate, 536
Naphthalene, 350
Narcotics, 463
Natural gas, 294, 308
Nature of reactants, 200
Nembutal, 457
Neomycin, 664
Neopentyl group, 303
**Nephron** The basic unit of the kidney, 189, 715
Nerve gases, 689
Nerve transmission (*See* Neurotransmitter)
Network solids, 156
**Neuron** A nerve cell, 686–7
**Neurotransmitter** A compound that communicates between two nerve cells or between a nerve cell and another cell, 685–95
  adrenergic, 690–3, 695
  cholinergic, 687–90, 695
  peptidergic, 694–5
**Neutralization** The reaction between an acid and a base to produce a salt and water, 227–8, 242–6
Neutral salts, 236–7
**Neutral solution** An aqueous solution whose pH = 7, 227, 231–7
**Neutron** A neutral particle that is a constituent of atoms, 41–2, 254
  in nuclear fission, 270–1
  role in holding nucleus together, 255
Newton, 11

Niacin, 597
Nicotinamide, 596–7
Nicotinamide adenine dinucleotide (*See* $NAD^+$)
Nicotinamide adenine dinucleotide, reduced form (*See* NADH)
Nicotine, 461, 597
Nicotinic acid, 597
Niemann-Pick disease, 531
Nirenberg, Marshall, 666
Nitration, 349
Nitric acid, 221, 349
Nitric oxide, as an intracellular messenger, 694
Nitrites, in meat, 450, 693
Nitrocellulose, 505
Nitro compounds, 349
Nitrogen, uses of, 131
  solubility in water, 165
Nitrogen catabolism in amino acids, 626–9
Nitrogen fixation, 213
Nitroglycerine, 373, 693
*N*-Nitrosamines, 450
Nitrous acid, 450
Nitrous oxide, 20, 376
Nmr scans, 266
Nobel, Alfred, 373
**Noble gas** An element belonging to group 0 of the periodic table, 47
Nomenclature
  of alcohols, 362–4
  of aldehydes, 392–4
  of aldoses, 493
  of alkanes, 298–304
  of alkenes, 323–7
  of alkyl halides, 302
  of alkynes, 341
  of amides, 451–3
  of amines, 443–6
  of amino acids, 549
  of benzene derivatives, 346–9
  of carboxylic acids, 416–7
  of carboxylic acid salts, 421
  of carboxylic esters, 422–4
  of cycloalkanes, 304–7
  of enzymes, 574–7
  of ethers, 375
  of inorganic compounds, 88–90
  of ketones, 392–4
  of phenols, 372, 374
**Noncompetitive inhibitor** A compound that decreases enzyme activity by binding to a site other

than the active site, changing the shape of the active site, 578, 582–3

**Nonelectrolyte**   A substance that does not conduct a current when dissolved in water, 179
in body fluids, 716

**Nonmetal**   An element that tends to accept electrons, 47

Nonpolar compounds, 85

Norepinephrine, 692, 695–6

Norethindrone, 539

Norethynodrel, 539

**Normal alkane**   An alkane that has no branching, 295

Normal alkyl groups, 302

**Normal boiling point**   The boiling point at 1 atm pressure, 152

**Normality**   Equivalents of solute per liter of solution, 244–6
relation to molarity, 245

Novocain, 449, 462

N-terminal, 555

Nuclear chemistry, 251–73

Nuclear energy, 270–3

Nuclear equations, balancing, 255–6

**Nuclear fission**   The cleavage of large nuclei into smaller ones, with release of energy, 270–3

**Nuclear fusion**   The combining of two or more small nuclei into one larger one, with release of energy, 269–70

Nuclear magnetic resonance scans, 266

Nuclear medicine, 265–9

Nuclear power plant accidents, 259, 271–3

Nuclear reactors, 271–3

Nuclear waste, 271–3
storage, 273

**Nucleic acid**   A polymer composed of nucleotides, 649–80 (*See also* DNA *and* RNA)

**Nucleon**   A name given to both protons and neutrons, 42

**Nucleoside**   A compound composed of a sugar (ribose or deoxyribose) and a base, 652–4

Nucleosome, 671

**Nucleotide**   A compound composed of a sugar, a base, and phosphate, 581, 651, 653–61

**Nucleus**   The cluster of protons and

neutrons located in the center of an atom, 42
density of, 60
mass of, 59
size of, 59

Nuprin, 452

NutraSweet, 553

**Nutrient**   A component of food and drink that provides growth, replacement, and energy, 725–6

Nutrition, 725–38

Nylon, 455

Nyos, Lake, 310

### O

Obesity, 728

Octane number, 309

Octet rule, 68–9

**Oil**   A liquid fat, 521

**Okazaki fragment**   A short nucleotide chain of about 1000 units, 661

**Olefin**   An older name for alkene, 323

Oleic acid, 518–9

Oligosaccharides, 489, 500

Olympic games, 536

**Oncogene**   A gene that participates in the development of cancer, 676

Operator sites, 672

Operon, 672

Opium, 376, 460, 463

Opsin, 330

**Optical activity**   Rotation of the plane of polarized light, 483–7
of carbohydrates, 491, 494–5

Oral contraceptives, 281–2, 539

Orbitals, 49–50

**Organelle**   Small structure inside cells, 525, 592

**Organic chemistry**   The study of compounds that contain carbon, 285

Organic compounds, 285–7
properties of, 286
sources of, 287–9
structures of, 289–91

Organic synthesis, 288–9, 309

Organ scans, 265–8

Orinase, 698

Ornithine, 574, 627–9

Ortho compounds, 347

**Osmolarity**   The molarity of a solution multiplied by the number

of particles produced by each mole of solute, 187

**Osmosis**   The passage of solvent molecules from the dilute to the concentrated side of a semipermeable membrane, 185–8

**Osmotic membrane**   A semipermeable membrane that allows only solvent molecules, and not solute molecules, to pass through, 185–8

**Osmotic pressure**   The amount of external pressure that must be applied to equalize the two levels when solutions of different concentrations are separated by a semipermeable membrane, 185–8

Osteomyelitis, 139

Osteoporosis, 718

Outer shell, 54, 68

Ovalbumin, 548, 567

Oxalic acid, 416–7

Oxaloacetate, 599–600, 602, 625

**Oxidation**   Loss of electrons, or gain of oxygen or loss of hydrogen, 115–8
of alcohols, 366–8, 395, 419
of aldehydes, 395–7, 419
of monosaccharides, 497

Oxidative deamination, 615, 627

Oxidative decarboxylation, 618–9

**Oxidative phosphorylation**   The conversion of ADP to ATP in the common catabolic pathway, 594, 603–6
schematic diagram of, 603

**Oxidizing agent**   The species that gets reduced, 117, 367

Oxides, reaction with acids, 227–8

Oxidizing antiseptics, 117

Oxidoreductases, 575–6

Oxygen
in blood, 78, 713–4
and breathing, 144
in electron transport chain, 603–6
in medicine, 131, 139, 144, 200–1, 310
reaction with hydrogen, 195, 198–9
solubility in water, 165
toxicity, 139
uses of, 131

Oxygenated gasoline, 377

Oxygen-carbon dioxide cycle, 480

Oxytocin, 548, 559
Ozone, 40, 117, 382
Ozone layer, 382

## P

Pain relievers, 429, 452
Palmitic acid, 519, 643–4
Palmitoyl CoA, 644
Pancreas, 585, 698
Pantothenic acid, 598
Papain, 577
Para compounds, 347, 382
Paraffin, 307, 524
Paraldehyde, 405
Parathion, 689
Parenteral nutrition, 729
Parent name, 299
Parkinson's disease, 692
Partial charges, 84
**Partial pressure**  The pressure a gas would exert if it were alone in the container, 143
Parts per million and per billion, 175
**Pascal**  A unit of pressure, 133
Pasteur, Louis, 374, 484
**Pathway (biochemical)**  A series of consecutive biochemical reactions, 592
**Pauli exclusion principle**  An orbital can hold a maximum of two electrons, and these must have opposite spin, 51
Pauling, Linus, 82, 562, 656
PCBs, 382
Pellagra, 730
Penicillin, 586
Penicillinamine, 632
Penicillinase, 586
Penicilloic acid, 586
$n$-Pentane, 152–3
Pentobarbital, 457
Pentothal, 457
Pep pills, 446
Pepsin, 574, 577, 614, 737
**Peptide**  A short chain of amino acids, connected by peptide linkages, 552–5
**Peptide linkage**  The linkage —C—NH— that holds amino
$$\overset{\|}{\underset{O}{}}$$
acids together in peptides and proteins, 553

Peptidergic neurotransmitters, 694
**Percent concentration**  A method of expressing concentration by dividing the weight or volume of solute by the weight or volume of solution and multiplying by 100, 169–70, 174–5
**Percentage yield**  The actual yield divided by the theoretical yield times 100, 108
Perchloroethylene, 382
**Periodic Table**  A chart in which elements, when placed in order of atomic number, fall into groups of similar elements, 44–8
and electronegativity, 83
and electronic configuration, 54–6
and ionization energy, 56
Permanent waves, 567
Peroxides, 377, 555
Petroleum, 288, 309
Petroleum jelly, 308
**pH**  A number that tells the acidity of aqueous solutions, 231–6
of blood, 234, 242
of buffer solutions, 239–41
effect on enzyme action, 579
meters, 236
values of common materials, 234
(*See also* Isoelectric point)
Phagocytes, 631
Phagocytosis, 711
Pharmacist, 443
Phase changes, 157
Phenanthrene, 350
**Phenol**  A compound in which an —OH group is connected to an aromatic ring, 347, 372, 374–5, 420
as antiseptics, 374
physical properties of, 374
reaction with formaldehyde, 406
Phenol-formaldehyde, 406
Phenolic esters, 424
Phenolphthalein, 235
Phenylalanine, 549, 632, 737
Phenylalanine hydroxylase, 632
Phenylethylamine, 693
**Phenyl group**  The benzene ring with one hydrogen gone, $C_6H_5$—, 348
Phenylketonuria (PKU), 632
$o$-Phenylphenol, 374

Phenylpyruvate, 632
**Pheromone**  A compound used by insects to transmit messages to each other, 329
Phosphatase, 574, 585
Phosphate buffer, 240–2, 714
Phosphate ion, structure of, 87
Phosphates, 221, 434–7, 595
in nucleic acids, 654
Phosphatidates, 643
Phosphatidyl choline, 527, 529, 644
Phosphatidylethanolamine, 528
Phosphatidylserine, 528
Phosphodiesterase, 691
Phosphoenolpyruvate (PEP), 380, 616–8
Phosphoglycerate, 616–8
**Phosphoglyceride**  An ester composed of glycerol, two fatty acid molecules, phosphoric acid, and another compound, 523, 526–8, 643
Phosphohexose isomerase, 576, 585
Phospholipase $A_2$, 541
**Phospholipid**  An ester composed of an alcohol, fatty acids, and a phosphate group, 523, 644
Phosphonates, 689
Phosphor, 260
Phosphoric acid, 221, 434
anhydrides of, 435–6
esters of, 434–7
Phosphorylase, 603–7, 621, 637
**Phosphorylation**  The transfer of a phosphate group from one molecule to another, 436, 603–6, 637
**Photon**  The smallest unit of electromagnetic radiation, 252, 262, 269
**Photosynthesis**  The process in which plants synthesize carbohydrates from $CO_2$ and $H_2O$, with the help of sunlight and chlorophyll, 479, 638–40
pH paper, 235
**Physical change**  A change in which a substance retains its identity, 39
**Physical property**  A property that does not involve a chemical change, 22, 39, 307–8
**Physiological saline solution**  A solution of NaCl that has the same

concentration of salts as normal human blood, 187
Phytol, 358
Pi bond, 328, 343–4
Picric acid, 374–5, 604
Pill, the, 281-2, 539
Pi orbital, 73, 328
Piperidine, 350
Pipet, 14, 245
Pituitary gland, 269, 697
$pK_a$, 226, 240–1, 559
    of organic compounds, 420
Planar triangular molecules, 80
Plasma (*See* Blood plasma)
Plasmids, 676–80
Plaster of Paris, 177
Plastics, 340, 406
Platelets, 429, 710–2
Platinum, 204, 333
Platinum blues, 475
Pleated sheet, 561–2
Plywood, 406
Pneumonia, 701
Polar covalent bond, 86
**Polarimeter**   A device for measuring the rotation of the plane of polarized light, 484–6
**Polarized light**   Light that vibrates in only one plane, 484–6
**Polar molecule**   A molecule that has a positive and a negative pole, a dipole, 84–5
    effect on solubility, 167
Polio, 670, 699
Pollution,
    environmental, 21
    oil spillage, 21
    thermal, 168
Polyacrylonitrile, 340
Polyamides, 455
Polyatomic elements, 40–1
**Polyatomic ion**   An ion containing more than one atom, 87
Polychlorinated biphenyls, 382
Polyesters, 430, 432
Polyethylene, 338–41
**Polymer**   A long chain of repeating units, 338–41, 406, 430, 452, 455, 500–8
Polymerase chain reaction (PCR), 679
Polymerases, 661
Polymerization, 406, 430, 432
    of alkenes, 338–41

**Polypeptide**   A chain of amino acids, connected by peptide linkages, 554–5
Polypropylene, 340
**Polysaccharide**   A polymer composed of monosaccharides, 489, 500–8
Polystyrene, 340
Polyunsaturated oils, 521
Polyvinyl chloride, 200, 338–40
**Positron**   A nuclear particle which has the same mass as an electron but a charge of +1, 256
Positron emission, 256
Postsynaptic nerve endings, 686–7, 691
Potassium acid tartrate, 229
Potassium dichromate, 367
Potassium hydroxide, 221
Potassium ions, 67
    balance in the body, 180, 580, 716
    loss of, 180
Potassium permanganate, 117, 367
**Potential energy**   Stored energy, 24, 130
Pott, Percival, 351
Power, in exponential notation, A.1
Ppb and ppm, 175
Precipitin, 703
Prefix
    in metric system, 13
    in nomenclature, 89, 300
Premarin (estrogens), 718
Preservatives, 421
Pressure
    atmospheric, 132–3
    effect on reaction rates, 200
    effect on solubility, 168
    of gases, 133–40
Pressure cooking, 152
Presynaptic nerve endings, 686–7, 693
**Primary alcohol**   An alcohol in which the —OH is connected to $CH_3$— or $RCH_2$—, 362
    oxidation of, 366–7, 395, 419
**Primary amine**   An amine in which only one carbon is connected to the nitrogen, 443–4
**Primary carbon**   One that is connected to one or no other carbons, 296
Primary structure
    of nucleic acids, 654–6
    of proteins, 557–60
Principal energy level, 50

Procaine, 462
**Product**   A substance formed in a chemical reaction, 95
**Proenzyme**   The biologically inactive form of an enzyme, 578
Progesterone, 535, 537–9, 718
Progesterone receptor, 537
Prokaryotes, 672
Proline, 548–9
Promoter sites, 672
**Proof**   Twice the percentage (v/v) of ethanol in water, 170, 369
Propane, 101–2, 293–5
2-Propanol, 367, 371
Propanolol, 720
Proportional counter, 260
Propyl group, 299
Proscar, 718
**Prostaglandins**   A group of fatty-acid-like substances that have a wide variety of effects on body chemistry, 540–2
Prostate gland, 540, 585, 718
Prostatic hyperplasia, 718
**Prosthetic group**   The non-amino acid portion of a conjugated protein, 565
Protease, 577
Protective colloid, 183
**Protein**   A long chain of amino acids connected by peptide linkages, 554–67
    biosynthesis of, 649–50, 663, 665, 667–9
    denaturation of, 405, 565–7
    in the diet, 729–30
    digestion of, 614–5, 737–8
    as enzymes, 547, 567
    fibrous, 548
    functions of, 547–8
    globular, 548
    metabolism of, 614–5, 626–31
    primary structure, 557–60
    quaternary structure, 557, 565
    secondary structure, 557, 560–3
    tertiary structure, 557, 563–5
    test for, 221
Protein coat, 670
Protein deficiency, 729
Protein kinase, 584, 691
Proteolytic enzymes, 585
Prothrombin, 712
Proto-oncogene, 676

**Proton**   A positively charged particle located in the nucleus of atoms, 41, 254
   transfer of, 223–4
Proton gradient, 605
Protonphore, 604
Proton translocating ATPase, 605
Protoplast, 679
Proust, Joseph, 40
Provera (progesterone analog), 718
Proximal tubule, 715, 717
Prozac, 695
P site, 668–9
Puffer fish, 690
Purine, 461
Purine bases, 651–4
Putrescine, 447
Pyramidal molecules, 81–2
**Pyranose ring**   A ring of five carbons and one oxygen, found in carbohydrates, 489–90, 495
Pyridine, 352
Pyridoxine, 728
Pyrimidine, 352
Pyrimidine bases, 651–4
Pyrite, 155
Pyrophosphoric acid, 435–6, 628, 690
Pyrrole, 352
Pyrrolidine, 350
Pyruvate, 580
   from amino acids, 630
   conversion to acetyl CoA, 616–9
   end product of glycolysis, 616–9
   reduction of, 404
Pyruvate dehydrogenase, 619
Pyruvate kinase, 580

**Q**

Q enzyme, 603–5
Quantized energy levels, 49
Quarter-stagger arrangement, 565
Quartz, 155
Quaternary ammonium salt, 450–1
**Quaternary carbon**   One that is connected to four other carbons, 296
Quaternary structure, of proteins, 565
Quicklime, 110
Quinine, 464

**R**

**Racemic mixture**   A mixture of equal amounts of two enantiomers; it does not rotate the plane of polarized light, 486, 509, 550
**Rad**   A unit of radiation, 262
Radiation, 251–73
   badges, 264
   in cancer therapy, 268–9
   characteristics of, 259–61
   detection of, 260
   dosimetry, 261–5
   effect on health, 261–5
   table of exposure sources, 264
Radiation necrosis, 139
**Radiation sickness**   A disease which may be fatal, caused by exposure to large amounts of radiation, 264, 268
Radioactive dating, 258
Radioactive fallout, 272
Radioactive materials, medical uses, 265–9
Radioactive nuclei, 255–9
Radioactive therapy, 268–9
Radioactive tracers, 267–8
Radioactive waste, 271–3
**Radioactivity**   Particles and waves given off by unstable isotopes, 251–73 (*See also* Radiation)
**Radioisotope**   A radioactive isotope, 255
Radium, 254, 265
Radon gas, 263
Radon-222, 263
Rancidity, 523
Random coil conformations, 502, 562
Ranitidine, 693
Rates of reaction, 195–204
   and equilibrium, 209
   factors affecting, 200–4
Rayon, 505
**Reactant**   A substance that takes part in a chemical reaction, 95
Real gases, 132, 141
Receptors, 685
Recessive gene, 675
Recognition site, 665
Recombinant DNA, 676–80
Recommended daily allowances (RDA), 728, 730, 732–5
Red blood cells, 631, 710–1
**Redox reaction**   An oxidation-reduction reaction, 115
**Reducing agent**   The species that gets oxidized, 115
**Reducing sugar**   Any sugar that gives a positive test with a mild oxidizing agent such as Benedict's solution, 496
**Reduction**   Gain of electrons, or loss of oxygen, or gain of hydrogen, 117
   of aldehydes and ketones, 403
Refrigerants, 382
Regulators, of enzyme action, 584
Regulatory gene, 672
**Regulatory proteins**   Proteins that control gene expression, 548
**Rem**   A unit of radiation effect, 262
Renin, 721
Repeating unit, 338
Repetitious coding sequences, 673
**Replication**   The duplication of DNA molecules, 660–1
**Representative element**   Any element in columns IA to VIIA or 0 in the periodic table, 47
Repressor protein, 672
Reserpine, 464, 695
Residues, in proteins, 555
**Resonance**   The representation of a real structure as an average of fictitious Lewis structures, 345–6, 560
Respiration, 118
   and Boyle's law, 136
   and Dalton's law, 144
Respiratory acidosis, 237
Respiratory alkalosis, 237
Respiratory chain (*See* Electron transport chain)
Restriction endonucleases, 676
Retina, 330
   detachment of, 507
Retinal, 330
**Reversible reaction**   One that can be made to go in either direction, 204–14
Rheumatism, 429
Rheumatoid arthritis, 506
Rhodopsin, 330
Rhubarb, 418
Riboflavin, 596, 604
Ribonuclease, 567
Ribonucleic acid (*See* RNA)
Ribose, 492–3, 495, 652
**Ribosomal RNA (rRNA)**   RNA molecules which, together with proteins, make up the ribosomes on

which proteins are synthesized, 663, 665

**Ribosome**  Small spherical bodies in the cells on which protein synthesis takes place, 663, 665, 668–9
A and P sites, 668–9
40S, 60S, 665, 668–9

**Ribozyme**  Enzyme made of ribonucleic acids, 573, 663

Rickets, 731–2

Rifabutin, 701

Rifamycin, 664

**RNA (Ribonucleic acid)**  A polymer made of nucleotide units that plays an essential role in protein synthesis, 650–6, 662–9
components of, 651–6
dialysis of, 188
as enzymes, 573
mutations and, 675
primary structure of, 654–6, 659
in protein synthesis, 663–5, 667–71, 675
three types of, 662–3
in viruses, 670–1

RNA polymerase, 664, 672–3

Roberts, Richard J., 669

**Roentgen**  A unit of radiation, 261–2

Roentgen, Wilhelm, 261

Rotation of plane of polarized light, 484

Rounding off rule, A.9

rRNA (*See* Ribosomal RNA)

Rubber, 339

Rubbing alcohol, 371

RU486, 537, 539

Ruby, 154

Rufen, 452

Ruminant mammals, 736

Rusting, 118

Rutherford, Ernest, 252, 254

**S**

Sabin polio vaccine, 699

Saccharin, 553

Salicylic acid, 429

Saline solution, 187–8

Salk polio vaccine, 699

*Salmonella,* 675

**Salt**  An ionic compound, 228
hydrolysis of, 236–7

**Salt bridge**  An attraction between ionized side chains on proteins, 563–4

Salt bridge, in a voltaic cell, 116

Sanger, Frederick, 560

Saphenous vein, 430

**Saponification**  The base-catalyzed hydrolysis of carboxylic esters, 433
of fats, 522

Saran, 339

Sarin, 689

Saturated fatty acids, 520–1

**Saturated hydrocarbon**  A hydrocarbon that contains only single bonds, 323, 333

**Saturated solution**  One in which the solvent holds all the solute it can hold at a given temperature, 167–8

Schwann cell, 529

Scientific method, 9–11
in medicine, 10

**Scientific notation**  A method for representing large and small numbers by using powers of ten, A.1

Scintillation counter, 260

Scurvy, 730, 734

Seaborg, Glenn, 269

Secobarbital, 457

Seconal, 457

**Second**  The basic unit of time, 16

**Secondary alcohol**  An alcohol in which the —OH is connected to $R_2CH$— or $RR'CH$—, 362
oxidation of, 367, 395

**Secondary amine**  An amine in which two carbons are connected to the nitrogen, 443–4

**Secondary carbon**  One that is connected to two other carbons, 296

Secondary messengers, 690–1, 698

Secondary structure
of DNA, 656–9
of proteins, 552, 560–3

Seeding, 168

Seldane, 695, 709

Self-ionization of water, 230–1

Semiconservative process, 661

**Semipermeable membrane**  A thin slice of some material that allows only small molecules, and not large ones, to pass through, 185–8

Senility, 688

Separatory funnels, 21, 167

Serine, 528, 549, 644

Serotonin, 695

Serum (*See* Blood serum)

Sesquiterpenes, 324

Sex attractants, 329

Sex hormones, 537–9, 718

Sharp, Philip A., 669

Shapes of molecules, 77–82
effect on boiling point, 152

Shell, electron, 50, 54–5

Shroud of Turin, 258

SI, 12–3

Sickle cell anemia, 560, 675

Side chains,
of alkanes, 297
of nucleotides, 655
of peptides and proteins, 555

Side reactions, 107, 368

**Sievert**  A unit of radiation, 262

Sigma bond, 292, 328, 343

Signals, initiation-termination, 666–9

**Significant figures**  The number of digits known with certainty, A.6–A.9

Silk, 562

Silkworm moth, 329

Silver chloride, 111, 195, 211

Silver mirror, 396

**Single bond**  A covalent bond consisting of one pair of electrons, 72–3

**Single displacement reaction**  A reaction in which one element displaces another element from a compound, 109–10

Sites, control, operator, promoter, 672

Sitz bath, 189

Skeletal muscles, 620

Skunks, 380

Sleeping pills, 398, 457, 606

Smallpox, 9, 670, 699

Smoker's lung, 351

Snow, melting of, 184

**Soap**  A mixture of sodium salts of fatty acids used as a cleaning agent, 522, 524

Sobrero, Ascanio, 373

Socrates, 460

Sodium acetate, 238, 242

Sodium benzoate, 421

Sodium bicarbonate, 228–9

Sodium chloride, 69–70, 179
crystal structure, 69–70
isotonic solutions of, 187–8
to melt ice, 184

Sodium citrate, 712
Sodium *para*-(2-dodecyl)benzenesul-
    fonate, 525
Sodium dodecyl sulfate, 525
Sodium hydroxide, 221
    in the eyes, 227
Sodium hypochlorite, 117
Sodium iodide, 383
Sodium ions, balance in the body, 67,
    607, 716, 719
Sodium lamp, 485
Sodium propionate, 421
Sodium urate, 178
Soil, 182
**Solid**    A state of matter with a definite
    shape and volume, 20, 130,
    154–6
    solutions of, 164
**Solubility**    The maximum amount of a
    solute that dissolves in a given
    amount of solvent at a given
    temperature, 166–8
    of alcohols and ethers, 378–80
    of aldehydes and ketones, 394
    of alkanes, 308
    of alkenes, 332
    of amines, 447
    of amino acids, 551
    of carboxylic acids, 418, 421
    of carboxylic esters, 424
    of ionic compounds, 113
    of monosaccharides, 494
**Solute**    The substance that is dis-
    solved in a medium, to form a
    solution, 164
**Solution**    A homogeneous mixture,
    163–81
    characteristics, 165–6
    types, 164
**Solvation**    The surrounding of solute
    molecules by solvent molecules,
    175, 183, 186
Solvation layer, 176, 183
**Solvent**    The medium in which the
    solute is dissolved, 164
Soman, 689
Soot, 155–6
Sorbitol, 497
*sp* hybrid orbitals, 342
*sp*² hybrid orbitals, 327
*sp*³ hybrid orbitals, 292
Spacers, 673
Species (*See* Chemical species)

**Specific gravity**    Ratio of the density
    of any substance to that of
    water, 22–3
**Specific heat**    The amount of heat, in
    calories, necessary to raise the
    temperature of 1 g of any sub-
    stance by 1°C, 26–8
    of water, 26
Specific rotation, 485
**Spectator ion**    An ion that does not
    participate in a reaction, 111
Spectra (stretched polyethylene fibers),
    341
Spectrum of radiation, 261
Speed of light, 252, 270
Sperm cells, 264
**Sphingolipid**    A complex lipid con-
    taining the amino alcohol sphin-
    gosine, 523, 528–9, 644
Sphingomyelin, 528–9, 644
Sphingosine, 523, 528–9, 644
**Sphygmomanometer**    A device for
    the measurement of blood pres-
    sure, 134
Spin, electron, 51
Spina bifida, 730
Spinach, 418
Spirits, 369
Spirits of ammonia, 221
Stable nuclei, 255
Stalactites, 310
Standard temperature and pressure,
    140
Staples, absorbable, 432
**Starch**    A polymer of D-glucose in
    which all linkages are $\alpha[1 \rightarrow 4]$
    or $\alpha[1 \rightarrow 6]$, 501–4
    in the diet, 728
    digestion of, 736
**Starting materials**    The substances
    that are converted to other sub-
    stances in a chemical reaction,
    95
Starvation, 728
States of matter, 20, 129–58
Static electricity, 200
Steam, 20
Stearic acid, 518–9, 624–5
Steel, 48
**Stereoisomers**    Compounds with the
    same structural formulas but
    different three-dimensional
    shapes, 307

of alkenes, 328–32
in biological organisms, 508–9
of cyclic compounds, 305–7
enantiomers, 480–9
Steroid hormones, 535–9
**Steroids**    A class of compounds all of
    which contain four rings fused
    in one particular pattern, 532–40
    anabolic, 536
Stethoscope, 134
Sticky ends of DNA, 677–9
**Stoichiometry**    Weight relationships
    in chemical reactions, 104–7
Stomach juice, 736–7
Stone Age, 48
**STP**    Standard temperature (0°C) and
    pressure (1 atm), 140
Strassman, Fritz, 270
Streptomycin, 664
**Strong acid**    An acid that reacts al-
    most completely with water, an
    acid that gives up its proton eas-
    ily, 221–7, 236–7
**Strong base**    One that is soluble in
    water and dissociates completely
    to OH⁻ ions, 221–7, 236–7
Strontium-90, 46, 257
**Structural formula**    A formula that
    shows all atoms in a molecule
    and all the bonds connecting
    them, 74, 289–91
    condensed, 294
Structural genes, 672
**Structural isomers**    Compounds with
    the same molecular formula, but
    different structural formulas, 289
Strychnine, 464
Sublevel, 50
**Sublimation**    The conversion of a
    solid directly to a vapor, 158
Subshell, electron, 50
Substitution reactions, 344
    of aromatic rings, 349
**Substrate**    A compound whose reac-
    tion is speeded up by an en-
    zyme, 577
**Subunit**    One polypeptide chain of a
    multichain protein, 565
Succinate, 601
Succinate dehydrogenase, 601, 606
Succinylcholine, 575, 689
Sucrose, 499, 618, 728
    heating of, 131

optical activity of, 494
solubility of, 167
**Sugar** A general term for monosac-
  charides and disaccharides, 494
  in nucleic acids, 652–3
Sugar, table (*See* Sucrose)
Sugar phosphates, 497
Sulfa drugs, 268, 581
Sulfanilamide, 581
Sulfapyridine, 581
Sulfate ions, balance in the body, 716
  Lewis structure, 76
Sulfathiazole, 581
Sulfides (*See* Thioethers)
Sulfonation, 349
Sulfonic acids, 349
Sulfur, 37, 154
Sulfur dioxide, in acid rain, 165, 177
Sulfuric acid, 221, 349
  in acid rain, 165
  in the eyes, 227
Sulfurous acid, 165
Sulfur trioxide, 165, 177, 180
Sun, energy of, 269–70
Sunette, 553
Sunflower oil, 517, 521
Sunglasses, 211
Sunlight, 211
Superimposability, 480–1, 483
**Supersaturated solution** One that
  holds more solute than the sol-
  vent normally can at a given
  temperature, 167–8
Surface area of molecules, effect on
  boiling point, 152
Surface charges, 183–4
**Surface tension** The pull exerted by
  molecules towards the center of
  a liquid; it causes the surface to
  act like a tough layer, 153
**Suspension** A mixture that separates
  into layers if allowed to stand,
  182
Svedberg units, 665
Sweating, 180
Sweetness, 494, 553
Swine flu, 529
**Synapse** The space between the axon
  end of one neuron and the cell
  body or dendrite end of the
  next, 686–7
Synergistic effect, 457
Synovial tissue, 506, 707

Synthase, 642
Synthetases, 575
Synthetic rubbers, 339
Système Internationale, 12
Systolic blood pressure, 134

T
2,4,5-T, 175
Table salt (*See* Sodium chloride)
Table sugar (*See* Sucrose)
Tabun, 689
**Tachycardia** Fast heart beat, 180, 720
Tagamet, 695
Tagging of a drug, 267
Tatum, Edward, 650
Taurine, 540, 693
Taurocholate, 540
Tautomerism, 404, 537, 618
Taxol, 314
Tay-Sachs disease, 531
T cells, 701
Tear gas, 406
Technetium-99m, 265, 267, 269
Teflon, 340–1, 430
**Temperature** The hotness or cold-
  ness of a substance; the property
  that is measured in °C or K, 16–7
  body, 16, 202
  contrast with heat, 25
  conversions, 16
  effect, on enzyme action, 579
    on equilibrium, 211–3
    on reaction rates, 201–3
    on solubility, 167–8
    on vapor pressure, 151
  and kinetic energy, 130
  scales, 17
Teratogen, 509
Terephthalic acid, 430
Terfenadine, 695, 709
Termination signal, 664, 666
Terpenes, 324–5
**Tertiary alcohol** An alcohol in which
  the —OH is connected to a carbon attached to three other carbons, 362
  oxidation of, 368
**Tertiary amine** A amine in which
  three carbons are connected to
  the nitrogen, 443–4
**Tertiary carbon** One that is connected to three other carbons,
  296

Tertiary structure,
  of proteins, 557, 563–5
Testosterone, 535, 537, 718
Tetrachloroethene, 382
Tetrachloromethane, 382
Tetracycline, 267
1,1,1,2-Tetrafluoroethane, 382
Tetrahedral molecules, 79–81, 289, 294
**Tetrahedron** A geometrical figure
  made up of four equilateral tri-
  angles, 79
Tetrahydrocannabinol, 379
Tetrahydrofuran, 375
Tetrahydropyran, 350
Tetrahydrourushiol, 374
Tetraterpenes, 324
Tetrodotoxin, 690
TFIIIA, 673
Thalidomide, 509
THC, 379
  in cancer and AIDS treatment, 379
Theoretical yield, 107
**Theory** A hypothesis supported by
  evidence; a model that explains
  many interrelated facts, 11
Thermal burns, 139
Thermal pollution, 168
Thermite reaction, 120
Thiazide diuretics, 728
Thioethers, 380
**Thiol** A compound containing the
  —SH group attached to an alkyl
  or aryl carbon, 380, 420, 436, 567
Thiolase, 623–4
Thompson, Benjamin, 22
Thorazine, 606
Three Mile Island, 259, 271
Threonine, 549, 729
Thrombin, 508, 712
Thrombocytes, 711–2
Thromboplastin, 712
Thrombosis, 712
Thudichum, Johann, 528
Thymidine, 653
Thymine, 651
Thymol, 374
Thyroglobulin, 383, 567
Thyroid gland, 267, 269, 272, 383
Thyroxine, 267, 383, 696
Time, 16
**Tincture** A solution in which ethanol
  is the solvent, 117, 370
*Tinea pedis*, 421

**Titration** The addition of a base to an acid (or vice versa) until it is neutralized, 242–6

TNT, 100, 604

Tobacco plant, 679

Tolbutamide, 698

**Tollens' test** A test for aldehydes that uses $NH_3$ and silver ions to form a silver mirror, 396–7

Toluene, 347

**Torr** A unit of pressure, 133

Torricelli, Evangelista, 132

Trace elements, 36

Tranquilizers, 459, 464

**Transamination** A reaction catalyzed by enzymes in which an amino group of an amino acid is transferred to a ketone, 626–7, 629–30, 646

**Transcription** The transfer of the information contained in a molecule of DNA to a molecule of RNA, 663–4

Transcription factors, 673
helix-turn-helix, 673
leucine zipper, 673

Transferases, 575–6, 668

**Transfer RNA (tRNA)** Relatively small molecules of RNA, each of which bears a site that recognizes only one amino acid and another (the anticodon) that corresponds to only one codon on mRNA, 662–9

**Trans isomer** A stereoisomer in which two similar or identical groups are on opposite sides
of alkenes, 329–32
of cycloalkanes, 307
in vision, 330

**Transition elements** Elements in the B columns of the periodic table, plus elements 58 to 71 and 90 to 103, 47, 57, 69

**Transition state** The top of the energy hill, 199

**Translation** The process by which the information carried in mRNA is converted to the sequence of amino acids in the newly synthesized protein, 663, 665–9

Translocation, 668–9

**Transmutation** The changing of one element to another
artificial, 269–70
natural, 256–7

Transuranium elements, 269–70

Tranxene, 459

1-Triacontyl palmitate, 524

Triacylglycerol, 518

Tricarboxylic acid cycle (*See* Citric acid cycle)

Trichloroacetic acid, 419

Trichloromethane, 382

Triclosan, 377

9-Tricosene, 329

**Triglyceride** An ester of glycerol in which all three —OH groups are esterified, 518–23

Trigonal molecules, 79–80

Triiodomethane, 382

Trinitrotoluene, 100, 604

Tripeptides, 554

Triphosphoric acid, 435

**Triple bond** A bond in which two atoms share six electrons, 74, 290, 341–3

Triple helix of collagen, 562–3

Triplets, of bases, 665

**Triprotic acid** An acid that can give up three protons, 224

Tritium, 255, 269–70

tRNA (*See* Transfer RNA)

Tropocollagen, 563, 565

Trypsin, 574, 578, 614, 738

Trypsinogen, 578

Tryptophan, 549, 645, 729, 737

Tubules, distal and proximal, 715, 717, 719

Tubulin, 702

Tylenol, 452

**Tyndall effect** The scattering of light by a colloidal system, 182

Tyramine, 693

Tyrosine, 549, 551, 564, 737

Tyrosine-tRNA ligase, 576

**U**

UDP, 641, 648

UDP glucose, 638, 641, 645

Ulcers, 429, 541, 585, 693

Uncoupling agent, 604

Units, 12–7
conversions of, 18–9

Universal gas constant, R, 141

Unpaired electrons, 51

**Unsaturated compound** A compound that contains one or more double or triple bonds, 323, 333
test for, 334, 343

Unsaturated fatty acids, 519–21

**Unsaturated solution** One that holds less solute than the solvent can hold at a given temperature, 167–8

Unwinding proteins, 660

**Uphill reaction** Endothermic reaction, 198

Uracil, 651, 653

Uranium, 251

Uranium-235, 270–1

Uranium-238, 259
dating of rocks, 258

Urea, 456
conversion to barbituric acid, 457
hydrolysis of, 106, 456
in protein denaturation, 566
manufacture of, 108
synthesis of, 287
in urea cycle, 629

**Urea cycle** The metabolic pathway by which the nitrogen of amino acids is converted to urea, 615, 627–9

Urea-formaldehyde, 406

Urease, 106, 574

Uridine, 653

Uridine diphosphate, 641

Urine, 707, 715, 717–9, 736
acetone in, 626
blue, 645
pH of, 234

Urobilin, 631

Uronic acid, 497

**V**

Vaccination, 670, 699, 701

Vagina, 432

**Valence shell** The outer shell of electrons in an atom, 68

Valence shell electron pair repulsion theory, 79–82

Valeric acid, 417–8

Valine, 549, 560, 675

Valium, 459

Vanilla beans, 391, 407

Vanillin, 374, 407

**Vapor pressure** The pressure of a

vapor in equilibrium with its liquid, 150–2
lowering, 185
Vaseline, 308
Vasoconstrictors, 721
Vasodilators, 720
Vasopressin, 559, 719
Vegetable oils, 517, 520–1
Vesicles, 687
Vidarabine, 671
Vinegar, 219, 221, 367, 416, 419
Vinyl chloride, 338
Vira-A, 671
Viruses, 670–1, 699–701
**Visual pigments** Colored compounds in the retina, 330
Vital force, 286
**Vitamin** An organic compound necessary to healthy life that humans cannot synthesize but must obtain in the diet, 726, 730
list of, 732–4
Vitamin A, 325, 330, 733
Vitamin B$_{12}$, 733
Vitamin B family, 577, 596–8, 730, 733
Vitamin C, 405, 733
Vitamin D, 731, 733
Vitamin K, 712, 733
Vitreous, 507
Volcanoes, 310
Voltaic cells, 117–8
**Volume** The space occupied by matter. For gases this is the volume of the container, 13–4, 131
Volumetric flask, 14, 169, 171
VSEPR, 79–82
v/v percentage, 170
Vulcanization, 339

**W**
Waage, Peter, 206
Water,
addition to alkenes, 335–6
as an amphoteric substance, 224
in body fluids, 175, 180–1
composition of, 40
density of, 22, 147
in the diet, 730
formation of, 195, 198–9
hydrogen bonding in, 147
phase changes in, 157
polarity of, 84–5
reaction,
with alkali metals, 46
with carbon monoxide, 204–6, 207
with ethyl acetate, 209–11
self-ionization of, 230–1
shape of molecule, 79
as a solvent, 167, 175–81
specific heat of, 26
structure of, 147
unusual properties of, 175
(*See also* Hydration; Hydrolysis)
Water bugs, 153
**Water of hydration** Water molecules present in a crystalline solid, 177
Watson, James, 656–7
**Wavelength** In a wave, the distance from one crest to the next, 252–3
**Wax** A naturally occurring mixture of long-chain fatty acids and long-chain alcohols, 524
**Weak acid** An acid that reacts with water, to form H$_3$O$^+$ ions, only to a limited extent, 221–3, 236–7
strengths of, 224–6, 419–20

Weak bases, 221–3, 236–7
**Weight** The force a mass experiences under the pull of gravity, 15
Weight relationships in chemical reactions, 104–7
White blood cells, 711
Wilkins, Maurice, 656
Willow bark, 429
Wine, 369, 693
oxidation to vinegar, 367, 419
Wintergreen, 374, 425
Wöhler, Friedrich, 287
Women in science, 476–7
Wood, 504
Wood alcohol, 370
Wristwatch dials, 265
w/v percentage, 169–70
w/w percentage, 170

**XYZ**
**X-ray** Electromagnetic radiation of very high frequency, 253–4
diagnostic, 266, 268
Xylocaine, 462
Yew tree, 314
Yttrium-90, 269
Zeno, 33
Zinc, 37, 115–6
Zinc sulfide, 37
Zinc 10-undecylenate, 421
**Zwitterion** A compound that has a positive charge on one atom and a negative charge on another, 550–2, 556
Zygote, 650
Zymase, 369
**Zymogen** The biologically inactive form of an enzyme, 578

# The Most Important Functional Groups in Organic Chemistry

| Group | Type of Compound |
|---|---|
| $\diagup C = C \diagdown$ | Alkene |
| $-C \equiv C-$ | Alkyne |
| R—F (or Cl, Br, I) | Alkyl halide |
| R—OH | Alcohol |
| R—O—R' | Ether |
| R—SH | Thiol (mercaptan) |
| R—NH$_2$ <br> R$_2$NH <br> R$_3$N | Amine |
| R—C—H $\;\;\underset{\parallel}{\phantom{C}}\;$ O | Aldehyde |
| R—C—R' $\;\;\underset{\parallel}{\phantom{C}}\;$ O | Ketone |
| R—C—OH $\;\;\underset{\parallel}{\phantom{C}}\;$ O | Carboxylic acid |
| R—C—OR' $\;\;\underset{\parallel}{\phantom{C}}\;$ O | Carboxylic ester |
| R—C—NH$_2$ $\;\;\underset{\parallel}{\phantom{C}}\;$ O <br> R—C—NHR' $\;\;\underset{\parallel}{\phantom{C}}\;$ O <br> R—C—NR'R'' $\;\;\underset{\parallel}{\phantom{C}}\;$ O | Amide |